Fundamentals of Transport Phenomena

To the memory of my parents,
Smt Karma-Bhari and Shri Khushal Chand

Contents

Part 2: Momentum Transport (Fluid Mechanics)

Fundamentals of Transport Phenomena

RAJINDER PAL

Professor of Chemical Engineering
University of Waterloo
Ontario, Canada

CWP
Central West Publishing

This edition has been published by Central West Publishing, Australia

For more information about the books published by Central West Publishing, please visit https://centralwestpublishing.com

Disclaimer

Every effort has been made by the publisher and author while preparing this book; however, no warranties are made regarding the accuracy and completeness of the content. The publisher and author disclaim without any limitation all warranties as well as any implied warranties about sales, along with fitness of the content for a particular purpose. Citation of any website and other information sources does not mean any endorsement from the publisher and author. For ascertaining the suitability of the contents contained herein for a particular lab or commercial use, consultation with the subject expert is needed. In addition, while using the information and methods contained herein, the practitioners and researchers need to be mindful for their own safety, along with the safety of others, including the professional parties and premises for which they have professional responsibility. To the fullest extent of law, the publisher and author are not liable in all circumstances (special, incidental, and consequential) for any injury and/or damage to persons and property, along with any potential loss of profit and other commercial damages due to the use of any methods, products, guidelines, procedures contained in the material herein.

NATIONAL LIBRARY OF AUSTRALIA

A catalogue record for this book is available from the National Library of Australia

ISBN (print): 978-1-922617-64-4

Contents

Part 2: Momentum Transport (Fluid Mechanics)

Part 4: Mass Transport

Preface

I have been teaching various aspects of transport processes (fluid mechanics, heat transfer, mass transfer) to our second and third year undergraduate chemical engineering students at the University of Waterloo for more than thirty years. I have also taught the complete transport phenomena course at a graduate level to our MASc, MEng, and PhD students for more than twenty years. Students find transport phenomena courses to be remarkably interesting and applied but difficult at the same time, especially at the graduate level. Although several excellent textbooks are available on various aspects and complete transport phenomena, I have felt the need for a new textbook where the material is presented in order and style such that it is easily comprehensible to undergraduate and graduate students.

This book is divided into four parts. Part 1 consists of two chapters. Chapter one introduces the subject of transport phenomena with important practical applications. The similarities and differences between different transport processes are discussed. Chapter two reviews the necessary mathematical background dealing with vector and tensor operations and important theorems required in the development of microscopic equations of transport phenomena. Part 2 deals with momentum transport (fluid mechanics). It consists of seven chapters (chapters three to nine). Chapter three presents introductory concepts and definitions of momentum transport (fluid mechanics). Chapter four develops the microscopic equations of motions. The applications of the microscopic equations of motion are illustrated in chapter five. Chapter six is devoted to friction factor, friction factor correlations, and friction loss calculations in practical systems. The settling behaviour of particles is covered along with drift flux theory in chapter seven. Chapter eight is devoted to non-Newtonian flow. The flow of power law, Bingham plastic, and general time-independent non-Newtonian fluids is covered. Chapter nine deals with the flow of Newtonian and non-Newtonian fluids through packed beds and consolidated porous media. Part 3 deals with heat transport. It consists of six chapters (chapters ten to fifteen). Chapter ten presents general introduction to heat transfer dealing with different modes of heat transfer (convection, conduction, and radiation). Chapter eleven is fully devoted to heat transfer in a stationary medium, that is, conductive heat transfer. Chapter twelve deals with heat transfer in a moving fluid. The microscopic equations of total energy and thermal energy are developed. The solutions of the thermal energy equation are illustrated for different situations. Chapter thirteen is devoted to heat transfer coefficient, heat transfer coefficient correlations, and applications of convective heat transfer including heat exchangers. Heat transfer with phase change such as in boiling and condensation is covered in chapter fourteen. The last chapter of part 3 is devoted to radiative heat transfer which covers the nature of thermal radiation and radiative heat exchange between surfaces at different temperatures. Part 4 deals with mass transport. It consists of five chapters (chapters sixteen to twenty). Chapter sixteen covers mass transfer in a stationary medium. Mass transfer in a moving fluid is presented in chapter seventeen. Chapter eighteen is devoted to mass transfer coefficient, mass transfer coefficient correlations, and interphase mass transfer. The analysis and design of packed towers for mass transfer are covered in detail in chapter nineteen. The last chapter of the book, that is, chapter twenty is devoted to multicomponent diffusion in gases and liquids. The available textbooks in transport phenomena rarely provide satisfactory coverage of multicomponent diffusion.

Each chapter presents solved examples to illustrate the applications of concepts and theory covered in the chapter. Notation is included separately in each chapter. For the benefit of students, practice problems are included at the end of each chapter.

It is hoped that the students will be able to understand the concepts and material presented in this textbook without difficulty. As transport phenomena is at the core of chemical engineering practice, it is important that students develop a liking for this subject. Understanding and liking of a subject go hand in hand. Students are inspired to pursue a particular field at a more advanced level only if they can develop a good understanding and liking for the subject.

Finally, I would like to thank my wife AP, daughter AP, and son AP for their love and constant support.

Waterloo, March 2024

Rajinder Pal

Author

Rajinder Pal is a Professor in the Department of Chemical Engineering at the University of Waterloo since 1991, and a Fellow of the Chemical Institute of Canada. He received degrees in Chemical Engineering from I.I.T. Kanpur (B.Tech. 1981) and the University of Waterloo (Ph.D. 1987), followed by a postdoctoral fellowship at the University of Alberta. His research interests include complex fluid rheology, interfacial and transport phenomena, physical properties of composite materials, and non-equilibrium thermodynamics. He has published close to 170 refereed journal articles, 5 books, and 5 book chapters on these and related subjects, including 8 papers in engineering education journals. He has taught over ten different undergraduate and graduate courses related to equilibrium staged separation processes, fluid mechanics, heat and mass transfer, transport phenomena, physical chemistry, and thermodynamics.

Professor Pal has been the recipient of several awards including the University of Waterloo prestigious Distinguished Teacher Award (2020), Sandford Fleming Foundation Teaching Excellence Award (twice 2001 and 2017), an Outstanding Performance Award from the University of Waterloo (2009), and the Syncrude Canada Innovation Award from the Canadian Society for Chemical Engineering (1998). He is a Fellow of the Chemical Institute of Canada since 1999. He is currently serving as an Associate Editor of the open access journal Fluids. He is also an Academic Editor of well-known international journals Foods, Nanomaterials, ChemEngineering, Colloids & Interfaces, Materials Research Express and Eng. He was an Associate Editor of the Canadian Journal of Chemical Engineering for twelve years. He is a registered Professional Engineer in the province of Ontario, Canada.

Chapter 1

Introduction to Transport Phenomena

Transport phenomena deals with the study of momentum, heat, and mass transport processes. One reason for lumping the three transport processes together is that the molecular level mechanisms for the transport of momentum, heat, and mass are similar in nature especially in the case of gases. For example, momentum, heat, and mass are transported from one part of the gas to another under the application of driving forces via random motion and collision of molecules. The governing laws of momentum, heat, and mass transport also have a close analogy. Furthermore, more than one transport process (such as momentum and heat, or heat and mass) occur simultaneously in many applications. It should also be noted that momentum transport in a fluid involves the study of motion and stresses in fluids. Thus, momentum transport in fluids and fluid mechanics are one and the same subjects.

1.1 MOMENTUM TRANSPORT (FLUID MECHANICS)

When different parts of a fluid move with different velocities, that is, velocity gradient exists in the fluid, momentum transfer takes place from a region at high velocity to a region at low velocity. The driving force for momentum transport is velocity gradient in the fluid and momentum flows in the direction of decreasing velocity. As an example, consider one-dimensional flow of fluid in x-direction as shown in Figure 1.1. The flow is in x-direction and the velocity increases in positive y-direction. Due to velocity gradient in the y-direction, adjacent layers of fluid are moving with different velocities. As shown in the figure, x-momentum flows in the direction of decreasing velocity (that is, negative y-direction). The x-momentum is transported from one layer of fluid to another due to random motion and collision of molecules. Fluid molecules jump from one layer to another and transfer momentum. Note that the fluid molecules jump in both upward (positive y direction) and downward (negative y-direction) directions but there is no net exchange of mass from one layer to another. Only x-momentum is transferred from one layer to another. The x-momentum flux in the y-direction (x-momentum flow in y-direction per unit area normal to y) is proportional to the negative of velocity gradient in y-direction, that is:

$$(x - momentum\ flux) \propto \left(-\frac{dV_x}{dy}\right) \tag{1.1}$$

The negative sign is there to indicate that the momentum flows in the direction of decreasing velocity. In the present example, the velocity gradient is positive ($dV_x/dy =$ positive) and therefore, x-momentum flux is negative, that is, x-momentum flows in the negative y-direction.

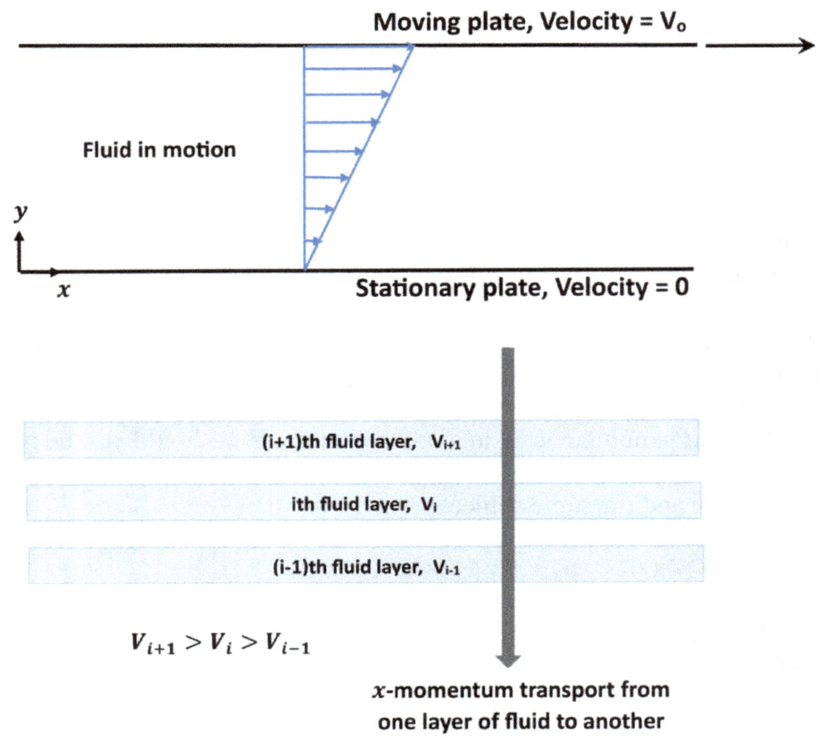

Figure 1.1 Momentum transport in a fluid subjected to velocity gradient.

The momentum flux in a fluid could also be interpreted as a shear stress (tangential force per unit area) acting on the fluid. For example, the x-momentum flux in the y-direction is equivalent to a shear stress τ_{yx}. The first subscript of τ_{yx} specifies the area on which the shear stress is acting, and the second subscript specifies the direction of stress. Thus, τ_{yx} is a shear stress acting on a y-surface, that is, x-z plane in the x-direction, as shown in Figure 1.2.

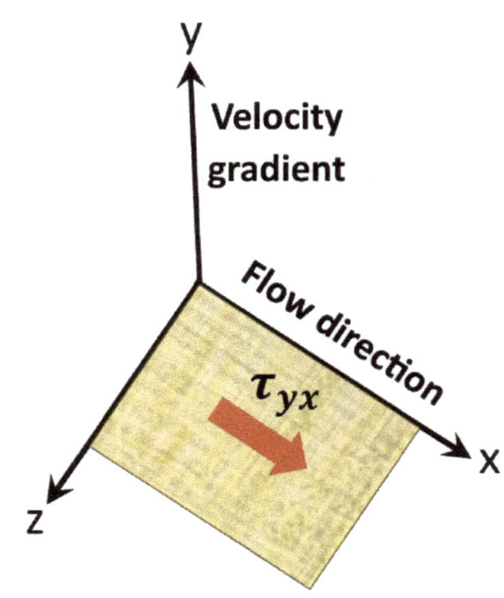

Figure 1.2 Shear stress acting on x-z plane when flow is in x-direction and velocity gradient is in y-direction (velocity increases with distance in y-direction).

Momentum flux and shear stress have the same units, that is, kg/m.s^2 (same as Pa). To explain their equivalence further, consider two adjacent thin layers of fluid moving past each other at slightly different velocities, as shown in Figure 1.3. The distance between the layers is a corresponding to approximately the mean free path of fluid molecules. The velocity of the upper fluid layer, that is, $V_x|_{y+a}$ is slightly higher than the velocity of the lower fluid layer, that is, $V_x|_y$.

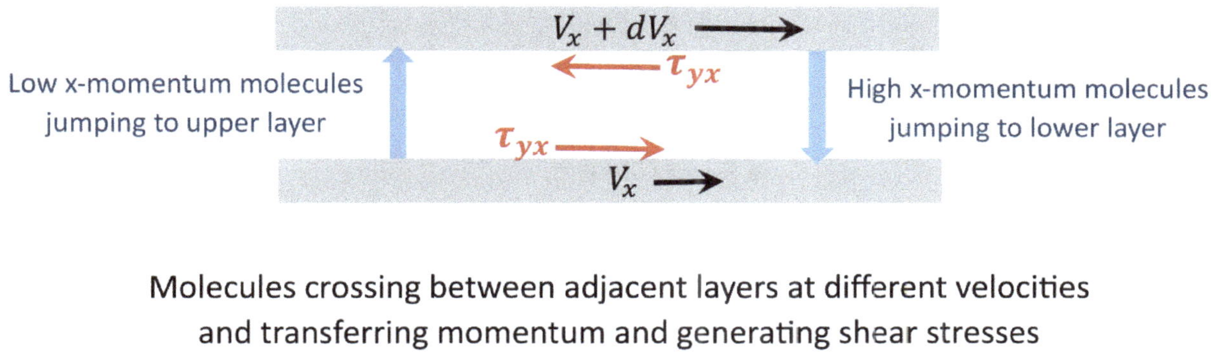

Figure 1.3 Equivalence of momentum flux and shear stress.

As noted earlier, due to random motion of molecules, the fluid molecules are crossing between the adjacent layers. The mass of any layer does not change but its momentum changes due to crossing of molecules between different layers moving at different velocities. As a result of crossing of molecules between adjacent layers, the x-momentum of the upper layer (see Figure 1.3) moving at a higher velocity $V_x|_{y+a}$ decreases whereas the x-momentum of the lower layer moving at a lower velocity $V_x|_y$ increases.

Let $\dot{m}_m$ be the molecular mass flow rate of fluid molecules crossing between the adjacent layers per unit contact area between layers. We assume that the molecular mass flow rate of fluid molecules per unit contact area is the same in both upward and downward directions so that the mass of each layer of fluid remains constant. The rate of increase of x-momentum of lower fluid layer per unit contact area is:

$$\text{Rate of increase of x} - \text{momentum of lower layer} = \dot{m}_m V_x|_{y+a} - \dot{m}_m V_x|_y \tag{1.2}$$

Therefore, the lower fluid layer experiences a tangential force per unit area (shear stress) in the forward direction (see Figure 1.3) in accordance with the Newton's second law of motion which states that the rate of change of momentum of a body is equal to net force acting on the body.

The upper fluid layer experiences a decrease in x-momentum due to crossing of molecules. The rate of decrease of x- momentum of upper fluid layer per unit contact area is:

$$\text{Rate of decrease of x} - \text{momentum of upper layer} = \dot{m}_m V_x|_y - \dot{m}_m V_x|_{y+a} \tag{1.3}$$

Consequently, the upper fluid layer experiences a tangential force per unit area (shear stress) in the backward direction (see Figure 1.3) in accordance with the Newton's second law of motion.

As the rate of increase of x-momentum of the lower fluid layer per unit contact area is equivalent to τ_{yx}, we can further write:

$$\tau_{yx} = \dot{m}_m \left(V_x|_{y+a} - V_x|_y\right) \tag{1.4}$$

The velocity gradient between the adjacent layers of fluid can be approximated as:

$$\frac{dV_x}{dy} \approx \frac{V_x|_{y+a} - V_x|_y}{a} \tag{1.5}$$

Combining the above equations, we get:

$$\tau_{yx} = \dot{m}_m a \frac{dV_x}{dy} \tag{1.6}$$

We can further re-write this equation as:

$$\tau_{yx} = (Zma)\frac{dV_x}{dy} \tag{1.7}$$

where Z is the frequency of collision of molecules per unit contact area and m is the mass of a fluid molecule. If τ_{yx} is interpreted as a shear stress, it is positive as given by the above equation, that is, the shear stress acts in the positive x-direction on a positive y-surface (see Figure 1.2). If τ_{yx} is interpreted as a momentum flux, we need to multiply the above equation with a negative sign as momentum flows in the direction of decreasing velocity. Thus, momentum flux τ_{yx} is given as:

$$\tau_{yx} = -(Zma)\frac{dV_x}{dy} \tag{1.8}$$

Finally, it should be noted that the mechanism of momentum transport described in the preceding discussion, that is, random motion and collision of molecules from one layer to another is generally valid for gases only. In liquids, the momentum exchange due to random motion and collision of molecules is usually small. In liquids, momentum transport between adjacent layers is mediated by attractive (cohesive) forces between the molecules. The cohesive forces tend to keep the adjacent molecules of a liquid in a fixed position relative to each other and to resist relative motion between different parts of the liquid. Thus, due to cohesive forces between the molecules, the upper layer moving at a higher velocity tends to drag the lower layer moving at a lower velocity in the forward direction and the lower layer tends to resist the motion of the upper layer. Consequently, the lower layer experiences a shear stress in the forward direction and the upper layer experiences a shear stress in the backward direction. The transmission of shear stress from a faster moving layer to a slower moving layer is equivalent of transport of momentum between adjacent layers.

1.2 HEAT TRANSPORT

When different parts of a material medium, are at different temperatures, that is, temperature gradient exists in the medium, energy flows from a region at high temperature to a region at low temperature. This energy which flows by virtue of temperature gradient is called heat. Heat flows in the direction of decreasing temperature. As an example, consider one-dimensional heat transfer in y-direction as shown in Figure 1.4. The material medium is a stationary fluid (no bulk motion) subjected to a temperature gradient as shown. The temperature in the medium increases in the positive y-direction. Due to temperature gradient in the y-direction, adjacent layers of fluid are at different temperatures. The heat is transported from one layer of fluid to another due to random motion and collision of molecules. The heat flows in the downward (negative y) direction. Fluid molecules jump from one layer to another and transfer heat. Once again, the fluid molecules jump in both upward and downward y-directions but there is no net exchange of mass from one layer to another. Only heat is transferred from one layer to another. The heat flux in the y-direction (heat transfer rate per unit area normal to y) is proportional to the negative of temperature gradient in y-direction, that is:

$$(heat\ flux) \propto \left(-\frac{dT}{dy}\right) \tag{1.9}$$

The negative sign is there to indicate that heat flows in the direction of decreasing temperature. In the present example, the temperature gradient is positive (dT/dy = positive) and therefore, heat flux is negative, that is, heat flows in the negative y-direction.

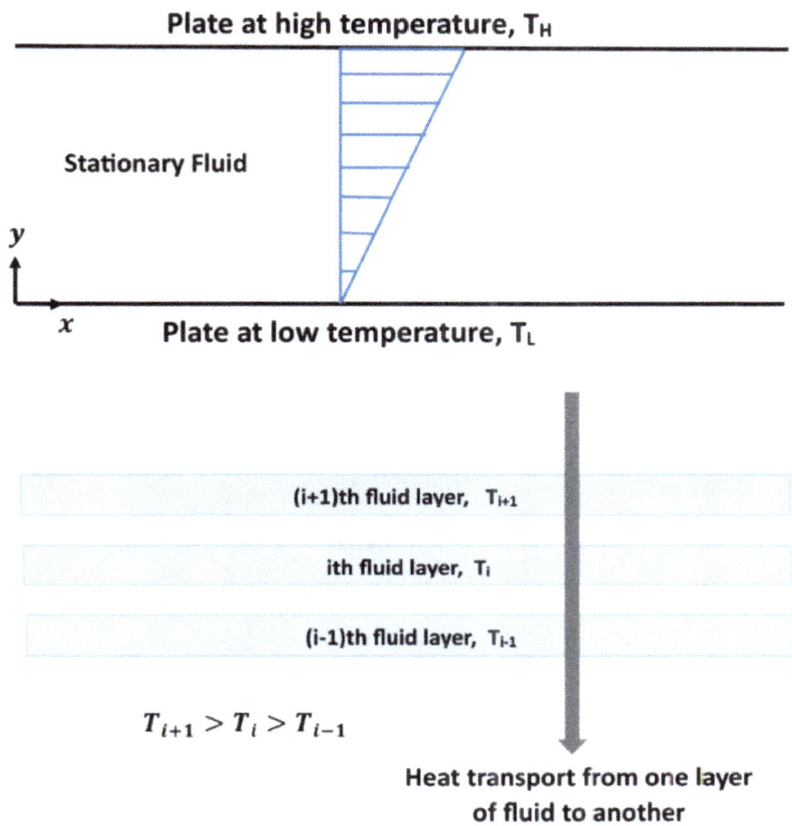

Figure 1.4 Heat transport in a fluid subjected to temperature gradient.

1.3 MASS TRANSPORT

When different parts of a material medium, say fluid, have a different concentration of some species A, that is, concentration gradient of species A exists in the medium, species A flows (diffuses) spontaneously from a region at high concentration to a region at low concentration. This molecular mass transport of species A occurs in the direction of decreasing concentration of A and the driving force for mass transport is the concentration gradient in A. As an example, consider one-dimensional mass transport of species A in y-direction as shown in Figure 1.5. The concentration of species A decreases in the downward direction. Due to concentration gradient in the y-direction, adjacent layers of fluid have a different concentration of species A. Molecules of A move randomly in both upward and downward directions. However, there occurs a net transport of species A in the downward direction due to concentration gradient. The number of molecules of species A moving in the downward direction is much larger than the number of molecules of A moving in the upward direction. The molar flux of species A in the y-direction (molar flow rate per unit area normal to y) is proportional to the negative of concentration gradient in y-direction, that is:

$$(molar\ flux) \propto \left(-\frac{dC_A}{dy}\right) \tag{1.10}$$

The negative sign is there to indicate that mass flows in the direction of decreasing concentration. In the present example, the concentration gradient is positive (dC_A/dy = positive) and therefore, molar flux is negative, that is, mass flows in the negative y-direction.

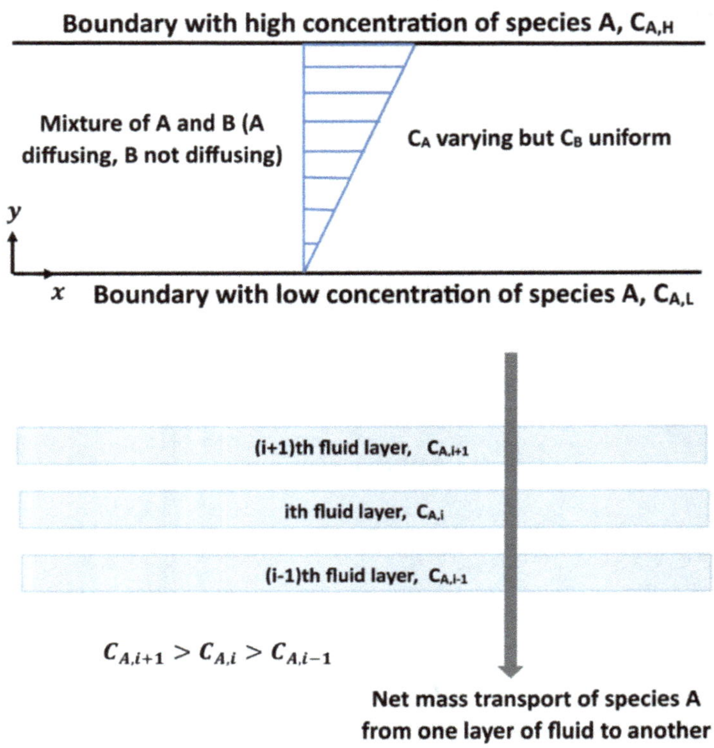

Figure 1.5 Mass transport of species A in a fluid subjected to concentration gradient in A.

1.4 DIFFERENCES BETWEEN THE THREE TRANSPORT PROCESSES

Although the molecular mechanisms of momentum, heat, and mass transport are similar, there are important differences between the three transport processes. Table 1.1 compares the three transport processes.

Table 1.1 Comparison of different transport processes

Type of transport process	Driving force	Quantities of practical interest
Momentum	$\nabla\vec{V}$ (velocity gradient, tensor)	Velocity distribution, momentum fluxes (viscous stresses), forces
Heat	∇T (temperature gradient, vector)	Temperature distribution, heat flux, heat transfer rates
Mass	∇C_A (concentration gradient, vector)	Concentration distribution, mass or molar fluxes, mass transfer rates

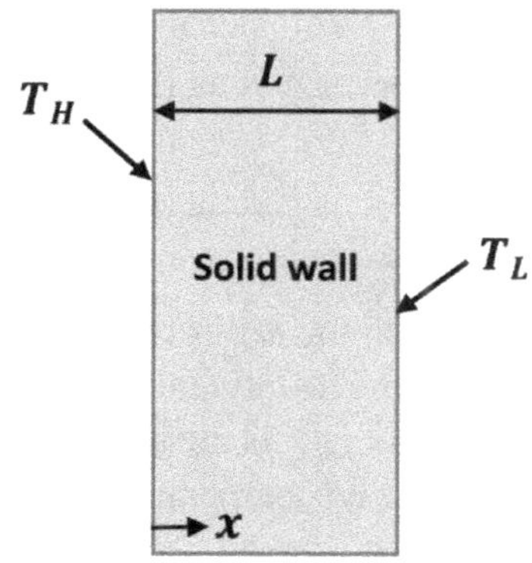

Solution

i) The temperature gradient is as follows:

$$\frac{dT}{dx} = -\left(\frac{T_H - T_L}{L}\right)$$

ii) Yes, the heat flux in the x-direction (q_x'') is constant. The heat flux in the x-direction (q_x'') is proportional to the negative of temperature gradient, that is, $-dT/dx$. As dT/dx is constant, the heat flux q_x'' is the same everywhere in the medium. Also note that the heat flux is positive as heat flows in the positive x direction (direction of decreasing temperature).

1.6.3 Example 3

Consider a solid plane wall of material B through which gaseous species A is diffusing. The molar concentration of species A at the left and right interfaces on the solid side are $C_{A,H}$ and $C_{A,L}$, respectively (see figure). The concentration of species A varies linearly with the position in the wall as given below:

$$C_A = C_{A,H} - \left(\frac{C_{A,H} - C_{A,L}}{L}\right)x$$

iii) Determine the concentration gradient of species A in the wall.
iv) Is the molar flux of species A in the x-direction ($J_{A,x}$) constant? Why?

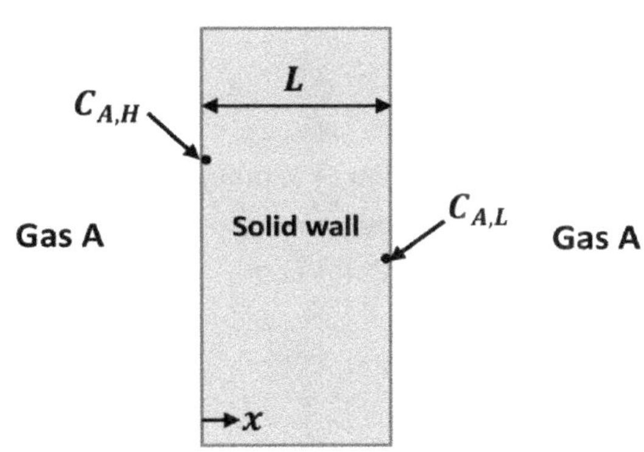

Solution

iii) The concentration gradient of species A is as follows:

$$\frac{dC_A}{dx} = -\left(\frac{C_{A,H} - C_{A,L}}{L}\right)$$

iv) Yes, the molar flux of species A in the x-direction ($J_{A,x}$) is constant. The molar flux in the x-direction ($J_{A,x}$) is proportional to the negative of concentration gradient, that is, $-d\,C_A/dx$. As $d\,C_A/dx$ is constant, the molar flux $J_{A,x}$ is the same everywhere in the medium. Also note that the molar flux of species A is positive as species A flows in the positive x direction (direction of decreasing C_A).

NOTATION

a	mean free path of molecules
C_A	molar concentration of species A
d	distance
$J_{A,r}$	molar flux of species A in r-direction
$J_{A,x}$	molar flux of species A in x-direction
L	length or thickness
m	mass
$\dot{m}$	mass flow rate
$\dot{m}_m$	mass flow rate of molecules
q''_x	x-component of heat flux
r	radial co-ordinate or distance in radial direction
R	radius of a pipe
T	temperature
V	velocity
$\vec{V}$	velocity vector
V_x	velocity in x-direction
V_z	velocity in z-direction
x	x co-ordinate or distance in x direction
y	y co-ordinate or distance in y direction
z	z co-ordinate or distance in z direction
Z	frequency of collision of molecules per unit contact area

Greek Letters

∇	mathematical del operator
τ_{yx}	shear stress on y-surface in the x -direction or x-momentum flux in y-direction
τ_{rz}	shear stress on r-surface in the z -direction or z-momentum flux in r-direction

PRACTICE PROBLEMS

Problem 1

Consider steady flow in a cylindrical pipe. The motion of the fluid in the pipe (z-direction) is caused by pressure-gradient applied to the fluid. The velocity distribution in the pipe is given as:

$$\frac{V_z}{V_{max}} = 1 - \left(\frac{r}{R}\right)^2$$

where V_z is the local velocity at any radial position, V_{max} is the maximum fluid velocity, r is any radial position in the pipe, and R is the radius of the pipe.

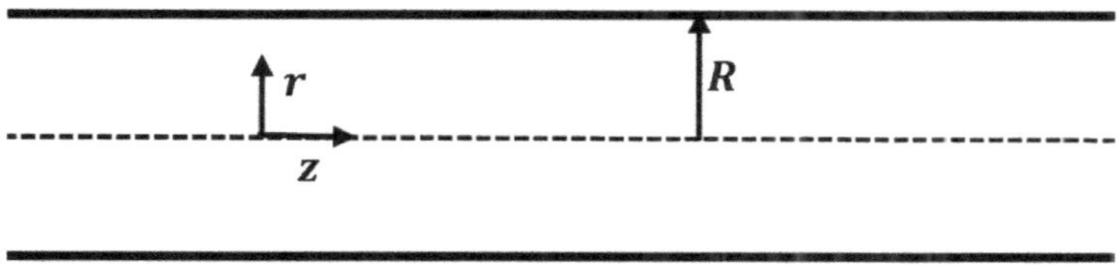

i) Determine the radial positions where the fluid velocities are zero and maximum.
ii) Plot the velocity distribution as V_z/V_{max} versus r/R.
iii) Determine the velocity gradient dV_z/dr at the pipe wall and centre of the pipe. Where is the velocity gradient highest?
iv) Where is the momentum flux τ_{rz} zero?
v) Where is the momentum flux τ_{rz} highest?

Note that τ_{rz} is z-momentum flux in r-direction.

Problem 2
Consider a solid cylindrical shell subjected to different temperatures at the two surfaces as shown in the figure below. The temperature varies with the radial position as follows:

$$\frac{T - T_1}{T_2 - T_1} = \frac{ln\left(\frac{r}{r_1}\right)}{ln\left(\frac{r_2}{r_1}\right)}$$

You can assume that $T_1 > T_2$.

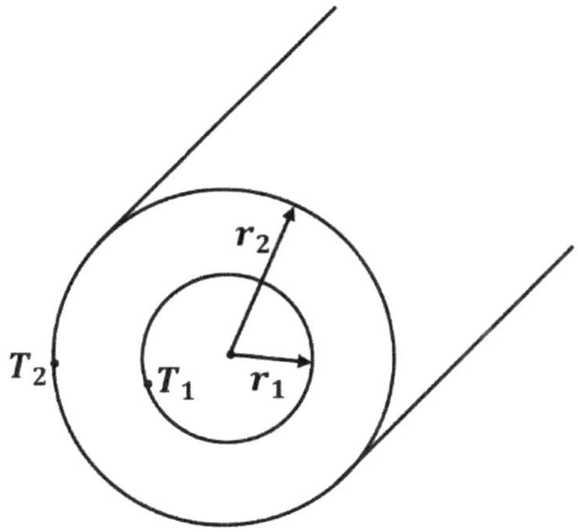

i) Determine the temperature gradient dT/dr in the shell.
ii) Is the temperature gradient constant? If not, where is the temperature gradient highest?
iii) Is the heat flux in the radial direction (q_r'') constant? If not, give explanation.

Problem 3

Consider a solid spherical shell of material B subjected to different concentrations at the two surfaces as shown in the figure below. The concentration of diffusing species A varies with the radial position as follows:

$$C_A = C_{A,1} + \frac{C_{A,2} - C_{A,1}}{\left(\frac{1}{r_1} - \frac{1}{r_2}\right)}\left(\frac{1}{r} - \frac{1}{r_1}\right)$$

where $C_{A,1}$ is the molar concentration of A within the shell at $r = r_1$ and $C_{A,2}$ is the molar concentration of A within the shell at $r = r_2$. You can assume that $C_{A,1} > C_{A,2}$.

Spherical shell

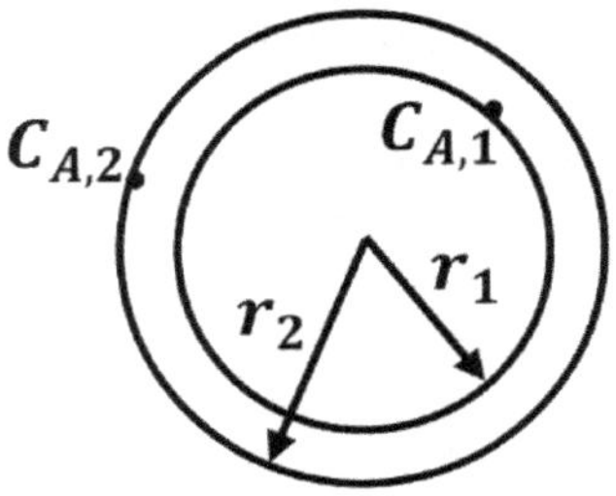

i) Determine the concentration gradient dC_A/dr in the shell.
ii) Is the concentration gradient constant? If not, where is the concentration gradient highest?
iii) Is the molar flux in the radial direction ($J_{A,r}$) constant? If not, give explanation.

FURTHER READING

1. R.B. Bird, W.E. Stewart, E.N. Lightfoot. *Transport Phenomena*. 2nd Edition, Wiley, New York, 2007.
2. C.J. Geankoplis, A.A. Hersel, D.H. Lepek. *Transport Processes and Separation Process Principles*. 5th Edition, Prentice Hall, New Jersey, 2018.
3. J.R. Welty, C.E. Wicks, R.E. Wilson, G. Rorrer. *Fundamentals of Momentum, Heat, and Mass Transfer*. 4th Edition, Wiley, New York, 2001.
4. F.P. Incropera, D.P. Dewitt, T.L. Bergman, A.S. Lavine. *Fundamentals of Heat and Mass Transfer*. 6th Edition, Wiley, New York, 2007.
5. J.P. Holland. *Heat Transfer*. 8th Edition, McGraw Hill, New York, 1997.
6. R.L. Street, G.Z. Watters, J.K. Vannard. *Elementary Fluid Mechanics*. 7th Edition, Wiley, New York, 1996.

Chapter 2

Review of Mathematical Background

2.1 SCALARS, VECTORS, AND VECTOR OPERATIONS

2.1.1 Scalars

Physical quantities which are specified by a single component (single number) are called scalar quantities. They do not possess any direction. Examples include mass, volume, energy, temperature, concentration, etc.

2.1.2 Vectors

Physical quantities which are specified by three components are called vector quantities. They possess both direction and magnitude. Examples include velocity, force, momentum, etc. In a cartesian coordinate system, vectors are represented as:

$$\vec{A} = A_x\hat{\imath} + A_y\hat{\jmath} + A_z\hat{k} \tag{2.1}$$

where $\vec{A}$ is a vector, A_x, A_y, and A_z are components of vector in $x, y,$ and z directions, and $\hat{\imath}, \hat{\jmath},$ and $\hat{k}$ are unit vectors in $x, y,$ and z directions. The magnitude of a vector is given as:

$$|\vec{A}| = \sqrt{A_x^2 + A_y^2 + A_z^2} \tag{2.2}$$

2.1.3 Vector Operations

Dot Product

The dot product between two vectors $\vec{A}$ and $\vec{B}$ is defined as:

$$\vec{A} \cdot \vec{B} = |\vec{A}||\vec{B}|cos\theta \tag{2.3}$$

where $|\vec{A}|$ and $|\vec{B}|$ are magnitudes of the vectors and θ is the angle between the two vectors. The dot product between two vectors is a scalar and therefore, the dot product is also referred to as scalar product. The dot product of a vector with itself is just the square of the magnitude of the vector:

$$\vec{A} \cdot \vec{A} = |\vec{A}|^2 \tag{2.4}$$

The dot product between vectors could be written in terms of vector components as:

$$\vec{A} \cdot \vec{B} = A_xB_x + A_yB_y + A_zB_z \tag{2.5}$$

There are nine dot products between unit vectors as given below:

$$\hat{\imath} \cdot \hat{\imath} = \hat{\jmath} \cdot \hat{\jmath} = \hat{k} \cdot \hat{k} = 1 \tag{2.6}$$

$$\hat{\imath} \cdot \hat{\jmath} = \hat{\jmath} \cdot \hat{\imath} = \hat{\jmath} \cdot \hat{k} = \hat{k} \cdot \hat{\jmath} = \hat{k} \cdot \hat{\imath} = \hat{\imath} \cdot \hat{k} = 0 \tag{2.7}$$

The following rules are applicable to dot product:

$$\vec{A} \cdot \vec{B} = \vec{B} \cdot \vec{A} \quad \text{(dot product is commutative)} \tag{2.8}$$

$$\vec{A} \cdot (\vec{B} + \vec{C}) = \vec{A} \cdot \vec{B} + \vec{A} \cdot \vec{C} \quad \text{(dot product is distributive)} \tag{2.9}$$

Cross Product

The cross product between two vectors $\vec{A}$ and $\vec{B}$ is defined as:

$$\vec{A} \times \vec{B} = |\vec{A}||\vec{B}| sin\theta \hat{n} \tag{2.10}$$

where $\hat{n}$ is a unit vector normal to the plane of $\vec{A}$ and $\vec{B}$ (see Figure 2.1), and θ is the angle between the two vectors. The direction of the cross product is determined by a right-hand rule. As the cross product of two vectors is a vector, the cross product is also referred to as vector product. From the definition of cross product, it follows that the cross product of a vector with itself is zero:

$$\vec{A} \times \vec{A} = 0 \tag{2.11}$$

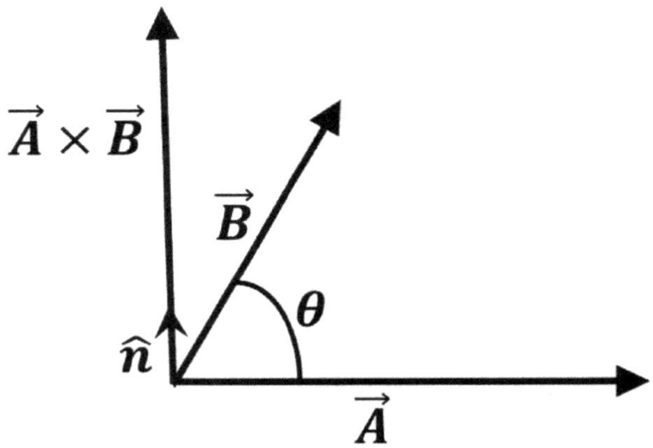

Figure 2.1 Cross product between two vectors.

The cross product could be written in component form as:

$$\vec{A} \times \vec{B} = \begin{vmatrix} \hat{\imath} & \hat{\jmath} & \hat{k} \\ A_x & A_y & A_z \\ B_x & B_y & B_z \end{vmatrix} = \hat{\imath} \begin{vmatrix} A_y & A_z \\ B_y & B_z \end{vmatrix} - \hat{\jmath} \begin{vmatrix} A_x & A_z \\ B_x & B_z \end{vmatrix} + \hat{k} \begin{vmatrix} A_x & A_y \\ B_x & B_y \end{vmatrix} =$$
$$\hat{\imath}(A_y B_z - A_z B_y) - \hat{\jmath}(A_x B_z - A_z B_x) + \hat{k}(A_x B_y - A_y B_x) \tag{2.12}$$

There are nine cross products between unit vectors as given below:

$$\hat{\imath} \times \hat{\jmath} = \hat{k} \; ; \; \hat{\jmath} \times \hat{k} = \hat{\imath} ; \; \hat{k} \times \hat{\imath} = \hat{\jmath} \quad \text{(unit vectors in clockwise direction)} \tag{2.13}$$

$$\hat{\imath} \times \hat{k} = -\hat{\jmath} \; ; \; \hat{k} \times \hat{\jmath} = -\hat{\imath} ; \; \hat{\jmath} \times \hat{\imath} = -\hat{k} \quad \text{(unit vectors in anti-clockwise direction)} \tag{2.14}$$

$$\hat{\imath} \times \hat{\imath} = 0 \ ; \ \hat{\jmath} \times \hat{\jmath} = 0 \ ; \ \hat{k} \times \hat{k} = 0 \tag{2.15}$$

The following rules are applicable to cross product:

$$\vec{A} \times \vec{B} \neq \vec{B} \times \vec{A} \quad \text{(cross product is non-commutative)} \tag{2.16}$$

$$\vec{A} \times (\vec{B} + \vec{C}) = \vec{A} \times \vec{B} + \vec{A} \times \vec{C} \quad \text{(cross product is distributive)} \tag{2.17}$$

$$(\vec{A} \times \vec{B}) \times \vec{C} \neq \vec{A} \times (\vec{B} \times \vec{C}) \quad \text{(cross product is non-associative)} \tag{2.18}$$

2.1.4 Vector Differential Operations

Del (Nabla) Operator

The Del or Nabla operator is defined as:

$$\nabla = \hat{\imath}\frac{\partial}{\partial x} + \hat{\jmath}\frac{\partial}{\partial x} + \hat{k}\frac{\partial}{\partial x} \tag{2.19}$$

Gradient of a Scalar

The gradient of a scalar *S* is defined as:

$$\nabla S = \hat{\imath}\frac{\partial s}{\partial x} + \hat{\jmath}\frac{\partial s}{\partial x} + \hat{k}\frac{\partial s}{\partial x} \tag{2.20}$$

∇S is normal to surface generated by the scalar function $S(x, y, z) = \text{constant}$.

Divergence of a Vector

The divergence of a vector $\vec{V}$ is defined as:

$$div\vec{V} = \nabla \cdot \vec{V} = \frac{\partial V_x}{\partial x} + \frac{\partial V_y}{\partial y} + \frac{\partial V_z}{\partial z} \tag{2.21}$$

Note that divergence of a fluid velocity vector can be interpreted as the rate of volumetric strain of fluid. For incompressible fluids, it is zero.

Curl of a Vector

The curl of a vector $\vec{V}$ is defined as

$$curl\,\vec{V} = \nabla \times \vec{V} = \begin{vmatrix} \hat{\imath} & \hat{\jmath} & \hat{k} \\ \frac{\partial}{\partial x} & \frac{\partial}{\partial y} & \frac{\partial}{\partial z} \\ V_x & V_y & V_z \end{vmatrix} = \hat{\imath}\begin{vmatrix} \frac{\partial}{\partial y} & \frac{\partial}{\partial z} \\ V_y & V_z \end{vmatrix} - \hat{\jmath}\begin{vmatrix} \frac{\partial}{\partial x} & \frac{\partial}{\partial z} \\ V_x & V_z \end{vmatrix} + \hat{k}\begin{vmatrix} \frac{\partial}{\partial x} & \frac{\partial}{\partial y} \\ V_x & V_y \end{vmatrix} =$$

$$\hat{\imath}\left(\frac{\partial V_z}{\partial y} - \frac{\partial V_y}{\partial z}\right) - \hat{\jmath}\left(\frac{\partial V_z}{\partial x} - \frac{\partial V_x}{\partial z}\right) + \hat{k}\left(\frac{\partial V_y}{\partial x} - \frac{\partial V_x}{\partial y}\right) \tag{2.22}$$

Note that the curl of fluid velocity vector is equal to local vorticity of fluid which is two times the local angular velocity of fluid. Thus, the curl of fluid velocity vector is a measure of local rotation of fluid.

Laplacian Operator

The Laplacian operator is defined as:

$$\nabla^2 = \frac{\partial^2}{\partial x^2} + \frac{\partial^2}{\partial y^2} + \frac{\partial^2}{\partial z^2} \tag{2.23}$$

Laplacian of a Scalar

The Laplacian of a scalar S is given as:

$$\nabla^2 S = \frac{\partial^2 S}{\partial x^2} + \frac{\partial^2 S}{\partial y^2} + \frac{\partial^2 S}{\partial z^2} \tag{2.24}$$

Laplacian of a Vector

The Laplacian of a vector $\vec{V}$ is given as:

$$\nabla^2 \vec{V} = \hat{\imath}(\nabla^2 V_x) + \hat{\jmath}(\nabla^2 V_y) + \hat{k}(\nabla^2 V_z) = \hat{\imath}\left(\frac{\partial^2 V_x}{\partial x^2} + \frac{\partial^2 V_x}{\partial y^2} + \frac{\partial^2 V_x}{\partial z^2}\right) + \hat{\jmath}\left(\frac{\partial^2 V_y}{\partial x^2} + \frac{\partial^2 V_y}{\partial y^2} + \frac{\partial^2 V_y}{\partial z^2}\right) + \hat{k}\left(\frac{\partial^2 V_z}{\partial x^2} + \frac{\partial^2 V_z}{\partial y^2} + \frac{\partial^2 V_z}{\partial z^2}\right) \tag{2.25}$$

2.2 INDEX NOTATION OF WRITING VECTORS AND VECTOR OPERATIONS

Thus far we have used Gibbs notation to represent vectors and vector operations. An equivalent way of representing vector and vector operations involves the usage of index notation. Any vector $\vec{A}$ in index notation can be expressed as:

$$\vec{A} = \sum_{i=1}^{3} A_i\,\hat{\delta}_i = A_1\hat{\delta}_1 + A_2\hat{\delta}_2 + A_3\hat{\delta}_3 \tag{2.26}$$

where A_i is the component associated with i-coordinate axis and $\hat{\delta}_i$ is a unit vector in the direction of i-coordinate axis. Thus, $A_1, A_2,$ and A_3 are the three components of $\vec{A}$ in the directions of coordinate axes 1,2, and 3, respectively. The corresponding unit vectors in the directions of coordinate axes 1,2, and 3 are $\hat{\delta}_1$, $\hat{\delta}_2$, and $\hat{\delta}_3$, respectively (see Figure 2.2). Index notation is very useful as it allows vector and vector operations to be represented in a compact form.

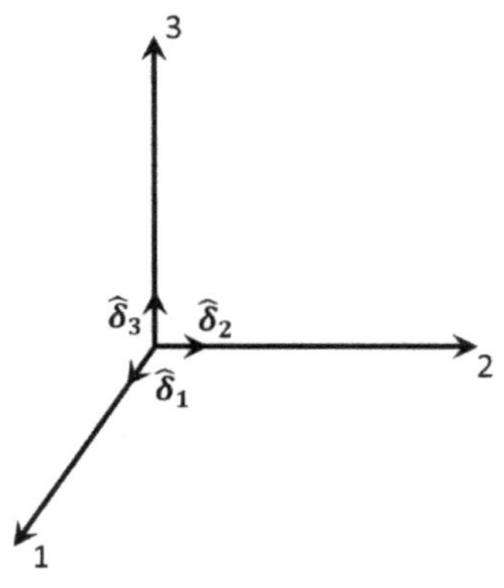

Figure 2.2 Unit vectors in the three coordinate directions.

Many relations in index notation are expressed in terms of two special symbols, namely Kronecker delta and permutation symbol. They are defined below.

2.2.1 Kronecker Delta (δ_{ij})

The Kronecker delta δ_{ij} is defined as:

$$\delta_{ij} = \begin{matrix} 1 & when\ i = j \\ 0 & when\ i \neq j \end{matrix} \tag{2.27}$$

Thus, $\delta_{11} = \delta_{22} = \delta_{33} = 1$. Note that the indices i and j vary from 1 to 3.

2.2.2 Permutation Symbol (ε_{ijk})

The permutation symbol ε_{ijk} is defined as:

$$\varepsilon_{ijk} = \begin{matrix} +1 & if\ ijk\ clockwise\ (123, 231, 312) \\ -1 & if\ ijk\ anti-clockwise\ (132, 321, 213) \\ 0 & if\ any\ two\ or\ three\ \ indices\ are\ alike \end{matrix} \tag{2.28}$$

Interestingly, a 3x3 determinant can be written in terms of the permutation symbol:

$$\begin{vmatrix} a_1 & a_2 & a_3 \\ b_1 & b_2 & b_3 \\ c_1 & c_2 & c_3 \end{vmatrix} = \sum_{i=1}^{3} \sum_{j=1}^{3} \sum_{k=1}^{3} \varepsilon_{ijk}\, a_i b_j c_k \tag{2.29}$$

2.2.3 Dot Product Between Unit Vectors

The dot product (or scalar product) of two unit vectors $\hat{\delta}_i$ and $\hat{\delta}_j$ is given as:

$$\hat{\delta}_i \cdot \hat{\delta}_j = \delta_{ij} \tag{2.30}$$

where δ_{ij} is Kronecker delta ($\delta_{ij} = 1$ when $i = j$; $\delta_{ij} = 0$ when $i \neq j$). Note that this relation covers all nine dot products of unit vectors if we vary the indices i and j from 1 to 3.

2.2.4 Cross Product of Two Unit Vectors

The cross product (or vector product) of two unit vectors $\hat{\delta}_i$ and $\hat{\delta}_j$ is given as:

$$\hat{\delta}_i \times \hat{\delta}_j = \sum_{k=1}^{3} \varepsilon_{ijk}\, \hat{\delta}_k \tag{2.31}$$

in which ε_{ijk} is the permutation symbol $\varepsilon_{ijk} = +1$ if ijk = 123, 231, or 312; $\varepsilon_{ijk} = -1$ if ijk = 321, 132, or 213; $\varepsilon_{ijk} = 0$ if any two or three indices are alike)

2.2.5 Dot Product of Two Vectors

Let vector $\vec{A}$ be $\sum_{i=1}^{3} A_i \hat{\delta}_i$ and vector $\vec{B} = \sum_{j=1}^{3} B_j \hat{\delta}_j$. The dot product of $\vec{A}$ and $\vec{B}$ is:

$$\vec{A} \cdot \vec{B} = \sum_{i=1}^{3} \sum_{j=1}^{3} A_i B_j \left(\hat{\delta}_i \cdot \hat{\delta}_j\right) = \sum_{i=1}^{3} \sum_{j=1}^{3} A_i B_j\, \delta_{ij} = \sum_{i=1}^{3} A_i B_i \tag{2.32}$$

2.2.6 Cross Product of Two Vectors

The cross product of two vectors $\vec{A}$ and $\vec{B}$ is given as:

$$\vec{A} \times \vec{B} = \{\sum_{i=1}^{3} \hat{\delta}_i A_i\} \times \{\sum_{j=1}^{3} \hat{\delta}_j B_j\} = \sum_{i=1}^{3} \sum_{j=1}^{3} A_i B_i \left(\hat{\delta}_i \times \hat{\delta}_j\right)$$
$$= \sum_{i=1}^{3} \sum_{j=1}^{3} \sum_{k=1}^{3} \varepsilon_{ijk}\, A_i B_j \hat{\delta}_k = \sum_{i=1}^{3} \sum_{j=1}^{3} \sum_{k=1}^{3} \varepsilon_{ijk}\, \hat{\delta}_i A_j B_k \tag{2.33}$$

Note that we can interchange the indices of $A_i B_j \hat{\delta}_k$ by multiplying the product with a negative sign.

2.2.7 Divergence of a Vector

The divergence of a vector $\vec{A}$ is given as:

$$div\ \vec{A} = \nabla \cdot \vec{A} = \sum_{i=1}^{3} \sum_{j=1}^{3} \left(\frac{\partial A_j}{\partial x_i}\right) \hat{\delta}_i \cdot \hat{\delta}_j = \sum_{i=1}^{3} \sum_{j=1}^{3} \left(\frac{\partial A_j}{\partial x_i}\right) \delta_{ij}$$
$$= \sum_{i=1}^{3} \left(\frac{\partial A_i}{\partial x_i}\right) \tag{2.34}$$

2.2.8 Curl of a Vector

The curl of a vector $\vec{A}$ is given as:

$$curl\ \vec{A} = \nabla \times \vec{A} = \sum_{i=1}^{3} \sum_{j=1}^{3} \left(\frac{\partial A_j}{\partial x_i}\right) \hat{\delta}_i \times \hat{\delta}_j = \sum_{i=1}^{3} \sum_{j=1}^{3} \sum_{k=1}^{3} \varepsilon_{ijk} \left(\frac{\partial A_j}{\partial x_i}\right) \hat{\delta}_k$$
$$= \sum_{i=1}^{3} \sum_{j=1}^{3} \sum_{k=1}^{3} \varepsilon_{ijk}\ \hat{\delta}_i \left(\frac{\partial A_k}{\partial x_j}\right) \tag{2.35}$$

2.2.9 Laplacian Operator

The Laplacian operator in index notation is:

$$\nabla^2 = \sum_{i=1}^{3} \frac{\partial^2}{\partial x_i^2} \tag{2.36}$$

2.2.10 Laplacian of a Scalar

The Laplacian of a scalar S in index notation is given as:

$$\nabla^2 S = \sum_{i=1}^{3} \frac{\partial^2 S}{\partial x_i^2} \tag{2.37}$$

2.2.11 Laplacian of a Vector

The Laplacian of a vector $\vec{V}$ is in index notation is given as:

$$\nabla^2 \vec{V} = \sum_{i=1}^{3} \hat{\delta}_i \nabla^2 V_i \tag{2.38}$$

2.3 TENSORS AND TENSOR OPERATIONS

2.3.1 Tensor

A tensor is a mathematical entity which has nine components, and which transforms according to certain transformation rules under a change of coordinates. While the components of a tensor vary with the coordinate system used for reference, the tensor is a definite entity with some properties which do not vary with the co-ordinate system used. Such properties are called invariants of a tensor. We cannot give any physical interpretation to a tensor as it is simply a collection of nine components. However, when tensor is operated with other quantities the resultant quantity may have a physical significance.

In physical problems where we need simultaneous specification of two directions to describe the situation requires tensors. For example, consider the velocity gradient $\nabla\vec{V}$. At each point in the flow field, there are three velocity components V_1, V_2, and V_3. Each of these velocity components may vary with distance in the three coordinate directions 1,2, and 3. Therefore, we need to specify nine velocity gradient components at each point in the flow field to describe the velocity gradient $\nabla\vec{V}$ at that point. Thus, $\nabla\vec{V}$ is a tensor quantity given as follows in index notation:

$$\nabla\vec{V} = \sum_{i=1}^{3} \sum_{j=1}^{3} \left(\frac{\partial V_j}{\partial x_i}\right) \hat{\delta}_i\hat{\delta}_j \tag{2.39}$$

Another example where tensor is required to describe the situation is stress. To specify stress vector at any point in the flow field, we must specify the surface vector (area vector). For a surface specified by a unit normal vector $\hat{n}$ (see Figure 2.3), we have a stress vector $\vec{\pi}_{(\hat{n})}$ acting on the surface. Since the surface vector has three components (three unit surfaces in the three coordinate directions), we have to specify stress vectors on each of three unit surfaces. Thus, to specify the state of stress at any point in the flow field, we need to specify nine components of stress, that is, the stress at any point in the flow field is a tensor, given as follows in index notation:

$$\bar{\bar{\pi}} = \sum_{i=1}^{3} \sum_{j=1}^{3} \pi_{ij}\, \hat{\delta}_i\hat{\delta}_j \tag{2.40}$$

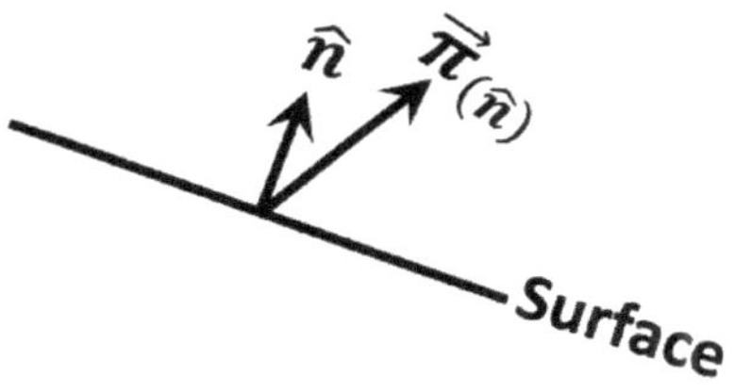

Figure 2.3 Stress vector acting on a surface.

2.3.2 Unit Tensor

The unit tensor is defined as:

$$\bar{\bar{\delta}} = \sum_{i=1}^{3} \sum_{j=1}^{3} \delta_{ij}\, \hat{\delta}_i\hat{\delta}_j \tag{2.41}$$

When a dot product is taken between a vector and unit tensor, the vector is recovered:

$$\vec{A} \cdot \bar{\bar{\delta}} = \vec{A} \tag{2.42}$$

2.3.3 Transpose of a Tensor

Consider a tensor $\bar{\bar{\sigma}}$ given as:

$$\bar{\bar{\sigma}} = \sum_{i=1}^{3} \sum_{j=1}^{3} \sigma_{ij}\, \hat{\delta}_i\hat{\delta}_j \tag{2.43}$$

The transpose of tensor $\bar{\bar{\sigma}}$, denoted as $\bar{\bar{\sigma}}^T$, is given as:

$$\bar{\bar{\sigma}}^T = \sum_{i=1}^{3} \sum_{j=1}^{3} \sigma_{ji} \, \hat{\delta}_i \hat{\delta}_j \quad (2.44)$$

Thus, the transpose of a tensor is obtained by interchanging subscripts from σ_{ij} to σ_{ji}.

2.3.4 Symmetric Tensor

A tensor $\bar{\bar{\sigma}}$ is symmetric if

$$\bar{\bar{\sigma}}^T = \bar{\bar{\sigma}} \quad (2.45)$$

$$\sigma_{ij} = \sigma_{ji} \quad (2.46)$$

2.3.5 Anti-Symmetric Tensor

A tensor $\bar{\bar{\sigma}}$ is anti-symmetric if

$$\bar{\bar{\sigma}}^T = -\bar{\bar{\sigma}} \quad (2.47)$$

$$\sigma_{ij} = -\sigma_{ji} \quad (2.48)$$

Any tensor $\bar{\bar{\sigma}}$ could be split into symmetric and anti-symmetric tensors as follows:

$$\bar{\bar{\sigma}} = \bar{\bar{u}} + \bar{\bar{w}} \quad (2.49)$$

where

$$\bar{\bar{u}} = \frac{1}{2}(\bar{\bar{\sigma}} + \bar{\bar{\sigma}}^T) \quad (2.50)$$

$$\bar{\bar{w}} = \frac{1}{2}(\bar{\bar{\sigma}} - \bar{\bar{\sigma}}^T) \quad (2.51)$$

Here $\bar{\bar{u}}$ is symmetric tensor and $\bar{\bar{w}}$ is anti-symmetric tensor.

2.3.6 Trace of a Tensor

The trace of a tensor $\bar{\bar{\sigma}}$ is defined as:

$$tr(\bar{\bar{\sigma}}) = \sum_{i=1}^{3} \sigma_{ii} \quad (2.52)$$

2.3.7 Invariants of a Tensor

While the components of a tensor vary with the coordinate system used for reference, its invariants do not vary with the coordinate systems. Tensors have three invariants, given below:

First Invariant (I_1)

The first invariant of a tensor $\bar{\bar{\sigma}}$ is as follows:

$$I_1 = \bar{\bar{\sigma}} : \bar{\bar{\delta}} = \sum_{i=1}^{3} \sigma_{ii} \quad (2.53)$$

Second Invariant (I_2)

The second invariant of a tensor $\bar{\bar{\sigma}}$ is as follows:

$$I_2 = \bar{\bar{\sigma}} : \bar{\bar{\sigma}} = \sum_{i=1}^{3} \sum_{j=1}^{3} \sigma_{ii}\, \sigma_{ji} \tag{2.54}$$

Third Invariant (I_3)

The third invariant of a tensor $\bar{\bar{\sigma}}$ is as follows:

$$I_3 = det(\bar{\bar{\sigma}}) = \sum_{i=1}^{3} \sum_{j=1}^{3} \sum_{i=1}^{3} \varepsilon_{ijk} \sigma_{1i} \sigma_{2j}\, \sigma_{3k} \tag{2.55}$$

The invariants of a tensor are important in modeling the flow behavior of materials such as non-Newtonian fluids. For example, the viscosity (a scalar) of a shear-thinning non-Newtonian fluid can be expressed as a function of the second invariant of the rate of deformation tensor.

2.3.8 Dyadic Product of Two Vectors

The dyadic product of two vectors $\vec{A}$ and $\vec{B}$ is a tensor given as:

$$\vec{A}\, \vec{B} = \sum_{i=1}^{3}\ \sum_{j=1}^{3}\ A_i\, B_j\, \hat{\delta}_i\ \hat{\delta}_j \tag{2.56}$$

Note that the two vectors are placed next to one another with no sign in between them. The products $\hat{\delta}_i\ \hat{\delta}_j$ are called unit dyads. There are nine unit dyads.

2.3.9 Operations for Unit Dyads

The operations between tensors involve operations for unit dyads described below:

Double dot product between two unit dyads

$$\hat{\delta}_i\ \hat{\delta}_j : \hat{\delta}_k\ \hat{\delta}_l = (\hat{\delta}_j \cdot \hat{\delta}_k)(\hat{\delta}_i \cdot \hat{\delta}_l) = \delta_{jk}\delta_{il} \tag{2.57}$$

Single dot product between two unit dyads

$$\hat{\delta}_i\ \hat{\delta}_j \cdot \hat{\delta}_k\ \hat{\delta}_l = (\hat{\delta}_j \cdot \hat{\delta}_k)(\hat{\delta}_i\hat{\delta}_l) = \delta_{jk}\hat{\delta}_i\hat{\delta}_l \tag{2.58}$$

Dot product between unit dyad and unit vector

$$\hat{\delta}_i\ \hat{\delta}_j \cdot \hat{\delta}_k\ = \hat{\delta}_i(\hat{\delta}_j \cdot \hat{\delta}_k) = \delta_{jk}\hat{\delta}_i \tag{2.59}$$

$$\hat{\delta}_i \cdot \hat{\delta}_j\hat{\delta}_k\ = (\hat{\delta}_i \cdot \hat{\delta}_j)(\hat{\delta}_k) = \delta_{ij}\hat{\delta}_k \tag{2.60}$$

Cross product between unit dyad and unit vector

$$\hat{\delta}_i\ \hat{\delta}_j \times \hat{\delta}_k\ = \hat{\delta}_i(\hat{\delta}_j \times \hat{\delta}_k) = \sum_{l=1}^{3} \varepsilon_{jkl}\, \hat{\delta}_i\hat{\delta}_l \tag{2.61}$$

$$\hat{\delta}_i \times \hat{\delta}_j\hat{\delta}_k\ = (\hat{\delta}_i \times \hat{\delta}_j)(\hat{\delta}_k) = \sum_{l=1}^{3} \varepsilon_{ijl}\, \hat{\delta}_l\hat{\delta}_k \tag{2.62}$$

2.3.10 Operations for Tensors

Addition of two tensors

The sum of two tensors is a tensor whose components are the additions of the corresponding components of the two tensors as indicated below:

$$\bar{\bar{\sigma}} + \bar{\bar{\pi}} = \sum_{i=1}^{3} \sum_{j=1}^{3} \sigma_{ij}\ \hat{\delta}_i\hat{\delta}_j + \sum_{i=1}^{3} \sum_{j=1}^{3} \pi_{ij}\ \hat{\delta}_i\hat{\delta}_j = \sum_{i=1}^{3} \sum_{j=1}^{3} (\sigma_{ij}+\pi_{ij})\ \hat{\delta}_i\hat{\delta}_j \quad (2.63)$$

Multiplication of a Tensor by a Scalar

The multiplication of a scalar S with a tensor $\bar{\bar{\sigma}}$ is given as:

$$S\bar{\bar{\sigma}} = S\sum_{j=1}^{3} \sigma_{ij}\ \hat{\delta}_i\hat{\delta}_j = \sum_{j=1}^{3} S\sigma_{ij}\ \hat{\delta}_i\hat{\delta}_j \quad (2.64)$$

Each component of a tensor is multiplied with a scalar.

Double Dot Product of Two Tensors

The double dot product between two tensors is given as:

$$\bar{\bar{\sigma}} : \bar{\bar{\pi}} = \sum_{i=1}^{3} \sum_{j=1}^{3} \sigma_{ij}\ \hat{\delta}_i\hat{\delta}_j : \sum_{k=1}^{3} \sum_{l=1}^{3} \pi_{kl}\ \hat{\delta}_k\hat{\delta}_l = \sum_{i=1}^{3} \sum_{j=1}^{3} \sum_{k=1}^{3} \sum_{l=1}^{3} \sigma_{ij}\pi_{kl}(\hat{\delta}_i\hat{\delta}_j : \hat{\delta}_k\hat{\delta}_l) = \sum_{i=1}^{3} \sum_{j=1}^{3} \sum_{k=1}^{3} \sum_{l=1}^{3} \sigma_{ij}\pi_{kl}\delta_{jk}\delta_{il} = \sum_{i=1}^{3} \sum_{j=1}^{3} \sigma_{ij}\pi_{ji} \quad (2.65)$$

Note that the double dot product between two tensors is a scalar product.

Single Dot Product of Two Tensors

$$\bar{\bar{\sigma}} \cdot \bar{\bar{\pi}} = \sum_{i=1}^{3} \sum_{j=1}^{3} \sigma_{ij}\ \hat{\delta}_i\hat{\delta}_j \cdot \sum_{k=1}^{3} \sum_{l=1}^{3} \pi_{kl}\ \hat{\delta}_k\hat{\delta}_l = \sum_{i=1}^{3} \sum_{j=1}^{3} \sum_{k=1}^{3} \sum_{l=1}^{3} \sigma_{ij}\pi_{kl}(\hat{\delta}_i\hat{\delta}_j \cdot \hat{\delta}_k\hat{\delta}_l) = \sum_{i=1}^{3} \sum_{j=1}^{3} \sum_{k=1}^{3} \sum_{l=1}^{3} \sigma_{ij}\pi_{kl}\delta_{jk}\hat{\delta}_i\hat{\delta}_l = \sum_{i=1}^{3} \sum_{j=1}^{3} \sum_{l=1}^{3} \sigma_{ij}\pi_{jl}\ \hat{\delta}_i\hat{\delta}_l \quad (2.66)$$

Dot Product of a Tensor with a Vector

$$\bar{\bar{\pi}} \cdot \vec{V} = \sum_{i=1}^{3} \sum_{j=1}^{3} \pi_{ij}\ \hat{\delta}_i\hat{\delta}_j \cdot \sum_{k=1}^{3} V_k\ \hat{\delta}_k = \sum_{i=1}^{3} \sum_{j=1}^{3} \sum_{k=1}^{3} \pi_{ij}V_k\ (\hat{\delta}_i\hat{\delta}_j \cdot \hat{\delta}_k) = \sum_{i=1}^{3} \sum_{j=1}^{3} \sum_{k=1}^{3} \pi_{ij}V_k\ \delta_{jk}\hat{\delta}_i = \sum_{i=1}^{3} \sum_{j=1}^{3} \pi_{ij}V_j\hat{\delta}_i \quad (2.67)$$

Cross Product of a Tensor with a Vector

$$\bar{\bar{\pi}} \times \vec{V} = \sum_{i=1}^{3} \sum_{j=1}^{3} \pi_{ij}\ \hat{\delta}_i\hat{\delta}_j \times \sum_{k=1}^{3} V_k\ \hat{\delta}_k = \sum_{i=1}^{3} \sum_{j=1}^{3} \sum_{k=1}^{3} \pi_{ij}V_k\ (\hat{\delta}_i\hat{\delta}_j \times \hat{\delta}_k) = \sum_{i=1}^{3} \sum_{j=1}^{3} \sum_{k=1}^{3} \sum_{l=1}^{3} \pi_{ij}V_k\ \varepsilon_{jkl}\hat{\delta}_i\hat{\delta}_l \quad (2.68)$$

Divergence of a Tensor

$$div(\bar{\bar{\pi}}) = \nabla \cdot \bar{\bar{\pi}} = \sum_{i=1}^{3} \hat{\delta}_i \frac{\partial}{\partial x_i} \cdot \sum_{j=1}^{3} \sum_{k=1}^{3} \pi_{jk}\hat{\delta}_j\ \hat{\delta}_k = \sum_{i=1}^{3} \sum_{j=1}^{3} \sum_{k=1}^{3} \frac{\partial \pi_{jk}}{\partial x_i} \hat{\delta}_i \cdot \hat{\delta}_j\hat{\delta}_k = \sum_{i=1}^{3} \sum_{j=1}^{3} \sum_{k=1}^{3} \frac{\partial \pi_{jk}}{\partial x_i} \delta_{ij}\ \hat{\delta}_k = \sum_{i=1}^{3} \sum_{k=1}^{3} \frac{\partial \pi_{ik}}{\partial x_i} \hat{\delta}_k \quad (2.69)$$

2.4 ORDER OF A PRODUCT

We have come across four kinds of vector or tensor products: Dyadic product (no sign), Cross product (×), Dot product (·), and Double dot product (:). The orders associated with different product signs are as follows:

Product type	Sign	Order
Dyadic	No sign	0
Cross	×	-1
Dot	·	-2
Double dot	:	-4

The order of any product between vectors and/or tensors is given as:

$$Order\ of\ a\ Product = \sum order\ of\ individual\ quantities + order\ associated\ with\ the\ product\ sign \tag{2.70}$$

The order of a scalar is zero, the order of a vector is one, and the order of a tensor is two. Thus, the order of a product of vectors and tensors can tell us whether a given product is a scalar, a vector, or a tensor. For example, consider the product $\bar{\bar{\pi}} \cdot \vec{V} \times \vec{A}$. Is this product a scalar, a vector, or a tensor? The sum of the order of individual quantities is 2+1+1=4. The order associated with the signs is -2-1=-3. Thus, the order of the product $\bar{\bar{\pi}} \cdot \vec{V} \times \vec{A}$ is 4-3=1. Consequently, the given product $\bar{\bar{\pi}} \cdot \vec{V} \times \vec{A}$ is a vector.

2.5 GAUSS DIVERGENCE THEOREM

Let V be some closed region bounded by surface S. Let $\vec{F}$ be some vector function which is continuous, and which has continuous first order partial derivatives in some domain containing V, the Gauss divergence theorem states that:

$$\iiint_V (\nabla \cdot \vec{F}) dV = \iint_S (\hat{n} \cdot \vec{F}) dS \tag{2.71}$$

The Gauss divergence theorem is used extensively to convert surface integrals to volume integrals in the derivations of microscopic equations of motion.

Proof

We can re-write the Gauss divergence theorem as:

$$\iiint_V \left(\frac{\partial F_x}{\partial x} + \frac{\partial F_y}{\partial y} + \frac{\partial F_z}{\partial z}\right) dV = \iint_S \left[F_x(\hat{n} \cdot \hat{\imath}) + F_y(\hat{n} \cdot \hat{\jmath}) + F_z(\hat{n} \cdot \hat{k})\right] dS \tag{2.72}$$

If we can prove that:

$$\iiint_V \left(\frac{\partial F_x}{\partial x}\right) dV = \iint_S [F_x(\hat{n} \cdot \hat{\imath})] dS \tag{2.73}$$

$$\iiint_V \left(\frac{\partial F_y}{\partial y}\right) dV = \iint_S [F_y(\hat{n} \cdot \hat{\jmath})] dS \tag{2.74}$$

$$\iiint_V \left(\frac{\partial F_z}{\partial z}\right) dV = \iint_S \left[F_z(\hat{n} \cdot \hat{k})\right] dS \tag{2.75}$$

Then obviously the Gauss divergence theorem is proven. Hence, we will prove only one of the above relations. The other two can be proven in the same way.

Let us prove:

$$\iiint_V \left(\frac{\partial F_z}{\partial z}\right) dV = \iint_S [F_z(\hat{n} \cdot \hat{k})] dS \tag{2.76}$$

Consider a closed spherical region of volume V as shown in Figure 2.4. Let $\vec{F}$ be the vector function which is continuous, and which has continuous first order partial derivatives in the domain containing V. We divide the spherical surface into two portions by a curve along which $\hat{n} \cdot \hat{k} = 0$. The upper hemispherical surface is expressed by an equation $z = h(x, y)$ and the bottom hemispherical surface is expressed by an equation $z = g(x, y)$. The projection of the spherical region on to xy plane is area R^*.

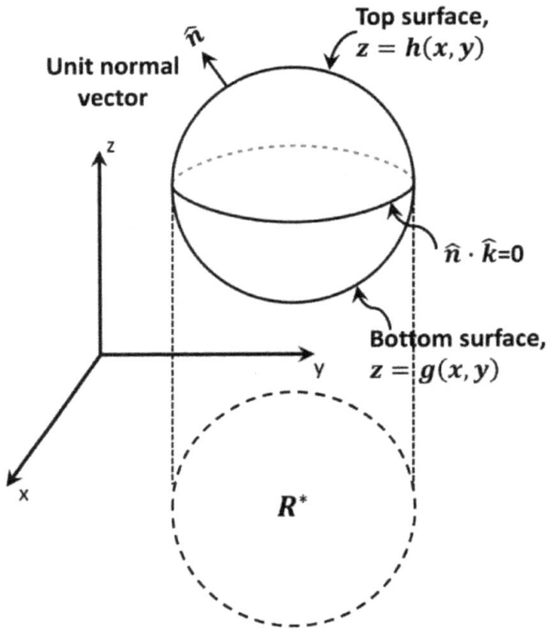

Figure 2.4 A closed region of volume V and its projection in the xy plane.

We now determine the volume integral on the left-hand side of the Gauss divergence theorem:

$$\iiint_V \left(\frac{\partial F_z}{\partial z}\right) dV = \iiint_V \left(\frac{\partial F_z}{\partial z}\right) dxdydz = \iint_{R^*} \left[\int_{z=g(x,y)}^{z=h(x,y)} \left(\frac{\partial F_z}{\partial z}\right) dz\right] dxdy = \iint_{R^*} F_z(x, y, h) dxdy - \iint_{R^*} F_z(x, y, g) dxdy \tag{2.77}$$

The area integrals on the right-hand side are to be evaluated over the projected area R^* in the xy plane. Now we express the projected areas in the xy plane in terms of the surface areas of the hemispheres as:

$$dxdy = (\hat{n} \cdot \hat{k}) dS \qquad \text{(top hemispherical surface)} \tag{2.78}$$

$$dxdy = -(\hat{n} \cdot \hat{k}) dS \qquad \text{(bottom hemispherical surface)} \tag{2.79}$$

Consequently, the volume integral becomes:

$$\iiint_V \left(\frac{\partial F_z}{\partial z}\right) dV = \iint_{S_{top}} F_z(x,y,h)(\hat{n}\cdot\hat{k})dS + \iint_{S_{bottom}} F_z(x,y,g)(\hat{n}\cdot\hat{k})dS = \iint_S [F_z(\hat{n}\cdot\hat{k})]dS \tag{2.80}$$

Hence the Gauss divergence theorem is proved.

Note that the volume region that we considered here is xy simple or xy projectable. A more general proof of the Gauss divergence theorem can be found in advanced engineering mathematics books.

2.6 REYNOLDS TRANSPORT THEOREM

The fundamental laws of mechanics are generally expressed for a material *system* defined as a fixed quantity of matter which maintains its identity as it undergoes a change in conditions. For example, Newton's second law of motion states that "the net force acting on a system is equal to the rate of change of momentum of the system". However, in practical applications, one is more interested in what goes on in a *control volume* defined as an arbitrarily selected volume in space through which mass, momentum, and energy flow. The control volume can be fixed or moving, rigid or deformable.

The Reynolds transport theorem allows the application of laws of mechanics originally expressed for systems to control volumes. It can be stated as follows: The rate of change of any quantity (mass, momentum, or energy) of the system at an instant when the system occupies the given control volume is equal to the instantaneous rate of accumulation of the quantity inside the control volume plus the net efflux of the quantity through the control surface. Thus,

$$\frac{D}{Dt}\iiint_{system}(\rho b)dV = \frac{d}{dt}\iiint_{CV}(\rho b)dV + \iint_{CS}(\rho b)\left(\hat{n}\cdot\vec{W}\right)dS \tag{2.81}$$

The left-hand side is a material time derivative of the system, the first term on the right-hand side is the time derivative of the control volume, the second term on the right-hand side is the net efflux of the quantity through the control surface (surface enclosing the control volume), b is the specific quantity of interest (mass, momentum, or energy per unit mass), ρ is the density of the fluid, $\hat{n}$ is the unit outward normal to the control surface, $\vec{W}$ is the fluid velocity vector relative to control surface, S is the area of the control surface. For fixed but deforming control volume, the Reynolds transport theorem becomes:

$$\frac{D}{Dt}\iiint_{system}(\rho b)dV = \frac{\partial}{\partial t}\iiint_{CV}(\rho b)dV + \iint_{CS}(\rho b)\left(\hat{n}\cdot\vec{W}\right)dS \tag{2.82}$$

If we further assume that the control volume is nondeforming, that is, the control volume is rigid and fixed, the Reynolds transport theorem reduces to:

$$\frac{D}{Dt}\iiint_{system}(\rho b)dV = \iiint_{CV}\frac{\partial}{\partial t}(\rho b)dV + \iint_{CS}(\rho b)\left(\hat{n}\cdot\vec{V}\right)dS \tag{2.83}$$

Note that for a rigid (nondeforming) and fixed control volume, the velocity of control surface is zero ($\vec{V}_{CS} = 0$) and hence, $\vec{W} = \vec{V} - \vec{V}_{CS} = \vec{V}$. It is this form of the Reynolds transport theorem (that is, Equation (2.83)) that is used in the derivation of the microscopic equations of motion in Chapter 4.

Proof

Let the material system be such that at any time t, it occupies the full control volume as shown in Figure 2.5. At time $t + \delta t$, a portion of the system leaves the control volume. Let V_I be the portion of the control volume vacated by the system in time δt. Let V_{II} be the portion of the control volume that still contains part of the system. Let V_{III} be the part of the system that has left the control volume in time δt.

Let B be the quantity of interest. It could be mass, momentum, or energy. Thus,

$$B = \iiint_V (\rho b) dV \tag{2.84}$$

where b is the specific value of B given as $b = B/\text{mass}$. The time rate of change of B for the system is given as:

$$\frac{DB_{system}}{Dt} = \lim_{\delta t \to 0}\left[\frac{B_{system,t+\delta t} - B_{system,t}}{\delta t}\right] = \lim_{\delta t \to 0}\left[\frac{\iiint_{V,system,t+\delta t} \rho b dV - \iiint_{V,system,t} \rho b dV}{\delta t}\right] \tag{2.85}$$

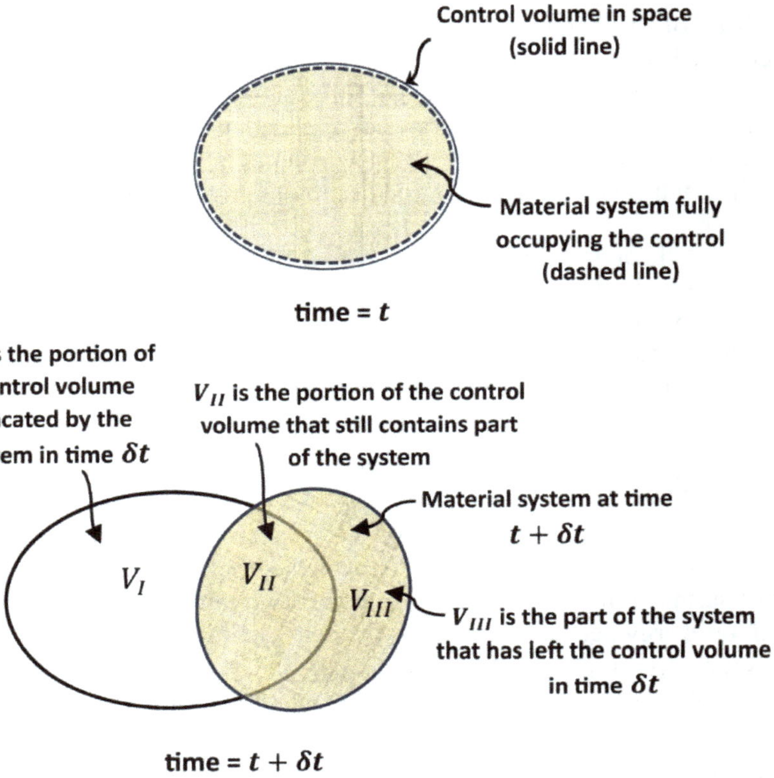

Figure 2.5 Control volume and material system at times t and t+δt.

At time $= t$, $V_{system,t} = V_{CV,t}$ where subscript CV refers to control volume. At time $= t + \delta t$, $V_{system,t+\delta t} = V_{CV,t+\delta t} - V_I + V_{III}$. Thus,

$$\frac{DB_{system}}{Dt} = \lim_{\delta t \to 0}\left[\frac{\iiint_{CV,t+\delta t} \rho b dV + \iiint_{V_{III}} \rho b dV - \iiint_{V_I} \rho b dV - \iiint_{CV,t} \rho b dV}{\delta t}\right] = \lim_{\delta t \to 0}\left[\frac{\iiint_{CV,t+\delta t} \rho b dV - \iiint_{CV,t} \rho b dV}{\delta t}\right] + \lim_{\delta t \to 0}\left[\frac{\iiint_{V_{III}} \rho b dV}{\delta t}\right] - \lim_{\delta t \to 0}\left[\frac{\iiint_{V_I} \rho b dV}{\delta t}\right] \tag{2.86}$$

Each term on the right-hand side of the above equation can be re-written as:

$$\lim_{\delta t \to 0}\left[\frac{\iiint_{CV,t+\delta t} \rho b dV - \iiint_{CV,t} \rho b dV}{\delta t}\right] = \frac{d}{dt}\iiint_{CV} (\rho b) dV \tag{2.87}$$

$\lim_{\delta t \to 0}\left[\frac{\iiint_{V_{III}} \rho b dV}{\delta t}\right] = \iint_{S_{out}} (\rho b)(\hat{n} \cdot \vec{W}) dS$ =Rate at which B leaves the control volume through the control surface portion S_{out} (2.88)

$\lim_{\delta t \to 0}\left[\frac{\iiint_{V_{I}} \rho b dV}{\delta t}\right] = -\iint_{S_{in}} (\rho b)(\hat{n} \cdot \vec{W}) dS$ = Rate at which B enters the control volume through the control surface S_{in} (2.89)

Thus, we can write:

$$\frac{D}{Dt}\iiint_{system} (\rho b) dV = \frac{d}{dt}\iiint_{CV} (\rho b) dV + \iint_{CS} (\rho b)\,(\hat{n} \cdot \vec{W}) dS \tag{2.90}$$

where the second term on the right-hand side is the net efflux of the quantity B through the total control surface CS $(= S_{out} + S_{in})$

2.7 EXAMPLES

2.7.1 Example 1

The vector $\vec{V}$ has the following initial point $P(x, y, z)$: $P(2,0,0)$ and terminal point $Q(x, y, z)$: $Q(3,2,0)$. Find the components and magnitude of $\vec{V}$.

Solution

Components of $\vec{V}$: $V_x = 3 - 2 = 1,\ V_y = 2 - 0 = 2,\ V_z = 0 - 0 = 0$

Magnitude of $\vec{V} = \sqrt{(1)^2 + (2)^2 + (0)^2} = \sqrt{5}$

2.7.2 Example 2

If the initial point of vector $\vec{V}$ in the above problem is $P(x, y, z)$: $P(1, -2, 1)$, find the corresponding terminal point $Q(x, y, z)$.

Solution

$x = -1 + V_x = -1 + 1 = 0, \quad y = -2 + V_y = -2 + 2 = 0, \quad z = 1 + V_z = 1 + 0 = 1$
Thus, $Q(x, y, z)$: $Q(0,0,1)$

2.7.3 Example 3

If the components of a vector $\vec{V}$, that is, V_x, V_y, and V_z, are $(2, -2, 0)$ and the initial point is $P(x, y, z)$: $P(2,0,1)$, find the corresponding terminal point $Q(x, y, z)$ and the magnitude of $\vec{V}$.

Solution

Terminal point $Q(x, y, z)$: $(2 + 2, 0 - 2, 1 + 0) = (4, -2, 1)$

Magnitude of $\vec{V}$: $\sqrt{(2)^2 + (-2)^2 + (0)^2} = \sqrt{8}$

2.7.4 Example 4

If $\vec{A} = 2\hat{\imath} + \hat{\jmath} + 3\hat{k}$, $\vec{B} = \hat{\imath} - 4\hat{k}$, and $\vec{C} = 3\hat{\imath} - \hat{\jmath} + 2\hat{k}$, determine $\vec{A} \cdot \vec{B}$, $\vec{A} + \vec{B}$, and $(\vec{A} + \vec{B}) \cdot \vec{C}$

Solution

$\vec{A} \cdot \vec{B} = 2 + 0 + 3(4) = -10$

$\vec{A} + \vec{B} = 3\hat{\imath} + \hat{\jmath} - \hat{k}$

$(\vec{A} + \vec{B}) \cdot \vec{C} = 9 - 1 - 2 = 6$

2.7.5 Example 5

If $\vec{A} = \hat{\imath} + \hat{\jmath} + \hat{k}$, $\vec{B} = -\hat{\imath} + \hat{\jmath}$, and $\vec{C} = 3\hat{\imath} + \hat{k}$, determine the cosine of the angle between the following vectors: $\vec{A}$ and $\vec{B}$, $(\vec{A} + \vec{B})$ and $\vec{C}$

Solution

Cosine of the angle between the vectors $\vec{A}$ and $\vec{B}$:

$$cos\theta = \frac{\vec{A} \cdot \vec{B}}{|\vec{A}||\vec{B}|}$$

$\vec{A} \cdot \vec{B} = -1 + 1 = 0$, $|\vec{A}| = \sqrt{(1)^2 + (1)^2 + (1)^2} = \sqrt{3}$, , $|\vec{B}| = \sqrt{(-1)^2 + (1)^2 + (0)^2} = \sqrt{2}$

Thus,

$$cos\theta = \frac{\vec{A} \cdot \vec{B}}{|\vec{A}||\vec{B}|} = \frac{0}{\sqrt{3} \times \sqrt{2}} = 0$$

Cosine of the angle between the vectors $(\vec{A} + \vec{B})$ and $\vec{C}$:

$(\vec{A} + \vec{B}) = 2\hat{\jmath} + \hat{k}$, $\vec{C} = 3\hat{\imath} + \hat{k}$, $(\vec{A} + \vec{B}) \cdot \vec{C} = 1$, $|\vec{A} + \vec{B}| = \sqrt{5}$, $|\vec{C}| = \sqrt{10}$

Thus,

$$cos\theta = \frac{(\vec{A} + \vec{B}) \cdot \vec{C}}{|\vec{A} + \vec{B}||\vec{C}|} = \frac{1}{\sqrt{5} \times \sqrt{10}} = \frac{1}{\sqrt{50}}$$

2.7.6 Example 6

Given $\vec{A} = \hat{\imath} + \hat{\jmath}$, $\vec{B} = -\hat{\imath} + 2\hat{\jmath}$, determine $\vec{A} \times \vec{B}$ and $\vec{B} \times \vec{A}$

Solution

$$\vec{A}\times\vec{B}=\begin{vmatrix}\hat{\imath} & \hat{\jmath} & \hat{k}\\ 1 & 1 & 0\\ -1 & 2 & 0\end{vmatrix}=\hat{\imath}\begin{vmatrix}1 & 0\\ 2 & 0\end{vmatrix}-\hat{\jmath}\begin{vmatrix}1 & 0\\ -1 & 0\end{vmatrix}+\hat{k}\begin{vmatrix}1 & 1\\ -1 & 2\end{vmatrix}=3\hat{k}$$

$$\vec{B}\times\vec{A}=\begin{vmatrix}\hat{\imath} & \hat{\jmath} & \hat{k}\\ -1 & 2 & 0\\ 1 & 1 & 0\end{vmatrix}=\hat{\imath}\begin{vmatrix}2 & 0\\ 1 & 0\end{vmatrix}-\hat{\jmath}\begin{vmatrix}-1 & 0\\ 1 & 0\end{vmatrix}+\hat{k}\begin{vmatrix}-1 & 2\\ 1 & 1\end{vmatrix}=-3\hat{k}$$

2.7.7 Example 7

Using index notation, prove the following:

i) $\bar{\bar{\delta}}\cdot\vec{V}=\vec{V}\cdot\bar{\bar{\delta}}=\vec{V}$

ii) $S\bar{\bar{\delta}}:\nabla\vec{V}=S(\nabla\cdot\vec{V})$

iii) $\nabla\cdot S\bar{\bar{\delta}}=\nabla S$

Solution

i) $\bar{\bar{\delta}}\cdot\vec{V}=\vec{V}\cdot\bar{\bar{\delta}}=\vec{V}$

LHS:

$$\bar{\bar{\delta}}\cdot\vec{V}=\sum_i\sum_j\delta_{ij}\,\hat{\delta}_i\hat{\delta}_j\cdot\sum_k V_k\hat{\delta}_k=\sum_i\sum_j\sum_k\delta_{ij}\,V_k\left(\hat{\delta}_i\hat{\delta}_j\cdot\hat{\delta}_k\right)=\sum_i\sum_j\sum_k\delta_{ij}\,\delta_{jk}\,V_k\hat{\delta}_i$$

For non-zero values, $i=j=k$. Thus,

$$\bar{\bar{\delta}}\cdot\vec{V}=\sum_i V_i\hat{\delta}_i$$

RHS:

$$\vec{V}=\sum_i V_i\hat{\delta}_i$$

LHS=RHS. Hence proved.

ii) $S\bar{\bar{\delta}}:\nabla\vec{V}=S(\nabla\cdot\vec{V})$

LHS:

$$S\bar{\bar{\delta}}:\nabla\vec{V}=S\sum_i\sum_j\delta_{ij}\,\hat{\delta}_i\hat{\delta}_j:\sum_k\frac{\partial}{\partial x_k}\hat{\delta}_k\sum_l V_l\,\hat{\delta}_l=S\sum_i\sum_j\sum_k\sum_l\delta_{ij}\left(\frac{\partial V_l}{\partial x_k}\right)\left(\hat{\delta}_i\hat{\delta}_j:\hat{\delta}_k\hat{\delta}_l\right)$$

$$= S\sum_i\sum_j\sum_k\sum_l \delta_{ij}\left(\frac{\partial V_l}{\partial x_k}\right)(\hat{\delta}_j\cdot\hat{\delta}_k)(\hat{\delta}_i\cdot\hat{\delta}_l) = S\sum_i\sum_j\sum_k\sum_l \delta_{ij}\left(\frac{\partial V_l}{\partial x_k}\right)(\delta_{jk})(\delta_{il})$$

For non-zero values, $i = j = k = l$. Thus,

$$S\bar{\bar{\delta}}:\nabla\vec{V} = S\sum_i\left(\frac{\partial V_i}{\partial x_i}\right)$$

RHS:

$$S(\nabla\cdot\vec{V}) = S\sum_i\sum_j\frac{\partial}{\partial x_i}\hat{\delta}_i\cdot V_j\hat{\delta}_j = S\sum_i\sum_j\frac{\partial V_j}{\partial x_i}(\hat{\delta}_i\cdot\hat{\delta}_j) = S\sum_i\sum_j\frac{\partial V_j}{\partial x_i}\delta_{ij}$$

For non-zero values, $i = j$. Thus,

$$S(\nabla\cdot\vec{V}) = S\sum_i\frac{\partial V_i}{\partial x_i}$$

LHS=RHS. Hence proved.

iii) $\nabla\cdot S\bar{\bar{\delta}} = \nabla S$

LHS:

$$\nabla\cdot S\bar{\bar{\delta}} = \sum_i\frac{\partial}{\partial x_i}\hat{\delta}_i\cdot S\sum_j\sum_k\delta_{jk}\hat{\delta}_j\hat{\delta}_k = \sum_i\sum_j\sum_k\frac{\partial S}{\partial x_i}\delta_{jk}\,\hat{\delta}_i\cdot\hat{\delta}_j\hat{\delta}_k = \sum_i\sum_j\sum_k\frac{\partial S}{\partial x_i}\delta_{jk}\,\delta_{ij}\hat{\delta}_k$$

For non-zero values, $i = j = k$. Thus,

$$\nabla\cdot S\bar{\bar{\delta}} = \sum_i\frac{\partial S}{\partial x_i}\hat{\delta}_i$$

RHS:

$$\nabla S = \sum_i\frac{\partial}{\partial x_i}\hat{\delta}_i S = \sum_i\frac{\partial S}{\partial x_i}\hat{\delta}_i$$

LHS=RHS. Hence proved.

2.7.8 Example 8

Expand the following double-dot products of dyads in terms of the components:

$\vec{U}\vec{V}:\vec{W}\vec{S}$

$\vec{W}\vec{S}:\vec{U}\vec{V}$

Is $\vec{U}\vec{V}:\vec{W}\vec{S} = \vec{W}\vec{S}:\vec{U}\vec{V}$?

Solution

Expansion of $\vec{U}\vec{V}:\vec{W}\vec{S}$

$$\vec{U}\vec{V}:\vec{W}\vec{S} = \sum_i\sum_j\sum_k\sum_l U_iV_j\,\hat{\delta}_i\hat{\delta}_j : W_kS_l\hat{\delta}_k\hat{\delta}_l = \sum_i\sum_j\sum_k\sum_l U_iV_jW_kS_l\,(\hat{\delta}_i\hat{\delta}_j : \hat{\delta}_k\hat{\delta}_l)$$
$$= \sum_i\sum_j\sum_k\sum_l U_iV_jW_kS_l\,\delta_{jk}\delta_{il}$$

For non-zero values, $j = k$ and $i = l$ Thus,

$$\vec{U}\vec{V}:\vec{W}\vec{S} = \sum_i\sum_j U_iV_jW_jS_i$$

Expansion of $\vec{W}\vec{S}:\vec{U}\vec{V}$

$$\vec{W}\vec{S}:\vec{U}\vec{V} = \sum_i\sum_j\sum_k\sum_l W_iS_j\,\hat{\delta}_i\hat{\delta}_j : U_kV_l\hat{\delta}_k\hat{\delta}_l = \sum_i\sum_j\sum_k\sum_l W_iS_j\,U_kV_l(\hat{\delta}_i\hat{\delta}_j : \hat{\delta}_k\hat{\delta}_l)$$
$$= \sum_i\sum_j\sum_k\sum_l W_iS_jU_kV_l\,\delta_{jk}\delta_{il}$$

For non-zero values, $j = k$ and $i = l$ Thus,

$$\vec{W}\vec{S}:\vec{U}\vec{V} = \sum_i\sum_j W_iS_j\,U_jV_i = \sum_i\sum_j W_jS_i\,U_iV_j$$

Is $\vec{U}\vec{V}:\vec{W}\vec{S} = \vec{W}\vec{S}:\vec{U}\vec{V}$?

Upon comparison of the expansions of $\vec{U}\vec{V}:\vec{W}\vec{S}$ and $\vec{W}\vec{S}:\vec{U}\vec{V}$, we find that they are the same. Thus,

$$\vec{U}\vec{V}:\vec{W}\vec{S} = \vec{W}\vec{S}:\vec{U}\vec{V}$$

2.7.9 Example 9

Given the vector:

$$\vec{A} = 2x_1\hat{\delta}_1 + (-2x_2 + 2x_2x_3)\hat{\delta}_2 - x_3^2\hat{\delta}_3$$

Find the divergence of $\vec{A}$

Solution

$$\vec{A} = A_1\hat{\delta}_1 + A_2\hat{\delta}_2 + A_3\hat{\delta}_3 = 2x_1\hat{\delta}_1 + (-2x_2 + 2x_2x_3)\hat{\delta}_2 - x_3^2\hat{\delta}_3$$

$$A_1 = 2x_1,\ A_2 = (-2x_2 + 2x_2x_3),\ A_3 = -x_3^2$$

$$\text{div}\vec{A} = \nabla\cdot\vec{A} = \frac{\partial A_1}{\partial x_1} + \frac{\partial A_2}{\partial x_2} + \frac{\partial A_3}{\partial x_3} = (2) + (-2 + 2x_3) + (-2x_3)$$

Thus,

$\text{div}\vec{A} = 0$

2.7.10 Example 10

Given the vector:

$\vec{A} = 2x_1\hat{\delta}_1 - 2x_2\hat{\delta}_2 + 2x_1x_2\hat{\delta}_3$

Find the curl of $\vec{A}$

Solution

$$\text{Curl}\,\vec{A} = \begin{vmatrix} \hat{\delta}_1 & \hat{\delta}_2 & \hat{\delta}_3 \\ \dfrac{\partial}{\partial x_1} & \dfrac{\partial}{\partial x_2} & \dfrac{\partial}{\partial x_3} \\ A_1 & A_2 & A_3 \end{vmatrix} = \hat{\delta}_1 \begin{vmatrix} \dfrac{\partial}{\partial x_2} & \dfrac{\partial}{\partial x_3} \\ A_2 & A_3 \end{vmatrix} - \hat{\delta}_2 \begin{vmatrix} \dfrac{\partial}{\partial x_1} & \dfrac{\partial}{\partial x_3} \\ A_1 & A_3 \end{vmatrix} + \hat{\delta}_3 \begin{vmatrix} \dfrac{\partial}{\partial x_1} & \dfrac{\partial}{\partial x_2} \\ A_1 & A_2 \end{vmatrix}$$

$$= \hat{\delta}_1 \left[\frac{\partial A_3}{\partial x_2} - \frac{\partial A_2}{\partial x_3}\right] - \hat{\delta}_2 \left[\frac{\partial A_3}{\partial x_1} - \frac{\partial A_1}{\partial x_3}\right] + \hat{\delta}_3 \left[\frac{\partial A_2}{\partial x_1} - \frac{\partial A_1}{\partial x_2}\right]$$

$A_1 = 2x_1,\ A_2 = (-2x_2),\ A_3 = 2x_1x_2$

$\frac{\partial A_1}{\partial x_2} = 0\ ,\ \frac{\partial A_1}{\partial x_3} = 0\ ,\ \frac{\partial A_2}{\partial x_1} = 0\ ,\ \frac{\partial A_2}{\partial x_3} = 0\ ,\ \frac{\partial A_3}{\partial x_2} = 2x_1\ ,\ \frac{\partial A_3}{\partial x_1} = 2x_2$

Thus,

$\text{Curl}\,\vec{A} = \hat{\delta}_1[2x_1 - 0] - \hat{\delta}_2[2x_2 - 0] + \hat{\delta}_3[0 - 0] = 2x_1\hat{\delta}_1 - 2x_2\hat{\delta}_2$

2.7.11 Example 11

Given

$\rho = 3x_2x_3$

$\vec{A} = 4x_1\hat{\delta}_1 - x_2\hat{\delta}_2 - x_3\hat{\delta}_3$

Find the surface integral:

$$\int_S (\rho\hat{n}\cdot\vec{A})\,dS$$

where $\hat{n}$ is a unit normal vector to the surface.

Solution

Using the Gauss divergence theorem

$$\int_S (\rho\hat{n}\cdot\vec{A})\,dS = \int_S (\hat{n}\cdot\rho\vec{A})\,dS = \int_V \nabla\cdot(\rho\vec{A})dV$$

Let us determine $\nabla\cdot(\rho\vec{A})$:

$$\rho\vec{A} = 12x_1x_2x_3\hat{\delta}_1 - 3x_2^2x_3\hat{\delta}_2 - 3x_2x_3^2\hat{\delta}_3$$

$$\nabla\cdot(\rho\vec{A}) = \frac{\partial}{\partial x_1}[12x_1x_2x_3] + \frac{\partial}{\partial x_2}[-3x_2^2x_3] + \frac{\partial}{\partial x_3}[-3x_2x_3^2]$$

$$= 12x_2x_3 - 6x_2x_3 - 6x_2x_3 = 0$$

Thus,

$$\int_S (\rho\hat{n}\cdot\vec{A})\,dS = \int_V \nabla\cdot(\rho\vec{A})dV = 0$$

2.7.12 Example 12

Given the following vectors and tensor:

$$\vec{A} = 3\hat{\delta}_1 + 4\hat{\delta}_3$$

$$\vec{B} = 2\hat{\delta}_2 - 6\hat{\delta}_3$$

$$\bar{\bar{D}} = 3\hat{\delta}_1\hat{\delta}_1 + 2\hat{\delta}_1\hat{\delta}_3 - 4\hat{\delta}_2\hat{\delta}_2 - 5\hat{\delta}_3\hat{\delta}_2$$

Compute $\vec{A}\cdot\bar{\bar{D}}\cdot\vec{B}$

Solution

Use index notation to expand $\vec{A}\cdot\bar{\bar{D}}\cdot\vec{B}$ as follows:

$$\vec{A}\cdot\bar{\bar{D}}\cdot\vec{B} = \sum_i A_i\hat{\delta}_i\cdot\sum_j\sum_k D_{jk}\,\hat{\delta}_j\hat{\delta}_k\cdot\sum_l B_l\hat{\delta}_l = \sum_i\sum_j\sum_k\sum_l A_iD_{jk}B_l(\hat{\delta}_i\cdot\hat{\delta}_j\hat{\delta}_k\cdot\hat{\delta}_l)$$
$$= \sum_i\sum_j\sum_k\sum_l A_iD_{jk}B_l\,\delta_{ij}\delta_{kl}$$

For non-zero values, $i = j$ and $k = l$ Thus,

$$\vec{A}\cdot\bar{\bar{D}}\cdot\vec{B} = \sum_i\sum_k A_iD_{ik}B_k = \sum_i A_iD_{i1}B_1 + A_iD_{i2}B_2 + A_iD_{i3}B_3$$
$$= [A_1D_{11}B_1 + A_2D_{21}B_1 + A_3D_{31}B_1] + [A_1D_{12}B_2 + A_2D_{22}B_2 + A_3D_{32}B_2]$$
$$+ [A_1D_{13}B_3 + A_2D_{23}B_3 + A_3D_{33}B_3]$$

Given that:

$$A_1 = 3,\ \ A_2 = 0,\ A_3 = 4$$

$B_1 = 0, \quad B_2 = 2, \quad B_3 = -6$

$D_{11} = 3, \quad D_{13} = 2, \quad D_{22} = -4, \quad D_{32} = -5$

Thus,

$$\vec{A} \cdot \bar{\bar{D}} \cdot \vec{B} = [0 + 0 + 0] + [0 + 0 + 4(-5)(2)] + [3(2)(-6) + 0 + 0] = -76$$

2.7.13 Example 13

An initially empty tank of diameter 8 ft is being filled with water from a tap, 1 inch in diameter. The rate of flow of water from the tap is steady at 8 ft^3/min. Estimate the time rate of change of the height of water in the tank, dh/dt, at any instant.

Solution

Refer to the figure given below.

This problem involves a fixed but deforming control volume. The fluid velocity and control surface velocity are given as:

$$\vec{V} = -\left(\frac{Q}{A_{tap}}\right)\hat{k}$$

$$\vec{V}_{CS} = \left(\frac{dh}{dt}\right)\hat{k}$$

where $Q = 8 \ \text{ft}^3/\text{min}$ and A_{tap} is the cross-sectional area of the water tap.

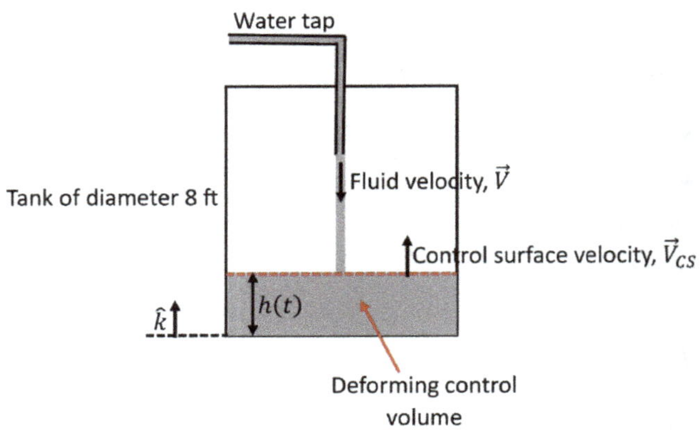

$$\vec{V} = -\left(\frac{Q}{A_{tap}}\right)\hat{k}$$

$$\vec{V}_{CS} = \left(\frac{dh}{dt}\right)\hat{k}$$

Mass balance using the Reynolds transport theorem for fixed but deforming control volume gives:

$\frac{D}{Dt}\iiint_{system}(\rho)dV = \frac{\partial}{\partial t}\iiint_{CV}(\rho)dV + \iint_{CS}(\rho)\left(\hat{n}\cdot\vec{W}\right)\mathrm{dS} = 0$

The fluid velocity relative to the control surface ($\vec{W}$) is:

$\vec{W} = \vec{V} - \vec{V}_{CS} = -\left(\frac{Q}{A_{tap}}\right)\hat{k} - \left(\frac{dh}{dt}\right)\hat{k} = -\left[\left(\frac{Q}{A_{tap}}\right) + \left(\frac{dh}{dt}\right)\right]\hat{k}$

We can now evaluate the volume and surface integrals as follows:

$$\frac{\partial}{\partial t}\iiint_{CV}(\rho)dV = \frac{\partial}{\partial t}(A_{tank}\rho h) = A_{tank}\rho\left(\frac{dh}{dt}\right)$$

$$\iint_{CS}(\rho)\left(\hat{n}\cdot\vec{W}\right)\mathrm{dS} = -\rho A_{tap}\left[\left(\frac{Q}{A_{tap}}\right) + \left(\frac{dh}{dt}\right)\right]$$

Thus, mass balance becomes:

$$\frac{D}{Dt}\iiint_{system}(\rho)dV = A_{tank}\rho\left(\frac{dh}{dt}\right) - \rho A_{tap}\left[\left(\frac{Q}{A_{tap}}\right) + \left(\frac{dh}{dt}\right)\right] = 0$$

Upon re-arrangement, we get:

$$\left(\frac{dh}{dt}\right) = \frac{Q}{A_{tank} - A_{tap}}$$

As $Q = 8\ \mathrm{ft}^3/\mathrm{min}$, $A_{tank} = \frac{\pi}{4}(8)^2 = 50.27\ \mathrm{ft}^2$, $A_{tap} = \frac{\pi}{4}\left(\frac{1}{12}\right)^2 = 5.45\times10^{-3}\ \mathrm{ft}^2$, we get:

$$\left(\frac{dh}{dt}\right) = \frac{8}{50.27 - 5.45\times10^{-3}} = 0.159\ \mathrm{ft/min}$$

Note that as $A_{tap} \ll A_{tank}$, we could have neglected A_{tap} in the estimation of $\left(\frac{dh}{dt}\right)$.

NOTATION

$\vec{A}$ any vector

A_x, A_y, A_z components of vector $\vec{A}$

A_i component of vector $\vec{A}$ in index notation

$|\vec{A}|$ magnitude of vector $\vec{A}$

$\vec{A}\vec{B}$ dyadic product between two vectors $\vec{A}$ and $\vec{B}$

$\vec{A}\cdot\vec{B}$ dot or scalar product between two vectors $\vec{A}$ and $\vec{B}$

$\vec{A}\times\vec{B}$ cross or vector product between two vectors $\vec{A}$ and $\vec{B}$

b specific quantity of interest (mass, momentum, or energy per unit mass)

CS control surface

CV control volume

$curl(\vec{A})$ curl of any vector $\vec{A}$
$div(\vec{A})$ divergence of any vector $\vec{A}$
$det(\bar{\bar{\sigma}})$ determinant of any tensor $\bar{\bar{\sigma}}$
$\vec{F}$ some vector function
I_1, I_2, I_3 three invariants of a tensor
$\hat{\imath}, \hat{\jmath}, \hat{k}$ unit vectors in cartesian coordinate system
$\hat{n}$ unit outward normal to the control surface
S area of control surface
$tr(\bar{\bar{\sigma}})$ trace of any tensor $\bar{\bar{\sigma}}$
V volume
$\vec{V}$ fluid velocity vector
$\vec{W}$ fluid velocity vector relative to control surface

Greek Letters

$\hat{\delta}_1, \hat{\delta}_2, \hat{\delta}_3$ unit vectors in directions 1, 2, and 3.
$\hat{\delta}_i$ unit vector in index notation
δ_{ij} Kronecker delta
$\hat{\delta}_i\hat{\delta}_j$ unit dyad, dyadic product between two unit vectors
$\bar{\bar{\delta}}$ unit tensor
ε_{ijk} permutation symbol
σ_{ij} component of a tensor $\bar{\bar{\sigma}}$
$\bar{\bar{\sigma}}$ any tensor
$\bar{\bar{\sigma}}^T$ transpose of a tensor $\bar{\bar{\sigma}}$

Mathematical symbols and Operators

∇ del operator
∇^2 Laplacian operator
$\times$ cross or vector product sign
$\cdot$ single dot product sign
$:$ double dot product sign

PRACTICE PROBLEMS

Problem 1

The initial point of a vector $\vec{A}$ is $(-1.-1.-1)$ and the terminal point is $(3,0,0)$. Find the components A_x, A_y, A_z of the vector and its magnitude.

Problem 2

The initial point of a vector $\vec{A}$ is $(-6.-2.-1)$. The components of the vector $\vec{A}$ are: $A_x = 6, A_y = 2, A_z = 1$. Find the corresponding terminal point of the vector and its magnitude.

Problem 3

Given that $\vec{A} = 2\hat{\imath} + \hat{\jmath} + 3\hat{k}$, $\vec{B} = \hat{\imath} - 4\hat{k}$, and $\vec{C} = 3\hat{\imath} - \hat{\jmath} + 2\hat{k}$, determine $(\vec{A} - \vec{C}) \cdot \vec{B}$ and $\vec{A} \cdot \vec{B} - \vec{C} \cdot \vec{B}$. Are they the same?

Problem 4

Given that $\vec{A} = \hat{\imath} + \hat{\jmath} + \hat{k}$, $\vec{B} = -\hat{\imath} + \hat{\jmath}$, and $\vec{C} = 3\hat{\imath} + \hat{k}$, determine the cosine of the angle between the vectors (i) $\vec{A}$ and $\vec{B} + \vec{C}$ and (ii) $\vec{A} + \vec{C}$ and $\vec{A} - \vec{C}$

Problem 5

Given that $\vec{A} = 4\hat{\imath} - \hat{k}$ and $\vec{B} = -2\hat{\imath} + \hat{\jmath} + 3\hat{k}$, determine $\vec{A} \times \vec{B}$

Problem 6

Given the vector:

$$\vec{A} = x_1\hat{\delta}_1 + (-x_2 + 5x_2x_3)\hat{\delta}_2 - 5x_3^2\hat{\delta}_3$$

Find the divergence of $\vec{A}$

Problem 7

Given the vector:

$$\vec{A} = x_1\hat{\delta}_1 + (-x_2 + 5x_2x_3)\hat{\delta}_2 + 3x_1x_2\hat{\delta}_3$$

Find the curl of $\vec{A}$

Problem 8

Determine the order of the following products:

$$\vec{A} \cdot \vec{B},\ \bar{\bar{A}} : \nabla\vec{B}, \bar{\bar{A}} \cdot \vec{B}, \vec{A} \cdot \nabla\vec{B}$$

Problem 9

Use the index notation to prove:

$$\vec{A}\vec{B} : \vec{C}\vec{D} = \vec{A}\vec{C} : \vec{B}\vec{D} = (\vec{A} \cdot \vec{D})(\vec{B} \cdot \vec{C})$$

Problem 10

Determine

$$\iint_S (\hat{n} \cdot \vec{F})dS$$

where $\vec{F} = (3x + z)\hat{\imath} + (y^2 - sinx^2z)\hat{\jmath} + (xz + ye^{x^5})\hat{k}$, $\hat{n}$ is a unit vector normal to the surface S, and $0 \le x \le 1,\ 0 \le y \le 3,\ 0 \le z \le 2$.

FURTHER READING

1. R.B. Bird, W.E. Stewart, E.N. Lightfoot. *Transport Phenomena*. 2nd Edition, Wiley, New York, 2007.
2. R. Pal. *Rheology of Particulate Dispersions and Composites*. CRC Press, Boca Raton, 2007.
3. R.L. Street, G.Z. Watters, J.K. Vannard. *Elementary Fluid Mechanics*. 7th Edition, Wiley, New York, 1996.
4. B.R. Munson, D.F. Young, T.H. Okiishi. *Fundamentals of Fluid Mechanics*. 4th Edition, Wiley, New York, 2002.

Part 2

Momentum Transport (Fluid Mechanics)

Chapter 3

Introduction to Fluid Mechanics

3.1 CONCEPT OF A FLUID

From fluid mechanics perspective, matter exists in two states: solid and fluid (liquid or gas). The distinction between the two states is based on their response to an applied shear (tangential) stress. Solids are materials which undergo a fixed amount of deformation (shear strain) under the application of shear stress (see Figure 3.1). Thus, solids do not flow under the application of shear stress. A fluid on the other hand is a material that undergoes continuous deformation (rate of shear strain) under the influence of shear stress. Thus, fluids flow under the application of shear stress, no matter how small the shear stress is (see Figure 3.2). For example, honey is a fluid that spreads (flows) readily on bread under the application of shear by a knife. Likewise, paint is a fluid that spreads (flows) readily on the wall under the shearing action of a brush.

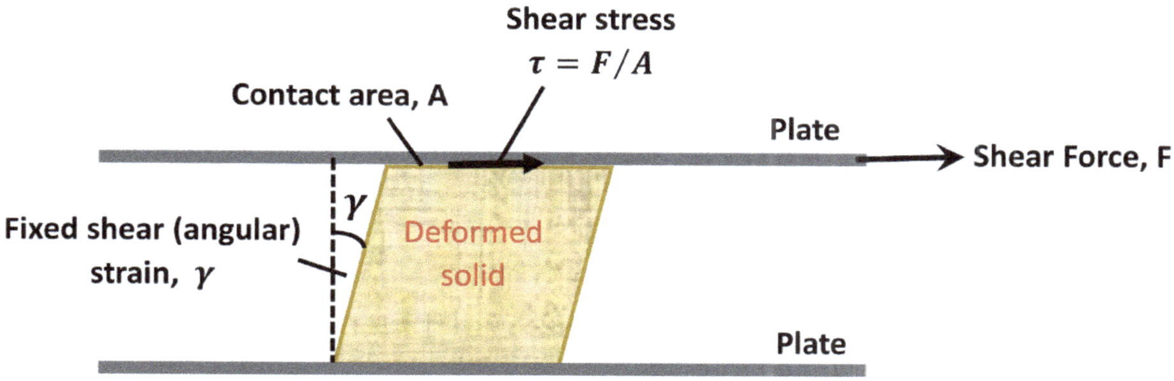

Figure 3.1 Solid undergoes a fixed amount of shear strain (angular deformation) under the application of shear force or shear stress.

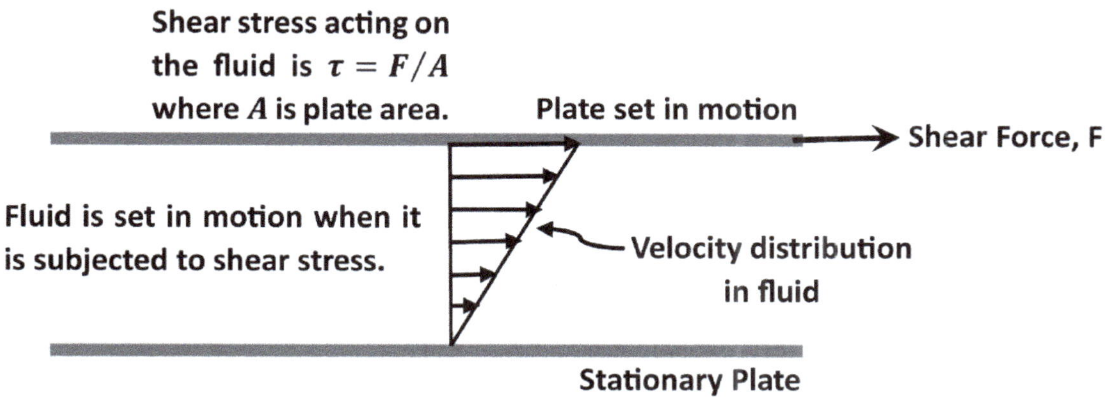

Figure 3.2 Fluid undergoes continuous deformation (flows) under the application of shear force or shear stress. Thus, it is the rate of shear strain $\dot{\gamma}$ that is important in fluids.

3.2 CONTINUUM HYPOTHESIS

Fluid like any other matter is discrete or discontinuous at the microscopic level. It is composed of molecules with empty space in between the molecules. Solids, liquids, and gases differ in the average spacing of molecules. The molecules of solids are more closely packed as compared with the molecules of liquids. In gases, the average spacing between the molecules is much larger as compared with solids and liquids.

In the analysis of fluid motion, the discrete molecular structure of fluid is ignored, and it is treated as a continuous medium. This hypothesis that fluid is a continuum is referred to as continuum hypothesis. At each point of this continuous medium, one can assign unique values of flow field variables. Thus, the flow field variables are continuous functions of the space co-ordinates and time. At any point in space, the values of the field variables can be regarded as the average values of the various molecules which occupy a small volume in about that point.

For continuum hypothesis to be valid, it is necessary that the average spacing between the molecules be very small in comparison with the characteristic size of the flow system. In most situations encountered in engineering and science, this condition is readily met.

3.3 LAGRANGIAN AND EULERIAN DESRIPTIONS OF FLUID FLOW

There are two basic methods of describing the motion of fluids: Lagrangian and Eulerian methods.
In the *Lagrangian* description (also referred to as material or system description), the motion of each fluid particle is tracked to observe changes in its properties. Each particle of the continuum is labeled, usually by specifying its spatial co-ordinates at some initial time. Then, the path, velocity etc. of each individual particle are traced as time passes.

Suppose that at some initial time $t = t_o$, the fluid particle occupies a position described by its position vector $\vec{R}$ measured from the origin of some fixed frame of reference (see Figure 3.3). Let the position vector of the fluid particle sometime later, at time t, be $\vec{r}$. Then the equation of the form:

$$\vec{r} = \vec{r}(\vec{R}, t) \quad \text{with} \quad \vec{r}(\vec{R}, t_o) = \vec{R} \tag{3.1}$$

describes the path of every fluid particle, which, at time $t = t_o$ is located at $\vec{r} = \vec{R}$. Obviously, $\vec{R}$ is different for different fluid particles.

Equation (3.1) can also be written in component form if we take $\vec{R}$ as:

$$\vec{R} = X\hat{\imath} + Y\hat{\jmath} + Z\hat{k} \tag{3.2}$$

and $\vec{r}$ as

$$\vec{r} = x\hat{\imath} + y\hat{\jmath} + z\hat{k} \tag{3.3}$$

where $\hat{\imath}, \hat{\jmath}, \hat{k}$ are unit vectors in the three coordinate directions. Thus,

$$x = x(X, Y, Z, t) \tag{3.4a}$$

$$y = y(X, Y, Z, t) \tag{3.4b}$$

$$z = z(X, Y, Z, t) \tag{3.4c}$$

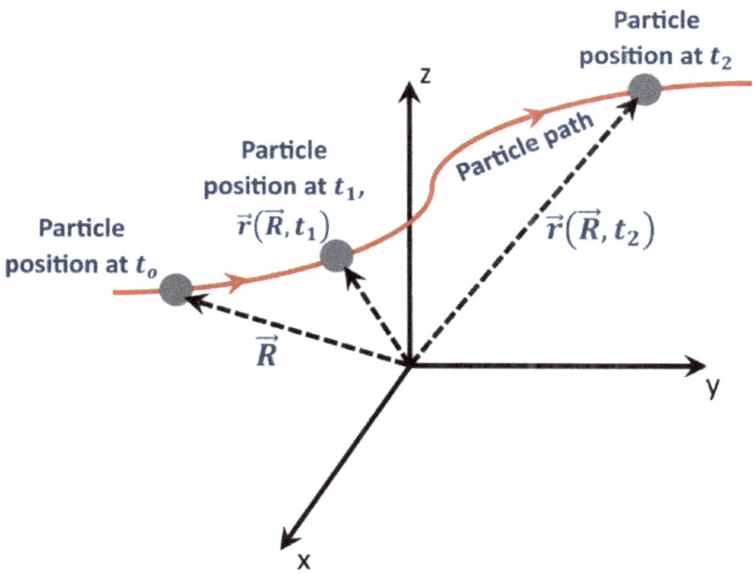

Figure 3.3 Lagrangian description of fluid motion.

In Equation (3.4), the co-ordinates (X, Y, Z) serve to identify the different particles of the fluid and are known as material co-ordinates. Equation (3.1) or Equation (3.4) is said to define the motion of a fluid. An important point to note is that in the Lagrangian or material description, the independent variables are: X, Y, Z, and t, where X, Y, Z are the co-ordinates of the particle at some reference time $t = t_o$.

As the fluid undergoes motion, physical quantities (other than the position vector already discussed) associated with the individual fluid particles change with time. All these physical quantities can be expressed as functions of time and material co-ordinates X, Y, Z of the individual particles. Thus,

$$F = F(X, Y, Z, t) \tag{3.5}$$

where F is any physical quantity (scalar, vector, or tensor) associated with the fluid particle.

In contrast to the Lagrangian approach, the *Eulerian* method does not trace individual fluid particles when the fluid undergoes motion. It rather describes the changes at every fixed point (or control volume) in the space occupied by the fluid as a function of time. For example, the velocity field in Eulerian description can be written as: $\vec{V} = \vec{V}(x, y, z, t)$. The independent variables here are space co-ordinates x, y, z and time t. As any given spatial position is occupied by different fluid particles at different times, the Eulerian description (also referred to as spatial description) does not provide direct information regarding the changes in particle properties as it moves about. It only describes the change of quantities at a fixed location or region as a function of time. The Eulerian approach is also referred to as control volume approach (see Figure 3.4).

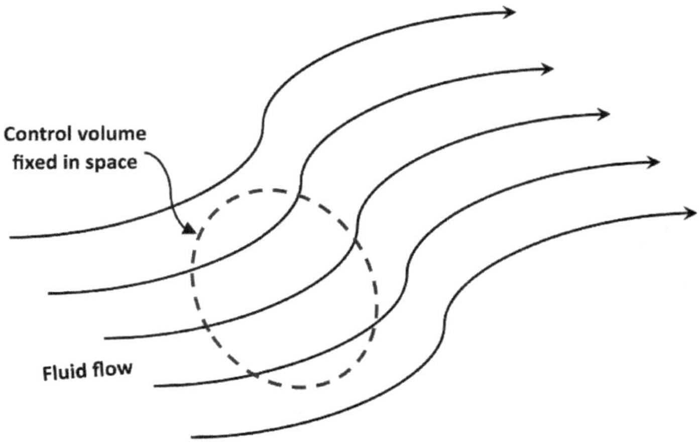

Figure 3.4 Eulerian description of fluid motion.

The advantage of the Lagrangian approach is that the laws of mechanics can be applied directly to each fluid particle. The disadvantage of this approach is that it is expensive computationally to keep track of changes of all the fluid particles. Therefore, this approach is seldom used in fluid mechanics applications. It is used only in some numerical simulations. The Eulerian approach is mathematically much simpler to apply. Hence it is more commonly used in solving fluid mechanics problems.

3.4 PARTIAL AND SUBSTANTIAL TIME DERIVATIVES

Let $F(x, y, z, t)$ be a spatial (Eulerian) function representing some flow field variable (density, velocity, etc.). The partial time derivative, denoted as $\partial F/\partial t$, describes the time rate of change F at a fixed point in space. Thus, it gives the local rate of change of F at a given location.

The substantial time derivative of the spatial function $F(x, y, z, t)$, denoted as DF/Dt, is defined as:

$$\frac{DF}{Dt} = \frac{\partial F}{\partial t} + \vec{V} \cdot \nabla F \tag{3.6}$$

where $\vec{V}$ is the fluid velocity vector. The substantial time derivative DF/Dt describes the time rate of change of F for a particle instantaneously located at (x, y, z). The term $\partial F/\partial t$ gives the local rate of change of F at a given location and is zero under steady-state conditions. The term $\vec{V} \cdot \nabla F$ gives the advective rate of change of F resulting from the movement of a particle from one position in space to another. The advective term is zero only if F is space invariant. Thus, the particle instantaneously located at (x, y, z) senses both local $(\partial F/\partial t)$ and spatial $(\vec{V} \cdot \nabla F)$ changes.

The substantial time derivative is also referred to material or particle time derivative. While the concept of this derivative is Lagrangian in that it gives the time rate of change of F for a material particle instantaneously located at (x, y, z), the derivative itself is Eulerian. Many important flow field variables require the determination of a material time derivative. For example, the acceleration of a material particle instantaneously located at (x, y, z) is simply the material time derivative of the velocity field $\vec{V}(x, y, z, t)$, that is, $D\vec{V}/Dt$,

3.5 PRINCIPLES OF CONSERVATION OF MASS, MOMENTUM, AND ENERGY

The microscopic equations of motion of fluid are derived by applying the principles of conservation of mass, momentum, and energy to fluid flow. The principles are discussed here briefly.

The principle of conservation of mass states that matter cannot be created or destroyed (in the absence of nuclear reactions). Thus, the mass of a material system (collection of matter of fixed identity) is constant over time and the material time derivative of system's mass is zero:

$$\frac{D}{Dt}\iiint_{system}\rho dV = 0 \tag{3.7}$$

The principle of conservation of momentum is equivalent to Newton's second law of motion. It states that the time rate of change of momentum of a material system (mass of fixed identity) is equal to vector sum of external forces acting on the system. Thus,

$$\frac{D}{Dt}\iiint_{system}\rho\vec{V}dV = \sum\vec{F} \tag{3.8}$$

The principle of conservation of energy is equivalent to the first law of thermodynamics. It states that the time rate of change of total energy of a material system (mass of fixed identity) is equal to the rate of heat addition to the system minus the rate of work done by the system on the surroundings. Thus,

$$\frac{D}{Dt}\iiint_{system}\rho e dV = \dot{Q} - \dot{W} \tag{3.9}$$

where e is total energy per unit mass, $\dot{Q}$ is rate of heat input to the system, and $\dot{W}$ is rate of work done by the system on its surroundings.

3.6 CONCEPT OF VISCOSITY

Consider two large horizontal parallel plates separated by a small distance y_o (see Figure 3.5). The space between the plates is filled with some fluid whose viscosity needs to be defined. A force F_o is applied on the top plate in the horizontal direction while the bottom plate is held stationary. Upon the application of force F_o on the top plate, the plate is set in motion. The plate accelerates initially and then moves at a steady velocity V_o. The motion of the top plate also sets the fluid in motion. Due to no-slip boundary conditions, the fluid adjacent to the top plate moves at the plate velocity of V_o and the fluid adjacent to the lower plate remains stationary. Thus, there exists a velocity gradient in the fluid. Provided that y_o is small and V_o is not too high, that is, the flow is laminar, the velocity profile of the fluid is linear, given as follows:

$$V_x = \frac{V_o}{y_o}y \tag{3.10}$$

The velocity gradient in the fluid is:

$$\frac{dV_x}{dy} = \frac{V_o}{y_o} = \text{constant} \tag{3.11}$$

The shear stress acting on the fluid adjacent to the top plate is given as:

$$\tau_{yx} = \frac{F_o}{A} \tag{3.12}$$

where A is the area of the top plate (contact area between the top plate and fluid).

Experience indicates that the shear force per unit area (F_o/A) required to move the top plate is directly proportional to the velocity gradient (V_o/y_o) in the fluid. In other words, F_o/A is doubled when either the plate velocity V_o is doubled or the gap between the plates y_o is reduced to one-half. Thus:

$$\tau_{yx}\alpha\frac{dV_x}{dy} \tag{3.13}$$

The proportionality constant between the shear stress τ_{yx} and velocity gradient dV_x/dy is a property of the fluid called *viscosity* η, that is:

$$\tau_{yx} = \eta \frac{dV_x}{dy} \tag{3.14}$$

This relationship which defines viscosity of a fluid is called *Newton's law of viscosity*. According to the Newton law of viscosity, the shear stress at any point in the fluid τ_{yx} is viscosity times the velocity gradient dV_x/dy in the fluid at that point. Thus,

$$\eta = \frac{\tau_{yx}}{dV_x/dy} \tag{3.15}$$

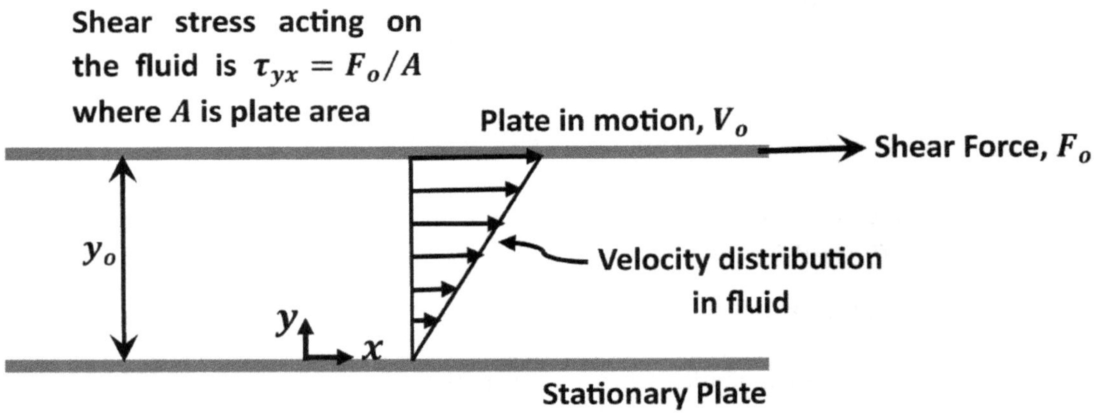

Figure 3.5 Flow between two large parallel plates.

In principle, the viscosity of a fluid can be measured using the parallel plate device of Figure 3.5 by varying the shear force and recording the corresponding top plate velocities at steady state. From the shearing force versus top plate velocity data, one can calculate the shear stress τ_{yx} (Equation (3.9)) versus dV_x/dy (Equation (3.8)) data and plot it as shown in Figure 3.6. The slope of the plot is the fluid viscosity.

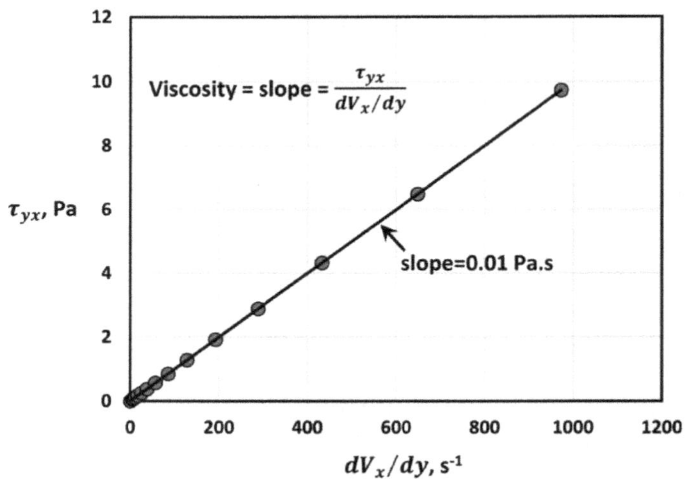

Figure 3.6 Shear stress τ_{yx} versus dV_x/dy plot.

Equivalence of Velocity Gradient dV_x/dy and Shear Rate $\dot{\gamma}$

It can be readily shown that shown the velocity gradient dV_x/dy is the same as shear rate $\dot{\gamma}$, that is, the rate of angular deformation of fluid. Figure 3.7 shows the angular deformation of a fluid element subjected to velocity gradient dV_x/dy. At some time t, the fluid element has a shape indicated by ABCD in Figure 3.7. At time $t + \Delta t$, the fluid element deforms to shape AB'CD'. The angular deformation of the fluid element in time Δt is:

$$tan(\Delta\gamma) = \frac{BB'}{AB} = \frac{V_o \Delta t}{y_o} \tag{3.16}$$

In the limit $\Delta t \to 0$, Equation (3.16) can be re-written as:

$$\frac{d\gamma}{dt} = \lim_{\Delta t \to 0} \frac{tan(\Delta\gamma)}{\Delta t} = \lim_{\Delta t \to 0} \left(\frac{\Delta\gamma}{\Delta t}\right) = \frac{V_o}{y_o} \tag{3.17}$$

As the velocity profile is linear, we can further re-write Equation (3.17) as:

$$\frac{d\gamma}{dt} = \frac{dV_x}{dy} \tag{3.18}$$

Thus, the velocity gradient dV_x/dy is equivalent to shear rate $\dot{\gamma}$.

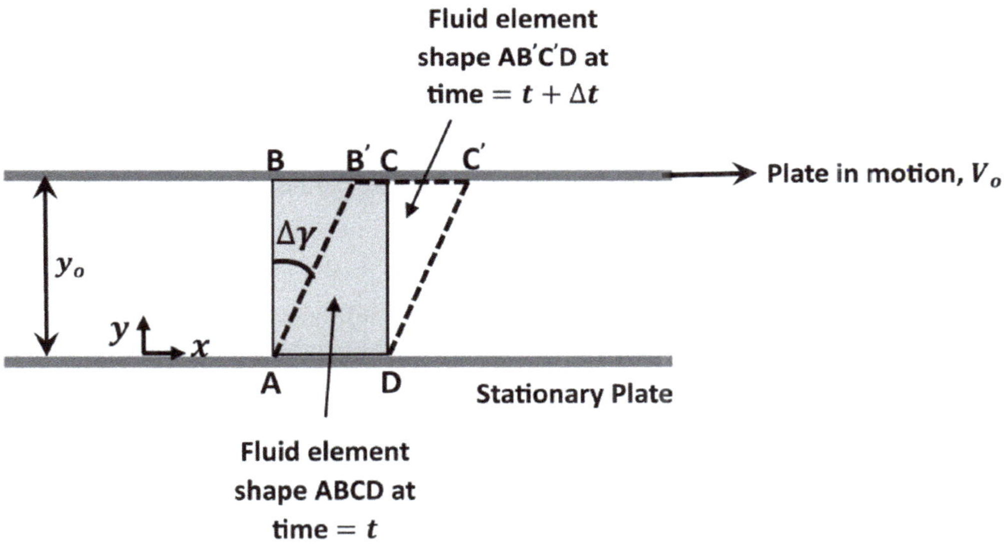

Figure 3.7 Shear strain or angular deformation of a fluid element.

3.7 VISCOSITY OF LIQUIDS AND GASES

3.7.1 Liquids

The viscosity of liquids is generally a strong function of temperature. It decreases with the increase in temperature. The viscosity of liquids has a weak dependence on pressure. It generally increases with the increase in pressure. However, the pressure effect is quite small and is generally neglected in most applications.

The temperature dependence of liquid viscosity is usually expressed in the form:

$$\eta = Aexp(B/T) \tag{3.19}$$

where A and B are constants for a given liquid and T is the absolute temperature. For example, $A = 1.732 \times 10^{-6}$ N.s/m^2 and B =1863 K for water. Table 3.1 gives the viscosities of some liquids at 20 °C and 101.3 kPa.

Table 3.1 Viscosities of liquids at 20 °C and 101.3 kPa

Liquid	Viscosity (mPa.s)
Ethyl alcohol	1.19
Gasoline	0.317
Carbon tetrachloride	0.958
Kerosene	1.92
Glycerin	1500
Mercury	1.58

3.7.2 Gases

The viscosity of gases is given by the following theoretical equation based on kinetic theory of gases (also referred to as Chapman-Enskog theory):

$$\eta = 2.6693 \times 10^{-5} \frac{\sqrt{MT}}{\sigma^2 \Omega} \tag{3.20}$$

where μ is in units of Poise, M is molecular weight of gas (g/mol), T is absolute temperature (K), σ is the characteristic diameter of the molecules (referred to as collision diameter) in units of angstroms, and Ω is a dimensionless quantity referred to as collision integral.

The collision integral Ω is a function of dimensionless temperature defined as:

$$T^* = T/(\varepsilon/k_B) \tag{3.21}$$

where ε is the characteristic energy of interaction between the molecules and k_B is the Boltzmann constant. Figure 3.8 shows the plot of Ω as a function of T^*. The figure also shows the plot of the following equation:

$$\Omega = 1.16145(T^*)^{-0.14874} + 0.52487exp(-0.7732T^*) + 2.1617exp(-2.43787T^*) \tag{3.22}$$

The equation describes the collision integral data very well.

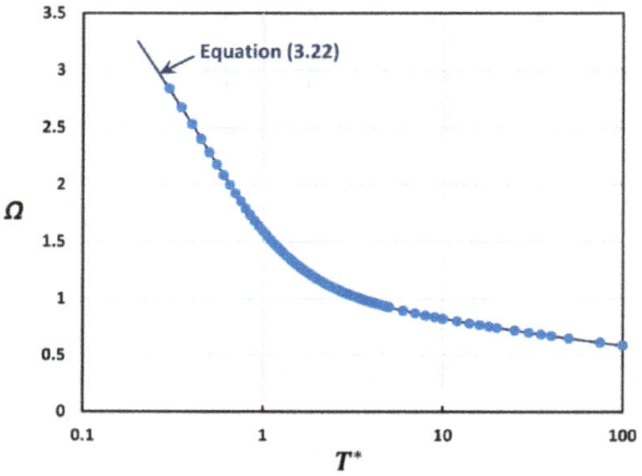

Figure 3.8 Collision integral as a function of T^*.

Table 3.2 gives the parameters needed to estimate the viscosity of some gases from Equation (3.20).

Table 3.2 Values of the parameters needed to estimate the viscosity of gases

Gas	Chemical formula	σ (angstroms)	ε/k_B (K)	M
Methane	CH_4	3.76	149	16
Carbon monoxide	CO	3.69	91.7	28
Carbon dioxide	CO_2	3.94	195	44
Ethane	C_2H_6	4.44	216	30
Propane	C_3H_8	5.12	237	44
n-Butane	C_4H_{10}	4.69	531	58
Hydrogen	H_2	2.83	59.7	2
Ammonia	NH_3	2.9	558	17
Nitrogen	N_2	3.8	71.4	28
Oxygen	O_2	3.47	107	32
Helium	He	2.58	10.2	4

As indicated by Equation (3.20), the viscosity of gases increases with the increase in temperature. Thus, the temperature dependence of viscosity of gases is very different from that of liquids. We observe just the opposite behaviour. However, the pressure dependence of viscosity of gas is negligible as in the case of liquids.

Another useful equation to predict the viscosity of gases is the Sutherland equation given as:

$$\eta = \frac{AT^{1.5}}{(B+T)} \tag{3.23}$$

where A and B are constants for a given gas and T is the absolute temperature. For example, $A = 1.458 \times 10^{-6}$ N.s/(m^2.K$^{1/2}$) and $B = 110.4$ K for air. Table 3.3 gives Sutherland equation constants for different gases.

Table 3.3 Sutherland equation constants for different gases

Gas	$A \times 10^6$, N.s/(m^2.K$^{1/2}$)	B (K)
CO_2	1.55	233
CO	1.40	109
Freon-12	1.48	317
He	1.52	97.8
H_2	0.649	70.6
CH_4	0.983	155
N_2	1.39	102
O_2	1.65	110

3.8 DENSITY OF LIQUIDS AND GASES

The density of a fluid is defined as its mass per unit volume:

$$\rho = \frac{m}{V} \tag{3.24}$$

The density of liquids varies little with pressure. The effect of temperature is also small, but temperature dependence of density is larger as compared with pressure dependence of density. It decreases with the increase in temperature. For example, water has a density of 1000 kg/m^3 at 4°C and 958.1 kg/m^3 at 100 °C.

The density of liquids can be considered constant if the temperature variation is not large. Table 3.4 gives the densities of several common liquids at 20 °C and 101.3 kPa.

Table 3.4 Densities of some common liquids at 20 °C and 101.3 kPa

Liquid	Density (kg/m^3)
Ethyl alcohol	789
Gasoline	726
Carbon tetrachloride	1,590
Kerosene	814
Glycerin	1,260
Mercury	13,600

Unlike liquids, the density of gases varies strongly with both pressure and temperature. It increases with the increase in pressure and decreases with the increase in temperature. For example, the density of air at 15 °C and 101.3 kPa is 1.23 kg/m^3. However, it is doubled to 2.46 kg/m^3 when the pressure is doubled at the same temperature.

The ideal gas law gives the density of gas as:

$$\rho = \frac{P}{RT} \tag{3.25}$$

where R is the specific gas constant in units of J/kg.K. Thus, the density of gas is directly proportional to pressure and inversely proportional to temperature. Table 3.5 gives the specific gas constant values for selected gases.

Table 3.5 Specific gas constant values for selected gases

Gas	R, J/(kg.K)
Air	287.05
CO_2	189
He	2077
H_2	4127
CH_4	518
N_2	297
O_2	260

Equation (3.25) closely approximates the behaviour of a real gas under normal conditions when the gas is not near liquefaction condition.

Incompressible flow

Incompressible fluids have constant density in space and time. Thus, the material (substantial) time derivative of density is zero in incompressible flow:

$$\frac{D\rho}{Dt} = 0 \tag{3.26}$$

The flow of liquids can be treated as incompressible flow. The flow of gases could also be treated as incompressible flow if the Mach number, defined below, is smaller than 0.3:

$$Ma = \frac{V}{c} \tag{3.27}$$

where V is fluid velocity and c is the speed of sound. The speed of sound in an ideal gas is given by:

$$c = \sqrt{kRT} \tag{3.28}$$

where k is the heat capacity ratio and R is the specific gas constant.

3.9 NON-NEWTONIAN FLUIDS

In section 3.6, the viscosity of fluid was defined as the ratio of shear stress to velocity gradient or shear rate (see Equation (3.15)). It is the slope of shear stress versus shear rate plot (see Figure 3.6). It could be further stated that the slope of shear stress versus shear rate plot (that is, viscosity) is constant independent of the shear rate and duration of shearing of fluid. Thus, the viscosity does not change with time when the fluid is subjected to a constant shear rate or with the change in shear rate. Fluids which exhibit these characteristics are called Newtonian fluids. However, many fluids of practical importance do not exhibit these characteristics. Such fluids are referred to as non-Newtonian fluids.

Homogeneous fluids consisting of small molecules generally behave as Newtonian fluids. All gases are Newtonian in nature. Most abundant fluids on our planet earth, that is, air and water, are also Newtonian fluids.

Non-Newtonian fluids can be classified into two groups: 1) time-independent non-Newtonian fluids and 2) time-dependent non-Newtonian fluids. The time-independent fluids include Shear-thinning (pseudoplastic) fluids, shear-thickening (dilatant) fluids, Bingham-plastic fluids, and yield pseudoplastic fluids. The time-dependent fluids include thixotropic and rheopectic fluids.

The apparent viscosity of non-Newtonian fluids is defined in the same way as the viscosity of Newtonian fluids, that is, the ratio of shear stress to shear rate:

$$\eta_a = \frac{\tau}{\dot{\gamma}} \tag{3.29}$$

However, the term "apparent viscosity" is used to describe the viscosity of non-Newtonian fluids as η_a is not a true property of the fluid. It depends on factors such as shear rate and duration of shearing, in addition to the nature of the fluid.

Fluids generally behave as non-Newtonian fluids when they possess some structural elements. It the microstructure of the fluid that is altered with the application of shear stress (or shear rate) resulting in changes in viscosity. For example, polymer solution (mixture of polymer molecules and solvent) behaves as shear-thinning fluid, that is, the viscosity decreases with the increase in shear rate. Polymer molecules are large molecules referred to as macromolecules. They exist in a coiled state in a solution at rest. The application of shear (shear stress or shear rate) to the polymer solution causes coiled macromolecules to straighten out and align with the flow. Consequently, the viscosity of the polymer solution decreases with the increase in shear rate.

3.9.1 Time-independent Non-Newtonian Fluids

Figure 3.9 shows the flow curves of a shear-thinning (also referred to as pseudoplastic) fluid. The flow curve of shear stress versus shear rate (Figure 3.9a) is not linear as in the case of Newtonian fluids (see Figure 3.6). It is concave downwards, that is, the slope decreases with the increase in shear rate. Figure 3.9b shows the same data as apparent viscosity versus shear rate. As can be seen, the viscosity decreases with the increase in shear rate. Hence the fluid is called shear-thinning. Examples of pseudoplastic fluids are polymer solutions, polymer melts, paints, oil/water emulsions, suspensions of solid particles, etc. Some of these fluids are deliberately designed to be shear-thinning. For example, consider a paint. When we are brushing a paint on the wall, the shear rate between the brush and the wall is high. To enable easier brushing of a paint, it is required that the paint have a low viscosity at high shear rates. But we also want that the paint should stay on the wall and not flow on the wall after brushing. Thus, it is desired that the viscosity of the paint be high at zero or low shear rates.

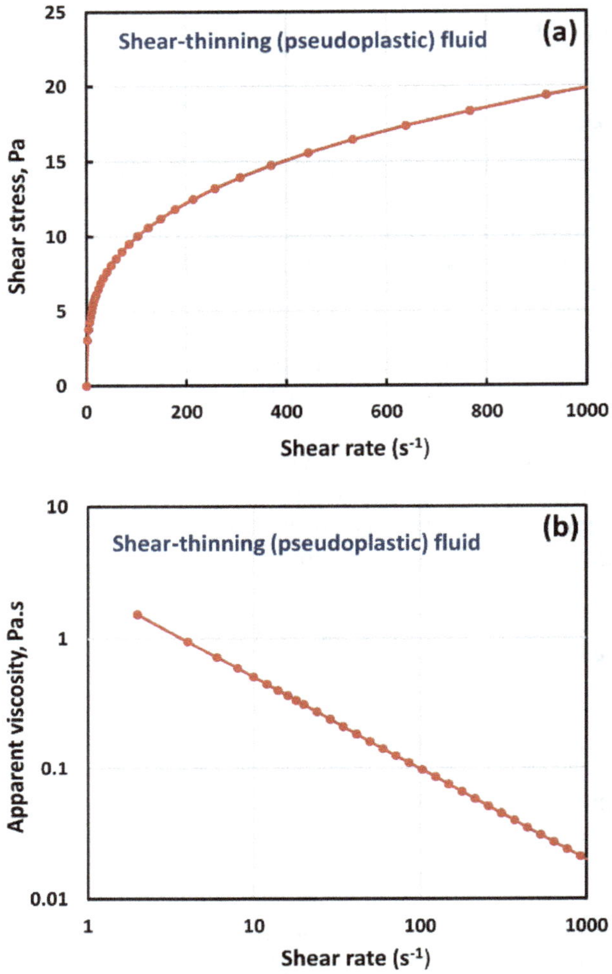

Figure 3.9 Flow curves of a shear-thinning (pseudoplastic) fluid.

Figure 3.10 shows the flow curves of a shear-thickening (also referred to as dilatant) fluid. The flow curve of shear stress versus shear rate (Figure 3.10a) is non-linear, concave upwards, that is, the slope increases with the increase in shear rate. Figure 3.10b shows the same data as apparent viscosity versus shear rate. As can be seen, the viscosity increases with the increase in shear rate. Hence the fluid is called shear-thickening. Examples of a dilatant fluid include suspensions of starch in water. For example, oobleck (see Figure 3.11) is a suspension of cornstarch and water. Oobleck behaves like a liquid at rest (zero shear rate) and like a solid when subjected to shear (tapped or rolled).

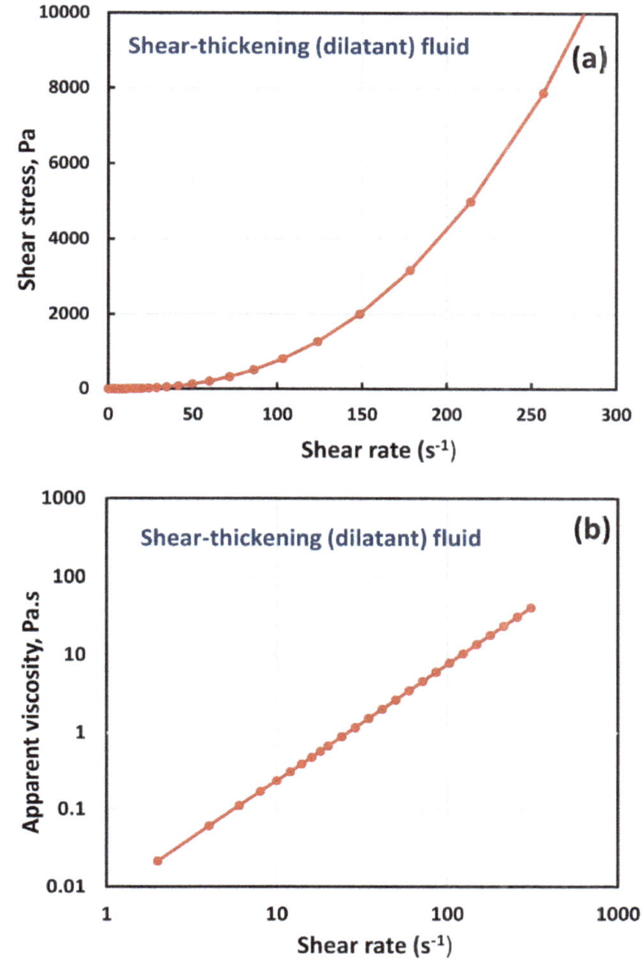

Figure 3.10 Flow curves of a shear-thickening (dilatant) fluid.

Figure 3.11 Suspension of cornstarch and water (shear-thickening fluid).

Figure 3.12 shows the flow curves of a Bingham-plastic fluid. The flow curve of shear stress versus shear rate (Figure 3.12a) does not pass through the origin. The y-intercept of the shear stress versus shear rate plot is the yield stress (τ_o) of the material. The material behaves like a solid (does not flow) at shear stresses below the yield stress. When the shear stress exceeds the yield stress, the material flows like a fluid with linear relationship between shear stress and shear rate. Figure 3.12b shows the same data as apparent viscosity versus shear stress. The apparent viscosity of a Bingham-plastic fluid is relatively very low at high shear stresses (or shear rates) but it shoots up as the shear stress is decreased towards the yield stress. Examples of Bingham-plastic fluids are ketchup, mayonnaise, butter, foams (for example, shaving foam), and toothpaste (see Figure 3.13). These fluids are deliberately designed to be Bingham-plastic for practical reasons. For example, consider ketchup. One would not like ketchup to be a thin fluidic material that flows like water. At the same time, one would like it to come out of the bottle readily when tapped (exceed the yield stress). Likewise, one would not like the toothpaste to be a thin watery liquid that flows out of the tube without any squeezing action. But when squeezed (exceed yield stress), it should come out of the tube readily. Similarly, mayonnaise, butter and shaving foam are designed to be Bingham-plastic materials. The desired characteristic of mayonnaise, butter, and shaving foam is the consistency of solid at rest. That is, they must possess some yield stress. But at the same time, one should be able to spread them when subjected to shear stress exceeding yield stress with the help of a knife or hand.

Yield-pseudoplastic fluids exhibit the combination of Bingham-plastic and pseudoplastic behaviours as shown in Figure 3.14. They possess a yield stress like Bingham-plastic fluids and the flow curve is concave downwards with the increase in shear rate like pseudoplastic fluids. Concentrated emulsions of oil and water, flocculated emulsions of oil and water, and suspensions of interacting particles (flocculated particles) often exhibit yield-pseudoplastic behaviour.

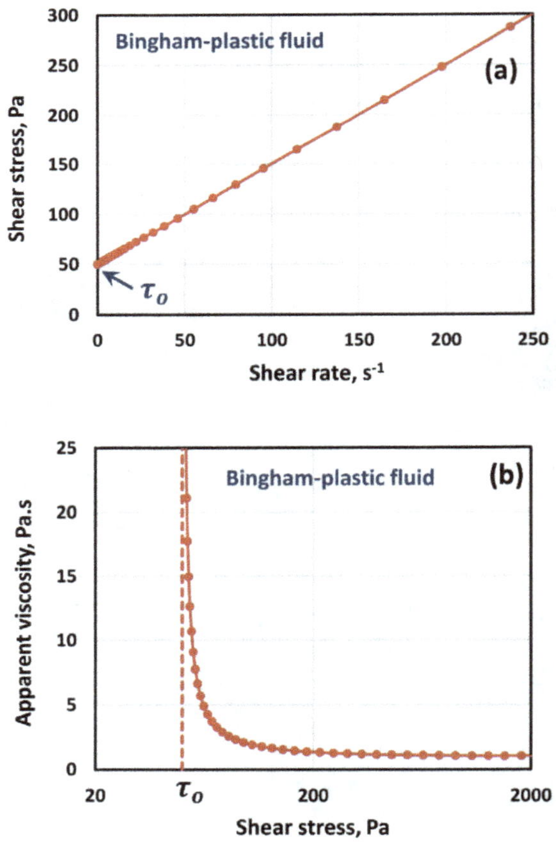

Figure 3.12 Flow curves of a Bingham-plastic fluid.

Figure 3.13 Examples of Bingham-plastic fluids.

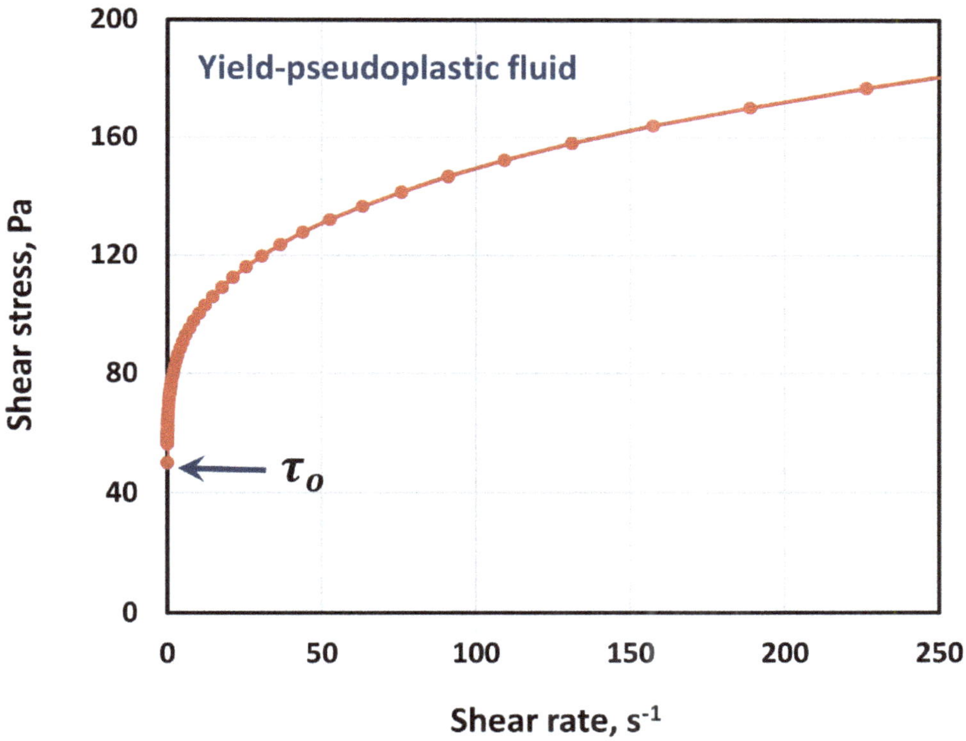

Figure 3.14 Flow curve of a yield-pseudoplastic fluid.

3.9.2 Time-dependent Non-Newtonian Fluids

Thixotropic fluids are time-dependent shear-thinning fluids. When subjected to a constant shear rate, their viscosity decreases with the duration of shear. When subjected to shear-rate sweep (see Figure 3.15(a)), they exhibit hysteresis loop. The shear stress in the forward sweep is higher than the shear stress in the backward sweep at the same shear rate. The hysteresis loop indicates that the fluid microstructure does not recover instantaneously to its original state. Examples of thixotropic fluids are multigrade oils, most greases, and some paints.

Rheopectic fluids (also referred to as anti-thixotropic fluids) are time-dependent shear-thickening fluids. When subjected to a constant shear rate, their viscosity increases with the duration of shear. When subjected to shear-rate sweep (see Figure 3.15(b)), they exhibit hysteresis loop like thixotropic fluids. However, the shear stress in the backward sweep is higher than the shear stress in the forward sweep at the same shear rate (opposite of thixotropic fluid). Examples of rheopectic fluids are printer inks, gypsum pastes, and body's synovial fluids.

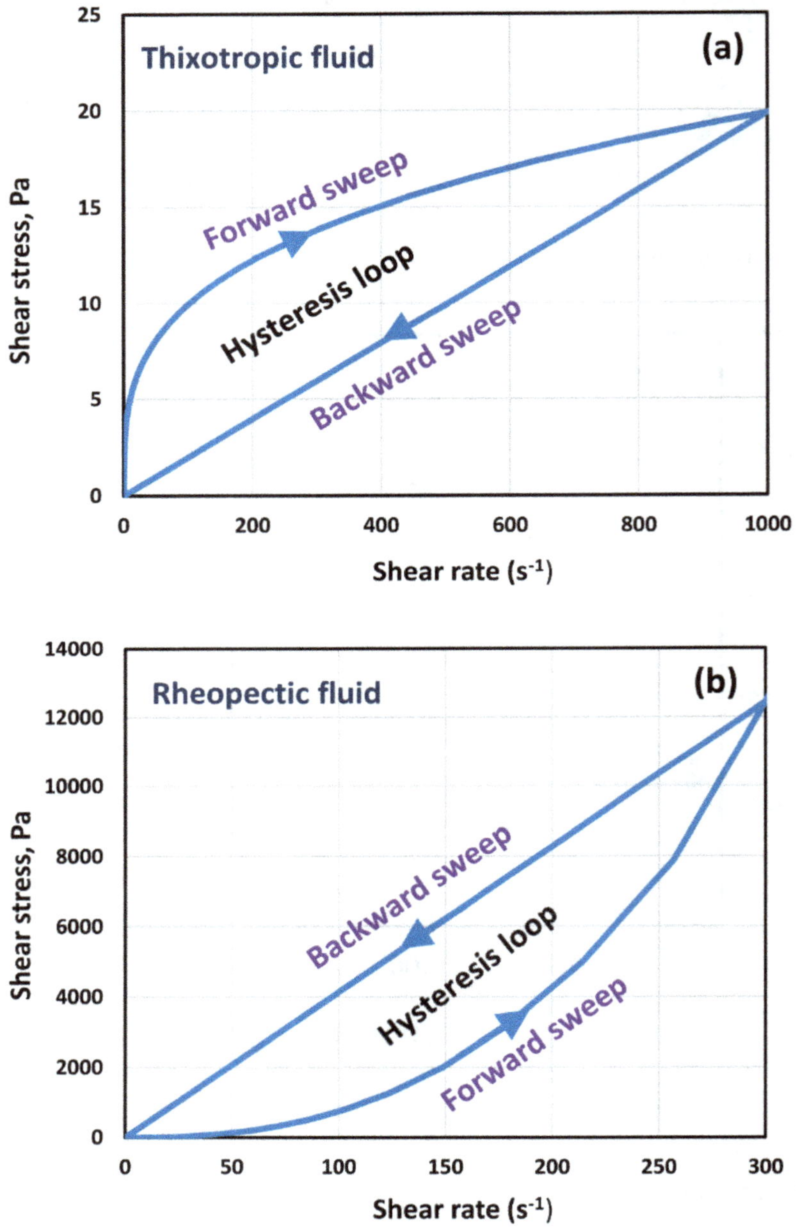

Figure 3.15 Hysteresis effects in thixotropic and rheopectic fluids.

3.10 LAMINAR AND TURBULENT FLOW REGIMES

Laminar flow is defined to be the flow where the fluid moves in smooth laminae, that is, adjacent layers of fluid slide past one another without any lateral mixing. There are neither crosscurrents nor eddies. The flow is very orderly.

Turbulent flow is characterized by irregular (disorderly, chaotic, random) motion of fluid particles (packets or lumps). Fluid no longer flows as smooth laminae. There occurs a strong lateral mixing action.

The distinction between laminar and turbulent flow regimes was first demonstrated experimentally by Osborne Reynolds in 1883. The experiment is now called Reynolds experiment. Consider steady flow of water in a pipe. We will inject a red dye through a thin needle into water flow at a very small flow rate to avoid any disturbance of water flow. At low water flow rate (see Figure 3.16) in the pipe, the stream of dye emerging from the needle follows a straight line, that is, the dye particles march along one behind the other. In this case, the water flow in the pipe is laminar. There is no lateral mixing action involved. The fluid (water) flows in smooth laminae.

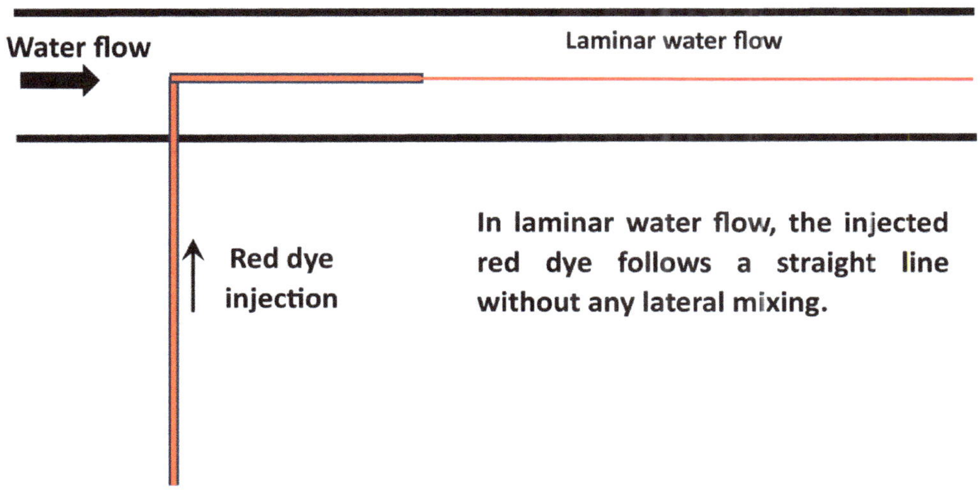

Figure 3.16 Reynolds experiment to demonstrate laminar regime.

We now perform the same experiment at high water flow rate (see Figure 3.17). A completely different behaviour is observed at high water flow rate. The injected dye no longer follows a straight line. Rather it mixes instantly across the entire cross-section of the pipe. This type of disorderly flow with intense mixing action is called turbulent flow. In turbulent flow, the fluid particles not only have a translational motion but also rotational motion. The rotational motion of fluid particles is superimposed over the bulk motion. The rotating fluid particles (packets) are referred to as eddies.

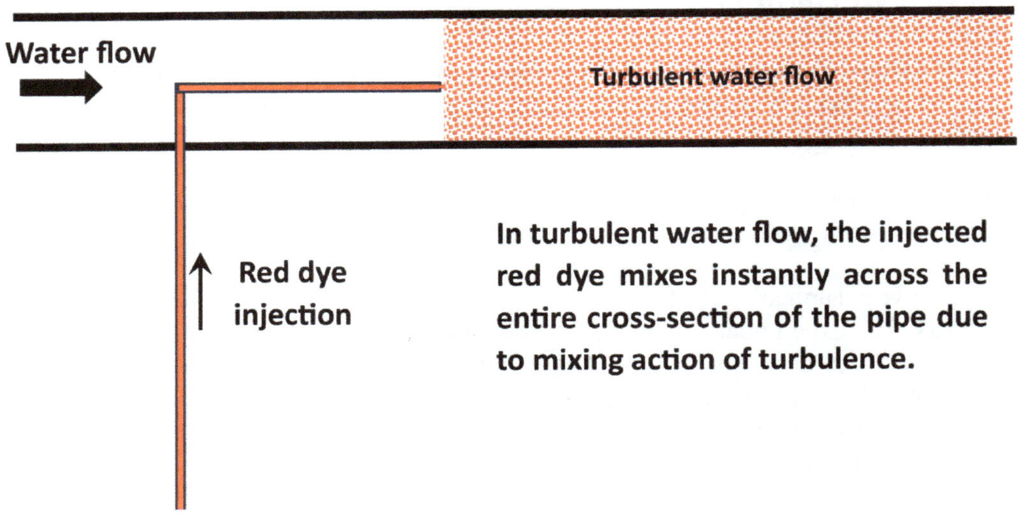

Figure 3.17 Reynolds experiment to demonstrate turbulent regime.

If the Reynolds experiment is carried out starting at a very low water flow rate and the flow rate is gradually increased, the flow is initially laminar and at some high flow rate, the flow becomes turbulent. Reynolds carried out such experiments under different conditions by varying the pipe diameter and fluid properties (density and viscosity) and observed the transition from laminar to turbulent regimes. He discovered that the transition from laminar to turbulent regime is governed by the following dimensionless group, now referred to as Reynolds number:

$$Re = \frac{\rho D \bar{V}}{\eta} \tag{3.30}$$

For a straight cylindrical pipe, the flow is always laminar when $Re < 2100$. The flow is almost always turbulent when $Re > 4000$. Between 2100 and 4000, the flow may be either laminar or turbulent or oscillate between laminar and turbulent regimes. In this transition region, the flow cannot be predicted.

Figure 3.18 shows the variation of pressure gradient $(\Delta P/L)$ with the volumetric flow rate (Q) in a hydraulically smooth horizontal pipe. In the laminar regime, the pressure gradient is proportional to the flow rate. In the turbulent regime, however, the pressure gradient is proportional to approximately $Q^{1.75}$. In the transition regime, there is no reproducible relation between pressure gradient and flow rate. Note that pressure gradient is independent of pipe roughness in laminar regime but increases with pipe roughness in turbulent regime.

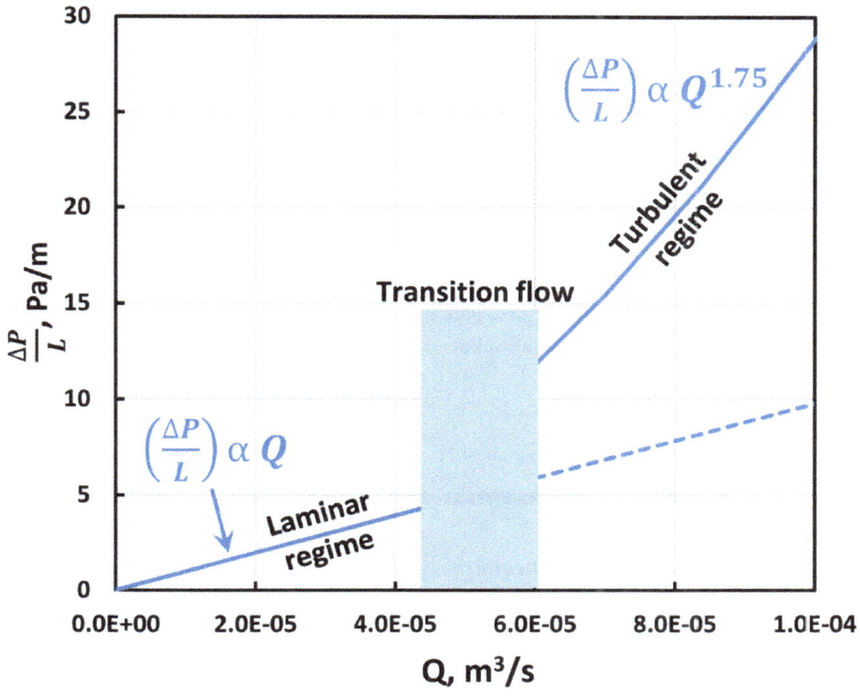

Figure 3.18 Pressure gradient versus volumetric flow rate in a hydraulically smooth horizontal pipe.

Laminar and Turbulent Flow Regimes in External Flows

Consider flow of fluid over a large flat plate (see Figure 3.19). The flow is uniform (flat velocity profile) before the plate. The motion of fluid is strongly influenced by the presence of the solid boundary, that is, the plate in the present case. The thin region of fluid moving adjacent to solid boundary where the velocity profile is influenced is called boundary layer. Above the boundary layer the velocity profile is uniform.

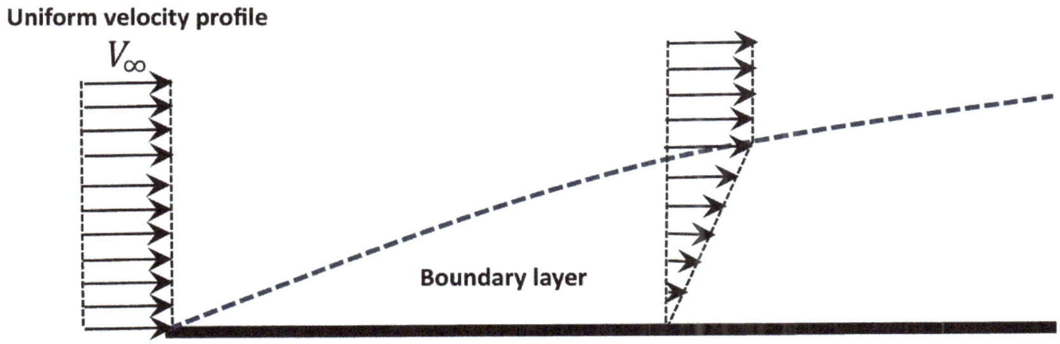

Figure 3.19 Formation of boundary layer for flow of fluid over a large flat plate.

Flow in the boundary layer could be laminar or turbulent. Flow is always laminar near the leading edge of the plate. At some distance downstream the plate, the flow behaviour in the boundary layer changes from laminar to turbulent (see Figure 3.20). When onset of turbulence occurs, the thickness of the boundary layer increases suddenly. Like in internal flows (pipe flow), the onset of turbulence in external flows is governed by Reynolds number. However, the Reynolds number is defined differently as:

$$Re = \frac{\rho x V_\infty}{\eta} \tag{3.31}$$

As x increases, the Reynolds number increases. For flow over a flat surface, the flow in the boundary layer is usually laminar for $Re_x < 2 \times 10^5$. The transition from laminar boundary layer to turbulent boundary layer begins at $Re_x = 2 \times 10^5$ and is completed at $Re_x \approx 3 \times 10^6$. The boundary layer is fully turbulent for $Re_x \geq 3 \times 10^6$. The transition occurs at a lower Re_x when the plate is rough and at a higher Re_x when the plate is hydraulically smooth.

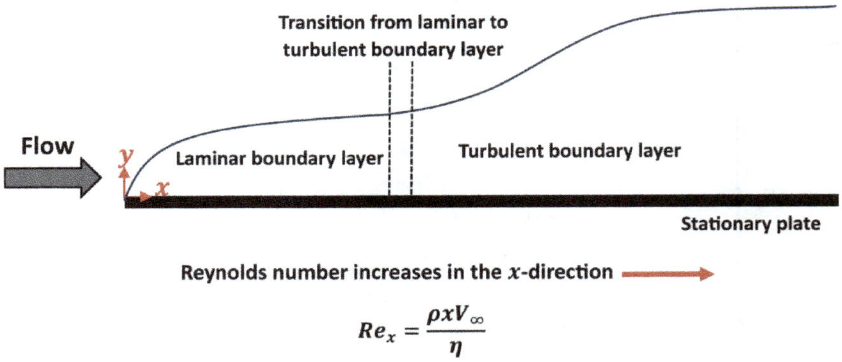

Figure 3.20 Formation of laminar and turbulent boundary layers over a flat plate.

3.11 EXAMPLES

3.11.1 Example 1

Given the motion of a body in Lagrangian description to be:

$x = X/(1 + tX)$; $y = Y$; $z = Z$

Find the spatial (Eulerian) and material (Lagrangian) descriptions of velocity.

Solution

The Lagrangian description of velocity can be expressed as

$$V_x = \left[\frac{\partial x}{\partial t}\right]_{(X,Y,Z-\text{fixed})} = -\frac{X^2}{(1+tX)^2}$$

$$V_y = \left[\frac{\partial y}{\partial t}\right]_{(X,Y,Z-\text{fixed})} = 0$$

$$V_z = \left[\frac{\partial z}{\partial t}\right]_{(X,Y,Z-\text{fixed})} = 0$$

Solving for X and substituting in the material description of velocity, one can obtain the spatial description of velocity. Thus,

$X = x/(1 - tx)$; $V_x = -x^2$; $V_y = 0$; $V_z = 0$

Note that the functional form of V_x in Eulerian description ($V_x = -x^2$) is different from the functional form in Lagrangian description, that is, $V_x = -X^2/(1+tX)^2$

3.11.2 Example 2

Given the velocity vector:

$$\vec{V} = x_1^2 t\hat{\delta}_1 + x_2 t^2 \hat{\delta}_2 + x_1 x_3 \hat{\delta}_3$$

Determine the expression for the substantial derivative $D\vec{V}/Dt$

Solution

By definition,

$$\frac{D\vec{V}}{Dt} = \frac{\partial \vec{V}}{\partial t} + \vec{V}\cdot\nabla\vec{V}$$

$$\frac{\partial \vec{V}}{\partial t} = x_1^2\hat{\delta}_1 + 2x_2 t\hat{\delta}_2$$

Use index notation to expand $\vec{V}\cdot\nabla\vec{V}$

$$\vec{V}\cdot\nabla\vec{V} = \sum_i V_i\hat{\delta}_i \cdot \sum_j \frac{\partial}{\partial x_j}\hat{\delta}_j \sum_k V_k\hat{\delta}_k = \sum_i\sum_j\sum_k V_i\left(\frac{\partial V_k}{\partial x_j}\right)(\hat{\delta}_i\cdot\hat{\delta}_j)\hat{\delta}_k = \sum_i\sum_j\sum_k V_i\left(\frac{\partial V_k}{\partial x_j}\right)\delta_{ij}\hat{\delta}_k$$

For non-zero values, $i = j$. Consequently,

$$\vec{V}\cdot\nabla\vec{V} = \sum_i\sum_k V_i\left(\frac{\partial V_k}{\partial x_i}\right)\hat{\delta}_k = \sum_i V_i\left[\left(\frac{\partial V_1}{\partial x_i}\right)\hat{\delta}_1 + \left(\frac{\partial V_2}{\partial x_i}\right)\hat{\delta}_2 + \left(\frac{\partial V_3}{\partial x_i}\right)\hat{\delta}_3\right]$$

$$= V_1\left[\left(\frac{\partial V_1}{\partial x_1}\right)\hat{\delta}_1 + \left(\frac{\partial V_2}{\partial x_1}\right)\hat{\delta}_2 + \left(\frac{\partial V_3}{\partial x_1}\right)\hat{\delta}_3\right] + V_2\left[\left(\frac{\partial V_1}{\partial x_2}\right)\hat{\delta}_1 + \left(\frac{\partial V_2}{\partial x_2}\right)\hat{\delta}_2 + \left(\frac{\partial V_3}{\partial x_2}\right)\hat{\delta}_3\right]$$
$$+ V_3\left[\left(\frac{\partial V_1}{\partial x_3}\right)\hat{\delta}_1 + \left(\frac{\partial V_2}{\partial x_3}\right)\hat{\delta}_2 + \left(\frac{\partial V_3}{\partial x_3}\right)\hat{\delta}_3\right]$$
$$= (x_1^2 t)\left[2x_1 t\hat{\delta}_1 + x_3\hat{\delta}_3\right] + (x_2 t^2)\left[t^2\hat{\delta}_2\right] + (x_1x_3)\left[x_1\hat{\delta}_3\right]$$
$$= 2x_1^3t^2\hat{\delta}_1 + x_2t^4\hat{\delta}_2 + (x_1^2x_3t + x_1^2x_3)\hat{\delta}_3$$

Thus, the substantial derivative is:

$$\frac{D\vec{V}}{Dt} = \frac{\partial \vec{V}}{\partial t} + \vec{V}\cdot\nabla\vec{V} = (x_1^2 + 2x_1^3t^2)\hat{\delta}_1 + (x_2t^4 + 2x_2t)\hat{\delta}_2 + (x_1^2x_3t + x_1^2x_3)\hat{\delta}_3$$

3.11.3 Example 3

Consider two large parallel plates separated by 0.55 cm and filled with a liquid of viscosity 1.8 mPa.s. The top plate is being pulled at a velocity of 0.12 m/s and the bottom plate is stationary. Calculate:

i) The velocity gradient in the fluid
ii) The shear stress in the fluid
iii) The drag force on the bottom plate if the plate area is 8 m^2.

iv) If another liquid of viscosity 9 mPa.s is used at the same plate spacing and same top velocity, what would be the shear stress in the fluid.

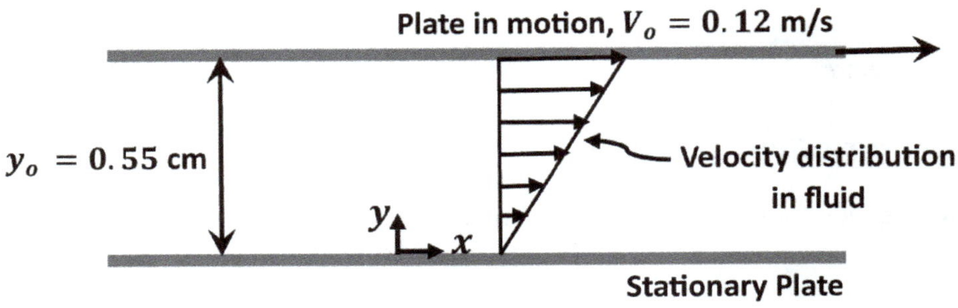

Solution

i) The velocity gradient in the fluid

$$\frac{dV_x}{dy} = \frac{V_o - 0}{y_o - 0} = \frac{0.12 - 0}{0.55 \times 10^{-3} - 0} = 21.82\ s^{-1}$$ (assuming linear velocity profile)

ii) The shear stress in the fluid

$$\tau_{yx} = \eta \frac{dV_x}{dy} = (1.8 \times 10^{-3}) \times 21.82 = 0.03928 \text{ Pa}$$

iii) The drag force on the bottom plate if the plate area is 8 m^2.

$$F_D = \tau_{yx} \times A = 0.03928 \times 8 = 0.3142 \text{ N}$$

iv) If another liquid of viscosity 9 mPa.s is used at the same plate spacing and same top velocity, what would be the velocity gradient and shear stress in the fluid.

The velocity gradient is the same, that is, $dV_x/dy = 21.82\ s^{-1}$

Thus, $\tau_{yx} = \eta \frac{dV_x}{dy} = (9 \times 10^{-3}) \times 21.82 = 0.196$ Pa

3.11.4 Example 4

Estimate the viscosity of CO_2 at 300 K and 1 atm using the Chapman-Enskog equation. You can take $M = 44.01$, $\sigma = 3.996\ A^o$, and $\Omega = 1.286$. Compare the value with Sutherland equation.

Solution

Chapman-Enskog equation:

$$\eta = 2.6693 \times 10^{-5} \frac{\sqrt{MT}}{\sigma^2 \Omega} = 2.6693 \times 10^{-5} \times \frac{\sqrt{44.01 \times 300}}{(3.996)^2 (1.286)} = 1.494 \times 10^{-4} \text{ Poise}$$

Sutherland equation:

$\eta = \frac{AT^{1.5}}{(B+T)}$ For CO_2, $A = 1.55 \times 10^{-6}, B = 233$

$$\eta = \frac{AT^{1.5}}{(B+T)} = \frac{1.55 \times 10^{-6}(300)^{1.5}}{(233+300)} = 1.511 \times 10^{-5}\text{Pa.s} = 1.511 \times 10^{-4}\text{Poise}$$

The values of the viscosity obtained through different equations are quite close.

3.11.5 Example 5

Calculate the Reynolds number for flow of oil at a flow rate of 9 ft^3/s in a 1 ft diameter pipe. The oil viscosity is 0.25 Poise and its density is 56 lb_m/ft^3. Is the flow laminar or turbulent?

Solution

$Q = 9\ ft^3/s,\ D = 1\ ft,\ \eta = 0.25\ Poise = 0.25 \times 6.72 \times 10^{-2}\ lb_m/ft.s$

$\bar{V} = Q/(\pi D^2/4) = 9/(\pi * 1^2/4) = 11.459\ ft/s$

$Re = \rho D\bar{V}/\eta = 56 \times 1 \times 11.459/(0.25 \times 6.72 \times 10^{-2}) = 3.82 \times 10^4$

As $Re \gg 2100$, the flow is highly turbulent.

3.11.6 Example 6

Atmospheric air at 25°C flows over a smooth plate that is 1m wide and 10m long. The air free stream velocity (V_∞) is 5 m/s parallel to the long side of the plate. The transition from laminar to turbulent flow occurs in the boundary layer at a Reynolds number of 5×10^5. Calculate the distance from the leading edge of the plate where the transition occurs. Given that:

$\rho = 1.18$ kg/m^3 and $\eta = 1.83 \times 10^{-5}$ Pa.s

Solution

$V_\infty = 5\ m/s$

$Re = \rho x V_\infty/\eta = 5 \times 10^5$

$x = Re \times \eta/\rho V_\infty = 5 \times 10^5 \times 1.83 \times 10^{-5}/(1.18 \times 5) = 1.55\ m$

NOTATION

A	area or constant in Equations (3.19) and (3.23)
B	constant in Equations (3.19) and (3.23)
c	velocity of sound
D	diameter
e	total energy per unit mass
F	force or some flow field variable
$\vec{F}$	force vector

k	heat capacity ratio
k_B	Boltzmann constant
L	length
m	mass
M	molecular weight
Ma	Mach number
P	pressure
Q	volumetric flow rate
$\dot{Q}$	rate of heat transfer
R	gas constant
$\vec{R}$	initial position vector of a fluid particle
$\vec{r}$	position vector
Re	Reynolds number
t	time
t_o	initial time
T	temperature
T^*	dimensionless temperature
V	volume or velocity
$\vec{V}$	velocity vector
$\bar{V}$	average velocity
V_o	velocity of the plate
V_x	velocity component in x- direction
V_∞	free stream velocity
$\dot{W}$	rate of work
x	space co-ordinate in x- direction
X	material co-ordinate in x- direction
y	space co-ordinate in y- direction
Y	material co-ordinate in y- direction
y_o	gap between parallel plates
z	space co-ordinate in z- direction
Z	material co-ordinate in z- direction

Greek Letters

γ	shear strain
$\dot{\gamma}$	shear strain rate (shear rate)
∇	del operator
Δ	change in variable
ε	characteristic energy of interaction of molecules
η	viscosity
η_a	apparent viscosity
ρ	density
σ	characteristic diameter of the molecules (referred to as collision diameter)
τ	shear stress
τ_o	yield-stress
τ_{yx}	shear stress on y-plane in the x-direction
Ω	collision integral

Derivatives

$\partial/\partial t$ partial time derivative
d/dy ordinary derivative
D/Dt substantial or material time derivative

PRACTICE PROBLEMS

Problem 1
Consider the motion $\vec{r} = \vec{R} + ktY\hat{\imath}$, that is, $x = X + ktY$, $y = Y$, $z = Z$ where the material co-ordinates (X, Y, Z) give the position of a particle at t = 0. A material body, at t = 0, has the shape of a cube of unit sides as shown in the figure. Sketch the configuration of the body at time t if it undergoes the given motion.

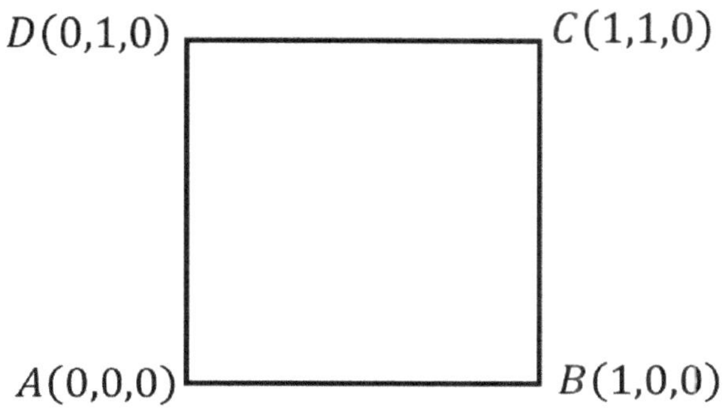

Configuration of the body at t=0

Problem 2
Consider the steady, two-dimensional velocity field given as:

$$\vec{V} = \left(\frac{V_o}{L}\right)(x\hat{\imath} - y\hat{\jmath})$$

Determine the substantial derivative of $\vec{V}$, that is, $D\vec{V}/Dt$.

Problem 3
In a parallel plate experiment, the spacing between the two plates is filled with some liquid of unknown viscosity. The temperature of the liquid is 20°C. If the force per unit contact area required to move the top plate at a constant velocity of 2.5 m/s is 43 Pa, Calculate the following:

i) The liquid viscosity
ii) The drag force per unit area on the bottom stationary plate

The spacing between the two plates is 1.8 cm.

Problem 4
Calculate the flow rate of a liquid of density 1000 kg/m^3 and viscosity 1 mPa.s in a 1 cm diameter pipe at a Reynolds number of 1000.

Problem 5

Estimate the viscosity of NH_3 at 493 K and 1 atm using the Chapman-Enskog equation. You can take $M = 17$, $\sigma = 2.9\ A^o$, and $\Omega = 1.7$.

Problem 6

Atmospheric air at 25°C flows over a smooth plate that is 1m wide and 12m long. The air free stream velocity (V_∞) is 6 m/s parallel to the long side of the plate. Calculate the Reynolds number at the following distances from the leading edge of the plate: $x = 1\ m$, $x = 8\ m$, and $x = 10\ m$. Is the flow laminar or turbulent at these locations. You can assume that the transition from laminar to turbulent flow occurs in the boundary layer at a Reynolds number of 5×10^5. Given that:

$\rho = 1.18$ kg/m^3 and $\eta = 1.83 \times 10^{-5}$ Pa. s

FURTHER READING

1. R. Pal. *Rheology of Particulate Dispersions and Composites*. CRC Press, Boca Raton, 2007.
2. R.B. Bird, W.E. Stewart, E.N. Lightfoot. *Transport Phenomena*. 2nd Edition, Wiley, New York, 2007.
3. C.J. Geankoplis, A.A. Hersel, D.H. Lepek. *Transport Processes and Separation Process Principles*. 5th Edition, Prentice Hall, New Jersey, 2018.
4. J.R. Welty, C.E. Wicks, R.E. Wilson, G. Rorrer. *Fundamentals of Momentum, Heat, and Mass Transfer*. 4th Edition, Wiley, New York, 2001.
5. H.A. Barnes, J.F. Hutton, K.Walters. *An Introduction to Rheology*. Elsevier, Amsterdam, 1989.
6. R.B. Bird, R.C. Armstrong, O. Hassager. *Dynamics of Polymeric Liquids*. Vol. 1, 2nd Edition, Wiley, New York, 1987.
7. R.G Larson. *The Structure and Rheology of Complex Fluids*. Oxford University Press, New York, 1999.
8. R.L. Street, G.Z. Watters, J.K. Vannard. *Elementary Fluid Mechanics*. 7th Edition, Wiley, New York, 1996.
9. B.R. Munson, D.F. Young, T.H. Okiishi. *Fundamentals of Fluid Mechanics*. 4th Edition, Wiley, New York, 2002.
10. J.O. Wilkes. *Fluid Mechanics for Chemical Engineers*. 3rd Edition. Prentice Hall, Boston, 2018.
11. W.L. McCabe, J.C. Smith, P. Harriott. *Unit Operations of Chemical Engineering*. 7th Edition, McGraw Hill, New York, 2005.

Chapter 4

Microscopic Equations of Motion

4.1 CONTINUITY EQUATION

The principle of conservation of mass states that matter cannot be created or destroyed (in the absence of nuclear reactions). Thus, the mass of a material system (collection of matter of fixed identity) is constant over time and the material time derivative of system's mass is zero:

$$\frac{D}{Dt}\iiint_{system}\rho dV = 0 \tag{4.1}$$

Using Reynolds Transport Theorem, the material time derivative of system's mass can be expressed as:

$$\frac{D}{Dt}\iiint_{system}\rho dV = \iiint_{CV}\left(\frac{\partial\rho}{\partial t}\right)dV + \iint_{CS}\rho(\hat{n}\cdot\vec{V})dS \tag{4.2}$$

where the integrals on the right-hand side are carried out over a nondeforming control volume fixed in space. The second integral on the right-hand side of Equation (4.1) can be converted into volume integral using the Gauss divergence theorem:

$$\iint_{CS}\rho(\hat{n}\cdot\vec{V})dS = \iiint_{CV}(\nabla\cdot\rho\vec{V})\,dV \tag{4.3}$$

The combination of Equations (4.1) - (4.3) gives:

$$\iiint_{CV}\left(\frac{\partial\rho}{\partial t} + \nabla\cdot\rho\vec{V}\right)dV = 0 \tag{4.4}$$

Since the control volume is arbitrarily chosen, the only way in which the above equation can be satisfied for all possible choices of control volume is for integrand to be zero. Thus,

$$\frac{\partial\rho}{\partial t} + \nabla\cdot\rho\vec{V} = 0 \tag{4.5}$$

This is the continuity equation. It can be written for the three coordinate systems as follows:

Cartesian Coordinates (x, y, z):

$$\frac{\partial\rho}{\partial t} + \left[\frac{\partial(\rho V_x)}{\partial x} + \frac{\partial(\rho V_y)}{\partial y} + \frac{\partial(\rho V_z)}{\partial z}\right] = 0 \tag{4.6}$$

Cylindrical Coordinates (r, θ, z):

$$\frac{\partial\rho}{\partial t} + \frac{1}{r}\frac{\partial}{\partial x}(\rho r V_r) + \frac{1}{r}\frac{\partial}{\partial\theta}(\rho V_\theta) + \frac{\partial}{\partial z}(\rho V_z) = 0 \tag{4.7}$$

Spherical Coordinates (r, θ, φ):

$$\frac{\partial\rho}{\partial t} + \frac{1}{r^2}\frac{\partial}{\partial r}(\rho r^2 V_r) + \frac{1}{r\sin\theta}\frac{\partial}{\partial\theta}(\rho V_\theta\sin\theta) + \frac{1}{r\sin\theta}\frac{\partial}{\partial\varphi}(\rho V_\varphi) = 0 \tag{4.8}$$

The continuity equation, Equation (4.5), could also be re-cast in terms of the substantial derivative of density as follows:

$$\frac{D\rho}{Dt} = -\rho(\nabla \cdot \vec{V}) \tag{4.9}$$

For incompressible flows, $D\rho/Dt = 0$. Thus, the continuity equation for incompressible flows reduces to:

$$\nabla \cdot \vec{V} = 0 \tag{4.10}$$

4.2 STRESS EQUATION OF MOTION

4.2.1 Stress Vector and Stress Tensor

Two types of forces act on a fluid. One of these is the body force that acts throughout a volume and is expressed as force per unit mass or force per unit volume. For example, gravitational force per unit mass of a fluid is $\vec{g}$ and force per unit volume is $\rho\vec{g}$, where $\vec{g}$ is the acceleration due to gravity and ρ is the fluid density. The body force acts at a distance and not because of direct contact. The other type of force acting on a fluid is the surface force (also referred to as contact force). Surface force acts on a surface, whether it is an arbitrary internal surface or the bounding surface of the continuum and is expressed as force per unit area.

Stress Vector

Consider a fluid occupying a volume V and imagine a closed surface S within the volume V, as shown in Figure 4.1. The material outside S exerts a surface force on the material inside S. Let $\Delta\vec{F}$ be the total surface force exerted on a small surface element of area ΔS, whose spatial orientation is given by the unit normal vector $\hat{n}$. The force $\Delta\vec{F}$ depends on the location, size of the area element, and the orientation of the area element specified by unit normal vector $\hat{n}$. The average surface force per unit area on ΔS is $\Delta\vec{F}/\Delta S$. As ΔS tends to zero, keeping $\hat{n}$ constant, the ratio $\Delta\vec{F}/\Delta S$ approaches a definite limit $d\vec{F}/dS$. The limiting vector $d\vec{F}/dS$ is called the stress vector or sometimes the traction vector. Thus,

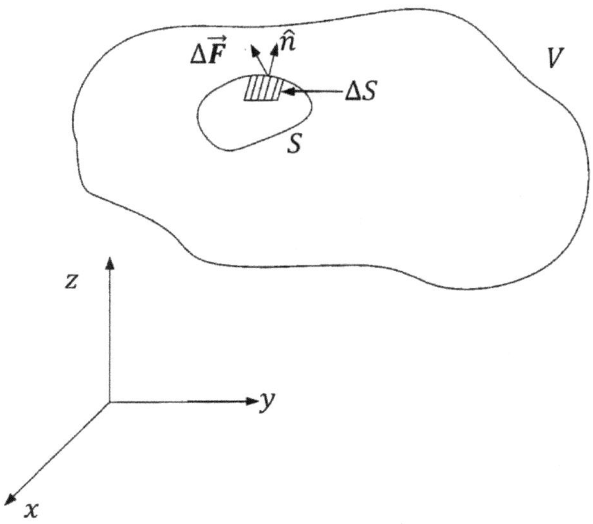

Figure 4.1 A material continuum occupying a volume V and subjected to surface forces.

$$\vec{\pi}_{(\hat{n})} = \lim_{\Delta S \to 0} \frac{\Delta \vec{F}}{\Delta S} = \frac{d\vec{F}}{dS} \tag{4.11}$$

where $\vec{\pi}_{(\hat{n})}$ is the stress vector which is a function of position, surface orientation, and time in general, that is:

$$\vec{\pi}_{(\hat{n})} = \vec{\pi}_{(\hat{n})}(\hat{n}, \vec{r}, t) \tag{4.12}$$

where $\vec{r}$ is the position vector and t is time.

Stress Tensor

Consider an arbitrary point Q in a fluid continuum. For each unit normal vector $\hat{n}$ at Q, there exists a corresponding stress vector $\vec{\pi}_{(\hat{n})}$. The state of stress at point Q is defined by the totality of all possible pairs of companion vectors $\vec{\pi}_{(\hat{n})}$ and $\hat{n}$ at that point. As an infinite number of unit normal vectors can be imagined at Q, each having a different orientation, there exist an infinite number of pairs of $\hat{n}$ and $\vec{\pi}_{(\hat{n})}$ at point Q. Fortunately every pair of $\hat{n}$ and associated $\vec{\pi}_{(\hat{n})}$ need not be specified. According to Cauchy's theorem, if the stress vectors acting on three mutually perpendicular planes at a given point are known, then all stress vectors at that point can be determined. Cauchy's formula, given below, allows us to express the stress vector at any given location as $\hat{n}$ varies:

$$\vec{\pi}_{(\hat{n})} = \bar{\bar{\pi}} \cdot \hat{n} \tag{4.13}$$

where $\bar{\bar{\pi}}$ is the stress tensor that transforms the unit normal vector $\hat{n}$ into the stress vector $\vec{\pi}_{(\hat{n})}$. The stress tensor is independent of $\hat{n}$ (orientation of the surface) and has nine components. It is a point function which completely describes the state of stress at a given point. The stress tensor is symmetric, that is:

$$\bar{\bar{\pi}} = (\bar{\bar{\pi}})^T \tag{4.14}$$

Consequently, $\bar{\bar{\pi}}$ has only six independent components. The stress tensor $\bar{\bar{\pi}}$ can be further re-written as:

$$\bar{\bar{\pi}} = -P\bar{\bar{\delta}} + \bar{\bar{\tau}} \tag{4.15}$$

where $\bar{\bar{\tau}}$ is known as the extra or deviatoric stress tensor (also referred to as viscous stress tensor), P is local pressure, and $\bar{\bar{\delta}}$ is the unit tensor. The stress vector could also be re-written as:

$$\vec{\pi}_{(\hat{n})} = -P\hat{n} + \vec{\tau}_{(\hat{n})} \tag{4.16}$$

where $\vec{\tau}_{(\hat{n})}$ is viscous stress vector.

4.2.2 Stress Equation of Motion

Newton's second law of motion states that the time rate of change of linear momentum of a body is equal to the resultant force (sum of forces) acting upon the body. When applied to fluid as a material system, Newton's second law of motion can be expressed as:

$$\frac{D}{Dt} \iiint_{system} \rho \vec{V} dV = \sum \vec{F} = \vec{F}_{body} + \vec{F}_{surface} \tag{4.17}$$

where the body force $\vec{F}_{body}$ is:

$$\vec{F}_{body} = \iiint_V \rho\vec{g}\, dV \tag{4.18}$$

and the surface force $\vec{F}_{surface}$ is:

$$\vec{F}_{surface} = \iint_S \vec{\pi}_{(\hat{n})}\, dS = \iint_S (\bar{\bar{\pi}} \cdot \hat{n}) dS \tag{4.19}$$

where V is the volume of the fluid material system and S is the bounding surface of the system. Using Gauss's divergence theorem, $\vec{F}_{surface}$ can be re-written as

$$\vec{F}_{surface} = \iiint_V (\nabla \cdot \bar{\bar{\pi}})\, dV \tag{4.20}$$

Combining Equations (4.17), (4.18), and (4.20) gives:

$$\frac{D}{Dt}\iiint_{system} \rho\vec{V} dV = \iiint_V \rho\vec{g}\, dV + \iiint_V (\nabla \cdot \bar{\bar{\pi}})\, dV \tag{4.21}$$

Using the Reynolds transport theorem, the substantial time derivative in Equation (4.21) can be written as:

$$\frac{D}{Dt}\iiint_{system} \rho\vec{V} dV = \iiint_{CV} \frac{\partial(\rho\vec{V})}{\partial t} dV + \iint_{CS} (\hat{n} \cdot \rho\vec{V}\vec{V}) dS \tag{4.22}$$

Applying the Gauss divergence theorem to the surface integral on the right-hand side of Equation (4.22), we get:

$$\frac{D}{Dt}\iiint_{system} \rho\vec{V} dV = \iiint_{CV} \frac{\partial(\rho\vec{V})}{\partial t} dV + \iiint_{CV} (\nabla \cdot \rho\vec{V}\vec{V}) dV \tag{4.23}$$

Substituting the expression for substantial derivative from Equation (4.23) into Equation (4.21) gives:

$$\iiint_{CV} \frac{\partial(\rho\vec{V})}{\partial t} dV + \iiint_{CV} (\nabla \cdot \rho\vec{V}\vec{V}) dV = \iiint_V \rho\vec{g}\, dV + \iiint_V (\nabla \cdot \bar{\bar{\pi}})\, dV \tag{4.24}$$

Noting that the volume of system and control volume are the same, we can re-write this equation as follows:

$$\iiint_{CV} \left[\frac{\partial(\rho\vec{V})}{\partial t} + (\nabla \cdot \rho\vec{V}\vec{V}) - \rho\vec{g} - (\nabla \cdot \bar{\bar{\pi}})\right] dV = 0 \tag{4.25}$$

Since this equation holds for an arbitrary volume control volume, the integrand must also be zero, that is:

$$\frac{\partial}{\partial t}(\rho\vec{V}) + \nabla \cdot \rho\vec{V}\vec{V} = \rho\vec{g} + \nabla \cdot \bar{\bar{\pi}} \tag{4.26}$$

This is the stress equation of motion.

Using the definition of substantial derivative and the continuity equation, it can be readily shown that:

$$\frac{\partial}{\partial t}(\rho\vec{V}) + \nabla \cdot \rho\vec{V}\vec{V} = \rho\frac{D\vec{V}}{Dt} \tag{4.27}$$

Thus, the stress equation of motion becomes:

$$\rho \frac{D\vec{V}}{Dt} = \rho\left[\frac{\partial \vec{V}}{\partial t} + \vec{V}\cdot\nabla\vec{V}\right] = \rho\vec{g} + \nabla\cdot\bar{\bar{\pi}} \tag{4.28}$$

Using Equation (4.15), the stress equation of motion could be further re-cast as:

$$\rho\left[\frac{\partial \vec{V}}{\partial t} + \vec{V}\cdot\nabla\vec{V}\right] = \rho\vec{g} - \nabla P + \nabla\cdot\bar{\bar{\tau}} \tag{4.29}$$

The stress equation of motion, Equation (4.29), could be interpreted as follows: The rate of increase of momentum of fluid per unit volume, that is, left hand side of Equation (4.29), is equal to the sum of gravitational, pressure, and viscous forces acting on the fluid per unit volume.

As the stress equation of motion (Equation (4.29)) is a vector equation, it has three components. It can be written for the three coordinate systems as follows:

Cartesian Coordinates (x, y, z):

x-direction

$$\rho\left[\frac{\partial V_x}{\partial t} + V_x\frac{\partial V_x}{\partial x} + V_y\frac{\partial V_x}{\partial y} + V_z\frac{\partial V_x}{\partial z}\right] = \rho g_x - \frac{\partial P}{\partial x} + \frac{\partial \tau_{xx}}{\partial x} + \frac{\partial \tau_{yz}}{\partial y} + \frac{\partial \tau_{zx}}{\partial z} \tag{4.30}$$

y-direction

$$\rho\left[\frac{\partial V_y}{\partial t} + V_x\frac{\partial V_y}{\partial x} + V_y\frac{\partial V_y}{\partial y} + V_z\frac{\partial V_y}{\partial z}\right] = \rho g_y - \frac{\partial P}{\partial y} + \frac{\partial \tau_{yy}}{\partial x} + \frac{\partial \tau_{xy}}{\partial y} + \frac{\partial \tau_{zy}}{\partial z} \tag{4.31}$$

z-direction

$$\rho\left[\frac{\partial V_z}{\partial t} + V_x\frac{\partial V_z}{\partial x} + V_y\frac{\partial V_z}{\partial y} + V_z\frac{\partial V_z}{\partial z}\right] = \rho g_z - \frac{\partial P}{\partial z} + \frac{\partial \tau_{zz}}{\partial x} + \frac{\partial \tau_{xz}}{\partial y} + \frac{\partial \tau_{yz}}{\partial z} \tag{4.32}$$

Cylindrical Coordinates (r, θ, z):

r-direction

$$\rho\left(\frac{\partial V_r}{\partial t} + V_r\frac{\partial V_r}{\partial r} + \frac{V_\theta}{r}\frac{\partial V_r}{\partial \theta} - \frac{V_\theta^2}{r} + V_z\frac{\partial V_r}{\partial z}\right) = -\frac{\partial P}{\partial r} + \rho g_r + \left[\frac{1}{r}\frac{\partial}{\partial r}(r\tau_{rr}) + \frac{1}{r}\frac{\partial \tau_{\theta r}}{\partial \theta} - \frac{\tau_{\theta\theta}}{r} + \frac{\partial \tau_{zr}}{\partial z}\right] \tag{4.33}$$

θ-direction

$$\rho\left(\frac{\partial V_\theta}{\partial t} + V_r\frac{\partial V_\theta}{\partial r} + \frac{V_\theta}{r}\frac{\partial V_\theta}{\partial \theta} + \frac{V_r V_\theta}{r} + V_z\frac{\partial V_\theta}{\partial z}\right) = -\frac{1}{r}\frac{\partial P}{\partial \theta} + \rho g_\theta + \left[\frac{1}{r^2}\frac{\partial}{\partial r}(r^2\tau_{r9}) + \frac{1}{r}\frac{\partial \tau_{\theta\theta}}{\partial \theta} + \frac{\partial \tau_{z\theta}}{\partial z}\right] \tag{4.34}$$

z-direction

$$\rho\left(\frac{\partial V_z}{\partial t} + V_r\frac{\partial V_z}{\partial r} + \frac{V_\theta}{r}\frac{\partial V_z}{\partial \theta} + V_z\frac{\partial V_z}{\partial z}\right) = -\frac{\partial P}{\partial z} + \rho g_z + \left[\frac{1}{r}\frac{\partial}{\partial r}(r\tau_{rz}) + \frac{1}{r}\frac{\partial \tau_{\theta z}}{\partial \theta} + \frac{\partial \tau_{zz}}{\partial z}\right] \tag{4.35}$$

Spherical Coordinates (r, θ, φ):

r-direction

$$\rho\left(\frac{\partial V_r}{\partial t} + V_r\frac{\partial V_r}{\partial r} + \frac{V_\theta}{r}\frac{\partial V_r}{\partial \theta} + \frac{V_\varphi}{r\sin\theta}\frac{\partial V_r}{\partial \varphi} - \frac{V_\theta^2+V_\varphi^2}{r}\right) = -\frac{\partial P}{\partial r} + \rho g_r + \left[\frac{1}{r^2}\frac{\partial}{\partial r}(r^2\tau_{rr}) + \frac{1}{r\sin\theta}\frac{\partial}{\partial\theta}(\tau_{\theta r}\sin\theta) + \frac{1}{r\sin\theta}\frac{\partial \tau_{\varphi r}}{\partial\varphi} - \frac{\tau_{\theta\theta}+\tau_{\varphi\varphi}}{r}\right] \tag{4.36}$$

θ-direction

$$\rho\left(\frac{\partial V_\theta}{\partial t} + V_r\frac{\partial V_\theta}{\partial r} + \frac{V_\theta}{r}\frac{\partial V_\theta}{\partial \theta} + \frac{V_\varphi}{r\sin\theta}\frac{\partial V_\theta}{\partial \varphi} + \frac{V_r V_\theta}{r} - \frac{V_\varphi^2\cot\theta}{r}\right) = -\frac{1}{r}\frac{\partial P}{\partial r} + \rho g_\theta + \left[\frac{1}{r^2}\frac{\partial}{\partial r}(r^2\tau_{r\theta}) + \frac{1}{r\sin\theta}\frac{\partial}{\partial\theta}(\tau_{\theta\theta}\sin\theta) + \frac{1}{r\sin\theta}\frac{\partial \tau_{\varphi\theta}}{\partial\varphi} + \frac{\tau_{r\theta}}{r} - \frac{\tau_{\varphi\varphi}\cot\theta}{r}\right] \tag{4.37}$$

φ-direction

$$\rho\left(\frac{\partial V_\varphi}{\partial t} + V_r\frac{\partial V_\varphi}{\partial r} + \frac{V_\theta}{r}\frac{\partial V_\varphi}{\partial \theta} + \frac{V_\varphi}{r\sin\theta}\frac{\partial V_\varphi}{\partial \varphi} + \frac{V_\varphi V_r}{r} + \frac{V_\theta V_\varphi\cot\theta}{r}\right) = -\frac{1}{r\sin\theta}\frac{\partial P}{\partial\varphi} + \rho g_\varphi + \left[\frac{1}{r^2}\frac{\partial}{\partial r}(r^2\tau_{r\varphi}) + \frac{1}{r}\frac{\partial \tau_{\theta\varphi}}{\partial\theta} + \frac{1}{r\sin\theta}\frac{\partial \tau_{\varphi\varphi}}{\partial\varphi} + \frac{\tau_{r\varphi}}{r} - \frac{2\tau_{\theta\varphi}\cot\theta}{r}\right] \tag{4.38}$$

4.3 CONSTITUTIVE EQUATION

The viscous stress in a fluid is directly proportional to the local rate of deformation of fluid. The relationship between viscous stress and rate of deformation of fluid is called constitutive equation. Consider a material continuum (fluid) undergoing flow. When the flow is uniform (zero velocity gradient), fluid material elements undergo only translational motion (see Figure 4.2). When subjected to velocity gradient, the fluid material elements in general undergo translation, rotation, and deformation (linear and shear strains). Figure 4.3 shows linear strain (stretching) of fluid element, Figure 4.4 shows shear strain (angular strain) of fluid element, and Figure 4.5 shows rotation of fluid element when fluid is subjected to velocity gradient.

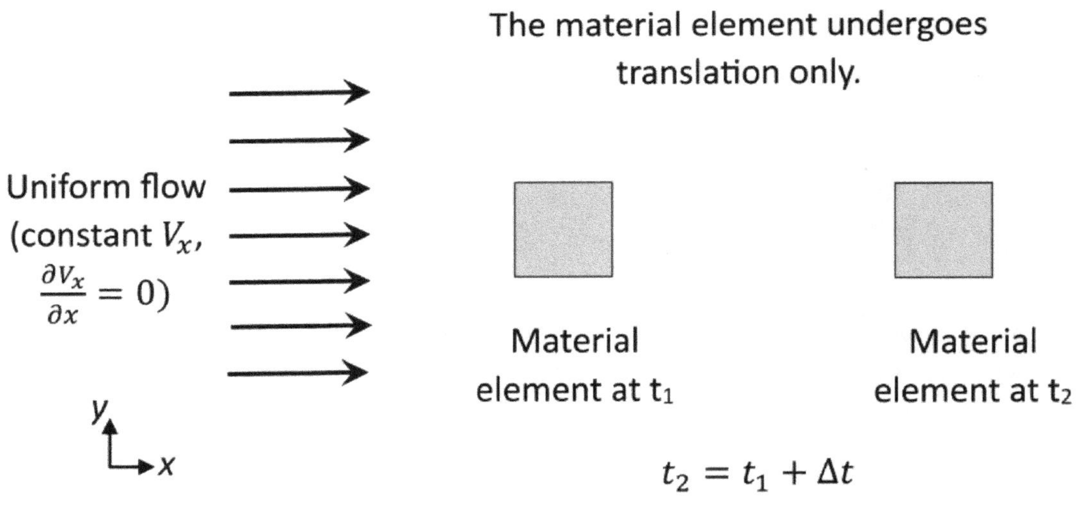

Figure 4.2 Translation of a fluid element when the fluid is subjected to uniform flow with zero velocity gradient.

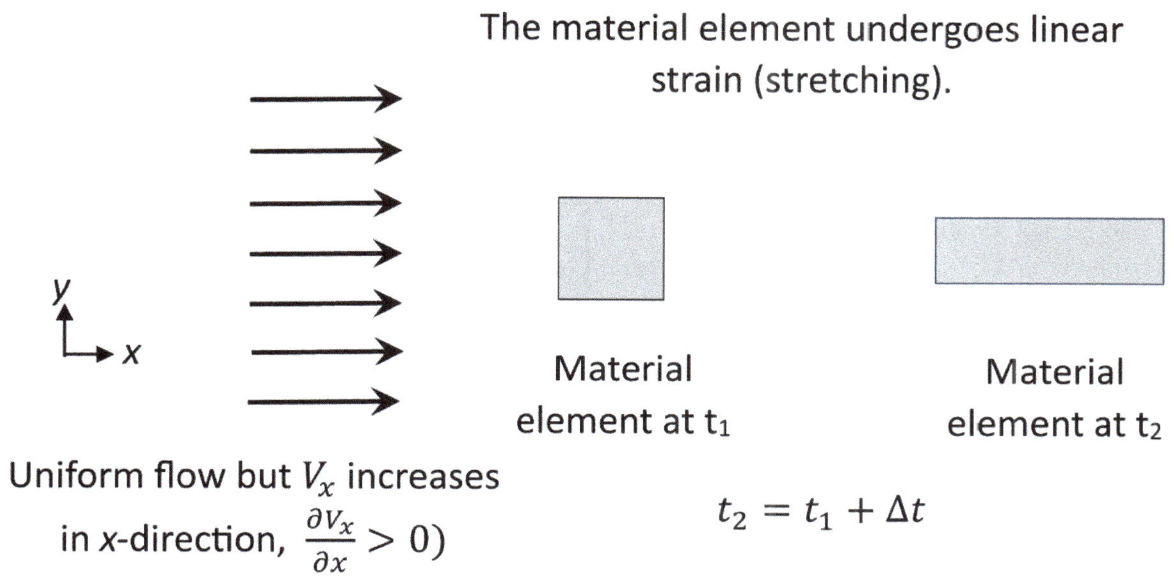

Figure 4.3 Linear strain (stretching) of a fluid element when the fluid is subjected to velocity gradient.

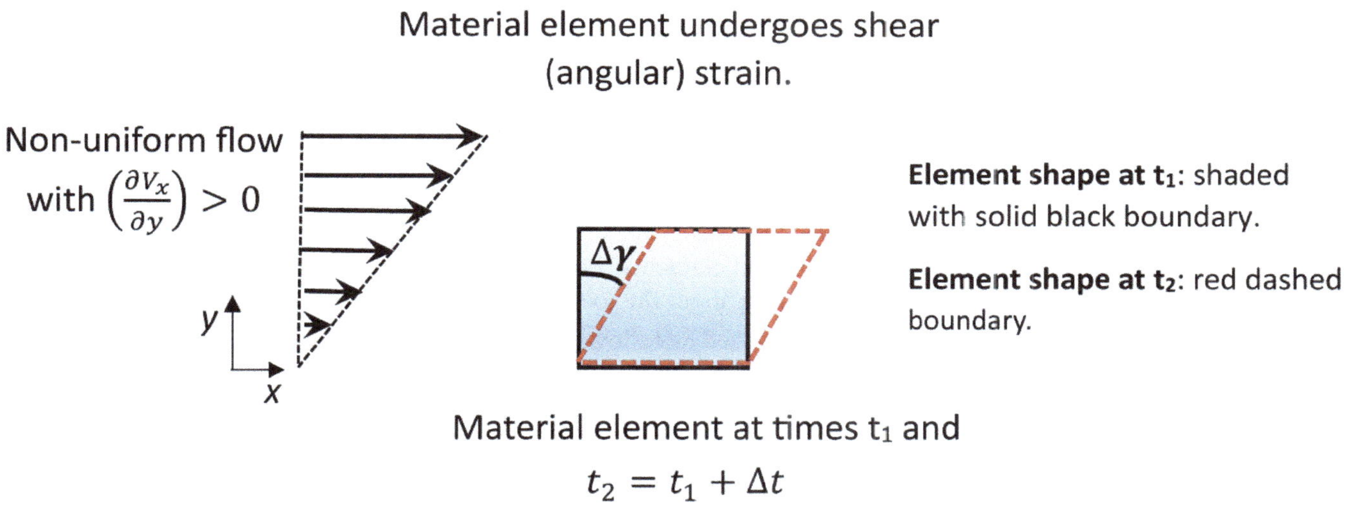

Figure 4.4 Shear strain (angular deformation) of a fluid element when the fluid is subjected to velocity gradient.

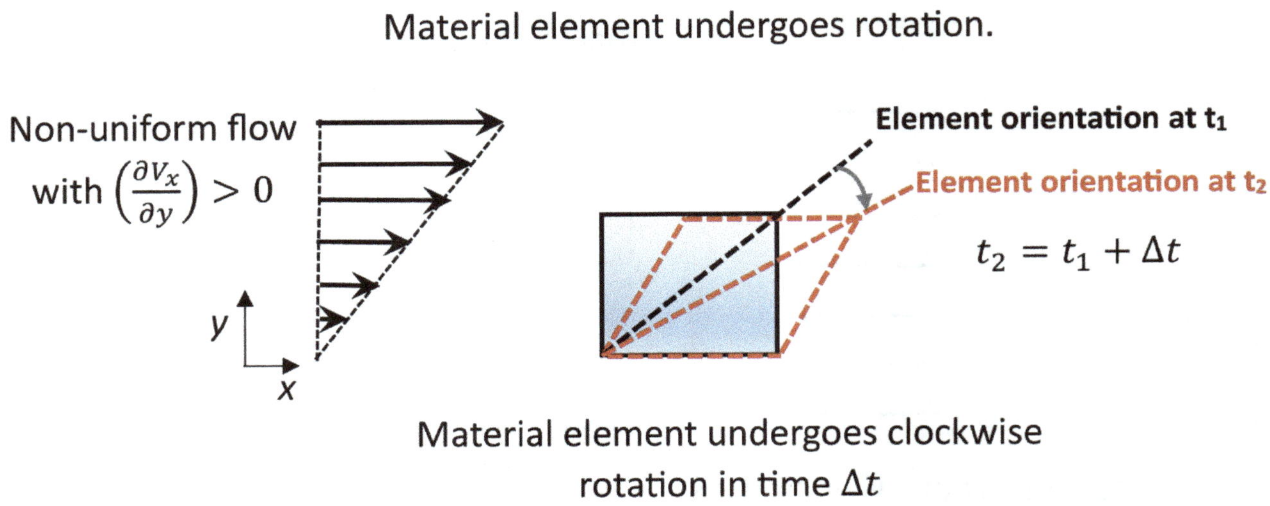

Figure 4.5 Rotation of a fluid element when the fluid is subjected to velocity gradient.

The velocity gradient tensor $\nabla\vec{V}$ contains all the information on rotation and deformation of the material elements. We can separate the rotation and deformation of fluid by splitting the velocity gradient tensor $\nabla\vec{V}$ into two tensors, namely the rate of deformation tensor $\bar{\bar{d}}$ and rate of rotation tensor $\bar{\bar{\omega}}$, as shown below:

$$\bar{\bar{d}} = \frac{1}{2}\left[\nabla\vec{V} + \left(\nabla\vec{V}\right)^T\right] \tag{4.39}$$

$$\bar{\bar{\omega}} = \frac{1}{2}\left[\nabla\vec{V} - \left(\nabla\vec{V}\right)^T\right] \tag{4.40}$$

The rate of deformation tensor $\bar{\bar{d}}$ is symmetric whereas rotation tensor $\bar{\bar{\omega}}$ is anti-symmetric tensor. All the information about deformation of material elements (linear and shear strains) is contained in $\bar{\bar{d}}$. The rotation tensor $\bar{\bar{\omega}}$ contains all the information about the local rates of rotation of material elements. Thus, the velocity gradient tensor $\nabla\vec{V}$ can be expressed as follows:

$$\nabla\vec{V} = \begin{pmatrix} \frac{\partial V_x}{\partial x} & \frac{\partial V_x}{\partial y} & \frac{\partial V_x}{\partial z} \\ \frac{\partial V_y}{\partial x} & \frac{\partial V_y}{\partial y} & \frac{\partial V_y}{\partial z} \\ \frac{\partial V_z}{\partial x} & \frac{\partial V_z}{\partial y} & \frac{\partial V_z}{\partial z} \end{pmatrix} = (\text{linear strain tensor} + \text{shear strain tensor}) + (\text{rotation tensor}) \tag{4.41}$$

where linear strain (stretching), shear strain, and rotation tensors are:

$$\text{Linear strain tensor} = \begin{pmatrix} \frac{\partial V_x}{\partial x} & 0 & 0 \\ 0 & \frac{\partial V_y}{\partial y} & 0 \\ 0 & 0 & \frac{\partial V_z}{\partial z} \end{pmatrix} \tag{4.42}$$

$$\text{Shearing strain tensor} = \begin{pmatrix} 0 & \frac{1}{2}\left(\frac{\partial V_x}{\partial y} + \frac{\partial V_y}{\partial x}\right) & \frac{1}{2}\left(\frac{\partial V_x}{\partial z} + \frac{\partial V_z}{\partial x}\right) \\ \frac{1}{2}\left(\frac{\partial V_x}{\partial y} + \frac{\partial V_y}{\partial x}\right) & 0 & \frac{1}{2}\left(\frac{\partial V_y}{\partial z} + \frac{\partial V_z}{\partial y}\right) \\ \frac{1}{2}\left(\frac{\partial V_x}{\partial z} + \frac{\partial V_z}{\partial x}\right) & \frac{1}{2}\left(\frac{\partial V_y}{\partial z} + \frac{\partial V_z}{\partial y}\right) & 0 \end{pmatrix} \tag{4.43}$$

$$\text{Rotation tensor} = \begin{pmatrix} 0 & \frac{1}{2}\left(\frac{\partial V_x}{\partial y} - \frac{\partial V_y}{\partial x}\right) & \frac{1}{2}\left(\frac{\partial V_x}{\partial z} - \frac{\partial V_z}{\partial x}\right) \\ -\frac{1}{2}\left(\frac{\partial V_x}{\partial y} - \frac{\partial V_y}{\partial x}\right) & 0 & \frac{1}{2}\left(\frac{\partial V_y}{\partial z} - \frac{\partial V_z}{\partial y}\right) \\ -\frac{1}{2}\left(\frac{\partial V_x}{\partial z} - \frac{\partial V_z}{\partial x}\right) & -\frac{1}{2}\left(\frac{\partial V_y}{\partial z} - \frac{\partial V_z}{\partial y}\right) & 0 \end{pmatrix} \tag{4.44}$$

Note that the rate of deformation tensor $\bar{\bar{d}}$ is the sum of linear and shearing strain tensors.

The viscous stress in a fluid is independent of rotation of fluid and depends only on the rate of deformation of fluid. Thus, the constitutive equation of a fluid is a relationship of the form:

$$\bar{\bar{\tau}} = f(\bar{\bar{d}}) \tag{4.45}$$

where $\bar{\bar{\tau}}$ is the viscous stress tensor. Note that not all materials follow the same form of a constitutive equation. These equations are particular to individual class of materials (Newtonian versus non-Newtonian), and they serve to distinguish one class of materials from another.

The Newtonian fluids obey the following form of a constitutive equation:

$$\bar{\bar{\tau}} = 2\eta\bar{\bar{d}} + \lambda(tr\bar{\bar{d}})\bar{\bar{\delta}} \tag{4.46}$$

where η and λ are the first and second viscosity coefficients, respectively of the fluid. Note that $tr\bar{\bar{d}} = \nabla \cdot \vec{V}$. The second viscosity coefficient λ is related to the first viscosity coefficient η (shear viscosity) and dilational (bulk) viscosity η^k as follows:

$$\lambda = \eta^k - \frac{2}{3}\eta \tag{4.47}$$

Consequently, the constitutive equation for Newtonian fluids can be written as:

$$\bar{\bar{\tau}} = 2\eta\bar{\bar{d}} + \left(\eta^k - \frac{2}{3}\eta\right)(tr\bar{\bar{d}})\bar{\bar{\delta}} \tag{4.48}$$

The bulk viscosity η^k is related to viscous stresses produced only because of volumetric (dilational) strain of fluid elements. It is zero for monoatomic gases. Even for polyatomic gases, it is rather small and can frequently be neglected. For incompressible fluids (liquids are generally assumed to be incompressible), $tr\bar{\bar{d}} = \nabla \cdot \vec{V} = 0$, and therefore, the constitutive equation (Equation (4.48)) reduces to the following form regardless of the value of second viscosity coefficient λ:

$$\bar{\bar{\tau}} = 2\eta\bar{\bar{d}} \tag{4.49}$$

The constitutive equation (Equation (4.49)) can be written in component form in different coordinates as follows:

Cartesian Coordinates (x, y, z):

$$\tau_{xx} = 2\eta \left(\frac{\partial V_x}{\partial x}\right) \tag{4.50}$$

$$\tau_{yy} = 2\eta \left(\frac{\partial V_y}{\partial y}\right) \tag{4.51}$$

$$\tau_{zz} = 2\eta \left(\frac{\partial V_z}{\partial z}\right) \tag{4.52}$$

$$\tau_{xy} = \tau_{yx} = \eta \left(\frac{\partial V_x}{\partial y} + \frac{\partial V_y}{\partial x}\right) \tag{4.53}$$

$$\tau_{yz} = \tau_{zy} = \eta \left(\frac{\partial V_y}{\partial z} + \frac{\partial V_z}{\partial y}\right) \tag{4.54}$$

$$\tau_{zx} = \tau_{xz} = \eta \left(\frac{\partial V_z}{\partial x} + \frac{\partial V_x}{\partial z}\right) \tag{4.55}$$

Cylindrical Coordinates (r, θ, z):

$$\tau_{rr} = 2\eta \left(\frac{\partial V_r}{\partial r}\right) \tag{4.56}$$

$$\tau_{\theta\theta} = 2\eta \left(\frac{1}{r}\frac{\partial V_\theta}{\partial \theta} + \frac{V_r}{r}\right) \tag{4.57}$$

$$\tau_{zz} = 2\eta \left(\frac{\partial V_z}{\partial z}\right) \tag{4.58}$$

$$\tau_{r\theta} = \tau_{\theta r} = \eta \left[r\frac{\partial}{\partial r}\left(\frac{V_\theta}{r}\right) + \frac{1}{r}\frac{\partial V_r}{\partial \theta}\right] \tag{4.59}$$

$$\tau_{\theta z} = \tau_{z\theta} = \eta \left(\frac{\partial V_\theta}{\partial z} + \frac{1}{r}\frac{\partial V_z}{\partial \theta}\right) \tag{4.60}$$

$$\tau_{zr} = \tau_{rz} = \eta \left(\frac{\partial V_z}{\partial r} + \frac{\partial V_r}{\partial z}\right) \tag{4.61}$$

Spherical Coordinates (r, θ, φ):

$$\tau_{rr} = 2\eta \left(\frac{\partial V_r}{\partial r}\right) \tag{4.62}$$

$$\tau_{\theta\theta} = 2\eta \left(\frac{1}{r}\frac{\partial V_\theta}{\partial \theta} + \frac{V_r}{r}\right) \tag{4.63}$$

$$\tau_{\varphi\varphi} = 2\eta \left(\frac{1}{r\sin\theta}\frac{\partial V_\varphi}{\partial \varphi} + \frac{V_r}{r}\frac{V_\theta \cot\theta}{r}\right) \tag{4.64}$$

$$\tau_{r\theta} = \tau_{\theta r} = \eta \left[r\frac{\partial}{\partial r}\left(\frac{V_\theta}{r}\right) + \frac{1}{r}\frac{\partial V_r}{\partial \theta}\right] \tag{4.65}$$

$$\tau_{\theta\varphi} = \tau_{\varphi\theta} = \eta \left[\frac{\sin\theta}{r}\frac{\partial}{\partial \theta}\left(\frac{V_\varphi}{\sin\theta}\right) + \frac{1}{r\sin\theta}\frac{\partial V_\theta}{\partial \varphi}\right] \tag{4.66}$$

$$\tau_{\varphi r} = \tau_{r\varphi} = \left[\eta \frac{1}{r\sin\theta}\frac{\partial V_r}{\partial \varphi} + r\frac{\partial}{\partial r}\left(\frac{V_\varphi}{r}\right)\right] \tag{4.67}$$

4.4 NAVIER-STOKES EQUATION

For incompressible Newtonian fluids, the combination of stress equation of motion (Equation (4.29)) and constitutive equation (Equation (4.49)) yields:

$$\rho\left[\frac{\partial \vec{V}}{\partial t}+\vec{V}\cdot\nabla\vec{V}\right]=-\nabla P+\rho\vec{g}+\eta\nabla^2\vec{V} \tag{4.68}$$

This is the celebrated Navier-Stokes equation. The Navier-Stokes equation together with the continuity equation describe the flow of Newtonian fluids of constant density and viscosity. The Navier-Stokes equation can be expressed in different coordinates as follows:

Cartesian Coordinates (x, y, z):

x-direction

$$\rho\left(\frac{\partial V_x}{\partial t}+V_x\frac{\partial V_x}{\partial x}+V_y\frac{\partial V_x}{\partial y}+V_z\frac{\partial V_x}{\partial z}\right)=-\frac{\partial P}{\partial x}+\rho g_x+\eta\left(\frac{\partial^2 V_x}{\partial x^2}+\frac{\partial^2 V_x}{\partial y^2}+\frac{\partial^2 V_x}{\partial z^2}\right) \tag{4.69}$$

y-direction

$$\rho\left(\frac{\partial V_y}{\partial t}+V_x\frac{\partial V_y}{\partial x}+V_y\frac{\partial V_y}{\partial y}+V_z\frac{\partial V_y}{\partial z}\right)=-\frac{\partial P}{\partial y}+\rho g_y+\eta\left(\frac{\partial^2 V_y}{\partial x^2}+\frac{\partial^2 V_y}{\partial y^2}+\frac{\partial^2 V_y}{\partial z^2}\right) \tag{4.70}$$

z-direction

$$\rho\left(\frac{\partial V_z}{\partial t}+V_x\frac{\partial V_z}{\partial x}+V_y\frac{\partial V_z}{\partial y}+V_z\frac{\partial V_z}{\partial z}\right)=-\frac{\partial P}{\partial z}+\rho g_z+\eta\left(\frac{\partial^2 V_z}{\partial x^2}+\frac{\partial^2 V_z}{\partial y^2}+\frac{\partial^2 V_z}{\partial z^2}\right) \tag{4.71}$$

Cylindrical Coordinates (r, θ, z):

r-direction

$$\begin{aligned}\rho\left(\frac{\partial V_r}{\partial t}+V_r\frac{\partial V_r}{\partial r}+\frac{V_\theta}{r}\frac{\partial V_r}{\partial \theta}-\frac{V_\theta^2}{r}+V_z\frac{\partial V_r}{\partial z}\right)&=-\frac{\partial P}{\partial r}\\+\rho g_r+\eta\left[\frac{\partial}{\partial r}\left(\frac{1}{r}\frac{\partial}{\partial r}(rV_r)\right)+\frac{1}{r^2}\frac{\partial^2 V_r}{\partial \theta^2}-\frac{2}{r^2}\frac{\partial V_\theta}{\partial \theta}+\frac{\partial^2 V_r}{\partial z^2}\right]&\end{aligned} \tag{4.72}$$

θ-direction

$$\begin{aligned}\rho\left(\frac{\partial V_\theta}{\partial t}+V_r\frac{\partial V_\theta}{\partial r}+\frac{V_\theta}{r}\frac{\partial V_\theta}{\partial \theta}+\frac{V_rV_\theta}{r}+V_z\frac{\partial V_\theta}{\partial z}\right)&=-\frac{1}{r}\frac{\partial P}{\partial \theta}\\+\rho g_\theta+\eta\left[\frac{\partial}{\partial r}\left(\frac{1}{r}\frac{\partial}{\partial r}(rV_\theta)\right)+\frac{1}{r^2}\frac{\partial^2 V_\theta}{\partial \theta^2}-\frac{2}{r^2}\frac{\partial V_r}{\partial \theta}+\frac{\partial^2 V_\theta}{\partial z^2}\right]&\end{aligned} \tag{4.73}$$

z-direction

$$\begin{aligned}\rho\left(\frac{\partial V_z}{\partial t}+V_r\frac{\partial V_z}{\partial r}+\frac{V_\theta}{r}\frac{\partial V_z}{\partial \theta}+V_z\frac{\partial V_z}{\partial z}\right)&=-\frac{\partial P}{\partial z}\\+\rho g_z+\eta\left[\frac{1}{r}\frac{\partial}{\partial r}\left(r\frac{\partial V_z}{\partial r}\right)+\frac{1}{r^2}\frac{\partial^2 V_z}{\partial \theta^2}+\frac{\partial^2 V_z}{\partial z^2}\right]&\end{aligned} \tag{4.74}$$

Spherical Coordinates (r, θ, φ):

r-direction

$$\rho\left(\frac{\partial V_r}{\partial t}+V_r\frac{\partial V_r}{\partial r}+\frac{V_\theta}{r}\frac{\partial V_r}{\partial \theta}+\frac{V_\varphi}{r\sin\theta}\frac{\partial V_r}{\partial \varphi}-\frac{V_\theta^2+V_\varphi^2}{r}\right)=-\frac{\partial P}{\partial r}$$
$$+\rho g_r+\eta\left(\nabla^2 V_r-\frac{2V_r}{r^2}-\frac{2}{r^2}\frac{\partial V_\theta}{\partial \theta}-\frac{2V_\theta\cot\theta}{r^2}-\frac{2}{r^2\sin\theta}\frac{\partial V_\varphi}{\partial \varphi}\right) \quad (4.75)$$

θ-direction

$$\rho\left(\frac{\partial V_\theta}{\partial t}+V_r\frac{\partial V_\theta}{\partial r}+\frac{V_\theta}{r}\frac{\partial V_\theta}{\partial \theta}+\frac{V_\varphi}{r\sin\theta}\frac{\partial V_\theta}{\partial \varphi}+\frac{V_rV_\theta}{r}-\frac{V_\varphi^2\cot\theta}{r}\right)=-\frac{1}{r}\frac{\partial P}{\partial \theta}$$
$$+\rho g_\theta+\eta\left(\nabla^2 V_\theta+\frac{2}{r^2}\frac{\partial V_r}{\partial}-\frac{V_\theta}{r^2\sin^2\theta}-\frac{2\cos\theta}{r^2\sin^2\theta}\frac{\partial V_\varphi}{\partial \varphi}\right) \quad (4.76)$$

φ-direction

$$\rho\left(\frac{\partial V_\varphi}{\partial t}+V_r\frac{\partial V_\varphi}{\partial r}+\frac{V_\theta}{r}\frac{\partial V_\varphi}{\partial \theta}+\frac{V_\varphi}{r\sin\theta}\frac{\partial V_\varphi}{\partial \varphi}+\frac{V_\varphi V_r}{r}+\frac{V_\theta V_\varphi}{r}\cot\theta\right)=-\frac{1}{r\sin\theta}\frac{\partial P}{\partial \varphi}$$
$$+\rho g_\varphi+\eta\left(\nabla^2 V_\varphi-\frac{V_\varphi}{r^2\sin^2\theta}+\frac{2}{r^2\sin^2\theta}\frac{\partial V_r}{\partial \varphi}+\frac{2\cos\theta}{r^2\sin^2\theta}\frac{\partial V_\theta}{\partial \varphi}\right) \quad (4.77)$$

Note that the Laplacian ∇^2 in spherical coordinates is:

$$\nabla^2=\frac{1}{r^2}\frac{\partial}{\partial r}\left(r^2\frac{\partial}{\partial r}\right)+\frac{1}{r^2\sin\theta}\frac{\partial}{\partial \theta}\left(\sin\theta\frac{\partial}{\partial \theta}\right)+\frac{1}{r^2\sin^2\theta}\left(\frac{\partial^2}{\partial \varphi^2}\right) \quad (4.78)$$

4.5 MECHANICAL ENERGY EQUATION

The microscopic mechanical energy equation can be derived from the stress equation of motion:

$$\rho\frac{D\vec{V}}{Dt}=\rho\vec{g}-\nabla P+\nabla\cdot\bar{\bar{\tau}} \quad (4.79)$$

We take the dot product of fluid velocity vector with the above stress equation of motion as shown below:

$$\rho\vec{V}\cdot\frac{D\vec{V}}{Dt}=\rho\vec{V}\cdot\vec{g}-\vec{V}\cdot\nabla P+\vec{V}\cdot\nabla\cdot\bar{\bar{\tau}} \quad (4.80)$$

Thus,

$$\rho\frac{D}{Dt}\left(\frac{1}{2}V^2\right)=\rho\vec{V}\cdot\vec{g}-\vec{V}\cdot\nabla P+\vec{V}\cdot\nabla\cdot\bar{\bar{\tau}} \quad (4.81)$$

where V is the magnitude of the velocity vector. Equation (4.81) is the microscopic mechanical energy equation. We can re-cast Equation (4.81) in another form. Let

$$\vec{g}=-\nabla\varphi \quad (4.82)$$

where φ is gravitational potential (potential energy per unit mass). Substituting Equation (4.82) into Equation (4.81), we get:

$$\rho \frac{D}{Dt}\left(\frac{1}{2}V^2\right) = -\rho\vec{V}\cdot\nabla\varphi - \vec{V}\cdot\nabla P + \vec{V}\cdot\nabla\cdot\bar{\bar{\tau}} \tag{4.83}$$

Using the definition of substantial derivative,

$$\frac{D\varphi}{Dt} = \frac{\partial\varphi}{\partial t} + \vec{V}\cdot\nabla\varphi = \vec{V}\cdot\nabla\varphi \tag{4.84}$$

Note that the gravitational potential does not vary with time. Combination of Equations (4.83) and (4.84) yield another form of the mechanical energy equation as:

$$\rho \frac{D}{Dt}\left(\frac{1}{2}V^2 + \varphi\right) = -\vec{V}\cdot\nabla P + \vec{V}\cdot\nabla\cdot\bar{\bar{\tau}} \tag{4.85}$$

We can interpret the mechanical energy equation, Equation (4.85), as follows: The rate of increase of mechanical energy (sum of kinetic and potential energies) of fluid per unit volume is equal to the contribution of the rate of work done on fluid per unit volume by pressure forces that goes into the increase of the mechanical energy plus the contribution of the rate of work done on fluid per unit volume by viscous forces that goes into the increase of the mechanical energy. Note that some of the work done by pressure and viscous forces on the fluid per unit volume goes to internal energy increase of the fluid.

For inviscid and incompressible fluid ($\eta = 0$, ρ=constant), the mechanical energy Equation (4.81) reduces to:

$$\frac{D}{Dt}\left(\frac{1}{2}\rho V^2\right) = \rho\vec{V}\cdot\vec{g} - \vec{V}\cdot\nabla P \tag{4.86}$$

Bernoulli Equation

To derive the Bernoulli equation, we need to define a streamline as the equation is valid along a streamline. A streamline is a curve in space drawn so that the fluid velocity vector is everywhere tangent to the curve. In steady flow, streamlines are the curves traced by the particles of fluid. In unsteady flow, the velocity vector at a given point will change both its magnitude and its direction with time. As the streamlines change with time in unsteady flows, particle trajectories are not equal to the streamlines.

Equation of Streamline

The arc length, S, along any curve in space is a function of spatial coordinates, that is:

$$S = S(x, y, z) \tag{4.87}$$

Alternatively,

$$x = x(S), \quad y = y(S), \quad z = z(S) \tag{4.88}$$

The position vector $\vec{r}$ (see Figure 4.5) can be expressed as:

$$\vec{r} = \vec{r}(S) = x(S)\hat{\imath} + y(S)\hat{\jmath} + z(S)\hat{k} \tag{4.89}$$

The unit tangent vector $\hat{\lambda}$ to the curve is then given as:

$$\hat{\lambda} = \lim_{\Delta S\to 0}\frac{\Delta\vec{r}}{\Delta S} = \frac{d\vec{r}}{dS} \tag{4.90}$$

Thus, a streamline is expressed as:

$$\hat{\lambda} = \frac{\vec{V}}{V} = \frac{d\vec{r}}{ds} \tag{4.91}$$

$$\frac{dx}{ds} = \frac{V_x}{V}\,, \quad \frac{dy}{ds} = \frac{V_y}{V}\,, \quad \frac{dz}{ds} = \frac{V_z}{V} \tag{4.92}$$

where V is the magnitude of the fluid velocity vector.

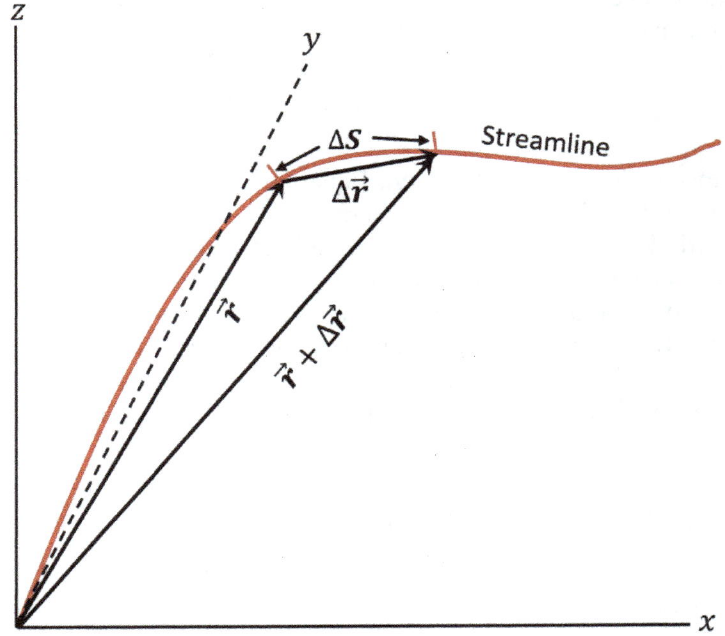

Figure 4.5 Depiction of a streamline.

We now apply the mechanical energy equation for inviscid, incompressible fluid (Equation (4.86)) along a streamline. At steady state, Equation (4.86) gives:

$$\vec{V} \cdot \left[\nabla\left(\frac{1}{2}\rho V^2\right) + \nabla P - \rho\vec{g}\right] = 0 \tag{4.93}$$

Applying Equation (4.93) along a streamline, we get:

$$\frac{d}{ds}\left(\frac{1}{2}\rho V^2\right) + \frac{dP}{ds} - \rho g_s = 0 \tag{4.94}$$

The component of gravity vector along the streamline is given as:

$$g_s = \hat{\lambda} \cdot \vec{g} = \left(\frac{dx}{ds}\hat{\imath} + \frac{dy}{ds}\hat{\jmath} + \frac{dz}{ds}\hat{k}\right) \cdot \left(g_x\hat{\imath} + g_y\hat{\jmath} + g_z\hat{k}\right) \tag{4.95}$$

Assuming gravity to be in the z-direction,

$$g_x = g_y = 0\,, \quad g_z = -g \tag{4.96}$$

Thus,

$$g_s = -g\left(\frac{dz}{ds}\right) \tag{4.97}$$

Combining Equations (4.94) and (4.97), we get:

$$\frac{d}{ds}\left[\frac{1}{2}\rho V^2 + P + \rho g z\right] = 0 \tag{4.98}$$

Equation (4.98) implies that:

$$\frac{1}{2}\rho V^2 + P + \rho g z = \text{constant} \tag{4.99}$$

Since ρ is constant, Equation (4.99) can be re-cast as:

$$\frac{P}{\rho} + gz + \frac{1}{2}V^2 = \text{constant} \tag{4.100}$$

This is the classical Bernoulli equation applicable along a streamline. It assumes steady state and the fluid to be ideal (inviscid and incompressible). According to the Bernoulli equation, the mechanical energy (pressure energy + kinetic energy + potential energy) of ideal fluid is conserved along a streamline.

4.6 EXAMPLES

4.6.1 Example 1

Liquid flows in a straight pipe of radius R as shown in the figure below. At the entrance (section 1), the velocity distribution is uniform with $V_z = U_o$. At section 2, the velocity distribution is given as: $V_z = V_{max}[1 - (r/R)^2]$ where V_{max} is the maximum velocity at the centre $r = 0$. Determine the relationship between V_{max} and U_o.

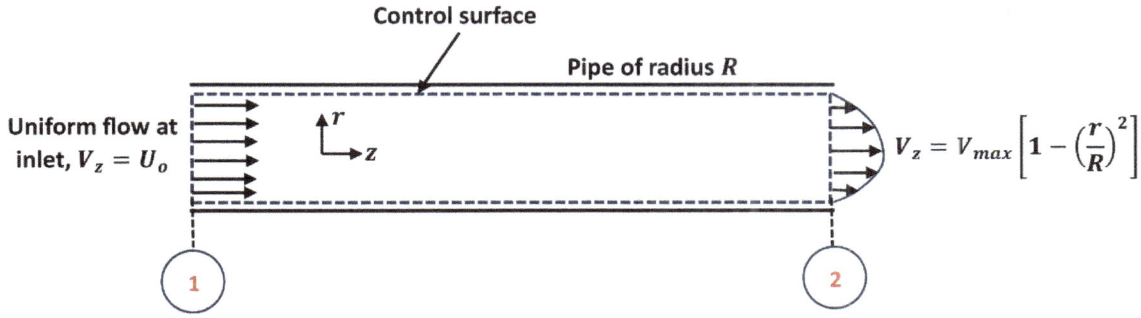

Solution

The control volume and the corresponding control surface are shown in the figure. From Equations (4.1) and (4.2):

$$\iiint_{CV} \left(\frac{\partial \rho}{\partial t}\right) dV + \iint_{CS} \rho(\hat{n} \cdot \vec{V}) dS = 0$$

As density is constant, the first integral on the left-hand side of the above equation is zero. Thus,

$$\iint_{CS} \rho(\hat{n}\cdot\vec{V})dS = 0$$

The above surface integral is evaluated as follows:

$$\iint_{A_1} \rho(\hat{n}\cdot\vec{V})dS + \iint_{A_2} \rho(\hat{n}\cdot\vec{V})dS = -\rho U_o A_1 + \iint_{A_2} \rho(\hat{n}\cdot\vec{V})dS = 0$$

where A_1 is cross-section area at section 1 and A_2 is cross-section area at section 2. To evaluate the surface integral over A_2, we take a thin circular strip of area dS as shown below:

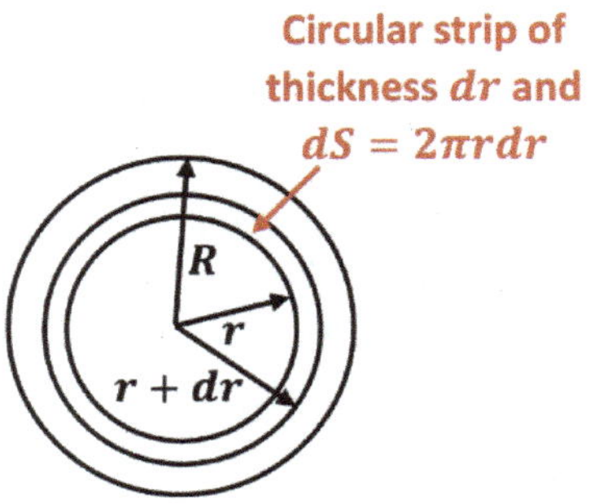

Thus,

$$-\rho U_o A_1 + \rho \int_0^R V_z\,(2\pi r)dr = 0$$

As density is constant and $V_z = V_{max}[1-(r/R)^2]$, we get:

$$-U_o\pi R^2 + 2\pi V_{max}\int_0^R [1-(r/R)^2]rdr = 0$$

Upon integration and simplification, we obtain:

$$V_{max} = 2U_o$$

4.6.2 Example 2

Water is flowing in a turbulent manner in a U-shaped horizontal bend (I.D.=10.16 cm) at 0.085 m^3/s (see figure). The pressure at the inlet is 144.69 kPa. The pressure at the outlet is 130.91 kPa. The density of water is 1000 kg/m^3. Calculate the horizontal components (F_x and F_y) of the anchoring force required to keep the bend in place.

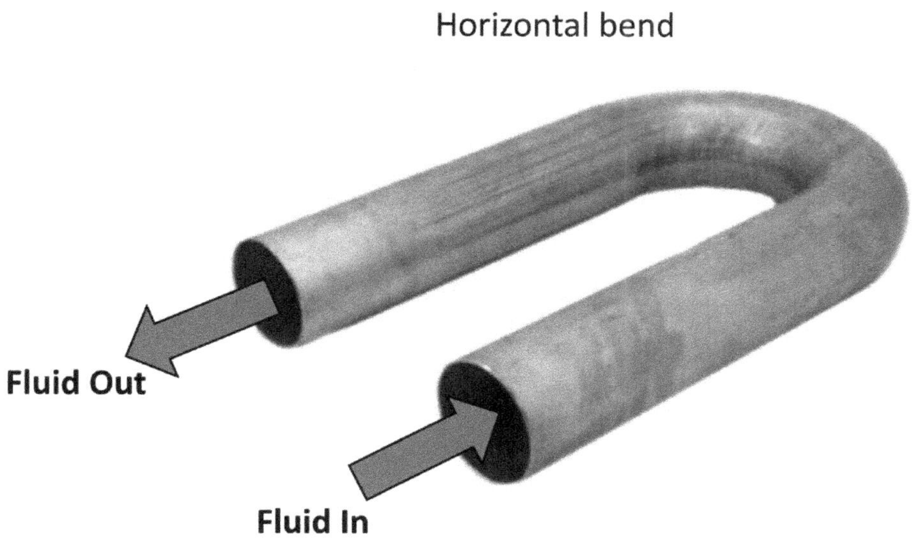

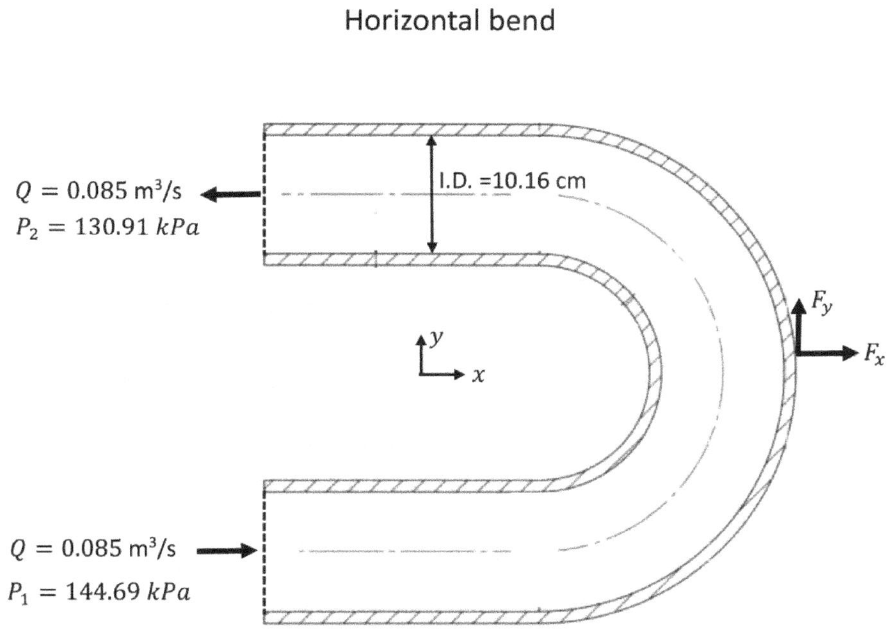

Solution

Momentum balance, Equation (4.17), gives:

$$\frac{D}{Dt}\iiint_{system}\rho\vec{V}dV = \sum\vec{F}$$

Using Equation (4.22), we obtain:

$$\iiint_{CV}\frac{\partial(\rho\vec{V})}{\partial t}dV + \iint_{CS}(\hat{n}\cdot\rho\vec{V}\vec{V})dS = \sum\vec{F}$$

Assuming steady state,

$$\iint_{CS} (\hat{n} \cdot \rho \vec{V}\vec{V}) dS = \sum \vec{F}$$

We apply this equation to the control volume consisting of the volume of the bend itself. The surface integral is evaluated as follows:

$$\iint_{CS} (\hat{n} \cdot \rho \vec{V}\vec{V}) dS = \iint_{CS} (\vec{V})(\hat{n} \cdot \rho \vec{V}) dS = \iint_{A_1} (\vec{V}_1)(\hat{n} \cdot \rho \vec{V}_1) dS + \iint_{A_1} (\vec{V}_2)(\hat{n} \cdot \rho \vec{V}_2) dS$$

where $\vec{V}_1 = |\vec{V}_1|\hat{\imath}$ and $\vec{V}_2 = -|\vec{V}_2|\hat{\imath}$. The right-hand side integrals in the above equation are:

$$\iint_{A_1} (\vec{V}_1)(\hat{n} \cdot \rho \vec{V}_1) dS = |\vec{V}_1|\hat{\imath}(-\dot{m}_1)$$

$$\iint_{A_2} (\vec{V}_2)(\hat{n} \cdot \rho \vec{V}_2) dS = -|\vec{V}_2|\hat{\imath}(\dot{m}_2)$$

As $\dot{m}_1 = \dot{m}_2 = \dot{m}$ and $|\vec{V}_1| = |\vec{V}_2| = V$, we get:

$$\iint_{CS} (\hat{n} \cdot \rho \vec{V}\vec{V}) dS = -2\dot{m}V\hat{\imath} = \sum \vec{F}$$

Let us now determine the required forces:

$$\sum \vec{F} = P_1 A_1 \hat{\imath} + P_2 A_2 \hat{\imath} + F_x \hat{\imath} + F_y \hat{\jmath} = (P_1 A_1 + P_2 A_2 + F_x)\hat{\imath} + F_y \hat{\jmath}$$

where F_x and F_y are the reaction forces acting on the fluid by the bend. Upon comparison with momentum balance, we get:

$$(P_1 A_1 + P_2 A_2 + F_x) = -2\dot{m}V$$

$$F_y = 0$$

Upon substitution of the given values, the following results are obtained:

$$V = 4Q/\pi D^2 = 4 \times 0.085/\pi(0.1016)^2 = 10.484 \text{ m/s}$$

$$\dot{m} = \rho Q = 1000 \times 0.085 = 85 \text{ kg/s}$$

$$F_x = -2\dot{m}V - (P_1 + P_2)A = -2 \times 85 \times 10.484 - (144.69 + 130.91) \times 10^3 \times \frac{\pi}{4}(0.1016)^2$$
$$= -4016.66\ N$$

Thus, the horizontal components of the anchoring force required to keep the bend in place are:

$F_x = -4016.66\ N$ and $F_y = 0$

4.6.3 Example 3

A liquid of density $\rho = 1000$ kg/m^3 is flowing at an unknown velocity V_1 m/s through a horizontal pipe of diameter $D_1 = 0.65$ m at a pressure of 2×10^6 Pa, and then it passes to a section of the pipe in which the diameter is gradually reduced to 0.13 m and the pressure of 1.2×10^4 Pa. Assuming frictionless and streamline flow, calculate the velocities V_1 and V_2

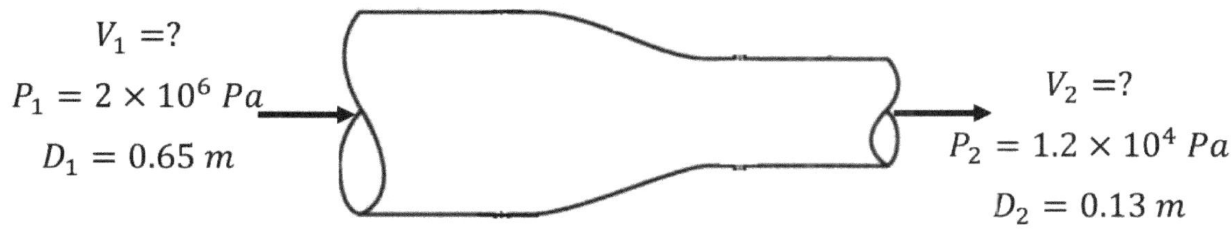

Solution

Apply Bernoulli equation, Equation (4.100), between entrance and exit:

$$\frac{P_1}{\rho} + gz_1 + \frac{V_1^2}{2} = \frac{P_2}{\rho} + gz_2 + \frac{V_2^2}{2}$$

As $z_1 = z_2$, the Bernoulli equation gives:

$$\frac{P_1}{\rho} + \frac{V_1^2}{2} = \frac{P_2}{\rho} + \frac{V_2^2}{2}$$

From mass balance,

$$A_1 V_1 \rho_1 = A_2 V_2 \rho_2$$

As $\rho_1 = \rho_2$, the mass balance equation gives:

$$V_1 = \left(\frac{D_2}{D_1}\right)^2 V_2 = \left(\frac{0.13}{0.65}\right)^2 V_2 = 0.04 V_2$$

Using the relationship between V_1 and V_2 and the Bernoulli equation, we get:

$$V_2^2 \left[\frac{1}{2} - \frac{(0.04)^2}{2}\right] = \frac{P_1 - P_2}{\rho} = \frac{2 \times 10^6 - 1.2 \times 10^4}{1000}$$

Thus,

$V_2 = 63.106$ m/s and $V_1 = 0.04 \times 63.106 = 2.524$ m/s

NOTATION

CS	control surface
CV	control volume

$\vec{F}$	force vector
$\vec{F}_{body}$	body force
$\vec{F}_{surface}$	surface force
F_x, F_y, F_z	force components in x, y, and z directions
$\Delta\vec{F}$	surface force on differential surface element
g_x, g_y, g_z	gravity components in x,y, and z directions, cartesian co-ordinates
g_r, g_θ, g_z	gravity components in r, θ, and z directions, cylindrical co-ordinates
g_r, g_θ, g_φ	gravity components in r, θ, and φ directions, spherical co-ordinates
$\vec{g}$	acceleration dur to gravity vector
$\hat{\imath}, \hat{\jmath}, \hat{k}$	unit vectors in x,y, and z directions in cartesian co-ordinates
$\dot{m}$	mass flow rate
$\hat{n}$	unit normal vector
P	pressure
Q	volumetric flow rate or any point in space
r	radial co-ordinate
$\vec{r}$	position vector
S	surface area or arc length
t	time
tr	trace of a tensor
V	volume or magnitude of velocity vector
$\vec{V}$	velocity vector
V_x, V_y, V_z	velocity components in cartesian co-ordinates
V_r, V_θ, V_z	velocity components in cylindrical co-ordinates
V_r, V_θ, V_φ	velocity components in spherical co-ordinates
V_{max}	maximum velocity
x, y, z	space co-ordinates

Greek Letters

∇	del operator
Δ	change in variable
$\bar{\bar{\delta}}$	unit tensor
η	shear viscosity
η^k	bulk viscosity
θ	cylindrical or spherical angle co-ordinate
λ	shear or angular strain
$\vec{\lambda}$	unit tangent vector
$\vec{\pi}_{(\hat{n})}$	total stress vector
$\bar{\bar{\pi}}$	total stress tensor
$\bar{\bar{\pi}}^T$	transpose of total stress tensor
ρ	density
$\vec{\tau}_{(\hat{n})}$	viscous stress vector
$\bar{\bar{\tau}}$	viscous stress tensor
τ_{ij}	viscous stress components
φ	spherical co-ordinate or gravitational potential
$\bar{\bar{\omega}}$	vorticity tensor

Derivatives

∂/∂ partial time derivative
d/d ordinary derivative
D/Dt substantial or material time derivative

PRACTICE PROBLEMS

Problem 1

A two-dimensional velocity field is given by

$$\vec{V} = a(x^2 - y^2)\hat{\imath} - 2axy\hat{\jmath}$$

where a is a constant. Does this velocity field satisfy the continuity equation for incompressible flow?

Problem 2

Consider a steady-state one-dimensional radial flow in the $r\theta$ plane of a cylindrical geometry, given by $V_r = f(r)$ and $V_\theta = 0$. Determine the form of the function $f(r)$ if the flow is incompressible.

Problem 3

Water flows in a horizontal cylindrical pipe at steady state. The water flow at the inlet is essentially inviscid. Therefore, the velocity at the inlet may be assumed to be constant at 7.25 m/s across the entire cross section; that is:

$$\vec{V}_1 = 7.25\hat{\imath}$$

The velocity at some point downstream the pipe is given by

$$\vec{V}_2 = V_{max}\left[1 - \frac{r}{R}\right]^{1/7}\hat{\imath}$$

where V_{max} is the velocity at the centreline and R is the pipe radius, 7.5 cm.

If the flow is steady and incompressible, what is V_{max} in m/s?

Problem 4

A crude oil having a density of 892 kg/m^3 is flowing through the piping arrangement shown in the figure below at a total volumetric flow rate of 1.388×10^{-3} m^3/s entering section 1. The flow divides equally between sections 2 and 3. Determine the following surface integrals:

i) $\iint_{A_1} \rho\vec{V}\cdot\hat{n}\,dS$ at section 1
ii) $\iint_{A_2} \rho\vec{V}\cdot\hat{n}\,dS$ at section 2
iii) $\iint_{A_3} \rho\vec{V}\cdot\hat{n}\,dS$ at section 3
iv) $\iint_{CS} \rho\vec{V}\cdot\hat{n}\,dS$ for the entire control surface corresponding to the given piping arrangement.

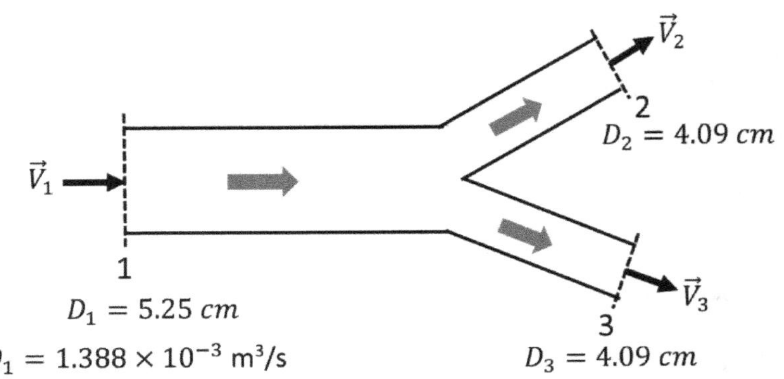

Problem 5

Water is flowing at steady state through a horizontal reducing elbow making a 60° turn in a horizontal plane. At the entering section, $D_1 = 20\ ft, V_1 = 50\ ft/s, P_1 = 40\ psi$. At the exit section, $D_2 = 16\ ft$, $P_2 = 15.8\ psi$. The density of water is 62.4 lb_m/ft^3. Calculate the horizontal components $(F_x$ and $F_y)$ of the anchoring force required to keep the elbow in place. What is the resultant anchoring force?

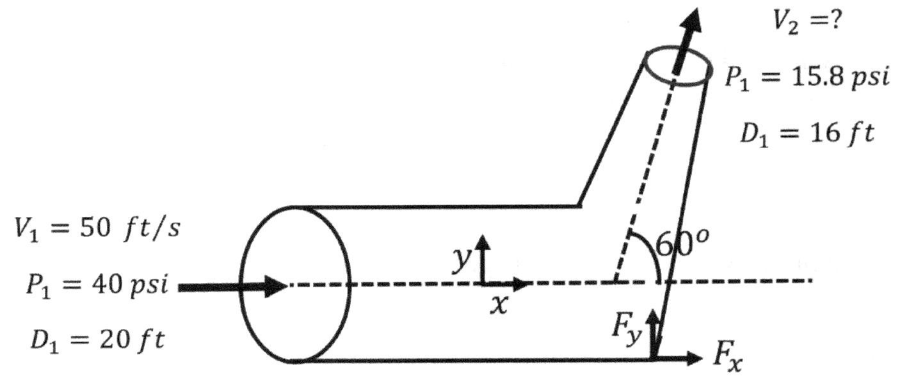

Problem 6

Brine is draining from the bottom of a large open tank through a 4.1 cm diameter pipe. The drainpipe ends at a point 5.15 m below the surface of the brine in the tank. Assuming frictionless and streamline flow, calculate the velocity of brine at the point of discharge from the pipe. The density of brine is 1155 kg/m³.

FURTHER READING

1. R.B. Bird, W.E. Stewart, E.N. Lightfoot. *Transport Phenomena*. 2nd Edition, Wiley, New York, 2007.
2. R. Pal. *Rheology of Particulate Dispersions and Composites*. CRC Press, Boca Raton, 2007.
3. R.L. Street, G.Z. Watters, J.K. Vannard. *Elementary Fluid Mechanics*. 7th Edition, Wiley, New York, 1996.
4. B.R. Munson, D.F. Young, T.H. Okiishi. *Fundamentals of Fluid Mechanics*. 4th Edition, Wiley, New York, 2002.

Chapter 5

Applications of Microscopic Equations of Motion

In this chapter, exact solutions of the equations of motion are developed for a variety of laminar flow problems including steady one non-vanishing velocity component problems, unsteady one non-vanishing velocity component problems, and steady two non-vanishing velocity component problems.

5.1 STEADY-STATE PROBLEMS WITH ONE NON-VANISHING VELOCITY COMPONENT

5.1.1 Boundary Generated Flow Between Large Parallel Plates

Problem:

Consider two infinitely large parallel plates, a distance of $2B$ apart as shown in Figure 5.1. The bottom plate is stationary. The top plate is moving at a constant velocity of V_o. Assuming steady state, laminar flow of a Newtonian fluid of constant properties, determine the velocity distribution and shear stress distribution in the fluid. Plot the velocity and shear stress distributions.

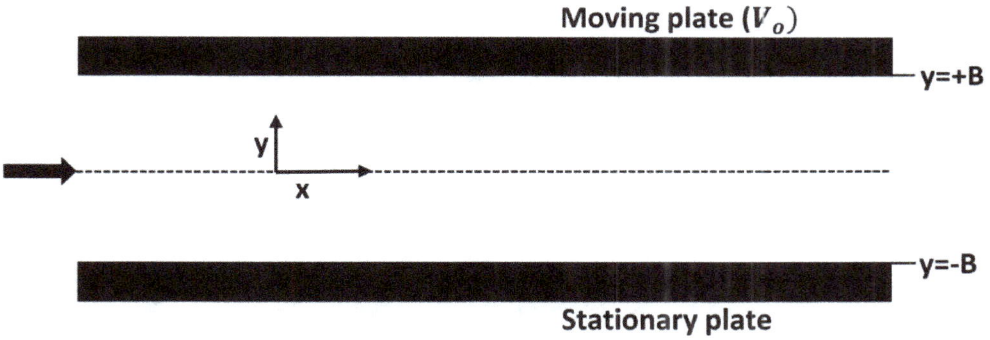

Figure 5.1 Boundary generated flow between two large parallel plates.

Solution:

As the flow is in x -direction, only V_x is non-zero. The continuity equation gives:

$$\frac{\partial V_x}{\partial x} = 0 \tag{5.1}$$

The x-direction Navier-Stokes equation in Cartesian coordinates gives:

$$0 = \eta\left(\frac{d^2 V_x}{dy^2}\right) \tag{5.2}$$

Upon integration, Equation (5.2) yields:

$$V_x = C_1 y + C_2 \tag{5.3}$$

The integration constants C_1 and C_2 can be determined from the following boundary conditions:

$$\text{B.C. 1:} \quad \text{at } y = -B,\ V_x = 0 \text{ (no-slip boundary condition)} \tag{5.4}$$

$$\text{B.C. 2:} \quad \text{at } y = +B,\ V_x = V_o \text{ (no-slip boundary condition)} \tag{5.5}$$

The solution is:

$$C_1 = \frac{V_o}{2B},\ C_2 = \frac{V_o}{2} \tag{5.6}$$

$$V_x = \left(\frac{V_o}{2}\right)\left(1 + \frac{y}{B}\right) \tag{5.7}$$

The shear stress distribution in the fluid can be obtained from Newton's law of viscosity (constitutive equation):

$$\tau_{yx} = \eta\left(\frac{dV_x}{dy}\right) = \eta\frac{V_o}{2B} \tag{5.8}$$

Thus, the shear stress is constant everywhere in the fluid.

The plots of velocity and shear stress distributions are shown in Figure 5.2.

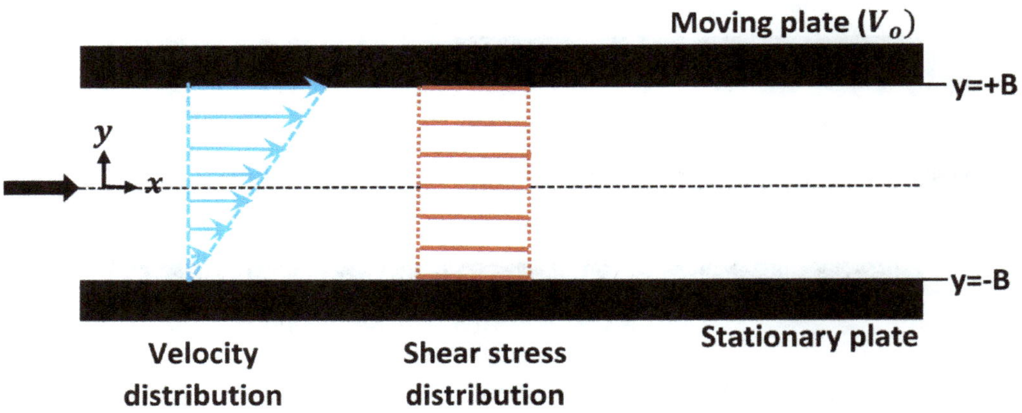

Figure 5.2 Velocity and shear stress distributions.

5.1.2 Pressure and Boundary Generated Flow Between Large Parallel Plates

Problem:

Consider two infinitely large parallel plates, a distance of $2B$ apart as shown in Figure 5.3. The bottom plate is stationary. The top plate is moving at a constant velocity of V_o, thus contributing to motion. There is an additional contribution to fluid motion caused by applying a constant pressure gradient dP/dx (negative value). Assuming steady state, laminar flow of a Newtonian fluid of constant properties, determine the velocity distribution and shear stress distribution in the fluid. Plot the velocity and shear stress distributions.

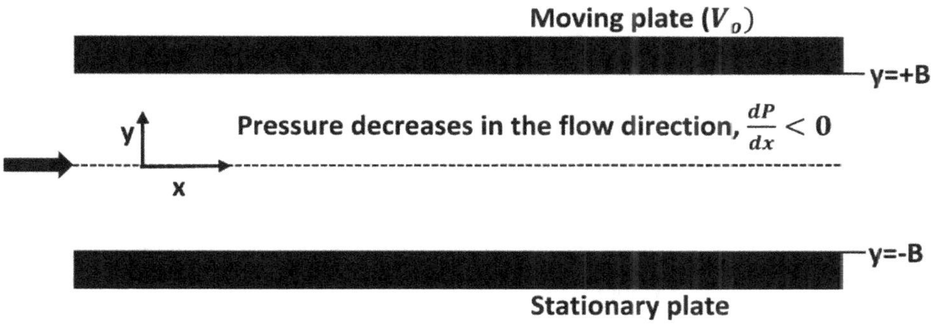

Figure 5.3 Flow generated by moving boundary and pressure gradient between two large parallel plates.

Solution:

As the flow is in x -direction, only V_x is non-zero. The continuity equation gives:

$$\frac{\partial V_x}{\partial x} = 0 \tag{5.9}$$

The x-direction Navier-Stokes equation in Cartesian coordinates gives:

$$0 = -\frac{\partial P}{\partial x} + \eta\left(\frac{\partial^2 V_x}{\partial y^2}\right) \tag{5.10}$$

As P is a function of just x and V_x is a function of just y, we can re-write Equation (5.10) as:

$$\left(\frac{d^2 V_x}{\partial d}\right) = \frac{1}{\eta}\left(\frac{dP}{dx}\right) \tag{5.11}$$

Upon integration, Equation (5.11) yields:

$$V_x = \frac{1}{\eta}\left(\frac{dP}{dx}\right)\frac{y^2}{2} + C_1 y + C_2 \tag{5.12}$$

The integration constants C_1 and C_2 can be determined from the following boundary conditions:

B.C. 1: at $y = -B,\ V_x = 0$ (no-slip boundary condition) (5.13)

B.C. 2: at $y = +B,\ V_x = V_o$ (no-slip boundary condition) (5.14)

The solution is:

$$C_1 = \frac{V_o}{2B},\ C_2 = -\frac{1}{\eta}\left(\frac{dP}{dx}\right)\frac{B^2}{2} \tag{5.15}$$

$$V_x = \left(\frac{V_o}{2}\right)\left(1 + \frac{y}{B}\right) + \frac{1}{\eta}\left(-\frac{dP}{dx}\right)\frac{B^2}{2}\left[1 - \left(\frac{y}{B}\right)^2\right] \tag{5.16}$$

The shear stress distribution in the fluid can be obtained from Newton's law of viscosity (constitutive equation):

$$\tau_{yx} = \eta\left(\frac{dV_x}{dy}\right) = \eta\frac{V_o}{2B} + \left(\frac{dP}{dx}\right)y \tag{5.17}$$

The shear stress is zero at:

$$y = \frac{\eta V_o}{2B}\frac{1}{\left(-\frac{dP}{dx}\right)} \tag{5.18}$$

The plots of velocity and shear stress distributions are shown in Figure 5.4. Note that the velocity is maximum and shear stress is zero at y given by Equation (5.18).

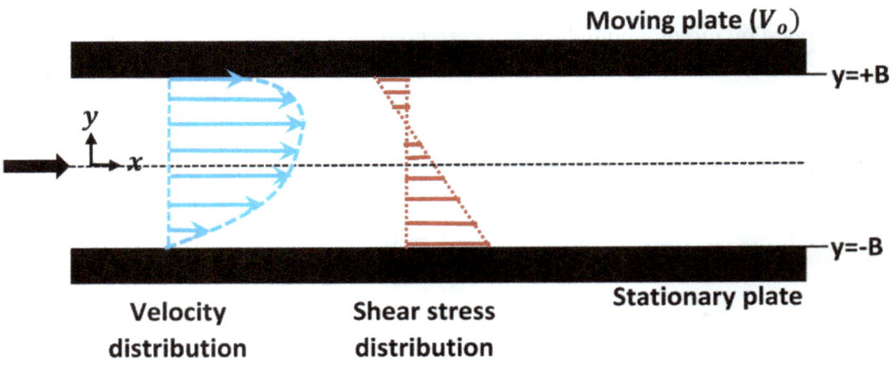

Figure 5.4 Velocity and shear stress distributions.

5.1.3 Thin Liquid Film Flow Over an Inclined Plate

Problem:

Consider steady laminar flow of a Newtonian liquid film flowing down a flat inclined plate due to gravity (see Figure 5.5). The plate is inclined at an angle α with the horizontal. The length and width of the plate are L and W, respectively. The thickness of the liquid film is δ. Determine the following:

i) Velocity distribution in the liquid film
ii) Maximum fluid velocity
iii) Average fluid velocity
iv) Volumetric flow rate of liquid in the film
v) Film thickness in terms of mass flow rate per unit width
vi) Drag force exerted by the fluid on the plate

Plot velocity and shear stress distributions

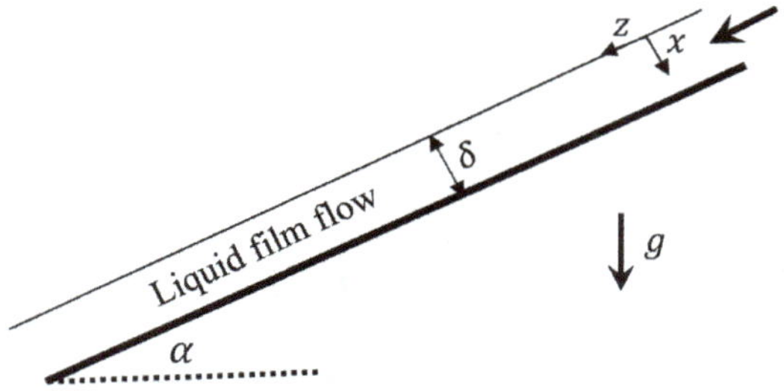

Figure 5.5 Liquid film flow over a flat inclined plate due to gravity.

Solution:

Velocity Distribution:

As the flow is in z-direction, only V_z is non-zero. From continuity equation:

$$\frac{\partial V_z}{\partial z} = 0 \tag{5.19}$$

The z -component of the Navier-Stokes equation in Cartesian coordinates gives:

$$\eta \frac{d^2 V_z}{dx^2} = -\rho g_z \tag{5.20}$$

As $g_z = g \sin\alpha$, Equation (5.20) becomes:

$$\eta \frac{d^2 V_z}{dx^2} = -\rho g \sin\alpha \tag{5.21}$$

Upon integration:

$$V_z = -\left(\frac{\rho g \sin\alpha}{\eta}\right)\frac{x^2}{2} + C_1 x + C_2 \tag{5.22}$$

The integration constants C_1 and C_2 can be determined from the following boundary conditions:

B.C. 1: at $x = 0, \tau_{xz} = 0, \frac{dV_z}{dx} = 0$ (shear stress at liquid surface is zero) (5.23)

B.C. 2: $at\ x = \delta, V_z = 0$ (no-slip boundary condition) (5.24)

The solution is:

$$C_1 = 0\,,\quad C_2 = \left(\frac{\rho g \sin\alpha}{\eta}\right)\frac{\delta^2}{2} \tag{5.25}$$

$$V_z(x) = \frac{\rho g(\sin\alpha)\delta^2}{2\eta}\left[1 - \left(\frac{x}{\delta}\right)^2\right] \tag{5.26}$$

Maximum Fluid Velocity:

The fluid velocity is maximum at $x = 0$:

$$V_{max} = \frac{\rho g(\sin\alpha)\delta^2}{2\eta} \tag{5.27}$$

Average Fluid Velocity:

The average velocity of the liquid film can be determined as follows:

$$Q = \bar{V}(\delta W) = \int_0^\delta V_z W dx \tag{5.28}$$

$$\bar{V} = \frac{1}{\delta}\int_0^\delta V_z dx \tag{5.29}$$

where Q is the volumetric flow rate, W is the width of the plate, and $\bar{V}$ is the average liquid velocity in the film. From Equations (5.26) and (5.29), we obtain:

$$\bar{V} = \frac{\rho g \sin\alpha \delta^2}{3\eta} \tag{5.30}$$

Note that:

$$\bar{V} = \frac{2}{3} V_{max} \tag{5.31}$$

Volumetric Flow Rate of Liquid in the Film:

$$Q = \bar{V}(\delta W) = \frac{\rho g \sin\alpha \delta^3}{3\eta} W \tag{5.32}$$

Film Thickness in Terms of Mass Flow Rate Per Unit Width:

From Equation (5.32),

$$\delta^3 = \frac{3\eta Q}{g\rho \sin\alpha W} = \frac{3\eta \dot{m}}{g\rho^2 \sin\alpha W} \tag{5.33}$$

where $\dot{m}$ is the liquid mass flow rate. From Equation (5.33), it follows that:

$$\delta = \left(\frac{3\eta \dot{\Gamma}}{g\rho^2 \sin\alpha}\right)^{1/3} \tag{5.34}$$

where $\dot{\Gamma}$ is the liquid mass flow rate per unit width of the plate.

Drag Force Exerted by the Fluid on the Plate:

The drag force on the plate exerted by the fluid is as follows:

$$F_D = (-\tau_{xz}|_{x=\delta})WL \tag{5.35}$$

where L is the length of the plate. To determine the drag force F_D, we need to calculate the shear stress τ_{xz} as follows:

$$\tau_{xz} = \eta\left(\frac{dV_z}{dx}\right) \tag{5.36}$$

From Equations (5.26) and (5.36), we get:

$$\tau_{xz} = -\rho g x \sin\alpha \tag{5.37}$$

Thus,

$$F_D = \rho g \sin\alpha\, \delta W L \tag{5.38}$$

Upon re-arrangement, Equation (5.38) gives:

$$F_D = mg \sin\alpha \tag{5.39}$$

where m is the mass of the liquid film. Thus, the drag force exerted by the liquid film on the inclined plate is equal to the weight of the liquid along the direction of flow.

Plot Velocity and Shear Stress Distributions:

The plots of velocity and shear stress distributions in a falling liquid film over an inclined plate are shown in Figure 5.6.

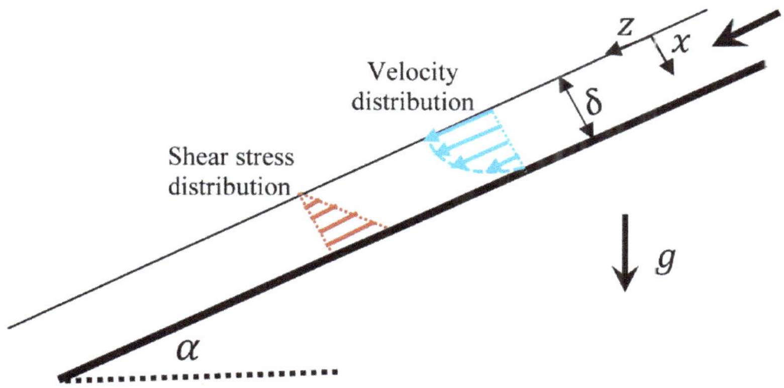

Figure 5.6 Velocity and shear stress distributions in a liquid film.

5.1.4 Flow in a Pipe

Problem:

Consider steady laminar flow of a Newtonian fluid of constant properties in a horizontal pipe of radius R and length L under the influence of pressure gradient $\partial P/\partial z$ (see Figure 5.7). Do the following:

- i) Determine velocity distribution in the pipe
- ii) Determine maximum fluid velocity
- iii) Determine average fluid velocity
- iv) Derive the Hagen-Poiseuille law
- v) Determine the drag force exerted by the fluid on the pipe wall
- vi) Plot velocity and shear stress distributions

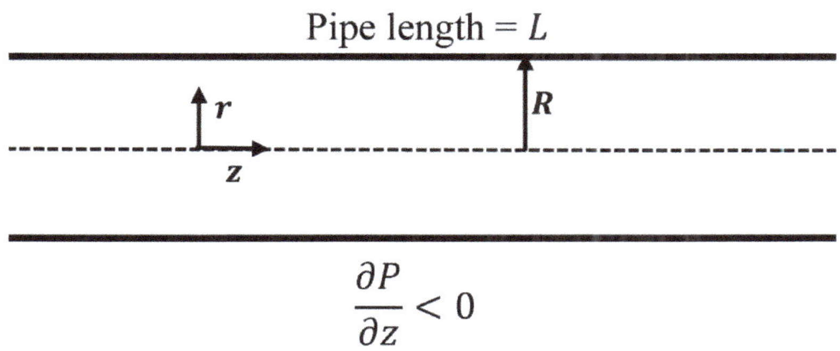

Figure 5.7 Fluid flow in a horizontal flow under the influence of pressure gradient.

Solution:

Velocity Distribution:
As the flow is in z-direction, only V_z is non-zero. From continuity equation:

$$\frac{\partial V_z}{\partial z} = 0 \tag{5.40}$$

The z -component of the Navier-Stokes equation in cylindrical coordinates gives:

$$-\frac{\partial P}{\partial z} + \rho g_z + \frac{\eta}{r}\frac{\partial}{\partial r}\left(r\frac{\partial V_z}{\partial r}\right) = 0 \tag{5.41}$$

P is a function of z only, V_z is a function of r only, and $g_z = 0$. Thus, Equation (5.41) can be re-written as:

$$\frac{\eta}{r}\frac{d}{dr}\left(r\frac{dV_z}{dr}\right) = \frac{dP}{dz} \tag{5.42}$$

The left-hand side of Equation (5.42) is a function of r only and the right-hand side of Equation (5.42) is a function of z only, and r and z are independent variables. The only way we can have a function of r and a function of z equal is for them to be a constant. Thus, dP/dz =constant. We can now integrate Equation (5.42) as:

$$r\frac{dV_z}{dr} = \left(\frac{dP}{dz}\right)\frac{r^2}{2\eta} + C_1 \tag{5.43}$$

Consequently,

$$\frac{dV_z}{dr} = \left(\frac{dP}{dz}\right)\frac{r}{2\eta} + \frac{C_1}{r} \tag{5.44}$$

We must take $C_1 = 0$ as otherwise the velocity gradient diverges at the centre of the pipe. Note that due to symmetry of flow, the velocity gradient is zero at the centre of the pipe. Thus,

$$\frac{dV_z}{dr} = \left(\frac{dP}{dz}\right)\frac{r}{2\eta} \tag{5.45}$$

Upon further integration, Equation (5.45) gives:

$$V_z = \left(\frac{dP}{dz}\right)\frac{r^2}{4\eta} + C_2 \tag{5.46}$$

The integration constant C_2 can be determined from the following boundary condition:

$$\text{B.C.:} \quad at\ r = R, V_z = 0 \text{ (no-slip boundary condition)} \tag{5.47}$$

The solution is:

$$C_2 = -\left(\frac{dP}{dz}\right)\frac{R^2}{4\eta} \tag{5.48}$$

$$V_z(r) = -\left(\frac{dP}{dz}\right)\frac{R^2}{4\eta}\left[1 - \left(\frac{r}{R}\right)^2\right] \tag{5.49}$$

Maximum Fluid Velocity:

The fluid velocity is maximum at $r = 0$:

$$V_{max} = \frac{R^2}{4\eta}\left(-\frac{dP}{dz}\right) \tag{5.50}$$

Thus, the velocity distribution given in Equation (5.49) can be re-written as:

$$\frac{V_z(r)}{V_{max}} = \left[1 - \left(\frac{r}{R}\right)^2\right] \tag{5.51}$$

Average Fluid Velocity:

The average velocity of fluid can be determined as follows:

$$Q = \bar{V}(\pi R^2) = \int_0^R V_z(2\pi r)dr \tag{5.52}$$

$$\bar{V} = \frac{2}{R^2}\int_0^R V_z dr \tag{5.53}$$

where Q is the volumetric flow rate, and $\bar{V}$ is the average fluid velocity in the pipe. From Equations (5.51) and (5.53), we obtain:

$$\bar{V} = \frac{1}{2}V_{max} \tag{5.54}$$

where V_{max} is given by Equation (5.50).

Hagen-Poiseuille law:

The Hagen-Poiseuille law relates the pressure gradient in pipeline laminar flow of Newtonian fluid to the volumetric flow rate of fluid. It can be derived using Equations (5.50) and (5.53) as follows:

$$\bar{V} = \frac{1}{2}V_{max} = \frac{R^2}{8\eta}\left(-\frac{dP}{dz}\right) \tag{5.55}$$

As $\bar{V} = Q/(\pi R^2)$, Equation (5.55) can be re-arranged as:

$$\left(-\frac{dP}{dz}\right) = \frac{\Delta P}{L} = \frac{8\eta Q}{\pi R^4} \tag{5.56}$$

where ΔP is pressure-drop in the pipe (positive value). Equation (5.56) is the Hagen-Poiseuille law. It could also be expressed in terms of pipe diameter as:

$$\frac{\Delta P}{L} = \frac{128\eta Q}{\pi D^4} \tag{5.57}$$

Thus, the pressure gradient in pipeline laminar flow of Newtonian fluid is directly proportional to the volumetric flow rate of fluid.

Drag Force Exerted by the Fluid on the Pipe Wall:

The drag force on the pipe wall exerted by the fluid is as follows:

$$F_D = (-\tau_{rz}|_{r=R})\pi DL \tag{5.58}$$

where L is the length of the pipe. To determine the drag force F_D, we need to calculate the shear stress τ_{rz} as follows:

$$\tau_{rz} = \eta\left(\frac{dV_z}{dr}\right) \tag{5.59}$$

From Equations (5.51) and (5.59), we get:

$$\tau_{rz} = -\frac{2\eta V_{max}}{R^2}r \tag{5.60}$$

Upon substitution of V_{max} from Equation (5.50) into Equation (5.60), we get:

$$\tau_{rz} = -\frac{1}{2}\left(\frac{\Delta P}{L}\right)r \tag{5.61}$$

Thus, the drag force on the pipe wall is:

$$F_D = (\Delta P)\pi R^2 \tag{5.62}$$

Plot Velocity and Shear Stress Distributions:

The plots of velocity and shear stress distributions in pipe flow are shown in Figure 5.8.

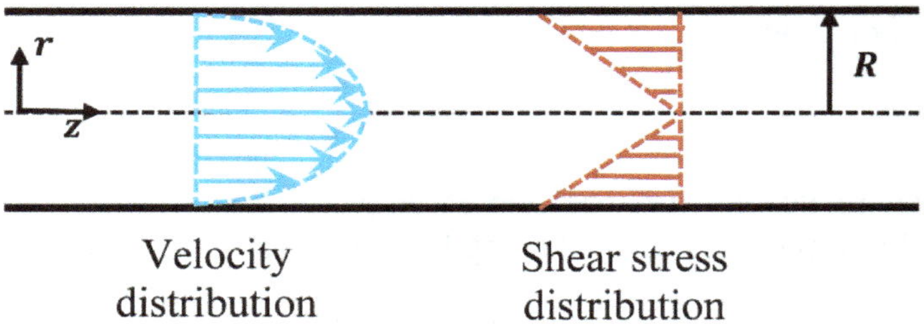

Figure 5.8 Velocity and shear stress distributions in pipe flow.

5.1.5 Flow in Annular Region of Concentric Pipes

Problem:

Consider steady laminar flow of a Newtonian fluid of constant properties in an annular region between two concentric horizontal pipes of radii kR and R under the influence of pressure gradient $\partial P/\partial z$ (see Figure 5.9). Do the following:

i) Determine velocity distribution in the annular region
ii) Determine maximum fluid velocity
iii) Determine average fluid velocity
iv) Determine volumetric flow rate
v) Determine the shear stress distribution
vi) Determine the drag force exerted by the fluid on the inner and outer pipes of the annulus
vii) Plot velocity and shear stress distributions

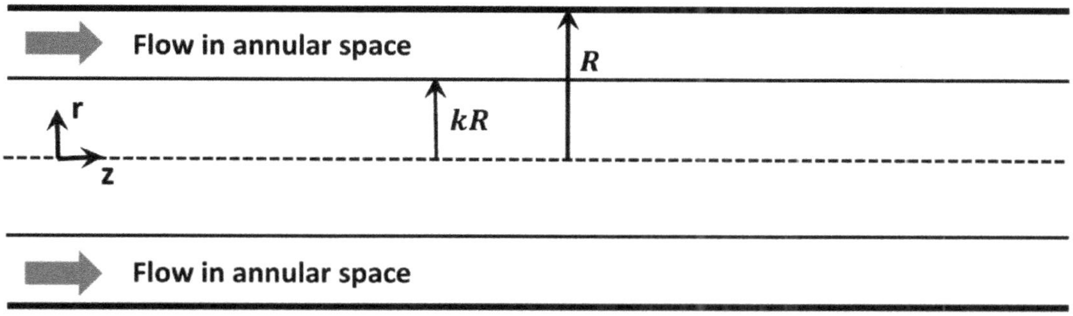

Figure 5.9 Fluid flow in the annular region of horizontal concentric pipes under the influence of pressure gradient.

Solution:

Velocity Distribution:

As the flow is in z-direction, only V_z is non-zero. From continuity equation:

$$\frac{\partial V_z}{\partial z} = 0 \tag{5.63}$$

The z -component of the Navier-Stokes equation in cylindrical coordinates gives:

$$-\frac{\partial P}{\partial z} + \rho g_z + \frac{\eta}{r}\frac{\partial}{\partial r}\left(r\frac{\partial V_z}{\partial r}\right) = 0 \tag{5.64}$$

P is a function of z only, V_z is a function of r only, and $g_z = 0$. Thus, Equation (5.64) can be re-written as:

$$\frac{\eta}{r}\frac{d}{dr}\left(r\frac{dV_z}{dr}\right) = \frac{dP}{dz} \tag{5.65}$$

The left-hand side of Equation (5.65) is a function of r only and the right-hand side of Equation (5.65) is a function of z only, and r and z are independent variables. The only way we can have a function of r and a function of z equal is for them to be a constant. Thus, dP/dz =constant. We can now integrate Equation (5.65) as:

$$r\frac{dV_z}{dr} = \left(\frac{dP}{dz}\right)\frac{r^2}{2\eta} + C_1 \tag{5.66}$$

Upon further integration, we get:

$$V_z = \left(\frac{dP}{dz}\right)\frac{r^2}{4\eta} + C_1 lnr + C_2 \tag{5.67}$$

The integration constants C_1 and C_2 can be determined from the following boundary conditions:

B.C. 1: at $r = kR, V_z = 0$ (no-slip boundary condition) (5.68)

B.C. 2: $at\ x = R, V_z = 0$ (no-slip boundary condition) (5.69)

The solution is:

$$C_1 = \left(-\frac{dP}{dz}\right)\frac{R^2}{4\eta}\left[\frac{(1-k^2)}{ln(1/k)}\right], \quad C_2 = \left(-\frac{dP}{dz}\right)\frac{R^2}{4\eta}\left[1 + (1-k^2)\frac{lnR}{lnk}\right] \tag{5.70}$$

$$V_z(r) = \left(-\frac{dP}{dz}\right)\frac{R^2}{4\eta}\left[1 - \left(\frac{r}{R}\right)^2 + \frac{(1-k^2)}{ln(1/k)} ln\left(\frac{r}{R}\right)\right] \tag{5.71}$$

Note that when $k = 0$, Equation (5.71) reduces to the velocity distribution for flow in a cylindrical pipe, that is, Equation (5.49).

Maximum Fluid Velocity:

The fluid velocity is maximum where the $dV_z/dr = 0$. Thus,

$$\frac{dV_z}{dr} = \left(-\frac{dP}{dz}\right)\frac{R^2}{4\eta}\left[-\frac{2r}{R^2} + \frac{(1-k^2)}{ln(1/k)}\left(\frac{1}{r}\right)\right] = 0 \tag{5.72}$$

Equation (5.72) gives the following value of r where the fluid velocity is maximum:

$$r = \left[\sqrt{\frac{(1-k^2)}{2ln(1/k)}}\right] R \tag{5.73}$$

Upon substituting the expression for r from Equation (5.73) into Equation (5.71), we get:

$$V_{max} = \left(-\frac{dP}{dz}\right)\frac{R^2}{4\eta}\left[1 - \left\{\frac{(1-k^2)}{2ln(1/k)}\right\}\left\{1 - ln\left(\frac{1-k^2}{2ln(1/k)}\right)\right\}\right] \tag{5.74}$$

Average Fluid Velocity:

The average velocity of fluid can be determined as follows:

$$Q = \bar{V}\pi R^2(1-k^2) = \int_{kR}^{R} V_z(2\pi r)dr \tag{5.75}$$

$$\bar{V} = \frac{2}{R^2(1-k^2)}\int_{kR}^{R} V_z dr \tag{5.76}$$

where Q is the volumetric flow rate, and $\bar{V}$ is the average fluid velocity in the annular region. From Equations (5.71) and (5.76), it can be readily shown that:

$$\bar{V} = \left(-\frac{dP}{dz}\right)\frac{R^2}{8\eta}\left[\left\{\frac{1-k^4}{1-k^2}\right\} - \left\{\frac{1-k^2}{ln(1/k)}\right\}\right] \tag{5.77}$$

Volumetric Flow Rate:

The volumetric flow rate Q is as follows:

$$Q = \bar{V}\pi R^2(1-k^2) = \left(-\frac{dP}{dz}\right)\frac{\pi R^4}{8\eta}\left[(1-k^4) - \frac{(1-k^2)^2}{ln(1/k)}\right] \tag{5.78}$$

Shear Stress Distribution:

From the constitutive equation for Newtonian fluids, the shear stress τ_{rz} is given as:

$$\tau_{rz} = \eta\frac{dV_z}{dr} \tag{5.79}$$

$$\tau_{rz} = -\left(-\frac{dP}{dz}\right)\frac{R}{2}\left[\frac{r}{R} - \frac{(1-k^2)}{2ln(1/k)}\left(\frac{R}{r}\right)\right] \tag{5.80}$$

Drag Force Exerted by the Fluid on the Pipe Wall:

The drag force on the annulus exerted by the fluid is as follows:

$$F_D = (-\tau_{rz}|_{r=R})(2\pi RL) + (\tau_{rz}|_{r=kR})(2\pi kRL) \tag{5.81}$$

The first term of on the right-hand side of Equation (5.81) is the drag force on the outer pipe and the second term of on the right-hand side of Equation (5.81) is the drag force on the inner pipe. The stresses at the walls are:

$$-\tau_{rz}|_{r=R} = \left(-\frac{dP}{dz}\right)\frac{R}{2}\left[1 - \frac{(1-k^2)}{2ln(1/k)}\right] \tag{5.82}$$

$$\tau_{rz}|_{r=kR} = -\left(-\frac{dP}{dz}\right)\frac{R}{2}\left[k - \frac{(1-k^2)}{2kln(1/k)}\right] \tag{5.83}$$

From Equations (5.81) to (5.83), it follows that:

$$F_D = (\Delta P)(1-k^2)\pi R^2 \tag{5.84}$$

Note that $\Delta P = -(dP/dz)L$

Plot Velocity and Shear Stress Distributions:

The plots of velocity and shear stress distributions in annular flow are shown in Figure 5.10.

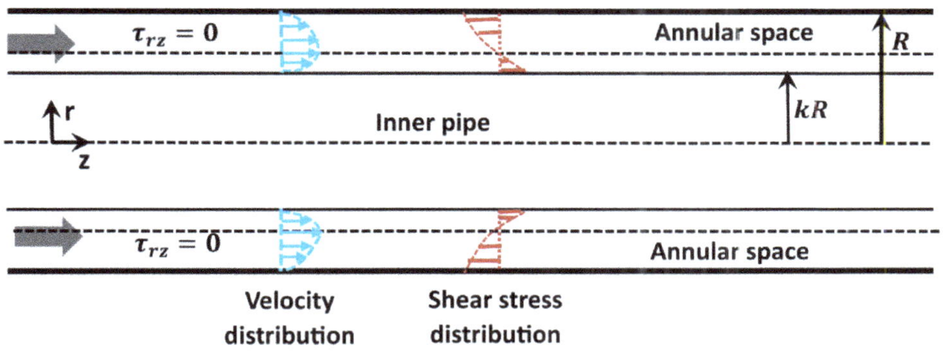

Figure 5.10 Velocity and shear stress distributions in annulus. Note that the velocity profile is skewed towards the inner pipe wall. The maximum in velocity occurs at r value where $\tau_{rz} = 0$. Also, the shear stress distribution is not linear.

5.1.6 Flow in a Co-axial Cylinder Viscometer

Problem:

A co-axial cylinder viscometer consists of two concentric cylinders with the annular space between the cylinders filled with a fluid of unknown viscosity. One cylinder is rotated at a known angular velocity while the other one is held stationary. From the measurement of the torque required to rotate the cylinder, the viscosity of the fluid is deduced. Consider a co-axial cylinder viscometer with annular space filled with a Newtonian liquid of constant properties (see Figure 5.11). The outer cylinder is rotated at an angular velocity of ω and the inner cylinder is held stationary.

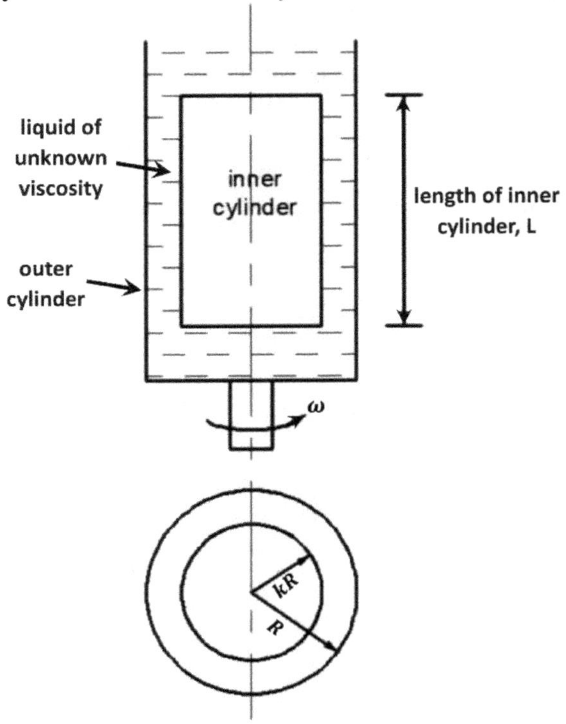

Figure 5.11 Co-axial cylinder viscometer.

Assume that the flow is steady and laminar and that the end effects are negligible. Let the radii of the inner and outer cylinders be kR and R, respectively where k is some constant in the range $0 < k < 1$. Determine the following:

i) Velocity distribution in the liquid
ii) Shear stress distribution in the liquid
iii) Torque required to turn the outer cylinder

Solution:

Velocity Distribution:

As the flow is in θ-direction, only V_θ is non-zero. From continuity equation:

$$\frac{\partial V_\theta}{\partial \theta} = 0 \tag{5.85}$$

Thus, V_θ is a function of r only. The θ-direction Navier-Stokes equation in cylindrical coordinates gives:

$$\frac{d}{dr}\left(\frac{1}{r}\frac{d}{dr}(rV_\theta)\right) = 0 \tag{5.86}$$

Upon integration, we get:

$$V_\theta = \frac{C_1 r}{2} + \frac{C_2}{r} \tag{5.87}$$

The integration constants C_1 and C_2 can be determined from the following boundary conditions:

B.C. 1: at $r = R, V_\theta = \omega R$ (no-slip boundary condition) (5.88)

B.C. 2: at r = k$R, V_\theta = 0$ (no-slip boundary condition) (5.89)

The solution is:

$$C_1 = \frac{2\omega}{k^2-1}, \quad C_2 = -\frac{k^2R^2\omega}{k^2-1} \tag{5.90}$$

$$V_\theta = \frac{\omega Rk}{k^2-1}\left[\frac{kR}{r} - \frac{r}{kR}\right] \tag{5.91}$$

Shear Stress Distribution:

Using the constitutive equation, the shear stress distribution in the fluid can be determined as follows:

$$\tau_{r\theta} = \eta\left[r\frac{d}{dr}\left(\frac{V_\theta}{r}\right)\right] \tag{5.92}$$

Upon substitution of V_θ from Equation (5.91) and carrying out the differentiation, we obtain:

$$\tau_{r\theta} = \frac{2\eta(kR)^2\omega}{(1-k^2)r^2} \tag{5.93}$$

Torque Required to Turn the Outer Cylinder:

The torque (T) required to rotate the outer cylinder is as follows:

$$T = (2\pi RL)R\tau_{r\theta}|_{r=R} = \frac{4\pi\eta(kR)^2\omega L}{1-k^2} \tag{5.94}$$

where L is the length of the inner cylinder. Equation (5.94) can be re-written as:

$$T = (G\eta)\omega \tag{5.95}$$

where G is a geometric constant, given as:

$$G = \frac{4\pi(kR)^2L}{1-k^2} \tag{5.96}$$

Thus, one can determine the unknown viscosity of a Newtonian liquid from the known value of G and the slope of torque (T) versus angular velocity (ω) plot.

5.1.7 Flow in a Cone and Plate Viscometer

Problem:

A cone and plate viscometer is widely used to measure the viscosity of liquids. It consists of a flat circular plate which is held stationary (see Figure 5.12). A cone rotates at the top of the plate with a known angular velocity ω. The apex of the cone just touches the bottom plate. The liquid of unknown viscosity is placed in the narrow gap between the cone and plate. The angle between the cone and the plate is typically of the order of one to two degrees. From the measurement of the torque transmitted to the bottom plate, the viscosity of the liquid is deduced.

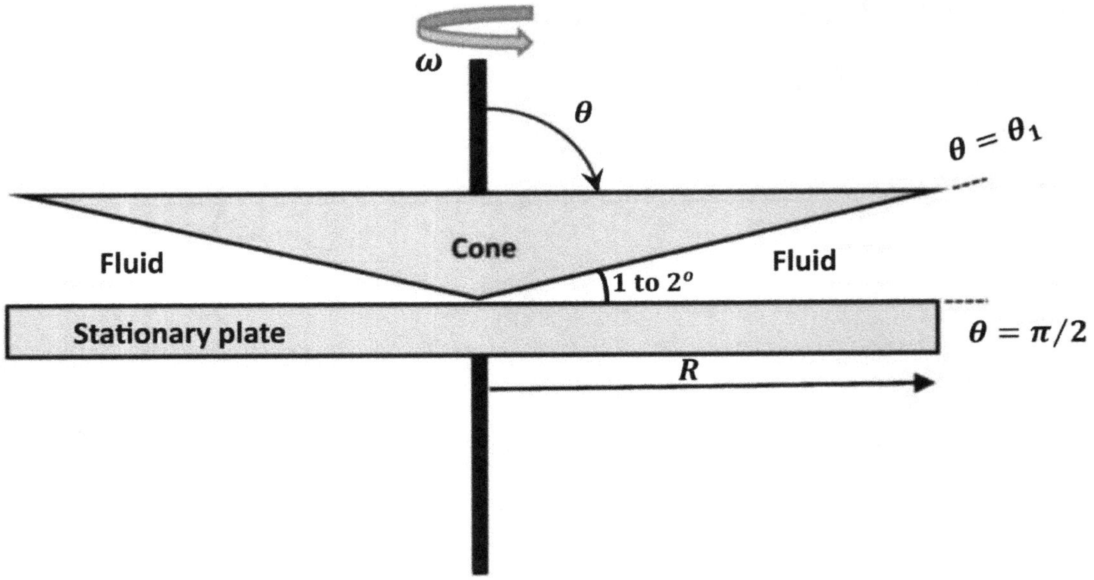

Figure 5.12 Cone and plate viscometer.

Consider a cone and plate viscometer filled with a Newtonian liquid of constant properties. Use a spherical coordinate system with origin fixed at the centre of the bottom plate (corresponding to the apex of the cone). The bottom plate of radius R is held stationary and the top cone is rotated at an angular velocity ω. Determine the following:

i) Velocity distribution in the liquid
ii) Shear stress distribution in the liquid
iii) Torque transmitted to the bottom plate

Solution:

Velocity Distribution:
As the motion is in φ-direction, only V_φ is non-zero. The other two components of velocity (V_r and V_θ) are zero.

From the continuity equation:

$$\frac{\partial V_\varphi}{\partial \varphi} = 0 \tag{5.97}$$

The φ -direction Navier-Stokes equation gives:

$$\nabla^2 V_\varphi - \frac{V_\varphi}{r^2 \sin^2 \theta} = 0 \tag{5.98}$$

Upon substitution of ∇^2operator in spherical coordinates and simplification, the above equation reduces to:

$$\frac{\partial}{\partial r}\left(r^2\frac{\partial V_\varphi}{\partial r}\right)+\frac{1}{\sin\theta}\frac{\partial}{\partial\theta}\left(\sin\theta\frac{\partial V_\varphi}{\partial\theta}\right)-\frac{V_\varphi}{\sin^2\theta}=0 \quad (5.99)$$

Or,

$$r^2\frac{\partial^2 V_\varphi}{\partial r^2}+\frac{\partial^2 V_\varphi}{\partial\theta^2}+2r\frac{\partial V_\varphi}{\partial r}+\cot\theta\frac{\partial V_\varphi}{\partial\theta}-\frac{V_\varphi}{\sin^2\theta}=0 \quad (5.100)$$

This linear, homogeneous, partial differential equation must be solved subject to the following boundary conditions:

B.C. 1: at $\theta=\frac{\pi}{2}$, $V_\varphi=0$ for all r (no-slip boundary condition) (5.101)

B.C. 2: at $\theta=\theta_1, V_\varphi=\omega r\sin\theta_1$ for all r (no-slip boundary condition) (5.102)

B.C. 3: at $r=0$, $V_\varphi=0$ (no-slip boundary condition) (5.103)

B.C. 4: V_φ is bounded for all r (5.104)

Using the separation of variables method, we can write the solution as:

$$V_\varphi=\Re(r)\Phi(\theta) \quad (5.105)$$

Substituting V_φ in the partial differential equation, we get:

$$\frac{r^2}{\Re}\frac{d^2\Re}{dr^2}+\frac{2r}{\Re}\frac{d\Re}{dr}=-\frac{1}{\Phi}\frac{d^2\Phi}{d\theta^2}-\frac{\cot\theta}{\Phi}\frac{d\Phi}{d\theta}+\frac{1}{\sin^2\theta} \quad (5.106)$$

The left side of the equation is a function of r only and the right side of the equation is a function of θ only. For the two sides to be equal for all r and θ, they must be equal to a constant, that is:

$$\frac{r^2}{\Re}\frac{d^2\Re}{dr^2}+\frac{2r}{\Re}\frac{d\Re}{dr}=-\frac{1}{\Phi}\frac{d^2\Phi}{d\theta^2}-\frac{\cot\theta}{\Phi}\frac{d\Phi}{d\theta}+\frac{1}{\sin^2\theta}=\beta \quad (5.107)$$

Thus, we obtain the following two ordinary differential equations:

$$r^2\frac{d^2\Re}{dr^2}+2r\frac{d\Re}{dr}-\beta\Re=0 \quad (5.108)$$

$$\frac{d^2\Phi}{d\theta^2}+\cot\theta\frac{d\Phi}{d\theta}-\frac{\Phi}{\sin^2\theta}+\beta\Phi=0 \quad (5.109)$$

Equation (5.108) is the Euler-Cauchy equation. Substituting the solution $\Re=r^\alpha$ into the equation, we obtain the following two roots of α:

$$\alpha_1=\frac{-1+\sqrt{1+4\beta}}{2} \quad \text{and} \quad \alpha_2=\frac{-1+\sqrt{1+4\beta}}{2} \quad (5.110)$$

Thus, the general solution is:

$$\Re=a_1r^{\alpha_1}+a_2r^{\alpha_2} \quad (5.111)$$

Using the boundary conditions that at $\theta = \theta_1$, $V_\varphi = \Omega r \sin\theta_1$ and that the solution is bounded for all r, we find that $\alpha_1 = 1, \beta = 2, \alpha_2 = -2$, and $a_2 = 0$. Consequently,

$$\Re = a_1 r \tag{5.112}$$

Consider now the other differential equation in Φ. If we substitute $x = cos\theta$, the differential equation in Φ can be re-cast as:

$$(1 - x^2)\frac{d^2\Phi}{dx^2} - 2x\frac{d\Phi}{dx} + \Phi\left(\beta - \frac{1}{1-x^2}\right) = 0 \tag{5.113}$$

This equation is a special case of the following Associated Legendre differential equation:

$$(1 - x^2)y'' - 2xy' + \left[n(n + 1) - \frac{m^2}{1-x^2}\right]y = 0 \tag{5.114}$$

where m and n are non-negative integers (m is called the order and n the degree of the equation). The general solution of the associated Legendre equation is as follows:

$$y = b_1 P_n^m(x) + b_2 Q_n^m(x) \tag{5.115}$$

where $P_n^m(x)$ is the associated Legendre function of first kind (order m, degree n) and $Q_n^m(x)$ is the associated Legendre function of second kind (order m, degree n).

The order and degree of the Associated Legendre differential equation in Φ are both unity (note that $\beta = 2$ gives $n = 1$). Consequently:

$$\Phi = b_1 P_1^1(x) + b_2 Q_1^1(x) \tag{5.116}$$

where the associated Legendre function of first kind $P_1^1(x)$ (order 1, degree1) and the associated Legendre function of second kind $Q_1^1(x)$ (order 1, degree1) are given as:

$$P_1^1(x) = \sqrt{1 - x^2} \tag{5.117}$$

$$Q_1^1(x) = \frac{x}{\sqrt{1-x^2}} + \frac{\sqrt{1-x^2}}{2} ln\left(\frac{1+x}{1-x}\right) \tag{5.118}$$

Thus,

$$\Phi = b_1\left(\sqrt{1 - x^2}\right) + b_2\left[\frac{x}{\sqrt{1-x^2}} + \frac{\sqrt{1-x^2}}{2} ln\left(\frac{1+x}{1-x}\right)\right] \tag{5.119}$$

Since $x = cos\theta$, Equation (5.119) becomes:

$$\Phi = b_1 \sin\theta + b_2\left[\cot\theta + \frac{\sin\theta}{2} ln\left(\frac{1+\cos\theta}{1-\cos\theta}\right)\right] \tag{5.120}$$

Thus, the velocity distribution of fluid is as follows:

$$V_\varphi = \Re(r)\Phi(\theta) = a_1 r\left[b_1 \sin\theta + b_2\left\{\cot\theta + \frac{\sin\theta}{2} ln\left(\frac{1+\cos\theta}{1-\cos\theta}\right)\right\}\right] \tag{5.121}$$

or,

$$V_\varphi = r\left[B_1 \sin\theta + B_2\left\{\cot\theta + \frac{\sin\theta}{2} ln\left(\frac{1+\cos\theta}{1-\cos\theta}\right)\right\}\right] \tag{5.122}$$

where B_1 $(= a_1 b_1)$ and B_2 $(= a_1 b_2)$ are new constants. These constants can be determined from the boundary conditions stated earlier as:

$$\text{At } \theta = \frac{\pi}{2}, V_\varphi = 0 \text{ for all } r. \text{ Thus, } B_1 = 0 \tag{5.123}$$

$$\text{At } \theta = \theta_1, V_\varphi = \omega r \sin\theta_1 \text{ for all } r. \text{ Thus, } B_2 = \frac{\omega \sin\theta_1}{\cot\theta_1 + \frac{\sin\theta_1}{2} ln\left(\frac{1+\cos\theta_1}{1-\cos\theta_1}\right)} \tag{5.124}$$

Consequently, the velocity distribution is as follows:

$$V_\varphi = \omega r \sin\theta_1 \left[\frac{\cot\theta + \frac{\sin\theta}{2} ln\left(\frac{1+\cos\theta}{1-\cos\theta}\right)}{\cot\theta_1 + \frac{\sin\theta_1}{2} ln\left(\frac{1+\cos\theta_1}{1-\cos\theta_1}\right)}\right] \tag{5.125}$$

Since $\theta_1 \approx \frac{\pi}{2}$ and $\theta \approx \frac{\pi}{2}$, this expression simplifies to:

$$V_\varphi = \omega r \frac{\cos\theta}{\cos\theta_1} \tag{5.126}$$

Shear Stress Distribution:

Now that the velocity distribution is known, we can determine shear stress $\tau_{\theta\varphi}$distribution in the fluid using the Newton's law of viscosity as:

$$\tau_{\theta\varphi} = \eta\left[\frac{\sin\theta}{r}\frac{\partial}{\partial\theta}\left(\frac{V_\varphi}{\sin\theta}\right) + \frac{1}{r\sin\theta}\frac{\partial V_\theta}{\partial\varphi}\right] = \eta\left[\frac{\sin\theta}{r}\frac{\partial}{\partial\theta}\left(\frac{V_\varphi}{\sin\theta}\right)\right] \tag{5.127}$$

Upon substituting the velocity distribution and carrying out the necessary differentiation, the shear stress distribution is as follows:

$$\tau_{\theta\varphi} = -\frac{\eta\omega}{\cos\theta_1 \sin\theta} = -\frac{\eta\omega}{\cos\theta_1} \tag{5.128}$$

Thus, the shear stress $\tau_{\theta\varphi}$is constant throughout the gap between the cone and plate.

Torque Transmitted to the Bottom Plate:

The torque (T) transmitted to the bottom plate can be calculated in the following manner:

$$T = \int_0^R -\tau_{\theta\varphi}\Big|_{\theta=\frac{\pi}{2}} r(2\pi r)dr = \frac{2\pi\eta\omega R^3}{3\cos\theta_1} \tag{5.129}$$

Equation (5.129), upon rearrangement, gives:

$$T = G\eta\omega \tag{5.130}$$

where G is a geometric constant, given as:

$$G = \frac{2\pi R^3}{3\cos\theta_1} \tag{5.131}$$

Thus, one can determine the unknown viscosity of a Newtonian liquid from the known value of G and the slope of torque (T) versus angular velocity (ω) plot.

5.1.8 Flow in a Parallel Plate Viscometer

Problem:

A parallel plate or disk viscometer is used to measure the viscosity of viscous liquids. It consists of a flat circular plate which is held stationary (see Figure 5.13). Another parallel plate rotates at the top of the plate with a known angular velocity ω. The liquid of unknown viscosity is placed in the narrow gap between the two plates (height H). From the measurement of the torque required to rotate the top plate, the viscosity of the liquid is deduced.

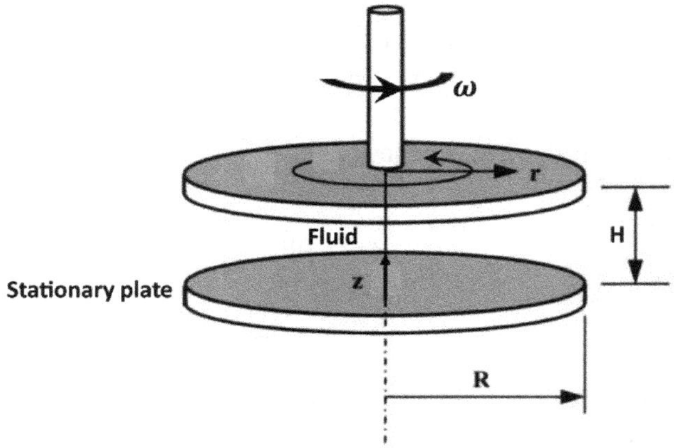

Figure 5.13 Parallel plate viscometer.

Consider a parallel plate viscometer filled with a Newtonian liquid of constant properties. Use a cylindrical coordinate system with origin fixed at the centre of the bottom plate. The bottom plate of radius R is held stationary and the top plate is rotated at an angular velocity ω. Determine the following:

i) Velocity distribution in the liquid
ii) Shear stress distribution in the liquid
iii) Torque required to rotate the top plate at an angular velocity ω

Solution:

Velocity Distribution:

As the motion is in θ-direction, only V_θ is non-zero. The other two components of velocity (V_r and V_z) are zero. From the continuity equation:

$$\frac{\partial V_\theta}{\partial \theta} = 0 \tag{5.132}$$

The θ-direction Navier-Stokes equation gives:

$$\left[\frac{\partial}{\partial r}\left(\frac{1}{r}\frac{\partial}{\partial r}(rV_\theta)\right) + \frac{\partial^2 V_\theta}{\partial z^2}\right] = 0 \tag{5.133}$$

We need to solve this partial differential equation for V_θ subject to the following boundary conditions:

B.C. 1: at $z = 0$, $V_\theta = 0$ for all r (no-slip boundary condition) (5.134)

B.C. 2: at $z = \mathrm{H}$, $V_\varphi = \omega r$ for all r (no-slip boundary condition) (5.135)

B.C. 3: at $r = 0$, $V_\theta = 0$ (5.136)

B.C. 4: V_θ is bounded for all r (5.137)

Clearly, V_θ is a function of r and z. The second boundary condition suggests the structure of the solution. As there is no restriction on r in this boundary condition, the solution is expected to be of the form:

$$V_\theta = rf(z) \tag{5.138}$$

Upon substitution of this solution into the above partial differential equation, we obtain:

$$\frac{d^2 f}{dz^2} = 0 \tag{5.139}$$

Upon integration and using the boundary conditions, the following expression is obtained for $f(z)$:

$$f(z) = \frac{\omega}{H} z \tag{5.140}$$

Thus, the velocity distribution is:

$$V_\theta = \frac{\omega}{H} rz \tag{5.141}$$

Shear Stress Distribution:

Now that the velocity distribution is known, we can determine shear stress $\tau_{z\theta}$ distribution in the fluid using the Newton's law of viscosity as:

$$\tau_{z\theta} = \eta \frac{\partial V_\theta}{\partial z} = \frac{\eta\omega}{H} r \tag{5.142}$$

Torque Required to Rotate the Top Plate:

The torque (T) required to rotate the top plate is:

$$T = \int_0^R \tau_{z\theta}|_{z=H}\, r(2\pi r)dr = \frac{\pi\eta\omega R^4}{2H} \tag{5.143}$$

Equation (5.143), upon rearrangement, gives:

$$T = G\eta\omega \tag{5.144}$$

where G is a geometric constant, given as:

$$G = \frac{\pi R^4}{2\mathrm{H}} \tag{5.145}$$

Thus, one can determine the unknown viscosity of a Newtonian liquid from the known value of G and the slope of torque (T) versus angular velocity (ω) plot.

5.1.9 Flow Around a Rotating Sphere

Problem:

A solid sphere of radius R is rotated slowly in a large pool of stagnant Newtonian fluid at a constant angular velocity of ω (see Figure 5.14) Determine the following:

i) Velocity distribution in the fluid
ii) Shear stress distribution in the fluid
iii) Torque required to rotate the sphere at an angular velocity ω

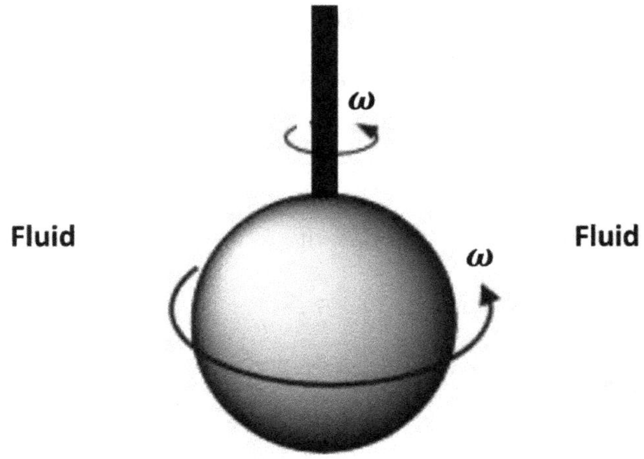

Figure 5.14 A solid sphere rotating in a large pool of stagnant Newtonian fluid.

Solution:

Velocity Distribution:

As the motion is in φ-direction, only V_φ is non-zero. The other two components of velocity (V_r and V_θ) are zero.

From the continuity equation:

$$\frac{\partial V_\varphi}{\partial \varphi} = 0 \tag{5.146}$$

The φ -direction Navier-Stokes equation gives:

$$\nabla^2 V_\varphi - \frac{V_\varphi}{r^2 \sin^2 \theta} = 0 \tag{5.147}$$

Upon substitution of ∇^2operator in spherical coordinates and simplification, the above equation reduces to:

$$\frac{\partial}{\partial r}\left(r^2 \frac{\partial V_\varphi}{\partial r}\right) + \frac{1}{\sin\theta}\frac{\partial}{\partial \theta}\left(\sin\theta \frac{\partial V_\varphi}{\partial \theta}\right) - \frac{V_\varphi}{\sin^2 \theta} = 0 \tag{5.148}$$

Or,

$$r^2\frac{\partial^2 V_\varphi}{\partial r^2}+\frac{\partial^2 V_\varphi}{\partial \theta^2}+2r\frac{\partial V_\varphi}{\partial r}+\cot\theta\frac{\partial V_\varphi}{\partial \theta}-\frac{V_\varphi}{\sin^2\theta}=0 \tag{5.149}$$

This linear, homogeneous, partial differential equation must be solved subject to the following boundary conditions:

B.C. 1: at $r = \mathrm{R}$, $V_\varphi = \omega R \sin\theta$ for all θ (no-slip boundary condition) (5.150)

B.C. 2: at $r \to \infty, V_\varphi = 0$ for all θ (5.151)

B.C. 3: at $\theta = 0$, $V_\varphi = 0$ for all r (5.152)

B.C. 4: V_φ is bounded for all θ (5.153)

The first boundary condition suggests the structure of the solution. As there is no restriction on θ in this boundary condition, the solution is expected to be of the form:

$$V_\varphi = f(r)\sin\theta \tag{5.154}$$

Upon substitution of this solution into the partial differential equation, Equation (5.149), we obtain:

$$r^2\frac{d^2 f}{dz^2}+2\mathrm{r}\frac{df}{dr}-2\mathrm{f}=0 \tag{5.155}$$

Equation (5.155) is the Euler-Cauchy equation. Substituting the solution $f = r^\alpha$ into the equation, we obtain the following two roots of α:

$$\alpha_1 = -2 \quad \text{and} \quad \alpha_2 = 1 \tag{5.156}$$

Thus, the general solution is:

$$f = C_1 r^{-2} + C_2 r \tag{5.157}$$

Using B.C. 2 (Equation (5.151),

$$\text{At } r \to \infty, V_\varphi = 0,\ C_2 = 0 \tag{5.158}$$

Using B.C. 1 (Equation (5.150),

$$\text{at } r = \mathrm{R},\ \ V_\varphi = \omega R\sin\theta,\ C_1 = \omega R^3 \tag{5.159}$$

Substituting C_1 and C_2 into Equation (5.157), we get:

$$f = \frac{\omega R^3}{r^2} \tag{5.160}$$

Thus, the velocity distribution is:

$$V_\varphi = \frac{\omega R^3 \sin\theta}{r^2} \tag{5.161}$$

Shear Stress Distribution:

Now that the velocity distribution is known, we can determine shear stress $\tau_{r\varphi}$ distribution in the fluid using the Newton's law of viscosity as:

$$\tau_{r\varphi} = \eta r \frac{\partial}{\partial r}\left(\frac{V_\varphi}{r}\right) \tag{5.162}$$

From Equations (5.161) and (5.162), it follows that:

$$\tau_{r\varphi} = -3\eta \frac{\omega R^3 \sin\theta}{r^3} \tag{5.163}$$

Torque Required to Rotate the Sphere:

The torque (T) required to rotate the sphere is:

$$T = \int_0^{2\pi}\int_0^{\pi} -\tau_{r\varphi}\big|_{r=R} R^2 sin\theta d\theta d\varphi \tag{5.164}$$

Upon substituting the expression for $\tau_{r\varphi}$ from Equation (5.163) into Equation (5.164) and integrating, we get:

$$T = 8\pi\eta R^3\omega \tag{5.165}$$

5.1.10 Flow in Non-Circular Cross-section Pipes

Triangular Cross-section Duct

A horizontal duct has a length L and a cross-section of triangular shape, bounded by surfaces $y = H$, $y = (\sqrt{3})x$, and $y = (-\sqrt{3})x$ (see Figure 5.15). Obtain the expression for the velocity profile for the steady laminar flow of a Newtonian fluid of constant properties in this duct.

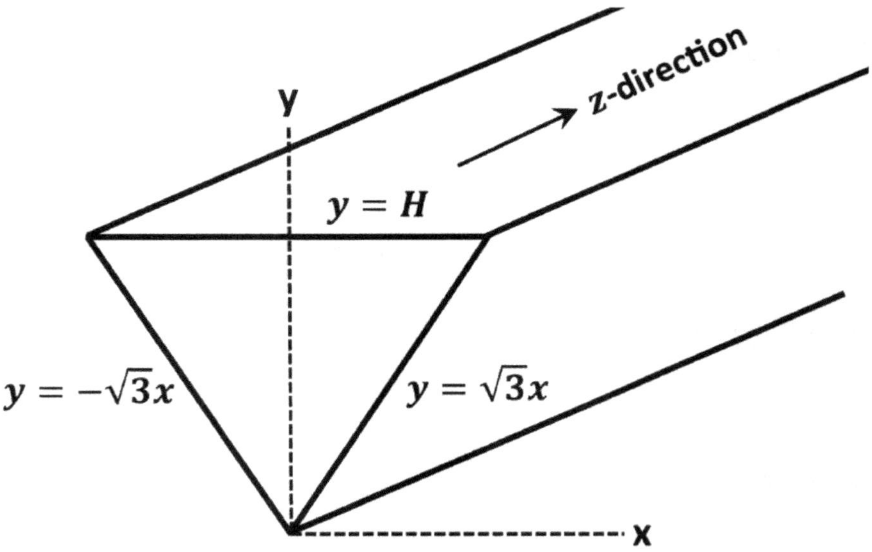

Figure 5.15 A horizontal duct with a cross-section of triangular shape.

Solution:

As the motion is in z-direction, only V_z is non-zero. The other two components of velocity (V_x and V_y) are zero.

From the continuity equation:

$$\frac{\partial V_z}{\partial z} = 0 \tag{5.166}$$

Thus,

$$V_z = V_z(x, y) \tag{5.167}$$

The z -direction Navier-Stokes equation in Cartesian coordinates gives:

$$-\frac{\partial P}{\partial z} + \eta\left(\frac{\partial^2 V_z}{\partial x^2} + \frac{\partial^2 V_z}{\partial y^2}\right) = 0 \tag{5.168}$$

One can often guess the solution from the boundary condition. To satisfy the boundary conditions (see Figure 5.15), we can guess the solution as:

$$V_z = A(y - \sqrt{3}x)(y + \sqrt{3}x)(y - H) \tag{5.169}$$

This solution satisfies the no-slip boundary condition, that is, $V_z = 0$ at the boundary. It also satisfies the continuity equation. It must also satisfy the Navier-Stokes equation (Equation (5.168):

$$\frac{\partial V_z}{\partial x} = -6Axy + 6AHx \tag{5.170}$$

$$\frac{\partial^2 V_z}{\partial x^2} = -6Ay + 6AH \tag{5.171}$$

$$\frac{\partial V_z}{\partial y} = 3Ay^2 - 3Ax^2 - 2AHy \tag{5.172}$$

$$\frac{\partial^2 V_z}{\partial y^2} = 6Ay + 2AH \tag{5.173}$$

$$\frac{\partial^2 V_z}{\partial x^2} + \frac{\partial^2 V_z}{\partial y^2} = 4AH \tag{5.174}$$

Substituting Equation (5.174) into the Navier-Stokes equation, Equation (5.168), we get:

$$\frac{\partial P}{\partial z} = \eta\left(\frac{\partial^2 V_z}{\partial x^2} + \frac{\partial^2 V_z}{\partial y^2}\right) = 4AH\eta \tag{5.175}$$

From Equation (5.175):

$$A = \frac{1}{4H\eta}\left(\frac{\partial P}{\partial z}\right) \tag{5.176}$$

Thus, the solution for velocity profile is:

$$V_z = \frac{1}{4H\eta}\left(\frac{\partial P}{\partial z}\right)(y^2 - 3x^2)(y - H) \tag{5.177}$$

The solution satisfies the continuity equation, the Navier-Stokes equation, and the boundary conditions.

Elliptical Cross-section Duct

Consider steady laminar flow of a Newtonian fluid of constant properties in a horizontal duct of length L and a cross-section of elliptical shape (see Figure 5.16):

$$\frac{x^2}{a^2} + \frac{y^2}{b^2} = 1$$

Obtain the expressions for velocity distribution and volumetric flow rate in this duct.

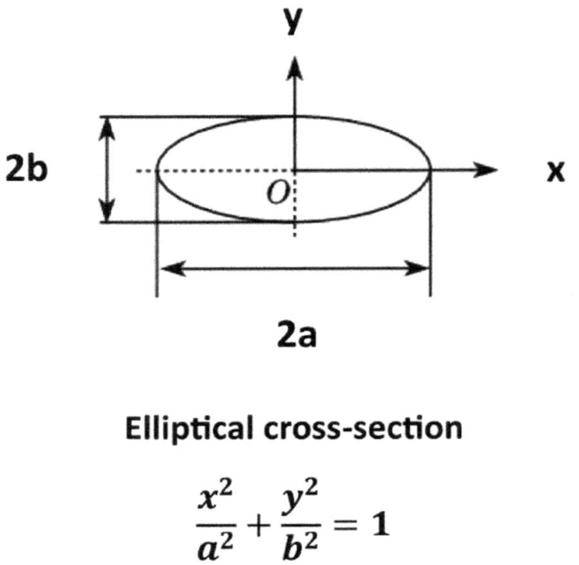

Figure 5.16 Horizontal duct with elliptical shape cross section.

Solution:

Velocity Distribution:

As the motion is in z-direction, only V_z is non-zero. The other two components of velocity (V_x and V_y) are zero.

From the continuity equation:

$$\frac{\partial V_z}{\partial z} = 0 \tag{5.178}$$

Thus,

$$V_z = V_z(x, y) \tag{5.179}$$

The z -direction Navier-Stokes equation in Cartesian coordinates gives:

$$-\frac{\partial P}{\partial z} + \eta\left(\frac{\partial^2 V_z}{\partial x^2} + \frac{\partial^2 V_z}{\partial y^2}\right) = 0 \tag{5.180}$$

One can guess the solution from the boundary condition. To satisfy the boundary conditions (see Figure 5.16), we can guess the solution as:

$$V_z = A\left[1 - \frac{x^2}{a^2} - \frac{y^2}{b^2}\right] \tag{5.181}$$

This solution satisfies the no-slip boundary condition, that is, $V_z = 0$ at the boundary. It also satisfies the continuity equation. It must also satisfy the Navier-Stokes equation (Equation (5.180):

$$\frac{\partial V_z}{\partial x} = A\left(-\frac{2x}{a^2}\right) \tag{5.182}$$

$$\frac{\partial^2 V_z}{\partial x^2} = A\left(-\frac{2}{a^2}\right) \tag{5.183}$$

$$\frac{\partial V_z}{\partial y} = A\left(-\frac{2y}{b^2}\right) \tag{5.184}$$

$$\frac{\partial^2 V_z}{\partial y^2} = A\left(-\frac{2}{b^2}\right) \tag{5.185}$$

$$\frac{\partial^2 V_z}{\partial x^2} + \frac{\partial^2 V_z}{\partial y^2} = -2A\left(\frac{1}{a^2} - \frac{1}{b^2}\right) \tag{5.186}$$

Substituting Equation (5.186) into the Navier-Stokes equation, Equation (5.180), we get:

$$\frac{\partial P}{\partial z} = \eta\left(\frac{\partial^2 V_z}{\partial x^2} + \frac{\partial^2 V_z}{\partial y^2}\right) = -2\eta A\left(\frac{1}{a^2} - \frac{1}{b^2}\right) \tag{5.187}$$

From Equation (5.187):

$$A = \frac{a^2 b^2}{2\eta(a^2+b^2)}\left(-\frac{\partial P}{\partial z}\right) \tag{5.188}$$

Thus, the solution for velocity distribution is:

$$V_z = \frac{a^2b^2}{2\eta(a^2+b^2)}\left(-\frac{\partial P}{\partial z}\right)\left[1 - \frac{x^2}{a^2} - \frac{y^2}{b^2}\right] \tag{5.189}$$

The solution satisfies the continuity equation, the Navier-Stokes equation, and the boundary conditions.

Volumetric Flow Rate:

The volumetric flow rate can be obtained as:

$$Q = \int_{-b}^{b}\int_{-\frac{a}{b}\sqrt{b^2-y^2}}^{\frac{a}{b}\sqrt{b^2-y^2}} V_z dxdy = 4\int_0^b\int_0^{\frac{a}{b}\sqrt{b^2-y^2}} V_z dxdy \tag{5.190}$$

Upon substitution of V_z from Equation (5.189) into Equation (5.190) and carrying out the integration, we get:

$$Q = \frac{\pi}{4\eta}\left(-\frac{\partial P}{\partial z}\right)\frac{a^3b^3}{(a^2+b^2)} \tag{5.191}$$

5.1.11 Simultaneous Flow of Two Immiscible Liquids in a Parallel Plate Geometry

Problem:

Consider steady flow of two immiscible fluids of different densities and viscosities in the z-direction in a horizontal thin slit (parallel plates) under the influence of pressure gradient (see Figure 5.17). The flow rates of the two fluids are so adjusted that the bottom half of the slit is filled with the more viscous and heavier fluid (fluid I) and the top half is filled with the less viscous and lighter fluid (fluid II). Determine the velocity and shear stress distributions in the two fluids. Discuss your results.

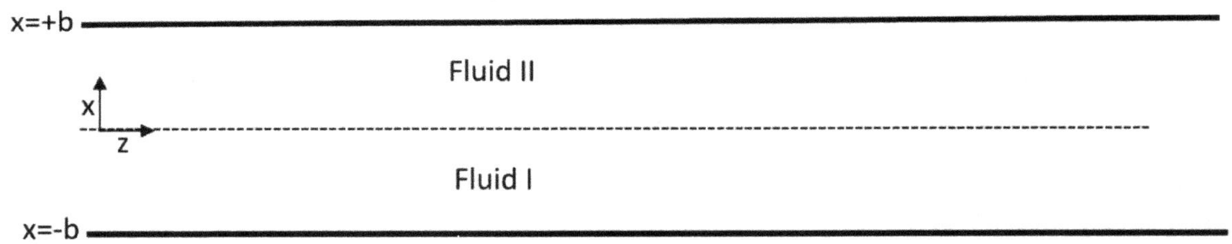

Figure 5.17 Simultaneous flow of two immiscible liquids between parallel plates.

Solution:

The continuity equation gives:

$$\frac{\partial V_z}{\partial z} = 0 \tag{5.192}$$

Thus, V_z is a function of x only. The z -direction Navier-Stokes equation in Cartesian coordinates gives:

$$\eta\left(\frac{d^2V_z}{dx^2}\right) = \frac{dP}{dz} \tag{5.193}$$

Upon integration

$$\left(\frac{dV_z}{dx}\right) = \frac{1}{\eta}\left(\frac{dP}{dz}\right)x + C_1 \quad (5.194)$$

$$V_z = \frac{1}{\eta}\left(\frac{dP}{dz}\right)\frac{x^2}{2} + C_1 x + C_2 \quad (5.195)$$

These equations are valid for both fluids I and II

Fluid II

Using the no-slip boundary condition (V_z=0 at $x = +b$), one can express velocity-gradient and velocity in fluid II as:

$$\left(\frac{dV_z^{II}}{dx}\right) = \frac{1}{\eta^{II}}\left(\frac{dP}{dz}\right)x + C_1^{II} \quad (5.196)$$

$$V_z^{II} = -\frac{b^2}{2\eta^{II}}\left(\frac{dP}{dz}\right)\left(1 - \left(\frac{x}{b}\right)^2\right) + C_1^{II}(x - b) \quad (5.197)$$

where superscript II refers to fluid II (less viscous and lighter fluid).

Fluid I

Using the no-slip boundary condition (V_z=0 at $x = -b$), one can express velocity-gradient and velocity in fluid I as:

$$\left(\frac{dV_z^{I}}{dx}\right) = \frac{1}{\eta^{I}}\left(\frac{dP}{dz}\right)x + C_1^{I} \quad (5.198)$$

$$V_z^{I} = -\frac{b^2}{2\eta^{I}}\left(\frac{dP}{dz}\right)\left(1 - \left(\frac{x}{b}\right)^2\right) + C_1^{I}(x + b) \quad (5.199)$$

where superscript I refers to fluid I (more viscous and heavier fluid).

We still have two unknowns: C_1^I and C_1^{II}. They can be determined using the following boundary conditions at the interface:

$$x = 0 \qquad V_z^I = V_z^{II} \qquad \text{(continuity of velocities)} \quad (5.200)$$

$$x = 0 \qquad \tau_{xz}^I = \tau_{xz}^{II} \qquad \text{(continuity of momentum fluxes)} \quad (5.201)$$

From Newton's law of viscosity:

$$\tau_{xz} = \eta\left(\frac{\partial V_z}{\partial x} + \frac{\partial V_x}{\partial z}\right) = \eta\left(\frac{\partial V_z}{\partial x}\right) \quad (5.202)$$

Thus, the boundary conditions at the interface can be expressed as:

$$x = 0 \qquad -\frac{b^2}{2\eta^I}\left(\frac{dP}{dz}\right) + C_1^I b = -\frac{b^2}{2\eta^{II}}\left(\frac{dP}{dz}\right) - C_1^{II} b \quad (5.203)$$

$$x = 0 \qquad \eta^I C_1^I = \eta^{II} C_1^{II} \tag{5.204}$$

Solving for C_1^I and C_1^{II} we obtain:

$$C_1^I = \left(\frac{\eta^{II}-\eta^I}{\eta^{II}+\eta^I}\right)\left(\frac{b}{2\eta^I}\right)\left(\frac{dp}{dz}\right) \tag{5.205}$$

$$C_1^{II} = \left(\frac{\eta^{II}-\eta^I}{\eta^{II}+\eta^I}\right)\left(\frac{b}{2\eta^{II}}\right)\left(\frac{dp}{dz}\right) \tag{5.206}$$

Thus, for fluid I:

$$V_z^I = -\frac{b^2}{2\eta^I}\left(\frac{dP}{dz}\right)\left[\frac{2\eta^I}{\eta^I+\eta^{II}} + \left(\frac{\eta^I-\eta^{II}}{\eta^I+\eta^{II}}\right)\left(\frac{x}{b}\right) - \left(\frac{x}{b}\right)^2\right] \tag{5.207}$$

$$\left(\frac{dV_z^I}{dx}\right) = \frac{b}{\eta^I}\left(\frac{dP}{dz}\right)\left[\left(\frac{x}{b}\right) - \frac{1}{2}\left(\frac{\eta^I-\eta^{II}}{\eta^I+\eta^{II}}\right)\right] \tag{5.208}$$

$$\tau_{xz}^I = b\left(\frac{dP}{dz}\right)\left[\left(\frac{x}{b}\right) - \frac{1}{2}\left(\frac{\eta^I-\eta^{II}}{\eta^I+\eta^{II}}\right)\right] \tag{5.209}$$

For fluid II:

$$V_z^{II} = -\frac{b^2}{2\eta^{II}}\left(\frac{dP}{dz}\right)\left[\frac{2\eta^{II}}{\eta^I+\eta^{II}} + \left(\frac{\eta^I-\eta^{II}}{\eta^I+\eta^{II}}\right)\left(\frac{x}{b}\right) - \left(\frac{x}{b}\right)^2\right] \tag{5.210}$$

$$\left(\frac{dV_z^{II}}{dx}\right) = \frac{b}{\eta^{II}}\left(\frac{dP}{dz}\right)\left[\left(\frac{x}{b}\right) - \frac{1}{2}\left(\frac{\eta^I-\eta^{II}}{\eta^I+\eta^{II}}\right)\right] \tag{5.211}$$

$$\tau_{xz}^{II} = b\left(\frac{dP}{dz}\right)\left[\left(\frac{x}{b}\right) - \frac{1}{2}\left(\frac{\eta^I-\eta^{II}}{\eta^I+\eta^{II}}\right)\right] \tag{5.212}$$

The plane of zero shear stress ($\tau_{xz} = 0$) is located at:

$$\left(\frac{x}{b}\right) = \frac{1}{2}\left(\frac{\eta^I-\eta^{II}}{\eta^I+\eta^{II}}\right) \tag{5.213}$$

Since $\eta^I > \eta^{II}$, x is positive. Thus, the plane of zero shear stress, and hence maximum velocity, is located in the less viscous fluid (fluid II).

The average velocities in the two fluids are as follows:

$$\bar{V}^I = \frac{1}{b}\int_{-b}^{0} V_z^I dx = \frac{b^2}{3\eta^I}\left(-\frac{dp}{dz}\right) \tag{5.214}$$

$$\bar{V}^{II} = \frac{1}{b}\int_{0}^{b} V_z^{II} dx = \frac{b^2}{3\eta^{II}}\left(-\frac{dp}{dz}\right) \tag{5.215}$$

Thus

$$\frac{\bar{V}^{II}}{\bar{V}^I} = \frac{\eta^I}{\eta^{II}} \tag{5.216}$$

Since $\eta^I > \eta^{II}$, the average velocity in the less viscous fluid ($\bar{V}^{II}$) is higher than the average velocity in the more viscous fluid ($\bar{V}^I$).

Figure 5.18 shows the velocity and shear stress distributions in the two fluids.

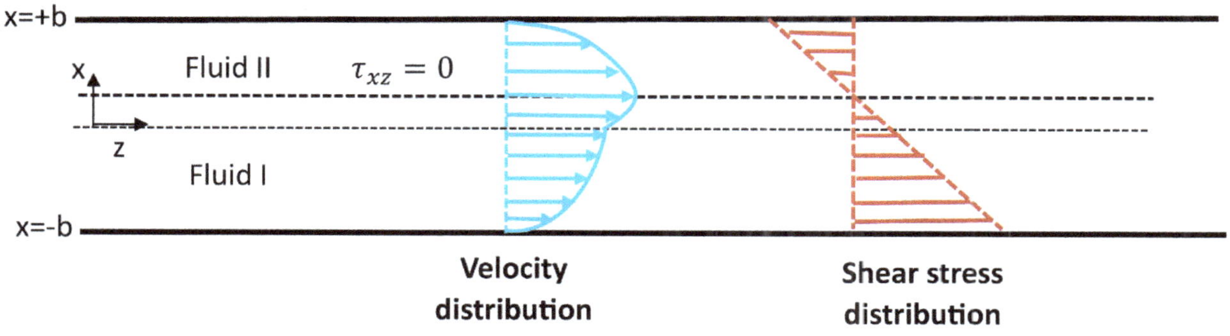

Figure 5.18 Velocity and shear stress distributions in simultaneous flow of two immiscible liquids between parallel plates.

Discussion:

This example illustrates some important concepts about two-phase flow, such as:

a) the less viscous fluid moves faster than the more viscous fluid under the same pressure gradient. As the average velocity of the less viscous fluid II is higher than that of the more viscous fluid ($\bar{V}^{II} > \bar{V}^{I}$), the less viscous fluid II slips past the more viscous fluid I although there is no actual slip at the interface,

b) the in-situ volume fractions of the two fluids are the same in the present example as the slit is half filled with fluid I and half filled with fluid II, that is, $\varphi^{I}_{in-situ} = \varphi^{II}_{in-situ}$, where $\varphi^{I}_{in-situ}$ is the in-situ volume fraction of fluid I and $\varphi^{II}_{in-situ}$ is the in-situ volume fraction of fluid II. However, the volume fractions of the two fluids in the feed (or discharge) are not the same, as indicated below:

$$\varphi^{II}_{feed} > \varphi^{I}_{feed} \tag{5.217}$$

where

$$\varphi^{I}_{feed} = \frac{Q^{I}}{Q^{I}+Q^{II}} \quad \text{and} \quad \varphi^{II}_{feed} = \frac{Q^{II}}{Q^{I}+Q^{II}} \tag{5.218}$$

where Q is the volumetric flow rate. Note that in the present example where the slit is half filled with fluid I and half filled with fluid II, the feed volume fractions are related to the average velocities in the slit as:

$$\varphi^{I}_{feed} = \frac{Q^{I}}{Q^{I}+Q^{II}} = \frac{\bar{V}^{I}}{\bar{V}^{I}+\bar{V}^{II}} \quad \text{and} \quad \varphi^{II}_{feed} = \frac{Q^{II}}{Q^{I}+Q^{II}} = \frac{\bar{V}^{II}}{\bar{V}^{I}+\bar{V}^{II}} \tag{5.219}$$

c) the faster moving fluid (fluid II) has the feed volume fraction higher than the in-situ volume fraction, that is:

$$\varphi^{II}_{feed} > \varphi^{II}_{in-situ} \tag{5.220}$$

d) the slower moving fluid (fluid I) has the feed volume fraction lower than the in-situ volume fraction, that is:

$$\varphi_{feed}^{I} < \varphi_{in-situ}^{I} \tag{5.221}$$

In a variation of the example under consideration, let the feed volume fractions of the two fluids be the same, that is:

$$\varphi_{feed}^{II} = \varphi_{feed}^{I} \tag{5.222}$$

Now the less viscous and faster moving fluid (fluid II) occupies less portion of the cross-section as compared with the more viscous and slower moving fluid (fluid I), that is:

$$\varphi_{in-situ}^{I} > \varphi_{in-situ}^{II} \tag{5.223}$$

Also:

$$\varphi_{feed}^{I} < \varphi_{in-situ}^{I} \tag{5.224}$$

$$\varphi_{feed}^{II} > \varphi_{in-situ}^{II} \tag{5.225}$$

5.2 UNSTEADY-STATE PROBLEMS WITH ONE NON-VANISHING VELOCITY COMPONENT

5.2.1 Large Flat Plate Suddenly Set in Motion

Problem:

A large flat plate immersed in a Newtonian fluid of constant properties is initially stationary (see Figure 5.19). The plate is suddenly set in motion at a constant velocity of V in the x-direction. Determine the velocity distribution in the fluid as a function of x and t.

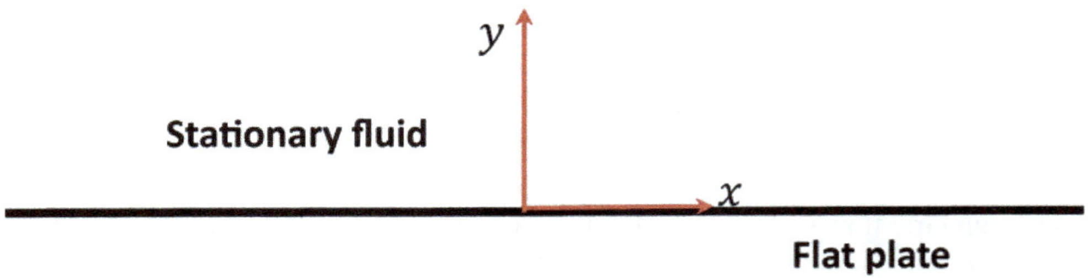

Figure 5.19 A large flat plate immersed in a Newtonian fluid suddenly set in motion.

Solution:

The continuity equation gives:

$$\frac{\partial V_x}{\partial x} = 0 \tag{5.226}$$

Thus, V_x is a function of y and t only. The x-direction Navier-Stokes equation in Cartesian coordinates gives:

$$\frac{\partial V_x}{\partial t} = \nu\left(\frac{\partial^2 V_z}{\partial y^2}\right) \tag{5.227}$$

where ν is the kinematic viscosity, defined as η/ρ.

To obtain the velocity distribution, $V_x(y,t)$, we need to solve the *PDE* (partial differential equation), Equation (5.227), subject to the following initial condition and boundary conditions:

$$\text{I.C.:} \quad V_x(y,0) = 0 \tag{5.228}$$

$$\text{B.C. 1:} \quad V_x(0,t) = V \tag{5.229}$$

$$\text{B.C. 2:} \quad V_x(\infty,t) = 0 \tag{5.230}$$

The *PDE,* Equation (5.227), can be solved using the Laplace transform. Take Laplace transform of Equation (5.227):

$$\mathcal{L}\left(\frac{\partial V_x}{\partial t}\right) = \nu\mathcal{L}\left(\frac{\partial^2 V_z}{\partial y^2}\right) \tag{5.231}$$

$$\mathcal{L}\left(\frac{\partial V_x}{\partial t}\right) = s\mathcal{L}(V_x) - V_x(y,0) = sV_x(y,s) \tag{5.232}$$

$$\mathcal{L}\left(\frac{\partial^2 V_z}{\partial y^2}\right) = \frac{d^2V_x(y,s)}{dy^2} \tag{5.233}$$

where s is a Laplace variable. Substituting the Laplace transforms in Equations (5.232) and (5.233) into Equation (5.231), we get:

$$sV_x(y,s) = \nu\frac{d^2V_x(y,s)}{dy^2} \tag{5.234}$$

Upon rearrangement, Equation (5.234) gives:

$$\frac{d^2V_x}{dy^2} - \frac{s}{\nu}V_x = 0 \tag{5.235}$$

This equation must be solved subject to the following boundary conditions:

$$\text{B.C. 1:} \quad V_x(0,s) = V/s \tag{5.236}$$

$$\text{B.C. 2:} \quad V_x(\infty,s) = 0 \tag{5.237}$$

Equation (5.235) is a linear, second order, homogeneous *ODE* (ordinary differential equation) with constant coefficients. To obtain the general solution, we take $V_x = e^{my}$ and substitute into Equation (5.235) to obtain:

$$m^2 - \frac{s}{\nu} = 0 \tag{5.238}$$

$$m = \pm\sqrt{\frac{s}{\nu}} \tag{5.239}$$

Thus, the general solution of the *ODE* (Equation (5.235)) is:

$$V_x = a_1 e^{\sqrt{\frac{s}{\nu}}y} + a_2 e^{-\sqrt{\frac{s}{\nu}}y} \tag{5.240}$$

For solution to be bounded, $a_1 = 0$. Consequently,

$$V_x = a_2 e^{-\sqrt{\frac{s}{\nu}}y} \tag{5.241}$$

Using the boundary equation (B.C. 1, Equation (5.236)), we get:

$$\frac{V_x(y,s)}{V} = \frac{1}{s} e^{-\sqrt{\frac{s}{\nu}}y} \tag{5.242}$$

Thus,

$$\frac{V_x(y,t)}{V} = \mathcal{L}^{-1}\left(\frac{1}{s} e^{-\sqrt{\frac{s}{\nu}}y}\right) \tag{5.243}$$

From the Laplace transform tables given in mathematical handbooks, we get:

$$erfc\left(\frac{a}{2\sqrt{t}}\right) = \mathcal{L}^{-1}\left(\frac{1}{s} e^{-a\sqrt{s}}\right) \tag{5.244}$$

where $erfc$ is the complementary error function defined as:

$$erfc(\lambda) = 1 - erf(\lambda)\,;\;\; erf(\lambda) = \frac{2}{\sqrt{\pi}} \int_0^{\lambda} e^{-\lambda^2}\, d\lambda \tag{5.245}$$

Upon comparison of Equation (5.244) with the Equation (5.243), we get:

$$\frac{V_x(y,t)}{V} = erfc\left(\frac{y}{2\sqrt{\nu t}}\right) = 1 - erf\left(\frac{y}{2\sqrt{\nu t}}\right) \tag{5.246}$$

This is the final solution of the transient problem. Figure 5.20 shows the plot of $\frac{V_x(y,t)}{V}$ as a function of $\frac{y}{2\sqrt{\nu t}}$.

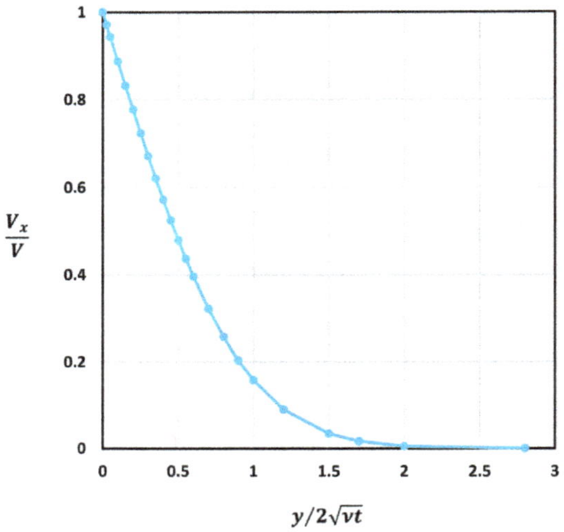

Figure 5.20 Velocity distribution in dimensionless form for plate suddenly set in motion.

5.2.2 Unsteady Pipe Flow

Problem:

Consider a long horizontal pipe of length L and radius R connected to a fluid reservoir at one end and on-off valve at the other end. The fluid is Newtonian of constant viscosity and density. The valve is opened suddenly and the fluid flows through the pipe under a constant pressure gradient $(P_o - P_L)/L$, where L is the length of the pipe. The fluid accelerates initially due to an imbalance between the pressure gradient and opposing viscous stresses. As the velocity of the fluid increases, the opposing viscous stresses also increase. The flow eventually becomes steady when the opposing viscous forces are balanced by the pressure gradient. Note that the acceleration in the fluid is uniform along the pipe as the fluid is incompressible. Determine the transient velocity distribution in the pipe.

Solution:

From the continuity equation:

$$\frac{\partial V_z}{\partial z} = 0 \tag{5.247}$$

The z-direction Navier-Stokes equation (cylindrical coordinates) gives:

$$\rho\frac{\partial V_z}{\partial t} = -\frac{\partial P}{\partial z} + \eta\left[\frac{1}{r}\frac{\partial}{\partial r}\left(r\frac{\partial V_z}{\partial r}\right)\right] \tag{5.248}$$

The initial and the boundary conditions are:

$$\text{I.C.: at } t = 0, V_z = 0 \tag{5.249}$$

$$\text{B.C. 1: at } r = 0, \frac{\partial V_z}{\partial r} = 0 \tag{5.250}$$

$$\text{B.C. 2: at } r = R, V_z = 0 \text{ (no-slip)} \tag{5.251}$$

The steady-state solution to this problem (see section 5.1.4) is as follows:

$$V_z = V_{max}\left[1 - \left(\frac{r}{R}\right)^2\right] \quad \text{where} \quad V_{max} = \left(-\frac{\partial p}{\partial z}\right)\frac{R^2}{4\eta} = \frac{(p_o - p_L)R^2}{4\eta L} \tag{5.252}$$

For the sake of convenience, the Navier-Stokes equation and the initial and boundary conditions are expressed in terms of the following dimensionless variables:

$$\xi = \frac{r}{R}, \quad \tau = \frac{t}{(R^2/\nu)}, \quad \varphi = \frac{V_z}{V_{max(S.S.)}} = \frac{4\eta L V_z}{(p_o - p_L)R^2} \tag{5.253}$$

$$\frac{\partial^2\varphi}{\partial\xi^2} + \frac{1}{\xi}\frac{\partial\varphi}{\partial\xi} - \frac{\partial\varphi}{\partial\tau} + 4 = 0 \tag{5.254}$$

Equation (2.254) is a linear, non-homogeneous partial differential equation that is to be solved subject to the following initial and the boundary conditions:

$$\text{I.C.: at } \tau = 0, \varphi(\xi, 0) = 0 \tag{5.255}$$

$$\text{B.C. 1: at } \xi = 0, \frac{\partial \varphi}{\partial \xi} = 0 \tag{5.256}$$

$$\text{B.C. 2: at } \xi = 1, \varphi(1, \tau) = 0 \text{ (no-slip)} \tag{5.257}$$

As the partial differential equation is non-homogeneous, the separation of variables technique of solving the partial differential equation will not work. However, this problem could be solved by expressing the solution as follows:

$$\varphi(\xi, \tau) = W(\xi, \tau) + \psi(\xi) \tag{5.258}$$

where the first term on the right-hand side of Equation (5.258) is the transient part of the solution and the second term on the right-hand side of Equation (5.258) is the steady state part of the solution. Upon substitution of this solution into the partial differential equation (Equation (5.254)), we obtain:

$$\frac{\partial^2 W}{\partial \xi^2} + \frac{1}{\xi}\frac{\partial W}{\partial \xi} - \frac{\partial W}{\partial \tau} = -\left[\frac{d^2\psi}{d\xi^2} + \frac{1}{\xi}\frac{d\psi}{d\xi} + 4\right] \tag{5.259}$$

If ψ is defined such that

$$\frac{d^2\psi}{d\xi^2} + \frac{1}{\xi}\frac{d\psi}{d\xi} + 4 = 0 \tag{5.260}$$

then the original partial differential equation, Equation (5.254), is satisfied if:

$$\frac{\partial^2 W}{\partial \xi^2} + \frac{1}{\xi}\frac{\partial W}{\partial \xi} - \frac{\partial W}{\partial \tau} = 0 \tag{5.261}$$

Thus, we need to solve two equations: one ordinary differential equation for $\psi(\xi)$ and one linear and homogeneous partial differential equation for $W(\xi, \tau)$.

From the initial condition, Equation (5.255), it follows that:

$$\varphi(\xi, 0) = W(\xi, 0) + \psi(\xi) = 0 \tag{5.262}$$

$$W(\xi, 0) = -\psi(\xi) \tag{5.263}$$

From the symmetry boundary condition, Equation (5.256, it follows that:

$$\frac{\partial W(0,\tau)}{\partial \xi} = 0 \text{ and } \left.\frac{d\psi}{d\xi}\right|_{\xi=0} = 0 \tag{5.264}$$

From the no-slip boundary condition, Equation (5.257), it follows that:

$$W(1, \tau) = -\psi(1) \tag{5.265}$$

If we set $\psi(1) = 0$, then

$$W(1, \tau) = 0 \tag{5.266}$$

The ordinary differential equation for $\psi(\xi)$, Equation (5.260), can be re-written as:

$$\frac{d}{d\xi}\left(\xi\frac{d\psi}{d\xi}\right) = -4\xi \quad (5.267)$$

Upon integration Equation (5.267), we get:

$$\frac{d\psi}{d\xi} = -2\xi + \frac{C_1}{\xi} \quad (5.268)$$

C_1 must be taken zero otherwise the symmetry condition (Equation (5.264)) at $\xi = 0$ is not satisfied. Upon further integration, and using the boundary condition, $\psi(1) = 0$, we get the following solution for ψ:

$$\psi = 1 - \xi^2 \quad (5.269)$$

Now we need to solve the partial differential equation, Equation (5.261):

$$\frac{\partial^2 W}{\partial\xi^2} + \frac{1}{\xi}\frac{\partial W}{\partial\xi} - \frac{\partial W}{\partial\tau} = 0$$

subject to the following boundary conditions: $W(1,\tau) = 0$, $\frac{\partial W(0,\tau)}{\partial\xi} = 0$, and the initial condition:

$$W(\xi,0) = -\psi(\xi) = -(1-\xi^2) \quad (5.270)$$

Using the separation of variables method, the solution is postulated to be:

$$W(\xi,\tau) = E(\xi)T(\tau) \quad (5.271)$$

Upon substitution this solution into the partial differential equation, Equation (5.261), we obtain

$$\frac{1}{E}\frac{d^2E}{d\xi^2} + \frac{1}{\xi E}\frac{dE}{d\xi} = \frac{1}{T}\frac{dT}{d\tau} \quad (5.272)$$

The left-hand side of Equation (5.272) is a function of ξ only and the right-hand side of Equation (5.272) is a function of τ only. For the two sides to be equal for all ξ and τ, they must be equal to a constant. Also, the constant must be negative for the T(τ) to be bounded. Thus:

$$\frac{1}{E}\frac{d^2E}{d\xi^2} + \frac{1}{\xi E}\frac{dE}{d\xi} = \frac{1}{T}\frac{dT}{d\tau} = -\alpha^2 \quad (5.273)$$

The solution of the differential equation dealing with T in Equation (5.273) is as follows:

$$T = C_o\, exp(-\alpha^2\tau) \quad (5.274)$$

where C_o is a constant. The differential equation dealing with E in Equation (5.273) is as follows:

$$\frac{d^2E}{d\xi^2} + \frac{1}{\xi}\frac{dE}{d\xi} + \alpha^2E = 0 \quad (5.275)$$

We can re-cast Equation (5.275) in terms of a new variable $z = \xi\alpha$ as:

$$z^2\frac{d^2E}{dz^2} + z\frac{dE}{d\xi} + z^2E = 0 \quad (5.276)$$

Equation (5.276) is the well-known Bessel's differential equation. The general form of the Bessel's differential equation is as follows:

$$x^2y'' + xy' + (x^2 - \nu^2)y = 0 \quad \text{where} \quad \nu \geq 0 \tag{5.277}$$

The order ν of the Bessel's equation could be integer or non-integer. The general solution of the Bessel equation is:

$$y = C_1 J_\nu(x) + C_2 Y_\nu(x) \tag{5.278}$$

where $J_\nu(x)$ is a Bessel function of first kind of order ν and $Y_\nu(x)$ is a Bessel function of second kind of order ν.

The order of the Bessel equation under consideration, Equation (5.276), is zero. Thus, the solution for E is as follows:

$$E(z) = C_1 J_0(z) + C_2 Y_0(z) \tag{5.279}$$

where $J_0(z)$ and $Y_0(z)$ are given as:

$$J_0(z) = \sum_{m=0}^{\infty} \frac{(-1)^m z^{2m}}{2^{2m}(m!)^2} = 1 - \frac{z^2}{2^2(1!)^2} + \frac{z^4}{2^4(2!)^2} - \ldots\ldots\ldots\ldots \tag{5.280}$$

$$Y_0(z) = \frac{2}{\pi}\left[\left(\gamma + ln\frac{z}{2}\right)J_0(z) + \sum_{m=0}^{\infty} \frac{(-1)^{m-1} h_m z^{2m}}{2^{2m}(m!)^2}\right] \tag{5.281}$$

where γ is Euler's constant (= 0.5772156) and

$$h_m = \sum_{m=1}^{m}\left(\frac{1}{m}\right) \tag{5.282}$$

Important points to note about $J_0(z)$ and $Y_0(z)$ are as follows: $J_0(z)$ is an oscillating function of z with many zeros (crosses the z-axis many times) and $Y_0(z)$ is also an oscillating function of z and is unbounded at $z = 0$ ($Y_0(0) \to -\infty$).

The solution for E in terms of ξ is as follows:

$$E(\xi) = C_1 J_0(\alpha\xi) + C_2 Y_0(\alpha\xi) \tag{5.283}$$

For bounded solution, $C_2 = 0$. Consequently,

$$E(\xi) = C_1 J_0(\alpha\xi) \tag{5.284}$$

Using the boundary condition, Equation (5.266), we get:

$$W(1,\tau) = E(1)T(\tau) = 0 \tag{5.285}$$

We cannot take $T(\tau) = 0$. Thus,

$$E(1) = C_1 J_0(\alpha) = 0, \; J_0(\alpha) = 0 \tag{5.286}$$

We have multiple solutions of $J_0(\alpha) = 0$, that is, there are many values of α for which $J_0(\alpha)$ is zero. As α has multiple values, the solutions for E, T and W are also multiple. Let the values of α for which $J_0(\alpha)$ is zero be denoted by α_n. Thus:

$$E_n(\xi) = C_{1n} J_0(\alpha_n \xi) \tag{5.287}$$

$$T_n(\tau) = C_{on}\, exp(-\alpha_n^2 \tau) \tag{5.288}$$

$$W_n(\xi, \tau) = B_n J_0(\alpha_n \xi)\, exp(-\alpha_n^2 \tau) \tag{5.289}$$

Since the partial differential equation dealing with W, Equation (5.261), is linear and homogeneous, the linear combination of solutions is also a solution. The general solution for W is as follows:

$$W(\xi, \tau) = \sum_{n=1}^{\infty} B_n J_0(\alpha_n \xi)\, exp(-\alpha_n^2 \tau) \tag{5.290}$$

where α's are zeros of $J_0(\alpha)$. However, we still need to determine the unknown constants B_n. Using the initial condition, Equation (5.270), $W(\xi, 0) = -\psi(\xi) = -(1 - \xi^2)$, we obtain:

$$\sum_{n=1}^{\infty} B_n J_0(\alpha_n \xi) = -(1 - \xi^2) \tag{5.291}$$

Multiply both sides by $J_0(\alpha_m \xi)\xi d\xi$ and integrate from 0 to 1:

$$\sum_{n=1}^{\infty} B_n \int_0^1 J_0(\alpha_n \xi) J_0(\alpha_m \xi)\xi d\xi = -\int_0^1 (1 - \xi^2) J_0(\alpha_m \xi)\xi d\xi \tag{5.292}$$

The Bessel functions $J_0(\alpha_m \xi)$ and $J_0(\alpha_n \xi)$ form an orthogonal set on the interval $0 \le \xi \le 1$with respect to the weight function ξ, that is:

$$\int_0^1 J_0(\alpha_n \xi) J_0(\alpha_m \xi)\xi d\xi = 0 \quad \text{when} \quad m \ne n \tag{5.293}$$

Thus:

$$B_m \int_0^1 J_0^2(\alpha_m \xi)\xi d\xi = -\int_0^1 (1 - \xi^2)\, \xi J_0(\alpha_m \xi) d\xi \tag{5.294}$$

From handbook of mathematical functions, the integrals in this equation are as follows:

$$\int_0^1 J_0^2(\alpha_m \xi)\xi d\xi = \frac{1}{2} J_1^2(\alpha_m) \tag{5.295}$$

$$\int_0^1 (1 - \xi^2)\, \xi J_0(\alpha_m \xi) d\xi = \frac{4}{\alpha_m^3} J_1(\alpha_m) \tag{5.296}$$

Hence the constants B_m are given as:

$$B_m = -\frac{8}{\alpha_m^3 J_1(\alpha_m)} \tag{5.297}$$

The solution for the velocity distribution is finally obtained as follows:

$$\varphi(\xi, \tau) = \psi(\xi) + W(\xi, \tau) = (1 - \xi^2) - 8\sum_{m=1}^{\infty} \frac{J_0(\alpha_m \xi)}{\alpha_m^3 J_1(\alpha_m)} exp(-\alpha_m^2 \tau) \tag{5.298}$$

where α_m are the roots of $J_0(\alpha_m) = 0$.

The first few terms of this solution are given below:

$$\varphi(\xi,\tau) = (1-\xi^2) - 8\left[\frac{J_0(\alpha_1\xi)}{\alpha_1^3 J_1(\alpha_1)} exp(-\alpha_1^2\tau) + \frac{J_0(\alpha_2\xi)}{\alpha_2^3 J_1(\alpha_2)} exp(-\alpha_2^2\tau) + \frac{J_0(\alpha_3\xi)}{\alpha_3^3 J_1(\alpha_3)} exp(-\alpha_3^2\tau) + \ldots\ldots\ldots\right] \tag{5.299}$$

where $\alpha_1 = 2.405$, $\alpha_2 = 5.52$, and $\alpha_3 = 8.654$.

Figure 5.21 shows the plots of velocity distributions generated from Equation (5.299) for different values of dimensionless time τ.

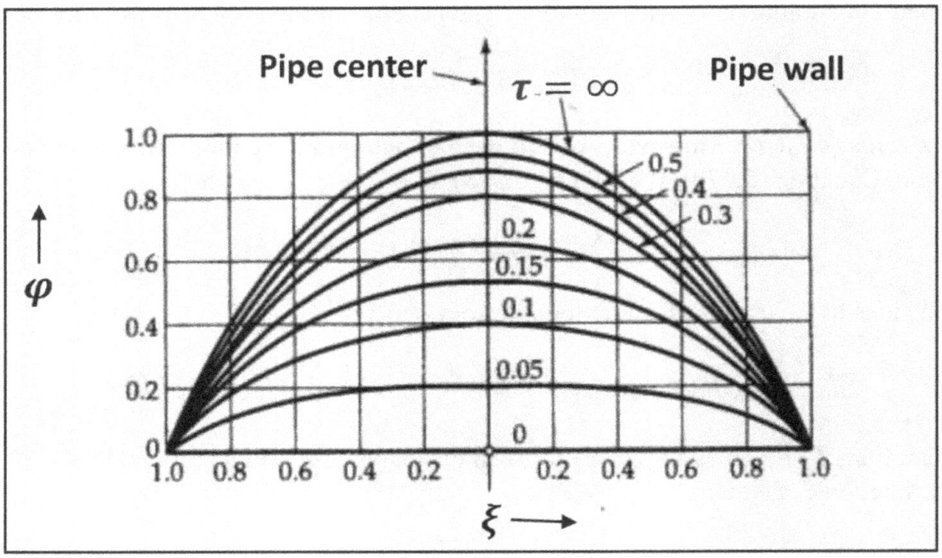

Figure 5.21 Velocity distribution in dimensionless form for unsteady pipe flow.

5.3 STEADY-STATE PROBLEMS WITH TWO NON-VANISHING VELOCITY COMPONENTS

Thus far, we have been dealing with viscous flow problems having one non-vanishing velocity component. When we have two non-vanishing velocity components, we must solve three coupled partial differential equations (two components of Navier-Stokes equation and one continuity equation). To solve such problems, the concept of stream function ψ is utilized. It should be noted that the concept of stream function works only for incompressible flow of Newtonian fluids involving two-directional (two non-vanishing velocity components) and two-dimensional (two space variables) flows. The concept of stream function allows the transformation of three coupled partial differential equations into a single fourth order partial differential equation. The general procedure is as follows:

- The velocity components are expressed as derivatives of stream function ψ such that the continuity equation is automatically satisfied.
- Appropriate derivatives are taken of the two components of the Navier-Stokes equation to eliminate the gravity terms. Note that the gravity is assumed to be constant.
- The equations obtained by taking the derivatives of the Navier-Stokes equations (gravity terms eliminated) are expressed in terms of the stream function ψ.

- The two stream function equations (equivalent to Navier-Stokes equations) obtained in the previous step are combined to eliminate the pressure terms.

- The resulting equation is a fourth order partial differential equation in terms of the stream function ψ. This single equation needs to be solved for the stream function subject to the given initial and boundary conditions.

- The stream function ψ obtained through the solution of the fourth order partial differential equation is finally used to determine the unknown velocity components.

5.3.1 Stream Function in Different Coordinate Systems

Cartesian Coordinates:

We consider two-dimensional (no z −dependence) and two directional laminar flow ($V_z = 0$) of a Newtonian fluid of constant properties. The continuity equation gives:

$$\frac{\partial V_x}{\partial x} + \frac{\partial V_y}{\partial y} = 0 \tag{5.300}$$

We now define the stream function such that the continuity equation, Equation (5.300), is automatically satisfied. The stream function is thus defined as:

$$V_x = -\frac{\partial \psi}{\partial y} \ ; \ V_y = \frac{\partial \psi}{\partial x} \tag{5.301}$$

If the velocity components in Equation (5.301) are substituted into the left-hand side of the continuity equation, Equation (5.300), the result is zero. Thus, the continuity equation is automatically satisfied and is no longer needed.

Cylindrical Coordinates:

We consider two-dimensional (no θ −dependence) and two directional axisymmetric laminar flow ($V_\theta = 0$) of a Newtonian fluid of constant properties. The continuity equation gives:

$$\frac{1}{r}\frac{\partial (rV_r)}{\partial r} + \frac{\partial V_z}{\partial z} = 0 \tag{5.302}$$

Upon rearrangement of Equation (5.302),

$$\frac{\partial (rV_r)}{\partial r} + \frac{\partial (rV_z)}{\partial z} = 0 \tag{5.303}$$

We now define the stream function such that the continuity equation, Equation (5.303), is automatically satisfied. The stream function is defined accordingly as:

$$rV_r = \frac{\partial \psi}{\partial z} \ ; \ rV_z = -\frac{\partial \psi}{\partial r} \tag{5.304}$$

If the velocity components in Equation (5.304) are substituted into the left-hand side of the continuity equation, Equation (5.303), the result is zero. Thus, the continuity equation is automatically satisfied and is no longer needed.

Spherical Coordinates:

We consider two-dimensional (no φ −dependence) and two directional axisymmetric laminar flow ($V_\varphi = 0$) of a Newtonian fluid of constant properties. The continuity equation gives:

$$\frac{1}{r^2}\frac{\partial(r^2V_r)}{\partial r}+\frac{1}{rsin\theta}\frac{\partial(V_\theta sin\theta)}{\partial\theta}=0 \tag{5.305}$$

Upon rearrangement of Equation (5.305),

$$\frac{\partial(r^2V_r sin\theta)}{\partial r}+\frac{\partial(rV_\theta sin\theta)}{\partial\theta}=0 \tag{5.306}$$

We now define the stream function such that the continuity equation, Equation (5.306), is automatically satisfied. The stream function is defined accordingly as:

$$r^2V_r sin\theta=-\frac{\partial\psi}{\partial\theta}\ ;\ rV_\theta sin\theta=\frac{\partial\psi}{\partial r} \tag{5.307}$$

If the velocity components in Equation (5.307) are substituted into the left-hand side of the continuity equation, Equation (5.306), the result is zero. Thus, the continuity equation is automatically satisfied and is no longer needed.

5.3.2 Navier-Stokes Equations in Terms of Stream Function in Different Coordinate Systems

We can now re-cast the Navier-Stokes equations in terms of the stream function for different coordinate systems.

Cartesian Coordinates:

Assuming $V_z = 0$, no z −dependence,

x-direction Navier-Stokes equation

$$\rho\left(\frac{\partial V_x}{\partial t}+V_x\frac{\partial V_x}{\partial x}+V_y\frac{\partial V_x}{\partial y}\right)=-\frac{\partial P}{\partial x}+\rho g_x+\eta\left(\frac{\partial^2V_x}{\partial x^2}+\frac{\partial^2V_x}{\partial y^2}\right) \tag{5.308}$$

y-direction Navier-Stokes equation

$$\rho\left(\frac{\partial V_y}{\partial t}+V_x\frac{\partial V_y}{\partial x}+V_y\frac{\partial V_y}{\partial y}\right)=-\frac{\partial P}{\partial y}+\rho g_y+\eta\left(\frac{\partial^2V_y}{\partial x^2}+\frac{\partial^2V_y}{\partial y^2}\right) \tag{5.309}$$

We differentiate x-direction equation, Equation (5.308), with respect to y:

$$\frac{\partial}{\partial y}\left(\frac{\partial V_x}{\partial t}\right)+\frac{\partial}{\partial y}\left(V_x\frac{\partial V_x}{\partial x}+V_y\frac{\partial V_x}{\partial y}\right)=-\frac{1}{\rho}\frac{\partial}{\partial y}\left(\frac{\partial P}{\partial x}\right)+\frac{\partial}{\partial y}g_x+\nu\frac{\partial}{\partial y}\left(\frac{\partial^2V_x}{\partial x^2}+\frac{\partial^2V_x}{\partial y^2}\right) \tag{5.310}$$

Substituting velocity components from Equation (5.301) into Equation (5.310) gives:

$$-\frac{\partial}{\partial t}\left(\frac{\partial^2\psi}{\partial y^2}\right)+\frac{\partial}{\partial y}\left[\left(\frac{\partial\psi}{\partial y}\right)\frac{\partial^2\psi}{\partial x\partial y}-\left(\frac{\partial\psi}{\partial x}\right)\frac{\partial^2\psi}{\partial y^2}\right]=-\frac{1}{\rho}\frac{\partial^2P}{\partial y\partial x}-\nu\frac{\partial}{\partial y}\left[\frac{\partial^3\psi}{\partial y\partial x^2}+\frac{\partial^3\psi}{\partial y^3}\right] \tag{5.311}$$

Note that the derivative of g_x is zero.

We now differentiate y-direction equation, Equation (5.309), with respect to x:

$$\frac{\partial}{\partial x}\left(\frac{\partial V_y}{\partial t}\right)+\frac{\partial}{\partial x}\left(V_x\frac{\partial V_y}{\partial x}+V_y\frac{\partial V_y}{\partial y}\right)=-\frac{1}{\rho}\frac{\partial}{\partial x}\left(\frac{\partial P}{\partial y}\right)+\frac{\partial}{\partial x}g_y+\nu\frac{\partial}{\partial x}\left(\frac{\partial^2 V_y}{\partial x^2}+\frac{\partial^2 V_y}{\partial y^2}\right) \tag{5.312}$$

Substituting velocity components from Equation (5.301) into Equation (5.312) gives:

$$-\frac{\partial}{\partial t}\left(\frac{\partial^2\psi}{\partial x^2}\right)+\frac{\partial}{\partial x}\left[\left(-\frac{\partial\psi}{\partial y}\right)\frac{\partial^2\psi}{\partial x^2}+\left(\frac{\partial\psi}{\partial x}\right)\frac{\partial^2\psi}{\partial x\partial y}\right]=-\frac{1}{\rho}\frac{\partial^2 P}{\partial x\partial y}+\nu\frac{\partial}{\partial x}\left[\frac{\partial^3\psi}{\partial x^3}+\frac{\partial^3\psi}{\partial x\partial y^2}\right] \tag{5.313}$$

Note that the derivative of g_y is zero.

Subtracting Equation (5.313) from Equation (5.311), we obtain:

$$-\frac{\partial}{\partial t}(\nabla^2\psi)+\frac{\partial}{\partial y}\left[\left(\frac{\partial\psi}{\partial y}\right)\frac{\partial^2\psi}{\partial x\partial y}-\left(\frac{\partial\psi}{\partial x}\right)\frac{\partial^2\psi}{\partial y^2}\right]-\frac{\partial}{\partial x}\left[\left(-\frac{\partial\psi}{\partial y}\right)\frac{\partial^2\psi}{\partial x^2}+\left(\frac{\partial\psi}{\partial x}\right)\frac{\partial^2\psi}{\partial x\partial y}\right]=-\nu\frac{\partial}{\partial y}\left[\frac{\partial^3\psi}{\partial y\partial x^2}+\frac{\partial^3\psi}{\partial y^3}\right]-\nu\frac{\partial}{\partial x}\left[\frac{\partial^3\psi}{\partial x^3}+\frac{\partial^3\psi}{\partial x\partial y^2}\right] \tag{5.314}$$

Note that the pressure terms cancel out. Upon simplification, Equation (5.324) reduces to:

$$-\frac{\partial}{\partial t}(\nabla^2\psi)+\left(\frac{\partial\psi}{\partial y}\right)\left[\frac{\partial^3\psi}{\partial x\partial y^2}+\frac{\partial^3\psi}{\partial x^3}\right]-\left(\frac{\partial\psi}{\partial x}\right)\left[\frac{\partial^3\psi}{\partial x^2\partial y}+\frac{\partial^3\psi}{\partial y^3}\right]=-\nu\nabla^4\psi \tag{5.315}$$

Upon further simplification of Equation (5.315), we get:

$$-\frac{\partial}{\partial t}(\nabla^2\psi)+\left(\frac{\partial\psi}{\partial y}\right)\left[\frac{\partial}{\partial x}(\nabla^2\psi)\right]-\left(\frac{\partial\psi}{\partial x}\right)\left[\frac{\partial}{\partial y}(\nabla^2\psi)\right]=-\nu\nabla^4\psi \tag{5.316}$$

where

$$\nabla^2=\frac{\partial^2}{\partial x^2}+\frac{\partial^2}{\partial y^2}\ ;\quad \nabla^4=\nabla^2(\nabla^2)=\left(\frac{\partial^2}{\partial x^2}+\frac{\partial^2}{\partial y^2}\right)\left(\frac{\partial^2}{\partial x^2}+\frac{\partial^2}{\partial y^2}\right) \tag{5.317}$$

Equation (5.316) could be re-written as:

$$\frac{\partial}{\partial t}(\nabla^2\psi)+\frac{\partial(\psi,\nabla^2\psi)}{\partial(x,y)}=\nu\nabla^4\psi \tag{5.318}$$

where $\frac{\partial(\psi,\nabla^2\psi)}{\partial(x,y)}$ is a Jacobian defined as:

$$\frac{\partial(\psi,\nabla^2\psi)}{\partial(x,y)}=\begin{vmatrix}\frac{\partial\psi}{\partial x} & \frac{\partial\psi}{\partial y}\\ \frac{\partial}{\partial x}(\nabla^2\psi) & \frac{\partial}{\partial y}(\nabla^2\psi)\end{vmatrix} \tag{5.319}$$

Equation (5.318) is the final fourth order partial differential equation that needs to be solved for the stream function subject to the given initial condition and boundary conditions.

One can derive the stream function fourth order partial differential equations like Equation (5.318) for axisymmetric motions in cylindrical and spherical coordinates using the same approach. They are summarized in Table 5.1.

Table 5.1 Stream function equations in different coordinate systems

Coordinate system	Velocity components	Fourth order partial differential equations for stream function
Cartesian with $V_z = 0$ and no z −dependence	$V_x = -\frac{\partial\psi}{\partial y}$; $V_y = \frac{\partial\psi}{\partial x}$	$\frac{\partial}{\partial t}(\nabla^2\psi) + \frac{\partial(\psi,\nabla^2\psi)}{\partial(x,y)} = \nu\nabla^4\psi$ $\nabla^2 = \frac{\partial^2}{\partial x^2} + \frac{\partial^2}{\partial y^2}$ $\nabla^4 = \nabla^2(\nabla^2) = \left(\frac{\partial^2}{\partial x^2} + \frac{\partial^2}{\partial y^2}\right)\left(\frac{\partial^2}{\partial x^2} + \frac{\partial^2}{\partial y^2}\right)$
Cylindrical with $V_\theta = 0$ and no θ −dependence (ax-isymmetric flow)	$V_r = \frac{1}{r}\frac{\partial\psi}{\partial z}$; $V_z = -\frac{1}{r}\frac{\partial\psi}{\partial r}$	$\frac{\partial}{\partial t}(E^2\psi) - \frac{1}{r}\frac{\partial(\psi,E^2\psi)}{\partial(r,z)} - \frac{2}{r^2}\left(\frac{\partial\psi}{\partial z}\right)E^2\psi = \nu E^4\psi$ $E^2 = \frac{\partial^2}{\partial r^2} - \frac{1}{r}\frac{\partial}{\partial r} + \frac{\partial^2}{\partial z^2}$ $E^4\psi = E^2(E^2\psi)$
Spherical with $V_\varphi = 0$ and no φ −dependence (ax-isymmetric flow)	$V_r = -\frac{1}{r^2 sin\theta}\frac{\partial\psi}{\partial\theta}$; $V_\theta = \frac{1}{r sin\theta}\frac{\partial\psi}{\partial r}$	$\frac{\partial}{\partial t}(E^2\psi) + \frac{1}{r^2 sin\theta}\frac{\partial(\psi,E^2\psi)}{\partial(r,\theta)} - \frac{2E^2\psi}{r^2 sin^2\theta}\left(\frac{\partial\psi}{\partial r}cos\theta - \frac{1}{r}\frac{\partial\psi}{\partial\theta}sin\theta\right) = \nu E^4\psi$ $E^2 = \frac{\partial^2}{\partial r^2} + \frac{sin\theta}{r^2}\frac{\partial}{\partial\theta}\left(\frac{1}{sin\theta}\frac{\partial}{\partial\theta}\right)$ $E^4\psi = E^2(E^2\psi)$

5.4 CREEPING FLOW

The general stream function equations (see last column of Table 5.1) are extremely difficult to solve, even in the simplest case of flow around a solid spherical particle. The main difficulty is the non-linear terms on the left-hand side of the fourth order stream function equations summarized in Table 5.1. Fortunately, the non-linear terms can be neglected if the flow is very slow, referred to as creeping flow. In what follows, the Navier-Stokes equation is written in dimensionless form and the condition defining creeping flow is determined.

Consider the Navier-Stokes equation given in the vector form as below:

$$\rho\left[\frac{\partial\vec{V}}{\partial t} + \vec{V}\cdot\nabla\vec{V}\right] = -\nabla P + \rho\vec{g} + \eta\nabla^2\vec{V} \quad (5.320)$$

As $\vec{g} = -\nabla\varphi$ where φ is the gravitational potential (potential energy per unit mass), Equation (5.320) can be re-written as:

$$\rho\left[\frac{\partial\vec{V}}{\partial t} + \vec{V}\cdot\nabla\vec{V}\right] = -\nabla P' + \eta\nabla^2\vec{V} \quad (5.321)$$

where $P' = P + \rho\varphi$ (referred to as modified pressure). The continuity equation is:

$$\nabla\cdot\vec{V} = 0 \quad (5.322)$$

Now we define all the relevant dimensionless variables as:

$$V^* = \vec{V}/U,\ t^* = tU/L,\ x^* = x/L, y^* = y/L, z^* = z/L,\ P^* = P'L/\eta U,\ \nabla^* = L\nabla \tag{5.323}$$

where the starred quantities are dimensionless variables and U, L are characteristic velocity and length of the system. Substituting the dimensionless variables into the Navier-Stokes and continuity equations, we get:

$$\left(\frac{\rho UL}{\eta}\right)\left[\frac{\partial V^*}{\partial t^*} + V^* \cdot \nabla^* V^*\right] = -\nabla^* P^* + \nabla^{*2} V^* \tag{5.324}$$

$$\nabla^* \cdot V^* = 0 \tag{5.325}$$

Equation (5.324) is a dimensionless form of the Navier-Stokes equation and Equation (5.325) is a dimensionless form of continuity equation. The term $\left(\frac{\rho UL}{\eta}\right)$ in Equation (5.324) is the characteristic Reynolds number (Re) of the system. Creeping flow can now be defined as the flow where $Re \to 0$. Therefore, for sufficiently small characteristic Reynolds number ($Re \ll 1$), the left-hand side of the Navier-Stokes equation, Equation (5.324), can be neglected. Consequently, Equations (5.324) and (5.325) become the following in dimensional form:

$$-\nabla P' + \eta \nabla^2 \vec{V} = 0 \tag{5.326}$$

$$\nabla \cdot \vec{V} = 0 \tag{5.327}$$

Equations (5.326) – (5.327) are called the creeping flow equations. Thus, the creeping flow approximation reduces the non-linear Navier-Stokes equation to a linear equation. This simplifies the mathematical problem greatly as the theory of linear partial differential equations is much better developed than the theory of non-linear partial differential equations. Table 5.2 summarizes the stream function equations valid for creeping flows.

Table 5.2 Stream function equations valid for creeping flows

Coordinate system	Velocity components	Fourth order partial differential equations for stream function under creeping flow
Cartesian with $V_z = 0$ and no z –dependence	$V_x = -\frac{\partial \psi}{\partial y}$; $V_y = \frac{\partial \psi}{\partial x}$	$\nabla^4 \psi = 0$ $\nabla^2 = \frac{\partial^2}{\partial x^2} + \frac{\partial^2}{\partial y^2}$ $\nabla^4 = \nabla^2(\nabla^2) = \left(\frac{\partial^2}{\partial x^2} + \frac{\partial^2}{\partial y^2}\right)\left(\frac{\partial^2}{\partial x^2} + \frac{\partial^2}{\partial y^2}\right)$
Cylindrical with $V_\theta = 0$ and no θ –dependence (axisymmetric flow)	$V_r = \frac{1}{r}\frac{\partial \psi}{\partial z}$; $V_z = -\frac{1}{r}\frac{\partial \psi}{\partial r}$	$E^4 \psi = 0$ $E^2 = \frac{\partial^2}{\partial r^2} - \frac{1}{r}\frac{\partial}{\partial r} + \frac{\partial^2}{\partial z^2}$ $E^4 \psi = E^2(E^2 \psi)$
Spherical with $V_\varphi = 0$ and no φ –dependence (axisymmetric flow)	$V_r = -\frac{1}{r^2 sin\theta}\frac{\partial \psi}{\partial \theta}$; $V_\theta = \frac{1}{r sin\theta}\frac{\partial \psi}{\partial r}$	$E^4 \psi = 0$ $E^2 = \frac{\partial^2}{\partial r^2} + \frac{sin\theta}{r^2}\frac{\partial}{\partial \theta}\left(\frac{1}{sin\theta}\frac{\partial}{\partial \theta}\right)$ $E^4 \psi = E^2(E^2 \psi)$

5.4.1 Creeping Flow Around Solid Spherical Particle

We now consider creeping flow around a stationary single solid spherical particle of radius R. The solution of creeping flow equations around a stationary solid spherical particle leads to the derivation of the Stokes law for drag force on the particle due to fluid motion. The Stokes law is important in the estimation of settling velocity of particles in the fluid.

To obtain the drag force on the particle, we need to first solve for the velocity field, the pressure distribution, and the shear-stress distribution around the stationary particle. From the pressure distribution and shear-stress distribution around the particle, the drag force on the particle in the direction of motion can be determined.

For axi-symmetric flow around a stationary solid sphere, $V_\varphi = 0$ and there is no φ - dependence of flow-field variables. The non-zero velocity components are V_r and V_θ. To solve for V_r and V_θ, we utilize the concept of stream function ψ. The continuity equation is automatically satisfied and the velocity components V_r and V_θ are given as:

$$V_r = -\frac{1}{r^2 \sin\theta}\frac{\partial\psi}{\partial\theta} \tag{5.328}$$

$$V_\theta = \frac{1}{r \sin\theta}\frac{\partial\psi}{\partial r} \tag{5.329}$$

The fourth-order partial differential equation for stream function applicable for axi-symmetric creeping flow around a solid sphere is as follows:

$$E^4\psi = E^2(E^2\psi) = 0 \tag{5.330}$$

where E^2 is given as:

$$E^2 = \frac{\partial^2}{\partial r^2} + \frac{\sin\theta}{r^2}\frac{\partial}{\partial\theta}\left(\frac{1}{\sin\theta}\frac{\partial}{\partial\theta}\right) = \frac{\partial^2}{\partial r^2} + \frac{1}{r^2}\frac{\partial^2}{\partial\theta^2} - \frac{\cot\theta}{r^2}\frac{\partial}{\partial\theta} \tag{5.331}$$

The boundary conditions are:

$$\text{B.C.1 : at } r = R\,,\ V_r = -\frac{1}{r^2 \sin\theta}\frac{\partial\psi}{\partial\theta} = 0 \tag{5.332}$$

$$\text{B.C.2 : at } r = R\,,\ V_\theta = \frac{1}{r \sin\theta}\frac{\partial\psi}{\partial r} = 0 \tag{5.333}$$

$$\text{B.C.3 : at } r \to \infty,\ V_r = -\frac{1}{r^2 \sin\theta}\frac{\partial\psi}{\partial\theta} = V_\infty \cos\theta \tag{5.334}$$

$$\text{B.C.4 : at } r \to \infty,\ V_\theta = \frac{1}{r \sin\theta}\frac{\partial\psi}{\partial r} = -V_\infty \sin\theta \tag{5.335}$$

Note that far away from the stationary particle, the flow is in z direction with free stream velocity of V_∞. From the B.C. 3 or B.C. 4, upon integration, it can readily be shown that:

$$\text{at } r \to \infty,\ \psi = -\frac{1}{2}V_\infty r^2 \sin^2\theta \tag{5.336}$$

This boundary condition suggests the structure of the solution. As there is no restriction on θ in this boundary condition, the solution is expected to be of the form:

$$\psi = f(r)\sin^2\theta \tag{5.337}$$

Upon substitution of this stream function into $E^2\psi$, we obtain:

$$E^2\psi = F(r)\, sin^2\,\theta \tag{5.338}$$

where

$$F(r) = \frac{d^2 f}{dr^2} - \frac{2f}{r^2} \tag{5.339}$$

Thus, it follows that:

$$E^2(E^2\psi) = sin^2\,\theta \left(\frac{d^2}{dr^2} - \frac{2}{r^2}\right)\left(\frac{d^2}{dr^2} - \frac{2}{r^2}\right) f \tag{5.340}$$

Since $E^4\psi = 0$, we obtain:

$$\left(\frac{d^2}{dr^2} - \frac{2}{r^2}\right)\left(\frac{d^2}{dr^2} - \frac{2}{r^2}\right) f = 0 \tag{5.341}$$

This ordinary differential equation can be re-written as:

$$r^4\frac{d^4 f}{dr^4} - 4r^2\frac{d^2 f}{df^2} + 8r\frac{df}{dr} - 8f = 0 \tag{5.342}$$

This is a fourth order Euler-Cauchy equation. It is a linear, homogeneous ordinary differential equation with variable coefficients. The solution of this equation is of the form $f = r^n$. Upon substituting this solution into the differential equation, Equation (5.342), we get:

$$(n-1)(n-2)(n-4)(n+1) = 0 \tag{5.343}$$

The roots of this equation are: $n = 1, n = 2, n = 4$, and $n = -1$.

We can now write the general solution of the Euler-Cauchy equation as follows:

$$f = \frac{A}{r} + Br + Cr^2 + Dr^4 \tag{5.344}$$

To satisfy the stream function boundary condition at $r \to \infty$, $D = 0$, and $C = -V_\infty/2$. The solution of the stream function equation $E^4\psi = 0$ thus becomes:

$$\psi = \left(\frac{A}{r} + Br - \frac{1}{2}V_\infty r^2\right) sin^2\,\theta \tag{5.345}$$

We can now express the unknown velocity components V_r and V_θ as follows:

$$V_r = -\frac{1}{r^2\, sin\,\theta}\frac{\partial\psi}{\partial\theta} = \left(V_\infty - \frac{2A}{r^3} - \frac{2B}{r}\right) cos\,\theta \tag{5.346}$$

$$V_\theta = \frac{1}{r\, sin\,\theta}\frac{\partial\psi}{\partial r} = \left(-V_\infty - \frac{A}{r^3} + \frac{B}{r}\right) sin\,\theta \tag{5.347}$$

The unknown constants A and B can be determined from the boundary conditions at the surface of the particle (Equations (5.332) and (5.333)). These boundary conditions give the following expressions for A and B :

$$A = -\frac{1}{4}V_\infty R^3 \tag{5.348}$$

$$B = \frac{3}{4}V_\infty R \tag{5.349}$$

Thus, the stream function and the velocity components are as follows:

$$\psi = f(r)\, sin^2\,\theta = \frac{V_\infty}{2}\left(\frac{3}{2}Rr - \frac{R^3}{2r} - r^2\right) sin^2\,\theta \tag{5.350}$$

$$\frac{V_r}{V_\infty} = \left[1 + \frac{1}{2}\left(\frac{R}{r}\right)^3 - \frac{3}{2}\left(\frac{R}{r}\right)\right] cos\,\theta \tag{5.351}$$

$$\frac{V_\theta}{V_\infty} = -\left[1 - \frac{1}{4}\left(\frac{R}{r}\right)^3 - \frac{3}{4}\left(\frac{R}{r}\right)\right] sin\,\theta \tag{5.352}$$

The pressure distribution in the fluid can now be determined using the following equations:

$$\frac{\partial P}{\partial r} - \rho g_r = -\frac{\eta}{r^2\, sin\,\theta}\frac{\partial(E^2\psi)}{\partial\theta} \tag{5.353}$$

$$\frac{\partial P}{\partial\theta} - r\rho g_\theta = \frac{\eta}{sin\,\theta}\frac{\partial(E^2\psi)}{\partial r} \tag{5.354}$$

These equations follow from the r and θ components of the Navier-Stokes equations under creeping flow condition. It can be readily shown that:

$$E^2\psi = -\frac{3}{2}V_\infty\left(\frac{R}{r}\right) sin^2\,\theta \tag{5.355}$$

$$\frac{\partial(E^2\psi)}{\partial\theta} = -3V_\infty\left(\frac{R}{r}\right) sin\,\theta\, cos\,\theta \tag{5.356}$$

$$\frac{\partial(E^2\psi)}{\partial r} = \frac{3}{2}V_\infty\left(\frac{R}{r^2}\right) sin^2\,\theta \tag{5.357}$$

$$\frac{\partial P}{\partial r} - \rho g_r = \frac{3\eta V_\infty R\, cos\,\theta}{r^3} \tag{5.358}$$

$$\frac{\partial P}{\partial\theta} - r\rho g_\theta = \frac{3\eta V_\infty R\, sin\,\theta}{2r^2} \tag{5.359}$$

$$dP = \left(\frac{\partial P}{\partial r}\right)dr + \left(\frac{\partial P}{\partial\theta}\right)d\theta = \left[\frac{3\eta V_\infty R\, cos\,\theta}{r^3} + \rho g_r\right]dr + \left[\frac{3\eta V_\infty R\, sin\,\theta}{2r^2} + r\rho g_\theta\right]d\theta \tag{5.360}$$

$$dP = \frac{3}{2}\eta V_\infty R\left[\left(\frac{2\, cos\,\theta}{r^3}\right)dr + \left(\frac{sin\,\theta}{r^2}\right)d\theta\right] - \rho g dz \tag{5.361}$$

Using the following relation

$$d\left(\frac{cos\,\theta}{r^2}\right) = -\left[\left(\frac{2\, cos\,\theta}{r^3}\right)dr + \left(\frac{sin\,\theta}{r^2}\right)d\theta\right] \tag{5.362}$$

we can express dP as follows:

$$dP = -\frac{3}{2}\eta V_\infty R d\left(\frac{cos\,\theta}{r^2}\right) - \rho g dz \tag{5.363}$$

Upon integration, we obtain:

$$P = -\frac{3}{2}\eta V_\infty R\left(\frac{\cos\theta}{r^2}\right) - \rho g z + C \tag{5.364}$$

Taking $P = P_o$ at $z = 0$ as $r \to \infty$, the pressure distribution around a solid spherical particle is given as:

$$P = P_o - \frac{3}{2}\eta V_\infty R\left(\frac{\cos\theta}{r^2}\right) - \rho g z \tag{5.365}$$

In addition to the pressure distribution, shear-stress distribution $\tau_{r\theta}$ on the particle surface is needed to determine the drag force on the particle. From Newton's law of viscosity, $\tau_{r\theta}$ is given as:

$$\tau_{r\theta} = \eta\left[r\frac{\partial}{\partial r}\left(\frac{V_\theta}{r}\right) + \frac{1}{r}\frac{\partial V_r}{\partial\theta}\right] \tag{5.366}$$

Upon substituting the expressions for V_r and V_θ and carrying out the required differentiation, the following result is obtained for $\tau_{r\theta}$:

$$\tau_{r\theta} = -\frac{3\eta V_\infty}{2R}\left(\frac{R}{r}\right)^4 \sin\theta \tag{5.367}$$

Force due to pressure distribution

The net force acting on the particle in the direction of motion (z –direction) due to pressure distribution around the particle can be determined as follows:

$$F_{z,p} = \int_0^{2\pi}\int_0^{\pi}(-P|_{r=R}\cos\theta)R^2\sin\theta\, d\theta d\varphi \tag{5.368}$$

Upon substituting the expression for pressure distribution and carrying out the integration, the following result is obtained:

$$F_{z,p} = \frac{4}{3}\pi R^3\rho g + 2\pi\eta R V_\infty \tag{5.369}$$

The first term on the right side is the buoyancy force acting on the particle. The second term is the drag force on the particle due to flow-induced pressure distribution around the particle surface. This second term is referred to as "form drag".

Force due to shear stress distribution

The net force acting on the particle in the direction of motion (z –direction) due to shear-stress $\tau_{r\theta}$ distribution around the particle surface can be determined as follows:

$$F_{z,\tau} = \int_0^{2\pi}\int_0^{\pi}(-\tau_{r\theta}|_{r=R}\sin\theta)R^2\sin\theta\, d\theta d\varphi \tag{5.370}$$

Upon substituting the expression for the shear-stress distribution and carrying out the integration, the following result is obtained:

$$F_{z,\tau} = 4\pi\eta R V_\infty \tag{5.371}$$

This is the net force acting on the particle in the direction of motion due to tangential stress distribution around the particle surface. This term is referred to as "skin drag".

Stokes Law

The total drag force acting on the particle due to fluid motion around the particle is:

$$F_D = 2\pi\eta R V_\infty + 4\pi\eta R V_\infty = 6\pi\eta R V_\infty \tag{5.372}$$

This equation is one form of the Stokes law. We can re-write the Stokes law in terms of the drag coefficient, defined as:

$$C_D = \frac{F_D}{A_p\left(\frac{1}{2}\rho V_\infty^2\right)} \tag{5.373}$$

where A_p is the projected area of the particle on the plane normal to the direction of flow. For a spherical particle, $A_p = \pi R^2$. Upon substituting the expression for F_D, we obtain:

$$C_D = \frac{6\pi\eta R V_\infty}{\pi R^2\left(\frac{1}{2}\rho V_\infty^2\right)} = \frac{24}{\left(\frac{\rho D_p V_\infty}{\eta}\right)} = \frac{24}{Re_p} \tag{5.374}$$

where Re_p is the particle Reynolds number. This equation is another form of the Stokes law. Note that the Stokes law is valid for creeping flow ($Re_p \ll 1$) which neglects the inertial terms ($\vec{V}\cdot\nabla\vec{V}$) of the Navier-Stokes equation. Modifications of the Stokes law have been carried out to consider the effect of the inertial terms. One such modification is the “Oseen’s approximation”, given as:

$$C_D = \frac{24}{Re_p}\left[1 + \frac{3}{16}Re_p\right] \tag{5.375}$$

Some authors [Proudman and Pearson, 1957; J. Fluid. Mech., V. 2, 237-262] have obtained even the next term in the expression for the drag coefficient:

$$C_D = \frac{24}{Re_p}\left[1 + \frac{3}{16}Re_p + \frac{9}{160}Re_p^2\, ln\left(\frac{Re_p}{2}\right)\right] \tag{5.376}$$

However, this and the Oseen expressions provide an over-correction of C_D when $Re_p > 1$. Several empirical equations have been proposed for C_D to cover a broad range of Re_p. One example of an empirical equation which fits the C_D data well up to $Re_p = 1000$ is as follows [Clift, R, Grace, J.R., Weber, M.E., Bubbles, Drops, and Particles, Academic Press, N.Y., 1972]:

$$C_D = \frac{24}{Re_p}\left[1 + 0.15 Re_p^{0.687}\right] \tag{5.377}$$

5.4.2 Creeping Flow Around a Spherical Gas Bubble

Consider a spherical bubble of radius R that is rising slowly in a Newtonian liquid in the z-direction at velocity V_∞. Let the origin of the coordinate system be fixed at the centre of the bubble. For axi-symmetric flow around a bubble, $V_\varphi = 0$ and there is no φ – dependence of flow-field variables. The non-zero velocity components are V_r and V_θ. To solve for V_r and V_θ, the procedure is the same as that used in the solution of creeping flow around a solid sphere. Equations (5.328) – (5.331) are valid. However, the boundary conditions are now different as compared with flow around a solid sphere. The boundary conditions are:

$$\text{B.C.1 : at } r = R\,,\ V_r = 0 \tag{5.378}$$

$$\text{B.C.2 : at } r = R\,,\ \tau_{r\theta} = 0 \quad (5.379$$

$$\text{B.C.3 : at } r \to \infty,\ V_r = -V_\infty cos\theta \quad (5.380)$$

$$\text{B.C.4 : at } r \to \infty,\ V_\theta = V_\infty sin\theta \quad (5.381)$$

Note that far away from the bubble, the flow is in negative z direction with free stream velocity of V_∞. From the B.C. 3 or B.C. 4, upon integration, it can readily be shown that:

$$\text{at } r \to \infty,\ \psi = \frac{1}{2} V_\infty r^2 \sin^2\theta \quad (5.382)$$

This boundary condition suggests the solution to be of the form:

$$\psi = f(r) \sin^2\theta \quad (5.383)$$

Upon substitution of this stream function into $E^4\psi = 0$, we obtain the same fourth order Euler-Cauchy equation as obtained in the case of solid sphere. Thus:

$$f = \frac{A}{r} + Br + Cr^2 + Dr^4 \quad (5.384)$$

To satisfy the stream function boundary condition at $r \to \infty$, $D = 0$ and $C = V_\infty/2$. The solution of the stream function equation $E^4\psi = 0$ thus becomes:

$$\psi = \left(\frac{A}{r} + Br + \frac{1}{2} V_\infty r^2\right) \sin^2\theta \quad (5.385)$$

We can now express the unknown velocity components V_r and V_θ as follows:

$$V_r = -\frac{1}{r^2 \sin\theta} \frac{\partial\psi}{\partial\theta} = -\left(V_\infty + \frac{2A}{r^3} + \frac{2B}{r}\right) \cos\theta \quad (5.386)$$

$$V_\theta = \frac{1}{r \sin\theta} \frac{\partial\psi}{\partial r} = \left(V_\infty - \frac{A}{r^3} + \frac{B}{r}\right) \sin\theta \quad (5.387)$$

From Newton's law of viscosity, $\tau_{r\theta}$ is given as:

$$\tau_{r\theta} = \eta\left[r \frac{\partial}{\partial r}\left(\frac{V_\theta}{r}\right) + \frac{1}{r} \frac{\partial V_r}{\partial\theta}\right] \quad (5.388)$$

Upon substituting the expressions for V_r and V_θ and carrying out the required differentiation, the following result is obtained for $\tau_{r\theta}$:

$$\tau_{r\theta} = \frac{6A\eta}{r^4} \sin\theta \quad (5.389)$$

Using the boundary conditions at the bubble surface, that is, B.C. 1 (Equation (5.378)) and B.C. 2 (Equation (5.379)), we get: $A = 0$ and $B = -V_\infty R/2$. Thus:

$$f = \left(\frac{V_\infty R^2}{2}\right)\left[\left(\frac{r}{R}\right)^2 - \left(\frac{r}{R}\right)\right] \quad (5.390)$$

$$\psi = \left(\frac{V_\infty R^2}{2}\right)\left[\left(\frac{r}{R}\right)^2 - \left(\frac{r}{R}\right)\right] sin^2\,\theta \tag{5.391}$$

$$V_r = V_\infty\left(\frac{R}{r} - 1\right) cos\,\theta \tag{5.392}$$

$$V_\theta = V_\infty\left(1 - \frac{R}{2r}\right) sin\,\theta \tag{5.393}$$

The pressure distribution in the matrix fluid can now be determined using the same approach as that used for solid sphere:

$$\frac{\partial P}{\partial r} - \rho g_r = -\frac{\eta}{r^2\,sin\,\theta}\frac{\partial(E^2\psi)}{\partial\theta} \tag{5.394}$$

$$\frac{\partial P}{\partial\theta} - r\rho g_\theta = \frac{\eta}{sin\,\theta}\frac{\partial(E^2\psi)}{\partial r} \tag{5.395}$$

where

$$E^2\psi = V_\infty\left(\frac{R}{r}\right) sin^2\,\theta \tag{5.396}$$

Thus:

$$\frac{\partial P}{\partial r} = -\left[\rho g + \frac{2\eta V_\infty R}{r^3}\right] cos\,\theta \tag{5.397}$$

$$\frac{\partial P}{\partial\theta} = \left[\rho g r - \frac{\eta V_\infty R}{r^2}\right] sin\,\theta \tag{5.398}$$

Note that $g_r = -gcos\theta$ and $g_\theta = gsin\theta$ where g is the acceleration due to gravity acting in the negative z direction. Using the above equations, one can write:

$$dP = \left(\frac{\partial P}{\partial r}\right) dr + \left(\frac{\partial P}{\partial\theta}\right) d\theta = -\left[\rho g + \frac{2\eta V_\infty R}{r^3}\right] cos\,\theta\,dr + \left[\rho g r - \frac{\eta V_\infty R}{r^2}\right] sin\,\theta\,d\theta = -d\left[(cos\,\theta)\left(\rho g r - \frac{\eta V_\infty R}{r^2}\right)\right] \tag{5.399}$$

Upon integration, we obtain:

$$P = P_o - (cos\,\theta)\left(\rho g r - \frac{\eta V_\infty R}{r^2}\right) \tag{5.400}$$

where $P = P_o$ at $z = 0$ as $r \to \infty$.

Thus, the pressure distribution at the bubble surface becomes:

$$P = P_o + (cos\,\theta)\left(\frac{\eta V_\infty}{R} - \rho g R\right) \tag{5.401}$$

Among the viscous stresses acting on the bubble surface, we need to consider only the normal stress τ_{rr} as the tangential stress $\tau_{r\theta}$ is zero.

$$\tau_{rr} = 2\eta\left(\frac{\partial V_r}{\partial r}\right) = -\frac{2\eta V_\infty R}{r^2} cos\,\theta \tag{5.402}$$

At the bubble surface:

$$\tau_{rr}|_{r=R} = -\frac{2\eta V_\infty}{R}\cos\theta \tag{5.403}$$

The net force acting on the bubble in the z-direction due to pressure and viscous stress distribution around the bubble can be determined as follows:

$$F_z = \int_0^{2\pi}\int_0^{\pi}((-P+\tau_{rr})|_{r=R}\cos\theta)R^2\sin\theta\, d\theta d\varphi \tag{5.404}$$

Upon substituting the expressions for pressure distribution and normal stress τ_{rr} and carrying out the integration, the following result is obtained:

$$F_z = \frac{4}{3}\pi R^3\rho g - 4\pi\eta R V_\infty \tag{5.405}$$

The first term on the right side is the buoyancy force acting on the bubble in the positive z-direction. The second term is the drag force on the bubble due to flow-induced pressure and normal-stress (τ_{rr}) distributions around the bubble surface. The drag force acts in the negative z-direction for the bubble rising in the positive z-direction. Note that the drag force on the bubble is entirely "form drag" as skin drag is zero.

The drag coefficient for the bubble is as follows:

$$C_D = \frac{F_D}{A_p\left(\frac{1}{2}\rho V_\infty^2\right)} = \frac{4\pi\eta R V_\infty}{\pi R^2\left(\frac{1}{2}\rho V_\infty^2\right)} = \frac{16}{\left(\frac{\rho D_b V_\infty}{\eta}\right)} = \frac{16}{Re_p} \tag{5.406}$$

where D_b is bubble diameter. Thus, the drag coefficient for a bubble is smaller than that of a solid particle when comparison is made at the same Re_p due to the absence of skin drag.

NOTATION

A_p	projected area of the particle on the plane normal to the direction of flow
C_D	drag coefficient
D	diameter
D_b	diameter of bubble
D_p	diameter of particle
erf	error function
$erfc$	complimentary error function
F_D	drag force
F_z	net force acting on the bubble in the z-direction due to pressure and viscous stress distribution around the bubble
$F_{z,p}$	net force acting on the particle in the direction of motion (z −direction) due to pressure distribution around the particle
$F_{z,\tau}$	net force acting on the particle in the direction of motion (z −direction) due to shear-stress $\tau_{r\theta}$ distribution around the particle surface
g	acceleration due to gravity
g_x	x-component of acceleration due to gravity
g_y	y-component of acceleration due to gravity

g_z z-component of acceleration due to gravity
$\vec{g}$ acceleration due to gravity vector
G geometric constant
H height
J_o Bessel function of first kind of order 0
J_1 Bessel function of first kind of order 1
J_ν Bessel function of first kind of order ν
k parameter specifying the radius of the inner pipe of a concentric double pipe geometry
L length
$\mathcal{L}$ Laplace transform
$\mathcal{L}^{-1}$ inverse of Laplace transform
P pressure
P^* dimensionless pressure
P' modified pressure
P_n^m $P_n^m(x)$ is the associated Legendre function of first kind (order m, degree n)
Q volumetric flow rate
Q_n^m $Q_n^m(x)$ is the associated Legendre function of second kind (order m, degree n)
r radial position or coordinate
R radius
Re Reynolds number
Re_p particle Reynolds number
s Laplace variable
t time
t^* dimensionless time
T temperature
V_x, V_y, V_z velocity components in cartesian coordinate system
V_r, V_θ, V_φ velocity components in spherical coordinate system
V_o constant velocity
V_{max} maximum velocity
V_∞ free stream velocity
$\vec{V}$ velocity vector
$\bar{V}$ average velocity
V^* dimensionless velocity
W width or some function
x, y, z space coordinates in cartesian coordinate system
x^*, y^*, z^* dimensionless space coordinates in cartesian coordinate system
Y_ν Bessel function of second kind of order ν

Greek Letters

α angle of inclination or constant
β constant
δ film thickness
η viscosity
θ space coordinate in cylindrical and spherical coordinates
ν kinematic viscosity
ξ dimensionless radius
ρ density
τ dimensionless time
τ_{rr} viscous stress component

τ_{rz}	viscous stress component
$\tau_{r\theta}$	viscous stress component
$\tau_{\theta\varphi}$	viscous stress component
φ	space coordinate in spherical coordinates or volume fraction or gravitation potential or dimensionless velocity
ψ	stream function
ω	angular velocity

Mathematical operators

∇	del operator or nabla
∇^*	dimensionless del operator
∇^2	Laplacian operator
∇^{*2}	dimensionless Laplacian operator
∇^4	operator defined in Table 5.1
E^2	operator defined in Table 5.1
E^4	operator defined in Table 5.1
$\frac{\partial(\psi,\nabla^2\psi)}{\partial(x,y)}$	Jacobian defined in Equation (5.319)

PRACTICE PROBLEMS

Problem 1

Consider steady laminar flow of a Newtonian fluid of constant properties between two large parallel plates 2.5 cm apart (see figure). The lower plate is stationary, and the top plate is moving at a velocity of V_o. If the pressure gradient in the direction of flow, that is, dP/dx, is $+48\ Pa/m$, at what velocity V_o will the net flow become zero at any station. The viscosity of the fluid is 0.12 mPa.s.

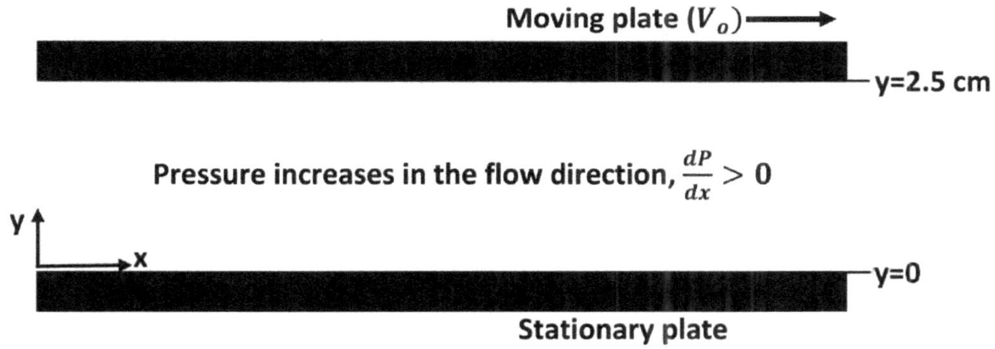

Problem 2

Consider an inclined plate moving with a constant velocity V_o. A liquid film is flowing over a moving plate (see figure). Determine the velocity distribution in the liquid film. You can assume that the shear stress τ_{yx} at the free surface is negligible. What is the net liquid flow rate per unit width of the plate? What is the film thickness δ if the net liquid flow rate is zero.

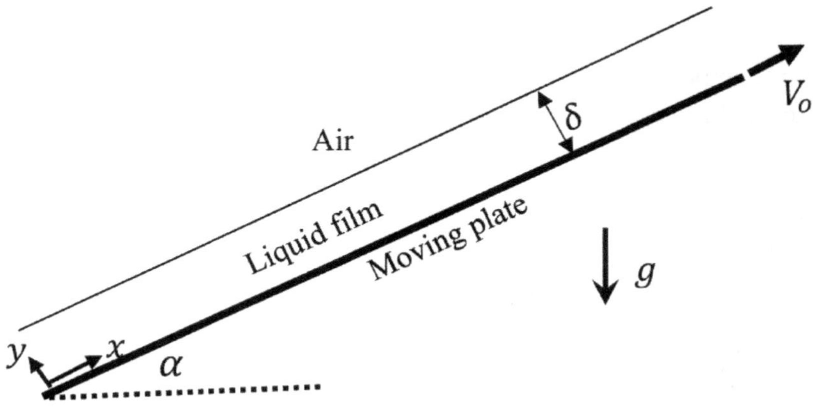

Problem 3
Consider steady laminar flow of a Newtonian fluid between two large parallel plates, a distance "d" apart. The upper plate is moving at a constant velocity of V_o and the bottom plate is stationary. What should be the pressure gradient if the shear stress at the lower stationary plate is zero?

Problem 4
Consider steady laminar flow of a Newtonian fluid between two large vertical parallel plates as shown in the figure. The left plate is stationary, and the right plate is moving at a constant velocity of V_o. The fluid is also subjected to a constant pressure gradient dP/dz. Determine the velocity distribution.

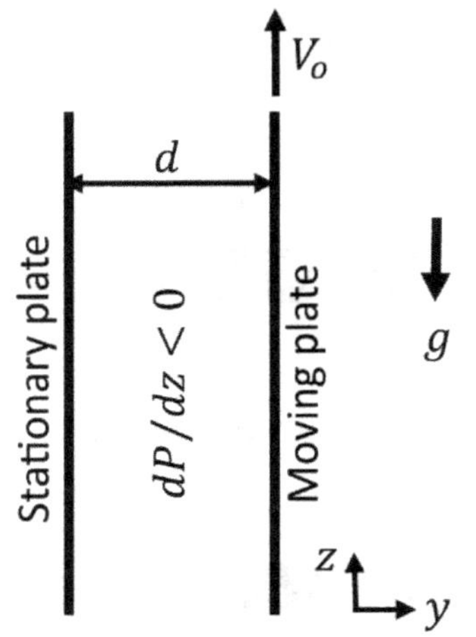

Problem 5
Consider steady laminar flow of a Newtonian fluid between two infinitely long horizontal concentric cylinders. The fluid motion is caused by the longitudinal motion of the inner cylinder at a constant velocity of V_o (see figure). The outer cylinder is stationary. There is no pressure-gradient applied. Determine the following:

i) Velocity distribution in the fluid
ii) Volumetric flow rate of fluid
iii) Drag force on the inner cylinder per unit length

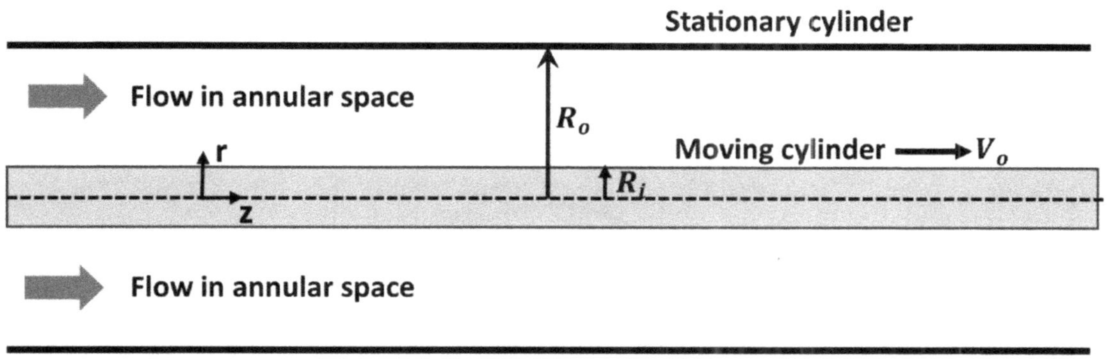

Problem 6
A long solid cylinder of radius R is placed vertically in a large pool of stagnant Newtonian fluid. The cylinder is rotated about its axis at a constant angular velocity of ω (see figure). Determine the following:

i) Velocity distribution in the fluid
ii) Shear stress distribution in the fluid
iii) Torque required to rotate the cylinder about its axis at a constant angular velocity of ω

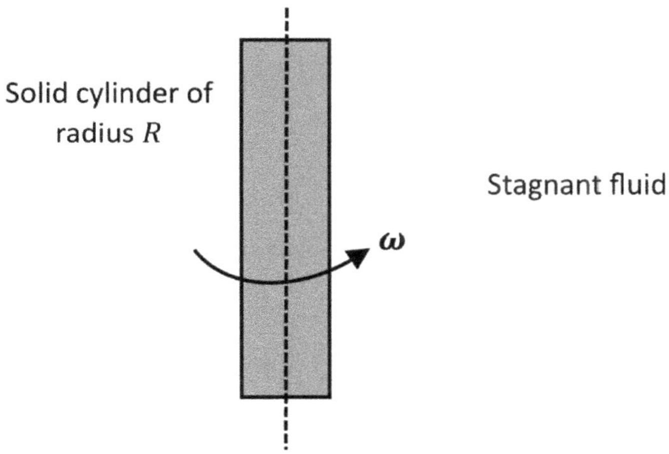

Problem 7
Consider the flow of a Newtonian fluid between two concentric rotating cylinders of radii R_i and R_o (see figure). The angular velocities of the inner and outer cylinders are ω_i and ω_o, respectively. Determine the velocity distribution in the fluid and the torques on the two cylinders.

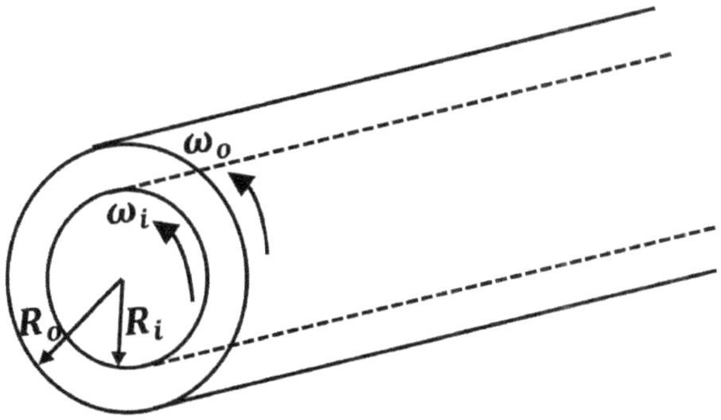

Problem 8

Consider the flow of a Newtonian fluid between two concentric rotating spheres of radii R_i and R_o (see figure). The angular velocities of the inner and outer spheres are ω_i and ω_o, respectively. Determine the velocity distribution in the fluid.

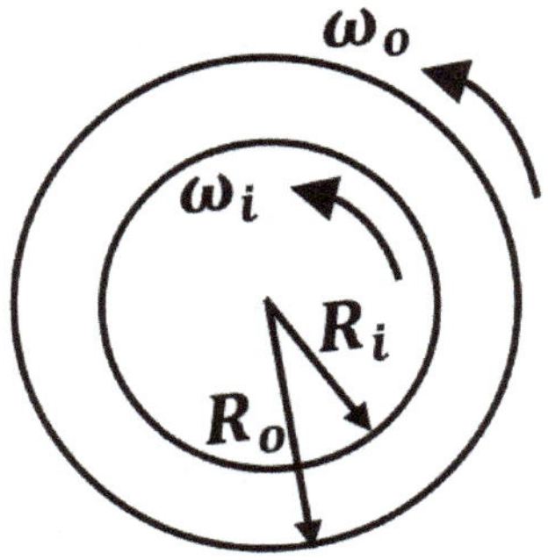

Problem 9

A straight duct has a length L and cross-section of square shape, bounded by plane surfaces $x = +B$, $x = -B$, $y = +B$, $y = -B$. Determine if the correct laminar velocity distribution for this problem is:

$$V_z = \left(\frac{\Delta P}{L}\right)\left(\frac{B^2}{4\mu}\right)\left[1-\left(\frac{x}{B}\right)^2\right]\left[1-\left(\frac{y}{B}\right)^2\right]$$

Problem 10

Consider the flow of a Newtonian fluid in the radial direction between parallel disks (see figure). The pressure is P_1 at $r = R_1$ and P_2 at $r = R_2$. Thus, the pressure differential driving the flow in the radial direction is $(P_1 - P_2)$. Assuming creeping flow between the disks, determine the velocity distribution $V_r(r, z)$.

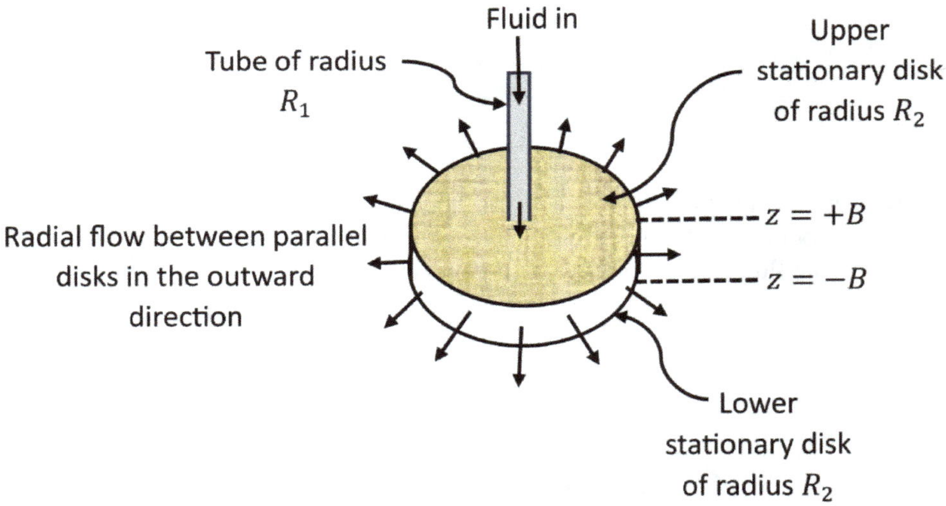

Problem 11

Consider the flow of a Newtonian fluid between two porous parallel plates at $y = h$ and $y = 0$. Let the main flow be generated by a constant pressure gradient $(-dP/dx)$. Let the porous walls be such that a uniform vertical crossflow is generated (see figure).

$$V_y = V_w = \text{constant}$$

Solve for the axial velocity profile, $V_x(y)$, in terms of the dimensionless variables:

$$\varphi = \frac{V_x}{(h^2/\eta)(-dP/dx)}, \qquad \xi = y/h, \qquad Re = V_w \rho h/\eta$$

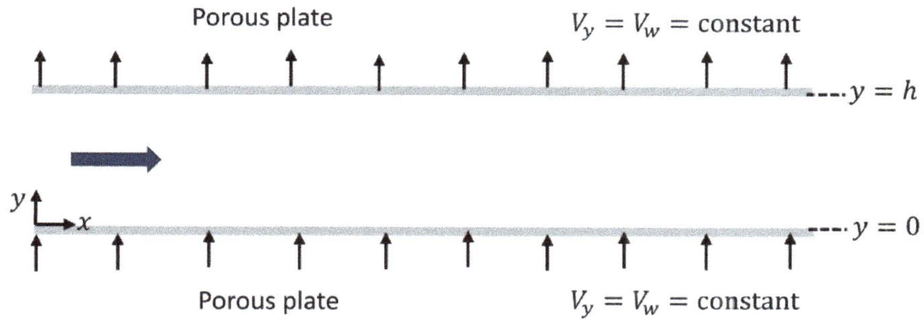

Problem 12

Re-do the unsteady-state pipe-flow problem (section 5.2.2) using the Laplace Transform Technique

Problem 13

Show that for axisymmetric creeping flow around a particle, the pressure distribution can be expressed in terms of the following equations:

$$\frac{\partial P'}{\partial r} = -\frac{\mu}{r^2 \sin\theta}\left(\frac{\partial (E^2\psi)}{\partial \theta}\right)$$

$$\frac{\partial P'}{\partial \theta} = \frac{\mu}{r \sin\theta}\left(\frac{\partial (E^2\psi)}{\partial r}\right)$$

where $P' = P + \rho g z$

FURTHER READING

1. R.B. Bird, W.E. Stewart, E.N. Lightfoot. *Transport Phenomena*. 2nd Edition, Wiley, New York, 2007.
2. C.J. Geankoplis, A.A. Hersel, D.H. Lepek. *Transport Processes and Separation Process Principles*. 5th Edition, Prentice Hall, New Jersey, 2018.
3. J.R. Welty, C.E. Wicks, R.E. Wilson, G. Rorrer. *Fundamentals of Momentum, Heat, and Mass Transfer*. 4th Edition, Wiley, New York, 2001.
4. R.L. Street, G.Z. Watters, J.K. Vannard. *Elementary Fluid Mechanics*. 7th Edition, Wiley, New York, 1996.
5. J.O. Wilkes, *Fluid Mechanics for Chemical Engineers*. 3rd Edition, Prentice Hall, Boston, 2018.
6. B.R. Munson, D.F. Young, T.H. Okiishi. *Fundamentals of Fluid Mechanics*. 4th Edition, Wiley, New York, 2002.

Chapter 6

Friction Factor and Friction Loss Calculations

In Chapter 5, analytical solutions of microscopic equations of motion were obtained for many problems involving laminar flow of Newtonian fluids. We could solve for velocity, shear stress, and pressure distributions in fluid flow. However, it is not always possible to obtain an analytical solution to the microscopic equations of motion even in the case of laminar flows when the geometry is complicated. Furthermore, turbulent flows are completely out of reach as it is not at all possible to obtain any analytical solutions of turbulent flows although the microscopic equations of motion are still valid. Fortunately, in many practical applications involving fluid flow, the detailed velocity, shear stress, and pressure distributions are not required. Often, the quantity of practical interest is the wall shear stress, that is, the shear stress exerted by the fluid on the wall on the flow geometry (see Figures 6.1 and 6.2).

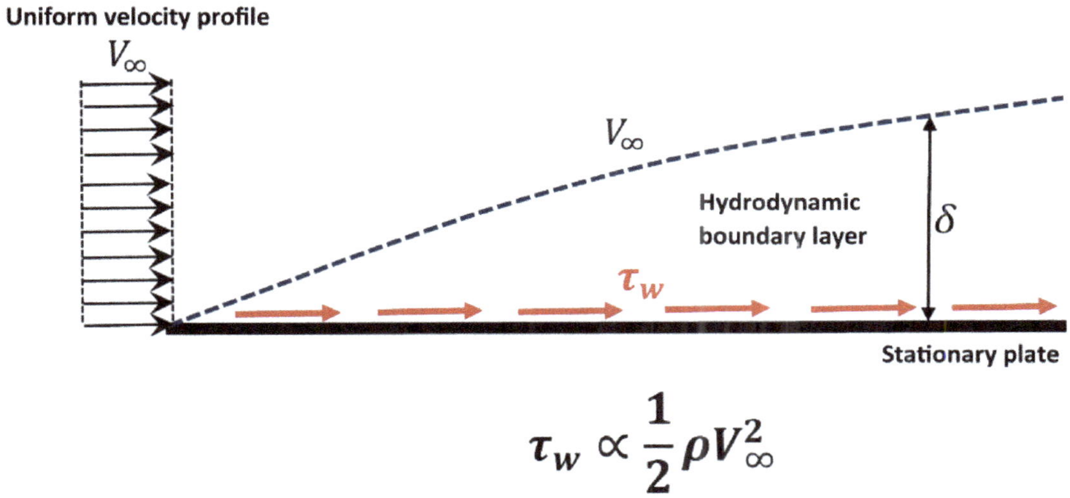

Figure 6.1 Flow over a solid flat boundary.

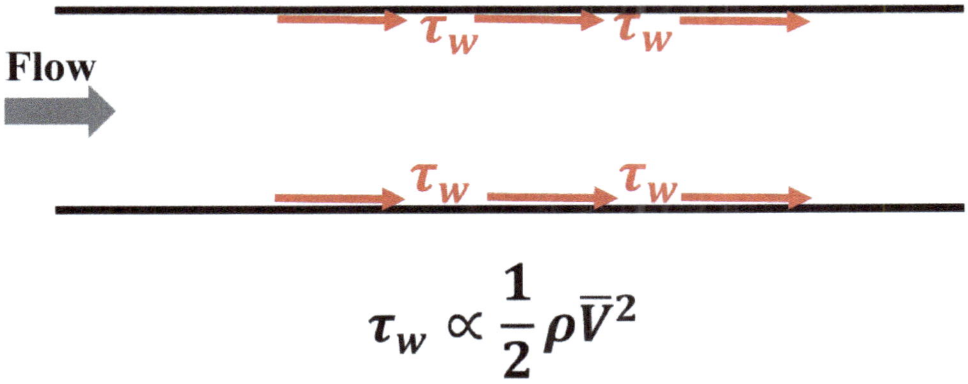

Figure 6.2 Flow in a cylindrical pipe.

There is enough experimental evidence to suggest that the wall shear stress τ_w is directly proportional to the characteristic kinetic energy per unit volume of fluid, that is,

$$\tau_w \propto \frac{1}{2}\rho V_c^2 \tag{6.1}$$

where ρ is fluid density and V_c is characteristic fluid velocity. For external flows (see Figure 6.1):

$$V_c = V_\infty \tag{6.2}$$

where V_∞ is the free-stream velocity. For internal flows (see Figure 6.2):

$$V_c = \bar{V} \tag{6.3}$$

where $\bar{V}$ is the average velocity in the internal flow geometry such as a pipe.

The friction factor is defined as the proportionality factor between τ_w and $\frac{1}{2}\rho V_c^2$. Thus, Equation (6.1) can be re-written as:

$$\tau_w = \frac{1}{2} f \rho V_c^2 \tag{6.4}$$

For external flows,

$$\tau_w = \frac{1}{2} f \rho V_\infty^2 \tag{6.5}$$

For internal flows,

$$\tau_w = \frac{1}{2} f \rho \bar{V}^2 \tag{6.6}$$

The wall stress, and hence the drag force exerted by the fluid on the solid boundary (plane wall or pipe wall), can be calculated from Equations (6.4) -(6.6) provided that the friction factor f for the given situation is known or could be estimated from correlations. The knowledge of the friction factor is also required in the calculation of friction loss, that is, mechanical energy dissipation into frictional heating during flow.

6.1 FRICTION FACTOR IN EXTERNAL FLOWS

Dimensional analysis shows that friction factor is a function of the Reynolds number. For external flows (see Figure 6.3), the local Reynolds number (Re_x) is defined as:

$$Re_x = \frac{\rho x V_\infty}{\eta} \tag{6.7}$$

where x is the distance from the leading edge of the solid boundary (plate in the present case). As x increases, the Reynolds number increases. For flow over a flat surface, the flow in the boundary layer is usually laminar for $Re_x < 2 \times 10^5$. The transition from laminar boundary layer to turbulent boundary layer begins at $Re_x = 2 \times 10^5$ and is completed at $Re_x \approx 3 \times 10^6$. The boundary layer is fully turbulent for $Re_x \geq 3 \times 10^6$. The transition occurs at a lower Re_x when the plate is rough and at a higher Re_x when the plate is hydraulically smooth. In performing calculations, however, it is generally assumed that the transition from laminar to turbulent flow occurs sharply at a critical Reynolds number. The commonly used value of critical Reynolds number is $Re_{x,critical} = 5 \times 10^5$.

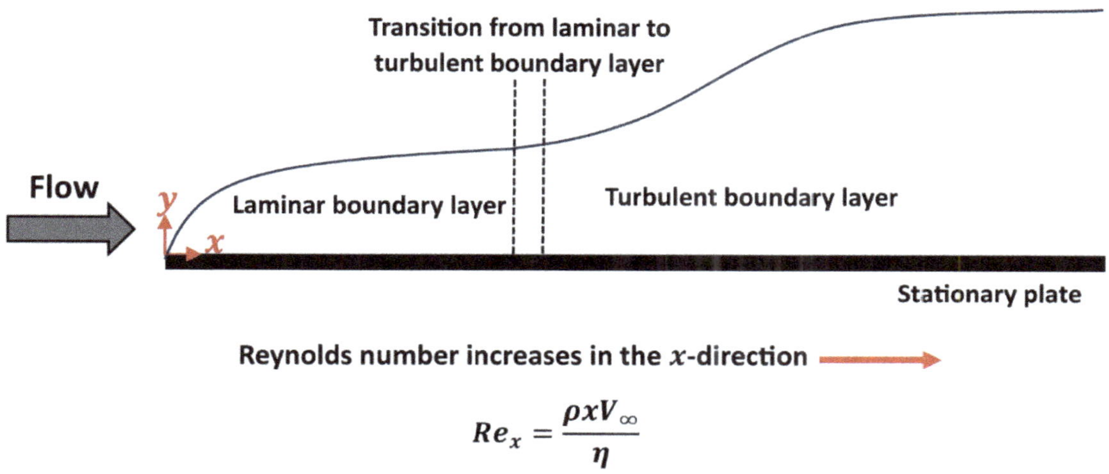

Figure 6.3 Boundary layer flow over a flat plate.

The following relations are valid for local friction factor in laminar and turbulent boundary layers over a smooth plate.

$$f_x = 0.664(Re_x)^{-1/2} \tag{6.8}$$

$$f_x = 0.0592(Re_x)^{-1/5} \tag{6.9}$$

Equation (6.8) is valid for laminar boundary layer and Equation (6.9) is valid for turbulent boundary layer. Equation (6.8) is an exact theoretical relationship obtained from the solution of boundary layer equations. The solution is given in Appendix A. Equation (6.9) is an empirical relation. Figure 6.4 shows the plots of local friction factor f_x versus local Reynolds number Re_x. The plots are generated from Equations (6.8) and (6.9). The friction factor decreases with the increase in Reynolds number. However, the decrease is steeper when the flow is laminar in the boundary layer. Upon transition from laminar to turbulent flow in the boundary layer, a sharp jump occurs in the friction factor.

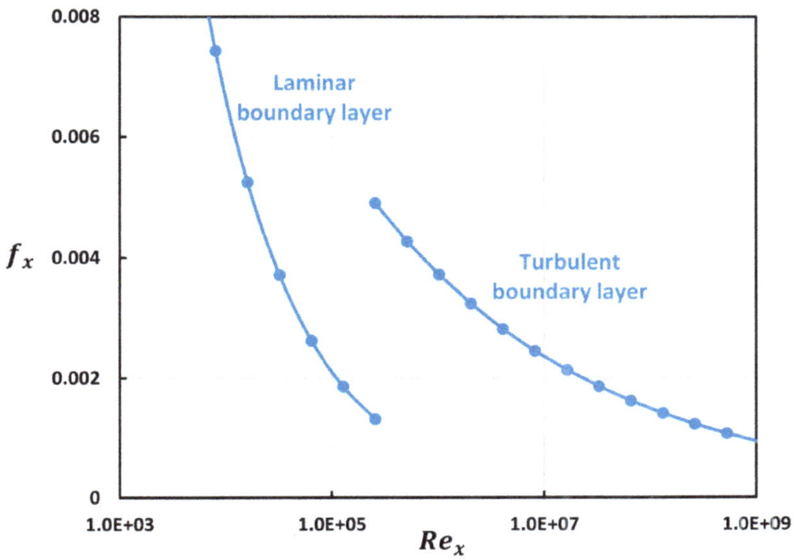

Figure 6.4 Local friction factor f_x versus local Reynolds number Re_x for flow over a flat plate.

The total drag force F_D exerted by the fluid flowing over the plate can be estimated as follows:

$$F_D = \bar{\tau}_w A \tag{6.10}$$

where $\bar{\tau}_w$ is the average wall stress over the entire plate and A is the area of the plate. Note that the wall stress, like friction factor, varies with the location x on the plate. The average wall stress $\bar{\tau}_w$ is given as:

$$\bar{\tau}_w = \frac{1}{2}\bar{f}\rho V_\infty^2 \tag{6.11}$$

where $\bar{f}$ is the average friction factor given as:

$$\bar{f} = \frac{1}{L}\int_0^L f_x\, dx \tag{6.12}$$

The evaluation of the integral in Equation (6.12) depends on the flow conditions over the plate. There are two possibilities: 1) the flow condition and length of the plate are such that the transition from laminar to turbulent flow does not occur over the plate, and hence, only laminar boundary layer flow occurs over the entire plate; 2) the transition from laminar to turbulent boundary layer flow occurs at some distance x on the plate. Thus, the flow over the plate is mixed boundary layer flow.

6.1.1 Only Laminar Boundary Layer Flow Over the Entire Plate

When only laminar boundary layer is present over the entire plate, the average friction factor $\bar{f}$ can be determined from Equations (6.8) and (6.12) as follows:

$$\bar{f} = \frac{1}{L}\int_0^L f_x\, dx = \frac{0.664}{L}\left(\frac{\rho V_\infty}{\eta}\right)^{-1/2}\int_0^L \frac{dx}{\sqrt{x}} = 2\left[0.664(Re_L)^{-1/2}\right] \tag{6.13}$$

where Re_L is the local Reynolds number at the end of the plate $\frac{\rho V_\infty L}{\eta}$. Equation (6.13) implies that:

$$\bar{f} = 2(f_x|_{x=L}) \tag{6.14}$$

where $f_x|_{x=L}$ is the local value of friction factor at $x = L$, given as

$$f_x|_{x=L} = 0.664(Re_L)^{-1/2} \tag{6.15}$$

From Equations (6.10), (6.11), and (6.14), it follows that:

$$F_D = \rho V_\infty^2 A(f_x|_{x=L}) \tag{6.16}$$

6.1.2 Mixed Boundary Layer Flow Over the Plate

When both laminar and turbulent boundary layers are present over the plate, the average friction factor $\bar{f}$ can be determined as follows:

$$\bar{f} = \frac{1}{L}\int_0^L f_x\, dx = \frac{1}{L}\left[\int_0^{x_c} f_{x,laminar}\, dx + \int_{x_c}^L f_{x,turbulent}\, dx\right] \tag{6.17}$$

where $f_{x,laminar}$ is given by Equation (6.8) and $f_{x,turbulent}$ is given by Equation (6.9). Upon substituting the expressions for $f_{x,laminar}$ and $f_{x,turbulent}$ into Equation (6.17) and integrating, the following result is obtained for the average friction factor:

$$\bar{f} = 0.074(Re_L)^{-1/5} - \frac{2A}{Re_L} \tag{6.18}$$

where

$$A = 0.037\left(Re_{x,c}\right)^{4/5} - 0.664\left(Re_{x,c}\right)^{1/2} \tag{6.19}$$

When $Re_{x,c} = 5 \times 10^5$, $A \approx 871$.

6.2 FRICTION FACTOR IN INTERNAL FLOWS

6.2.1 Circular Cross Section Pipes

For laminar flow of a Newtonian fluid in a circular cross section pipe, the friction factor is given by the following relation:

$$f = \frac{16}{Re} \tag{6.20}$$

This is an exact theoretical relationship. The friction factor in laminar pipe flow is independent of the roughness of the pipe wall. The flow in a pipe is laminar if Re is less than 2100. The flow is turbulent when Re is greater than about 4000. The flow oscillates randomly between laminar and turbulent regimes in the transition region corresponding to about $2100 < Re < 4000$.

Unlike laminar flow, the friction factor in turbulent flow depends on the pipe roughness, in addition to the Reynolds number. Thus:

$$f = f(Re, \varepsilon/D\,) \tag{6.21}$$

where ε/D is the relative roughness and ε is the roughness parameter (average height of the protrusions on the pipe surface). For hydraulically smooth pipes (ε/D approaches zero), the following equation, often referred to as Prandtl-vonKarman-Nikuradse, represents the experimental data well:

$$\frac{1}{\sqrt{f}} = 4.0\, log_{10}\left(Re\sqrt{f}\right) - 0.4 \quad 2.3 \times 10^3 < Re < 4 \times 10^6 \tag{6.22}$$

This equation is not explicit in f. The following equation, referred to as Barenblatt equation, is explicit in f and represents the experimental data well:

$$f = \frac{2}{\psi^{\frac{2}{\alpha+1}}} \qquad 3 \times 10^3 < Re < 3 \times 10^6 \tag{6.23}$$

where

$$\psi = \frac{e^{\frac{3}{2}}(\sqrt{3}+5\alpha)}{2^{\alpha}\alpha(\alpha+1)(\alpha+2)} \qquad \alpha = \frac{3}{2 lnRe} \tag{6.24}$$

The following simple forms of friction factor relations are also popular:

$$f = \frac{0.0791}{Re^{\frac{1}{4}}} \qquad 2.1 \times 10^3 < Re < 10^5 \tag{6.25}$$

$$f = \frac{0.046}{Re^{\frac{1}{5}}} \qquad 5 \times 10^4 < Re < 10^6 \tag{6.26}$$

$$4f = [0.79 \ln(Re) - 1.64]^{-2} \quad 3 \times 10^3 < Re < 5 \times 10^6 \tag{6.27}$$

Equation (6.25) is referred to as Blasius equation, Equation (6.25) as Knudsen and Katz equation, and Equation (6.27) as Filonenko equation.

For rough pipes, the following Colebrook's equation is most widely used:

$$\frac{1}{\sqrt{f}} = -4.0\, log_{10}\left(\frac{1.255}{Re\sqrt{f}} + \frac{\frac{\varepsilon}{D}}{3.7}\right) \tag{6.28}$$

The Colebrook equation is implicit in f and requires trial and error approach to solve for f. Haaland has proposed the following equation, explicit in f, as an approximation to the Colebrook equation:

$$\frac{1}{\sqrt{f}} = -3.6\, log_{10}\left[\frac{6.9}{Re} + \left(\frac{\frac{\varepsilon}{D}}{3.7}\right)^{\frac{10}{9}}\right] \qquad 4 \times 10^4 < Re < 10^8 \text{ and } 0 < \frac{\varepsilon}{D} < 0.05 \tag{6.29}$$

Another explicit equation which is very accurate for turbulent flow of Newtonian fluids in rough pipes is as follows:

$$1/\sqrt{f} = -4 log_{10}\left[\frac{\frac{\varepsilon}{D}}{3.7} - \frac{5.02}{Re} log_{10}\left(\frac{\frac{\varepsilon}{D}}{3.7} - \frac{5.02A}{Re}\right)\right] \tag{6.30}$$

$$A = log_{10}\left(\frac{\frac{\varepsilon}{D}}{3.7} + \frac{13}{Re}\right) \tag{6.31}$$

This equation is referred to as Zigrang and Sylvester equation.

Figure 6.5 shows the friction factor versus Reynolds number chart generated from Equation (6.20) for laminar flow and Equation (6.30) for turbulent flow. As expected, there is no dependence of friction factor on pipe roughness in laminar flow. However, in turbulent flow, the friction factor increases with the increase in relative roughness ε/D of the pipe. Consequently, we observe a family of friction factor versus Reynolds number curves in the turbulent region corresponding to different values of relative roughness ε/D.

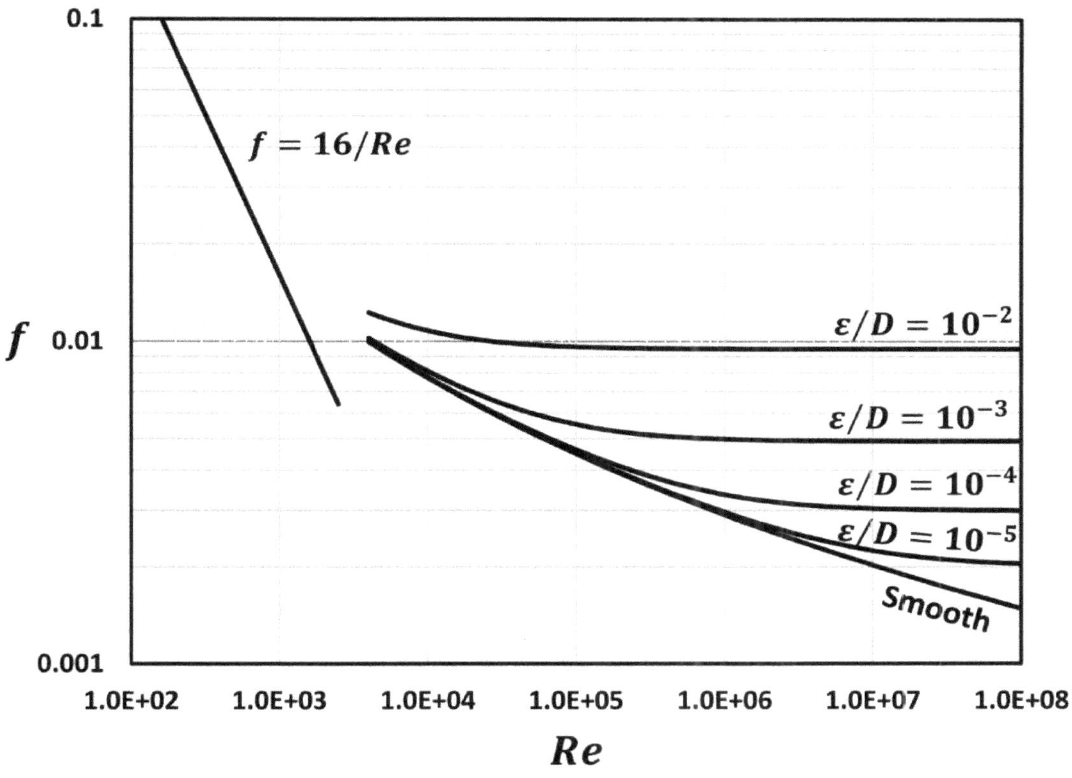

Figure 6.5 Friction factor versus Reynolds number for flow in a pipe.

6.2.2 Non-Circular Cross Section Pipes

So far, we have considered f versus Re relationships for pipes with circular cross section. For non- circular cross section pipes, the Reynolds number is defined based on the hydraulic diameter. The hydraulic diameter D_H is defined as:

$$D_H = 4\left[\frac{\text{cross sectional area of flow}}{\text{wetted perimeter}}\right] \tag{6.32}$$

The Reynolds number is defined as:

$$Re = \frac{\rho D_H \bar{V}}{\eta} \tag{6.33}$$

As an example, consider the annular geometry of two concentric pipes (see Figure 5.9). The hydraulic diameter of an annular geometry is:

$$D_H = 4\left[\frac{\pi(D_o^2 - D_i^2)/4}{\pi(D_o - D_i)}\right] = D_o - D_i = 2R(1-k) \tag{6.34}$$

where D_o is the diameter of the outer pipe, D_i is the diameter of the inner pipe, R is the radius of the outer pipe, and kR is the radius of the inner pipe. Thus, the Reynolds number for annular (concentric pipes) geometry is:

$$Re = \frac{2R\rho\bar{V}(1-k)}{\eta} \tag{6.35}$$

The friction factor is defined as:

$$f = \frac{\left(\frac{\Delta p D_H}{4L}\right)}{\frac{1}{2}\rho\bar{V}^2} \tag{6.36}$$

The average velocity for laminar flow in the annular region is given as (see Equation (5.77)):

$$\bar{V} = \frac{\Delta p R^2}{8\eta L}\left[\left(\frac{1-k^4}{1-k^2}\right) - \frac{(1-k^2)}{ln(1/k)}\right] \tag{6.37}$$

From Equations (6.35) to (6.37), we obtain:

$$f = \frac{16}{Re}\psi \tag{6.38}$$

where

$$\psi(k) = \frac{(1-k)^2}{\left[\frac{1-k^4}{1-k^2} - \frac{1-k^2}{ln(1/k)}\right]} \tag{6.39}$$

For a circular cross-section pipe, k is zero and ψ is unity. ψ increases with the increase in k. In the limit of $k \rightarrow 1$ (corresponding to flow between parallel plates), ψ reaches the highest value of 1.5 (see Figure 6.6). Note that Equation (6.38) is restricted to laminar flow in annular geometry.

For turbulent flow in pipes of non-circular cross section, one can use the f versus Re relationships for turbulent flow in pipes with circular cross section by replacing the diameter D with the hydraulic diameter D_H in these relationships. Thus, Re is given by Equation (6.33) and $\frac{\varepsilon}{D}$ is replaced by $\frac{\varepsilon}{D_H}$.

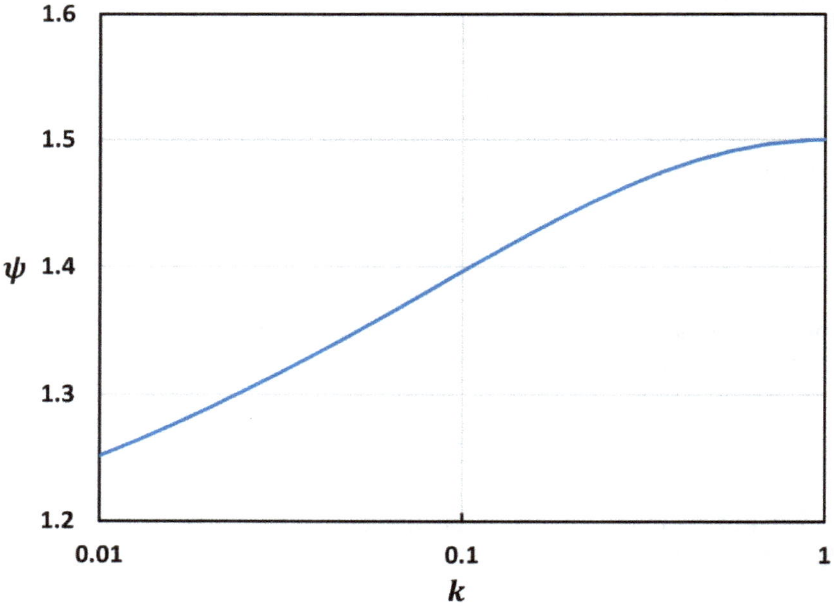

Figure 6.6 ψ variation with k.

6.3 ENGINEERING BERNOULLI EQUATION

The total energy balance over a control volume with one inlet and one outlet assuming steady state is as follows:

$$\dot{m}\left(\Delta h + \frac{1}{2}\Delta(V^2) + g\Delta z\right) = \dot{Q} - \dot{W}_{sh} \tag{6.40}$$

Here $\dot{m}$ is the mass flow rate, h is the specific enthalpy, V is the fluid velocity, z is the elevation with respect to some reference, $\dot{Q}$ is the rate of heat input to the control volume, and $\dot{W}_{sh}$ is the rate of shaft work output from the control volume. From thermodynamics, we know that:

$$dh = Tds + vdp \tag{6.41}$$

where s is the entropy per unit mass and v is the specific volume ($= 1/\rho$). Assuming frictionless flow, we can use the relation:

$$Tds = \delta Q \tag{6.42}$$

where Q is heat input per unit mass of the fluid. Combining Equations (6.41) and (6.42):

$$\Delta h = Q + \int_1^2 vdp \tag{6.43}$$

Thus:

$$-W_{sh} = \int_1^2 vdp + \frac{1}{2}\Delta(V^2) + g\Delta z \tag{6.44}$$

Equation (6.44) is a mechanical energy balance. It simply states that in frictionless flow, the increase in mechanical energy (kinetic, potential, and pressure) of the fluid is equal to the shaft work supplied to the fluid.

For real (viscous) fluids, we need to incorporate the frictional-loss term h_f (mechanical energy dissipation into frictional heating per unit mass of the fluid) as follows:

$$-W_{sh} = \int_1^2 vdp + \frac{1}{2}\Delta(V^2) + g\Delta z + h_f \tag{6.45}$$

For incompressible fluid $v = 1/\rho$ =constant and therefore:

$$\frac{p_1}{\rho} + gz_1 + \frac{1}{2}V_1^2 = \frac{p_2}{\rho} + gz_2 + \frac{1}{2}V_2^2 + h_f + W_{sh} \tag{6.46}$$

This equation assumes that the velocities at the inlet and outlet of the control volume are uniform, that is, the velocity profiles are flat at the entrance and discharge. For a flat profile, the kinetic energy per unit mass of the fluid is simply $V^2/2$. The velocity profile is not flat in the case of real viscous fluids and therefore, a correction is needed in the above mechanical energy balance equation. The kinetic energy per unit mass of the fluid can be determined from the given velocity distribution as follows:

$$\frac{K.E.}{mass} = \alpha\left(\frac{1}{2}\bar{V}^2\right) = \frac{\int_A \frac{1}{2}\rho V^3 dA}{(\rho\bar{V}A)} \tag{6.47}$$

where V is the local velocity, $\bar{V}$ is the average velocity and α is the kinetic-energy correction factor. From the above relationship, α is given as:

$$\alpha = \frac{1}{A}\int_A \left(\frac{V}{\bar{V}}\right)^3 dA = \frac{2}{R^2}\int_0^R \left(\frac{V}{\bar{V}}\right)^3 rdr \tag{6.48}$$

Thus, α can be determined from the velocity distribution. For a parabolic velocity distribution obtained in laminar flow of Newtonian fluids, $\alpha = 2$. For turbulent flows, the time-averaged velocity profile is relatively flat and therefore, α is only slightly larger than unity. As an approximation, α can be taken as unity for turbulent flows.

To consider the non-uniform velocity distributions at the inlet and outlet of the control volume, the mechanical energy balance equation needs to be modified as follows:

$$\frac{p_1}{\rho} + gz_1 + \frac{1}{2}\alpha_1\bar{V}_1^2 = \frac{p_2}{\rho} + gz_2 + \frac{1}{2}\alpha_2\bar{V}_2^2 + h_f + W_{sh} \tag{6.49}$$

This is the engineering Bernoulli equation used widely in the design and operation of pipeline systems. Note that α_1 at inlet may be different from α_2 at outlet due to a change in the flow regime (laminar to turbulent or vice versa) from inlet to outlet.

The friction loss h_f in the engineering Bernoulli equation is the sum of major loss, due to frictional effects in constant cross-sectional area pipes present in the pipeline system, and minor loss, resulting from frictional effects in pipe components such as: expansions and contractions, valves, and fittings. Thus,

$$h_{f,total} = h_{f,major} + h_{f,minor} \tag{6.50}$$

The major loss term can be evaluated from the following relation for pipeline flows:

$$h_{f,major} = 4f\frac{L}{D}\frac{\bar{V}^2}{2} \tag{6.51}$$

Thus, knowledge of friction factor is required to perform h_f calculation.

6.4 FLOW IN PIPE COMPONENTS

In a typical pipeline system, there are present various pipe components in addition to straight pipes of constant cross-section. The components consist of sudden enlargement of cross-section, sudden contraction of cross-section, and various types of fittings and valves. The flow patterns in the pipe components are generally quite complex and therefore, theoretical prediction of minor loss is difficult in most cases. However, sudden expansions and contractions are accessible to theoretical analysis.

6.4.1 Sudden Expansion

Consider a sudden expansion of the cross section of a pipe (see Figure 6.7). The fluid stream from a smaller pipe emerges as a jet into the larger pipe. The jet expands until it fills the full cross-section of the larger pipe. The velocity of the fluid decreases due to an increase in the cross-sectional area of flow. The pressure increases in compliance with the Bernoulli equation. In a decelerating flow under adverse pressure gradient, vortices are formed in the region between the expanding jet and pipe wall. A considerable loss of mechanical energy occurs due to frictional heating resulting from vortices.

Let the diameter of the pipe before expansion be D_1 (cross sectional area A_1) and the diameter after expansion be D_2 (cross sectional area A_2). Let the corresponding average velocities be $\bar{V}_1$ and $\bar{V}_2$. We will first do momentum balance on the fluid between sections "a" and "2", assuming steady state turbulent flow of a Newtonian fluid.

$$\begin{Bmatrix} \text{Rate at which momentum} \\ \text{enters the CV} \end{Bmatrix} = \beta_a \dot{m} \bar{V}_a \tag{6.52}$$

where β_a is the momentum correction factor at section "a" that accounts for a non-uniform velocity profile.

$$\begin{Bmatrix} \text{Rate at which momentum} \\ \text{leaves the CV} \end{Bmatrix} = \beta_2 \dot{m} \bar{V}_2 \tag{6.53}$$

$$\begin{Bmatrix} \text{Rate at which momentum is} \\ \text{generated within the CV} \end{Bmatrix} = \sum F \tag{6.54}$$

From momentum balance under steady state condition:

$$\beta_a \dot{m} \bar{V}_a - \beta_2 \dot{m} \bar{V}_2 + \sum F = 0 \tag{6.55}$$

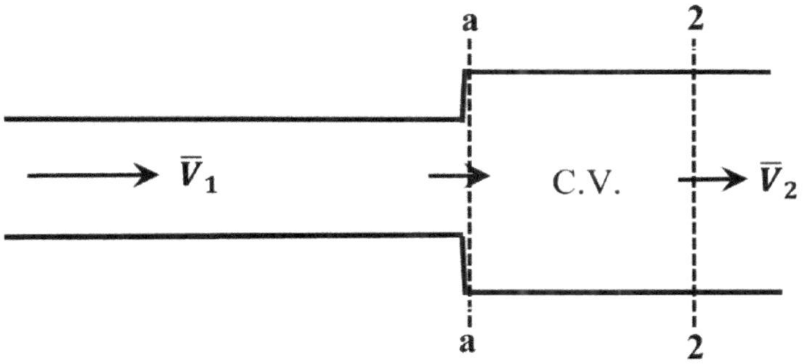

Figure 6.7 Schematic of sudden expansion.

As the flow is assumed to be turbulent at both sections "a" and "2", the momentum correction factors β_a and β_2 are approximately one (velocity profile nearly uniform). The forces acting on the fluid are due to pressure at the inlet and outlet (assuming negligible viscous stresses from the wall), that is,

$$\sum F = p_a A_2 - p_2 A_2 = (p_a - p_2) A_2 \tag{6.56}$$

Combining Equations (6.55) and (6.56), we get:

$$\dot{m}(\bar{V}_a - \bar{V}_2) + (p_a - p_2) A_2 = 0 \tag{6.57}$$

As $\bar{V}_a = \bar{V}_1$ and $p_a = p_1$ (where subscript 1 refers to pipe 1), we can re-write the momentum balance as:

$$\dot{m}(\bar{V}_1 - \bar{V}_2) + (p_1 - p_2)A_2 = 0 \tag{6.58}$$

Application of the Bernoulli equation, Equation (6.49), between sections "a" and "2" assuming $z_a = z_2$, $\alpha_a = \alpha_2$, and no shaft work, yields:

$$\frac{p_1}{\rho} + \frac{1}{2}\bar{V}_1^2 = \frac{p_2}{\rho} + \frac{1}{2}\bar{V}_2^2 + h_f \tag{6.59}$$

From mass balance:

$$\dot{m} = \rho A_1 \bar{V}_1 = \rho A_2 \bar{V}_2 \tag{6.60}$$

From Equations (6.59) and (6.60), we get:

$$\frac{p_2 - p_1}{\rho} = \frac{\bar{V}_1^2}{2}\left[1 - \left(\frac{A_1}{A_2}\right)^2\right] - h_f \tag{6.61}$$

From Equations (6.58) and (6.60), it follows that:

$$\frac{p_2 - p_1}{\rho} = \bar{V}_1^2\left[\frac{A_1}{A_2} - \left(\frac{A_1}{A_2}\right)^2\right] \tag{6.62}$$

Combining Equations (6.61) and (6.62), we get upon simplification:

$$h_f = \frac{\bar{V}_1^2}{2}\left[1 - \left(\frac{A_1}{A_2}\right)\right]^2 \tag{6.63}$$

It is customary to express the friction loss in the expansion as:

$$h_f = K_{exp}\left(\frac{\bar{V}_1^2}{2}\right) \tag{6.64}$$

where K_{exp} is the loss coefficient of expansion. Upon comparison of Equations (6.63) and (6.64), the loss coefficient of a sudden expansion is as follows:

$$K_{exp} = \left[1 - \left(\frac{A_1}{A_2}\right)\right]^2 = \left[1 - \left(\frac{D_1}{D_2}\right)^2\right]^2 \tag{6.65}$$

This relation is valid when the flow is turbulent in both small and large pipes.

6.4.2 Sudden Contraction

In a contraction, an abrupt reduction in the cross-section of the pipe occurs. As the streamlines cannot follow an abrupt change of cross-section, the fluid stream separates from the wall of the large diameter pipe and issues as a jet into the small diameter pipe. The jet cross-sectional area decreases up to a certain point called vena contracta. After the vena contracta, the jet expands until it fills the full cross-section of the small diameter pipe. Most of the mechanical energy loss in a contraction occurs in the region after the vena contracta where we have decelerating flow with adverse pressure gradient. In the region before the vena contracta where the jet contracts and we have accelerating flow with favourable pressure gradient, negligible loss of mechanical energy into frictional heating occurs.

In an abrupt contraction, the dissipation of mechanical energy is mainly due to the expansion followed by the contraction. Thus, the loss coefficient for a contraction can be determined from the loss coefficient expression for an expansion, Equation (6.65), as follows:

$$h_f = K_{con}\left(\frac{\bar{V}_2^2}{2}\right) = K_{con}C_c^2\left(\frac{\bar{V}_c^2}{2}\right) = [1 - C_c]^2\left(\frac{\bar{V}_c^2}{2}\right) \tag{6.66}$$

where C_c is the contraction coefficient, defined as the ratio of jet cross-sectional area at vena contracta to cross-sectional area of small diameter pipe, K_{con} is the loss coefficient of the contraction, $\bar{V}_2$ is average velocity in small diameter pipe, and $\bar{V}_c$ is the average velocity at vena contracta. From the above relation, the loss coefficient of a sudden contraction is given as:

$$K_{con} = \left(\frac{1}{C_c} - 1\right)^2 \tag{6.67}$$

The value of the contraction coefficient C_c itself depends on A_2/A_1, the ratio of cross-sectional area of small diameter pipe to cross-sectional area of large diameter pipe. In the limit of A_2/A_1 equal to unity (no change in cross-sectional area), C_c is unity and K_{con} is zero. As C_c is not usually known, empirical expressions have been proposed for K_{con} in terms of A_2/A_1. One such expression is as follows:

$$K_{con} = 0.55\left(1 - \frac{A_2}{A_1}\right) \tag{6.68}$$

6.4.3 Fittings and Valves

The mechanical energy loss due to friction in fittings and valves is expressed in the same manner as in sudden expansions and contractions, that is:

$$h_f = K_f\left(\frac{\bar{V}^2}{2}\right) \tag{6.69}$$

where h_f is the frictional mechanical energy loss per unit mass of fluid, K_f is the loss coefficient of the fitting or valve, and $\bar{V}$ is the average velocity in the pipe leading to the fitting. The flow patterns in fittings and valves are too complex to be analyzed theoretically and therefore, K_f values are determined experimentally for various fittings and valves. Table 6.1 gives the typical values of K_f for turbulent flow through various valves and fittings.

Table 6.1 Typical values of K_f for turbulent flow through various valves and fittings

Type of Fitting or Valve	Loss coefficient, K_f
Elbow, 45°	0.35
Elbow, 90°	0.75
Return bend	1.5
Coupling	0.04
Union	0.04
Gate valve (fully open)	0.17
Gate valve (half open)	4.5
Globe valve (fully open)	6.0
Globe valve (half open)	9.5
Angle valve (fully open)	2.0
Ball valve	70.0
Swing valve	2.0

6.5 EXAMPLES

6.5.1 Example 1

Consider the flow of water over a flat plate 1.5 m in length and 1.5 m in width. The free stream velocity of water is 0.15 m/s. Calculate:

i) The local friction factor and shear stress on the plate at $x = 0.25\ m$
ii) The average friction factor and shear stress over the entire plate
iii) The drag force on the plate exerted by water flow

The transition from laminar to turbulent flow occurs at a Reynolds number of 5×10^5. The properties of water are: $\rho = 998$ kg/m^3 and $\eta = 0.9$ mPa.s

Solution

Refer to the figure below.

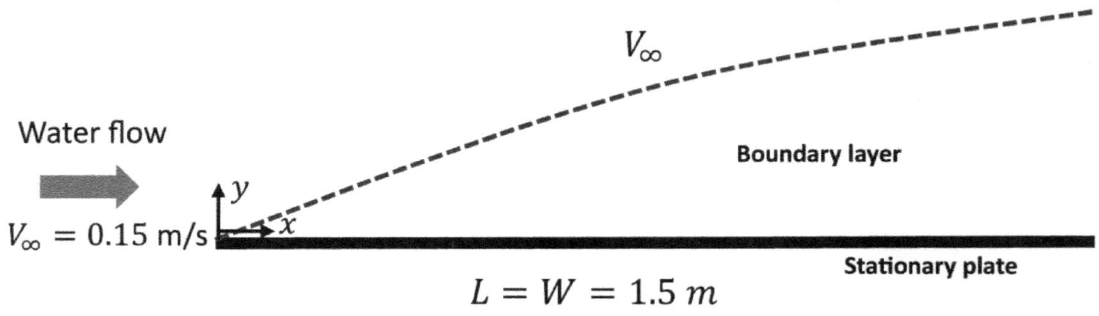

i) The local friction factor and shear stress on the plate at $x = 0.25\ m$

$$Re_x = \frac{\rho V_\infty x}{\eta} = \frac{998 \times 0.15 \times 0.25}{0.9 \times 10^{-3}} = 4.16 \times 10^4$$

Since $Re_x < 5 \times 10^5$, the flow is laminar at $x = 0.25\ m$. Thus,

$$f_x = 0.664(Re_x)^{-1/2} = 0.664(4.16 \times 10^4)^{-1/2} = 3.256 \times 10^{-3}$$

ii) The average friction factor and shear stress over the entire plate

$$Re_x|_{x=L=1.5\ m} = \frac{\rho V_\infty L}{\eta} = \frac{998 \times 0.15 \times 1.5}{0.9 \times 10^{-3}} = 2.495 \times 10^5$$

Since $Re_L < 5 \times 10^5$, the flow is laminar in the entire boundary layer. Thus,

$$\bar{f} = 2(f_x|_{x=L}) = 2 \times 0.664(Re_L)^{-1/2} = 2 \times 0.664(2.495 \times 10^5)^{-1/2} = 2.66 \times 10^{-3}$$

$$\bar{\tau}_w = \frac{1}{2}\bar{f}\rho V_\infty^2 = \frac{1}{2} \times 2.66 \times 10^{-3} \times 998 \times (0.15)^2 = 0.02986\ Pa$$

iii) The drag force on the plate exerted by water flow

$$F_D = \bar{\tau}_w A = 0.02986\ \times 1.5 \times 1.5 = 0.0672\ N$$

6.5.2 Example 2

A flat square plate of length and width of 1 m each is being pulled at a velocity of 9 m/s through water. Determine the power required to pull the plate. The transition from laminar to turbulent flow occurs at a Reynolds number of 5×10^5. The properties of water are: $\rho = 998$ kg/m^3 and $\eta = 0.9$ mPa.s

Solution

Refer to the figure below.

The plate is being pulled through water at 9 m/s. This problem is equivalent to a boundary layer problem where the plate is stationary, and water flows over the plate at a free stream velocity of 9 m/s as shown in the figure. The boundary layer is formed on both sides of the plate, top and bottom. We need to determine the total drag force on top and bottom sides of the plate. The power required is the product of drag force and velocity of 9 m/s.

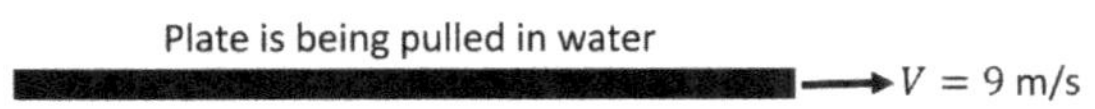

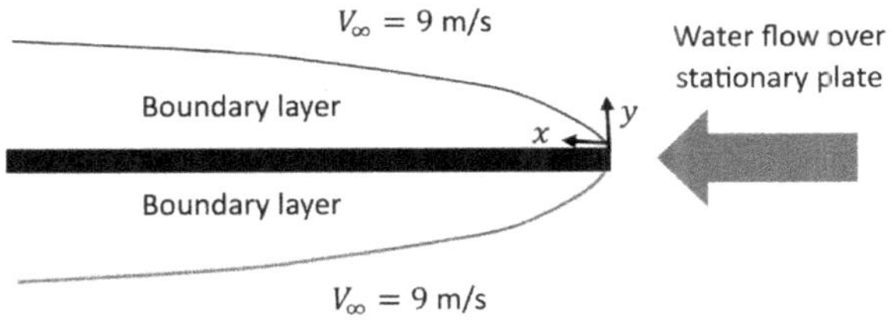

Let us calculate the Reynolds number at the end of the plate:

$$Re_L = \frac{\rho V_\infty L}{\eta} = \frac{998 \times 9 \times 1}{0.9 \times 10^{-3}} = 9.98 \times 10^6$$

Since $Re_L \gg 5 \times 10^5$, the flow is turbulent at the end of the plate. Thus, we have mixed boundary layer. The flow is laminar in the boundary layer at the leading portion of the plate and turbulent at the end portion of the plate.

For mixed boundary layer flow over a plate,

$$\bar{f} = 0.074(Re_L)^{-1/5} - \frac{2A}{Re_L}$$

where $A = 871$ if the critical Reynolds number is 5×10^5. Thus,

$$\bar{f} = 0.074(Re_L)^{-1/5} - \frac{2A}{Re_L} = 0.074(9.98 \times 10^6)^{-1/5} - \frac{2 \times 871}{9.98 \times 10^6} = 2.7726 \times 10^{-3}$$

$$\bar{\tau}_w = \frac{1}{2}\bar{f}\rho V_\infty^2 = \frac{1}{2} \times 2.7726 \times 10^{-3} \times 998 \times (9)^2 = 112.066\ Pa$$

$$F_D = \bar{\tau}_w \times A = 112.066 \times (2 \times 1 \times 1) = 224.131 \text{ N}$$

$$\text{Power} = F_D V_\infty = 224.131 \times 9 = 2017.183\ W$$

6.5.3 Example 3

Water is being transported by a horizontal 0.15 m diameter galvanized iron pipe. The relative roughness of the pipe ε/D is 0.001. The water flow rate is 0.03 m^3/s. The water properties are: $\rho = 1000$ kg/m^3 and $\eta = 1$ mPa.s. Determine the following:

i) Friction factor
ii) Wall shear stress
iii) Pressure gradient

Solution

i) Friction factor

$$Re = \frac{\rho \bar{V} D}{\eta} = \frac{4\rho Q}{\pi D \eta} = \frac{4 \times 1000 \times 0.03}{\pi(0.15)(1.0 \times 10^{-3})} = 2.546 \times 10^5$$

Flow is highly turbulent as $Re \gg 2100$. Use Zigrang and Sylvester equation to predict f:

$$1/\sqrt{f} = -4log_{10}\left[\frac{\frac{\varepsilon}{D}}{3.7} - \frac{5.02}{Re} log_{10}\left(\frac{\frac{\varepsilon}{D}}{3.7} - \frac{5.02A}{Re}\right)\right]$$

$$A = log_{10}\left(\frac{\frac{\varepsilon}{D}}{3.7} + \frac{13}{Re}\right)$$

$$A = log_{10}\left(\frac{0.001}{3.7} + \frac{13}{2.546 \times 10^5}\right) = -3.493$$

$$1/\sqrt{f} = -4log_{10}\left[\frac{0.001}{3.7} - \frac{5.02}{2.546 \times 10^5} log_{10}\left(\frac{0.001}{3.7} + \frac{5.02 \times (3.493)}{2.546 \times 10^5}\right)\right] = 13.881$$

$$f = 5.19 \times 10^{-3}$$

ii) Wall shear stress

$$\bar{V} = \frac{4Q}{\pi D^2} = \frac{4 \times 0.03}{\pi \times 0.15^2} = 1.698 \text{ m/s}$$

$$\tau_w = \frac{1}{2} f \rho \bar{V}^2 = \frac{1}{2} \times 5.19 \times 10^{-3} \times 1000 \times 1.698^2 = 7.482 \text{ Pa}$$

iii) Pressure gradient

From force balance,

$$\Delta P \left(\frac{\pi D^2}{4}\right) = \tau_w (\pi D L)$$

$$\frac{\Delta P}{L} = \frac{4\tau_w}{D} = \frac{4 \times 7.482}{0.15} = 199.52 \text{ Pa/m}$$

6.5.4 Example 4

Consider the flow of crude oil in the pipeline system shown in the figure. The flow rate of 2 m³/s splits between the pipes 1 and 2. The dimensions of the pipes are as follows:

Pipe 1: $L = 1000\ m$, $D = 0.6\ m$

Pipe 2: $L = 1000\ m$, $D = 0.4\ m$

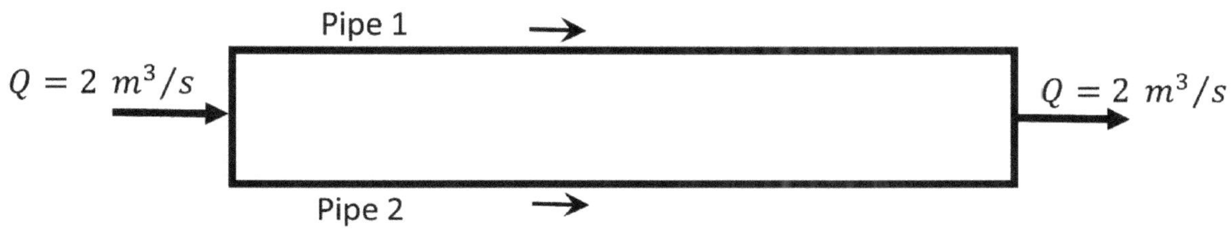

Assuming that the flow is laminar in the pipes, calculate the flow rates in the pipes.

Solution

For laminar flow in a pipe,

$$f = \frac{16}{Re}$$

It can be readily shown that this friction factor relationship is equivalent to:

$$\Delta P = \frac{128\eta QL}{\pi D^4}$$

The ΔP is the same in two pipes. Thus,

$$\frac{Q_1}{D_1^4} = \frac{Q_2}{D_2^4}$$

From mass balance:

$$Q = Q_1 + Q_2$$

Consequently,

$$Q_1 = \frac{Q}{\left[1 + \left(\frac{D_2}{D_1}\right)^4\right]} = \frac{2}{\left[1 + \left(\frac{0.4}{0.6}\right)^4\right]} = 1.67\ m^3/s$$

$$Q_2 = \frac{Q(D_2/D_1)^4}{\left[1 + \left(\frac{D_2}{D_1}\right)^4\right]} = \frac{2 \times (0.4/0.6)^4}{\left[1 + \left(\frac{0.4}{0.6}\right)^4\right]} = 0.33\ m^3/s$$

6.5.5 Example 5

Consider the flow of water in the pipeline system shown in the figure. The water flow rate of 0.03 m^3/s splits between the pipes 1 and 2. The dimensions of the pipes made of galvanized iron are as follows:

Pipe 1: $L = 3\ m,\ D = 0.05\ m$, relative roughness of pipe ε/D is 0.003

Pipe 2: $L = 3\ m,\ D = 0.1$, relative roughness of pipe ε/D is 0.0015

The water properties are: $\rho = 1000\ \text{kg/m}^3$ and $\eta = 1\ \text{mPa.s}$

i) Determine the flow rate in each pipe

ii) Calculate the pressure drop in the system.

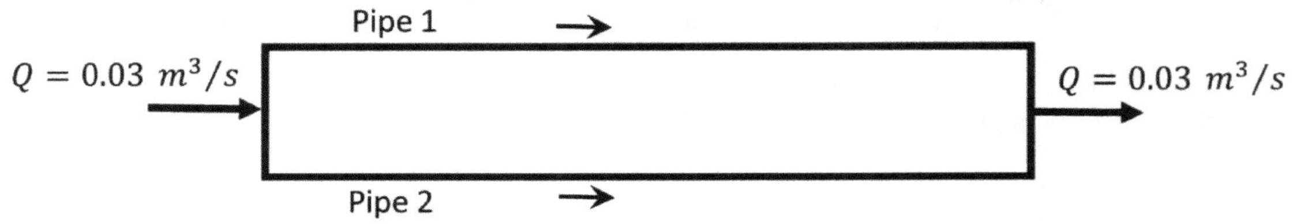

Solution

i) Determine the flow rate in each pipe

From mass balance:

$$Q = Q_1 + Q_2$$

$$0.03 = \frac{\pi \times 0.05^2}{4} V_1 + \frac{\pi \times 0.1^2}{4} V_2$$

$$15.279 = V_1 + 4V_2$$

The friction loss in the two pipes is the same, that is,

$$h_{f,pipe1} = h_{f,pipe2}$$

$$4f_1 \left(\frac{L_1}{D_1}\right) \frac{V_1^2}{2} = 4f_2 \left(\frac{L_2}{D_2}\right) \frac{V_2^2}{2}$$

$$f_1 V_1^2 = f_2 \left(\frac{D_1}{D_2}\right) V_2^2 = f_2 \left(\frac{0.05}{0.1}\right) V_2^2 = 0.5 f_2 V_2^2$$

Thus, the following equations need to be solved for V_1 and V_2 by trial-and-error approach:

$$15.279 = V_1 + 4V_2 \quad \text{(a)}$$

$$f_1 V_1^2 = 0.5 f_2 V_2^2 \quad \text{(b)}$$

As a first guess, we take $f_1 = f_2$. Thus, $V_1 = 2.2195$ m/s and $V_2 = 3.246$ m/s. Estimate new values of f_1 and f_2 as shown below.

f_1 Estimation:

$Re_1 = \frac{\rho V_1 D_1}{\eta} = \frac{1000 \times 2.295 \times 0.05}{10^{-3}} = 114{,}750$ Flow is highly turbulent.

Use Zigrang and Sylvester equation to predict f_1:

$$A = log_{10}\left(\frac{0.003}{3.7} + \frac{13}{114750}\right) = -3.03428$$

$$1/\sqrt{f} = -4log_{10}\left[\frac{0.003}{3.7} - \frac{5.02}{114750} log_{10}\left(\frac{0.003}{3.7} - \frac{5.02 \times (-3.03428)}{114750}\right)\right] = 12.10166$$

$$f_1 = 6.828 \times 10^{-3}$$

f_2 Estimation:

$Re_2 = \frac{\rho V_2 D_2}{\eta} = \frac{1000 \times 3.246 \times 0.10}{10^{-3}} = 324{,}600$ Flow is highly turbulent.

Use Zigrang and Sylvester equation to predict f_2:

$$A = log_{10}\left(\frac{0.0015}{3.7} + \frac{13}{324600}\right) = -3.3512$$

$$1/\sqrt{f} = -4log_{10}\left[\frac{0.0015}{3.7} - \frac{5.02}{324600}log_{10}\left(\frac{0.0015}{3.7} - \frac{5.02 \times (-3.3512)}{324600}\right)\right] = 13.36012$$

$$f_2 = 5.602 \times 10^{-3}$$

New values of V_1 and V_2 from Equations (a) and (b): $V_1 = 2.1089$ m/s and $V_2 = 3.2925$ m/s.

Next iteration:

$V_1 = 2.1089$ m/s and $V_2 = 3.2925$ m/s. Estimate new values of f_1 and f_2 following the same procedure as shown above. The next iteration gives:

$f_1 = 6.852 \times 10^{-3}$; $f_2 = 5.6 \times 10^{-3}$

From Equations (a) and (b), we get:

$V_1 = 2.105$ m/s and $V_2 = 3.293$ m/s

These values are very close to the previous iteration values.

Thus,

$V_1 = 2.105$ m/s and $Q_1 = V_1(\pi D_1^2/4) = 2.105 \times \pi \times 0.05^2/4 = 4.133 \times 10^{-3}\, m^3/s$

$V_2 = 3.293$ m/s and $Q_2 = V_2(\pi D_2^2/4) = 3.293 \times \pi \times 0.1^2/4 = 0.02586\, m^3/s$

ii) Calculate the pressure drop in the system.

$f_1 = 6.852 \times 10^{-3}$; $\tau_w = \frac{1}{2} f_1 \rho {V_1}^2 = \frac{1}{2} \times 6.852 \times 10^{-3} \times 1000 \times 2.105^2 = 15.1807$ Pa

$$\Delta P = \frac{4L\tau_w}{D} = \frac{4 \times 3 \times 15.1807}{0.05} = 3.643\ kPa$$

$f_2 = 5.6 \times 10^{-3}$; $\tau_w = \frac{1}{2} f_2 \rho {V_2}^2 = \frac{1}{2} \times 5.6 \times 10^{-3} \times 1000 \times 3.293^2 = 30.363$ Pa

$$\Delta P = \frac{4L\tau_w}{D} = \frac{4 \times 3 \times 30.363}{0.1} = 3.643\ kPa$$

Note that the pressure drop is the same for the two pipes as expected.

6.5.6 Example 6

Water is being discharged at a flow rate of 0.1 m^3/s from a large tank open to atmosphere through a series of three horizontal cast iron pipes as shown in the figure. The dimensions of the pipes are as follows:

Pipe 1: $L_1 = 25\ m$, $D_1 = 25\ cm$, $\varepsilon/D_1 = 0.001$

Pipe 2: $L_2 = 35\ m$, $D_2 = 20\ cm$, $\varepsilon/D_2 = 0.0013$

Pipe 3: $L_3 = 35\ m$, $D_3 = 15\ cm$, $\varepsilon/D_3 = 0.0017$

The water properties are: $\rho = 1000\ \text{kg/m}^3$ and $\eta = 1\ \text{mPa.s}$

Estimate the height of the water level H in the tank.

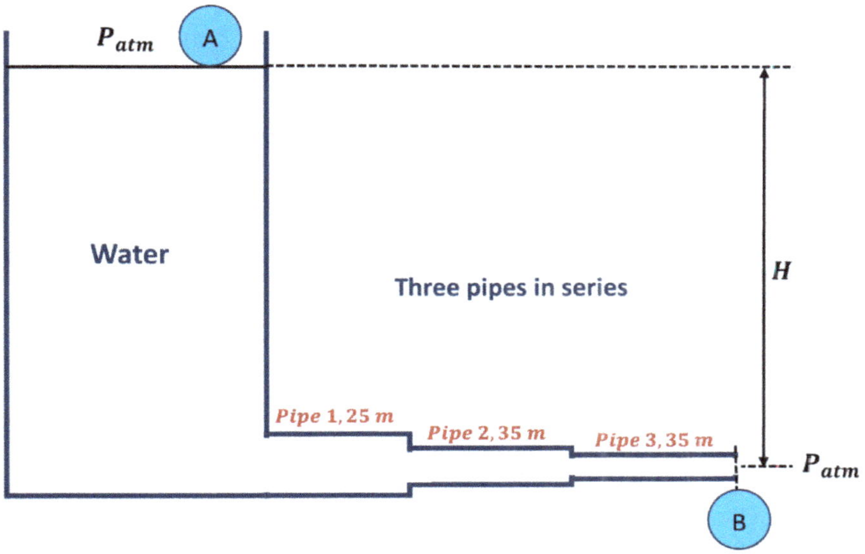

Solution

Apply engineering Bernoulli equation, Equation (6.49), between sections A and B (see figure):

$$\frac{p_A}{\rho} + gz_A + \frac{1}{2}\alpha_A \bar{V}_A^2 = \frac{p_B}{\rho} + gz_B + \frac{1}{2}\alpha_B \bar{V}_B^2 + h_f + W_{sh}$$

Since $p_A = p_B = p_{atm}$, $\bar{V}_A = 0$, $\alpha_B \cong 1, W_{sh} = 0$, the Bernoulli equation reduces to:

$$g(z_A - z_B) = gH = \frac{1}{2}\bar{V}_B^2 + h_f$$

Thus,

$$H = \frac{1}{g}\left[h_f + \frac{1}{2}\bar{V}_B^2\right]$$

$$\bar{V}_B = \frac{4Q}{\pi D_3^2} = \frac{4 \times 0.1}{\pi \times 0.15^2} = 5.66\ m/s$$

$$H = \frac{1}{g}\left[h_f + \frac{1}{2}\bar{V}_B^2\right] = \frac{1}{9.81}\left[h_f + \frac{1}{2} \times 5.66^2\right] = \frac{1}{9.81}\left[h_f + 16.02\right]\quad m$$

Estimation of friction loss, h_f:

The friction loss consists of the following: (1) contraction loss at the tank exit, (2) friction loss in pipe 1, (3) contraction loss from pipe 1 to pipe 2, (4) friction loss in pipe 2, (5) contraction loss from pipe 2 to pipe 3, and (6) friction loss in pipe 3.

(1) contraction loss at the tank exit:

$$K_{con} = 0.55\left(1 - \frac{A_2}{A_1}\right) = 0.55$$

$$\bar{V}_1 = \frac{4Q}{\pi D_1^2} = \frac{4 \times 0.1}{\pi \times 0.25^2} = 2.0372\ m/s$$

$$h_f = K_{con}\left(\frac{\bar{V}_1^2}{2}\right) = 0.55 \times \frac{2.0372^2}{2} = 1.1413\ J/kg$$

(2) friction loss in pipe 1

$$\bar{V}_1 = 2.0372\ m/s,\ Re_1 = \frac{\rho D_1 \bar{V}_1}{\eta} = \frac{1000 \times 0.25 \times 2.0372}{10^{-3}} = 5.093 \times 10^5$$

Flow is highly turbulent as $Re \gg 2100$. Use the Zigrang and Sylvester equation to predict f:

$$1/\sqrt{f} = -4log_{10}\left[\frac{\frac{\varepsilon}{D}}{3.7} - \frac{5.02}{Re}log_{10}\left(\frac{\frac{\varepsilon}{D}}{3.7} - \frac{5.02A}{Re}\right)\right] \qquad \text{where } A = log_{10}\left(\frac{\frac{\varepsilon}{D}}{3.7} + \frac{13}{Re}\right)$$

$$\varepsilon/D_1 = 0.001,\ A = -3.529,\ f = 5.056 \times 10^{-3}$$

$$h_f = 4f\frac{L}{D}\frac{\bar{V}^2}{2} = 4(5.056 \times 10^{-3})\left(\frac{25}{0.25}\right)\frac{2.0372^2}{2} = 4.1967\ J/kg$$

(3) contraction loss from pipe 1 to pipe 2

$$K_{con} = 0.55\left(1 - \frac{A_2}{A_1}\right) = 0.55\left[1 - \left(\frac{D_2}{D_1}\right)^2\right] = 0.55\left[1 - \left(\frac{0.20}{0.25}\right)^2\right] = 0.198$$

$$\bar{V}_2 = \frac{4Q}{\pi D_2^2} = \frac{4 \times 0.1}{\pi \times 0.20^2} = 3.1831\ m/s$$

$$h_f = K_{con}\left(\frac{\bar{V}_2^2}{2}\right) = 0.198 \times \frac{3.1831^2}{2} = 1.003\ J/kg$$

(4) friction loss in pipe 2

$$\bar{V}_2 = 3.1831\ m/s,\ Re_2 = \frac{\rho D_2 \bar{V}_2}{\eta} = \frac{1000 \times 0.20 \times 3.1831}{10^{-3}} = 6.3662 \times 10^5$$

Flow is highly turbulent as $Re \gg 2100$. Use the Zigrang and Sylvester equation to predict f:

$$1/\sqrt{f} = -4log_{10}\left[\frac{\frac{\varepsilon}{D}}{3.7} - \frac{5.02}{Re}log_{10}\left(\frac{\frac{\varepsilon}{D}}{3.7} - \frac{5.02A}{Re}\right)\right] \qquad \text{where } A = log_{10}\left(\frac{\frac{\varepsilon}{D}}{3.7} + \frac{13}{Re}\right)$$

$$\varepsilon/D_2 = 0.0013, A = -3.42972,\ f = 5.337 \times 10^{-3}$$

$$h_f = 4f\frac{L}{D}\frac{\bar{V}^2}{2} = 4(5.337 \times 10^{-3})\left(\frac{35}{0.20}\right)\frac{3.1831^2}{2} = 18.926\ \ J/kg$$

(5) contraction loss from pipe 2 to pipe 3

$$K_{con} = 0.55\left(1 - \frac{A_3}{A_2}\right) = 0.55\left[1 - \left(\frac{D_3}{D_2}\right)^2\right] = 0.55\left[1 - \left(\frac{0.15}{0.20}\right)^2\right] = 0.2406$$

$$\bar{V}_3 = \frac{4Q}{\pi D_3^2} = \frac{4 \times 0.1}{\pi \times 0.15^2} = 5.66\ m/s$$

$$h_f = K_{con}\left(\frac{\bar{V}_3^2}{2}\right) = 0.2406 \times \frac{5.66^2}{2} = 3.854\ J/kg$$

(6) friction loss in pipe 3

$$\bar{V}_3 = 5.66\ m/s,\ Re_3 = \frac{\rho D_3 \bar{V}_3}{\eta} = \frac{1000 \times 0.15 \times 5.66}{10^{-3}} = 8.49 \times 10^5$$

Flow is highly turbulent as $Re \gg 2100$. Use the Zigrang and Sylvester equation to predict f:

$$1/\sqrt{f} = -4log_{10}\left[\frac{\frac{\varepsilon}{D}}{3.7} - \frac{5.02}{Re}log_{10}\left(\frac{\frac{\varepsilon}{D}}{3.7} - \frac{5.02A}{Re}\right)\right] \qquad \text{where } A = log_{10}\left(\frac{\frac{\varepsilon}{D}}{3.7} + \frac{13}{Re}\right)$$

$$\varepsilon/D_3 = 0.0017, A = -3.32352,\ f = 5.672 \times 10^{-3}$$

$$h_f = 4f\frac{L}{D}\frac{\bar{V}^2}{2} = 4(5.672 \times 10^{-3})\left(\frac{35}{0.15}\right)\frac{5.66^2}{2} = 84.796\ \ J/kg$$

Total friction loss:

$$h_f = 1.1413 + 4.1967 + 1.003 + 18.926 + 3.854 + 84.796 = 113.917\ \ J/kg$$

Note that the minor loss ($= 1.1413 + 1.003 + 3.854 = 5.9983\ J/kg$) is just 5.26% of the total friction loss.

Thus, the height of the water level H in the tank is:

$$H = \frac{1}{9.81}\left[h_f + 16.02\right] = \frac{1}{9.81}[113.917 + 16.02] = 13.245\ m$$

6.5.7 Example 7

Water is being pumped from a large feed tank to a large, elevated tank as shown in the figure. The flow rate of water is 0.01 m^3/s. The pipe diameter is 0.1 m, and its total length is 200 m. The relative roughness of the pipe ε/D is 0.0005. The pump efficiency is 0.70. Calculate the kW of the pump required to do the job. The water properties are: $\rho = 1000\ kg/m^3$ and $\eta = 1\ mPa.s$.

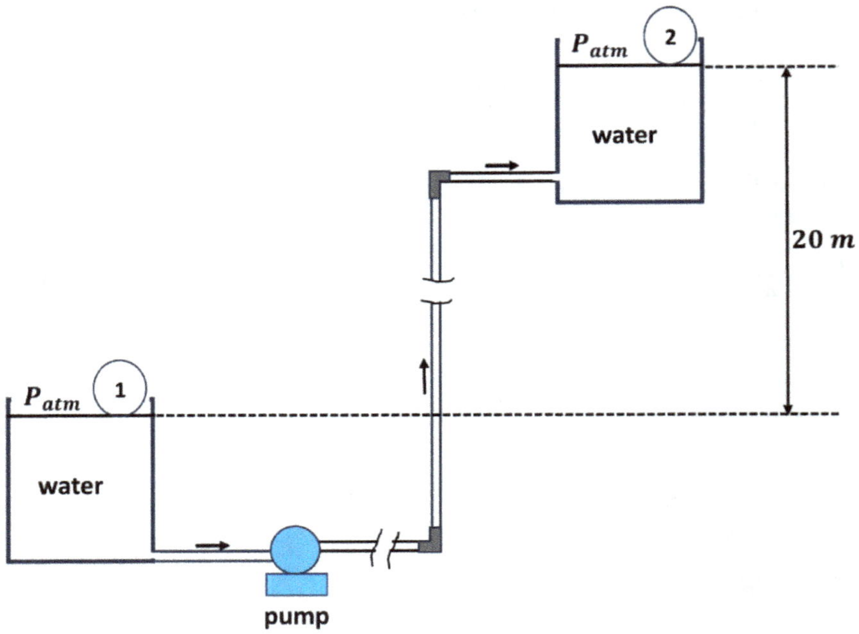

Solution

Apply engineering Bernoulli equation, Equation (6.49), between sections 1 and 2 (see figure):

$$\frac{p_1}{\rho} + gz_1 + \frac{1}{2}\alpha_1\bar{V}_1^2 = \frac{p_2}{\rho} + gz_2 + \frac{1}{2}\alpha_2\bar{V}_2^2 + h_f + W_{sh}$$

Since $p_1 = p_2 = p_{atm}$, $\bar{V}_1 \cong 0$, $\bar{V}_2 \cong 0$, the Bernoulli equation reduces to:

$$W_{sh} = g(z_1 - z_2) - h_f = -20g - h_f$$

Estimation of friction loss, h_f*:*

The friction loss consists of the following: (1) contraction loss at the feed tank exit, (2) friction loss in pipe of length 200 m, (3) friction loss in two 90° elbows, and (4) expansion loss at the exit of elevated tank.

(1) contraction loss at the feed tank exit:

$$K_{con} = 0.55\left(1 - \frac{A_2}{A_1}\right) = 0.55$$

$$\bar{V} = \frac{4Q}{\pi D^2} = \frac{4 \times 0.01}{\pi \times 0.1^2} = 1.2732\ m/s$$

$$h_f = K_{con}\left(\frac{\bar{V}^2}{2}\right) = 0.55 \times \frac{1.2732^2}{2} = 0.4458\ J/kg$$

(2) friction loss in a pipe

$$\bar{V} = 1.2732\ m/s,\ Re = \frac{\rho D\bar{V}}{\eta} = \frac{1000\times0.1\times1.2732}{10^{-3}} = 1.2732\times10^{5}$$

Flow is highly turbulent as $Re \gg 2100$. Use the Zigrang and Sylvester equation to predict f:

$$1/\sqrt{f} = -4log_{10}\left[\frac{\frac{\varepsilon}{D}}{3.7} - \frac{5.02}{Re}log_{10}\left(\frac{\frac{\varepsilon}{D}}{3.7} - \frac{5.02A}{Re}\right)\right] \qquad \text{where } A = log_{10}\left(\frac{\frac{\varepsilon}{D}}{3.7} + \frac{13}{Re}\right)$$

$\varepsilon/D = 0.0005,\ A = -3.62481,\ f = 4.931\times10^{-3}$

$$h_f = 4f\frac{L}{D}\frac{\bar{V}^2}{2} = 4(4.931\times10^{-3})\left(\frac{200}{0.1}\right)\frac{1.2732^2}{2} = 31.973\ J/kg$$

(3) friction loss in two 90° elbows

$$h_f = K_f\left(\frac{\bar{V}^2}{2}\right) = 0.75\frac{1.2732^2}{2} = 0.6079\ J/kg$$

For two elbows, $h_f = 2\times0.6079 = 1.258\ J/kg$

(4) expansion loss in elevated tank

$$K_{exp} = \left[1 - \left(\frac{A_1}{A_2}\right)\right]^2 = 1$$

$$h_f = K_{exp}\left(\frac{\bar{V}^2}{2}\right) = 1\times\frac{1.2732^2}{2} = 0.8105\ J/kg$$

Total friction loss:

$$h_f = 0.4458 + 31.973 + 1.2158 + 0.8105 = 34.4451\ J/kg$$

Note that the minor loss ($= 0.4458 + 1.258 + 0.8105 = 2.5143\ J/kg$) is just 7.3% of the total friction loss.

Thus, the shaft work for the pump is:

$$W_{sh} = -20g - h_f = -20\times9.81 - 34.4451 = -230.645\ J/kg$$

As the pump efficiency=0.70,

$$W_{sh,pump} = \frac{W_{sh}}{0.70} = -\frac{230.645}{0.70} = -329.493\ J/kg$$

Note that the pump shaft work is negative as it is the work supplied to the fluid. The shaft work is positive if it is produced by the fluid as in the case of turbines.

To calculate the kW of the pump required, we need to calculate the mass flow rate through the pump:

$$\dot{m} = \rho Q = 1000 \times 0.01 = 10\ kg/s$$

$$\dot{W}_{sh,pump} = \dot{m}W_{sh,pump} = 10 \times (-329.493\,) = -3294.93\ W = -3.295\ kW$$

Thus, the kW of the pump required is 3.295.

NOTATION

A	area or factor in Equations (6.18), (6.19), (6.30), (6.31)
C_c	contraction coefficient, Equation (6.66)
D	diameter
D_H	hydraulic diameter
f	friction factor
F	force
F_D	drag force
f_x	local friction factor
$\bar{f}$	average friction factor
g	acceleration due to gravity
h	specific enthalpy
h_f	friction loss
k	parameter specifying the radius of the inner pipe of a concentric double pipe system
K_{con}	loss coefficient of a sudden contraction
K_{exp}	loss coefficient of a sudden expansion
K_f	loss coefficient of a fitting or valve
L	length
$\dot{m}$	mass flow rate
p	pressure
Q	heat or volumetric flow rate
$\dot{Q}$	rate of heat input
r	radial co-ordinate
Re	Reynolds number
Re_x	local value of Reynolds number
Re_{x_c}	critical Reynolds number
Re_L	Reynolds number at the end of the plate at $x = L$
s	specific entropy
T	absolute temperature
V	velocity
$\bar{V}$	average velocity
V_c	characteristic fluid velocity
V_∞	free-stream velocity
W_{sh}	shaft work
$\dot{W}_{sh}$	rate of shaft work
x	distance in x-direction
z	elevation

Greek Letters

α	kinetic energy correction factor or parameter in Equation (6.24)
β	momentum correction factor

Δ	change in variable
δ	boundary layer thickness
ε	roughness of a pipe
η	viscosity
υ	specific volume
ρ	density
τ_w	wall shear stress
$\bar{\tau}_w$	average wall shear stress
ψ	parameter in Equations (6.24) and (6.39)

PRACTICE PROBLEMS

Problem 1

Compare the drag force exerted by flow of air with that of flow of water over a flat plate 0.48 m in length and 0.22 m in width at the same free stream velocity of 0.45 m/s. The transition from laminar to turbulent flow occurs at a Reynolds number of 5×10^5. The properties of air and water are:

Air: $\rho = 1.18$ kg/m^3 and $\eta = 1.85 \times 10^{-2}$ mPa.s

Water: $\rho = 1000$ kg/m^3 and $\eta = 1.0$ mPa.s

Problem 2

Re-do example 2 (section 6.5.2) with the plate being pulled at double the velocity, that is, 18 m/s.

Problem 3

Consider the flow of water in the pipeline system shown in the figure. The water flow rate of 0.05 m^3/s splits between the pipes 1 and 2. The dimensions of the pipes made of galvanized iron are as follows:

Pipe 1: $L = 5\ m$, $D = 0.05\ m$, relative roughness of pipe ε/D is 0.003

Pipe 2: $L = 5\ m$, $D = 0.1$, relative roughness of pipe ε/D is 0.0015

The water properties are: $\rho = 1000\ \mathrm{kg/m^3}$ and $\eta = 1\ \mathrm{mPa.s}$

(1) Determine the flow rate in each pipe

(2) Calculate the pressure drop in the system.

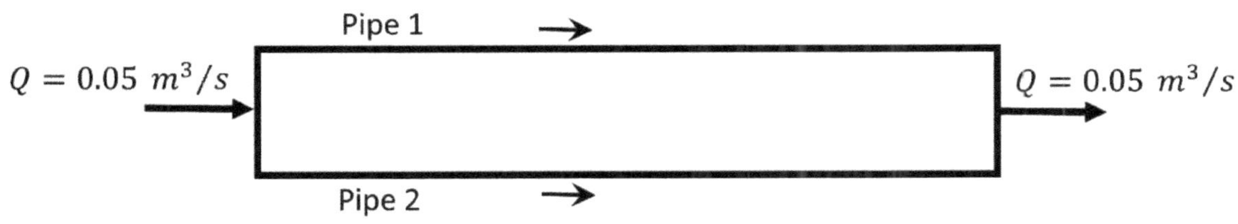

Problem 4
Consider the flow of water in the pipeline system shown in the figure. The water flow rate of 0.09 m^3/s splits between the pipes 1, 2, and 3. The dimensions of the pipes made of galvanized iron are as follows:

Pipe 1: $L = 5\ m$, $D = 0.05\ m$, relative roughness of pipe ε/D is 0.003

Pipe 2: $L = 5\ m$, $D = 0.1$, relative roughness of pipe ε/D is 0.0015

Pipe 3: $L = 5\ m$, $D = 0.15$, relative roughness of pipe ε/D is 0.001

The water properties are: $\rho = 1000\ \text{kg/m}^3$ and $\eta = 1\ \text{mPa. s}$

(1) Determine the flow rate in each pipe

(2) Calculate the pressure drop in the system.

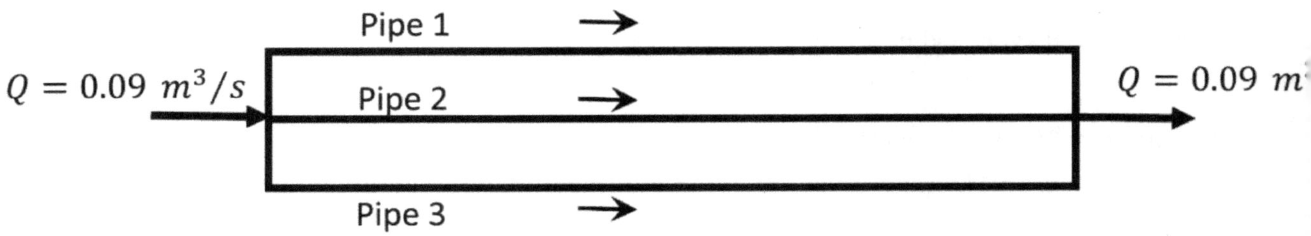

Problem 5
Water is being discharged at a flow rate of 0.05 m^3/s from a large tank open to atmosphere through a series of three horizontal cast iron pipes as shown in the figure. The dimensions of the pipes are as follows:

Pipe 1: $L_1 = 50\ m$, $D_1 = 25\ cm$, $\varepsilon/D_1 = 0.001$

Pipe 2: $L_2 = 70\ m$, $D_2 = 20\ cm$, $\varepsilon/D_2 = 0.0013$

Pipe 3: $L_3 = 70\ m$, $D_3 = 15\ cm$, $\varepsilon/D_3 = 0.0017$

The water properties are: $\rho = 1000\ \text{kg/m}^3$ and $\eta = 1\ \text{mPa. s}$

Estimate the height of the water level H in the tank.

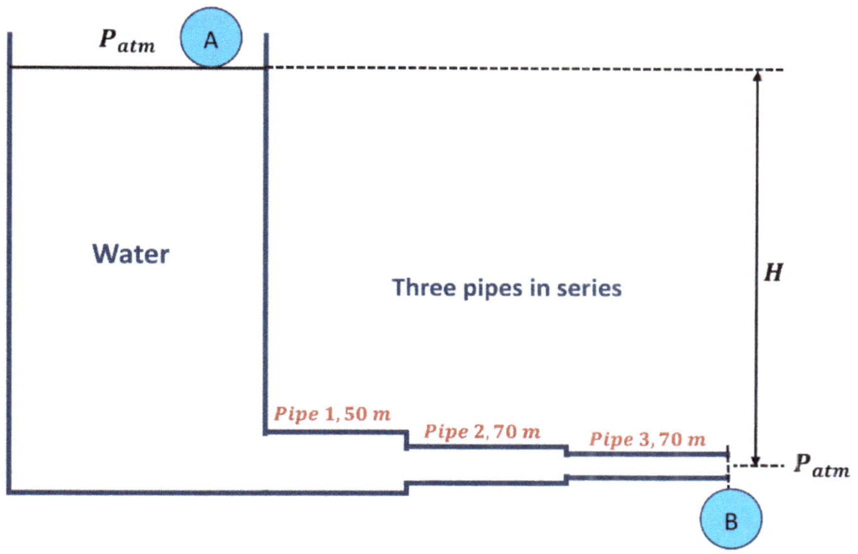

Problem 6

Water is being pumped from a large feed tank to a large, elevated tank as shown in the figure. The flow rate of water is 0.02 m^3/s. The pipe diameter is 0.1 m, and its total length is 500 m. The relative roughness of the pipe ε/D is 0.0005. The gate valves are fully open. The pump efficiency is 0.65. Calculate the kW of the pump required to do the job. The water properties are: $\rho = 1000\ kg/m^3$ and $\eta = 1\ mPa.s$.

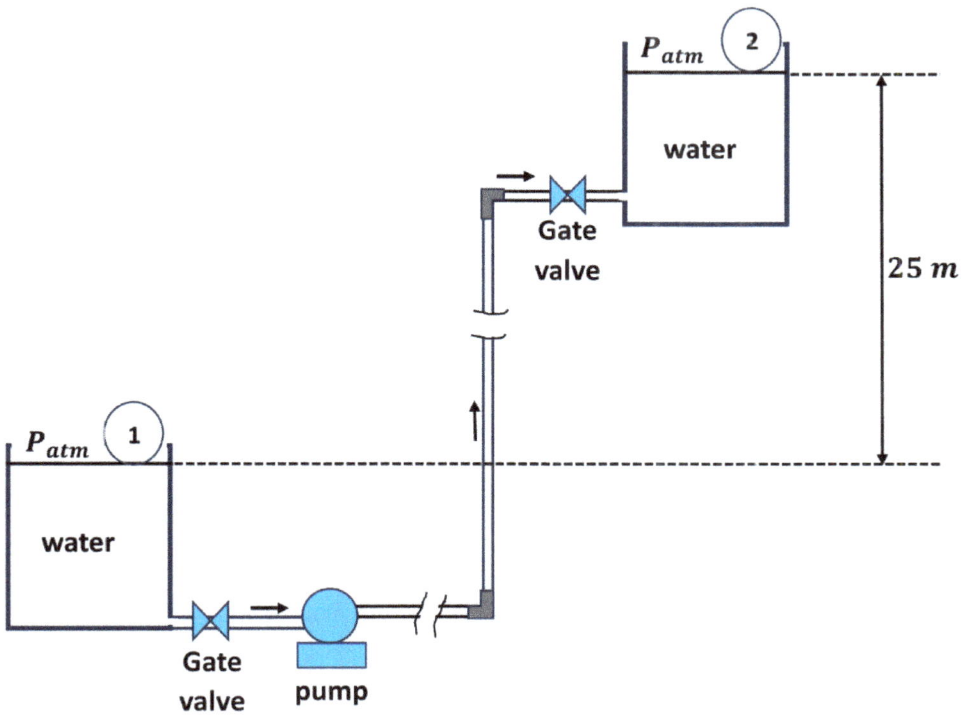

Problem 7

Consider a concentric double pipe heat exchanger with the inner pipe diameter of 2.54 cm and outer pipe diameter of 5.08 cm. The length of the heat exchanger is 30.48 m. Water is flowing in the annulus region at an average velocity of 3.048 m/s. Estimate the pressure drop in annular flow of water. The water properties are: $\rho = 1000\ kg/m^3$ and $\eta = 1\ mPa.s$. The relative roughness (ε/D) of the annulus is 0.0018.

FURTHER READING

1. R. Pal. Teaching fluid mechanics and thermodynamics simultaneously through pipeline flow experiments. *Fluids* 2019, 4, 103.
2. R.B. Bird, W.E. Stewart, E.N. Lightfoot. *Transport Phenomena.* 2nd Edition, Wiley, New York, 2007.
3. C.J. Geankoplis, A.A. Hersel, D.H. Lepek. *Transport Processes and Separation Process Principles*. 5th Edition, Prentice Hall, New Jersey, 2018.
4. J.R. Welty, C.E. Wicks, R.E. Wilson, G. Rorrer. *Fundamentals of Momentum, Heat, and Mass Transfer*. 4th Edition, Wiley, New York, 2001.
5. R.L. Street, G.Z. Watters, J.K. Vannard. *Elementary Fluid Mechanics*. 7th Edition, Wiley, New York, 1996.
6. J.O. Wilkes, *Fluid Mechanics for Chemical Engineers*. 3rd Edition, Prentice Hall, Boston, 2018.
7. B.R. Munson, D.F. Young, T.H. Okiishi. *Fundamentals of Fluid Mechanics*. 4th Edition, Wiley, New York, 2002.
8. R.C. Hibbeler. *Fluid Mechanics*. Pearson, Boston, 2015.
9. D.R. Gaskell. *An Introduction to Transport Phenomena in Materials Engineering*. Macmillan, New York, 1992.
10. D.J. Zigrang, N.D. Sylvester. Explicit approximations to the solution of Colebrook's friction factor equation. *AIChE J.* 1982, 28, 514–515.

Chapter 7

Settling Behaviour of Particles

7.1 SETTLING OF A SINGLE SOLID PARTICLE

Consider a solid particle of diameter D_p and mass m settling in a liquid matrix under the action of a gravitational force. The forces acting on the particle are gravity force (F_G), drag force (F_D), and buoyancy force (F_B). Thus, the net force acting on the particle in the downward direction is:

$$F_{net,downward} = m\frac{dU}{dt} = F_G - F_B - F_D \tag{7.1}$$

The gravitational force F_G on the particle is equal to mg, the drag force F_D on the particle is equal to $C_D A_p(\rho U^2/2)$, where C_D is the drag coefficient, A_p is the projected area of the particle measured in a plane normal to direction of motion of particle (for spherical particle, $A_p = \pi D_p^2/4$), U is the particle velocity, and the buoyancy force F_B on the particle is equal to weight of the fluid displaced by the particle, that is, $F_B = m\rho g/\rho_p$ where ρ is fluid density and ρ_p is particle density. Thus:

$$m\frac{dU}{dt} = mg - \frac{mg\rho}{\rho_p} - \frac{C_D A_p \rho U^2}{2} \tag{7.2}$$

or,

$$\frac{dU}{dt} = g\left(\frac{\rho_p - \rho}{\rho_p}\right) - \frac{C_D A_p \rho U^2}{2m} \tag{7.3}$$

Under steady state condition ($dU/dt = 0$), the particle settles at a constant velocity called the terminal velocity (U_t). Thus:

$$U_t = \sqrt{\frac{2gm(\rho_p - \rho)}{A_p \rho_p C_D \rho}} \tag{7.4}$$

For spherical particles:

$$m = \frac{\pi D_p^3 \rho_p}{6} \quad \text{and} \quad A_p = \frac{\pi D_p^2}{4} \tag{7.5}$$

This leads to the following settling velocity of spherical particles:

$$U_t = \sqrt{\frac{4g(\rho_p - \rho)D_p}{3C_D \rho}} \tag{7.6}$$

The drag coefficient is a function of the particle Reynolds number (Re_p), defined as:

$$\text{Re}_p = \frac{\rho D_p U_t}{\eta} \tag{7.7}$$

where η is fluid viscosity. For creeping flow around a spherical particle ($\mathrm{Re}_p \ll 1$), C_D is given by the Stokes law as:

$$C_D = \frac{24}{\mathrm{Re}_p} \tag{7.8}$$

Upon substitution of the expression for C_D into Equation (7.6), the following expression for the terminal settling velocity of spherical solid particles is obtained:

$$U_t = \frac{gD_p^2(\rho_p - \rho)}{18\eta} \qquad \mathrm{Re}_p \ll 1 \tag{7.9}$$

Equation (7.9) is also referred to as Stokes law. It is applicable when $\mathrm{Re}_p \leq 0.1$

At higher values of Re_p, C_D can be described by the following empirical equation:

$$C_D = \left[\sqrt{\frac{24}{Re_p}} + 0.5407\right]^2 \tag{7.10}$$

This equation is accurate up to Re_p of about 6000. It reduces to Stokes law at low Re_p and therefore covers the creeping flow regime. Also note that C_D is nearly constant over a Re_p range of 10^3 to 10^5:

$$C_D = 0.44 \qquad 10^3 < \mathrm{Re}_p \leq 10^5 \tag{7.11}$$

This result is often referred to as Newton's Law.

Upon substitution of $C_D = 0.44$ into Equation (7.6), the following expression for the terminal settling velocity of spherical solid particles in Newton's regime is obtained:

$$U_t = 1.74\sqrt{\frac{gD_p(\rho_p - \rho)}{\rho}} \qquad 10^3 < \mathrm{Re}_p \leq 10^5 \tag{7.12}$$

The terminal velocity of a particle in the intermediate Re_p regime ($0.1 < \mathrm{Re}_p \leq 10^3$) can be calculated using Equations (7.6) and (7.10) together.

To estimate the terminal velocity of a particle, we need to identify the Reynolds number range in which the motion of the particle lies. As U_t is not known, Re_p cannot be calculated directly. However, the Reynolds number range can be identified by defining a parameter K as:

$$K = D_p\left[\frac{g\rho(\rho_p - \rho)}{\eta^2}\right]^{1/3} \tag{7.13}$$

This parameter does not contain the terminal velocity. It can be readily shown that:

$$\text{Stokes regime } (\mathrm{Re}_p \leq 0.1) \text{ corresponds to } K \leq 1.22 \tag{7.14}$$

$$\text{Intermediate regime } (0.1 < \mathrm{Re}_p \leq 10^3) \text{ corresponds to } 1.22 < K \leq 69 \tag{7.15}$$

$$\text{Newton's regime } (10^3 < \mathrm{Re}_p \leq 10^5) \text{ corresponds to } 69 < K \leq 1489 \tag{7.16}$$

Figure 7.1 shows the plot of C_D versus Re_p covering all the three regimes: Stokes regime, Intermediate regime, and Newton's regime. The intermediate regime portion of the plot is generated from Equation (7.10).

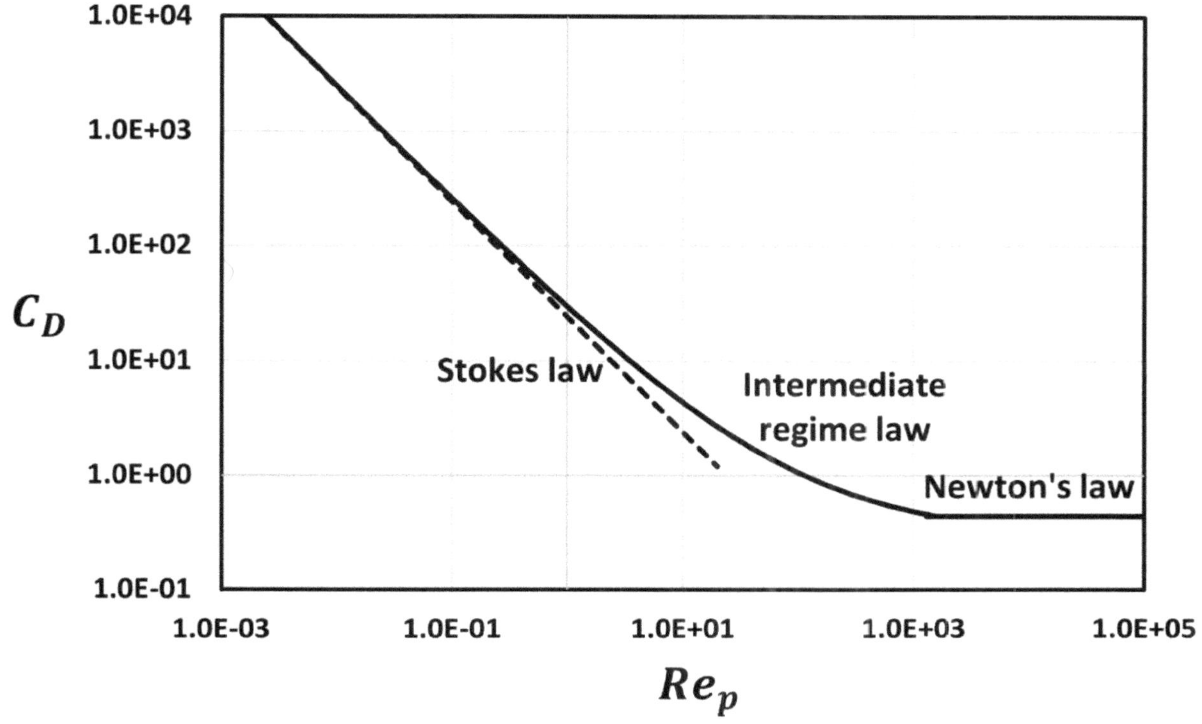

Figure 7.1 C_D versus Re_p plot covering all the three regimes: Stokes regime, Intermediate regime, and Newton's regime.

7.2 TERMINAL RISE VELOCITY OF DROPLETS AND BUBBLES

Droplets of a lighter liquid rise in another immiscible liquid of higher density. Assuming creeping flow regime, the drag coefficient C_D for a droplet is given as:

$$C_D = \frac{8}{Re_p}\left[\frac{2\eta_m+3\eta_d}{\eta_m+\eta_d}\right] \tag{7.17}$$

where η_m is the viscosity of the matrix liquid and η_d is the viscosity of the droplet liquid. For solid particles $\eta_d \to \infty$ and consequently Equation (7.17) reduces to the Stokes law, Equation (7.8).

Interestingly the drag force on a droplet is always smaller than the drag force on the rigid particle at the same Re_p. The tangential stresses acting on the droplet set up internal circulation within the droplets. The mobility of the interface makes the drag force less in comparison with rigid interface (solid particles).

The terminal rise velocity of a droplet is given by:

$$U_t = \frac{gD_p^2(\rho_m-\rho_d)}{6\eta_m}\left[\frac{\eta_m+\eta_d}{2\eta_m+3\eta_d}\right] \tag{7.18}$$

where D_p is droplet diameter, ρ_m is matrix fluid density, and ρ_d is droplet fluid density. This expression, Equation (7.18), is often referred to as the Hadamard-Rybcznski equation. It should be noted that the Hadamard-Rybcznski equation is valid only for clean interfaces. In the presence of impurities or surfactant at the interface, the mobility of the interface is decreased significantly, and the droplets behave more like rigid particles following the Stokes law.

For gas bubbles rising in a liquid matrix, the drag coefficient and the terminal rise velocity are given as:

$$C_D = \frac{16}{Re_p} \tag{7.19}$$

$$U_t = \frac{g D_p^2 (\rho_m - \rho_d)}{12 \eta_m} \tag{7.20}$$

These equations follow from the droplet equations, Equations (7.17) and (7.18), when $\eta_d \rightarrow 0$. Note that $Re_p = \rho_m D_p U_t / \eta_m$.

7.3 SETTLING OF A SWARM OF SOLID PARTICLES (HINDERED SETTLING)

In section 7.1, the settling behavior of a single solid particle in quiescent fluid was discussed. In practical applications, sedimentation of a swarm of particles is important. Due to hydrodynamic interactions between the neighboring particles, the settling velocity of a swarm of particles is hindered or slower. When a swarm of particles settle, the suspending medium fluid moves upward to compensate for the settling of particles (see Figure 7.2). Due to the backflow drag on the particles, the particles settle at a velocity lower than that of an isolated article.

The relative decrease in the settling velocity of particles in a *dilute* suspension is proportional to ϕ, the volume fraction of particles, and is given as:

$$\frac{U_p - U_o}{U_0} = -\alpha\phi \tag{7.21}$$

where U_p is the settling velocity of particles in a suspension of concentration ϕ. Equation (10) can be rewritten as:

$$U_p = U_0(1 - \alpha\phi) \tag{7.22}$$

Assuming the spatial distribution of particles to be random, Batchelor theoretically found the value of α to be 6.55.

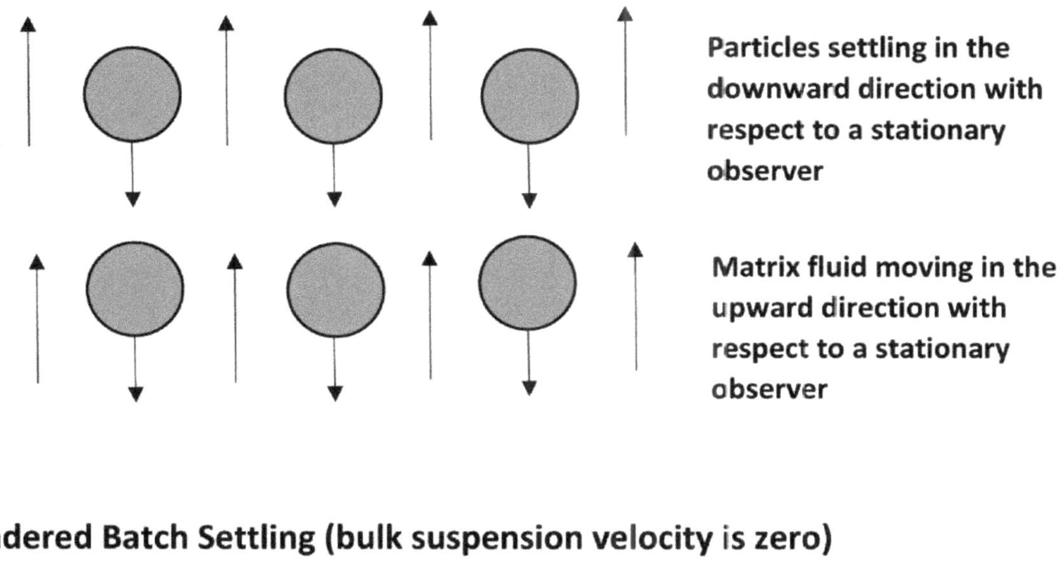

Figure 7.2 Hindered batch settling of particles.

The linear relationship, Equation (7.22), is valid only at low concentrations ($\phi < 0.05$). At high particle concentrations, the hindrance effect becomes more pronounced, and the relationship becomes non-linear. Several empirical and semi-empirical correlations have been proposed in the literature to describe hindered settling of particles at high particle concentrations. The most popular ones are due to Richardson and Zaki and Garside and Al-Dibouni. The Richardson and Zaki correlation is given as:

$$U_p = U_0(1-\phi)^n \tag{7.23}$$

where n is a function of particle Reynolds number Re_p as given below:

$$n = 4.65 \quad (\mathrm{Re}_p < 0.2) \tag{7.23a}$$

$$n = 4.35Re_p^{-0.03} \quad (0.2 < \mathrm{Re}_p < 1) \tag{7.23b}$$

$$n = 4.45Re_p^{-0.1} \quad (1 < \mathrm{Re}_p < 500) \tag{7.23c}$$

$$n = 2.39 \quad (\mathrm{Re}_p > 500) \tag{7.23d}$$

The Garside and Al-Dibouni correlation is given as:

$$\frac{U_R - A}{B - U_R} = 0.06(\mathrm{Re}_p)^{\varepsilon+0.2} \tag{7.24}$$

where $A = \varepsilon^{4.14}$, $B = 0.8\varepsilon^{1.28}$ (when $\varepsilon \leq 0.85$), $B = \varepsilon^{2.65}$ (when $\varepsilon > 0.85$), ε is volume fraction of fluid, and U_R is defined as:

$$U_R = \frac{U_p}{\varepsilon U_0} \tag{7.25}$$

In the limit $\mathrm{Re}_p \to 0$, Equation (7.24) reduces to:

$$U_p = U_0(1-\phi)^{5.14} \tag{7.26}$$

This equation has the same form as that of the Richardson and Zaki equation, Equation (7.23), with a different value of n. Al-Naafa and Sami-Selim proposed the following equation to describe their data on sedimentation of monodisperse suspensions:

$$U_p = U_0(1-\phi)^{6.55} \tag{7.27}$$

In the limit $\phi \to 0$, Equation (7.27) reduces to Batchelor's equation, Equation (7.22) with $\alpha = 6.55$.

7.4 DRIFT FLUX THEORY AND INTERPRETATION OF HINDERED SETTLING

The drift flux theory is used widely in the analysis of two-phase flows when the two phases move with different velocities. For example, it can be used to predict the in-situ gas hold up in bubbly gas-liquid two-phase flows from the knowledge of superficial gas and liquid velocities, bubble size, and fluid properties. The drift flux theory can be readily applied to sedimentation of particles in non-dilute suspensions. The advantage of the drift flux theory is that it can applied to both batch sedimentation and continuous sedimentation of suspensions. Note that in batch sedimentation there is no net motion of suspension, that is, the particles and the matrix fluid move in opposite directions such that the net suspension velocity is zero with respect to a stationary observer. In continuous sedimentation, the suspension has a net motion with a non-zero net velocity with respect to a stationary observer.

Consider a suspension of solid particles in Newtonian liquid. The drift velocity of solid particles, denoted as $\vec{U}_{pm}$, is defined as:

$$\vec{U}_{pm} = \vec{U}_p - \vec{U}_m \tag{7.28}$$

where $\vec{U}_p$ is the particle velocity and $\vec{U}_m$ is the mixture (suspension) velocity with respect to a stationary observer. Thus, the drift velocity of a particle is the velocity of a particle relative to an observer moving with the suspension velocity $\vec{U}_m$. The suspension velocity $\vec{U}_m$ can be expressed as:

$$\vec{U}_m = \vec{J}_p + \vec{J}_l \tag{7.29}$$

where $\vec{J}_p$ is the volumetric flux of particles and $\vec{J}_l$ is the volumetric flux of liquid phase. The volumetric fluxes are related to particle and liquid velocities as follows:

$$\vec{J}_p = \phi\, \vec{U}_p \tag{7.30}$$

$$\vec{J}_l = (1-\phi)\, \vec{U}_l \tag{7.31}$$

where $\vec{U}_l$ is the liquid phase velocity. From Equations (7.28) and (7.29), it follows that:

$$\vec{U}_{pm} = \vec{U}_p - (\vec{J}_p + \vec{J}_l) \tag{7.32}$$

Substituting the relations for $\vec{J}_p$ (Equation (7.30)) and $\vec{J}_l$ (Equation (7.31)) into Equation (7.32), we obtain:

$$\vec{U}_{pm} = (1 - \phi)(\vec{U}_p - \vec{U}_l) = (1 - \phi)\vec{U}_s \tag{7.33}$$

where $\vec{U}_s = (\vec{U}_p - \vec{U}_l)$ is the slip velocity between the particles and the fluid (liquid).

The drift flux of particles, denoted as $\vec{J}_{pm}$, is related to the drift velocity of particles $\vec{U}_{pm}$ as:

$$\vec{J}_{pm} = \phi\, \vec{U}_{pm} \tag{7.34}$$

From Equations (7.30), (7.32), and (7.34), it follows that:

$$\vec{J}_{pm} = (1 - \phi)\vec{J}_p - \phi\, \vec{J}_l \tag{7.35}$$

From the knowledge of volumetric fluxes $\vec{J}_p$, $\vec{J}_l$ and volume fraction of particles ϕ, one can calculate the drift flux using Equation (7.35). Once the drift flux is known, the drift velocity $\vec{U}_{pm}$ can be calculated from Equation (7.34) and the slip velocity $\vec{U}_s$ can be calculated from Equation (7.33).

The drift velocity can be expressed as:

$$\vec{U}_{pm} = \vec{U}_0 f(\phi) \tag{7.36}$$

where $\vec{U}_0$ is the settling velocity of a single isolated particle in quiescent liquid and $f(\phi)$ is the hindered settling function of the form of Richardson and Zaki correlation. Thus,

$$\vec{U}_{pm} = \vec{U}_0(1 - \phi)^n \tag{7.37}$$

$$\vec{J}_{pm} = \phi\, \vec{U}_{pm} = \vec{U}_0 \phi(1 - \phi)^n \tag{7.38}$$

where $n = 4.65$ assuming Re_p to be small.

Figure 7.3 shows the plots of normalized drift velocity (U_{pm}/U_0) and normalized drift flux (J_{pm}/U_0) versus particle concentration ϕ. The normalized drift velocity decreases with the increase in ϕ whereas the normalized drift flux goes through a maxima at some intermediate particle concentration.

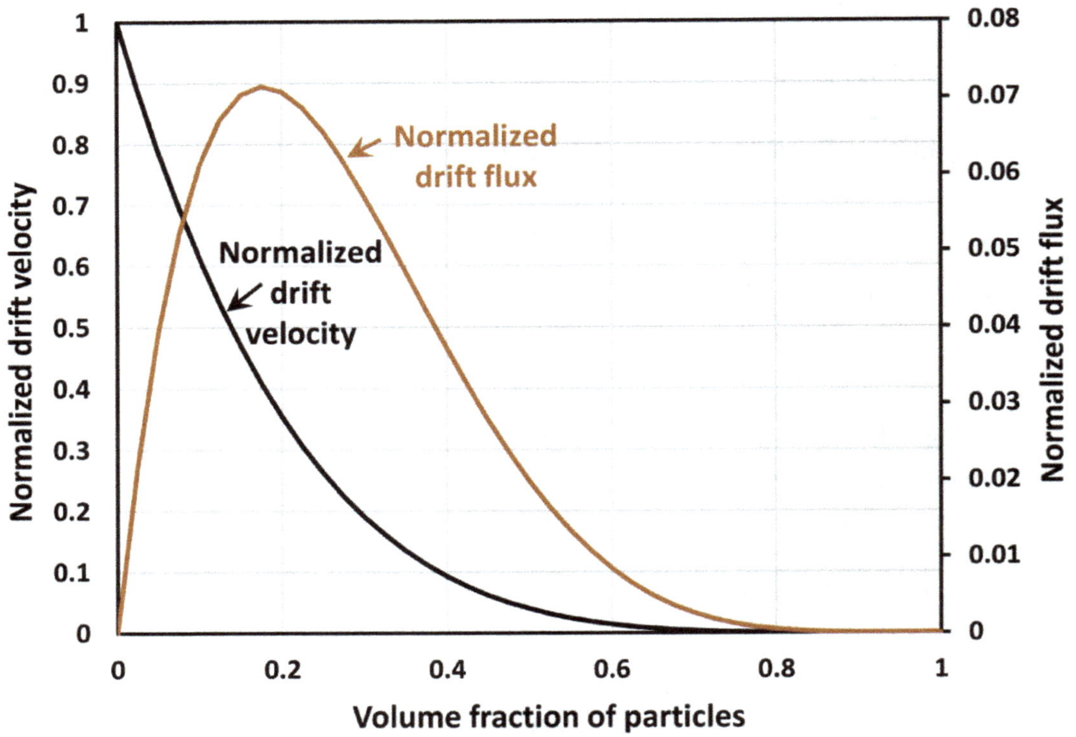

Figure 7.3 Normalized drift velocity (U_{pm}/U_0) and normalized drift flux (J_{pm}/U_0) versus particle concentration ϕ generated from Equations (7.37) and (7.38) using $n = 4.65$.

In the absence of any relative motion or slip between the particles and fluid, we have:

$$\vec{U}_{pm} = 0\,;\quad \vec{J}_{pm} = 0\,;\quad \vec{U}_s = 0 \tag{7.39}$$

Under these conditions, the suspension flow can be treated as a homogeneous single-phase flow with average properties.

In the present topic of discussion, that is, batch sedimentation of particles where there is no net flow of suspension, the following relations are applicable:

$$\vec{U}_m = 0 \tag{7.40}$$

$$\vec{U}_{pm} = \vec{U}_p \tag{7.41}$$

$$\vec{U}_s = \vec{U}_p/(1-\phi) \tag{7.42}$$

$$\vec{J}_{pm} = \phi\,\vec{U}_p \tag{7.43}$$

$$\vec{U}_p = \vec{U}_0(1-\phi)^n \tag{7.44}$$

$$\vec{J}_p = -\vec{J}_l = \vec{J}_{pm} = \vec{U}_0\phi(1-\phi)^n \tag{7.45}$$

7.5 EXAMPLES

7.5.1 Example 1

A fluid flows over a solid spherical object of diameter 5 cm at a free stream velocity of 18 m/s. Estimate the drag coefficient and drag force acting on the sphere. The properties of the fluid are: $\rho = 1.137$ kg/m^3 and $\eta = 1.9 \times 10^{-5}$ Pa.s.

Solution

Refer to the figure.

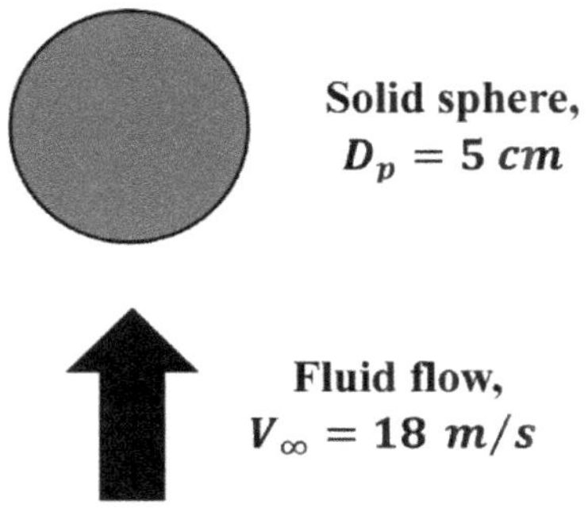

$$\mathrm{Re}_p = \frac{\rho D_p V_\infty}{\eta} = \frac{1.137 \times 18 \times (5 \times 10^{-2})}{1.9 \times 10^{-5}} = 5.39 \times 10^4$$

Thus, the flow regime is Newton's regime. Hence,

$$C_D = 0.44 \qquad 10^3 < \mathrm{Re}_p \le 10^5$$

$$F_D = C_D A_p (\rho V_\infty^2/2) = C_D\left(\pi D_p^2/4\right)(\rho V_\infty^2/2) = 0.44(\pi \times 0.05^2/4)(1.137 \times 18^2/2) = 0.159\ N$$

7.5.2 Example 2

A solid particle having a diameter of 0.01 mm is settling in air. The density of the particle is 1000 kg/m^3. Determine the terminal velocity of the particle. The properties of air are: $\rho = 1.137$ kg/m^3 and $\eta = 1.9 \times 10^{-5}$ Pa.s.

Solution

Refer to the figure.

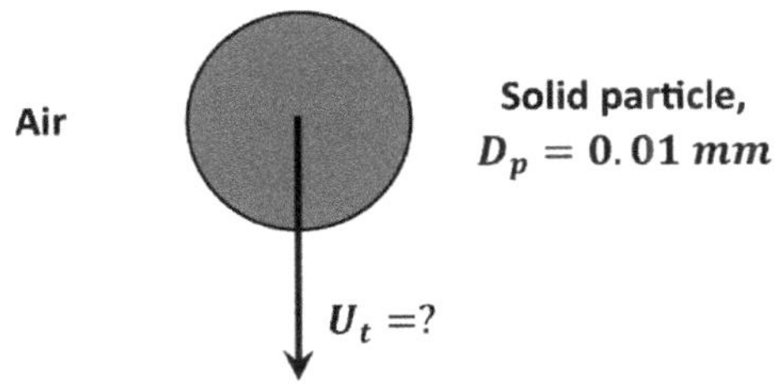

First determine the flow regime:

$$K = D_p\left[\frac{g\rho(\rho_p-\rho)}{\eta^2}\right]^{1/3} = (0.01\times10^{-3})\left[\frac{1.137\times9.81\times(1000-1.137)}{(1.9\times10^{-5})^2}\right]^{1/3} = 0.3137$$

Since $K < 2.6$, the flow regime is Stokes regime. Use Stokes law to estimate the terminal velocity.

$$U_t = \frac{gD_p^2(\rho_p-\rho)}{18\eta} = \frac{9.81\times(0.01\times10^{-3})^2(1000-1.137)}{18\times1.9\times10^{-5}} = 2.865\times10^{-3}\ m/s$$

Check Reynolds number

$$\mathrm{Re}_p = \frac{\rho D_p U_t}{\eta} = \frac{1.137\times(0.01\times10^{-3})\times2.865\times10^{-3}}{1.9\times10^{-5}} = 1.71\times10^{-3} \ll 1$$

7.5.3 Example 3

A hollow steel sphere, 5 mm in diameter, with a mass of 0.05 gm, is released in a column of liquid and attains a terminal velocity of 0.5 cm/s. The liquid density is 0.9 g/cc. The local acceleration of gravity is 980.7 cm/s^2.

(a) Will the sphere rise, or will it settle (fall) in the liquid?

(b) In which direction will the drag force act? Upward or Downward?

(c) Compute the drag force.

(d) Compute the drag coefficient

(e) Determine the viscosity of the liquid

Solution

(a) Will the sphere rise, or will it settle (fall) in the liquid?

Calculate density of the hollow sphere particle:

$$\rho_p = \frac{m}{V} = \frac{0.05\times10^{-3}}{4\pi(2.5\times10^{-3})^3/3} = 764\ \ kg/m^3$$

$$\rho_f = 900\ kg/m^3$$

Since the particle is lighter than the liquid, it would rise in the liquid.

(b) In which direction will the drag force act? Upward or Downward?

The drag force acts in the downward direction (see figure below).

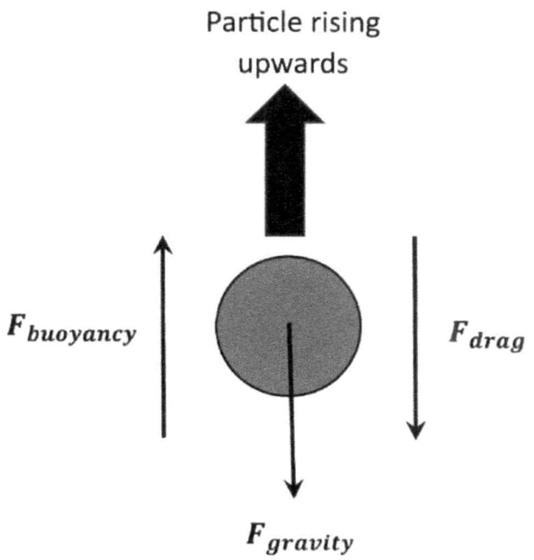

(c) Compute the drag force.

From force balance (see figure above),

$$F_{buoyancy} = F_{gravity} + F_{drag}$$

Thus,

$$F_{drag} = F_{buoyancy} - F_{gravity} = \frac{4}{3}\pi R^3(\rho_f - \rho_p)g = \frac{4}{3}\pi(2.5 \times 10^{-3})^3(900 - 764) \times 9.81$$
$$= 8.73 \times 10^{-5}\ N$$

(d) Compute the drag coefficient

$$C_D = \frac{F_D/A_p}{(\rho U_t^2/2)} = \frac{8.73 \times 10^{-5}/(\pi \times (2.5 \times 10^{-3})^2)}{900 \times (0.5 \times 10^{-2})^2/2} = 395$$

(e) Determine the viscosity of the liquid

Assume Stokes regime:

$$U_t = \frac{gD_p^2(\rho - \rho_p)}{18\eta}$$

$$\eta = \frac{9.81 \times (5 \times 10^{-3})^2(900 - 764)}{18 \times 0.5 \times 10^{-2}} = 0.3706\ Pa.s$$

Check if the assumed flow regime is correct:

$$\mathrm{Re}_p = \frac{\rho D_p U_t}{\eta} = \frac{900 \times (5 \times 10^{-3}) \times 0.5 \times 10^{-2}}{0.3706} = 0.0607 \ll 1$$

Thus, the assumed Stokes regime is correct.

7.5.4 Example 4

Consider hindered settling of particles. Solid particles of diameter of 0.01 mm are settling in a concentrated suspension with particle volume fraction of 0.30. The density of the particles is 2500 kg/m^3. Determine the settling velocity of the particles in the suspension. Use the Richardson and Zaki correlation for hindered settling. The properties of the matrix liquid are: $\rho = 1000$ kg/m^3 and $\eta = 1$ mPa.s.

Solution

For hindered settling, the Richardson and Zaki correlation gives:

$$U_p = U_0(1-\phi)^{4.65}$$

where U_p is the settling velocity of particles in the suspension and U_0 is the settling velocity of a single isolated particle. Assuming Stokes regime,

$$U_o = \frac{gD_p^2(\rho - \rho_p)}{18\eta} = \frac{9.81 \times (10 \times 10^{-6})^2(2500 - 1000)}{18 \times 10^{-3}} = 8.175 \times 10^{-5}\ m/s$$

Check if the assumed Stokes regime is correct:

$$\text{Re}_p = \frac{\rho D_p U_t}{\eta} = \frac{1000 \times (10 \times 10^{-6}) \times 8.175 \times 10^{-5}}{10^{-3}} = 8.175 \times 10^{-4} \ll 1$$

The settling velocity of particles in the suspension is

$$U_p = U_0(1-\phi)^{4.65} = 8.175 \times 10^{-5}\,(1 - 0.30)^{4.65} = 1.557 \times 10^{-5}\ m/s$$

Thus, the settling velocity of particles is much lower in a concentrated suspension as compared unhindered settling of a single particle.

7.5.4 Example 5

Consider the rise of air bubbles in a column as shown in the figure below.

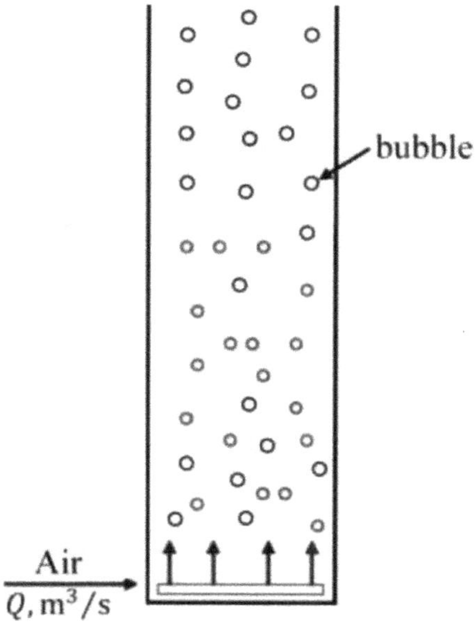

The superficial velocity of the air is 1.5 cm/s. The upward terminal rise velocity of a single isolated air bubble is 13 cm/s. The air bubbles are of average size of $D_p = 1$ mm. Apply the drift flux theory to estimate the gas hold-up (volume fraction of bubbles) of the column. The properties of the liquid are: $\rho = 1000$ kg/m^3 and $\eta = 1$ mPa.s.

Solution

The gas hold up can be obtained from the intersection of the operating line (Equation (7.35)) and drift flux characteristic curve (Equation (7.38)) given as follows:

$$\vec{J}_{pm} = (1-\phi)\vec{J}_p - \phi\,\vec{J}_l$$

$$\vec{J}_{pm} = \vec{U}_0\phi(1-\phi)^n$$

The normalized operating line and drift flux curve are as follows:

$$J_{pm}/U_o = (1-\phi)\big(J_p/U_o\big) - \phi(J_l/U_o)$$

$$J_{pm}/U_o = \phi(1-\phi)^n$$

Given: $J_p = 1.5$ cm/s, $J_l = 0$, $U_o = 13$ cm/s, $Re = \rho D_p U_o/\eta = 1000 \times 0.13 \times 0.001/10^{-3} = 130$. From Equation (7.23c), $n = 2.735$. The plots of operating line and drift flux curve are shown below. The gas hold-up in the column is given by the intersection of the operating line and drift-

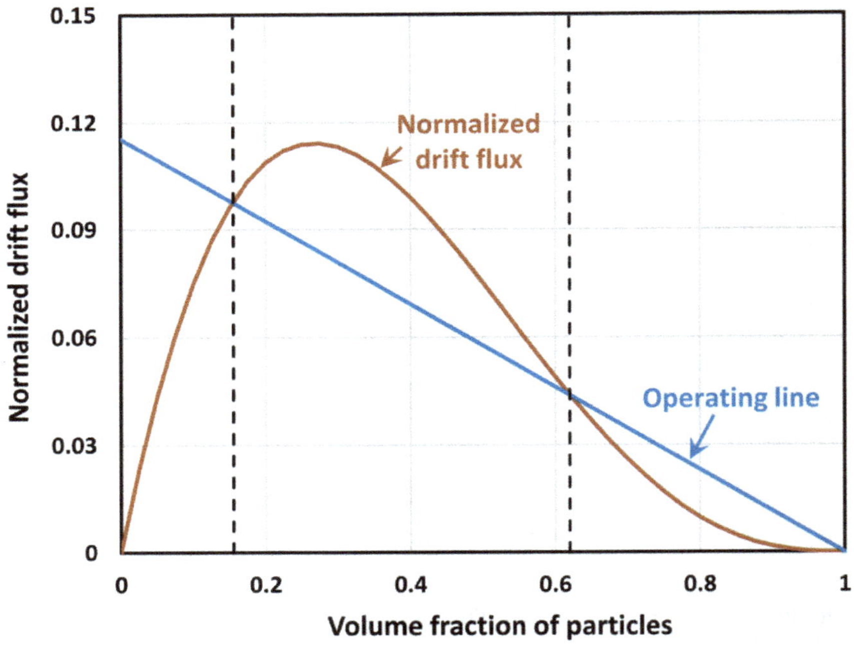

flux curve. Clearly, there are two intersection points and solutions: the dilute-phase solution with volume fraction of bubbles of 0.155 and the dense-phase solution with volume fraction of bubbles of 0.62. In a bubble column, the solution of $\phi = 0.155$ is expected. Two solutions are normally encountered in froth flotation columns where gas bubbles collect at the top of the column to form a more concentrated (dense) foamy phase.

NOTATION

A	parameter in Equation (7.24)
A_p	projected area of a particle
B	parameter in Equation (7.24)
C_D	drag coefficient
D_p	diameter of particle
F_B	buoyancy force
F_D	drag force
F_G	gravity force
g	acceleration due to gravity
$\vec{j}_l$	volumetric flux of liquid
$\vec{j}_p$	volumetric flux of particles
$\vec{j}_{pm}$	drift flux of particles
K	parameter defined in Equation (7.13)
n	exponent in Richardson and Zaki correlation, Equation (7.23)
Re_p	particle Reynolds number
t	time
U	velocity of a particle
U_o	settling velocity of a single isolated particle
U_p	settling velocity of particles in a suspension

U_R	defined in Equation (7.25)
U_t	terminal velocity of a single isolated particle. Note that $U_t = U_o$.
$\vec{U}_l$	liquid velocity vector
$\vec{U}_m$	mixture velocity vector
$\vec{U}_o$	settling velocity vector of a single isolated particle in quiescent liquid
$\vec{U}_p$	particle velocity vector
$\vec{U}_s$	slip velocity vector
$\vec{U}_{pm}$	drift velocity of particles

Greek Letters

α	proportionality factor in Equation (7.21)
ε	volume fraction of fluid, see Equation (7.24).
η	viscosity
η_d	viscosity of dispersed phase (droplets or bubbles)
η_m	viscosity of matrix (continuous phase)
ρ	density of fluid
ρ_p	density of particle
ϕ	volume fraction of particles

PRACTICE PROBLEMS

Problem 1
A solid sphere of 0.02 mm diameter and density of 2500 kg/m^3 is falling through a liquid matrix. The matrix density and viscosity are: 1000 kg/m^3 and 1 mPa.s, respectively. Calculate the terminal settling velocity of the particle.

Problem 2
A glass sphere of density 2620 kg/m^3 is falling through carbon tetrachloride as a liquid matrix. The matrix density and viscosity are: 1590 kg/m^3 and 0.958 mPa.s, respectively. The terminal velocity of the glass particle is 0.65 m/s. Estimate the diameter of the glass particle.

Problem 3
A steel ball of diameter 1 mm is released from rest at the surface of oil in a glass column (see figure). The ball reaches a constant terminal velocity at a location marked as 1. Estimate the terminal velocity. Also determine the time that the ball takes in falling from position 1 to position 2. Steel has a density of 7800 kg/m^3 and the properties of oil are: $\eta = 0.1$ Pa.s and $\rho = 850$ kg/m^3.

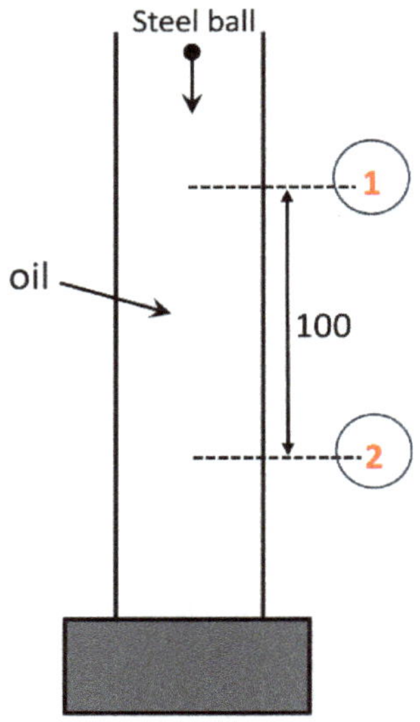

Problem 4

Re-do example 5 with the following changes:

Superficial velocity of air = 2.5 cm/s; The average diameter of air bubbles is 1.4 mm; upward terminal rise velocity of a single isolated air bubble is 15 cm/s.

The properties of the liquid are the same, that is, $\rho = 1000$ kg/m^3 and $\eta = 1$ mPa.s.

FURTHER READING

1. R. Pal. Modelling of sedimentation and creaming in suspensions and Pickering emulsions. *Fluids* 2019, 4, 186.
2. R. Pal, J.H. Masliyah. Flow characterization of a flotation column. *Can. J. Chem. Eng.* 1989, 67, 916-923.
3. R.B. Bird, W.E. Stewart, E.N. Lightfoot. *Transport Phenomena.* 2nd Edition, Wiley, New York, 2007.
4. C.J. Geankoplis, A.A. Hersel, D.H. Lepek. *Transport Processes and Separation Process Principles*. 5th Edition, Prentice Hall, New Jersey, 2018.
5. W.L. McCabe, J.C. Smith, P. Harriott. *Unit Operations of Chemical Engineering.* 6th Edition, McGraw Hill, New York, 2001.
6. G.K. Batchelor. Sedimentation in a dilute dispersion of spheres. *J. Fluid Mech.* 1972, 52, 245–268.
7. J.F. Richardson, W.N. Zaki. Sedimentation and fluidization. *Trans. Inst. Chem. Eng.* 1954, 32, 35–53.
8. J. Garside, M.R. Al-Dibouni. Velocity-voidage relationships for fluidization and sedimentation in solid-liquid systems. *Ind. Eng. Chem. Process Des. Dev.* 1977, 16, 206–214.
9. M.A. Al-Naafa, M.S. Selim. Sedimentation of monodisperse and bidisperse hard-sphere colloidal suspensions. *AIChE J.* 1992, 38, 1618–1630.
10. G.B. Wallis. *One-dimensional Two-phase Flow*. Dover publications, Inc., Garden city, New York, 2020.

Chapter 8

Non-Newtonian Flow

In this chapter, pipeline flow of time-independent non-Newtonian fluids is described. The discussion is initially restricted to two important classes of time-independent non-Newtonian fluids: Power-law fluids and Bingham-plastic fluids. The rheological behaviour of a majority of time-independent, purely viscous non-Newtonian fluids could be approximated by the power-law and Bingham-plastic models. A more general approach applicable to any time-independent, purely viscous non-Newtonian fluids is presented later in the chapter.

8.1 POWER-LAW FLUIDS

8.1.1 Laminar Flow

Consider steady state, fully developed, laminar flow of an incompressible non-Newtonian fluid in a horizontal, circular cross-section pipe of length L and radius R. The only velocity component that is non-zero is V_z. The other two velocity components V_x and V_y are both zero. The fluid obeys the following power-law model in pipe flow:

$$-\tau_{rz} = K\left(-\frac{dV_z}{dr}\right)^n \tag{8.1}$$

where K is the consistency index and n is the flow behaviour index. Both K and n are constant, independent of the velocity gradient. Pseudoplastic (shear-thinning) and dilatant (shear-thickening) fluids discussed in Chapter 3 can be described by a power-law model. For pseudoplastic fluids, $n < 1$ and for dilatant fluids, $n > 1$. A Newtonian fluid is a special case of the power-law fluid with $n = 1$ and $K = \eta$ where η is the fluid viscosity.

According to the continuity equation:

$$\frac{\partial V_z}{\partial z} = 0 \tag{8.2}$$

Thus, the velocity V_z, at any given radial position, remains constant in the axial direction. As the flow is axisymmetric, the velocity V_z is also independent of angular position θ. Thus, the axial velocity is a function of only radial position, that is, $V_z = V_z(r)$.

From force balance over a cylindrical, differential control volume of radius r and length dz, it follows that:

$$p(\pi r^2) + \tau_{rz}(2\pi r dz) = (p + dp)(\pi r^2) \tag{8.3}$$

$$\tau_{rz} = \frac{r}{2}\left(\frac{dp}{dz}\right) = -\frac{r}{2}\left(\frac{\Delta p}{L}\right) \tag{8.4}$$

As in the case of Newtonian fluids, the shear stress varies linearly with r. The magnitude of the shear-stress decreases toward the centre of the pipe. It is zero at the centreline and highest at the wall. Also,

$$\frac{\tau_{rz}}{\tau_w} = -\frac{r}{R} \tag{8.5}$$

where τ_w is wall shear stress, given as:

$$\tau_w = -\tau_{rz}|_{r=R} = \frac{R}{2}\left(\frac{\Delta p}{L}\right) \tag{8.6}$$

Upon substitution of the power-law model, Equation (8.1), into the force balance, Equation (8.4), we obtain:

$$-\tau_{rz} = K\left(-\frac{dV_z}{dr}\right)^n = \frac{r}{2}\left(\frac{\Delta p}{L}\right) \tag{8.7}$$

Equation (8.7) gives the following expression for the velocity gradient:

$$\frac{dV_z}{dr} = -\left(\frac{1}{2K}\right)^{1/n}\left(\frac{\Delta p}{L}\right)^{1/n} r^{1/n} \tag{8.8}$$

Upon integration and using the no-slip boundary condition that at $r = 0, V_z = 0$, the following velocity distribution is obtained:

$$V_z = \left(\frac{1}{2K}\frac{\Delta p}{L}\right)^{1/n}\left(\frac{n}{1+n}\right)(R)^{(1+n)/n}\left[1-\left(\frac{r}{R}\right)^{\frac{1+n}{n}}\right] \tag{8.9}$$

According to the velocity distribution, the fluid velocity is maximum at $r = 0$ and is given as:

$$V_{max} = \left(\frac{1}{2K}\frac{\Delta p}{L}\right)^{1/n}\left(\frac{n}{1+n}\right)(R)^{(1+n)/n} \tag{8.10}$$

When $n = 1$ and $K = \eta$, this expression reduces to the following expression for Newtonian fluids:

$$V_{max} = \frac{\Delta p R^2}{4\eta L} \tag{8.11}$$

The velocity distribution for power-law fluids can be re-written as:

$$\frac{V_z}{V_{max}} = \left[1-\left(\frac{r}{R}\right)^{\frac{1+n}{n}}\right] \tag{8.12}$$

If we plot V_z/V_{max} as a function of r/R, we find that for pseudoplastic fluids ($n < 1$), the velocity profile is flatter as compared with the Newtonian fluid case ($n = 1$), and for dilatant fluids ($n > 1$), the velocity profile is more pointed as compared with the parabolic profile of Newtonian fluid.

The average velocity of fluid in the pipe is:

$$\bar{V} = \frac{Q}{\pi R^2} = \frac{\int_0^R 2\pi r V_z dr}{\pi R^2} = \frac{2}{R^2} V_{max}\int_0^R r\left[1-\left(\frac{r}{R}\right)^{\frac{1+n}{n}}\right] = \left(\frac{1+n}{1+3n}\right)V_{max} \tag{8.13}$$

where Q is the volumetric flow rate of fluid. Thus, the velocity distribution could also be expressed as:

$$\frac{V_z}{\bar{V}} = \left(\frac{1+n}{1+3n}\right)\left[1-\left(\frac{r}{R}\right)^{\frac{1+n}{n}}\right] \tag{8.14}$$

Figure 8.1 shows the plots of $V_z/\bar{V}$ versus r/R for different values of n. As noted earlier, the velocity profile is flatter for pseudoplastic fluids ($n < 1$) as compared with the Newtonian fluid case ($n = 1$), and for dilatant fluids ($n > 1$), the velocity profile is more pointed as compared with the parabolic profile of Newtonian fluid.

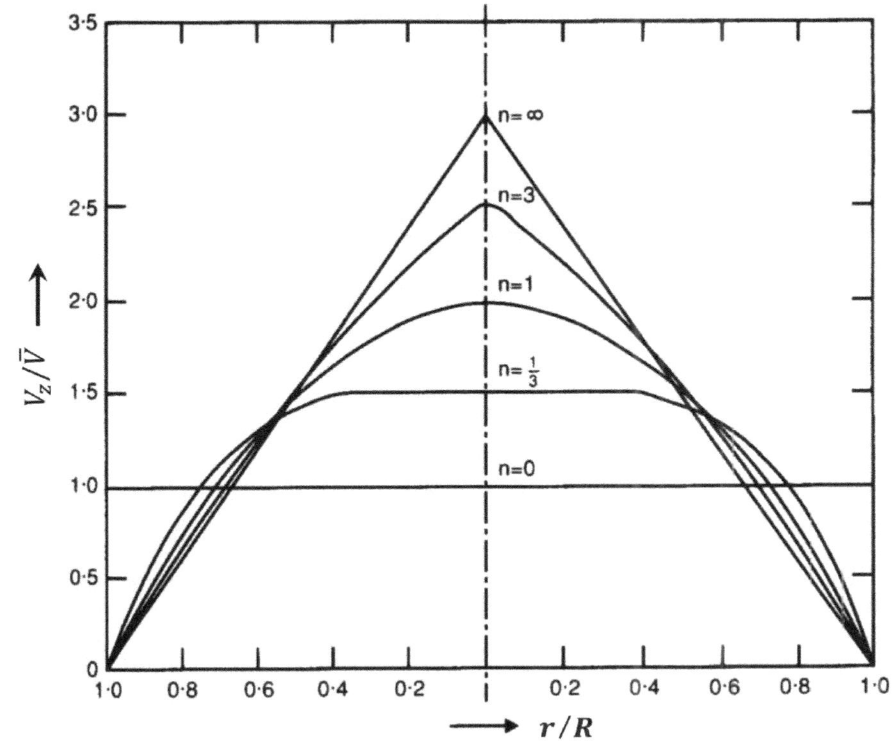

Figure 8.1 Plots of $V_z/\bar{V}$ versus r/R for different values of n.

Upon substitution of the expression for V_{max}, the following flow rate versus pressure drop relationship for the power-law fluid is obtained:

$$Q = \left(\frac{n}{1+3n}\right) \pi R^{(1+3n)/n} \left[\frac{1}{2K}\frac{\Delta p}{L}\right]^{1/n} \tag{8.15}$$

When $n = 1$ and $K = \eta$, this expression reduces to the following Newtonian case:

$$Q = \frac{\pi R^4}{8\eta}\left(\frac{\Delta p}{L}\right) = \frac{\pi D^4}{128\eta}\left(\frac{\Delta p}{L}\right) \tag{8.15a}$$

where D is the pipe diameter. The wall stress is given as follows:

$$\tau_w = -\tau_{rz}|_{r=R} = \frac{R}{2}\left(\frac{\Delta p}{L}\right) \tag{8.16}$$

Using the flow rate versus pressure drop relationship (Equation (8.15)) and eliminating the pressure-gradient, the wall stress can be expressed as:

$$\tau_w = K\left(\frac{1+3n}{4n}\right)^n \left(\frac{8\bar{V}}{D}\right)^n \tag{8.17}$$

Upon substitution of this wall stress relation into the definition of friction factor (f), we obtain:

$$f = \frac{\tau_w}{\frac{1}{2}\rho\bar{V}^2} = \frac{K\left(\frac{1+3n}{4n}\right)^n\left(\frac{8\bar{V}}{D}\right)^n}{\frac{1}{2}\rho\bar{V}^2} = \frac{16}{\left[\frac{\rho D^n\bar{V}^{2-n}}{(8)^{n-1}K\left(\frac{1+3n}{4n}\right)^n}\right]} \quad (8.18)$$

The generalized Reynolds number for power-law fluids Re_n is defined as:

$$Re_n = \frac{\rho D^n\bar{V}^{2-n}}{(8)^{n-1}K\left(\frac{1+3n}{4n}\right)^n} \quad (8.19)$$

Thus, the friction factor versus generalized Reynolds number law for laminar flow of incompressible, power-law non-Newtonian fluids is as follows:

$$f = \frac{16}{Re_n} \quad (8.20)$$

When $n = 1$ and $K = \eta$, this relationship reduces to the one for Newtonian fluids.

8.1.2 Turbulent Flow

Power-law fluids, like Newtonian fluids, flow in a laminar manner only at low to moderate values of the Reynolds number. The flow becomes turbulent at high values of the Reynolds number. The critical Reynolds number below which the flow is laminar is dependent on the power law index n. The critical Reynolds number for power-law fluids is given as:

$$Re_{n,critical} = \frac{2100(4n+2)(5n+3)}{3(3n+1)^2} \quad (8.21)$$

Another expression for the critical Reynolds number for power-law fluids recommended by American Petroleum Institute is: $Re_{n,critical} = 3470 - 1370n$. For Newtonian fluids ($n = 1$), Equation (8.21) gives the critical Reynolds number of 2100, as expected. Pseudoplastic fluids ($n < 1$) exhibit a delay in the transition from laminar to turbulent regime as compared with Newtonian fluids, that is, $Re_{n,critical} > 2100$ when $n < 1$. Also, $Re_{n,critical}$ increases with the decrease in the value of n. Dilatants fluids ($n > 1$), on the other hand, exhibit early transition to turbulent regime as compared with Newtonian fluids; the critical Reynolds number is less than 2100.

For fully developed turbulent flow of power-law fluids in smooth pipes, the following Dodge-Metzner correlation is widely used:

$$\frac{1}{\sqrt{f}} = \frac{4}{(n)^{0.75}}\log_{10}\left[Re_n\, f^{(1-0.5n)}\right] - \frac{0.4}{(n)^{1.2}} \quad (8.22)$$

For Newtonian fluids ($n = 1$), Equation (8.22) reduces to the Prandtl-VonKarman-Nikuradse equation.

Figure 8.2 shows the Dodge-Metzner friction factor-Reynolds number chart for non-Newtonian power-law fluids. In laminar flow, a single line is obtained given by Equation (8.20). In turbulent flow, a family of curves is obtained from Equation (8.22) corresponding to different values of n.

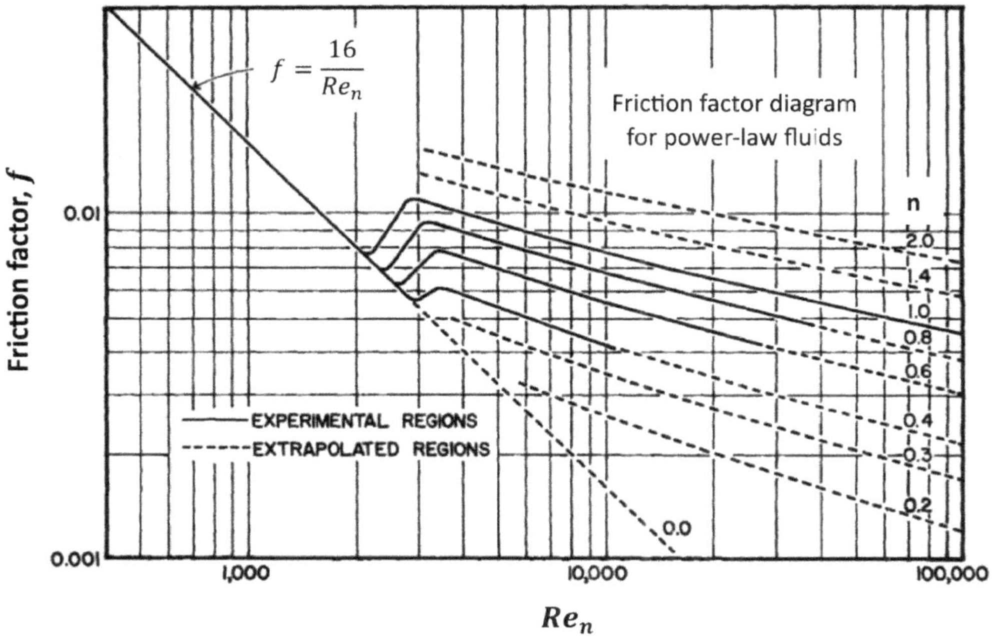

Figure 8.2 Dodge-Metzner friction factor-Reynolds number chart for power-law fluids.

The Dodge-Metzner correlation is not explicit in friction factor. Iterative procedure is needed to solve for f. Equations explicit in friction factor have been proposed for power-law fluids but they are somewhat less accurate. For example, Dodge and Metzner proposed the following Blasius-type equation, explicit in friction factor:

$$f = \frac{\alpha}{Re_n^{\beta}} \tag{8.23}$$

where α and β were given in the form of plots. Pal has proposed the following expressions for α and β:

$$\alpha = 0.0077\ln(n) + 0.078 \tag{8.24}$$

$$\beta = 0.25\,(n)^{-0.22} \tag{8.25}$$

When $n = 1$, Equation (8.23) reduces to the following equation which is very close to the Blasius equation for turbulent flow of Newtonian fluids:

$$f = \frac{0.078}{Re^{1/4}} \tag{8.26}$$

Irvine has proposed another friction factor equation for turbulent flow of power-law fluids in smooth pipes as follows:

$$f = \left[\frac{A(n)}{Re_n}\right]^{1/(3n+1)} \tag{8.27}$$

where $A(n)$ is given as:

$$A(n) = \frac{2^{n+4}}{7^{7n}}\left[\frac{4n}{3n+1}\right]^{3n^2} \tag{8.28}$$

When $n = 1$, Equation (8.27) reduces to the following form of the Blasius equation:

$$f = \frac{0.07895}{Re^{1/4}} \tag{8.29}$$

8.2 BINGHAM-PLASTIC FLUIDS

8.2.1 Laminar Flow

Bingham-plastic fluids are characterized by the following relationship between shear stress and velocity gradient in pipe flow:

$$-\tau_{rz} = \tau_0 + \eta_B\left(-\frac{dV_z}{dr}\right) \tag{8.30}$$

where τ_0 is the yield-stress and η_B is the Bingham viscosity of the fluid. Bingham-plastic fluids flow only when the applied shear stress exceeds the yield stress. For flow in a pipe, the magnitude of the shear stress decreases toward the centre of the pipe. At some radial location R_p near the centre of the pipe, the magnitude of the shear stress τ_{rz} becomes equal to the yield stress. Consequently, there occurs a solid plug-like flow in the middle of the pipe.

The relationship between the plug radius R_p and pipe radius R can be determined from force balance relations:

$$\tau_{rz} = \frac{r}{2}\left(\frac{\Delta p}{L}\right) \tag{8.31}$$

$$\tau_0 = -\tau_{rz}|_{r=R_p} = \frac{R_p}{2}\left(\frac{\Delta p}{L}\right) \tag{8.32}$$

$$\tau_w = -\tau_{rz}|_{r=R} = \frac{R}{2}\left(\frac{\Delta p}{L}\right) \tag{8.33}$$

Combining the above two relations, we obtain:

$$\frac{\tau_0}{\tau_w} = \frac{R_p}{R} \tag{8.34}$$

Now consider the region $R_p < r < R$ where the shear stress is larger than the yield-stress. In this region, the fluid velocity decreases from the plug velocity at $r = R_p$ to zero at the pipe wall ($r = R$). The velocity distribution in this region can be determined as follows:

$$-\tau_{rz} = \tau_0 + \eta_B\left(-\frac{dV_z}{dr}\right) = \frac{r}{2}\left(\frac{\Delta p}{L}\right) \tag{8.35}$$

Upon re-arrangement:

$$\frac{dV_z}{dr} = \frac{1}{\eta_B}\left[\tau_0 - \frac{r}{2}\left(\frac{\Delta p}{L}\right)\right] \tag{8.36}$$

Upon integration and using the no-slip boundary condition, we obtain the following velocity distribution:

$$V_z = \frac{\Delta p R^2}{4\eta_B L}\left[1 - \left(\frac{r}{R}\right)^2\right] - \frac{\tau_0 R}{\eta_B}\left(1 - \frac{r}{R}\right) \tag{8.37}$$

When the yield-stress is zero, this expression reduces to the parabolic velocity distribution for Newtonian fluids. The plug velocity can be determined from this distribution by taking $r = R_p$:

$$V_p = V_z|_{r=R_p} = \frac{\Delta p R^2}{4\eta_B L}\left[1 - \left(\frac{R_p}{R}\right)^2\right] - \frac{\tau_0 R}{\eta_B}\left(1 - \frac{R_p}{R}\right) \tag{8.38}$$

Upon substitution of the following equation for τ_0 into Equation (8.38),

$$\tau_0 = \frac{R_p}{2}\left(\frac{\Delta p}{L}\right) \tag{8.39}$$

the expression for V_p becomes:

$$V_p = \frac{\Delta p R^2}{4\eta_B L}\left[1 + \left(\frac{R_p}{R}\right)^2 - \frac{2R_p}{R}\right]^2 = \frac{\Delta p R^2}{4\eta_B L}\left[1 - \frac{R_p}{R}\right]^2 \tag{8.40}$$

Now that the velocity distribution is known, one can determine the flowrate versus pressure-drop relationship as follows:

$$Q = \int_0^R 2\pi r V_z dr = \int_0^{R_p} 2\pi r V_p dr + \int_{R_p}^R 2\pi r V_z dr \tag{8.41}$$

Upon substituting the expressions for V_p and V_z and integrating, the following result is obtained:

$$Q = \frac{\pi R^2}{2\eta_B}\left(\frac{\Delta p}{L}\right)\left[1 - \frac{R_p}{R}\right]^2\left(\frac{R_p^2}{2}\right) + \frac{\pi R^2}{2\eta_B}\left(\frac{\Delta p}{L}\right)\left[\frac{r^2}{2} - \frac{r^4}{4R^2}\Big|_{R_p}^{R}\right] - \frac{2\pi\tau_0 R}{\eta_B}\left[\frac{r^2}{2} - \frac{r^3}{3R}\Big|_{R_p}^{R}\right] \tag{8.42}$$

If we now substitute the expression for τ_0 (Equation (8.39)) and take the limits as indicated above, the following relationship is obtained upon simplification:

$$Q = \frac{\pi R^4}{8\eta_B}\left(\frac{\Delta p}{L}\right)\left[1 - \frac{4}{3}\left(\frac{R_p}{R}\right) + \frac{1}{3}\left(\frac{R_p}{R}\right)^4\right] \tag{8.43}$$

As noted earlier, the ratio R_p/R is equal to τ_0/τ_w. Thus, the flowrate versus pressure-gradient relationship for Bingham-plastic fluids is as follows:

$$Q = \frac{\pi R^4}{8\eta_B}\left(\frac{\Delta p}{L}\right)\left[1 - \frac{4}{3}\varphi + \frac{\varphi^4}{3}\right] \tag{8.44}$$

where $\varphi = \tau_0/\tau_w$. The average velocity in the pipe is:

$$\bar{V} = \frac{Q}{\pi R^2} = \frac{R^2}{8\eta_B}\left(\frac{\Delta p}{L}\right)\left[1 - \frac{4}{3}\varphi + \frac{\varphi^4}{3}\right] \tag{8.45}$$

Therefore, the wall stress can be expressed as:

$$\tau_w = -\tau_{rz}|_{r=R} = \frac{R}{2}\left(\frac{\Delta p}{L}\right) = \frac{\eta_B}{\left(1 - \frac{4}{3}\varphi + \frac{\varphi^4}{3}\right)}\left(\frac{8\bar{V}}{D}\right) \tag{8.46}$$

Using the definition of friction factor and this relationship for wall stress, the following equation is obtained:

$$f = \frac{\tau_w}{\frac{1}{2}\rho \bar{V}^2} = \frac{16\eta_B}{(\rho \bar{V} D)\left(1 - \frac{4}{3}\varphi + \frac{\varphi^4}{3}\right)} \tag{8.47}$$

Thus, the friction factor in steady laminar flow of Bingham-plastic fluids is given as:

$$f = \frac{16}{Re_B}\left[\frac{1}{1 - \frac{4}{3}\varphi + \frac{\varphi^4}{3}}\right] \tag{8.48}$$

where Re_B is the Reynolds number based on Bingham viscosity η_B. This relationship can be re-cast in terms of the Hedstrom number (He) as follows:

$$f = \frac{16}{Re_B}\left[1 + \frac{1}{6}\frac{He}{Re_B} - \frac{1}{3}\frac{He^4}{f^3 Re_B^7}\right] \tag{8.49}$$

where $He = \rho D^2 \tau_0 / \eta_B$. Thus, friction factor can be calculated from the knowledge of flow rate, pipe diameter, density, yield stress, and Bingham viscosity. However, an iterative procedure is required as the equation is not explicit in friction factor.

An alternative form of friction factor relation for Bingham plastic fluids is as follows:

$$f = \frac{16}{Re_B}\left[1 + \frac{1}{6}Bi - \frac{1}{3}\frac{Bi^4}{(f\ Re_B)^3}\right] \tag{8.50}$$

where Bi is the Bingham number, defined as $\tau_0 D/(\eta_B \bar{V})$. Note that $Bi = He/Re_B$.

8.2.2 Turbulent Flow

The critical Reynolds number below which the flow is laminar in pipeline flow of Bingham-plastic fluid is given as follows:

$$Re_{B,critical} = \left(\frac{\rho D \bar{V}}{\eta_B}\right)_{critical} = \left(\frac{1 - \frac{4}{3}\varphi_c + \frac{\varphi_c^4}{3}}{8\varphi_c}\right) He \tag{8.51}$$

where φ_c is the ratio of yield-stress to critical wall stress, given by the following expression:

$$\frac{\varphi_c}{(1-\varphi_c)^3} = \frac{He}{16800} \tag{8.52}$$

To estimate friction factor in fully developed turbulent flow of Bingham-plastic fluid in smooth pipes, the following expression proposed by Darby et al. can be used:

$$f = (f_L^\alpha + f_T^\alpha)^{1/\alpha} \tag{8.53}$$

where f_L is determined from the laminar flow relation, Equation (8.49), and f_T is given as:

$$f_T = 10^\beta Re_B^{-0.193} \tag{8.54}$$

The parameters α and β are given as follows:

$$\alpha = 1.7 + \frac{40000}{Re_B} \tag{8.55}$$

$$\beta = -1.47\left[1 + 0.146e^{-2.9\times10^{-5}He}\right] \quad (8.56)$$

8.3 GENERAL APPROACH TO ANALYZE PIPELINE FLOW OF TIME-INDEPENDENT NON-NEWTONIAN FLUIDS

8.3.1 Laminar Flow of Any Time-Independent Non-Newtonian Fluid

Consider any time-independent, purely viscous non-Newtonian fluid characterized by the following rheological model:

$$\dot{\gamma} = f(\tau) \quad (8.57)$$

where $\dot{\gamma}$ is the shear rate and τ is the shear stress. For steady, fully developed laminar flow in a cylindrical pipe, the shear-rate $\dot{\gamma}$ and shear stress τ are given as follows:

$$\dot{\gamma} = -\frac{dV_z}{dr} \quad \text{and} \quad \tau = -\tau_{rz} \quad (8.58)\text{-}(8.59)$$

Some examples of the rheological function $f(\tau)$ are given below:

$$\text{Newtonian fluid:} \quad \dot{\gamma} = f(\tau) = \frac{\tau}{\eta} \quad (8.60)$$

$$\text{Power-law fluid:} \quad \dot{\gamma} = f(\tau) = \left(\frac{\tau}{K}\right)^{1/n} \quad (8.61)$$

$$\text{Bingham-plastic:} \quad \dot{\gamma} = f(\tau) = \left(\frac{\tau-\tau_0}{\eta_B}\right) \quad (8.62)$$

$$\text{Yield-pseudoplastic:} \quad \dot{\gamma} = f(\tau) = \left(\frac{\tau-\tau_0}{K_y}\right)^{1/n} \quad (8.63)$$

where K_y is the consistency index of the yield-pseudoplastic fluid. We can now derive a general relationship between volumetric flow rate Q and wall stress (τ_w), valid for different types of non-Newtonian fluids characterized by the rheological function $f(\tau)$. The volume rate of flow Q in a pipe is:

$$Q = \int_0^R 2\pi r V_z dr \quad (8.64)$$

where V_z is a local velocity, r is radial location, and R is the pipe radius. Using integration by parts, Equation (8.64) can be expressed as:

$$Q = \pi r^2 V_z \Big|_0^R + \int_0^R \pi r^2 \left(-\frac{dV_z}{dr}\right) dr = \int_0^R \pi r^2 \left(-\frac{dV_z}{dr}\right) dr \quad (8.65)$$

Note that $V_z(R) = 0$ due to no-slip boundary condition. The shear stress τ varies with the radial position as follows:

$$\frac{\tau}{\tau_w} = \frac{r}{R} \quad (8.66)$$

Upon differentiation, Equation (8.66) gives:

$$d\tau = \frac{\tau_w}{R} dr \tag{8.67}$$

Also note that:

$$\left(-\frac{dV_z}{dr}\right) = \dot{\gamma} = f(\tau) \tag{8.68}$$

Thus, the flow rate Q given in Equation (8.65) can be expressed as:

$$Q = \frac{\pi R^3}{\tau_w^3} \int_0^{\tau_w} \tau^2 f(\tau)\, d\tau \tag{8.69}$$

Equation (8.68) is the general relationship between flow rate and wall stress. Upon rearrangement, another useful form of flow rate versus wall stress relationship is obtained:

$$\frac{4Q}{\pi R^3} = \frac{8\bar{V}}{D} = \frac{4}{\tau_w^3} \int_0^{\tau_w} \tau^2 f(\tau)\, d\tau \tag{8.70}$$

Equations (8.69) and (8.70) can be integrated directly to obtain the flow rate versus τ_w relationship for a non-Newtonian fluid of known rheological function $f(\tau)$. For example, consider a Newtonian fluid characterized by $f(\tau) = \tau/\eta$. Upon substitution of $f(\tau)$ in Equation (8.70) and integration, the following well-known result for Newtonian fluids is obtained:

$$\tau_w = \eta\left(\frac{8\bar{V}}{D}\right) \tag{8.71}$$

where $8\bar{V}/D$ is the true shear rate at the pipe wall.

8.3.2 Concept of Generalized Reynolds Number

For pipeline flow of a Newtonian fluid, the Reynolds number is defined as:

$$Re = \frac{\rho D\bar{V}}{\eta} \tag{8.72}$$

where η, the viscosity, is a fluid property that is constant, independent of the flow condition. In the case of non-Newtonian fluids, however, the viscosity is not constant. It varies with the shear rate, that is, it depends on the flow condition. This raises the issue of how to define the Reynolds number in pipeline flow of non-Newtonian fluids.

As wall region is most important in pipeline flows, it is appropriate to use some viscosity reflecting the flow condition in the wall region. One such viscosity is the effective viscosity defined as:

$$\eta_{eff} = \frac{\tau_w}{(8\bar{V}/D)} \tag{8.73}$$

For Newtonian fluids, the effective viscosity is the true viscosity of the fluid and $8\bar{V}/D$ is the true shear rate at the pipe wall. For non-Newtonian (time-independent) fluids, the effective viscosity is not constant; it varies with $8\bar{V}/D$. Also note that $8\bar{V}/D$ is no longer the true shear rate at the pipe wall. This term $(8\bar{V}/D)$ is, therefore, often called nominal wall shear rate.

In steady laminar flow of a time-independent non-Newtonian fluid, the wall stress can be expressed in the form of a power-law expression as:

$$\tau_w = K'\left(\frac{8\bar{V}}{D}\right)^{n'} \tag{8.74}$$

where K' and n' are not generally constant; they may vary with $8\bar{V}/D$. Combining the above two relations, Equations (8.73) and (8.74), gives:

$$\eta_{eff} = K'\left(\frac{8\bar{V}}{D}\right)^{n'-1} \tag{8.75}$$

Upon substitution of this effective viscosity in the Reynolds number, the following defining expression for the generalized Reynolds number ($Re_{n'}$) is obtained:

$$Re_{n'} = \frac{\rho D\bar{V}}{K'\left(\frac{8\bar{V}}{D}\right)^{n'-1}} = \frac{\rho D^{n'}\bar{V}^{2-n'}}{8^{n'-1}K'} \tag{8.76}$$

As this form of Reynolds number was first suggested by Metzner and Reed, it is often called Metzner-Reed Reynolds number. Note that for Newtonian fluids $K' = \eta$, $n' = 1$, and as expected, the generalized Reynolds number reduces to the conventional Reynolds number.

One important advantage of using the generalized Reynolds number, as defined in Equation (8.76), is that the friction factor relation for laminar flow of non-Newtonian fluids remains the same as that of Newtonian fluids. This can be shown as follows:

$$f = \frac{\tau_w}{\frac{1}{2}\rho\bar{V}^2} = \frac{K'\left(\frac{8\bar{V}}{D}\right)^{n'}}{\frac{1}{2}\rho\bar{V}^2} = \frac{16}{Re_{n'}} \tag{8.77}$$

It is important to realize that K' and $n'$are not generally constant. They may vary with wall stress or $8\bar{V}/D$. In evaluating the generalized Reynolds number, K' and n' values at the existing wall stress τ_w should be used. It should also be pointed out that the same definition of generalized Reynolds can be extended to turbulent flows provided that $K'$and n'values are determined from the laminar flow curve ($\tau_w$versus $8\bar{V}/D$ plot) at the existing wall stress τ_w in the turbulent regime.

Finally note that in the case of power-law fluids, the power-law constants K and n are related to the parameters $K'$and n'. For power-law fluids:

$$\tau_w = K\left(\frac{1+3n}{4n}\right)^n\left(\frac{8\bar{V}}{D}\right)^n \tag{8.78}$$

Thus:

$$\tau_w = K'\left(\frac{8\bar{V}}{D}\right)^{n'} = K\left(\frac{1+3n}{4n}\right)^n\left(\frac{8\bar{V}}{D}\right)^n \tag{8.79}$$

Equation (8.79) implies that:

$$K' = K\left(\frac{1+3n}{4n}\right)^n \quad \text{and} \quad n' = n \tag{8.80}$$

Thus, K' and $n'$are constant, independent of wall stress or $8\bar{V}/D$, for power-law fluids.

8.3.3 Turbulent Flow of Any Time-Independent Non-Newtonian Fluid

For fully developed turbulent flow of any time-independent, purely viscous non-Newtonian fluid in smooth pipes, the following Dodge-Metzner correlation is applicable:

$$\frac{1}{\sqrt{f}} = \frac{4}{(n')^{0.75}} log_{10}\left[Re_{n'} f^{(1-0.5n')}\right] - \frac{0.4}{(n')^{1.2}} \tag{8.81}$$

where $Re_{n'}$ is the generalized Reynolds number is defined in Equation (8.76).

The following explicit form of friction factor equation is also applicable:

$$f = \frac{\alpha}{Re_{n'}^{\beta}} \tag{8.82}$$

where α and β are approximately as follows:

$$\alpha = 0.0077\ln(n') + 0.078 \tag{8.83}$$

$$\beta = 0.25\,(n')^{-0.22} \tag{8.84}$$

8.3.4 Rheological Characterization of Any Time-Independent Non-Newtonian Fluid

Thus far the focus of discussion has been to develop relations to predict pressure gradient (wall stress) given the flow rate or vice versa. The rheological curve ($\tau_w$versus $8\bar{V}/D$ plot) was assumed to be known. In this section, we consider the problem of determining the true rheological curve ($\tau_w$versus wall shear-rate $\dot{\gamma}_w$) from the measurement of pressure-drop versus flow rate data. This approach, that is, tube-viscometry is widely used for rheological characterization of purely viscous, time-independent non-Newtonian fluids.

Upon differentiation of the Equation (8.70), we obtain:

$$\frac{d}{d\tau_w}\left(\tau_w^3 \frac{8\bar{V}}{D}\right) = \tau_w^3 \frac{d}{d\tau_w}\left(\frac{8\bar{V}}{D}\right) + \left(\frac{8\bar{V}}{D}\right)(3\tau_w^2) = 4\frac{d}{d\tau_w}\int_0^{\tau_w} \tau^2 f\,(\tau)\,d\tau \tag{8.85}$$

Applying the Leibnitz rule to the last term of the Equation (8.85):

$$4\frac{d}{d\tau_w}\int_0^{\tau_w} \tau^2 f\,(\tau)\,d\tau = 4\tau_w^2 f(\tau_w) \tag{8.86}$$

From Equations (8.85) and (8.86), it can be readily shown that:
\

$$f(\tau_w) = \left(\frac{8\bar{V}}{D}\right)\left[\frac{3}{4} + \frac{1}{4}\frac{d\,ln\left(\frac{8\bar{V}}{D}\right)}{d\,ln\,\tau_w}\right] \tag{8.87}$$

As $d\,ln\,\tau_w/d(8\bar{V}/D)$ is the slope (n') of τ_w versus $8\bar{V}/D$ plot on a log-log scale, we can re-write Equation (8.87) as:

$$f(\tau_w) = \left(\frac{8\bar{V}}{D}\right)\left[\frac{3n'+1}{4n'}\right] \tag{8.88}$$

The true shear rate at the pipe wall is:

$$\dot{\gamma}_W = f(\tau_W) = -\frac{dV_z}{dr}\bigg|_{r=R} = \left(\frac{8\bar{V}}{D}\right)\left[\frac{3n'+1}{4n'}\right] \quad (8.89)$$

Thus pressure-drop versus flow rate data obtained from steady laminar flow in tubes can be transformed into true rheological curve ($\tau_W$versus $\dot{\gamma}_W$).

8.4 EXAMPLES

8.4.1 Example 1

Consider two large parallel plates separated by 0.55 cm and filled with a power-law fluid of power-law constant: $K = 1.0\ Pa.s^n$ and $n = 0.5$. The top plate is being pulled at a velocity of 0.12 m/s and the bottom plate is stationary. Calculate:

i) The velocity gradient in the fluid
ii) The shear stress in the fluid
iii) The drag force on the bottom plate if the plate area is 10 m².

Solution

Refer to the figure below.

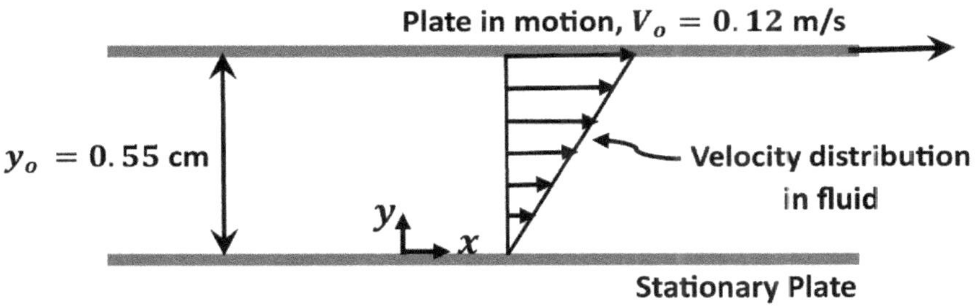

From continuity equation:

$$\frac{\partial V_x}{\partial x} = 0$$

From stress equation of motion in x-direction:

$$\rho\left[\frac{\partial V_x}{\partial t} + V_x\frac{\partial V_x}{\partial x} + V_y\frac{\partial V_x}{\partial y} + V_z\frac{\partial V_x}{\partial z}\right] = \rho g_x - \frac{\partial P}{\partial x} + \frac{\partial \tau_{xx}}{\partial x} + \frac{\partial \tau_{yx}}{\partial y} + \frac{\partial \tau_{zx}}{\partial z}$$

Only τ_{yx} term is relevant. Thus,

$$\frac{\partial \tau_{yx}}{\partial y} = 0$$

Upon integration,

$$\tau_{yx} = C_0$$

Using the power-law model,

$$\tau_{yx} = K\left(\frac{dV_x}{dy}\right)^n = C_0$$

Upon integration,

$$V_x = \left(\frac{C_0}{K}\right)^{1/n} y + C_1$$

Thus, the velocity profile is linear. Using the boundary conditions:

B.C. 1: $y = 0, V_x = 0$

B.C. 2: $y = 0.55 \text{ cm}, V_x = V_0 = 0.12 \text{ m/s}$

$C_1 = 0, \left(\frac{C_0}{K}\right)^{1/n} = 21.818$

Consequently, the velocity distribution is:

$$V_x = 21.818y$$

i) The velocity gradient in the fluid

$\frac{dV_x}{dy} = 21.818\ s^{-1}$

ii) The shear stress in the fluid

$\tau_{yx} = K\left(\frac{dV_x}{dy}\right)^n = (1) \times (21.818\,)^{0.5} = 4.671 \text{ Pa}$

The shear stress everywhere in the fluid is the same.

iii) The drag force on the bottom plate if the plate area is 10 m^2.

$F_D = \tau_{yx} \times A = 4.671 \times 10 = 46.71 \text{ N}$

8.4.2 Example 2

Consider two infinitely large parallel plates, a distance of $2B$ apart as shown below. The plates are stationary. A constant pressure gradient dP/dx (negative value) is applied to generate the flow. Assuming steady state, laminar flow of a power law fluid, determine the velocity distribution.

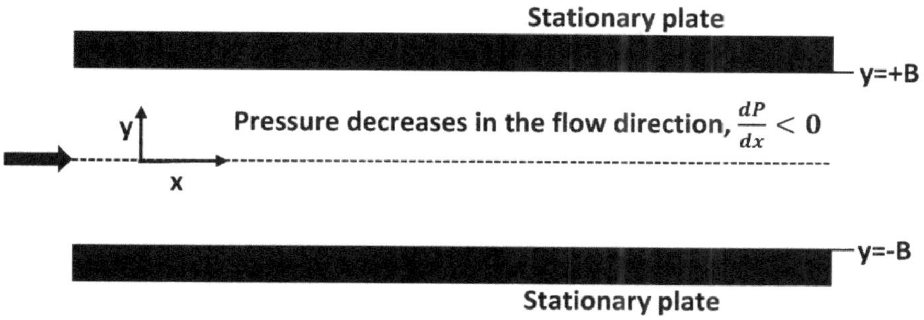

Solution:

As the flow is in x -direction, only V_x is non-zero. The continuity equation gives:

$$\frac{\partial V_x}{\partial x} = 0$$

The x-direction stress equation of motion in Cartesian coordinates is:

$$\rho\left[\frac{\partial V_x}{\partial t} + V_x\frac{\partial V_x}{\partial x} + V_y\frac{\partial V_x}{\partial y} + V_z\frac{\partial V_x}{\partial z}\right] = \rho g_x - \frac{\partial P}{\partial x} + \frac{\partial \tau_{xx}}{\partial x} + \frac{\partial \tau_{yx}}{\partial y} + \frac{\partial \tau_{zx}}{\partial z}$$

Upon simplification,

$$\frac{d\tau_{yx}}{dy} = \left(\frac{dP}{dx}\right)$$

The integration gives:

$$\tau_{yx} = \left(\frac{dP}{dx}\right)y + C_1$$

Due to symmetry of flow, τ_{yx} is zero at $y = 0$. Thus, $C_1 = 0$. Using the power-law model for τ_{yx}, we can write:

$$K\left(-\frac{dV_x}{dy}\right)^n = \left(\frac{dP}{dx}\right)y$$

$$-\frac{dV_x}{dy} = \left[\frac{1}{K}\left(\frac{dP}{dx}\right)\right]^{1/n} y^{1/n}$$

Upon integration,

$$V_x = -\left[\frac{1}{K}\left(\frac{dP}{dx}\right)\right]^{1/n} \frac{y^{\frac{n+1}{n}}}{\left(\frac{n+1}{n}\right)} + C_2$$

Using the boundary condition that $V_x = 0$ at 0 $y = \pm B$, we get

$$V_x = \left[\frac{1}{K}\left(\frac{dP}{dx}\right)\right]^{1/n}\left(\frac{n}{n+1}\right)\left[1-\left(\frac{y}{B}\right)^{\frac{1+n}{n}}\right]$$

The velocity is maximum at the centreline, $y = 0$:

$$V_{max} = \left[\frac{1}{K}\left(\frac{dP}{dx}\right)\right]^{1/n}\left(\frac{n}{n+1}\right)$$

Thus,

$$\frac{V_x}{V_{max}} = \left[1-\left(\frac{y}{B}\right)^{\frac{1+n}{n}}\right]$$

8.4.3 Example 3

A Bingham plastic fluid has a yield stress of 15 Pa and Bingham viscosity of 150 mPa.s. It is flowing under laminar condition in a 40 mm diameter pipe that is 200 m long. The pressure gradient applied is 3200 Pa/m. Calculate the flow rate of the fluid.

Solution:

The wall stress is:

$$\tau_w = \frac{R}{2}\left(\frac{\Delta p}{L}\right) = \frac{20\times10^{-3}}{2}\times 3200 = 32\ Pa$$

Thus,

$$\varphi = \tau_0/\tau_w = 15/32 = 0.46875$$

$$Q = \frac{\pi R^4}{8\eta_B}\left(\frac{\Delta p}{L}\right)\left[1-\frac{4}{3}\varphi+\frac{\varphi^4}{3}\right]$$

$$Q = \frac{\pi(20\times10^{-3})^4}{8\times150\times10^{-3}}(3200)\left[1-\frac{4}{3}(0.46875)+\frac{(0.46875)^4}{3}\right] = 5.2423\times10^{-4}\ m^3/s$$

8.4.4 Example 4

A power law fluid is flowing in a smooth horizontal tube having an inside diameter of 50 mm and a length of 3 m at an average velocity of 1 m/s. The power law constants are: $K = 0.5\ \mathrm{Pa.s^n}$ and $n = 0.645$. The density of the fluid is 1100 kg/m^3. Calculate the friction loss in J/kg and pressure drop in Pa.

Solution:

Given: $D = 50\ mm,\quad L = 3m,\quad \bar{V} = 1\ m/s,\quad K = 0.5\ \mathrm{Pa.s^n},\quad n = 0.645,\quad \rho = 1100\ kg/m^3$

$$Re_n = \frac{\rho D^n \bar{V}^{2-n}}{(8)^{n-1}K\left(\frac{1+3n}{4n}\right)^n} = \frac{1100\left(50\times10^{-3}\right)^{0.645}(1)^{2-0.645}}{(8)^{0.645-1}\times0.5\times\left(\frac{1+3\times0.645}{4\times0.645}\right)^{0.645}} = 613.4$$

$$Re_{n,critical} = \frac{2100(4n+2)(5n+3)}{3(3n+1)^2} = \frac{2100 \times (4 \times 0.645 + 2)(5 \times 0.645 + 3)}{3(3 \times 0.645 + 1)^2} = 2316.8$$

Since $Re_n < Re_{n,critical}$, the flow is laminar. Thus,

$$f = \frac{16}{Re_n} = \frac{16}{2316.8} = 0.0261$$

The friction loss h_f is

$$h_f = 4f\frac{L}{D}\left(\frac{\bar{V}^2}{2}\right) = 4 \times 0.0261 \times \frac{3}{50 \times 10^{-3}} \times \frac{1^2}{2} = 3.132\ J/kg$$

For horizontal tube,

$$h_f = \frac{\Delta P}{\rho};\quad \Delta P = h_f \rho = 3.132 \times 1100 = 3445.2\ Pa$$

8.4.5 Example 5

A polymeric solution following power-law behaviour is being discharged at a flow rate of 0.3 m^3/s from a large tank open to atmosphere through a series of three horizontal hydraulically smooth pipes as shown in the figure. The dimensions of the pipes are as follows:

Pipe 1: $L_1 = 25\ m$, $D_1 = 25\ cm$

Pipe 2: $L_2 = 35\ m$, $D_2 = 20\ cm$

Pipe 3: $L_3 = 35\ m$, $D_3 = 15\ cm$

The power-law constants are: $K = 0.05\ Pa.s^n$ and $n = 0.65$. The density of the fluid is 1000 kg/m^3. Estimate the height of the water level H in the tank. You can ignore minor losses.

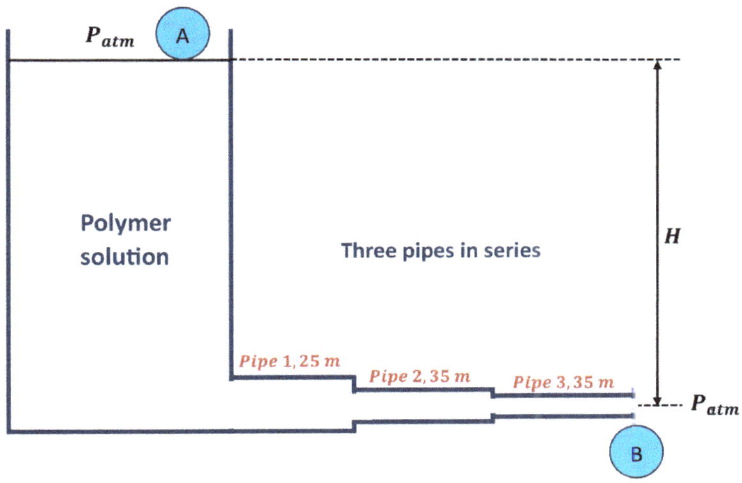

Solution:

Apply engineering Bernoulli equation, Equation (6.49), between sections A and B (see figure):

$$\frac{p_A}{\rho} + gz_A + \frac{1}{2}\alpha_A \bar{V}_A^2 = \frac{p_B}{\rho} + gz_B + \frac{1}{2}\alpha_B \bar{V}_B^2 + h_f + W_{sh}$$

Since $p_A = p_B = p_{atm}$, $\bar{V}_A = 0$, $\alpha_B \cong 1, W_{sh} = 0$, the Bernoulli equation reduces to:

$$g(z_A - z_B) = gH = \frac{1}{2}\bar{V}_B^2 + h_f$$

Thus,

$$H = \frac{1}{g}\left[h_f + \frac{1}{2}\bar{V}_B^2\right]$$

$$\bar{V}_B = \frac{4Q}{\pi D_3^2} = \frac{4 \times 0.3}{\pi \times 0.15^2} = 16.9765\ m/s$$

$$H = \frac{1}{g}\left[h_f + \frac{1}{2}\bar{V}_B^2\right] = \frac{1}{9.81}\left[h_f + \frac{1}{2} \times 16.9765^2\right] = \frac{1}{9.81}\left[h_f + 144.10\right]\ m$$

Estimation of friction loss, h_f:

The major friction loss consists of the following: (1) friction loss in pipe 1, (2) friction loss in pipe 2, and (3) friction loss in pipe 3.

(1) friction loss in pipe 1

$$\bar{V}_1 = \frac{4Q}{\pi D_1^2} = \frac{4 \times 0.3}{\pi \times 0.25^2} = 6.11\ m/s$$

$$Re_{n,1} = \frac{\rho D_1^n \bar{V}_1^{2-n}}{(8)^{n-1} K \left(\frac{1+3n}{4n}\right)^n} = \frac{1000(0.25)^{0.65}(6.11)^{2-0.65}}{(8)^{0.65-1} \times 0.05 \times \left(\frac{1+3 \times 0.65}{4 \times 0.65}\right)^{0.65}} = 41{,}610.93$$

$$Re_{n,critical} = \frac{2100(4n+2)(5n+3)}{3(3n+1)^2} = \frac{2100 \times (4 \times 0.65 + 2)(5 \times 0.65 + 3)}{3(3 \times 0.65 + 1)^2} = 2312.55$$

Since $Re_{n,1} > Re_{n,critical}$, the flow is turbulent in pipe 1.

Use the following explicit form of friction factor equation, that is, Equation (8.23), to estimate friction factor:

$$f = \frac{\alpha}{Re_n^\beta}$$

where α and β are given as:

$$\alpha = 0.0077 \ln(n) + 0.078 = 0.0077 ln(0.65) + 0.078 = 0.0747$$

$$\beta = 0.25\,(n)^{-0.22} = 0.25(0.65)^{-0.22} = 0.275$$

$$f = \frac{\alpha}{Re_n^\beta} = \frac{0.0747}{(41610.93)^{0.275}} = 4.01 \times 10^{-3}$$

$$h_f = 4f\frac{L}{D}\left(\frac{\bar{V}^2}{2}\right) = 4(4.01\times 10^{-3})\left(\frac{25}{0.25}\right)\left(\frac{6.11^2}{2}\right) = 29.94\ \ J/kg$$

(2) friction loss in pipe 2

$$\bar{V}_2 = \frac{4Q}{\pi D_2^2} = \frac{4\times 0.3}{\pi\times 0.20^2} = 9.549\ \ m/s$$

$$Re_{n,2} = \frac{\rho D_2^n \bar{V}_2^{2-n}}{(8)^{n-1}K\left(\frac{1+3n}{4n}\right)^n} = \frac{1000(0.20)^{0.65}(9.549)^{2-0.65}}{(8)^{0.65-1}\times 0.05\times\left(\frac{1+3\times 0.65}{4\times 0.65}\right)^{0.65}} = 281{,}885.38$$

$$Re_{n,critical} = \frac{2100(4n+2)(5n+3)}{3(3n+1)^2} = \frac{2100\times(4\times 0.65+2)(5\times 0.65+3)}{3(3\times 0.65+1)^2} = 2312.55$$

Since $Re_{n,1} > Re_{n,critical}$, the flow is turbulent in pipe 2.

Use the following explicit form of friction factor equation, that is, Equation (8.23), to estimate friction factor:

$$f = \frac{\alpha}{Re_n^{\beta}}$$

where α and β are given as:

$$\alpha = 0.0077\ln(n) + 0.078 = 0.0077ln(0.65) + 0.078 = 0.0747$$

$$\beta = 0.25\,(n)^{-0.22} = 0.25(0.65)^{-0.22} = 0.275$$

$$f = \frac{\alpha}{Re_n^{\beta}} = \frac{0.0747}{(281885.38)^{0.275}} = 2.369\times 10^{-3}$$

$$h_f = 4f\frac{L}{D}\left(\frac{\bar{V}^2}{2}\right) = 4(2.369\times 10^{-3})\left(\frac{35}{0.20}\right)\left(\frac{9.549^2}{2}\right) = 75.602\ J/kg$$

(3) friction loss in pipe 3

$$\bar{V}_3 = \frac{4Q}{\pi D_3^2} = \frac{4\times 0.3}{\pi\times 0.15^2} = 16.9765\ \ m/s$$

$$Re_{n,3} = \frac{\rho D_3^n \bar{V}_3^{2-n}}{(8)^{n-1}K\left(\frac{1+3n}{4n}\right)^n} = \frac{1000(0.15)^{0.65}(16.9765)^{2-0.65}}{(8)^{0.65-1}\times 0.05\times\left(\frac{1+3\times 0.65}{4\times 0.65}\right)^{0.65}} = 508{,}410.13$$

$$Re_{n,critical} = \frac{2100(4n+2)(5n+3)}{3(3n+1)^2} = \frac{2100\times(4\times 0.65+2)(5\times 0.65+3)}{3(3\times 0.65+1)^2} = 2312.55$$

Since $Re_{n,1} > Re_{n,critical}$, the flow is turbulent in pipe 3.

Use the following explicit form of friction factor equation, that is, Equation (8.23), to estimate friction factor:

$$f = \frac{\alpha}{Re_n^{\beta}}$$

where α and β are given as:

$$\alpha = 0.0077\ln(n) + 0.078 = 0.0077ln(0.65) + 0.078 = 0.0747$$

$$\beta = 0.25\,(n)^{-0.22} = 0.25(0.65)^{-0.22} = 0.275$$

$$f = \frac{\alpha}{Re_n^{\beta}} = \frac{0.0747}{(508410.13)^{0.275}} = 2.014 \times 10^{-3}$$

$$h_f = 4f\frac{L}{D}\left(\frac{\bar{V}^2}{2}\right) = 4(2.014 \times 10^{-3})\left(\frac{35}{0.15}\right)\left(\frac{16.9765^{\,2}}{2}\right) = 270.9\,J/kg$$

Total friction loss, h_f:

$$h_f = 29.94 + 75.602 + 270.9 = 376.442\ J/kg$$

Height of the water level H in the tank:

$$H = \frac{1}{9.81}\left[h_f + 144.10\right] = \frac{1}{9.81}[376.442 + 144.10] = 53.06\ \ m$$

NOTATION

$A(n)$	function of n, defined in Equation (8.28)
Bi	Bingham number, see Equation (8.50)
D	pipe diameter
f	friction factor
f_L	friction factor in laminar flow of Bingham plastic fluid
f_T	friction factor given by Equation (8.54)
He	Hedstrom number
H	height
h_f	friction loss
K	consistency index of a power-law fluid
K'	consistency index, generalized power law fluid
K_y	consistency index, yield-pseudoplastic fluid
L	length
n	flow behaviour index of a power-law fluid
n'	flow behaviour index of a generalized power-law fluid, see Equation (8.74)
p	pressure
Q	volumetric flow rate
r	radial distance or position
R	radius of pipe

R_p	plug radius, Bingham plastic fluid
Re	Reynolds number
Re_B	Reynolds number based on Bingham viscosity η_B
$Re_{B,critical}$	critical Reynolds number for Bingham plastic fluid
Re_n	generalized Reynolds number for power-law fluids
$Re_{n,critical}$	critical generalized Reynolds number
$Re_{n'}$	generalized Reynolds number for any time-independent non-Newtonian fluid
$\bar{V}$	average velocity
V_x	x-component of velocity
V_y	y-component of velocity
V_z	z-component of velocity
V_{max}	maximum velcocity
V_p	velocity of plug
z	distance in z-direction, z-coordinate, or elevation

Greek Letters

α	parameter in Equations (8.24), (8.55), and (8.83)
β	parameter in Equations (8.25), (8.56), and (8.84)
$\dot{\gamma}$	shear rate
$\dot{\gamma}_w$	shear rate at the pipe wall
Δ	change in variable
η	viscosity
η_B	Bingham viscosity
ρ	density
τ	shear stress
τ_o	yield stress
τ_{rz}	shear stress
τ_w	wall shear stress
φ	ratio of yield stress to wall stress
φ_c	ratio of yield stress to critical wall stress

PRACTICE PROBLEMS

Problem 1

Consider two infinitely large parallel plates, a distance of $2B$ apart as shown in the figure below. The bottom plate is stationary. The top plate is moving at a constant velocity of V_o, thus contributing to motion. There is an additional contribution to fluid motion caused by applying a constant pressure gradient dP/dx (negative value). Assuming steady state, laminar flow of a power-law fluid, determine the velocity distribution and shear stress distribution in the fluid.

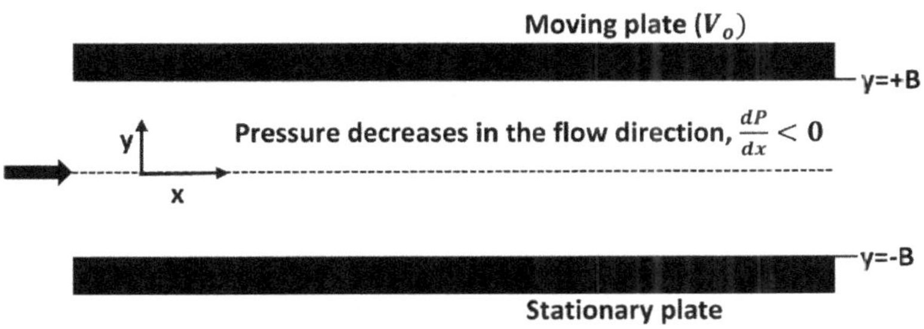

Problem 2

Consider steady laminar flow of a power-law liquid flowing down a flat inclined plate due to gravity, as shown in the figure below. The plate is inclined at an angle α with the horizontal. The length and width of the plate are L and W, respectively. The thickness of the liquid film is δ.

Determine the following:

i) Velocity distribution in the liquid film
ii) Maximum fluid velocity
iii) Average fluid velocity
iv) Volumetric flow rate of liquid in the film
v) Film thickness in terms of mass flow rate per unit width
vi) Drag force exerted by the fluid on the plate

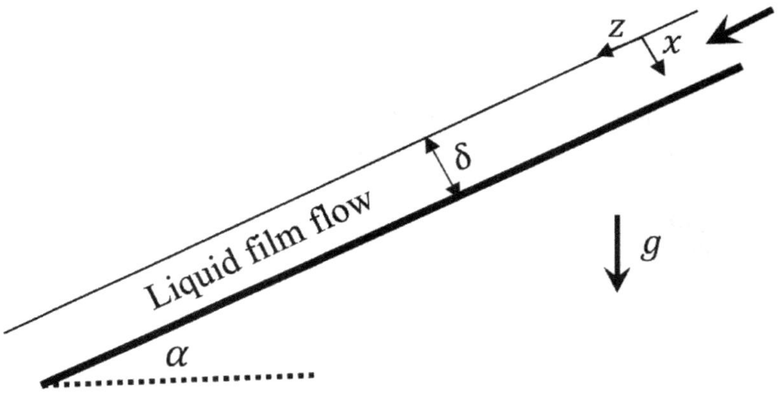

Problem 3

Consider steady laminar flow of a power-law fluid between two infinitely long horizontal concentric cylinders. The fluid motion is caused by the longitudinal motion of the inner cylinder at a constant velocity of V_o (see figure). The outer cylinder is stationary. There is no pressure-gradient applied.

Determine the following:

i) Velocity distribution in the fluid
ii) Volumetric flow rate of fluid
iii) Drag force on the inner cylinder per unit length

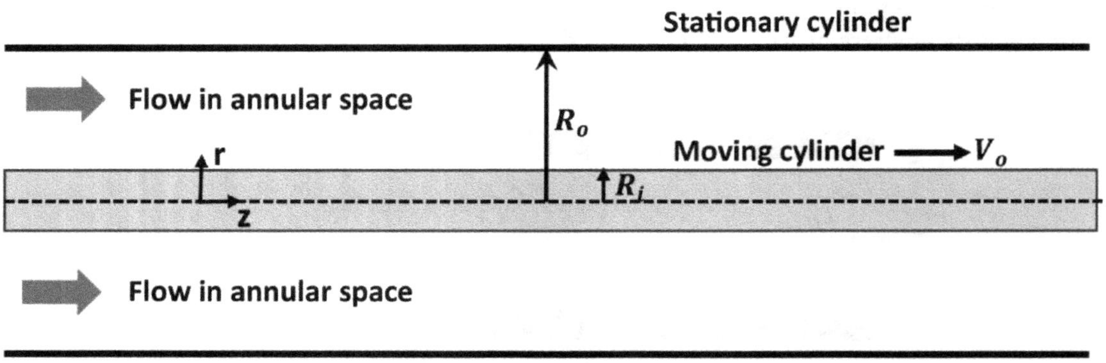

Problem 4

A power law fluid is being pumped from a large feed tank to a large, elevated tank as shown in the figure below. The flow rate of the fluid is 0.01 m^3/s. The pipe diameter is 0.1 m, and its total length is 200 m. The pipe is hydraulically smooth. The pump efficiency is 0.70. Calculate the kW of the pump required to do the job. The power law constants are: $K = 0.03\ Pa.s^n$ and $n = 0.70$. The fluid density is $\rho = 1000\ kg/m^3$. You can ignore minor losses.

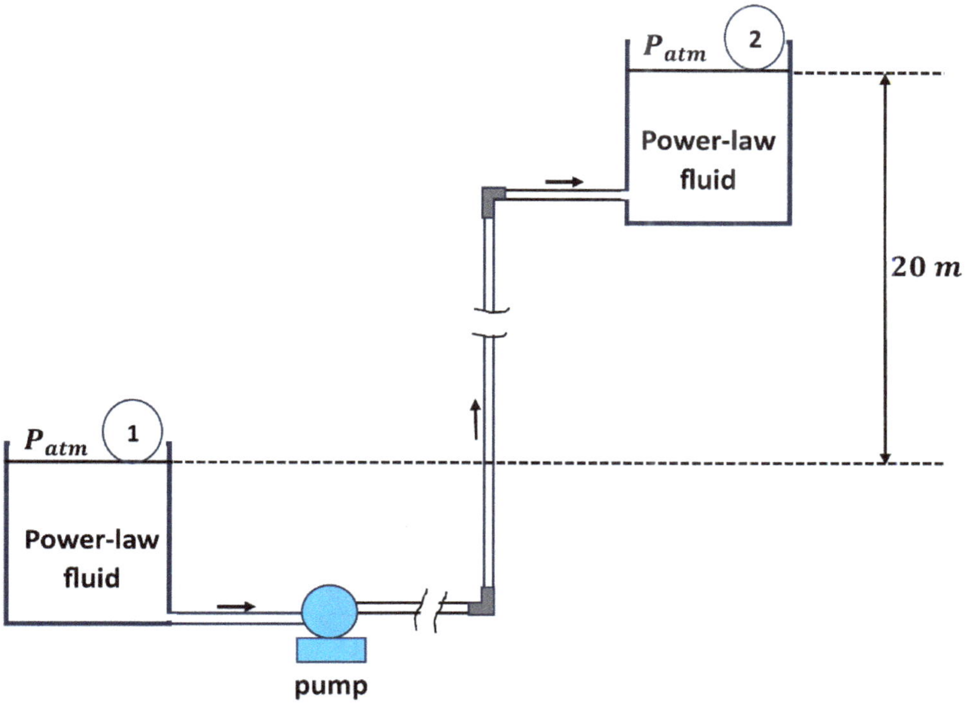

Problem 5

Consider the flow of a power-law fluid in the pipeline system shown in the figure below. The fluid flow rate of 0.03 m^3/s splits between the pipes 1 and 2. The dimensions of the pipes (hydraulically smooth) are as follows:

Pipe 1: $L = 3\ m$, $D = 0.05\ m$

Pipe 2: $L = 3\ m$, $D = 0.1$

The power-law constants are: $K = 0.015\ Pa.s^n$ and $n = 0.70$ and the fluid density is: $\rho = 1000\ \mathrm{kg/m^3}$.

i) Determine the flow rate in each pipe

ii) Calculate the pressure drop in the system.

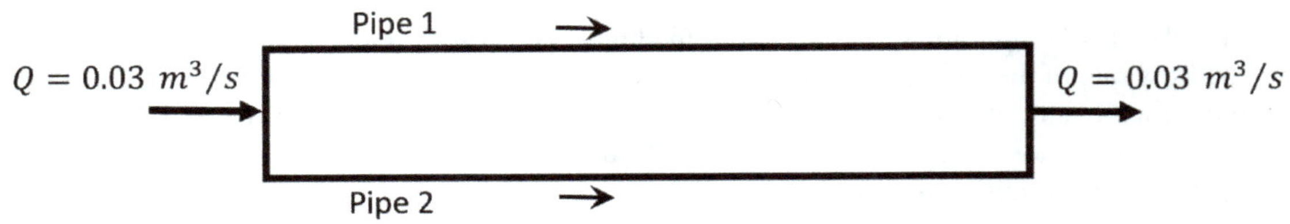

FURTHER READING

1. R.P. Chhabra, J.F. Richardson. *Non-Newtonian Flow in the Process Industries.* Butterworth Heinemann, Oxford, 1999.
2. R.B. Bird, W.E. Stewart, E.N. Lightfoot. *Transport Phenomena.* 2nd Edition, Wiley, New York, 2007.
3. C.J. Geankoplis, A.A. Hersel, D.H. Lepek. *Transport Processes and Separation Process Principles*. 5th Edition, Prentice Hall, New Jersey, 2018.
4. W.L. McCabe, J.C. Smith, P. Harriott. *Unit Operations of Chemical Engineering.* 6th Edition, McGraw Hill, New York, 2001.
5. R. Pal. Entropy generation in flow of highly concentrated non-Newtonian emulsions in smooth tubes. *Entropy* 2014, 16, 5178–5197.
6. S. Kinra, R. Pal. Pipe flow of suspensions of cellulose nanocrystals. *Fluids* 2023, 8, 275.
7. D.W. Dodge, A.B. Metzner. Turbulent flow of non-Newtonian systems. *AIChE. J.* 1959, 5, 189-204.
8. T.F. Irvine. A generalized Blasius equation for power-law fluids. *Chem. Eng. Communications* 1988, 65, 39-47.
9. P. Mishra, G. Tripathi. Transition from laminar to turbulent flow of purely viscous non-Newtonian fluids in tubes. *Chem. Eng. Sci.* 1971, 26, 915–921.

Chapter 9

Flow in Porous Media

9.1 INTRODUCTION

Porous medium is a material that consists of pores (voids, free space pockets) and solid phase. The pores may be occupied by a fluid (gas, oil, water etc.). A large variety of natural and synthetic materials are porous in nature. Examples include underground oil reservoirs, ceramics, solid foams, sand filters, wood, packed bed of particles used widely in chemical engineering applications. The pores of a porous medium usually form a three-dimensional inter-connected network and therefore, fluids can flow through the porous medium. If all pores of a porous medium are inter-connected, then the porosity (ε) of the porous medium is simply the fraction of the total volume of the medium that is occupied by the pores. The fraction of the total volume that is occupied by the solid phase is $(1-\varepsilon)$. When some pores are isolated or disconnected or have dead ends, then the effective porosity, defined as the ratio of connected void volume to total volume of the medium, is lower than the total porosity.

Porous media could be classified as consolidated or unconsolidated. In consolidated porous medium, the solid phase is basically a single piece of material, or the grains of the solid phase are cemented or fused together to form a single piece of solid phase. In unconsolidated porous medium, on the other hand, the grains or the particles of the solid phase are not cemented together and therefore, the porous medium is a multi-particle system like a packed bed of individual (un-cemented, un-glued) particles. Flow of single-phase Newtonian fluids (gas, water, oil) through packed beds and consolidated porous media is important from a practical point of view.

9.2 PRESSURE-LOSS IN FLOW THROUGH PACKED BEDS (UNCONSOLIDATED POROUS MEDIA)

Consider flow of a Newtonian fluid through a horizontal cylindrical packed bed of particles with bed diameter D_b and bed length L_b. As flow through a packed bed is quite complex, a rigorous theoretical derivation of pressure drop versus flow rate relationship is not possible. Only approximate models have been developed. In one approach, used widely to model flow through packed bed of particles, the packed bed is visualized as a bundle of identical capillary tubes. In its simplest form, the capillary tube bundle model assumes that the capillary tubes are straight, cylindrical of constant cross-section (uniform radius), and parallel (all oriented in the same direction), as shown in Figure 9.1. The average velocity in any capillary tube is the same as that in the packed bed. The capillary diameter is equal to the hydraulic diameter of the bed, d_H, defined as:

$$d_H = 4\left(\frac{volume\ of\ voids}{total\ wetted\ surface\ area}\right) = \left(\frac{4\varepsilon}{1-\epsilon}\right)\left(\frac{V_p}{S_p}\right) \tag{9.1}$$

where ε is the void fraction or porosity, V_p is the volume of a single particle, and S_p is the surface area of a single particle. In writing the above expression for d_H, it is assumed that all the particles are identical and that the wetted surface of the cylindrical container walls of the bed is negligible as compared with the total wetted surface area of the particles. For a bed of identical spherical particles, $V_p/S_p = d_p/6$, where d_p is the particle diameter. Thus, the hydraulic diameter becomes:

$$d_H = \left(\frac{2}{3}\right)\left(\frac{\varepsilon}{1-\epsilon}\right) d_p \tag{9.2}$$

The superficial velocity of fluid (V_s) is defined is:

$$V_s = \frac{Q}{A_b} \tag{9.3}$$

where Q is the volumetric flow rate of fluid and A_b is the total cross-sectional area of the bed. Thus V_s is the velocity of fluid in the bed if no particles were present in the bed. The average velocity $\bar{V}$of fluid in the bed, also called interstitial velocity, is defined as:

$$\bar{V} = \frac{Q}{A_{flow}} \tag{9.4}$$

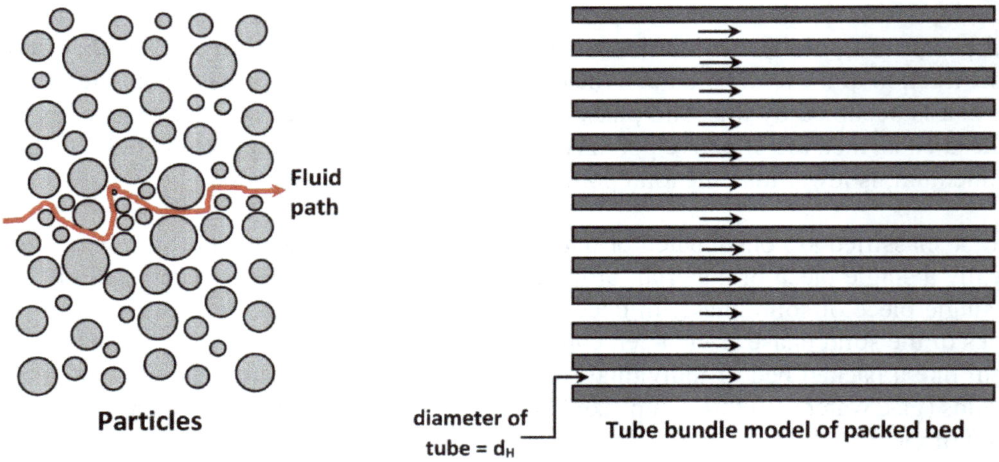

Figure 9.1 Path taken by fluid in a bed of particles and tube bundle model of packed bed.

where A_{flow} is the cross-sectional area of bed through which the fluid flows. The porosity of the bed ε is given as:

$$\varepsilon = \frac{A_{flow}}{A_b} \tag{9.5}$$

Therefore, the average or interstitial velocity of fluid in the bed is:

$$\bar{V} = \frac{V_s}{\varepsilon} \tag{9.6}$$

This expression of average velocity does not consider the tortuous path taken by fluid in the bed. Due to tortuosity of the bed, the actual average velocity of fluid in the bed is larger than that given by Equation (9.6). The tortuosity τ of a bed is defined as:

$$\tau = \frac{L_e}{L_b} \tag{9.7}$$

where L_e is average length of the tortuous path taken by the fluid and L_b is the straight length of the bed. To account for the tortuosity, the actual average velocity of the fluid in the bed can be expressed as follows:

$$\bar{V} = \tau \frac{V_s}{\varepsilon} \tag{9.8}$$

In laminar flow of a Newtonian fluid through a cylindrical tube, the pressure gradient in the direction of flow is given as:

$$-\left(\frac{dp}{dx}\right) = 32\left(\frac{\eta \bar{V}}{D^2}\right) \tag{9.9}$$

where η is the fluid viscosity and D is the tube diameter. Replacing D with hydraulic diameter d_H and the constant factor of 32 with C, the pressure-gradient in a capillary tube model of the packed-bed model can be expressed as:

$$-\left(\frac{dp}{dx}\right) = C\left(\frac{\eta \bar{V}}{d_H^2}\right) \tag{9.10}$$

Note that flow in the bed is assumed to be *laminar* here. The constant C is expected to be larger than 32, as the path of fluid in the bed is not straight. The fluid follows a tortuous path in the bed and consequently, the pressure drop over a certain straight length is expected to be more than that observed over the same length if the fluid path was non-tortuous. Therefore, the constant C is taken as 32τ where τ is the tortuosity. Upon substitution of the expressions for d_H and $\bar{V}$, and taking $C = 32\tau$, Equation (9.10) gives:

$$-\left(\frac{dp}{dx}\right) = 72\,\tau^2\left(\frac{\eta V_s}{d_p^2}\right)\frac{(1-\varepsilon)^2}{\varepsilon^3} \tag{9.11}$$

This equation assumes that the cross-section of the representative flow passage (capillary tube) in the porous medium is circular and constant. In reality the fluid moves through converging-diverging flow passages of non-uniform cross-sections. In converging-diverging flows, fluid experiences stretching or extensional deformation. The elongation or stretching of fluid elements results in additional dissipation of energy and pressure drop. Thus, Equation (9.11) needs to be modified further as follows:

$$-\left(\frac{dp}{dx}\right) = 72\,\tau^2 K_o\left(\frac{\eta V_s}{d_p^2}\right)\frac{(1-\varepsilon)^2}{\varepsilon^3} \tag{9.12}$$

where K_o is an empirical factor that considers the influence of non-constant cross-section of flow passage on pressure-gradient. The tortuosity τ is often taken to be $\sqrt{2}$ for random packing of uniform spheres. It is generally a function of the porosity of the porous medium. For example, the following model, based on the Maxwell equation for electrical conductivity of composite, is often used to describe the relationship between tortuosity and porosity:

$$\tau = 1.5 - 0.5\varepsilon \tag{9.13}$$

When $\tau = \sqrt{2}$ and $K_o = 25/24$, the following well-known Blake-Kozeny equation for laminar flow through randomly packed bed of uniform spheres is obtained:

$$-\left(\frac{dp}{dx}\right) = 150\left(\frac{\eta V_s}{d_p^2}\right)\frac{(1-\varepsilon)^2}{\varepsilon^3} \tag{9.14}$$

When $\tau = \sqrt{2}$ and $K_o = 5/4$, the following Carman-Kozeny equation, another well-known equation for laminar flow through randomly packed bed of uniform spheres, is obtained:

$$-\left(\frac{dp}{dx}\right) = 180\left(\frac{\eta V_s}{d_p^2}\right)\frac{(1-\varepsilon)^2}{\varepsilon^3} \tag{9.15}$$

Both Blake-Kozeny and Carman-Kozeny equations are popular in the literature although they use different values of the factor K_o. The observed difference in K_o values are probably related to differences in particle-shape, surface roughness of particles, and porosity of bed. Note that these equations are restricted to only laminar flow through bed of nearly uniform spheres. The packed-bed Reynolds number Re_b , defined below, should be less than 10 for flow to be laminar in the bed.

$$Re_b = \left(\frac{1}{1-\varepsilon}\right)\frac{\rho d_p V_s}{\eta} \tag{9.16}$$

For turbulent flow through packed beds ($Re_b > 1000$), the following Burke-Plummer equation is often used:

$$-\left(\frac{dp}{dx}\right) = 1.75\left(\frac{\rho V_s^2}{d_p}\right)\frac{(1-\varepsilon)}{\varepsilon^3} \tag{9.17}$$

In the transition region, the following equation, obtained by superposition of Blake-Kozeny and Burke-Plummer equations, proposed by Ergun is widely used:

$$-\left(\frac{dp}{dx}\right) = 150\left(\frac{\eta V_s}{d_p^2}\right)\frac{(1-\varepsilon)^2}{\varepsilon^3} + 1.75\left(\frac{\rho V_s^2}{d_p}\right)\frac{(1-\varepsilon)}{\varepsilon^3} \tag{9.18}$$

As the Ergun equation, Equation (9.18), is obtained by the superposition of laminar and turbulent expressions, it is valid over the full range of the packed-bed Reynolds number Re_b. The Ergun equation could also be re-cast in the following form:

$$f_b = \frac{150}{Re_b} + 1.75 \tag{9.19}$$

where f_b is the packed-bed friction factor defined as:

$$f_b = \frac{-\left(\frac{dp}{dx}\right)}{\left(\frac{\rho V_s^2}{d_p}\right)\frac{(1-\varepsilon)}{\varepsilon^3}} \tag{9.20}$$

The Ergun equation, Equation (9.18) or (9.19), is used extensively in the literature to describe pressure loss in packed beds. However, the following points should be kept in mind when using the Ergun equation: a) it is applicable to unconsolidated beds of nearly uniform-size spherical particles with appreciable roughness. For smooth spheres, it tends to over-predict f_b in the high Re_b range ($Re_b > 700$); b) if the bed particles are non-uniform in sizes, the Sauter mean diameter of particles should be used in the application of the Ergun equation; c) when particle shape deviates significantly from a sphere, the Ergun equation tends to under-predict f_b ; and d) wall effects can be important when $D_b/d_p < 10$. When wall effects are important, the experimental data show deviation from the predictions of the Ergun equation.

Figure 9.2 shows the prediction of the packed bed friction factor as a function of packed bed Reynolds number using the Ergun equation, Equation (9.19). In the limits of low Re_b and high Re_b , the Ergun equation predictions overlap with the predictions of the Blake-Kozeny equation (low Re_b) and the Burke-Plummer equation (high Re_b), as expected.

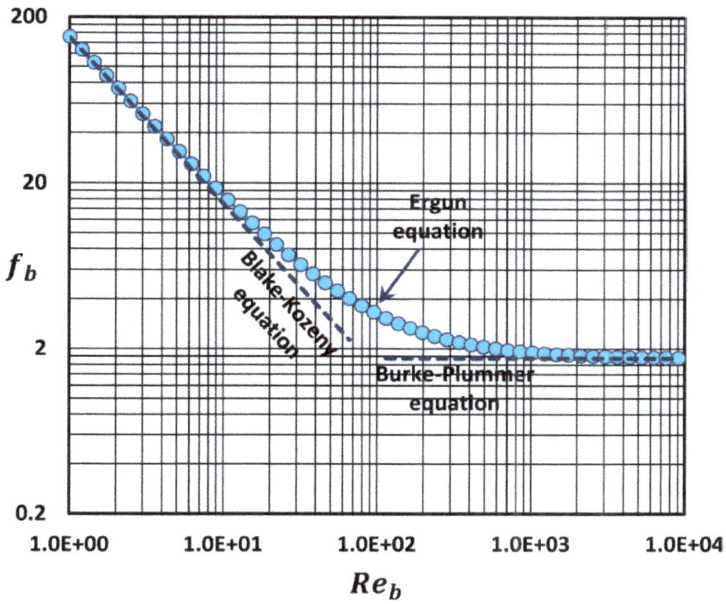

Figure 9.2 Prediction of packed bed friction factor as a function of packed bed Reynolds number. The data points are generated using the Ergun equation (Equation (9.19)).

9.3 PRESSURE-LOSS IN FLOW THROUGH CONSOLIDATED POROUS MEDIA

Laminar flow in consolidated porous media (see Figure 9.3) is usually described by Darcy's law given as follows:

$$-\frac{dp}{dx} = \frac{\eta}{k} V_s \tag{9.21}$$

where k is the permeability of the porous medium. Permeability is a measure of the ability of porous medium to conduct fluid. Higher k means lower resistance to flow and consequently, larger flow rate for the same pressure gradient. k has the dimensions of $(\text{length})^2$. It is usually expressed in darcies. For example, 1 darcy = 1 (cm/s)(centipoise)/(atm/cm) = 9.87×10^{-13} m^2.

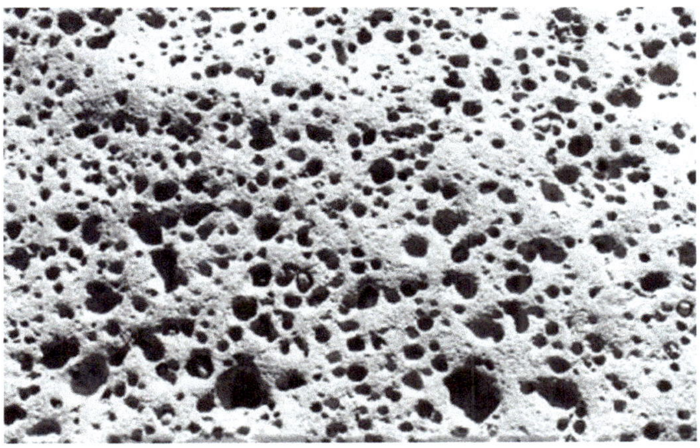

Figure 9.3 Consolidated porous medium (adapted from www.fotosearch.com.au).

Darcy's law is equally applicable to unconsolidated porous medium such as packed beds of free (un-cemented/un-fused) particles. Thus, the Blake-Kozeny equation or the Carman-Kozeny equation are special cases of the Darcy law. Upon comparison of the Darcy law with Blake-Kozeny/Carman-Kozeny equations, the permeability of a packed bed of uniform size spherical particles can be expressed as follows:

$$k = \frac{d_p^2 \varepsilon^3}{150(1-\varepsilon)^2} \tag{9.22}$$

$$k = \frac{d_p^2 \varepsilon^3}{180(1-\varepsilon)^2} \tag{9.23}$$

Equation (9.22) gives permeability based on Blake-Kozeny equation whereas Equation (9.23) gives permeability based on Carman/Kozeny equation.

Darcy's law is restricted to flows where viscous forces dominate over the inertial forces. At high flow rates, the flow in the porous medium becomes turbulent and Darcy's law is no longer valid. The pressure-gradient in the turbulent regime is much higher compared with that in the laminar regime at the same superficial velocity V_s. To extend the Darcy law to turbulent regime, Forchheimer modified the Darcy law by adding a quadratic term as follows:

$$-\frac{dp}{dx} = \frac{\eta}{k} V_s + \beta \rho V_s^2 \tag{9.24}$$

where the first term on the right side of the equation reflects the viscous effects and the second term reflects the inertial effects. β is called the non-Darcy flow coefficient with a dimension of 1/length. Equation (9.24) could be re-written in dimensionless form as:

$$f_{pm} = \frac{-\frac{dp}{dx}}{\beta \rho V_s^2} = 1 + \frac{1}{Re_{pm}} \tag{9.25}$$

where f_{pm} is the friction factor for flow in consolidated porous medium and Re_{pm} is the porous-medium Reynolds number, also referred to as Forchheimer number, defined as:

$$Re_{pm} = \frac{\rho V_s \beta k}{\eta} \tag{9.26}$$

At low Reynolds number ($Re_{pm} < 0.1$), the first term on the right side of the dimensionless Forchheimer equation, Equation (9.25), can be neglected and it reduces to Darcy's law. At high Reynolds number ($Re_{pm} > 10$), the second term on the right side of the dimensionless Forchheimer equation, Equation (9.25), can be neglected and it reduces to $f_{pm} = 1$. Figure 9.4 shows the prediction of f_{pm} as a function of Re_{pm} using the Forchheimer equation, Equation (9.25).

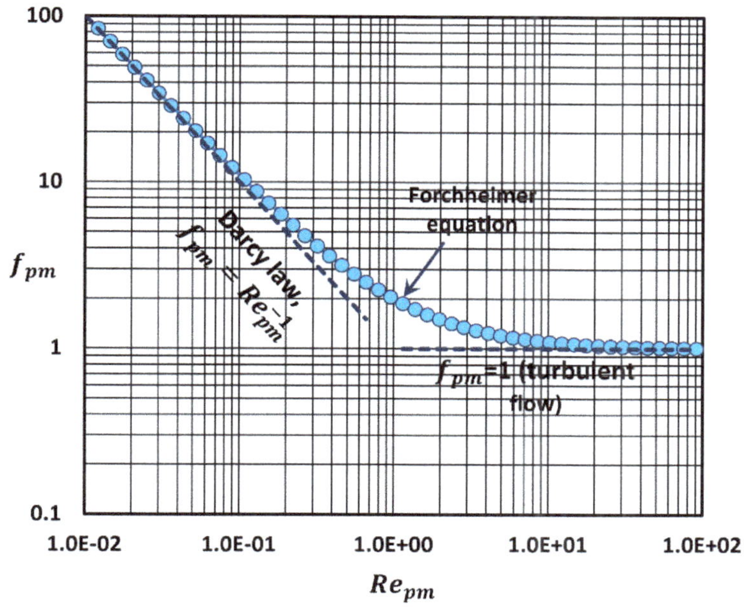

Figure 9.4 Prediction of porous medium friction factor as a function of porous medium Reynolds number. The data points are generated using the Forchheimer equation (Equation (9.25)).

To apply the Forchheimer equation, the value of non-Darcy flow coefficient is needed. It can be determined experimentally. The experimental procedure to determine β involves two steps: In the first step, the permeability k of the porous sample is determined from Darcy's law by restricting the experiments to low flow rates, and in the second step, the experiments are conducted at high flow rates and β is evaluated directly from the Forchheimer equation from the knowledge of pressure drop versus flow rate. The non-Darcy flow coefficient decreases with the increase in the permeability of the porous medium.

The Forchheimer equation is equally applicable to unconsolidated porous medium such as a packed bed of particles. Ergun equation is a special case of the Forchheimer equation. The Forchheimer equation, Equation (9.24) reduces to the Ergun equation, Equation (9.18), when permeability k and non-Darcy flow coefficient β are replaced by the following expressions:

$$k = \frac{d_p^2 \varepsilon^3}{150(1-\varepsilon)^2} \text{ and } \beta = \frac{7}{4}\left(\frac{1-\varepsilon}{d_p \varepsilon^3}\right) \tag{9.27}$$

The relationships between f_{pm} and f_b, and Re_{pm} and Re_b are as follows:

$$f_b = 1.75\, f_{pm} \tag{9.28}$$

$$Re_b = \left(\frac{600}{7}\right) Re_{pm} \tag{9.29}$$

9.4 FLOW OF NON-NEWTONIAN FLUIDS IN POROUS MEDIA

Consider a time-independent, purely viscous non-Newtonian fluid characterized by the following rheological model:

$$\dot{\gamma} = f(\tau) \tag{9.30}$$

where $\dot{\gamma}$ is the shear rate and τ is the shear stress.

To model the flow of such non-Newtonian fluids through a porous medium such as packed bed of particles, the packed bed is, once again, visualized as a bundle of identical capillary tubes. The capillary diameter is equal to the "hydraulic diameter of the bed", defined earlier.

9.4.1 Laminar Flow of Non-Newtonian Fluids in Porous Media

In steady laminar flow, the shear stress at the capillary wall can be expressed in the form of a power-law expression as:

$$\tau_w = K'(\Gamma_w)^{n'} \tag{9.31}$$

where Γ_w is the nominal shear rate at the capillary wall, and K' and n' are power-law parameters that may vary with the nominal wall shear rate Γ_w. The nominal shear rate at the capillary wall is given as:

$$\Gamma_w = \frac{8\bar{V}}{d_H} \tag{9.32}$$

This expression of nominal shear rate at the wall of the capillary assumes that the capillary has a circular cross-section. To account for deviation of the cross-section shape from circular, the nominal wall shear rate is expressed as:

$$\Gamma_w = K_o \frac{8\bar{V}}{d_H} \tag{9.33}$$

where $K_o = 1$ for circular cross-section. For packed-bed of uniform-size spherical particles, K_o is 5/4. From force balance, the average shear-stress at the wall of a capillary tube is as follows:

$$\tau_w = \frac{d_H}{4}\left(\frac{\Delta p}{L_e}\right) = \frac{d_H}{4\tau}\left(\frac{\Delta p}{L}\right) \tag{9.34}$$

Substituting the expression for d_H (Equation (9.2)) into the expressions for nominal wall shear rate and wall stress, and using Equation (9.8) relating $\bar{V}$ to V_s, we obtain:

$$\Gamma_w = 12K_o\tau\left(\frac{1-\varepsilon}{\varepsilon^2}\right)\left(\frac{V_s}{d_p}\right) \tag{9.35}$$

$$\tau_w = \frac{1}{6}\left(\frac{\varepsilon}{1-\varepsilon}\right)\frac{d_p}{\tau}\left(\frac{\Delta p}{L}\right) \tag{9.36}$$

Thus, the power-law expression gives:

$$\frac{1}{6}\left(\frac{\varepsilon}{1-\varepsilon}\right)\frac{d_p}{\tau}\left(\frac{\Delta p}{L}\right) = K'\left[12K_o\tau\left(\frac{1-\varepsilon}{\varepsilon^2}\right)\left(\frac{V_s}{d_p}\right)\right]^{n'} \tag{9.37}$$

Upon re-arrangement, this expression could be re-cast as:

$$f_b = \frac{\left(\frac{\Delta p}{L}\right)}{\left(\frac{\rho V_s^2}{d_p}\right)\left(\frac{1-\varepsilon}{\varepsilon^3}\right)} = \frac{72K_o\tau^2}{\left[\frac{\rho d_p^{n'} V_s^{2-n'}}{K'(1-\varepsilon)^{n'}}\right]\left[\frac{12K_o\tau}{\varepsilon^2}\right]^{1-n'}} \tag{9.38}$$

If we assume that K_o is 5/4 and $\tau = \sqrt{2}$, Equation (9.38) reduces to

$$f_b = \frac{180}{Re_{n',b}} \tag{9.39}$$

where the generalized Reynolds number for the packed bed is as follows:

$$Re_{n',b} = \left[\frac{\rho d_p^{n'} V_S^{2-n'}}{K'(1-\varepsilon)^{n'}}\right]\left[\frac{15\sqrt{2}}{\varepsilon^2}\right]^{1-n'} \tag{9.40}$$

This friction factor relation is valid for steady, laminar flow of any time-independent, purely viscous non-Newtonian fluid through packed bed of particles.

For truly power-law fluids characterized by the following rheological model:

$$\tau = K\dot{\gamma}^n \tag{9.41}$$

where K and n are constants independent of the shear rate, the following relationships hold:

$$n' = n \tag{9.42}$$

$$K' = K\left[\frac{3n+1}{4n}\right]^n \tag{9.43}$$

Consequently, the friction factor relation becomes:

$$f_b = \frac{180}{Re_{n,b}} \tag{9.44}$$

where the $Re_{n,b}$ is the generalized Reynolds number for power-law fluids, defined as follows:

$$Re_{n,b} = \left[\frac{\rho d_p^n V_S^{2-n}}{K(1-\varepsilon)^n}\right]\left(\frac{4n}{3n+1}\right)^n\left[\frac{15\sqrt{2}}{\varepsilon^2}\right]^{1-n} \tag{9.45}$$

For Bingham-plastic fluids, the nominal wall shear rate Γ_w is related to the wall stress as follows:

$$\Gamma_w = \frac{\tau_w}{\eta_B}\left(1 - \frac{4}{3}\varphi + \frac{\varphi^4}{3}\right) \tag{9.46}$$

Substituting the expressions for Γ_w and τ_w for flow in a packed bed of particles, we obtain:

$$12K_o\tau\left(\frac{1-\varepsilon}{\varepsilon^2}\right)\left(\frac{V_s}{D_p}\right) = \frac{1}{6}\left(\frac{\varepsilon}{1-\varepsilon}\right)\frac{d_p}{\tau}\left(\frac{\Delta p}{L}\right)\frac{\left(1-\frac{4}{3}\varphi+\frac{\varphi^4}{3}\right)}{\eta_B} \tag{9.47}$$

Using K_o= 5/4 and $\tau = \sqrt{2}$ and upon re-arrangement, Equation (9.470 gives:

$$\frac{\left(\frac{\Delta p}{L}\right)}{\left(\frac{\rho V_S^2}{d_p}\right)\left(\frac{1-\varepsilon}{\varepsilon^3}\right)} = \frac{180}{\left(1-\frac{4}{3}\varphi+\frac{\varphi^4}{3}\right)\left(\frac{\rho d_p V_S}{\eta_B(1-\varepsilon)}\right)} \tag{9.48}$$

As the left-hand side of Equation (9.48) is f_b, we can re-cast Equation (9.48) as follows:

$$f_b = \left[\frac{180}{Re_{B,b}\, F(\varphi)}\right] \tag{9.49}$$

where $Re_{B,b}$ is the Bingham Reynolds number for packed bed of particles, defined as:

$$Re_{B,b} = \frac{\rho d_p V_s}{\eta_B(1-\varepsilon)} \tag{9.50}$$

and $F(\varphi)$ is as follows:

$$F(\varphi) = \left(1 - \frac{4}{3}\varphi + \frac{\varphi^4}{3}\right) \tag{9.51}$$

Note that $\varphi = \tau_0/\tau_w$.

9.4.2 Turbulent Flow of Non-Newtonian Fluids in Porous Media

For turbulent flow of time-independent, purely viscous, non-Newtonian fluids through unconsolidated porous media, the following Ergun-type empirical correlation has been proposed in the literature:

$$f_b = \frac{\alpha}{Re_{n',b}} + \beta \tag{9.52}$$

where

$$f_b = \frac{\left(\frac{\Delta p}{L}\right)}{\left(\frac{\rho V_s^2}{d_p}\right)\left(\frac{1-\varepsilon}{\varepsilon^3}\right)} \tag{9.53}$$

and the generalized Reynolds number is defined as follows:

$$Re_{n',b} = \frac{\rho d_p V_s}{\eta_{eff}(1-\varepsilon)} \tag{9.54}$$

where

$$\eta_{eff} = K'\left[\frac{12V_s(1-\varepsilon)}{d_p\varepsilon^2}\right]^{n'-1} \tag{9.55}$$

Upon substituting the expression for effective viscosity in the expression for Reynolds number, we get:

$$Re_{n',b} = \left[\frac{\rho d_p^{n'} V_s^{2-n'}}{K'(1-\varepsilon)^{n'}}\right]\left[\frac{12}{\varepsilon^2}\right]^{1-n'} \tag{9.56}$$

Based on the experimental data, the following values of α and β are obtained: α=150 and β=1.75 (same values as in Ergun equation). Some authors have obtained slightly different values: α=160 and β=1.75. It should be noted that the definition of the generalized Reynolds number given in Equation (9.56) is a special case of the more general expression obtained earlier (see Equation (9.38)) as:

$$Re_{n',b} = \left[\frac{\rho d_p^{n'} V_s^{2-n'}}{K'(1-\varepsilon)^{n'}}\right]\left[\frac{12K_o\tau}{\varepsilon^2}\right]^{1-n'} \tag{9.57}$$

When $K_o = 1$ and $\tau = 1$, this expression reduces to Equation (9.56).

For power-law fluids of constant K and n, the following expression for generalized Reynolds number, obtained from the general expression Equation (9.57) with $K_o = 5/4$ and $\tau = \sqrt{2}$, is recommended:

$$Re_{n,b} = \left[\frac{\rho d_p^n V_s^{2-n}}{K(1-\varepsilon)^n}\right]\left[\frac{4n}{1+3n}\right]^n\left[\frac{15\sqrt{2}}{\varepsilon^2}\right]^{1-n} \tag{9.58}$$

The friction factor relation recommended for power-law fluids is:

$$f_b = \frac{150}{Re_{n,b}} + 1.75 \tag{9.59}$$

9.5 EXAMPLES

9.5.1 Example 1

A gas flows through a horizontal packed bed of spheres having a diameter of 1.25 cm. The bed void volume fraction is 0.38. The bed has a diameter of 1.25 m and a length of 2.5 m. The gas enters the bed at a mass flow rate of 1 kg/s. The properties of the gas are: $\rho = 1.1\ kg/m^3$ and $\eta = 1.9 \times 10^{-5} Pa.s$. Calculate the pressure drop of the gas in the bed.

Solution:

Refer to the figure below.

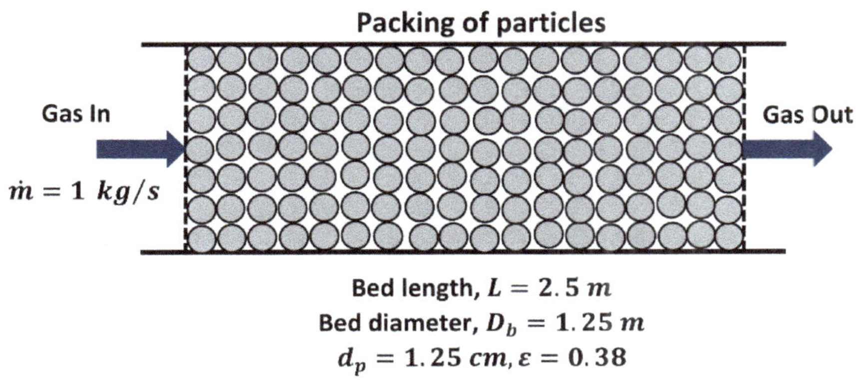

$$V_s = \frac{Q}{A_b} = \frac{\dot{m}/\rho}{\frac{\pi}{4}D_b^2} = \frac{1/1.1}{\frac{\pi}{4}(1.25^2)} = 0.7408\ m/s$$

$$Re_b = \left(\frac{1}{1-\varepsilon}\right)\frac{\rho d_p V_s}{\eta} = \frac{1}{(1-0.38)}\left[\frac{1.1 \times (1.25 \times 10^{-2}) \times 0.7408}{1.9 \times 10^{-5}}\right] = 864.68$$

Since $10 < Re_b < 1000$, we have a transition regime. Using Ergun equation,

$$f_b = \frac{150}{Re_b} + 1.75 = \frac{150}{864.68} + 1.75 = 1.9235$$

$$f_b = \frac{-\left(\frac{dp}{dx}\right)}{\left(\frac{\rho V_s^2}{d_p}\right)\frac{(1-\varepsilon)}{\varepsilon^3}} = \left(\frac{\Delta p}{L}\right)\left(\frac{\varepsilon^3}{1-\varepsilon}\right)\left(\frac{d_p}{\rho V_s^2}\right) = 1.9235$$

Note that Δp is pressure drop in the bed (positive value). Thus,

$$\left(\frac{\Delta p}{L}\right) = 1.9235\left(\frac{\rho V_s^2}{d_p}\right)\frac{(1-\varepsilon)}{\varepsilon^3} = 1.9235\left[\frac{1.1 \times 0.7408^2}{1.25 \times 10^{-2}}\right]\left[\frac{1-0.38}{0.38^3}\right] = 1049.55\ Pa/m$$

Therefore,

$$\Delta p = 1049.55\ \times 2.5 = 2623.88\ Pa$$

9.5.2 Example 2

Ion-exchange is commonly used for water softening. A water treatment unit consists of a packed bed filled with ion-exchange resin particles of diameter 2 mm ($d_p = 2\ mm$). The bed diameter (D_b) is 0.2 m and its length is 1.5 m. The void fraction ε of the bed is 0.38. The available pump limits the pressure drop over the bed to a maximum of 372.21 kPa. Determine the maximum flow rate of water that can be treated. The properties of water are: $\rho = 1000\ kg/m^3$ and $\eta = 1$ mPa. s.

Solution:

Refer to the figure below.

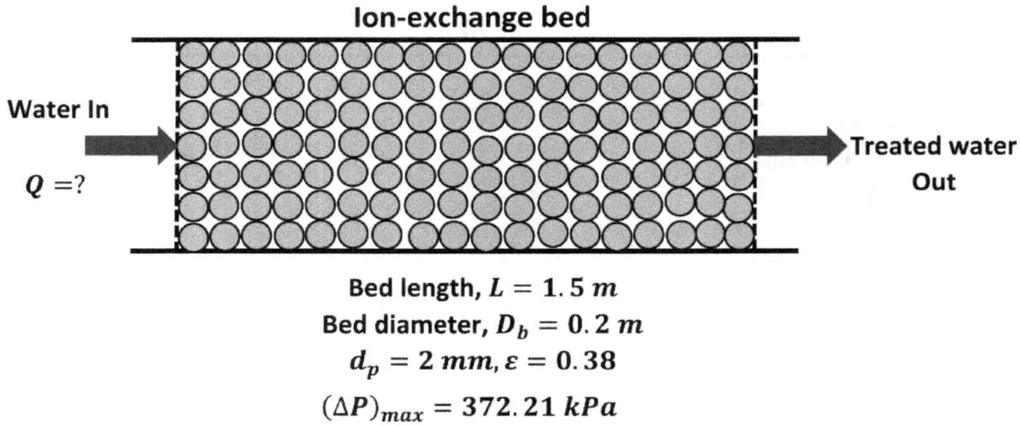

The following Ergun equation is applicable:

$$-\left(\frac{dp}{dx}\right) = 150\left(\frac{\eta V_s}{d_p^2}\right)\frac{(1-\varepsilon)^2}{\varepsilon^3} + 1.75\left(\frac{\rho V_s^2}{d_p}\right)\frac{(1-\varepsilon)}{\varepsilon^3}$$

Upon re-arrangement, we get:

$$1.75V_s^2 + \frac{150\eta(1-\varepsilon)}{\rho d_p}V_s - \left(\frac{\Delta p}{L}\right)\frac{d_p\varepsilon^3}{\rho(1-\varepsilon)} = 0$$

where Δp is pressure drop (upstream pressure – downstream pressure) in the bed. Thus, the unknown V_s can be calculated as:

$$V_s = \frac{-\frac{150\eta(1-\varepsilon)}{\rho d_p} \pm \sqrt{\left[\frac{150\eta(1-\varepsilon)}{\rho d_p}\right]^2 - 4 \times 1.75\left[-\left(\frac{\Delta p}{L}\right)\frac{d_p \varepsilon^3}{\rho(1-\varepsilon)}\right]}}{2 \times 1.75}$$

Substituting the given data, we get:

$$V_s = \frac{-\frac{150 \times 10^{-3}(1-0.38)}{1000 \times 2 \times 10^{-3}} \pm \sqrt{\left[\frac{150 \times 10^{-3}(1-0.38)}{1000 \times 2 \times 10^{-3}}\right]^2 - 4 \times 1.75\left[-\left(\frac{372.21 \times 10^3}{1.5}\right)\frac{2 \times 10^{-3} \times 0.38^3}{100(1-0.38)}\right]}}{2 \times 1.75}$$
$$= 0.1457\ m/s$$

Thus, the superficial velocity at a maximum pressure-drop of 372.21 kPa is 0.1457 m/s. Consequently,

$$Q = \frac{\pi}{4} D_b^2 V_s = \frac{\pi(0.2)^2 \times 0.1457}{4} = 0.00458\ \ m^3/s$$

9.5.3 Example 3

A core from an underground oil reservoir 11 cm long and 2.54 cm in diameter is placed in a core holder. Water flows through the core at a rate of 0.5 cm^3/s under a pressure drop of 0.9 atm. Determine the permeability of the porous core. The water viscosity is 1 mPa.s.

Solution:

The Darcy's law gives:

$$V_s = \frac{Q}{A} = \frac{k}{\eta}\left(\frac{\Delta p}{L}\right)$$
$$Q = 0.5\ cm^3/s\ ,\ \Delta p = 0.9\ atm,\ \ A = \pi(2.54)^2/4 = 5.067\ cm^2,\ \eta = 1\ cP, L = 11\ cm$$

Thus,

$$k = \eta\left(\frac{Q/A}{\Delta p/L}\right) = 1 \times \left(\frac{0.5/5.067}{0.9/11}\right) = 1.206 \text{ darcies}$$

9.5.4 Example 4

A power-law fluid (liquid) flows through a horizontal packed bed of spheres having a diameter of 1.25 cm. The bed void volume fraction is 0.38. The bed has a diameter of 1.25 m and a length of 2.5 m. The fluid enters the bed at a mass flow rate of 100 kg/s. The power-law constants of the fluid are: $K = 2.5\ Pa.s^n$ and $n = 0.60$. The density of the fluid is 1100 kg/m^3. Calculate the pressure drop of the fluid in the bed.

Solution:

Refer to the figure below.

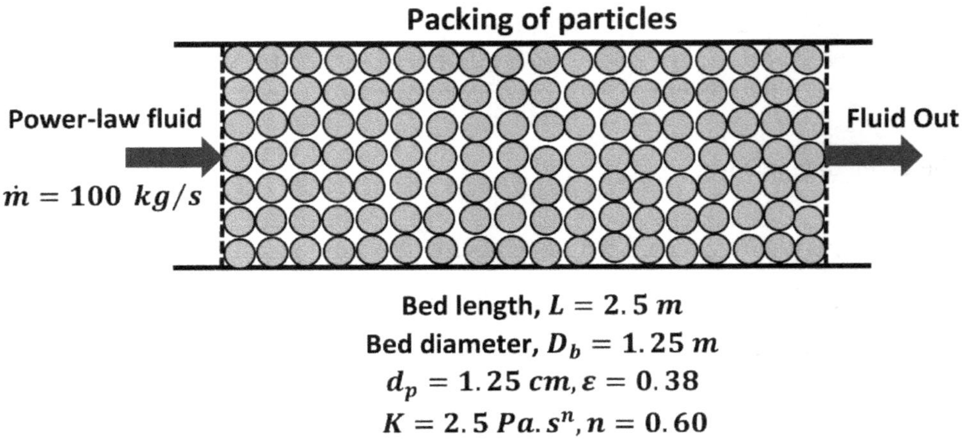

$$V_s = \frac{Q}{A_b} = \frac{\dot{m}/\rho}{\frac{\pi}{4}D_b^2} = \frac{100/1100}{\frac{\pi}{4}(1.25^2)} = 0.07408\ m/s$$

Using Equation (9.58), we get:

$$Re_{n,b} = \left[\frac{\rho d_p^n V_s^{2-n}}{K(1-\varepsilon)^n}\right]\left[\frac{4n}{1+3n}\right]^n\left[\frac{15\sqrt{2}}{\varepsilon^2}\right]^{1-n} = \left[\frac{1100\times(1.25\times10^{-2})^{0.60}(0.07408)^{2-0.60}}{2.5\times(1-0.38)^{0.60}}\right]\left[\frac{4\times0.60}{1+3\times0.60}\right]^{0.60}\left[\frac{15\sqrt{2}}{0.38^2}\right]^{1-0.60} = 7.42$$

The flow through the packed bed is laminar. The friction factor can be calculated as follows:

$$f_{bed} = \frac{150}{Re_{n,bed}} + 1.75 = \frac{150}{7.42} + 1.75 = 21.97$$

$$f_b = \frac{-\left(\frac{dp}{dx}\right)}{\left(\frac{\rho V_s^2}{d_p}\right)\frac{(1-\varepsilon)}{\varepsilon^3}} = \left(\frac{\Delta p}{L}\right)\left(\frac{\varepsilon^3}{1-\varepsilon}\right)\left(\frac{d_p}{\rho V_s^2}\right) = 21.97$$

Thus,

$$\left(\frac{\Delta p}{L}\right) = 21.97\left(\frac{\rho V_s^2}{d_p}\right)\frac{(1-\varepsilon)}{\varepsilon^3} = 21.97\left[\frac{1100\times0.07408^2}{1.25\times10^{-2}}\right]\left[\frac{1-0.38}{0.38^3}\right] = 119882.44\ Pa/m$$

Therefore,

$$\Delta p = 119882.44\ \times 2.5 = 299706.1\ Pa = 299.706\ kPa$$

NOTATION

A_b	cross-sectional area of the bed
C	constant in Equation (9.10)
d_H	hydraulic diameter
d_p	particle diameter

D tube diameter
D_b bed diameter
f_b packed-bed friction factor
f_{pm} friction factor for flow in consolidated porous medium
k permeability of the porous medium
K consistency index of a power-law fluid
K_o empirical factor in Equation (9.12)
K' consistency index of a generalized power-law fluid
L length
L_b length of a bed
L_e average length of the tortuous path taken by the fluid in the bed
$\dot{m}$ mass flow rate
n flow behaviour index of a power-law fluid
n' flow behaviour index of a generalized power law fluid
p pressure
Q volumetric flow rate
Re_b packed bed Reynolds number of a Newtonian fluid
$Re_{B,b}$ packed bed Reynolds number of a Bingham plastic fluid
$Re_{n,b}$ bed Reynolds number of a power-law fluid
$Re_{n',b}$ bed Reynolds number of a generalized power-law fluid
Re_{pm} Reynolds number of a consolidated porous medium
S_p surface area of a single particle
$\bar{V}$ average velocity
V_p volume of a single particle
V_s superficial velocity

Greek Letters

α parameter in Equation (9.52)
β non-Darcy flow coefficient or parameter in Equation (9.52)
$\dot{\gamma}$ shear rate
Γ_w nominal wall shear rate
Δ change in variable
ε porosity of porous medium
η viscosity
η_B Bingham viscosity
η_{eff} effective viscosity, Equation (9.55)
ρ density
τ tortuosity
τ shear stress
τ_w wall shear stress
φ ratio of yield stress to wall shear stress

PRACTICE PROBLEMS

Problem 1

A gas flows through a horizontal packed bed of spheres having a diameter of 10 mm. The bed void volume fraction is 0.40. The bed has a diameter of 0.70 m and a length of 2.5 m. The gas enters the bed at a mass flow rate of 0.5 kg/s. The properties of the gas are: $\rho = 1.2\ kg/m^3$ and $\eta = 2.0 \times 10^{-5} Pa.s$. Calculate the pressure drop of the gas in the bed.

Problem 2

Water flows through a horizontal packed bed filled with ion-exchange resin particles of diameter 4 mm ($d_p = 4\ mm$). The bed diameter (D_b) is 0.2 m and its length is 1.5 m. The void fraction ε of the bed is 0.38. Determine the flow rate of water if the pressure drop over the bed is 372.21 kPa. The properties of water are: $\rho = 1000\ kg/m^3$ and $\eta = 1$ mPa. s.

Problem 3

A fluid flows through a horizontal packed bed filled with spherical particles of diameter 2 mm. The pressure drop in the bed at two different volumetric flow rates is given as follows:

Q (ft^3/hr)	Δp (psi)
10	9
20	23

What would be the flow rate if the pressure drop is 45 psi?

If the particles of diameter 2 mm are now replaced with smaller particles of diameter 1mm, what would be the flow rate if the pressure drop is still 45 psi. Assume that the bed void fraction does not change with the change in particle size.

Problem 4

A power-law fluid (liquid) flows through a horizontal packed bed of spheres having a diameter of 1 mm. The bed void volume fraction is 0.40. The bed has a diameter of 0.1 m and a length of 2.5 m. The fluid enters the bed at a mass flow rate of 50 kg/s. The power-law constants of the fluid are: $K = 3.0\ Pa.s^n$ and $n = 0.55$. The density of the fluid is 1100 kg/m^3. Calculate the pressure drop of the fluid in the bed.

FURTHER READING

1. R. Pal. Teaching second law of thermodynamics via analysis of flow through packed beds and consolidated porous media. *Fluids* 2019, 4, 116.
2. R.P. Chhabra, J.F. Richardson. *Non-Newtonian Flow in the Process Industries.* Butterworth Heinemann, Oxford, 1999.
3. R.B. Bird, W.E. Stewart, E.N. Lightfoot. *Transport Phenomena.* 2nd Edition, Wiley, New York, 2007.
4. C.J. Geankoplis, A.A. Hersel, D.H. Lepek. *Transport Processes and Separation Process Principles.* 5th Edition, Prentice Hall, New Jersey, 2018.
5. W.L. McCabe, J.C. Smith, P. Harriott. *Unit Operations of Chemical Engineering.* 6th Edition, McGraw Hill, New York, 2001.
6. Y.F. Chen, J.Q. Zhou, S.H. Hu, R. Hu, C.B. Zhou. Evaluation of Forchheimer equation coefficients for non-Darcy flow in deformable rough-walled fractures. *J. Hydrology* 2015, 529, 993-1006.

Part 3

Heat Transport

Chapter 10

General Introduction to Heat Transfer

When two objects at different temperatures are brought in thermal contact (not necessarily physical contact) with each other, energy flows from one object to the other. This energy which flows by virtue of temperature difference between the two objects or due to temperature gradient within the same object is called "heat". So, heat is a form of energy that comes into play only in the presence of temperature gradient. Heat has meaning only during the transfer process. Once it has flowed into a body, it manifests as internal energy of the body. Because heat is relevant only during the transfer process, it is often referred to as "energy-in-transit" (energy in the process of being transferred). In accordance with the second law of thermodynamics, heat flows in the direction of decreasing temperature (from high temperature to low temperature).

10.1 DIFFERENT MODES OF HEAT TRANSFER

Heat transfer from one body to another (or from one location to another within the same medium) can take place by either conduction or convection or radiation or any combination of these three modes of heat transfer.

10.1.1 Conductive Heat Transfer

The flow of heat in a stationary medium_(solid or fluid) when there exists a temperature gradient in the medium is referred to as "heat conduction" or "conductive heat transfer". As an example, heat transfer in a plane solid wall when the two sides of the wall are exposed to different temperatures is due to conduction. The important features of conductive heat transfer are presence of material medium, the material medium is stationary (no bulk motion of matter is involved), and there exists a temperature gradient in the medium.

Rate Equation for Heat Conduction

For one-dimensional heat conduction in the x-direction, the rate of heat transfer per unit area perpendicular to the direction of heat flow is given by the following Fourier law of heat conduction:

$$q_x'' = -k\frac{dT}{dx} \tag{10.1}$$

where q_x'' is the heat flux (W/m^2) defined as the rate of heat transfer per unit area perpendicular to the direction of heat transfer, dT/dx is the temperature gradient (K/m) and k is the material property called thermal conductivity (W/m.K). The thermal conductivity is a measure of the ability of the material to conduct heat. For a prescribed temperature gradient, the conduction heat flux increases with the increase in the thermal conductivity. Thus, good conductors of heat have high values of k.

The total rate of heat transfer q is given by:

$$q = -kA\frac{dT}{dx} \tag{10.2}$$

where A is the area of heat transfer, area perpendicular to the direction of heat flow. The unit of q is W or J/s.

The thermal conductivity varies widely from one material to another (see Table 10.1). Gases are poor conductors of heat with low values of k whereas metallic solids are excellent conductors of heat conductors with high values of k. In general, the thermal conductivity of a solid is larger than that of a liquid which in turn has a larger k than that of a gas. The thermal conductivity of a solid may be four orders of magnitude larger than that of a gas.

The thermal conductivity of materials (solids, liquids, and gases) varies significantly with temperature. For example, the thermal conductivity of gases is proportional to the square root of temperature. However, the pressure dependence of thermal conductivity is usually weak.

Table 10.1 Thermal conductivity (W/m.K) of some materials at room temperature

Material	k (W/m.K)
Metallic solids	Silver: 420; Copper: 390
Non-metallic Solids	Glass:1; Limestone:1.5
Liquids	Water: 0.6; Engine oil: 0.15
Gases	Air: 0.026; Steam: 0.018

Mechanisms of Heat Conduction

In gases, heat is conducted by random motion and collision of molecules. The molecules of a gas are in ceaseless random motion even though there is no bulk motion. In a high temperature region, the gas molecules possess higher kinetic energy as compared with the molecules in a low temperature region. Due to random motion of molecules, some molecules from a high temp region reach a region of low temperature and give up part of their kinetic energy through collisions with lower energy molecules. In a similar random process, some molecules from a low temp region reach a region of high temperature and receive energy through collisions with high energy molecules. Thus, there occurs a net transfer of energy from a high temperature region to a low temperature region due to a diffusion process (random motion and collision of molecules). The mechanism of heat conduction in liquids is like that of gases, that is, random motion and collision of molecules. However, the molecules in liquids are more closely spaced and as a result intermolecular forces exert a strong influence on random motion and energy exchange in the collision process.

Heat conduction in solids occurs by two different mechanisms: (1) random motion and collision of free electrons, and (2) transmission of vibrational energy in the lattice structure. The "free" electrons (also called "electron gas") are electrons loosely bound to the atoms and are relatively free to move through the lattice structure of the solid. In the high temperature region, the free electrons possess higher kinetic energy as compared with electrons in a low temperature region. The free electrons transfer energy from a high temperature region to a low temperature region much like the gas molecules, that is, by random motion and collisions. This mechanism of heat conduction is dominant in metals (solids and liquid). In non-metallic solids, heat conduction occurs mainly due to lattice vibrations. The atoms or molecules of a solid are bound to each other by several bonds which act like springs. The atoms/molecules in the high temperature region of the solid vibrate more vigorously as compared with the atoms/molecules in the low temperature region. The vibrations are passed from the high temperature region to a low temperature region of the lattice structure through the spring-like motion of the bonds. The energy associated with lattice vibrations (lattice waves) is quantized and the quantized unit of vibrational energy is called "phonons". Thus, one can consider lattice waves as a continuous stream of packets of energy called phonons.

The thermal conductivity of isotropic solids can be expressed as:

$$k = k_e + k_{ph} \tag{10.3}$$

where k_e is the contribution of electron gas and k_{ph} is the contribution of lattice vibrations (phonons). In metallic solids, the dominant mechanism of heat conduction is electron gas and $k \approx k_e$. In non-metallic solids, heat conduction is mainly due to transmission of phonons through the lattice and $k \approx k_{ph}$.

10.1.2 Convective Heat Transfer

Convective heat transfer involves fluid motion in or around a solid object which is at a temperature different from that of the fluid. Thus, the flow of heat between a solid surface and a fluid moving adjacent to the solid surface, when the solid surface and the fluid are at different temperatures, is referred to as "heat convection" or "convective heat transfer".

Heat transfer by convection involves two mechanisms operating simultaneously – energy transfer due to random motion and collision of molecules (diffusion of energy) and superimposed upon the molecular motion is energy transfer by advection (advection means transport of a quantity, in the present case energy, by bulk fluid motion). Near the contact (solid) surface, the fluid is nearly stationary. Consequently, the heat is transferred by conduction only (random motion and collision of molecules). Away from the solid boundary, both mechanisms are important, that is, energy transfer occurs due to bulk motion (advection) as well as molecular phenomenon (diffusion).

Convection heat transfer can be classified into two broad groups based on the driving means for fluid flow. When the fluid motion is caused by external means, such as by a fan or a pump, we speak of forced convection. When fluid flow is induced by buoyancy forces, which arise from density differences caused by temperature variations in the fluid, we have "free" or "natural" convection.

Convection heat transfer could further involve external and internal flows. In external flows, fluid flows around a body that is completely submerged in it. The flow is unbounded flow. In internal flows, fluid flows in tubes or channels. Fluid is confined within the channel boundaries. The flow is bounded flow, surrounded by the solid channel boundaries.

Rate Equation for Convection

The rate equation for convective heat transfer is known as Newton's law of cooling. For external flows, the Newton's law of cooling is given as:

$$q_s'' = h(T_s - T_\infty) \tag{10.4}$$

where q_s'' is the local heat flux normal to the solid surface (also called surface heat flux), h is the local heat transfer coefficient (W/m^2.K), T_s is the local surface (solid) temperature, and T_∞ is the free stream temperature of the fluid. The heat transfer coefficient depends on several factors such as: geometry of the solid surface, nature of flow, and thermodynamic properties.

For internal flows, the Newton law of cooling is given as:

$$q_s'' = h(T_s - T_m) \tag{10.5}$$

where T_m is the mean or average temperature of the fluid (averaged over the cross-section at any axial location).

The total rate of heat transfer q may be obtained by integrating the local flux over the entire surface as:

$$q = \int_{A_s} q_s''\, dA_s = \int_{A_s} h(T_s - T_c)\, dA_s \tag{10.6}$$

where T_c is characteristic temperature of the fluid (T_∞ in the case of external flow and T_m in the case of internal flow) and A_s is the area of heat transfer (contact area between the fluid and solid).

The typical values of the heat transfer coefficient h (W/m^2.K) are summarized in the Table 10.2.

Table 10.2 Typical values of the heat transfer coefficient

Free convection, typical h values (W/m^2.K)	**Forced convection, typical h values (W/m^2.K)**	**Boiling/condensation (convection with phase-change), typical h values (W/m^2.K)**
Gases 2-25 Liquids 50-100	Gases 25-250 Liquids 50 – 20,000	2500 – 100,000

10.1.3 Radiative Heat Transfer

Nature of Thermal Radiation

All matter emits energy by virtue of having a non-zero absolute temperature. This energy, emitted in the form of electro-magnetic waves or photons due to non-zero absolute temperature, is called thermal radiation. Thermal radiation is emitted by matter due to changes in the vibrational, rotational, and electronic states of constituent molecules or atoms.

Thermal radiation covers a broad range of wavelengths: approximately 0.1 to 1000 μm. It includes part of UV region (0.1 to 0.38 μm), visible region (0.38 to 0.76 μm), and infra-red region (0.76 to 1000 μm). Thermal radiation follows the Planck Law of electromagnetic radiation:

$$e = hv = hc/\lambda \tag{10.7}$$

where e is the energy associated with photon, h is Planck's constant (= 6.626 x 10^{-34} J.s), v is frequency of radiation ($v = c/\lambda$, where λ is wavelength and c is speed of light, 3x10^8 m/s in vacuum). Higher the temperature of the object, smaller is the wavelength of the emitted thermal radiation and more energetic is the radiation. Thus, thermal radiation effect is important at high temperatures.

The rate at which energy is emitted by a surface per unit area is called the emissive power (E). The emissive power of an ideal (black) surface is given by the Stefan-Boltzmann Law:

$$E_b = \sigma T_s^4 \tag{10.8}$$

where E_b is the emissive power of a black surface, T_s is the absolute temperature of the surface, and σ is the Stefan-Boltzmann constant (= 5.67 x 10^{-8} W/(m^2. K^4). Note that E_b is the total emissive power of a black surface, integrated over all the wavelengths.

Black surfaces are surfaces which absorb all the incident thermal radiation (none of the incident radiation is reflected or transmitted; all is absorbed). Thus, black surfaces are perfect absorbers of thermal radiation. Any material which absorbs visible light appears as "visually black". However, this concept is extended to a much broader thermal band in heat transfer. A "thermally black" surface is one which absorbs all the incident thermal radiation. A surface which appears black to the eye may not be thermally black. For example, a surface may absorb all visible radiation and may reflect thermal radiation outside the visible band. Interestingly, snow is thermally black as it absorbs a major portion of the thermal radiation band and reflects the visible band.

Black surfaces are also perfect emitters of thermal radiation. The emissive power of a black surface is the highest as compared with any non-black surface at the same temperature. No real surface behaves like an ideal black surface. The emissivity (ε) of a real surface is defined as:

$$\varepsilon = E/E_b \tag{10.9}$$

where E is the emissive power of the real surface and E_b is the emissive power of a black (ideal) surface. For ideal (black) surface, $\varepsilon = 1$.

Thermal Radiation Exchange

When two objects (surfaces) at different temperatures are placed in view of each other, net transfer of energy occurs between the objects due to exchange of thermal radiation. The object at a higher temperature cools down and the object at a lower temperature gets heated up. This phenomenon whereby energy transfer between objects at different temperatures occurs due to exchange of thermal radiation is called radiative heat transfer.

As an example, consider two large black parallel surfaces of equal area (A) facing each other (see Figure 10.1). Let the absolute temperature of surface 1 be T_1 and absolute temperature of surface 2 be T_2. Let the rate at which energy leaving surface 1 and intercepted (seen) by surface 2 be $q_{1\to 2}$. In general, the radiation leaving surface 1 will include both emitted and reflected radiations. Also, not all the radiation leaving surface 1 will reach surface 2. However, in the present example, the radiation leaving surface 1 includes only the emitted radiation (black surface) and all the radiation leaving surface 1 is intercepted by surface 2 (parallel surfaces). From the Stefan-Boltzmann law:

$$q_{1\to 2} = \sigma T_1^4 A \tag{10.10}$$

The rate at which energy leaving surface 2 and intercepted (seen) by surface 1, that is, $q_{2\to 1}$, can be determined in a similar manner. It is given as:

$$q_{2\to 1} = \sigma T_2^4 A \tag{10.11}$$

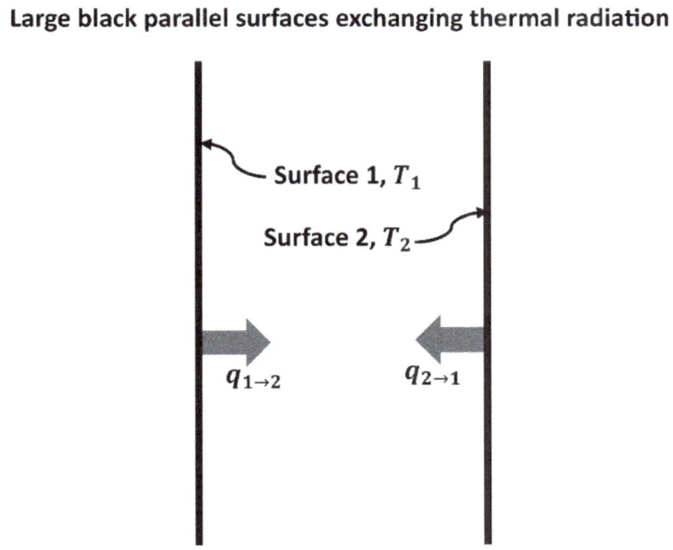

Figure 10.1 Two large black parallel surfaces exchanging thermal radiation.

Thus, the net rate of radiation heat transfer from black surface 1 to black surface 2 is:

$$q_{12} = q_{1\to 2} - q_{2\to 1} = \sigma(T_1^4 - T_2^4)A \tag{10.12}$$

10.2 SUMMARY

In summary, Table 10.3 compares the key features of the three modes of heat transfer. Convection is generally considered to be the most efficient mode of heat transfer. It is followed by conduction whereas radiation is the least efficient and slowest among the three modes of heat transfer.

Table 10.3 Comparison of the three modes of heat transfer

Conduction	Convection	Radiation
Needs matter.	Needs matter.	Does not need matter.
Energy transfer - a molecular phenomenon (diffusion process)	Energy transfer by bulk motion (advection) as well as diffusion	Energy transfer by photons (electromagnetic waves)
No bulk motion of matter.	Bulk motion of matter needed.	No bulk motion of matter.
Temperature difference /gradient needed.	Temperature difference /gradient needed.	Temperature difference between surfaces needed.

10.3 EXAMPLES

10.3.1 Example 1

Consider the concrete slab of a basement (see figure below). The slab is 15 m long, 10 m wide, and 10 inches thick. The temperatures during the winter are 19 °C at the top surface and 8 °C at the bottom surface. The thermal conductivity of the concrete is 1.48 W/(m.K). Calculate the rate of heat loss from the slab to the ground.

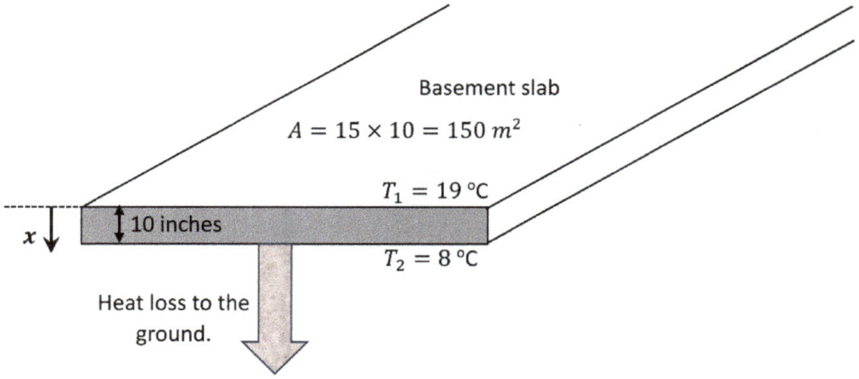

Solution:

From Fourier law of heat conduction:

$$q_x'' = -k\frac{dT}{dx} = -k\left[\frac{T_2 - T_1}{x_2 - x_1}\right] = -1.48\left[\frac{8 - 19}{0.254}\right] = 64.09\ W/m^2$$

The rate of heat loss q is:

$$q = q_x''A = 64.09 \times 150 = 9613.5\ W$$

10.3.2 Example 2

Consider heat transfer through a large plane wall of thickness 254 mm (see figure below). The heat flux through the wall is $q_x'' = 32\ W/m^2$. The surface temperature on the left side of the wall is 45^oC. Determine the surface temperature on the right side of the wall. The thermal conductivity of the wall material is 1.5 W/(m.K).

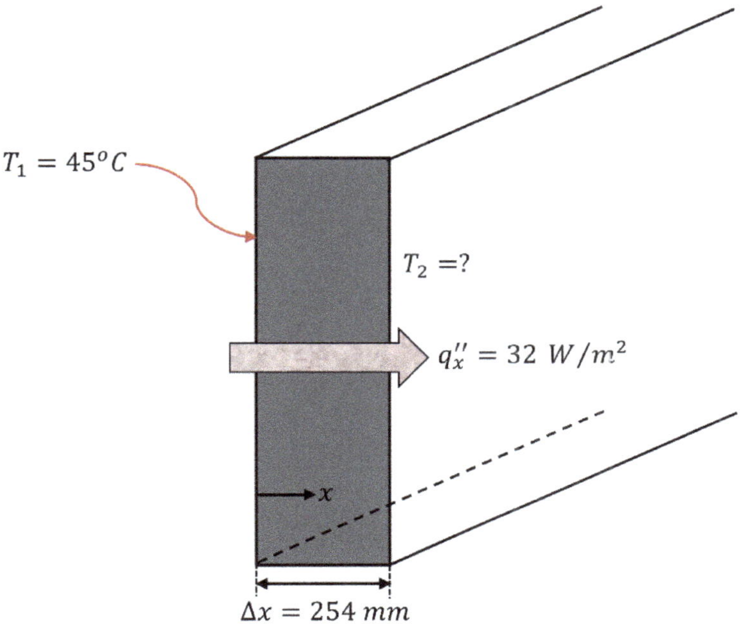

Solution:

From Fourier law of heat conduction:

$$q_x'' = -k\frac{dT}{dx} = -k\left[\frac{T_2 - T_1}{\Delta x}\right]$$

Upon rearrangement, we get:

$$T_2 = T_1 - q_x''\left(\frac{\Delta x}{k}\right) = 45 - 32\left(\frac{254\times10^{-3}}{1.5}\right) = 39.58^oC$$

10.3.3 Example 3

Consider heat loss from the human body (see figure below). The core temperature of the human body is 36.5^oC and the temperature of the human skin is 30^oC on a particular day. The drop in the temperature from 36.5^oC to 30^oC occurs across a sub skin layer of thickness approximately 1 cm. This sub skin layer of thermal conductivity about 0.45 W/(m.K) acts as an insulating layer for the human body.

i) Estimate the rate of heat loss from the skin surface to the surroundings per unit area

ii) Assuming that the ambient air temperature is 22^oC, estimate the value of the heat transfer coefficient at the skin surface.

You can neglect radiation heat loss from the human body to the surroundings.

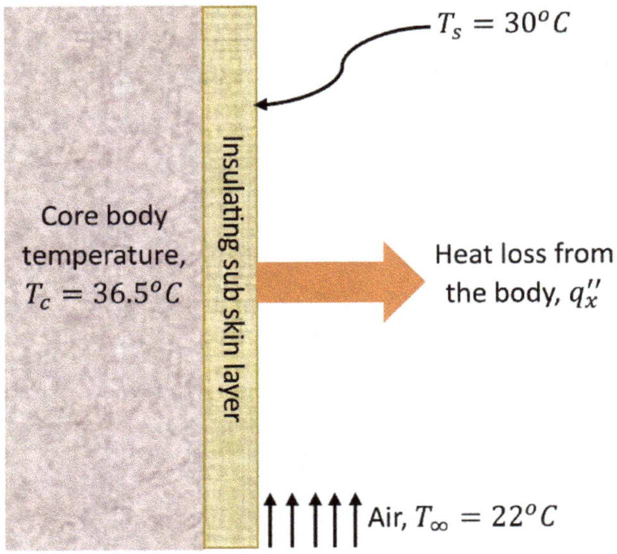

Solution:

i) Estimate the rate of heat loss from the skin surface to the surroundings per unit area

Apply Fourier's law of heat conduction to sub layer skin:

$$q_x'' = -k\frac{dT}{dx} = -k\left[\frac{T_s - T_c}{\Delta x}\right] = -0.45\left[\frac{30 - 36.5}{10^{-2}}\right] = 292.5 \;\; W/m^2$$

ii) Assuming that the ambient air temperature is 22^oC, estimate the value of the heat transfer coefficient at the skin surface.

Under steady state condition,

$$q_{x,conduction}'' = q_{x,convection}'' = 292.5 \;\; W/m^2$$

From Newton's law of cooling,

$$q_{x,convection}'' = h(T_s - T_\infty) = 292.5$$

Thus,

$$h = \frac{292.5}{(T_s - T_\infty)} = \frac{292.5}{(30 - 22)} = 39.56\ W/(m^2.K)$$

10.3.4 Example 4

Two very large parallel black plates (surfaces) at 500 °C and 100 °C exchange thermal radiation, as shown in the figure below.

(a) Calculate the net radiative heat transfer rate between the plates per unit area.

(b) If another black plate is placed between the 500 °C and 100 °C surfaces (see figure below), what would be its temperature at steady state?

(c) By how much is the heat transfer rate reduced due to placement of the centre plate?

Diagram for part (a)

Large black parallel surfaces exchanging thermal radiation

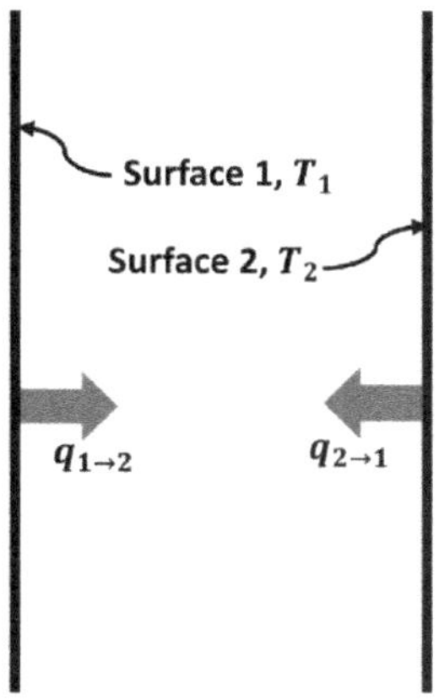

Diagram for parts (b) and (c)

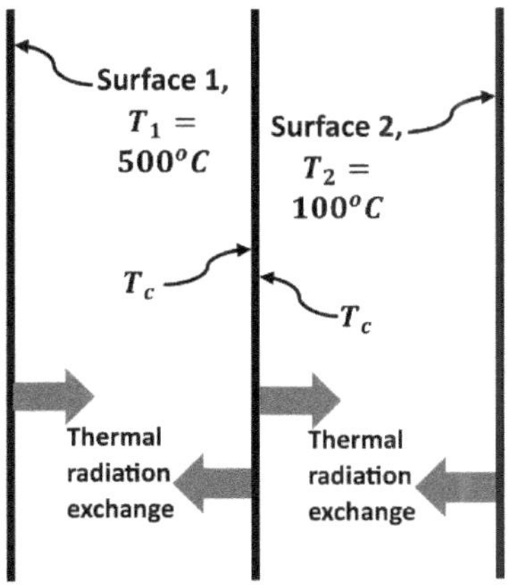

Solution:

(a) Calculate the net radiative heat transfer rate between the plates per unit area.

The net rate of radiation heat transfer from black surface 1 to black surface 2 is:

$$q_{12} = q_{1\to 2} - q_{2\to 1} = \sigma(T_1^4 - T_2^4)A$$

$$T_1 = 500 + 273.15 = 773.15\ K,\ \ T_2 = 100 + 273.15 = 373.15\ K$$

$$\frac{q_{12}}{A} = \sigma(T_1^4 - T_2^4) = (5.67 \times 10^{-8})[773.15^4 - 373.15^4] = 19160\ W/m^2$$

(b) If another black plate is placed between the 500 °C and 100 °C surfaces (see figure below), what would be its temperature at steady state?

$$q_{1c} = q_{1\to c} - q_{c\to 1} = \sigma(T_1^4 - T_c^4)A$$

$$q_{c2} = q_{c\to 2} - q_{2\to c} = \sigma(T_c^4 - T_2^4)A$$

At steady state,

$$q_{1c} = q_{c2}\ ;\ \ \sigma(T_1^4 - T_c^4)A = \sigma(T_c^4 - T_2^4)A$$

Thus,

$$T_c = \left[\frac{T_1^4 + T_2^4}{2}\right]^{1/4} = \left[\frac{773.15^4 + 373.15^4}{2}\right]^{1/4} = 658.784\ K = 385.634^oC$$

(c) By how much is the heat transfer rate reduced due to placement of the centre plate?

$$\frac{q_{1c}}{A} = \frac{q_{c2}}{A} = \sigma(T_1^4 - T_c^4) = (5.67 \times 10^{-8})[773.15^4 - 658.784^4] = 9580.318\ W/m^2$$

The reduction in heat transfer rate due to placement of the centre plate is:

$$= \frac{q_{without\ centre\ plate-}q_{with\ centre\ plate}}{q_{without\ centre\ plate-}} = \frac{19160 - 9580.318}{19160} = 0.50$$

Thus, the heat transfer rate is reduced to 50% of the value in the absence of the centre plate.

NOTATION

A	area of heat transfer
A_s	contact area between fluid and solid (same as heat transfer area)
c	velocity of electromagnetic radiation
E	emissive power of a body
E_b	emissive power of a black body
h	Planck's constant or heat transfer coefficient
k	thermal conductivity
k_e	contribution of electron gas to thermal conductivity
k_{ph}	contribution of lattice vibrations (phonons) to thermal conductivity

q heat transfer rate
$q_{1\to2}$ rate at which thermal radiation energy leaves surface 1 and is intercepted (seen) by surface 2
$q_{2\to1}$ rate at which thermal radiation energy leaves surface 2 and is intercepted (seen) by surface 1
q_{12} net rate of radiation heat transfer from surface 1 to surface 2
q''_x heat flux in x direction
q''_s local heat flux normal to the solid surface (also called surface heat flux)
T absolute temperature
T_c characteristic temperature of fluid
T_m mean temperature of fluid
T_s temperature of the surface
T_∞ free stream temperature of fluid
x distance in x direction, x- coordinate

Greek Letters

ε emissivity of a surface
λ wavelength of radiation
υ frequency of radiation
σ Stefan-Boltzmann constant

PRACTICE PROBLEMS

Problem 1

The wall of an industrial furnace is constructed from 0.18 m thick fireclay brick having a thermal conductivity of 1.75 W/m.K. The temperatures of the inner and outer surfaces are 1450 and 1175 K, respectively. Determine the rate of heat loss through a wall which is 0.6 m by 3.5 m on a side?

Problem 2

Consider a 3 m deep layer of a reactive material that generates heat volumetrically at a uniform rate of $\dot{q} = 25\ W/m^3$. The layer is exposed to ambient air at a temperature of $T_\infty = 25^oC$. The heat transfer coefficient at the surface of the layer is $h = 6.5\ W/m^2.K$. Calculate the temperature at the surface of the layer at steady state.

Problem 3

Two very large parallel plates are spaced 1.5 m apart. The temperatures of the plates are 750 °C and 250 °C. The plates exchange heat by thermal radiation. Calculate the net rate of radiation heat transfer between the plates assuming that the surfaces can be treated as black surfaces.

Problem 4

During summer, if you stand in front of an open refrigerator, you tend to feel cooler even though there is no fan present in the refrigerator to blow the cold air over you. Can you explain this?

Problem 5

During summer, you are quite comfortable in light clothes in your living room where the room temperature is set at 20 °C. However, in winter, you feel much cooler even though the room temperature is set at 20 °C and you are wearing the same light clothes. Can you explain this?

FURTHER READING

1. A. Bejan. *Heat Transfer*. Wiley, New York, 1993.
2. L.C. Thomas. *Heat Transfer*. Prentice Hall, New Jersey, 1992.
3. F.P. Incropera, D.P. Dewitt, T.L. Bergman, A.S. Lavine. *Fundamentals of Heat and Mass Transfer*. 6th Edition, Wiley, New York, 2007.
4. Holland, J.P. *Heat Transfer*. 8th Edition, McGraw Hill, New York, 1997.
5. W.S. Janna. *Engineering Heat Transfer*. PWS Publishers, Boston, 1986.
6. R.B. Bird, W.E. Stewart, E.N. Lightfoot. *Transport Phenomena*. 2nd Edition, Wiley, New York, 2007.
7. C.J. Geankoplis, A.A. Hersel, D.H. Lepek. *Transport Processes and Separation Process Principles*. 5th Edition, Prentice Hall, New Jersey, 2018.
8. J.R. Welty, C.E. Wicks, R.E. Wilson, G. Rorrer. *Fundamentals of Momentum, Heat, and Mass Transfer*. 4th Edition, Wiley, New York, 2001.
9. W.L. McCabe, J.C. Smith, P. Harriott. *Unit Operations of Chemical Engineering*. 6th Edition, McGraw Hill, New York, 2001.
10. F. Kreith, M.S. Bohn. *Principles of Heat Transfer*. 5th Edition, West Publishing Co., St. Paul, MN, USA, 1993.
11. R. Pal. *Electromagnetic, Mechanical, and Transport Properties of Composite Materials*. CRC Press, Boca Raton, Florida, 2015.

Chapter 11

Heat Transfer in a Stationary Medium (Conductive Heat Transfer)

11.1 HEAT FLUX VECTOR AND TEMPERATURE GRADIENT

The heat flux vector $\vec{q}''$ gives the rate of heat transfer per unit area (W/m^2) perpendicular to the direction of heat flow. It is expressed as:

$$\vec{q}'' = q_x''\hat{\imath} + q_y''\hat{\jmath} + q_z''\hat{k} \tag{11.1}$$

where q_x'',q_y'',q_z'' are the components of heat flux vector in x, y and z directions, respectively. q_x'' is the rate of heat transfer in x-direction per unit area normal to x, q_y'' is the rate of heat transfer in y-direction per unit area normal to y, and q_z'' is the rate of heat transfer in z-direction per unit area normal to z.

The temperature gradient ∇T at any point in the medium can be expressed as:

$$\nabla T = \frac{\partial T}{\partial x}\hat{\imath} + \frac{\partial T}{\partial y}\hat{\jmath} + \frac{\partial T}{\partial z}\hat{k} \tag{11.2}$$

where $\partial T/\partial x$ is the temperature gradient in x-direction, $\partial T/\partial y$ is the temperature gradient in y-direction, and $\partial T/\partial z$ is the temperature gradient in z-direction.

For one-dimensional heat conduction in x-direction:

$$q_x'' \neq 0;\; q_y'' = 0;\; q_z'' = 0;\; T = T(x);\; \frac{\partial T}{\partial x} \neq 0;\; \frac{\partial T}{\partial y} = 0;\; \frac{\partial T}{\partial y} = 0 \tag{11.3}$$

For two-dimensional heat conduction in x and y directions:

$$q_z'' = 0;\; q_x'' \neq 0;\; q_y'' \neq 0;\; T = T(x, y);\; \frac{\partial T}{\partial x} \neq 0;\; \frac{\partial T}{\partial y} \neq 0;\; \frac{\partial T}{\partial z} = 0 \tag{11.4}$$

For three-dimensional heat conduction, temperature is a function of all three space co-ordinates and all three heat flux components are non-zero:

$$q_x'' \neq 0;\; q_y'' \neq 0;\; q_z'' \neq 0;\; T = T(x, y, z);\; \frac{\partial T}{\partial x} \neq 0;\; \frac{\partial T}{\partial y} \neq 0;\; \frac{\partial T}{\partial z} \neq 0 \tag{11.5}$$

11.2 FOURIER'S LAW OF HEAT CONDUCTION

The Fourier Law of heat conduction states that the local heat flux vector $\vec{q}''$ at any point in the material is proportional to the local temperature gradient ∇T and is given as:

$$\vec{q}'' = -\bar{\bar{k}} \cdot \nabla T \tag{11.6}$$

where $\bar{\bar{k}}$ is the thermal conductivity tensor of the material. The minus sign in the Fourier's Law of heat conduction is a consequence of the second law of thermodynamics which requires that heat must flow in the direction of decreasing temperature. The thermal conductivity tensor is given as:

$$\bar{\bar{k}} = \sum_i \sum_j k_{ij}\, \hat{\delta}_i \hat{\delta}_j = \begin{pmatrix} k_{11} & k_{12} & k_{13} \\ k_{21} & k_{22} & k_{23} \\ k_{31} & k_{32} & k_{33} \end{pmatrix} \tag{11.7}$$

In the case of an isotropic medium, $k_{ij} = 0$ if $i \neq j$ and $k_{ii} = k$, that is, the non-diagonal components are zero and the diagonal components are all equal. Thus, for isotropic media, $\bar{\bar{k}} = k\bar{\bar{\delta}}$ ($\bar{\bar{\delta}}$ is a unit tensor) and the Fourier Law of heat conduction simplifies to:

$$\vec{q}'' = -k\bar{\bar{\delta}} \cdot \nabla T = -k\nabla T \tag{11.8}$$

where k is the thermal conductivity of the material. Note that for isotropic materials, the heat flux vector $\vec{q}''$ at any location is parallel to ∇T and hence, normal to the isothermal surface passing through that location. In the case of anisotropic materials, $\vec{q}''$ at any location is not parallel to ∇T and therefore, the direction of $\vec{q}''$ is not generally normal to the isothermal surface passing through that location. Table 11.1 presents the Fourier's law of heat conduction in rectangular, cylindrical, and spherical coordinate systems.

Table 11.1 Fourier's law of heat conduction in different co-ordinate systems

Co-ordinate system	**General form**	**Component form for the given co-ordinate system**
Cartesian	$\vec{q}'' = -k\nabla T$	$\vec{q}'' = q''_x\hat{\imath} + q''_y\hat{\jmath} + q''_z\hat{k} = -k\left[\left(\frac{\partial T}{\partial x}\right)\hat{\imath} + \left(\frac{\partial T}{\partial y}\right)\hat{\jmath} + \left(\frac{\partial T}{\partial z}\right)\hat{k}\right]$
Cylindrical	$\vec{q}'' = -k\nabla T$	$\vec{q}'' = q''_r\hat{r} + q''_\theta\hat{\theta} + q''_z\hat{z} = -k\left[\left(\frac{\partial T}{\partial r}\right)\hat{r} + \frac{1}{r}\left(\frac{\partial T}{\partial \theta}\right)\hat{\theta} + \left(\frac{\partial T}{\partial z}\right)\hat{z}\right]$
Spherical	$\vec{q}'' = -k\nabla T$	$\vec{q}'' = q''_r\hat{r} + q''_\theta\hat{\theta} + q''_\varphi\hat{\varphi} = -k\left[\left(\frac{\partial T}{\partial r}\right)\hat{r} + \frac{1}{r}\left(\frac{\partial T}{\partial \theta}\right)\hat{\theta} + \frac{1}{r\sin\theta}\left(\frac{\partial T}{\partial \varphi}\right)\hat{\varphi}\right]$

11.3 HEAT CONDUCTION EQUATION

The Fourier's law of heat conduction allows the calculation of heat flux provided that the temperature distribution in the medium is known. We need another equation, the heat conduction equation, to determine the temperature distribution in the medium. The heat conduction equation is obtained by applying energy balance to a control volume. Control volume (CV) is a region of space (within the material medium) bounded by a control surface. For non-flow systems, only energy and no mass can enter or leave the control volume. Energy can enter the control volume through the control surface in two ways: heat transfer and work transfer. However, for flow systems both mass and energy can enter or leave the control volume. Now in addition to heat and work transfer, energy can enter or leave the control volume in a direct manner by advection process whereby the mass entering and leaving the control volume carries energy with it.

Consider a stationary medium with non-uniform temperature distribution where only heat enters or leaves the control volume through conduction and there is no work transfer involved. For a cubical control volume (see Figure 11.1), heat enters the control volume through the x, y and z faces and leaves through the $x + \Delta x, y + \Delta y,$ and $z + \Delta z$ faces. Heat is also generated inside the control volume through means such as passage of electric current through the medium. Thus, energy balance can be written as:

$$\begin{Bmatrix}\text{Rate at which heat}\\ \text{enters the CV}\end{Bmatrix} - \begin{Bmatrix}\text{Rate at which heat}\\ \text{leaves the CV}\end{Bmatrix} + \begin{Bmatrix}\text{Rate at which heat is generated}\\ \text{within the CV}\end{Bmatrix} = \begin{Bmatrix}\text{Rate at which internal energy is stored or}\\ \text{accumulated inside the CV}\end{Bmatrix} \tag{11.9}$$

Let us now determine each term of the energy balance equation:

$$\begin{Bmatrix}\text{Rate at which heat}\\ \text{enters the CV}\end{Bmatrix} = q''_x|_x(\Delta y)(\Delta z) + q''_y\big|_y(\Delta x)(\Delta z) + q''_z|_z(\Delta y)(\Delta x) \tag{11.10}$$

$$\begin{Bmatrix}\text{Rate at which heat}\\ \text{leaves the CV}\end{Bmatrix} = q''_x|_{x+\Delta x}(\Delta y)(\Delta z) + q''_y\big|_{y+\Delta y}(\Delta x)(\Delta z) + q''_z|_{z+\Delta z}(\Delta y)(\Delta x) \tag{11.11}$$

$$\begin{Bmatrix}\text{Rate at which heat is generated}\\ \text{within the CV}\end{Bmatrix} = \dot{q}(\Delta x \Delta y \Delta z) \tag{11.12}$$

$$\begin{Bmatrix}\text{Rate at which internal energy is stored or}\\ \text{accumulated inside the CV}\end{Bmatrix} = \rho(\Delta x \Delta y \Delta z)\frac{\partial U}{\partial t} \tag{11.13}$$

where q''_x, q''_y and q''_z are the $x, y,$ and z components of heat flux vector, $\dot{q}$ is the rate of heat generation per unit volume, ρ is density of the medium, and U is internal energy per unit mass.

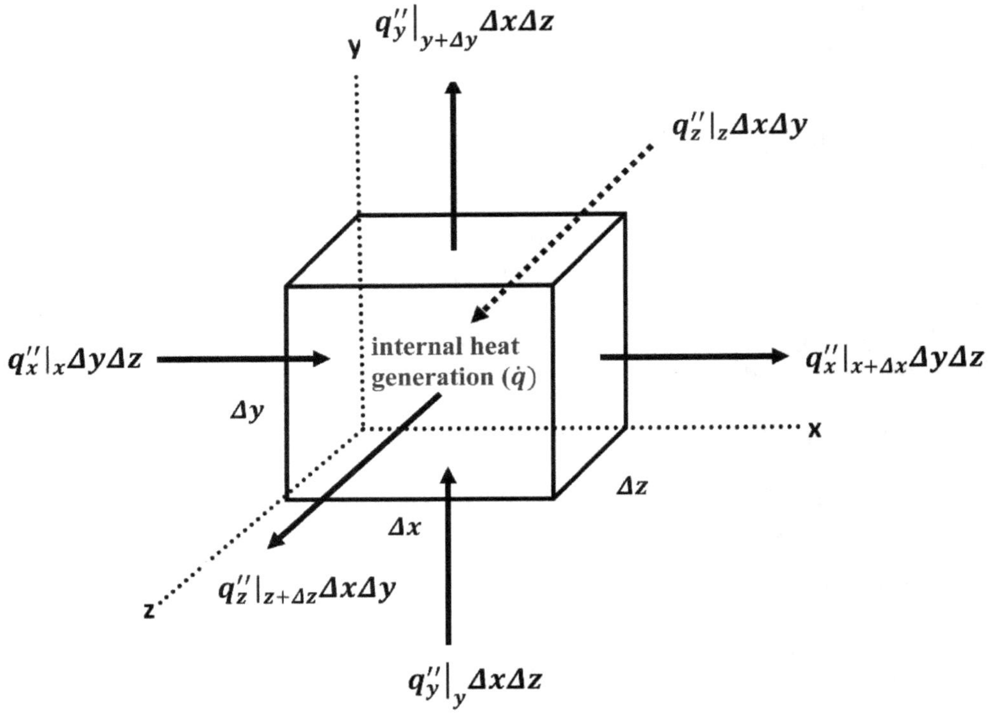

Figure 11.1 Energy balance over a cubical control volume.

Thus, energy balance could be written as:

$$-[(q''_x|_{x+\Delta x} - q''_x|_x)(\Delta y)(\Delta z)] - \left[(q''_y\big|_{y+\Delta y} - q''_y\big|_y)(\Delta x)(\Delta z)\right] - [(q''_z|_{z+\Delta z} - q''_z|_z)(\Delta y)(\Delta x)] + [\dot{q}(\Delta x\Delta y\Delta z)] = \rho(\Delta x\Delta y\Delta z)\frac{\partial U}{\partial t} \quad (11.14)$$

If we divide this equation by $(\Delta x\Delta y\Delta z)$ and take the limits $\Delta x \to 0, \Delta y \to 0, \Delta z \to 0$, the following result is obtained:

$$-\left[\frac{\partial q''_x}{\partial x} + \frac{\partial q''_y}{\partial y} + \frac{\partial q''_z}{\partial z}\right] + \dot{q} = \rho\frac{\partial U}{\partial t} \quad (11.15)$$

This energy balance is applicable to stationary systems, with or without chemical reactions. In the absence of chemical reactions (constant composition) and assuming constant density (incompressible material), one can express the time derivative of U as:

$$\frac{\partial U}{\partial t} = c_v\frac{\partial T}{\partial t} = c_p\frac{\partial T}{\partial t} \quad (11.16)$$

where c_p and c_v are constant-pressure and constant-volume heat capacities, respectively. Note that $c_p = c_v$ for incompressible materials. Thus, energy balance gives:

$$-\nabla\cdot\vec{q}'' + \dot{q} = \rho c_p\frac{\partial T}{\partial t} \quad (11.17)$$

where $\vec{q}''$is the heat flux vector. Substitution of the expression for heat flux vector $\vec{q}''$ from the Fourier law of heat conduction into Equation (11.17) gives:

$$\nabla \cdot (k\nabla T) + \dot{q} = \rho c_p \frac{\partial T}{\partial t} \tag{11.18}$$

where it is assumed that the medium is isotropic. Equation (11.18) is the most general form of the heat conduction equation. It describes the temperature variation with space co-ordinates and time. When the thermal conductivity k is constant, Equation (11.18) simplifies to:

$$\nabla^2 T + \frac{\dot{q}}{k} = \frac{1}{\alpha}\frac{\partial T}{\partial t} \tag{11.19}$$

where α is the thermal diffusivity (m^2/s) of material, defined as $k/\rho c_p$. In the absence of thermal energy generation, the heat conduction equation, Equation (11.19), reduces to:

$$\nabla^2 T = \frac{1}{\alpha}\frac{\partial T}{\partial t} \tag{11.20}$$

Table 11.2 presents the heat conduction equation, Equation (11.18), in rectangular, cylindrical, and spherical coordinate systems.

Table 11.2 Heat conduction equation, Equation (11.18), in different coordinate systems

Co-ordinate system	Heat conduction equation (general form)	Heat conduction equation for the given co-ordinate system
Cartesian	$\nabla \cdot (k\nabla T) + \dot{q} = \rho c_P \frac{\partial T}{\partial t}$ When k = constant: $\nabla^2 T + \frac{\dot{q}}{k} = \frac{1}{\alpha}\left(\frac{\partial T}{\partial t}\right)$	$\frac{\partial}{\partial x}\left(k\frac{\partial T}{\partial x}\right) + \frac{\partial}{\partial y}\left(k\frac{\partial T}{\partial y}\right) + \frac{\partial}{\partial z}\left(k\frac{\partial T}{\partial z}\right) + \dot{q} = \rho c_P \frac{\partial T}{\partial t}$ When k = constant: $\frac{\partial^2 T}{\partial x^2} + \frac{\partial^2 T}{\partial y^2} + \frac{\partial^2 T}{\partial z^2} + \frac{\dot{q}}{k} = \frac{1}{\alpha}\frac{\partial T}{\partial t}$
Cylindrical	$\nabla \cdot (k\nabla T) + \dot{q} = \rho c_P \frac{\partial T}{\partial t}$ When k = constant: $\nabla^2 T + \frac{\dot{q}}{k} = \frac{1}{\alpha}\left(\frac{\partial T}{\partial t}\right)$	$\frac{1}{r}\frac{\partial}{\partial r}\left(kr\frac{\partial T}{\partial r}\right) + \frac{1}{r^2}\frac{\partial}{\partial \theta}\left(k\frac{\partial T}{\partial \theta}\right) + \frac{\partial}{\partial z}\left(k\frac{\partial T}{\partial z}\right) + \dot{q} = \rho c_P \frac{\partial T}{\partial t}$ When k = constant: $\frac{1}{r}\frac{\partial}{\partial r}\left(r\frac{\partial T}{\partial r}\right) + \frac{1}{r^2}\frac{\partial^2 T}{\partial \theta^2} + \frac{\partial^2 T}{\partial z^2} + \frac{\dot{q}}{k} = \frac{1}{\alpha}\frac{\partial T}{\partial t}$
Spherical	$\nabla \cdot (k\nabla T) + \dot{q} = \rho c_P \frac{\partial T}{\partial t}$ When k = constant: $\nabla^2 T + \frac{\dot{q}}{k} = \frac{1}{\alpha}\left(\frac{\partial T}{\partial t}\right)$	$\frac{1}{r^2}\frac{\partial}{\partial r}\left(kr^2\frac{\partial T}{\partial r}\right) + \frac{1}{r^2 sin^2 \theta}\frac{\partial}{\partial \varphi}\left(k\frac{\partial T}{\partial \varphi}\right)$ $+ \frac{1}{r^2 sin\,\theta}\frac{\partial}{\partial \theta}\left(k\, sin\,\theta\frac{\partial T}{\partial \theta}\right) + \dot{q} = \rho c_P \frac{\partial T}{\partial t}$ When k = constant: $\frac{1}{r^2}\frac{\partial}{\partial r}\left(r^2\frac{\partial T}{\partial r}\right) + \frac{1}{r^2 sin^2 \theta}\frac{\partial^2 T}{\partial \varphi^2} + \frac{1}{r^2 sin\,\theta}\frac{\partial}{\partial \theta}\left(sin\,\theta\frac{\partial T}{\partial \theta}\right)$ $+ \frac{\dot{q}}{k} = \frac{1}{\alpha}\frac{\partial T}{\partial t}$

11.3.1 Boundary and Initial Conditions

To solve the heat conduction equation, the boundary and initial conditions that characterize the given heat transfer configuration must be specified. Because the heat conduction equation is second order in spatial co-ordinates, two boundary conditions are required for each co-ordinate. Only one initial condition is required as the equation is first order in time. Some examples of the boundary conditions in one-dimensional heat conduction problems are as follows:

a) The temperature at the boundary (located at $x = x_0$) of the conducting medium is held constant, that is, at $x = x_0, T = T_0$. This type of boundary condition is called Dirichlet boundary condition or boundary condition of first kind.

b) The heat flux at the boundary (located at $x = x_0$) of the conducting medium is held constant, that is, at $x = x_0$, $q''_x|_{x=x_0} = q''_o$. This type of boundary condition is called Newmann boundary condition or boundary condition of second kind. A special case of this boundary condition is $q''_x|_{x=x_0} = q''_o = 0$. No heat passes through the boundary surface located at $x = x_0$. This type of boundary condition is often referred to as adiabatic boundary condition.

c) Convection occurs at the boundary (located at $x = x_0$) of the conducting medium. Thus:

$$q''_x|_{x=x_0} = h\left(T|_{x=x_0} - T_\infty\right) \tag{11.21}$$

where h is the heat transfer coefficient and T_∞ is the free stream temperature of the fluid. This type of boundary condition is called boundary condition of third kind.

An example of the initial condition is the specification of temperature of the medium at the start of the time interval of interest, such as: at $t = t_0, T = T_0$. Another initial condition of interest is the specification of heat flux at the boundary of the medium at the start of the time interval of interest.

11.4 STEADY STATE HEAT CONDUCTION

Under steady state condition, the temperature distribution in an isotropic medium of constant k is governed by the following form of the heat conduction equation, Equation (11.19), referred to as Poisson equation:

$$\nabla^2 T + \frac{\dot{q}}{k} = 0 \tag{11.22}$$

In the absence of thermal energy generation, Equation (11.22) reduces to the following equation, referred to as Laplace equation:

$$\nabla^2 T = 0 \tag{11.23}$$

11.4.1 One-dimensional Steady State Heat Conduction

Assuming no thermal energy generation and the material to be isotropic of constant k, the Laplace form of the heat conduction equation, Equation (11.23), can be expressed in the three co-ordinate systems as follows:

Cartesian coordinates: For heat flow in the x-direction, Equation (11.23) gives:

$$\frac{d^2T}{dx^2} = 0 \tag{11.24}$$

Cylindrical coordinates: For heat flow in the radial direction, Equation (11.23) gives:

$$\frac{d}{dr}\left(r\frac{dT}{dr}\right) = 0 \tag{11.25}$$

Spherical coordinates: For heat flow in the radial direction, Equation (11.23) gives:

$$\frac{d}{dr}\left(r^2\frac{dT}{dr}\right) = 0 \tag{11.26}$$

For a slab (solid wall) of uniform thickness (see Figure 11.2), Equation (11.24) gives the following linear temperature profile:

$$\frac{T-T_1}{T_2-T_1} = \frac{x}{L} \tag{11.27}$$

where T_1 is the temperature of the slab at $x = 0$, T_2 is the temperature of the slab at $x = L$, and L is the slab thickness. The heat flux and the heat transfer rate through the slab are given as follows:

$$q''_x = -k\frac{dT}{dx} = \frac{k}{L}(T_1 - T_2) \tag{11.28}$$

$$q = q''_x A = \frac{kA}{L}(T_1 - T_2) \tag{11.29}$$

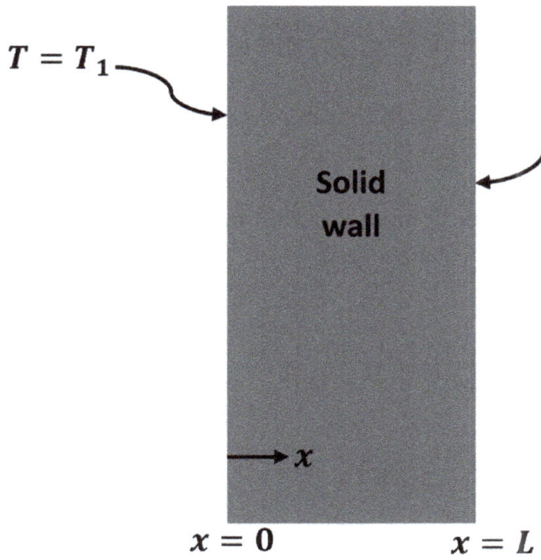

Figure 11.2 Conductive heat transfer through a solid plane wall.

For a cylindrical shell (see Figure 11.3) with boundary conditions: $T = T_1$ at $r = r_1$ (inner cylindrical surface) and $T = T_2$ at $r = r_2$ (outer cylindrical surface), Equation (11.25) gives the following temperature profile:

$$\frac{T - T_1}{T_2 - T_1} = \frac{ln(r/r_1)}{ln(r_2/r_1)} \tag{11.30}$$

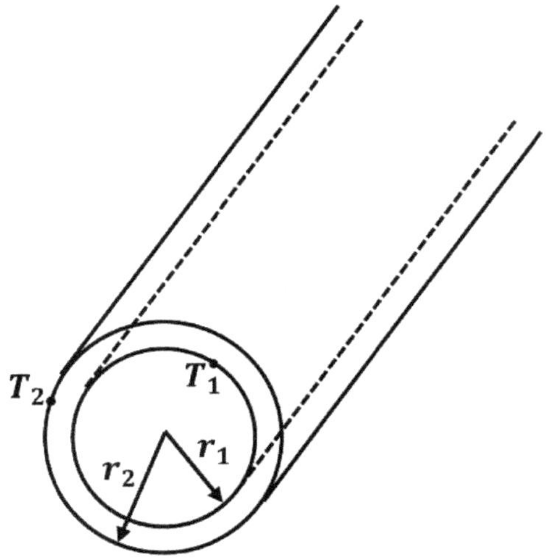

Figure 11.3 Conductive heat transfer through a cylindrical shell.

Thus, the temperature distribution is logarithmic in cylindrical shells. The heat flux and the heat transfer rate through the cylindrical shell are given as follows:

$$q_r'' = -k\frac{dT}{dr} = -\frac{k}{r}\frac{(T_2 - T_1)}{ln(r_2/r_1)} \tag{11.31}$$

$$q = q_r''(2\pi rL) = \frac{2\pi kL}{ln(r_2/r_1)}(T_1 - T_2) \tag{11.32}$$

where L is the length of the cylindrical shell. Note that the heat flux in a cylindrical shell is not constant. The heat flux decreases with the increase in the radial position r as the area of heat transfer increases.

For a spherical shell (see Figure 11.4) with boundary conditions: $T = T_1$ at $r = r_1$ (inner spherical surface) and $T = T_2$ at $r = r_2$ (outer spherical surface), Equation (11,26) gives the following temperature profile:

$$\frac{T - T_1}{T_2 - T_1} = \frac{1}{\left(\frac{1}{r_2} - \frac{1}{r_1}\right)}\left(\frac{1}{r} - \frac{1}{r_1}\right) \tag{11.33}$$

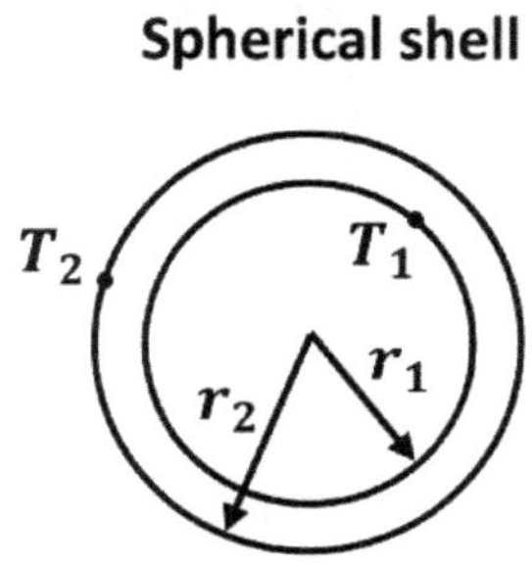

Figure 11.4 Conductive heat transfer through a spherical shell.

The heat flux and the heat transfer rate through the spherical shell are given as follows:

$$q_r'' = -k\frac{dT}{dr} = -\frac{k}{r^2}\frac{(T_2-T_1)}{\left(\frac{1}{r_1}-\frac{1}{r_2}\right)} \tag{11.34}$$

$$q = q_r''(4\pi r^2) = \frac{4\pi k}{\left(\frac{1}{r_2}-\frac{1}{r_1}\right)}(T_2 - T_1) \tag{11.35}$$

Note that at steady state the total heat transfer rate is constant, but the heat flux decreases with the increase in the radial position r due to an increase in the heat transfer area.

Table 11.3 summarizes the results for one-dimensional steady-state heat conduction in different geometries.

Table 11.3 One-dimensional Steady State Heat Conduction in Different Geometries

System with Boundary conditions	Temperature Distribution	Heat Flux (q'')	Heat Transfer Rate (q)
Plane Wall at $x = x_1, T = T_1$ at $x = x_2, T = T_2$	$T = c_1x + c_2$ $\frac{T-T_1}{T_2-T_1} = \frac{x-x_1}{x_2-x_1}$	q_x'' =constant $q_x'' = -k\frac{(T_2-T_1)}{(x_2-x_1)}$	q = constant $q = -kA\frac{(T_2-T_1)}{(x_2-x_1)}$
Cylindrical Shell at $r = r_1, T = T_1$ at $r = r_2, T = T_2$	$T = c_1 ln r + c_2$ $\frac{T-T_1}{T_2-T_1} = \frac{ln\left(\frac{r}{r_1}\right)}{ln\left(\frac{r_2}{r_1}\right)}$	q_r'' decreases with r $q_r'' = -\frac{k}{r}\frac{(T_2-T_1)}{ln\left(\frac{r_2}{r_1}\right)}$	q = constant $q = -\frac{2\pi kL(T_2-T_1)}{ln\left(\frac{r_2}{r_1}\right)}$
Spherical Shell at $r = r_1, T = T_1$ at $r = r_2, T = T_2$	$T = \frac{c_1}{r} + c_2$ $\frac{T-T_1}{T_2-T_1} = \frac{\left(\frac{1}{r}-\frac{1}{r_1}\right)}{\left(\frac{1}{r_2}-\frac{1}{r_1}\right)}$	q_r'' decreases with r $q_r'' = -\frac{k}{r^2}\frac{(T_2-T_1)}{\left(\frac{1}{r_1}-\frac{1}{r_2}\right)}$	q = constant $q = -\frac{4\pi k(T_2-T_1)}{\left(\frac{1}{r_1}-\frac{1}{r_2}\right)}$

11.4.2 Analogy Between Heat Conduction and Electric Current Flow

There exists a close analogy between one-dimensional steady state heat conduction (without thermal energy generation) and electric current flow. According to Ohm's law:

$$\frac{I}{A} = \sigma\left(\frac{\Delta V}{L}\right) \tag{11.36}$$

where I is current, A is cross-section area of conductor, σ is electrical conductivity, ΔV is voltage drop across the conductor, and L is the length/thickness of the conductor. Compare this Ohm's law with the following Fourier law of heat conduction for planar objects:

$$\frac{q}{A} = k\left(\frac{\Delta T}{L}\right) \tag{11.37}$$

Thus, the heat flow rate q is analogous to electric current I, thermal conductivity k is analogous to electric conductivity σ, temperature drop ΔT is analogous to voltage drop ΔV. The two laws could also be expressed in terms of resistances as:

$$I = \frac{\Delta V}{R} \quad \text{where } R = L/\sigma A \tag{11.38}$$

$$q = \frac{\Delta T}{R} \quad \text{where } R = L/kA \tag{11.39}$$

Thus, thermal resistance of an object is given as follows:

$$R = \frac{\text{Thickness of object}}{(\text{Thermal conductivity})(\text{Characteristic area of heat transfer})} \tag{11.40}$$

Figure 11.5 shows the electric analog of one-dimensional steady state heat conduction.

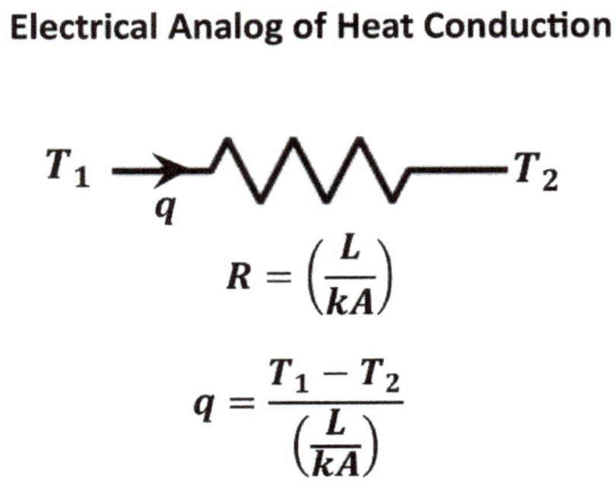

Figure 11.5 Electric analog of one-dimensional steady state heat conduction.

For cylindrical shells, the thickness is $(r_2 - r_1)$ and the characteristic area of heat transfer is the log mean area of inner and outer cylindrical surfaces, given as:

$$A_{lm} = \frac{A_2 - A_1}{ln\left(\frac{A_2}{A_1}\right)} = \frac{2\pi L(r_2 - r_1)}{ln\left(\frac{r_2}{r_1}\right)} \tag{11.41}$$

Therefore, the thermal resistance of cylindrical shells is:

$$R = \frac{(r_2 - r_1)}{(k)\left(\frac{2\pi L(r_2 - r_1)}{ln\left(\frac{r_2}{r_1}\right)}\right)} = \frac{ln\left(\frac{r_2}{r_1}\right)}{2\pi kL} \tag{11.42}$$

For spherical shells, the thickness is $(r_2 - r_1)$ and the characteristic area of heat transfer is the geometric mean area of inner and outer surfaces, given as:

$$A_{gm} = \sqrt{A_1 A_2} = 4\pi r_1 r_2 \tag{11.43}$$

Therefore, the thermal resistance of spherical shells is:

$$R = \frac{(r_2 - r_1)}{(k)(4\pi r_1 r_2)} = \frac{(r_2 - r_1)}{4\pi k r_1 r_2} \tag{11.44}$$

11.4.3 Temperature-Dependent Thermal Conductivity

The thermal conductivity of a material is generally a weak function of pressure but can vary significantly with temperature. Thus, we need to account for temperature variation of k. As k is not constant, the appropriate heat conduction equation is Equation (11.18):

$$\nabla \cdot (k\nabla T) + \dot{q} = \rho c_p \frac{\partial T}{\partial t} \tag{11.18}$$

Assuming steady state and no thermal energy generation, this equation simplifies to:

$$\nabla \cdot (k\nabla T) = 0 \tag{11.45}$$

Consider one-dimensional steady state heat conduction in a planar object in x direction. Equation (11.45) reduces to:

$$\frac{d}{dx}\left(k\frac{dT}{dx}\right) = 0 \tag{11.46}$$

Upon integration:

$$k\frac{dT}{dx} = C_1 \tag{11.47}$$

With further integration:

$$\int_{T_1}^{T_2} k(T)dT = C_1 \int_{x_1}^{x_2} dx \tag{11.48}$$

$$C_1 = \frac{\int_{T_1}^{T_2} k(T)dT}{(x_2 - x_1)} \tag{11.49}$$

Thus, the heat flux becomes:

$$q_x'' = -k\frac{dT}{dx} = -\frac{\int_{T_1}^{T_2} k(T)dT}{(x_2 - x_1)} \tag{11.50}$$

For constant k case, the heat flux is:

$$q_x'' = -k\frac{dT}{dx} = -\frac{k(T_2 - T_1)}{(x_2 - x_1)} \tag{11.51}$$

Upon comparing the above two expressions of heat flux, one can define a mean thermal conductivity as:

$$k_m = \frac{\int_{T_1}^{T_2} k(T)dT}{(T_2 - T_1)} \tag{11.52}$$

The heat flux in terms of mean thermal conductivity k_m is:

$$q_x'' = -\frac{k_m(T_2-T_1)}{(x_2-x_1)} \quad (11.53)$$

This is the same expression as that obtained in the constant k case, with k replaced by k_m. Thus, one can apply the same formulae for heat flux and thermal resistance obtained earlier for one-dimensional steady state heat conduction in different coordinate systems provided that the mean thermal conductivity, as defined in Equation (11.52), is used. However, the temperature distribution expressions obtained for constant k cases are no longer valid. For example, the temperature distribution in a planar object with constant k is linear but with variable k, say $k = a + bT$, it is non-linear as given below:

$$aT + (b/2)T^2 = C_1 x + C_2 \quad (11.54)$$

where C_1 and C_2 are integration constants.

11.4.4 Composite Systems with and without Contact Resistance

In many applications involving conductive heat transfer, conduction takes place in a multi-layer composite system consisting of several parallel layers of different materials joined together. If thermal contact between two layers of different materials is perfect, the interfacial boundary conditions are:

a) continuity of temperature at the interface, that is:

$$T_I = T_{II} \quad \text{at the interface} \quad (11.55)$$

b) continuity of heat flux at the interface, that is:

$$q_I'' = q_{II}'' \quad \text{at the interface} \quad (11.56)$$

$$-k_I \frac{dT_I}{dx}\bigg|_{\text{interface}} = -k_{II} \frac{dT_{II}}{dx}\bigg|_{\text{interface}} \quad (11.57)$$

The total thermal resistance for two layers in series with perfect thermal contact is given as follows:

$$R_{total} = R_I + R_{II} = \frac{L_I}{k_I A} + \frac{L_{II}}{k_{II} A} \quad (11.58)$$

where L_I and L_{II} are the thicknesses of the two layers.

The thermal contact between two layers of different materials is imperfect when the materials do not fit tightly together. Consequently, some voids or pockets of air/gas are formed at the interface between the two layers (see Figure 11.6). An imperfect contact between layers results in an interfacial resistance to heat transfer called thermal contact resistance, R_C'', defined as:

$$R_C'' = \frac{T_I - T_{II}}{q_{\text{interface}}''} \quad (11.59)$$

where $T_I - T_{II}$ is the temperature drop across the interface and $q_{\text{interface}}''$ is the heat flux at the interface. Note that the temperature distribution is not continuous at the interface although the heat flux is still continuous, that is:

$$T_I \neq T_{II} \quad \text{at the interface} \quad (11.60)$$

$$q_I'' = q_{II}'' = q_{\text{interface}}'' \quad \text{at the interface} \tag{11.61}$$

The interfacial or contact resistance is mainly a function of surface roughness, pressure keeping the two surfaces in contact, interfacial fluid material, and interface temperatures. At the interface, heat transfer occurs through conduction at the contact points and through convection and radiation across the trapped interfacial fluid.

The contact resistance R_C'' is defined on a unit interfacial-area basis. It has the SI units of m^2.K/W and is typically of the order of 10^{-4} m^2.K/W. The total contact resistance R_C is defined as:

$$R_C = \frac{R_C''}{A_{\text{interface}}} \tag{11.62}$$

where $A_{\text{interface}}$ is the interface area and R_C has SI units of K/W.

The total thermal resistance for two layers in series with imperfect thermal contact is as follows:

$$R_{total} = R_I + R_C + R_{II} = \frac{L_I}{k_I A} + \frac{R_C''}{A_{interface}} + \frac{L_{II}}{k_{II} A} \tag{11.63}$$

Figure 11.6 shows the electric analog of one-dimensional steady state heat conduction in composite system with imperfect thermal contact.

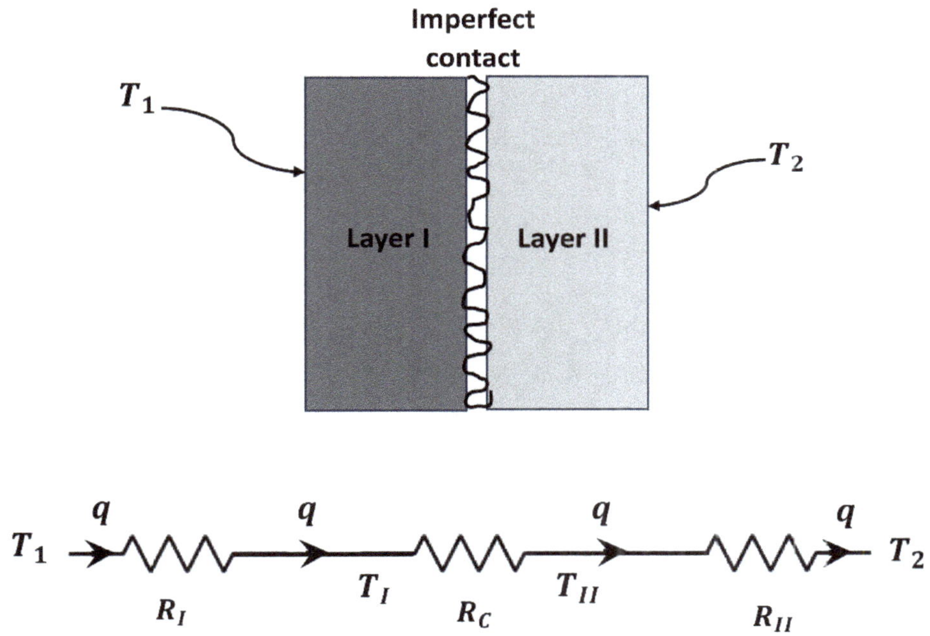

Figure 11.6 Electric analog of one-dimensional steady state heat conduction in two-layer composite system with imperfect thermal contact.

11.4.5 Conduction with Convection at the Boundaries

From Newton's law of cooling, the surface heat flux q_s'' at the solid boundary is given as:

$$q_s'' = h(T_s - T_C) \tag{11.64}$$

where T_C is the characteristic temperature of the fluid. For example, $T_C = T_\infty$ in external flows. From Equation (11.64) it follows that:

$$q = \frac{(T_s - T_C)}{\frac{1}{hA}} \tag{11.65}$$

Upon comparison with the Ohm's law (see Equations (11.38) and (11.39)), Equation (11.65) gives the convective heat transfer resistance as:

$$R = \frac{1}{hA} \tag{11.66}$$

Consider now a solid plane wall of thickness L and thermal conductivity k (see Figure 11.7). On one side of the wall, there is present a hot fluid with a free stream temperature of $T_{\infty,1}$ and on other side of the wall, there is present a cold fluid with free stream temperature of $T_{\infty,2}$. This problem has three thermal resistances in series as shown in Figure 11.7: convective resistance on the hot fluid side $(1/h_1A)$, conductive wall resistance (L/kA), and convective resistance on the cold fluid side $(1/h_2A)$. Thus, the total thermal resistance is:

$$R_{total} = \frac{1}{h_1A} + \frac{L}{kA} + \frac{1}{h_2A} \tag{11.67}$$

where A is the heat transfer area normal to the direction of heat flow, h_1 and h_2 are heat transfer coefficients on the two sides. The heat transfer rate q through this system is as follows:

$$q = \frac{T_{\infty,1} - T_{\infty,2}}{R_{total}} = \frac{T_{\infty,1} - T_{\infty,2}}{\frac{1}{h_1A} + \frac{L}{kA} + \frac{1}{h_2A}} \tag{11.68}$$

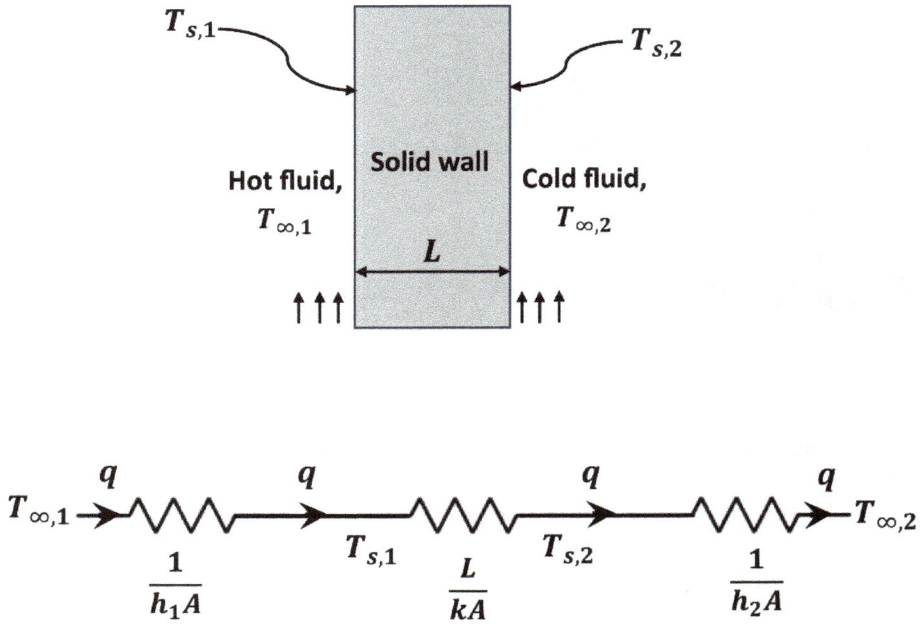

Figure 11.7 Electric analog of one-dimensional steady state heat conduction with convection at the boundaries.

Once q is known, the surface temperatures at the two sides of the solid wall can be calculated as follows:

$$T_{\infty,1} - T_{s,1} = q\left(\frac{1}{h_1 A}\right) \tag{11.69}$$

$$T_{s,2} - T_{\infty,2} = q\left(\frac{1}{h_2 A}\right) \tag{11.70}$$

In such combined conductive-convective problems, one can also define an overall heat transfer coefficient (U) as:

$$q = UA(T_{\infty,1} - T_{\infty,2}) = \frac{T_{\infty,1}-T_{\infty,2}}{R_{total}} \tag{11.71}$$

Thus, the overall heat transfer coefficient is given as:

$$U = \frac{1}{R_{total}A} = \frac{1}{\left(\frac{1}{h_1}+\frac{L}{k}+\frac{1}{h_2}\right)} \tag{11.72}$$

Consider another example where a hot fluid is flowing in a cylindrical pipe. Let the pipe wall be composite consisting of two layers, the inner layer of metal with $k = k_I$ and the outer layer of insulating material with $k = k_{II}$. The mean temperature of the fluid inside the pipe at any given axial position is T_m. The free stream temperature of the cold fluid outside the composite pipe wall is T_∞. Now there are four thermal resistors in series as shown in Figure 11.8: inside convective resistance, conductive metal-wall resistance, conductive insulation resistance, and outside convective resistance. Thus, the total resistance is:

$$R_{total} = \frac{1}{h_i A_i} + \frac{ln(r_2/r_1)}{2\pi k_I L} + \frac{ln(r_3/r_2)}{2\pi k_{II} L} + \frac{1}{h_o A_o} \tag{11.73}$$

where h_i and h_o are the inside and outside heat transfer coefficients, respectively, A_i is the inside area of the cylindrical geometry (equal to $2\pi r_1 L$), and A_o is the outside area of the cylindrical geometry (equal to $2\pi r_3 L$). The heat transfer rate q for this system is:

$$q = \frac{T_m - T_\infty}{R_{total}} = \frac{T_m - T_\infty}{\frac{1}{h_i A_i} + \frac{ln(r_2/r_1)}{2\pi k_I L} + \frac{ln(r_3/r_2)}{2\pi k_{II} L} + \frac{1}{h_o A_o}} \tag{11.74}$$

where T_m is the mean temperature of the fluid inside the pipe.

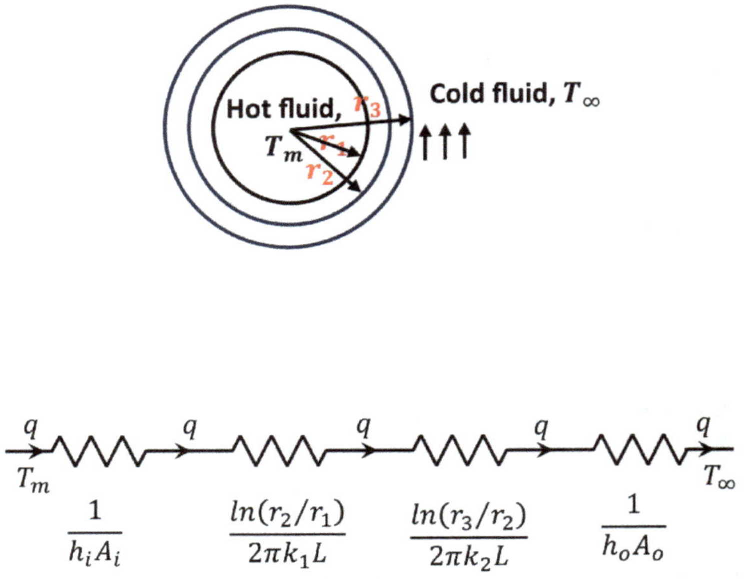

Figure 11.8 Composite pipe wall system with inside and outside convective resistances.

For a planar geometry, there is no ambiguity about the heat transfer area in the definition of the overall heat transfer coefficient. For radial geometries (cylindrical or spherical), however, one must specify the area on which the overall heat transfer coefficient is based. One can define the overall heat transfer coefficient either based on the inside heat transfer area (U_i) or based on the outside heat transfer area (U_o). Thus:

$$q = \frac{T_m - T_\infty}{R_{total}} = U_i A_i (T_m - T_\infty) = U_o A_o (T_m - T_\infty) \tag{11.75}$$

Upon comparison of different terms of Equation (11.75), we get:

$$U_o A_o = U_i A_i = \frac{1}{R_{total}} \tag{11.76}$$

Thus,

$$U_o = \frac{1}{\frac{1}{h_i}\left(\frac{A_o}{A_i}\right) + A_o \frac{ln(r_2/r_1)}{2\pi k_I L} + A_o \frac{ln(r_3/r_2)}{2\pi k_{II} L} + \frac{1}{h_o}} \tag{11.77}$$

$$U_i = \frac{1}{\frac{1}{h_i} + A_i \frac{ln(r_2/r_1)}{2\pi k_I L} + A_i \frac{ln(r_3/r_2)}{2\pi k_{II} L} + \frac{1}{h_o}\left(\frac{A_i}{A_o}\right)} \tag{11.78}$$

11.4.6 Heat Conduction with Thermal Energy Generation

Heat conduction with thermal energy generation is important industrially. In many applications, thermal energy is generated by passage of electric current through the medium.

Consider a plane solid wall of thickness $2L$, maintained at same temperatures on both sides, that is,

$$T = T_s \text{ at } x = +L \tag{11.79}$$

$$T = T_s \text{ at } x = -L \tag{11.80}$$

where x is the distance from the centre of the wall (see Figure 11.9). The rate of thermal energy generation within the wall per unit volume is $\dot{q}$. The temperature distribution in the wall can be obtained by solving the following Poisson form of heat conduction equation (assuming k is constant):

$$\frac{d^2T}{dx^2} + \frac{\dot{q}}{k} = 0 \tag{11.81}$$

The solution of this equation subject to the given boundary conditions is:

$$T - T_s = \frac{\dot{q}L^2}{2k}\left[1 - \left(\frac{x}{L}\right)^2\right] \tag{11.82}$$

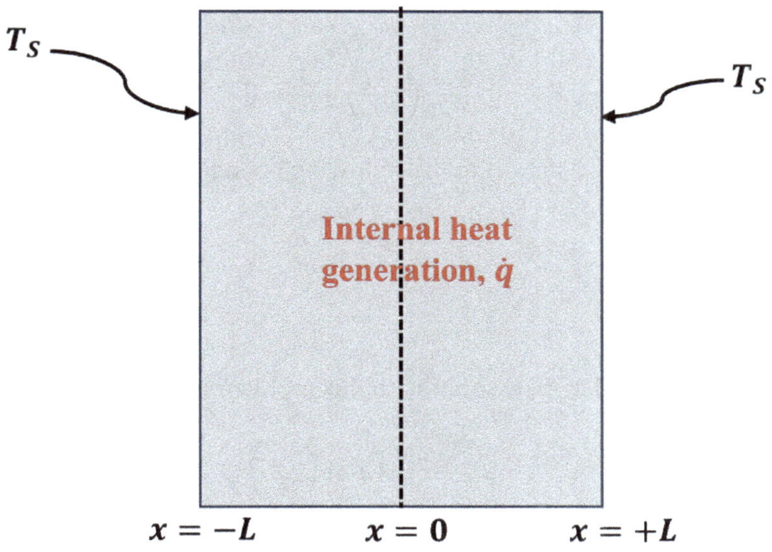

Figure 11.9 Heat conduction with internal thermal energy generation.

The temperature distribution is symmetric about the mid-plane with a maximum at $x = 0$ (mid-plane). The maximum temperature is:

$$T_{max} = T_s + \frac{\dot{q}L^2}{2k} \tag{11.83}$$

Thus, the temperature distribution can be expressed in dimensionless form as:

$$\frac{T-T_S}{T_{max}-T_S} = 1 - \left(\frac{x}{L}\right)^2 \tag{11.84}$$

The heat flux at any location is:

$$q_x'' = -k\frac{dT}{dx} = k\left[\frac{2x}{L^2}(T_{max} - T_s)\right] = \dot{q}x \tag{11.85}$$

The heat transfer rates at the two sides of the wall are:

$$q|_{x=-L} = -\dot{q}\left(\frac{V}{2}\right) \tag{11.86}$$

$$q|_{x=+L} = \dot{q}\left(\frac{V}{2}\right) \tag{11.87}$$

where V is the volume of the wall. Thus, the total heat generated within the wall leaves the wall from the two sides equally.

A similar analysis could be carried out for radial systems with heat generation. For example, consider a long, solid cylinder with heat generation. Heat generation in the cylinder could be due to passage of current. Heat generation within a solid cylindrical object is also encountered in nuclear reactors consisting of cylindrical fuel elements.

The Poisson form of heat conduction equation for this case of heat conduction in radial direction is as follows:

$$\frac{1}{r}\frac{d}{dr}\left(r\frac{dT}{dr}\right) + \frac{\dot{q}}{k} = 0 \tag{11.88}$$

Equation (11.88) needs to be solved subject to the following boundary conditions:

$$T = T_s \quad \text{at}\ r = R \tag{11.89}$$

$$\left.\frac{dT}{dr}\right|_{r=0} = 0 \tag{11.90}$$

where R is the radius of the cylinder. The solution to this problem is as follows:

$$T - T_s = \frac{\dot{q}R^2}{4k}\left(1 - \left(\frac{r}{R}\right)^2\right) \tag{11.91}$$

The temperature distribution is symmetric about the centre line with a maximum at $r = 0$ (centre line). The maximum temperature is:

$$T_{max} = T_s + \frac{\dot{q}R^2}{4k} \tag{11.92}$$

Thus, the temperature distribution can be expressed in dimensionless form as:

$$\frac{T - T_s}{T_{max} - T_s} = 1 - \left(\frac{r}{R}\right)^2 \tag{11.93}$$

The heat flux at any radial location is:

$$q_r'' = -k\frac{dT}{dr} = k\left[\frac{2r}{R^2}(T_{max} - T_s)\right] = \dot{q}r/2 \tag{11.94}$$

Thus,

$$q|_{r=R} = (2\pi RL)q_r''|_{r=R} = (\pi R^2 L)\dot{q} = V\dot{q} \tag{11.95}$$

where V is the volume of the cylinder.

11.4.7 Heat Transfer from Extended Surfaces

Extended surfaces are used to enhance heat transfer between a solid surface and a fluid moving adjacent to the solid surface. For example, consider a plane solid wall at a uniform temperature T_b. The adjoining fluid is moving at a free stream velocity of V_∞ with a free stream temperature of T_∞. Assuming $T_b > T_\infty$, heat will flow from the solid surface to the fluid. The heat flow rate q is given by:

$$q = hA(T_b - T_\infty) \tag{11.96}$$

where h is the heat transfer coefficient and A is the heat transfer area (area in contact with the moving fluid).

There are three ways to increase the rate of heat transfer between the plane solid surface and the moving fluid: 1) increase the fluid velocity to increase the heat transfer coefficient, 2) increase the temperature difference $T_b - T_\infty$ by reducing the fluid free stream temperature, and 3) increase the contact area A between the fluid and the solid across which convection is taking place. In many practical applications, one has little or no control on fluid velocity and fluid temperature. The only viable option to enhance the heat transfer rate in such applications is to increase the fluid/solid contact area across which convection occurs. The surface area exposed to the fluid can be "extended" by employing fins. Fins are protrusions from the base surface into the cooling fluid. They are made from high thermal conductivity material to provide maximum possible enhancement of heat transfer. Fins come in a variety of shapes and forms, such as: straight or longitudinal fins with different shape profiles (rectangular, trapezoidal, parabolic), annular fins, and cylindrical pin fins.

To analyse heat transfer from a fin, consider a long cylindrical pin fin of constant cross-section attached to a plane wall at a uniform surface temperature T_b, as shown in Figure 11.10. Let x be the axial coordinate of the fin. The surrounding fluid is at a free stream temperature of T_∞ and the heat transfer coefficient is h. Assuming that the transverse temperature gradients in the fin are negligible (the temperature at any given cross-section of the fin is uniform), the temperature in the fin is a function of x only.

Energy balance on a small cylindrical element (length Δx) of the fin shown in Figure 11.11 at steady state can be expressed as follows:

$$\begin{Bmatrix}\text{Rate at which heat enters}\\ \text{the element by conduction}\end{Bmatrix} = \begin{Bmatrix}\text{Rate at which heat leaves}\\ \text{the element by conduction}\end{Bmatrix} + \begin{Bmatrix}\text{Rate at which heat leaves}\\ \text{the element by convection}\end{Bmatrix} \tag{11.97}$$

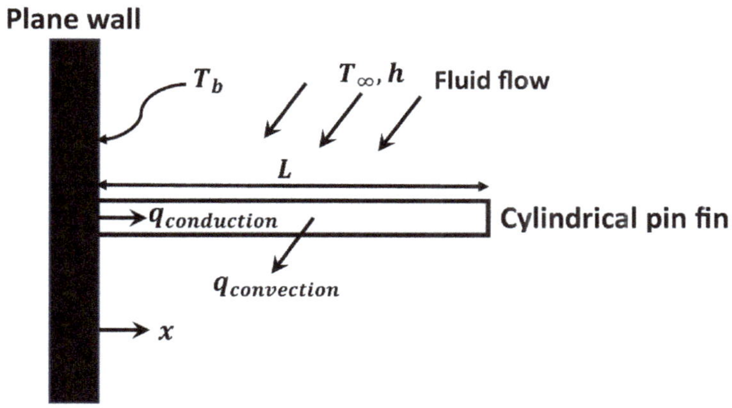

Figure 11.10 Heat transfer from a cylindrical pin fin.

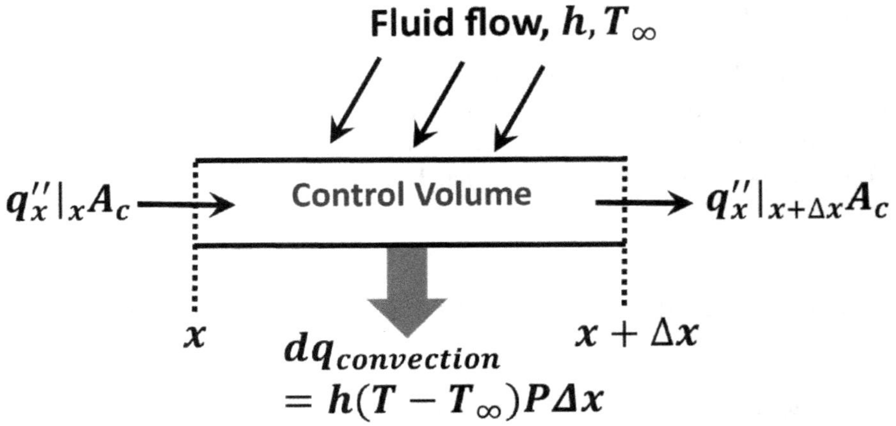

Figure 11.11 Energy balance over a differential control volume of a fin.

Each term of the energy balance (Equation (11.97)) is now determined (see Figure 11.11).

$$\begin{Bmatrix} \text{Rate at which heat} \\ \text{enters by conduction} \end{Bmatrix} = q''_x|_x A_c \tag{11.98}$$

$$\begin{Bmatrix} \text{Rate at which heat} \\ \text{leaves by conduction} \end{Bmatrix} = q''_x|_{x+\Delta x} A_c \tag{11.99}$$

$$\begin{Bmatrix} \text{Rate at which heat} \\ \text{leaves by convection} \end{Bmatrix} = h(T - T_\infty)P\Delta x \tag{11.100}$$

where A_c is the cross-sectional area and P is the perimeter of the fin. Thus, energy balance could be written as:

$$\frac{q''_x|_{x+\Delta x} - q''_x|_x}{\Delta x} = -\frac{h(T-T_\infty)P}{A_c} \tag{11.101}$$

In the limit $\Delta x \to 0$, this gives:

$$\frac{dq''_x}{dx} = \frac{-h(T-T_\infty)P}{A_c} \tag{11.102}$$

Using Fourier's law of heat conduction, Equation (11.102) can be re-written as:

$$\frac{d^2T}{dx^2} - \frac{hP(T-T_\infty)}{kA_c} = 0 \tag{11.103}$$

The boundary conditions are:

$$x = 0, \;\; T = T_b \tag{11.104}$$

$$x = L, \;\; \left.\frac{dT}{dx}\right|_{x=L} = 0 \tag{11.105}$$

where L is the length of the fin. The second boundary condition follows from the assumption of adiabatic tip. As the tip area of the fin is small, we can neglect heat transfer from the tip. The solution of this problem is as follows:

$$\frac{T-T_\infty}{T_b-T_\infty} = \frac{\cosh m(L-x)}{\cosh mL} \tag{11.106}$$

where m is defined as:

$$m = \sqrt{\frac{hP}{kA_c}} \tag{11.107}$$

The heat transfer rate from the fin can be determined by noting that all the heat lost by the fin must be conducted into the fin at the root or base ($x = 0$). Thus:

$$q = A_c q''_x|_{x=0} = -kA_c\left(\frac{dT}{dx}\right)_{x=0} = kA_c m(T_b - T_\infty)\tanh m\,L \tag{11.108}$$

The effectiveness (ε) of a fin is defined as the ratio of heat transfer rate from the fin to heat transfer rate without fin. Heat transfer without fin refers to heat transfer that would have occurred from the base area in the absence of the fin. Thus:

$$\varepsilon = \frac{q_{fin}}{A_c h(T_b-T_\infty)} \tag{11.109}$$

The effectiveness must be larger than unity for fin to provide additional heat transfer. For a cylindrical pin fin with adiabatic tip, ε is given as:

$$\varepsilon = \sqrt{\frac{kP}{hA_c}}\tanh m\,L \tag{11.110}$$

For long fins with $T \rightarrow T_\infty$ at the tip, this expression reduces to:

$$\varepsilon = \sqrt{\frac{kP}{hA_c}} \tag{11.111}$$

Fins are more effective if k of fin is high, diameter of fin is small (P/A_c is large), and the convection coefficient h is low (fins are more effective when the fluid is gas).

Another important measure of fin performance is the fin efficiency, defined as the ratio of actual rate of heat transfer from the fin to maximum possible rate of heat transfer from the fin:

$$\eta = \frac{q_{\text{fin,actual}}}{q_{\text{fin,maximum}}} \tag{11.112}$$

The heat transfer rate from the fin is maximum when the entire surface of the fin is at the base temperature T_b. Therefore,

$$\eta = \frac{q_{\text{fin,actual}}}{hA_f(T_b-T_\infty)} \tag{11.113}$$

where A_f is the surface are of the fin exposed to the fluid. The efficiency of a fin depends on several factors. It increases with the increase in thermal conductivity of the fin material. It decreases with the increase in fin length. It also depends on the geometry of the fin.

For cylindrical pin fin with adiabatic tip:

$$\eta = \frac{\tanh mL}{mL} \tag{11.114}$$

In most practical applications involving the use of fins, an array of fins is used to increase the heat transfer area. Let A_t be the total surface area in contact with the fluid. If there are N fins and the exposed surface area of each fin is A_f. Then:

$$A_t = NA_f + A_b \tag{11.115}$$

where A_b is the exposed portion of the base area (area not occupied by the fins). Thus, the total rate of heat transfer (q_t) from an array of fins is:

$$q_t = Nq_{fin} + A_b h(T_b - T_\infty) = N\eta h A_f(T_b - T_\infty) + A_b h(T_b - T_\infty) \tag{11.116}$$

Upon substitution of A_t and re-arrangement, this expression can be re-written as:

$$q_t = hA_t\left[1 - \frac{NA_f}{A_t}(1-\eta)\right](T_b - T_\infty) \tag{11.117}$$

The overall efficiency of an array of fins is defined as

$$\eta_o = \frac{q_t}{q_{max}} = \frac{q_t}{hA_t(T_b - T_\infty)} \tag{11.118}$$

where q_{max} is the maximum possible rate of heat transfer from the array assembly. Upon substitution of the expression for q_t, the overall efficiency can be written as:

$$\eta_o = \left[1 - \frac{NA_f}{A_t}(1-\eta)\right] \tag{11.119}$$

11.5 TRANSIENT HEAT CONDUCTION

When a solid body is suddenly subjected to a change in thermal environment, the temperature inside the body varies with both location and time. Such unsteady or transient heat conduction problems are commonly encountered in engineering applications. To obtain time dependence of the temperature distribution within an object, one needs to solve the heat conduction equation subject to the given initial and boundary conditions. As an example, consider transient heat conduction in a semi-infinite solid. Such a solid has a single plane surface and extends to infinity in all but one direction. The heat conduction equation for this situation can be written as:

$$\frac{\partial^2 T}{\partial x^2} = \frac{1}{\alpha}\frac{\partial T}{\partial t} \tag{11.120}$$

The object is initially at a uniform temperature of T_i. The exposed surface is suddenly subjected to a constant temperature of T_s. Thus, the boundary conditions are:\

$$T(0,t) = T_s \text{ and } T(\infty,t) = T_i \quad (11.121)$$

and the initial condition is:

$$T(x,0) = T_i \quad (11.122)$$

This problem can be solved by two different techniques: similarity technique and Laplace transform technique. According to the similarity technique, the two independent variables (x and t) are combined into a single variable η (called the similarity variable) such that the original partial differential equation is converted into an ordinary differential equation. This technique usually works if the T versus x profiles are similar at different times. For the problem at hand, the similarity variable is:

$$\eta = \frac{x}{2\sqrt{\alpha t}} \quad (11.123)$$

The partial differential equation (Equation (11.120)) can now be converted into an ordinary differential equation using η. Thus, Equation (11.120) becomes:

$$\frac{d^2T}{d\eta^2} = -2\eta\frac{dT}{d\eta} \quad (11.124)$$

and the new boundary conditions are:

$$T|_{\eta=0} = T_s \text{ and } T|_{\eta=\infty} = T_i \quad (11.125)$$

The solution of Equation (11.124) subjected to the given boundary conditions is as follows:

$$\frac{T-T_i}{T_s-T_i} = 1 - erf\left(\frac{x}{2\sqrt{\alpha t}}\right) \quad (11.126)$$

where erf (z) is the error function defined as:

$$erf(z) = \frac{2}{\sqrt{\pi}}\int_0^z exp(-u^2)du \quad (11.127)$$

The Laplace transform technique could also be applied to solve this problem. The Laplace transform of any function $f(t)$ is defined as:

$$\mathcal{L}[f(t)] = \int_0^\infty f(t)e^{-st}dt \quad (11.128)$$

where s is the Laplace variable. The given partial differential equation, Equation (11.120), can be converted into an ordinary differential equation using Laplace transformation. Thus,

$$\alpha\frac{d^2T}{dx^2} - sT = -T_i \quad (11.129)$$

where T is now a function of x and s. The new boundary conditions are:

$$T(x=0,s) = \frac{T_S}{s} \text{ and } T(x=\infty,s) = \frac{T_i}{s} \quad (11.130)$$

The solution of Equation (11.129) subjected to the given boundary conditions is:

$$T(x,s) = \frac{T_i}{s} + \left(\frac{T_s - T_i}{s}\right) e^{-(\sqrt{s/\alpha})x} \tag{11.131}$$

The same solution (Equation (11.126)) is obtained upon taking the inverse Laplace transform of Equation (11.131).

From the solution of $T(x,t)$, one can easily determine the instantaneous heat flux at the surface as:

$$q_x''(t) = -k\left(\frac{\partial T}{\partial x}\right)_{x=0} = \frac{k(T_s - T_i)}{\sqrt{\pi \alpha t}} \tag{11.132}$$

11.5.1 Lumped Capacity Analysis

In many transient heat conduction problems of practical interest, the object is initially at a uniform temperature. It is suddenly immersed in a fluid with a different temperature. Heat transfer occurs between the object and the fluid. This problem involves both conduction and convection, conduction within the object and convection at the boundaries of the object.

In general, the temperature of the object varies with both spatial coordinates and time. However, the problem is easy to solve if there is no spatial variation of temperature, that is, the temperature of the object is uniform spatially but varies with time. This can occur if the internal conductive resistance is negligible in comparison with the external convective resistance. The ratio of internal conductive resistance to external convective resistance is called the Biot number (Bi):

$$Bi = \frac{(L_c/kA)}{(1/hA)} = \frac{hL_c}{k} \tag{11.133}$$

where L_c is the characteristic length of the object.

The lumped capacity analysis assumes that the temperature of the object is uniform spatially but varies with time. This is a safe assumption when $Bi < 0.1$. For low Bi, the internal conductive resistance is quite small compared with the external convective resistance.

From energy balance:

$$\text{Rate of heat loss by convection} = \text{Rate of decrease of internal energy} \tag{11.134}$$

$$hA(T - T_\infty) = -mc_p \frac{dT}{dt} \tag{11.135}$$

where m is the mass of the object, A is its surface area, c_p is the heat capacity, T is the temperature at any time t, and T_∞ is the free-stream temperature of the fluid.

Upon integration, Equation (11.135) gives:

$$\frac{T - T_\infty}{T_i - T_\infty} = exp\left(-\frac{hA}{\rho V c_p} t\right) \tag{11.136}$$

where T_i is the initial temperature of the object and V is its volume. The instantaneous rate of heat transfer is:

$$q = hA(T - T_\infty) \tag{11.137}$$

The total amount of heat transferred to the fluid from $t = 0$ to time t is:

$$Q = mc_p(T_i - T) \tag{11.138}$$

The maximum amount of heat that can be transferred to the fluid is:

$$Q_{max} = mc_p(T_i - T_\infty) \tag{11.139}$$

11.5.2 Transient Heat Conduction with Non-negligible Internal Resistance

The lumped capacity analysis assumes negligible internal conductive resistance. It can only be applied if the Biot number is less than 0.1. When Bi is larger than 0.1, the internal conductive resistance is significant in comparison with the external convective resistance. Both spatial and time variations of temperature become important. In such situations, one needs to solve the heat conduction equation subject to the given initial and boundary conditions.

Consider transient heat conduction in a rectangular slab of thickness $2L$. The slab is initially at a uniform temperature of T_i. It is suddenly immersed in a cold fluid with a free stream temperature of T_∞. The internal conductive resistance is non-negligible as compared with external convective resistance. For this problem, the heat conduction equation along with the initial and boundary conditions are as follows:

$$\frac{\partial T}{\partial t} = \alpha \frac{\partial^2 T}{\partial x^2} \tag{11.140}$$

Initial condition:

$$T(x, t = 0) = T_i \tag{11.141}$$

Boundary conditions:

$$-k\left(\frac{\partial T}{\partial x}\right)_{x=L} = h(T|_{x=L} - T_\infty) \tag{11.142}$$

$$\left(\frac{\partial T}{\partial x}\right)_{x=0} = 0 \tag{11.143}$$

Note that at mid-plane ($x = 0$), the temperature gradient is zero due to symmetry of the problem.

This mathematical problem can be solved using the separation of variables technique. The solution $T(x, t)$ turns out to be an infinite series that can be expressed as:

$$\frac{T(x,t) - T_\infty}{T_i - T_\infty} = f\left(\frac{x}{L}, Fo, Bi\right) \tag{11.144}$$

where Fo is the Fourier number, defined as:

$$Fo = \frac{t}{(L^2/\alpha)} = \frac{\alpha t}{L^2} \tag{11.145}$$

The solution is available in the literature in the form of charts, referred to as "Heisler Charts". The Heisler charts are often presented in two parts. The first chart gives the dimensionless temperature at the centre or mid-plane ($x = 0$) of the object $(T_c - T_\infty)/(T_i - T_\infty)$, as a function of Fo and Bi. Once the mid-plane temperature is determined, the second chart is read to obtain the temperature $T(x,t)$ at any other location. The second chart presents $T(x,t)$ in dimensionless form as $(T(x,t) - T_\infty)/(T_c - T_\infty)$, as a function of Bi and x/L.

Figure 11.12 shows the Heisler charts for transient heat conduction problem involving a rectangular slab of thickness $2L$. The Heisler charts corresponding to radial geometries (cylindrical and spherical objects) are given in Appendix B. Note that Biot and Fourier numbers for radial geometries are defined as follows:

$$Bi = \frac{hr_o}{k} \tag{11.146}$$

$$Fo = \frac{\alpha t}{r_o^2} \tag{11.147}$$

where r_o is the radius of the cylinder or sphere.

In addition to local temperature, the other quantity of interest in transient heat conduction problems is the amount of heat exchanged between the solid and the ambient fluid during the time interval $0 - t$. The ratio of Q/Q_{max} for transient heat conduction problems are given in the form of dimensionless charts, referred to as Grober charts, where Q is the amount of heat exchanged during the time interval $0 - t$ and Q_{max} is the maximum amount of heat exchange possible between the object and the ambient fluid. Figure 11.13 shows the Grober chart for transient heat conduction problem involving a rectangular slab of thickness $2L$. The ratio Q/Q_{max} is plotted as a function of Bi^2Fo for different values of Biot number. Thus, the ratio Q/Q_{max} can be readily obtained from the chart. Note that Q_{max} is given as:

$$Q_{max} = \rho c_p V(T_i - T_\infty) \tag{11.148}$$

The Grober charts for transient heat conduction problems involving radial geometries (cylindrical and spherical objects) are given in Appendix B.

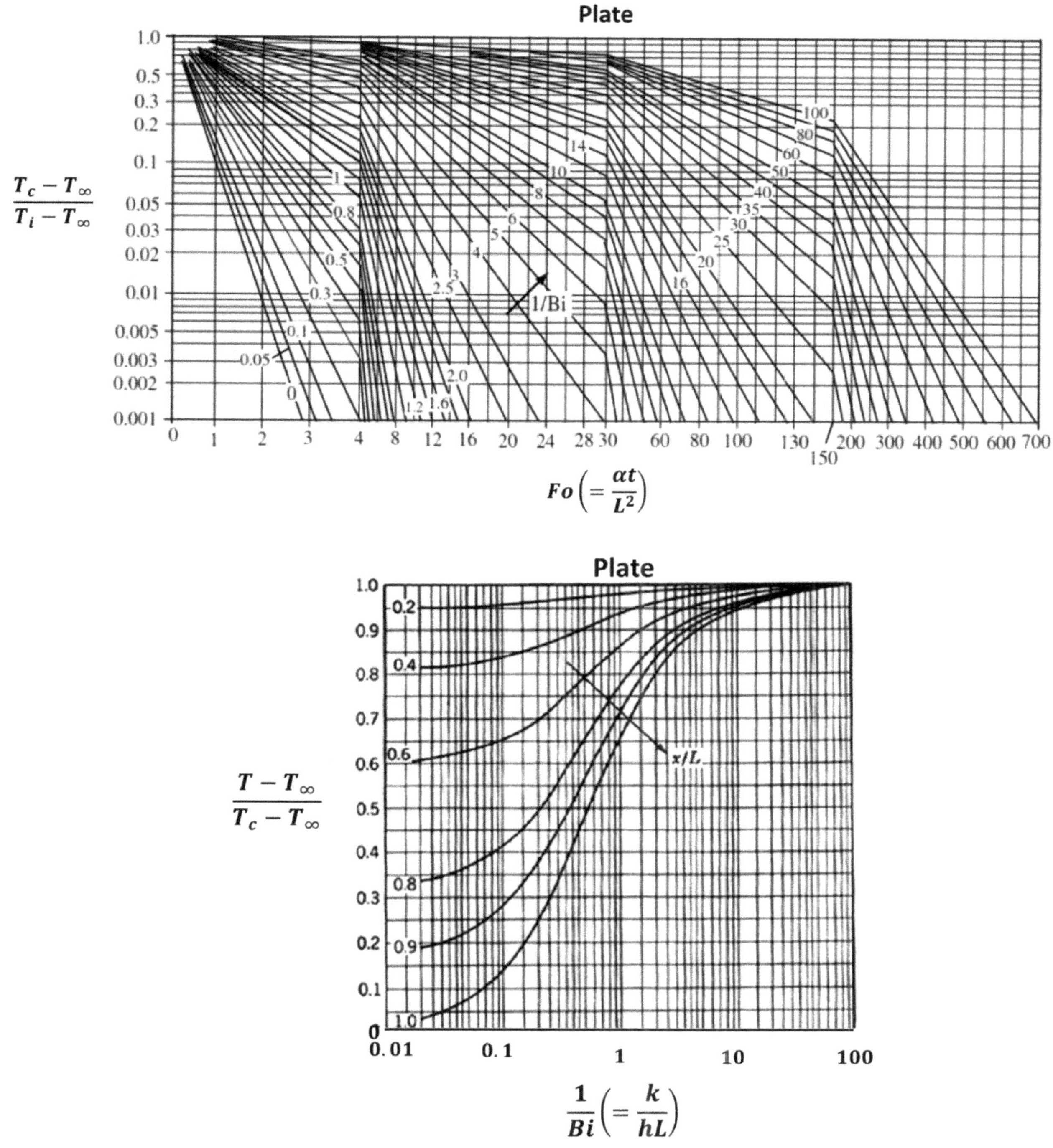

Figure 11.12 Heisler charts for transient heat conduction problem involving a rectangular slab of thickness $2L$.

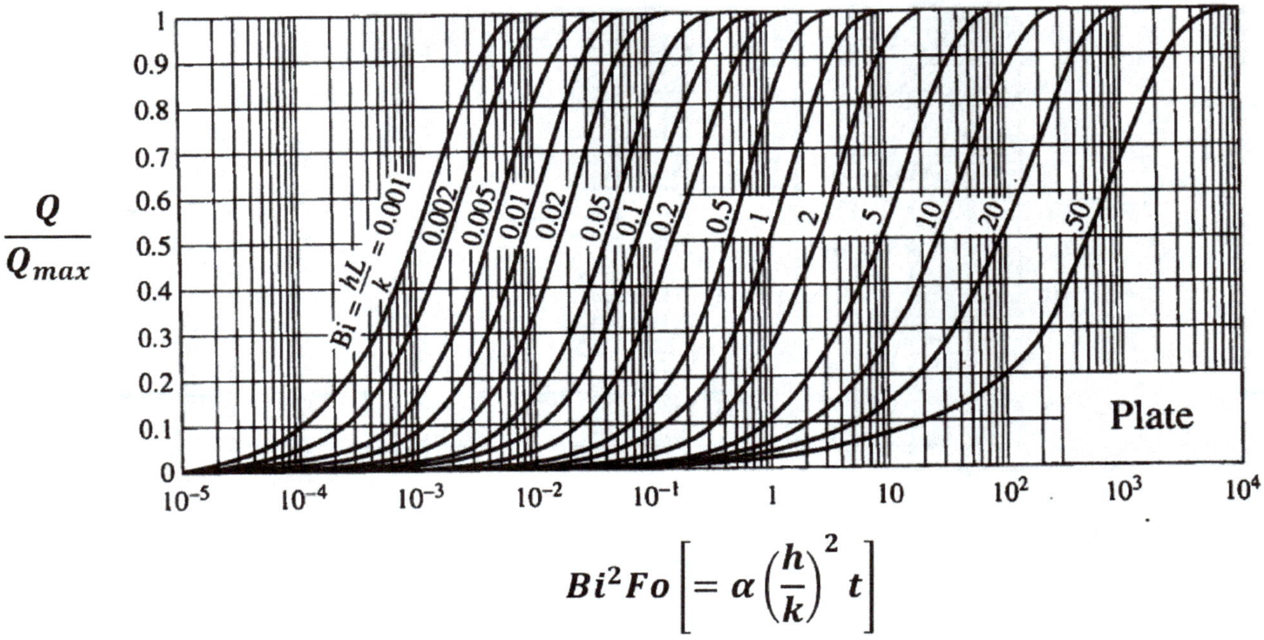

Figure 11.13 Grober chart for transient heat conduction problem involving a rectangular slab of thickness $2L$.

11.6 TWO-DIMENSIONAL HEAT CONDUCTION USING SHAPE FACTOR APPROACH

In two-dimensional steady state heat conduction problems where only two isothermal boundaries are involved, the heat transfer rate between the isothermal boundaries can be calculated from the following equation:

$$q = kS(\Delta T) \tag{11.149}$$

where k is the thermal conductivity of the medium between the isothermal boundaries, S is the shape factor of the system, and ΔT is the temperature difference between the isothermal boundaries. Note that the shape factor depends only on the shape or geometry of the system and has the units of length.

The shape factors for several two-dimensional heat conduction problems encountered in practice are given in Table 11.4. The shape factors for many more practical problems are available in the literature. The shape factors are determined using several different approaches such as graphical, conformal mapping, analytical, numerical, and empirical methods.

Table 11.4 Shape factors for sample two-dimensional heat conduction problems

Geometry	Diagram	Shape factor, S
Isothermal cylinder buried horizontally in a semi-infinite medium	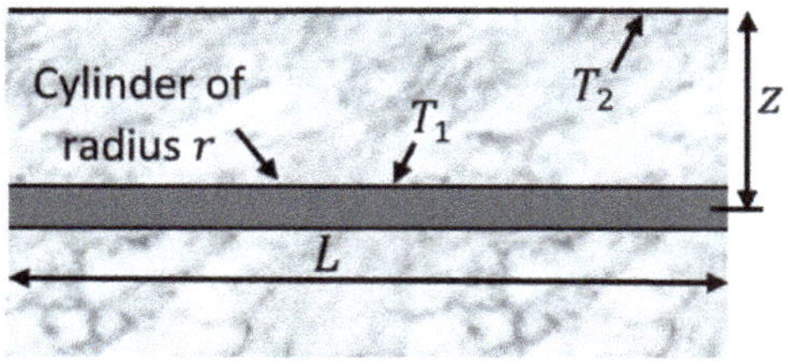	$S = \dfrac{2\pi L}{\cosh^{-1}(z/r)}$ $L \gg r$
Two isothermal cylinders buried horizontally in infinite medium	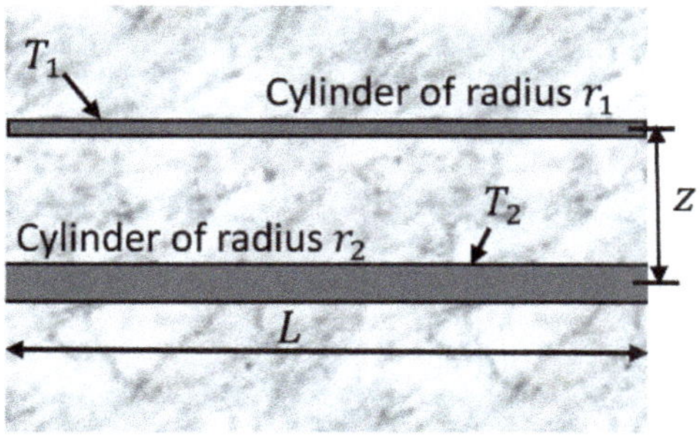	$S = \dfrac{2\pi L}{\cosh^{-1}\left[\frac{z^2-r_2^2-r_1^2}{2r_1r_2}\right]}$ $L \gg r_1, r_2, z$
Isothermal cylinder buried vertically in a semi-infinite medium	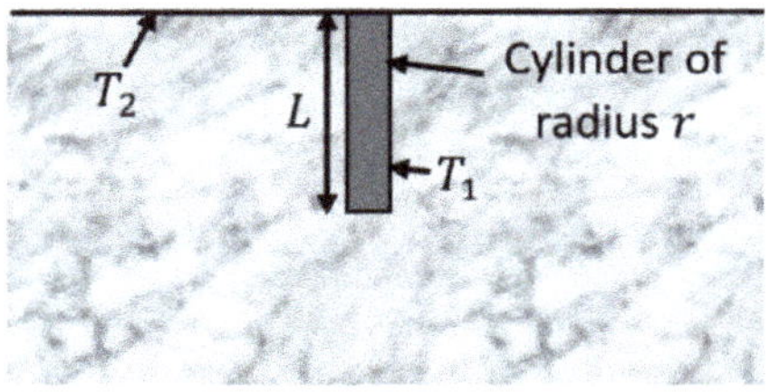	$S = \dfrac{2\pi L}{\ln(2L/r)}$ $L \gg 2r$
Isothermal sphere buried in a semi-infinite medium	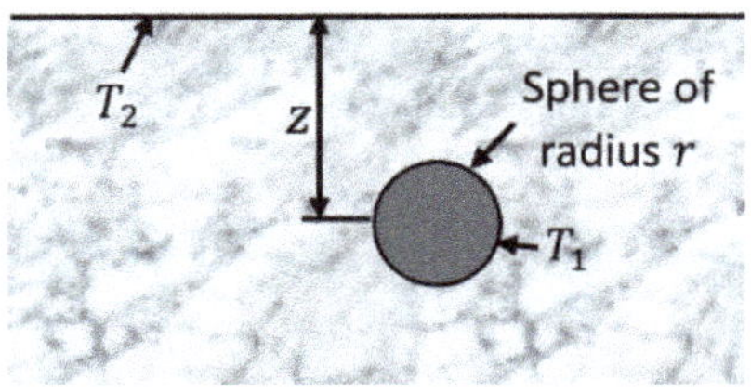	$S = \dfrac{4\pi r}{1-(r/2z)}$ $z > r$
Two isothermal spheres buried in infinite medium	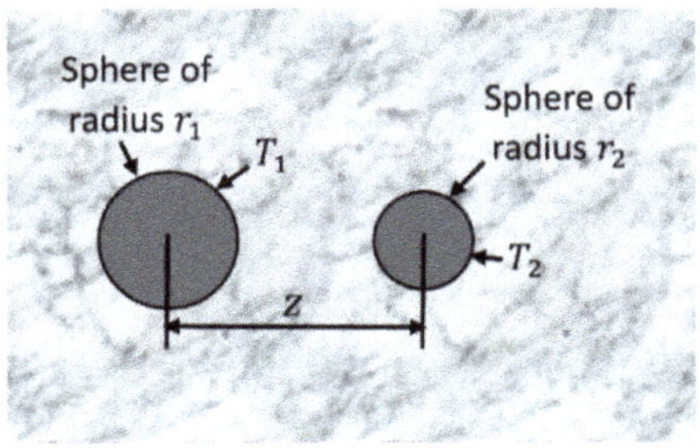	$S = \dfrac{4\pi r_2}{\left(\frac{r_2}{r_1}\right)\left[1-\frac{(r_1/z)^4}{1-(r_2/z)^2}\right]-\frac{2r_2}{z}}$ $z > 5r_1$

11.7 EXAMPLES

11.7.1 Example 1

Consider the composite wall as shown in the figure below. Determine the rate of heat transfer through the composite wall on a per-square-foot basis. Also calculate the unknown interface temperatures. Assume that the contact between the layers is perfect.

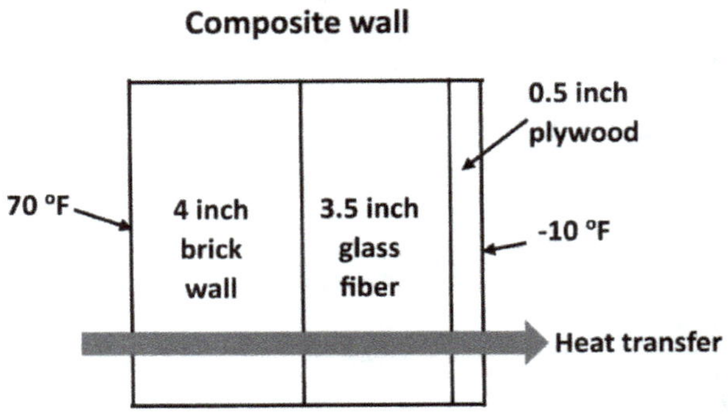

$K_{brick} = 0.38$ Btu/(hr.ft.°F)

$K_{glass\ fiber} = 0.02$ Btu/(hr.ft.°F)

$K_{plywood} = 0.063$ Btu/(hr.ft.°F)

Solution:

The electric analog of the composite wall is shown in the figure below. There are three resistors in series.

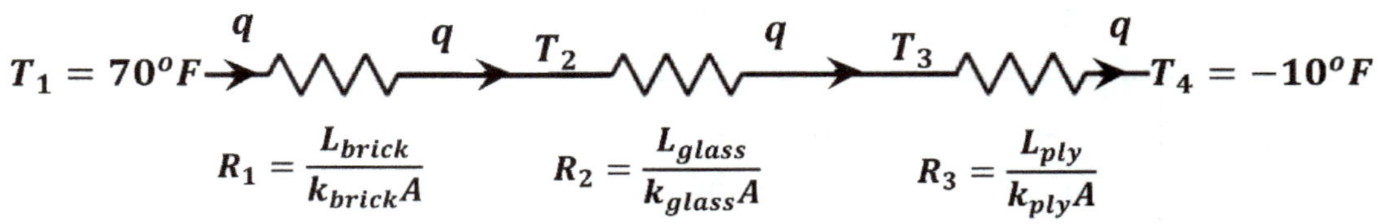

The total resistance is:

$R_{total} = R_1 + R_2 + R_3 = \frac{L_{brick}}{k_{brick}A} + \frac{L_{glass}}{k_{glass}A} + \frac{L_{ply}}{k_{ply}A} = \frac{4/12}{0.38A} + \frac{3.5/12}{0.02A} + \frac{0.5/12}{0.063A} = \frac{1}{A}[0.877 + 14.583 + 0.661] = \frac{16.12}{A}$ °F/(Btu/hr)

$q = \frac{\Delta T_{overall}}{R_{total}} = \frac{70-(-10)}{16.12}A\ ;\ \ \frac{q}{A} = 4.962\ \frac{Btu}{hr.ft^2}$

Interface temperatures:

$T_1 - T_2 = qR_1 = 4.962 \times 0.877 = 4.35^oF\ ;\ \ T_2 = T_1 - 4.35 = 70 - 4.35 = 65.65^oF$

$T_2 - T_3 = qR_2 = 4.962 \times 14.583 = 72.36^oF\ ;\ \ T_3 = T_2 - 72.36 = 70 - 72.36 = -6.71^oF$

11.7.2 Example 2

A thermopane window consists of two layers of glass separated by a layer of stagnant air. The glass layers of a given thermopane window are 6.5 mm thick and they are separated by a stagnant layer of 6.0 mm thick air. The thermal conductivities of glass and air are 0.88 and 0.024 $W/m.K$, respectively over the temperature range of interest. The area of the window is 1.75 m^2. The convective heat transfer coefficients on the inside and outside surfaces of the composite window are 13 $W/m^2.K$. The fluid free-stream temperature difference between the two sides of the composite window system is 29 K.

Calculate the following:

(a) the total thermal resistance
(b) the overall heat transfer coefficient
(c) the rate of heat loss

Solution:

Refer to the figures below.

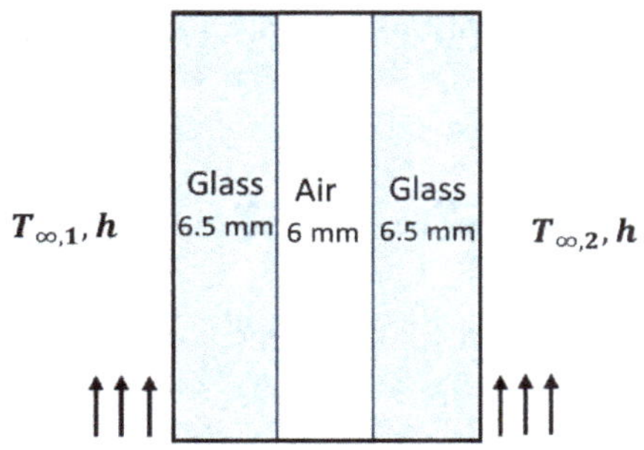

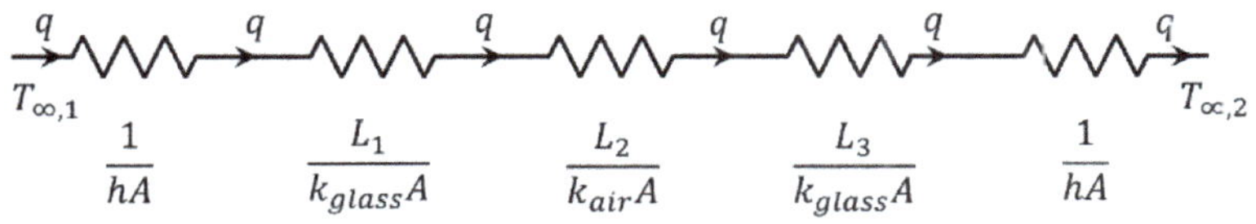

(a) the total thermal resistance

Thermopane window has five resistors in series as shown in the electric analog above. The total thermal resistance is:

$$R_{total} = R_1 + R_2 + R_3 + R_4 + R_5 = \frac{1}{hA} + \frac{L_1}{k_{glass}A} + \frac{L_2}{k_{air}A} + \frac{L_1}{k_{glass}A} + \frac{1}{hA}$$
$$= \frac{1}{A}\left[\frac{1}{13} + \frac{6.5 \times 10^{-3}}{0.88} + \frac{6.0 \times 10^{-3}}{0.024} + \frac{6.5 \times 10^{-3}}{0.88} + \frac{1}{13}\right] = \frac{0.4186}{A} = \frac{0.4186}{1.75}$$
$$= 0.2392\ K/W$$

Note that the air layer offers the greatest resistance to heat loss.

(b) the overall heat transfer coefficient

$$U = \frac{1}{AR_{total}} = \frac{1}{0.4186} = 2.389\ W/m^2.K$$

(c) the rate of heat loss

$$q = UA\left(T_{\infty,1} - T_{\infty,1}\right) = 2.389 \times 1.75 \times 29 = 121.24\ W$$

11.7.3 Example 3

A metal pipe of inside and outside diameters of 15 cm and 17 cm, respectively carries hot air at 60°C in the drying section of a car wash. The pipe is insulated with a 2.5 cm thick layer of glass wool. The heat transfer coefficient between the pipe wall and hot air is 68 $W/m^2.K$. The ambient air temperature is 20°C and the heat transfer coefficient between the glass wool and ambient air is 8.5 $W/m^2.K$. Calculate the following:

a) inside convective resistance per unit length of the pipe

b) total conductive resistance per unit length of the pipe

c) outside convective resistance per unit length of the pipe

d) overall heat transfer coefficient based on the inside area of the pipe

e) rate of heat loss per unit length of the pipe

The thermal conductivities of metal pipe and glass wool are 43 $W/m.K$ and 0.04 $W/m.K$, respectively.

Solution:

Refer to the figures below.

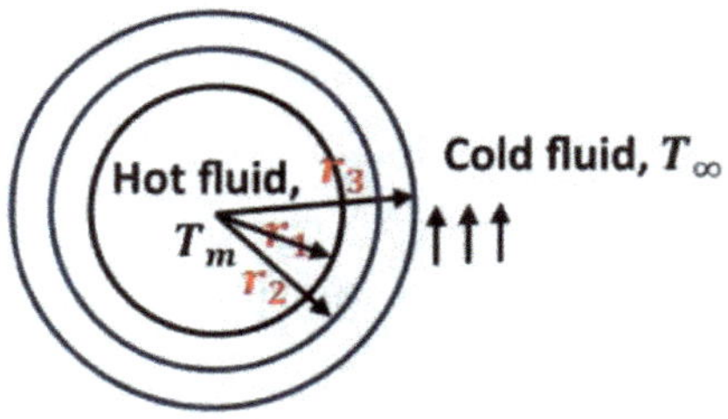

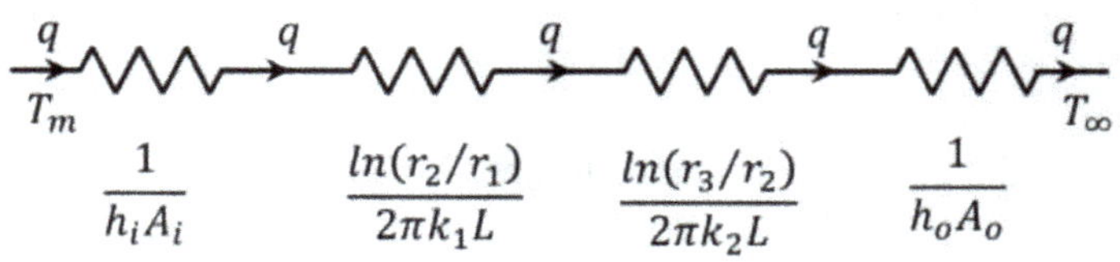

a) inside convective resistance per unit length of the pipe

$$R_1 = \frac{1}{h_i A_i} = \frac{1}{68 \times (2\pi r_1 L)} = \frac{1}{68 \times (2\pi)(7.5 \times 10^{-2})(1)} = 0.0312\ K/W$$

b) total conductive resistance per unit length of the pipe

$$R_2 = \frac{ln(r_2/r_1)}{2\pi k_1 L} = \frac{ln(8.5/7.5)}{2\pi(43)L} = \frac{4.63 \times 10^{-4}}{1}\ K/W = 4.63 \times 10^{-4}\ K/W$$

$$R_3 = \frac{ln(r_3/r_2)}{2\pi k_2 L} = \frac{ln(11/8.5)}{2\pi(0.04)L} = \frac{1.026}{1}\ K/W = 1.026\ K/W$$

$$R_{conductive} = R_2 + R_3 = 4.63 \times 10^{-4} + 1.026 = 1.02646\ K/W$$

c) outside convective resistance per unit length of the pipe

$$R_4 = \frac{1}{h_o A_o} = \frac{1}{8.5 \times (2\pi r_3 L)} = \frac{1}{8.5 \times (2\pi)(11 \times 10^{-2})L} = \frac{0.1702}{1}\ K/W = 0.1702\ K/W$$

d) overall heat transfer coefficient based on the inside area of the pipe

$$U_i A_i = \frac{1}{R_{total}} = \frac{1}{R_1 + R_2 + R_3 + R_4} = \frac{1}{0.0312 + 4.63 \times 10^{-4} + 1.02646 + 0.1702} = \frac{1}{1.2283} = 0.8141\ W/K$$

Thus,

$$U_i = \frac{0.8141}{2\pi r_1 L} = \frac{0.8141}{2\pi(7.5 \times 10^{-2}) \times 1} = 1.728\ W/m^2.K$$

e) rate of heat loss per unit length of the pipe

$$q = U_i A_i (T_m - T_\infty) = 0.8141(60 - 20) = 32.56\ W$$

11.7.4 Example 4

Consider a hot metal wire of diameter 2.0 mm. The temperature of the wire is maintained constant at 390 K. The wire is losing heat by convection due to flow of air across the wire. The free stream temperature of air is 295 K and the convection coefficient is 17 W/m^2.K.

a) Calculate the rate of heat loss per meter of wire length
b) If the wire is covered with a plastic insulation of thickness 5.0 mm and thermal conductivity of 0.5 W/m.K, what would be the rate of heat loss per meter of wire length?
c) With the addition of insulation, does the rate of heat loss increase or decrease? Explain.

Solution:

a) Calculate the rate of heat loss per meter of wire length

There is only one thermal resistance, that is, convective resistance:

Electrical Analog of Heat Transfer

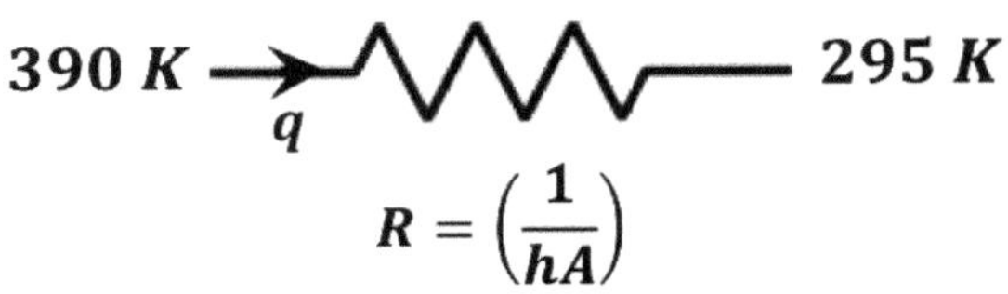

$$R = \frac{1}{hA} = \frac{1}{17\times(2\pi r L)} = \frac{1}{17\times(2\pi)(10^{-3})L} = \frac{9.362}{L}\ K/W$$

$$q = \frac{390-295}{\frac{9.362}{L}}\ ;\ \frac{q}{L} = \frac{390-295}{9.362} = 10.147\ W/m$$

b) If the wire is covered with a plastic insulation of thickness 5.0 mm and thermal conductivity of 0.5 W/m.K, what would be the rate of heat loss per meter of wire length?

Now there are two thermal resistors (conductive and convective) in series as shown below:

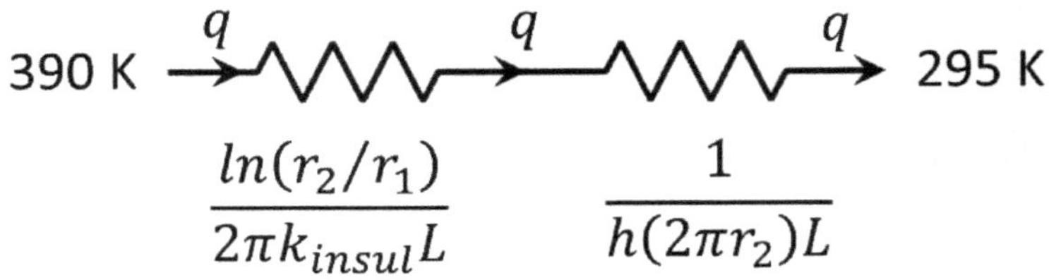

$$R_{conductive} = \frac{ln(r_2/r_1)}{2\pi k_{insul}L} = \frac{ln(6/1)}{2\pi(0.5)L} = \frac{0.57023}{L}\ K/W$$

$$R_{convective} = \frac{1}{hA} = \frac{1}{17\times(2\pi r_2 L)} = \frac{1}{17\times(2\pi)(6\times10^{-3})L} = \frac{1.5603}{L}\ K/W$$

$$R_{total} = R_{conductive} + R_{convective} = \frac{0.57023}{L} + \frac{1.5603}{L} = \frac{2.13053}{L}\ K/W$$

Thus,

$$q = \frac{390-295}{\frac{2.13053}{L}}\ ;\ \frac{q}{L} = \frac{390-295}{2.13053} = 44.59\ W/m$$

c) With the addition of insulation, does the rate of heat loss increase or decrease? Explain.

Interestingly, heat loss rate has increased by more than 4 times by adding a thick layer (5 mm) of insulation on a 2 mm diameter wire. This is counter intuitive as insulation should generally reduce the rate of heat loss. However, in the present situation, there are two opposing effects coming into play. The addition of insulation increases the conductive resistance but decreases the convective resistance due an increase in the convective heat transfer area. Overall, there is decrease in the total resistance and hence an increase in the heat transfer rate.

11.7.5 Example 5

A long solid cylindrical rod of radius 98 mm and thermal conductivity 0.49 W/m.K is generating heat at a uniform volumetric rate of 23,500 W/m^3. The rod is covered with a cylindrical sleeve of inner radius 98 mm and outer radius of 196 mm. The thermal conductivity of the sleeve material is 4.2 W/m.K. Convective heat transfer is taking place at the outside surface of the sleeve due to air flow at 25 °C. The outside convective heat transfer coefficient is 31 W/m^2.K.

Determine the following:

a) Interface temperatures between the rod and the sleeve and at the outer surface of the sleeve

b) Temperature at the centre of the rod.

Solution:

Refer to the figure below.

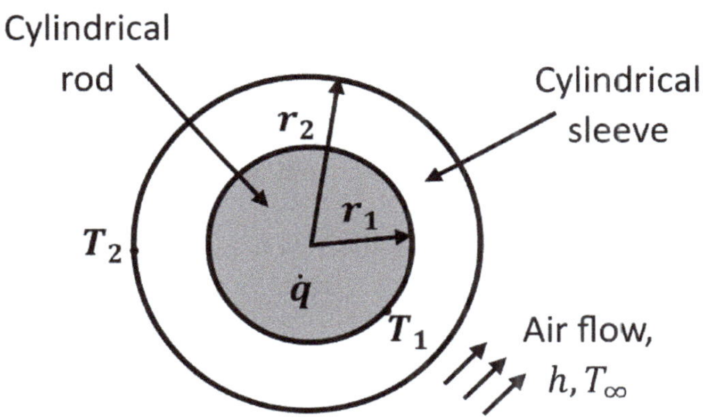

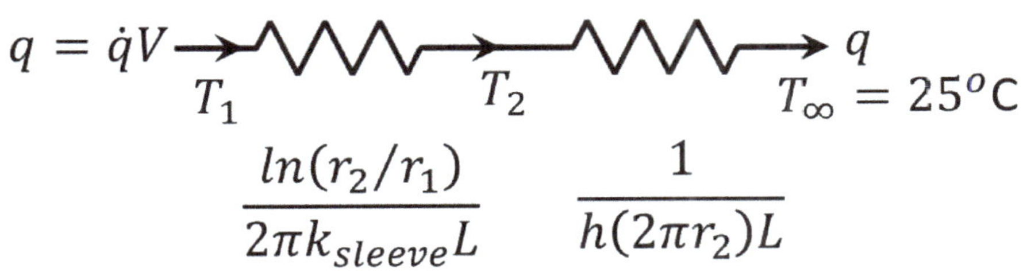

a) Interface temperatures between the rod and the sleeve and at the outer surface of the sleeve

$$q = \dot{q}V = 23500 \times \pi r_1^2 L = 23500 \times \pi \times (98 \times 10^{-3})^2 L = 709.039L$$

$$R_{total} = R_1 + R_2 = \frac{ln(r_2/r_1)}{2\pi k_{sleeve}L} + \frac{1}{h(2\pi r_2 L)} = \frac{ln(196/98)}{2\pi(4.2)L} + \frac{1}{31 \times 2\pi(196 \times 10^{-3})L} = \frac{0.026266}{L} + \frac{0.026194}{L} = \frac{0.05246}{L}$$

$q = \frac{T_1 - T_\infty}{R_{total}} = 709.039L$

$$T_1 = T_\infty + qR_{total} = 25 + (709.039L)\left(\frac{0.05246}{L}\right) = 62.196^oC$$

$$T_2 = T_1 - qR_1 = 62.196 - (709.039L)\left(\frac{0.026266}{L}\right) = 43.5724^oC$$

b) Temperature at the centre of the rod.

$$T_{centre} = T_{max} = T_1 + \frac{\dot{q}r_1^2}{4k_{rod}} = 62.196 + \frac{23500\times(98\times10^{-3})^2}{4\times0.49} = 177.346^oC$$

11.7.6 Example 6

The amount of heat generated in a single integrated circuit chip is a serious issue as there is a limit to the maximum allowable operating temperature of the chip. A dielectric liquid is used to remove the heat generated in the chip by way of convective heat transfer. One very effective way to enhance the dissipation of heat generated in the chip by convection is to extend the heat transfer area of the chip by mounting an array of copper fins on the chip. Consider a chip as shown in the figure below. The following information is given:

- Square chip of length and width 15 mm.
- Array of 7 × 7 copper fins are mounted on the chip
- The pin diameter (D_p) and length (L) are 1.25 mm and 12.5 mm, respectively. The thermal conductivity of the fin is 398 W/(m.K).
- The maximum allowable chip temperature T_c is 74 °C.
- The heat transfer coefficient on the dielectric liquid side h_o is 950 W/m^2.K
- The heat transfer coefficient on the ambient air side h_i is 35 W/m^2.K
- The free stream temperature of the fluids is 20 °C.
- The thermal contact between the chip and the board is perfect.
- The board thickness and thermal conductivity are 6 mm and 1.2 W/(m.K), respectively.

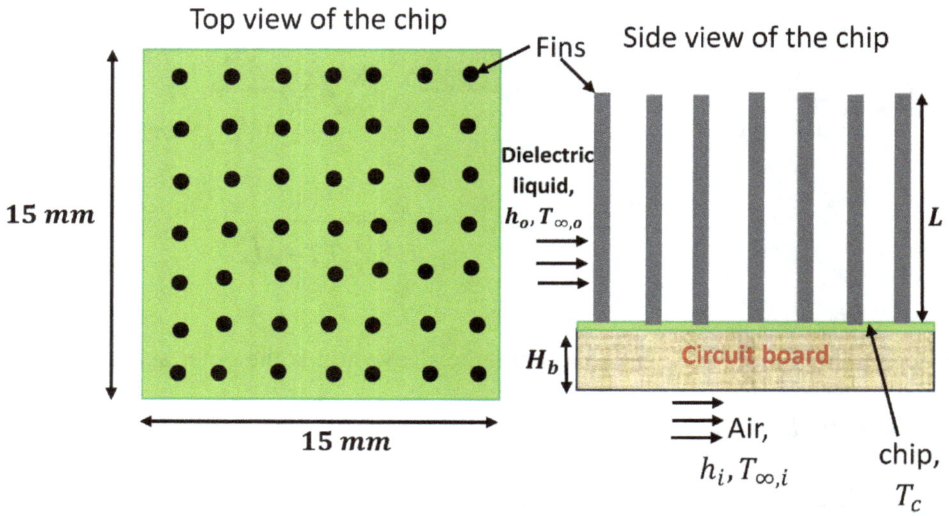

Estimate: (a) the effectiveness of a fin, (b) the efficiency of a fin, (c) the overall efficiency of the array of fins, (d) the rate of heat removal from the chip when the fins are not used, and (e) the rate of heat removal from the chip when the pins are in place. In your calculations, you can assume that the chip is at a maximum allowable temperature of 74 °C.

Solution:

(a) the effectiveness of a fin

$$\varepsilon = \frac{q_{fin}}{q_{no\,fin}} = \frac{kA_c m(T_b - T_\infty)\tanh mL}{A_c h(T_b - T_\infty)} = \frac{km\tanh mL}{h} \quad \text{where} \quad m = \sqrt{\frac{hP}{kA_c}}$$

$$m = \sqrt{\frac{hP}{kA_c}} = \sqrt{\frac{950 \times \pi D_p}{k\pi D_p{}^2/4}} = \sqrt{\frac{4 \times 950}{kD_p}} = \sqrt{\frac{4 \times 950}{398 \times 1.25 \times 10^{-3}}} = 87.397\ m^{-1}$$

$$\varepsilon = \frac{km\tanh mL}{h} = \frac{398 \times 87.397 \times \tanh(87.397 \times 12.5 \times 10^{-3})}{950} = 29.21$$

Thus, the fin is very effective. The rate of heat transfer from a fin is 29.21 times the rate of heat transfer without the fin.

(b) the efficiency of a fin

$$\eta = \frac{q_{fin,actual}}{q_{fin,max\,imum}} = \frac{q_{fin,actual}}{hA_f(T_b - T_\infty)} = \frac{\tanh m\,L}{mL} = \frac{\tanh(87.397 \times 12.5 \times 10^{-3})}{87.397\ \times 12.5 \times 10^{-3}} = 0.73025$$

(c) the overall efficiency of the array of fins

$$\eta_o = \frac{q_t}{q_{max}} = \frac{q_t}{hA_t(T_b - T_\infty)} = \left[1 - \frac{NA_f}{A_t}(1-\eta)\right]$$

$A_t = NA_f + A_b$ where A_f is the surface area of the fin and A_b is the exposed area of the base (not occupied by the fins)

$$A_f = \pi D_p L = \pi \times 1.25 \times 10^{-3} \times 12.5 \times 10^{-3} = 49.0874 \times 10^{-6}\ m^2$$

$$N = 7 \times 7 = 49$$

$$A_b = (15 \times 10^{-3})^2 - 49\left(\frac{\pi}{4}\right)(1.25 \times 10^{-3})^2 = 1.6487 \times 10^{-4}\ m^2$$

$$A_t = 49 \times 49.0874 \times 10^{-6} + 1.6487 \times 10^{-4} = 2.57 \times 10^{-3}\ m^2$$

$$\eta_o = \left[1 - \frac{NA_f}{A_t}(1-\eta)\right] = \left[1 - \frac{49 \times 49.0874 \times 10^{-6}}{2.57 \times 10^{-3}}(1 - 0.73025)\right] = 0.7446$$

(d) the rate of heat removal from the chip when the fins are not used

$$q_c = q_{board} + h_o A_{chip}(T_c - T_{\infty,o})$$

$$q_{board} = \frac{T_c - T_{\infty,i}}{R}$$

$$R = R_{conductive} + R_{convective} = \frac{H_b}{k_b A_{chip}} + \frac{1}{h_i A_{chip}} = \frac{6\times10^{-3}}{1.2\times15\times15\times10^{-6}} + \frac{1}{35\times15\times15\times10^{-6}} = 14.815 + 126.984 = 141.799 \quad K/W$$

$$q_{board} = \frac{T_c - T_{\infty,i}}{R} = \frac{74-20}{141.799} = 0.3808\ W$$

$$q_c = 0.3808 + 950 \times (15 \times 15 \times 10^{-6})(74-20) = 11.923\ W$$

(e) the rate of heat removal from the chip when the pins are in place

$$q_c = q_{board} + \eta_o h_o A_t (T_c - T_\infty) = 0.3808 + 0.7446 \times 950 \times 2.57 \times 10^{-3} \times (74-20) = 0.3808 + 98.1635 = 98.5443\ W$$

Thus, the addition of an array of fins increases the rate of removal from the chip by more than 8 times.

11.7.7 Example 7

The water mains should be buried under the ground surface well below a certain minimum depth to avoid freezing during cold periods in winter. What is the minimum depth to which the water mains should be buried to avoid freezing if the worst anticipated conditions during the winter are: ground surface temperature of -15^oC for full 60 days. You can assume that the ground is at a uniform temperature of 20^oC before the cold spell. The properties of the soil are as follows:

$$\rho = 2050\ kg/m^3\ ;\ k = 0.52\ W/m.K\ ;\ c_p = 1840\ J/kg.K\ ;\ \alpha = 0.138 \times 10^{-6}\ m^2/s$$

Solution:

Refer to the figure below.

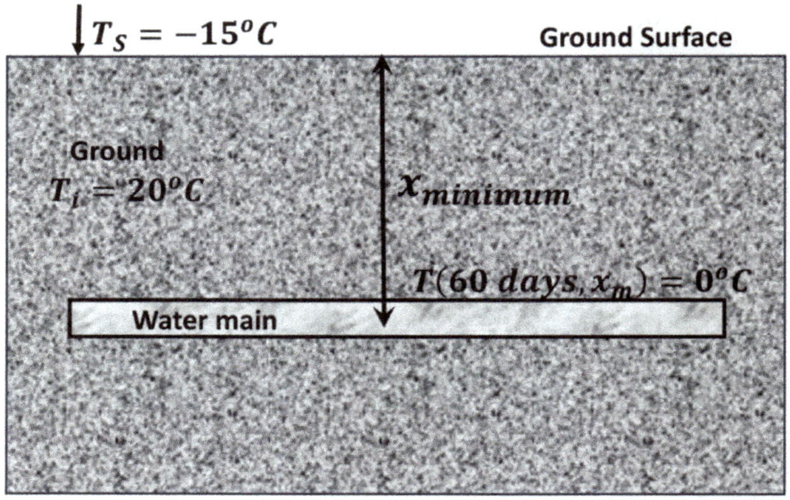

$$\frac{T-T_i}{T_S-T_i} = 1 - erf\left(\frac{x}{2\sqrt{\alpha t}}\right)$$

Upon rearrangement, we get:

$$\frac{T-T_S}{T_i-T_S} = erf\left(\frac{x}{2\sqrt{\alpha t}}\right)$$

$$\frac{T-T_S}{T_i-T_S} = \frac{0-15}{20-(-15)} = 0.42857 \quad ; \quad \frac{x}{2\sqrt{\alpha t}} = 0.40$$

$$x_m = 0.40 \times 2\sqrt{\alpha t} = 0.80\sqrt{0.138 \times 10^{-6} \times 60 \times 24 \times 3600} = 0.68\, m$$

Thus, the minimum depth to which the water mains must be buried is 0.68 m.

11.7.8 Example 8

A long solid cylinder of aluminium, 3 cm in radius, is initially at a uniform temperature of 220 °C. The cylinder is suddenly subjected to a convection environment of 82 °C with a heat transfer coefficient of 498 W/m^2.K. Estimate the following: a) the temperature of the cylinder after 45 seconds of exposure to the convection environment, b) the heat lost by the cylinder per unit length cylinder after 45 seconds of exposure to the convection environment. The properties of aluminium are as follows: $k = 237\ W/m.K$ and $\alpha = 97\, mm^2/s$.

Solution:

Refer to the figure below.

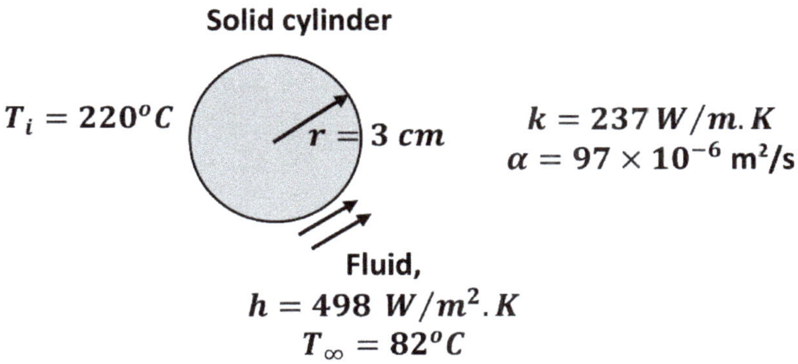

Calculate Biot number:

$$Bi = \frac{hL_c}{k} = \frac{hr}{k} = \frac{498 \times (3 \times 10^{-2})}{237} = 0.063 < 0.1$$

We can apply Lumped Capacity Analysis as $Bi < 0.1$. Thus,

$$\frac{T-T_\infty}{T_i-T_\infty} = exp\left(-\frac{hA}{\rho V c_p}t\right)$$

$$\frac{hA}{\rho V c_p} = \frac{2h\alpha}{rk} = \frac{2\times498\times97\times10^{-6}}{(3\times10^{-2})\times237} = 1.3588 \times 10^{-2}$$

$\frac{T-T_\infty}{T_i-T_\infty} = exp\left(-\frac{hA}{\rho V c_p}t\right) = exp(-1.3588 \times 10^{-2}t)$

a) the temperature of the cylinder after 45 seconds of exposure to the convection environment,

$t = 45\ s$

$T = T_\infty + (T_i - T_\infty)exp(-1.3588 \times 10^{-2}t) = 82 + (220 - 82)exp(-1.3588 \times 10^{-2} \times 45) = 156.873\ ^oC$

b) the heat lost by the cylinder per unit length cylinder after 45 seconds of exposure to the convection environment

The total amount of heat transferred to the fluid from $t = 0$ to time t is:

$Q = mc_p(T_i - T) = \rho V c_p(T_i - T) = \rho(\pi r^2 L)c_p(T_i - T)$

Thus,

$\frac{Q}{L} = \rho(\pi r^2)c_p(T_i - T) = \left(\frac{k}{\alpha}\right)(\pi r^2)(T_i - T) = \left(\frac{237}{97\times10^{-6}}\right) \times \pi \times (3 \times 10^{-2})^2(220 - 156.873) = 4.361 \times 10^5\ \ J/m$

11.7.9 Example 9

A large metal plate of thickness 5 cm initially at 24 °C is suddenly subjected to convection with T_∞ = 121 °C and heat transfer coefficient h of 17,000 W/(m^2.K). Estimate the temperature of the metal plate at the surface after 10 s. How much heat is transferred to the metal plate during this 10 s period on a unit area basis?

The metal properties are: $k = 235\ W/m.K$ and $\alpha = 97 \times 10^{-6}\ m^2/s$

Solution:

Calculate Biot number:

$$Bi = \frac{hL}{k} = \frac{17000 \times 2.5 \times 10^{-2}}{235} = 1.809\ > 0.1$$

Since $Bi\ > 0$, we cannot treat the object to be at a uniform temperature. Thus, the lumped capacity analysis is not valid. We need to use the Heisler and Grober charts.

$Fo = \frac{\alpha t}{L^2} = \frac{97\times10^{-6}\times10}{(2.5\times10^{-2})^2} = 1.552\ \ ;\ \ 1/Bi = 1/1.809 = 0.5528$

From Heisler chart (Figure 11.12):

$\frac{T_c-T_\infty}{T_i-T_\infty} = 0.23\ \ ;\ \ T_c = T_\infty + 0.23 \times (T_i - T_\infty) = 121 + 0.23 \times (24 - 121) = 98.69^oC$

From Heisler chart (Figure 11.12):

$\frac{x}{L} = 1$; $\frac{T-T_\infty}{T_c-T_\infty} = 0.5$; $T = T_\infty + 0.50 \times (T_c - T_\infty) = 121 + 0.50 \times (98.69 - 121) = 109.845^oC$

From Grober chart (Figure 11.12):

$Bi^2Fo = 1.809 \times 0.5528 = 1.0$; $Bi = 1.809$; $Q/Q_{max} = 0.27$

$Q_{max} = \rho c_p V(T_i - T_\infty) = \left(\frac{k}{\alpha}\right) A(2L)(T_i - T_\infty)$

$Q_{max}/A = \left(\frac{k}{\alpha}\right)(2L)(T_i - T_\infty) = \left(\frac{235}{97\times10^{-6}}\right)(5 \times 10^{-2})(24 - 121) = -11.75 \times 10^6\ J/m^2 = -11.75\ MJ/m^2$

Thus,

$Q = 0.27 \times Q_{max} = -0.27 \times 11.75 = -3.1725\ MJ/m^2$

The negative sign appears as heat is transferred to the object.

11.7.10 Example 10

It is proposed to store radioactive waste in a spherical vessel and bury the vessel 15 m below the earth's surface. The outside diameter of the spherical container is 1.5 m and the rate of heat generation in the radioactive waste is 550 W. Estimate the outside surface temperature of the container at steady state. You can take the surface temperature of the ground to be 22°C. The thermal conductivity of the soil is about 1.5 W/m.K.

Solution:

This problem can be solved using the shape factor approach. Refer to the figure below.

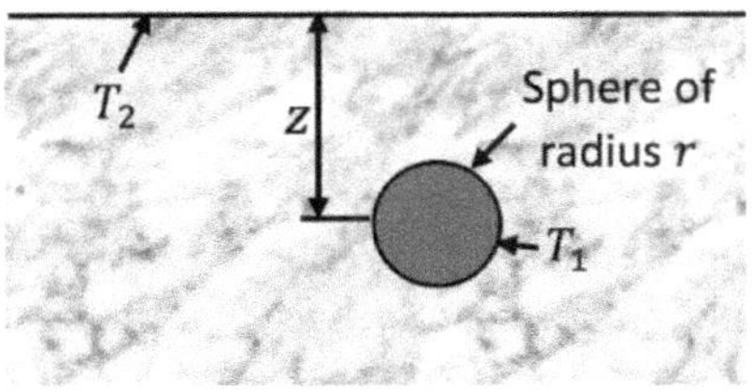

For isothermal sphere buried in a semi-infinite medium, the shape factor is:

$$S = \frac{4\pi r}{1-(r/2z)} \qquad z > r$$

Given that $z = 15\ m,\ r = 0.75\ m$, $S = \frac{4\pi\times0.75}{1-(0.75/2\times15)} = \frac{9.4248}{0.975} = 9.67\ m$

$q = kS(\Delta T) = 1.5 \times 9.67 \times (T - 22) = 550$; $\qquad T = 59.92^oC$

Thus, the outside surface temperature of the container is 59.92 °C.

NOTATION

A area of heat transfer or cross-section area
A_b exposed portion of the base area (area not occupied by the fins)
A_c cross-sectional area of the fin
A_f exposed surface area of a fin
A_{gm} geometric mean area of a spherical shell
A_i inside area of pipe
$A_{interface}$ interface area between two layers of materials
A_{lm} log-mean area of a cylindrical shell
A_o outside area of pipe
A_t total surface area of the fin and its base in contact with the fluid
Bi Biot number
CV control volume
c_p constant pressure specific heat capacity
erf error function
Fo Fourier number
h heat transfer coefficient
h_i heat transfer coefficient based on inside area of pipe
h_o heat transfer coefficient based on outside area of pipe
$\hat{\imath}$ unit vector in x-direction
$\hat{\jmath}$ unit vector in y-direction
$\hat{k}$ unit vector in z-direction
k thermal conductivity
$\bar{\bar{k}}$ thermal conductivity tensor
k_m mean thermal conductivity
k_{ij} component of thermal conductivity tensor
L length of an object
L_c characteristic length of an object
$\mathcal{L}$ Laplace transform
m mass or fin parameter defined in Equation (11.107)
N total number of fins
P perimeter of a fin
Q Q is the amount of heat exchanged during the time interval $0-t$
Q_{max} maximum amount of heat exchange possible between the object and the ambient fluid
q heat transfer rate
q_{fin} heat transfer rate from the fin
q_{max} maximum possible rate of heat transfer from the array assembly
q_t the total rate of heat transfer from an array of fins
$\dot{q}$ rate of heat generation per unit volume
$\vec{q}''$ heat flux vector
q_x'' heat flux in x- direction
q_y'' heat flux in y- direction
q_z'' heat flux in z- direction
q_r'' heat flux in r- direction
q_θ'' heat flux in θ- direction
q_φ'' heat flux in φ- direction
q_s'' surface heat flux
r radial coordinate or distance in radial direction
r_o radius of a sphere or a cylindrical object

$\hat{r}$	unit vector in radial direction
R	resistance
R_c	total contact resistance
R_c''	contact resistance on a unit interfacial-area basis
s	Laplace variable
S	shape factor
t	time
T	temperature
T_b	temperature at the base of the fin
T_c	characteristic temperature of fluid
T_i	initial temperature of an object
T_m	mean temperature of fluid
T_{max}	maximum temperature in an object
T_s	surface temperature
T_∞	free stream temperature of a fluid
U	overall heat transfer coefficient
U_i	overall heat transfer coefficient based on inside area
U_o	overall heat transfer coefficient based on outside area
V	volume of an object
x, y, z	space coordinates in cartesian coordinate system
$\hat{z}$	unit vector in z-direction

Greek Letters

α	thermal diffusivity
Δ	change in variable
∇	mathematical del operator
$\bar{\bar{\delta}}$	unit tensor
ε	effectiveness of a fin
η	efficiency of a fin
η_o	overall efficiency of an array of fins
θ	space coordinate in cylindrical and spherical coordinates
$\hat{\theta}$	unit vector in θ-direction
φ	space coordinate in spherical coordinates
$\hat{\varphi}$	unit vector in φ-direction

PRACTICE PROBLEMS

Problem 1

Consider a composite wall as shown in the figure below.

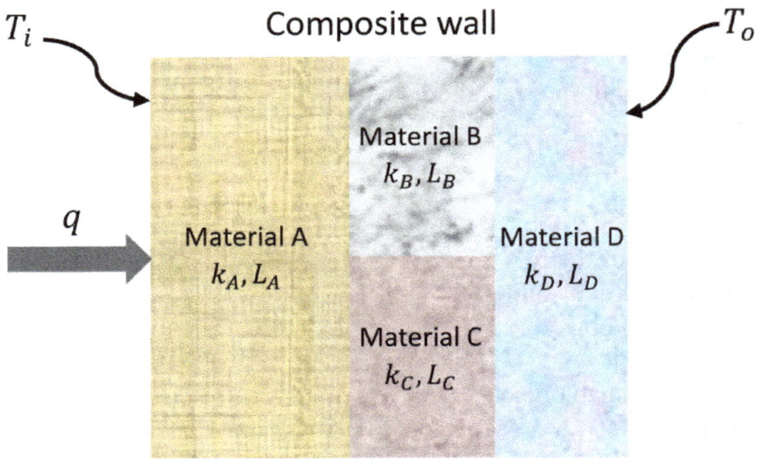

You are given the following information:

$$T_i = 355^oC, \quad T_o = 55^oC$$

Material	Thermal conductivity	Thickness of material	Heat transfer area
A	$k_A = 75\ W/m.K$	$L_A = 7\ cm$	1 m^2
B	$k_B = 45\ W/m.K$	$L_B = 5\ cm$	0.5 m^2
C	$k_C = 35\ W/m.K$	$L_C = 5\ cm$	0.5 m^2
D	$k_D = 25\ W/m.K$	$L_D = 4\ cm$	1 m^2

Note that the heat transfer area is the lateral area normal to the direction of heat flow.

Determine the following:

a) The rate of heat loss (q) from the composite wall

b) The heat flux (q'') in each material

c) The interface temperatures

Problem 2

The thermal conductivity of the wall of an industrial furnace constructed from fireclay brick varies with temperature as follows:

$$k = a + bT$$

where $a = 0.12$ and $b = 4.9 \times 10^{-5}$, T is in °F and k is in Btu/hr.ft.°F. The furnace wall is 5 inches thick. The heat transfer area of the wall is 15 ft^2. The inside and outside temperatures of the furnace wall are 1450 °F and 95 °F, respectively.

Determine the following:

a) The average thermal conductivity of the wall

b) The thermal resistance of the wall

c) The rate of heat loss through the wall

Problem 3

A chemical reactor of spherical shape is constructed from stainless steel ($k = 18\ W/m.K$). The inside diameter of the reactor is 0.75 m, and the thickness is 15 mm. The reactor is being used to produce pharmaceuticals. The inner surface temperature of the reactor is 53 °C and the outside surface temperature is 35 °C. Estimate the rate of heat loss from the reactor under these conditions.

Problem 4

A metal pipe (k = 28 Btu/hr.ft.°F) of inside and outside diameters of 3.2 inch and 3.65 inch respectively is transporting a fluid at a temperature of 345 °F. The inside convection coefficient is 35 Btu/hr.ft^2.°F. The pipe is insulated with a 1.2-inch-thick insulation material of low thermal conductivity of 0.11 Btu/hr.ft.°F. The pipe is exposed to ambient air at 78 °F and the outside convection coefficient is 2.2 Btu/hr.ft^2.°F. Calculate the following:

a) The overall heat transfer coefficient based on outside area of the composite pipe

b) The rate of heat loss per foot of the pipe

c) The temperatures at the inner surface of the pipe, at the interface between the pipe and the insulation, and at the outside surface of the insulation.

Problem 5

Consider two large planar slabs of same metal with rough surfaces in contact with each other under pressure, as shown in the figure below. The thickness of each slab is 1.5 cm. The surface roughness of the slab is 12 μm. The contact between the slabs is imperfect with contact resistance R''_C of 2.82×10^{-4} m^2.K/W. The temperatures of the free surfaces of the slabs are 408 and 398 °C (see figure below). The thermal conductivity of the metal is 245 W/m.K. Calculate the following:

a) the thermal resistances of the slab and the interface

b) the heat flux

c) the temperature drop at the contact between the slabs

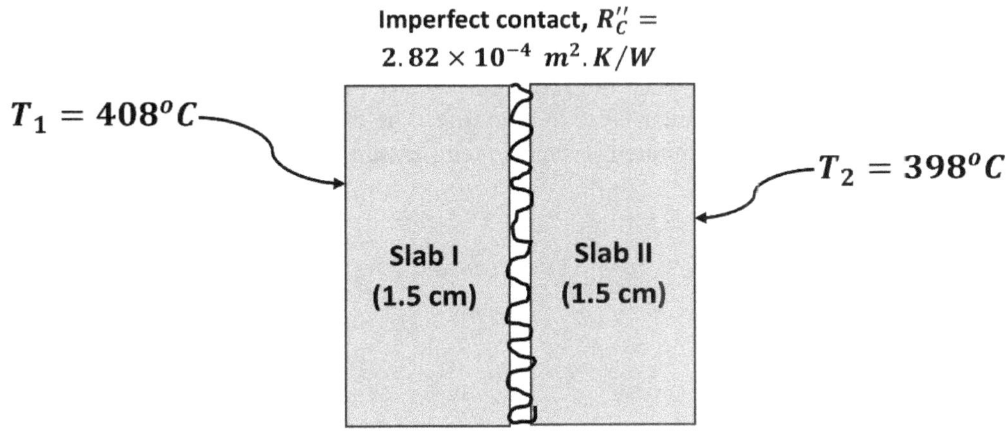

Problem 6

A composite plane wall consists of layers of materials A and B as shown in the figure below. The left side of the composite wall is perfectly insulated with no heat loss. Heat is being generated in material A at a volumetric rate of $\dot{q} = 1.25 \times 10^6 \ \ W/m^3$. There is no heat generation in material B. Convection is taking place on the right side of the composite wall as shown in the figure. The convection coefficient is $h = 997 \ W/m^2.K$ and the free stream temperature of the fluid is $T_\infty = 25^oC$. You are given the following additional information: $k_A = 69 \ W/m.K$, $L_A = 60 \ mm$, $k_B = 135 \ W/m.K$, $L_B = 25 \ mm$.

Determine the following temperatures (see figure): T_o, T_1, and T_2. You can assume steady state.

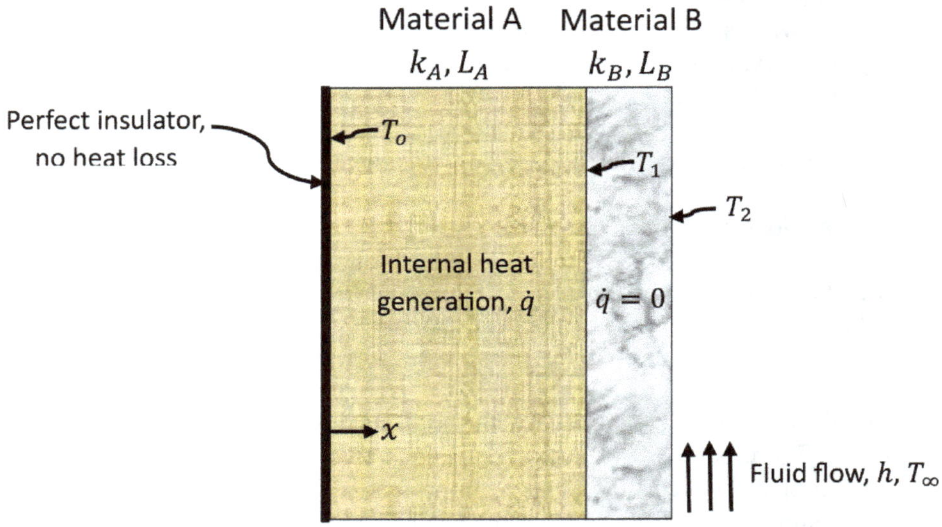

Problem 7

Heat is being generated in a cylindrical nuclear fuel rod (see figure below). The local rate of heat generation in the rod per unit volume is given by the following relation:

$$\dot{q} = \dot{q}_o \left[1 - \left(\frac{r}{r_1}\right)^2\right]$$

where $\dot{q}_o$ is a constant, r_1 is the radius of the rod, and r is any radial position in the rod from the centre. The fuel rod is surrounded by an annular layer of cladding. The cladding material is exposed to a coolant. Given the following information, estimate the surface temperatures T_1 and T_2:

$\dot{q}_o = 6.5 \times 10^8 \ \ W/m^3$, $r_1 = 11.5 \ mm$, $r_2 = 14.5 \ mm$, $h = 6800 \ W/m^2.K$, $T_\infty = 93^oC$, $k_{cladding} = 235 \ W/m.K$

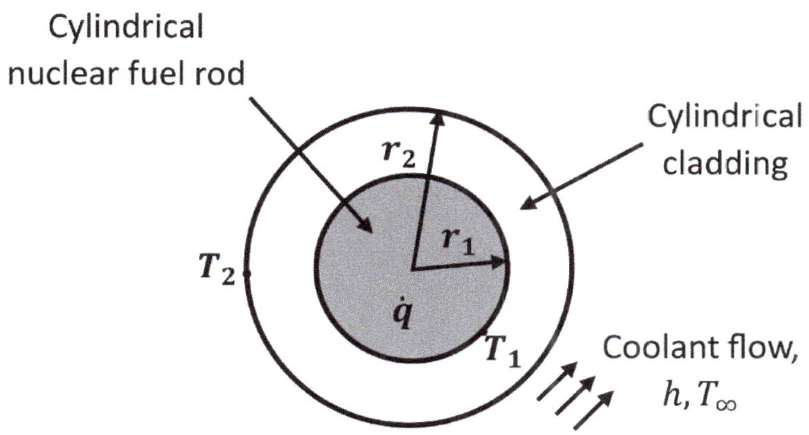

Problem 8

A long cylindrical concrete column of diameter 3.2 ft is used in the construction of a bridge. The thermal conductivity of the concrete material is 0.865 W/m.K. The heat of hydration, that is, the rate of heat released per unit mass of concrete, is 0.8 W/kg. The density of concrete is 2350 kg/m^3. Estimate the temperature at the centre of the concrete column when the outside surface temperature is 80 °C.

Problem 9

Consider a chip as shown in the figure below. The following information is given:

- Square chip of length and width 16 mm.
- Array of 6 × 6 copper fins are mounted on the chip
- The pin diameter (D_p) and length (L) are 1.35 mm and 14 mm, respectively. The thermal conductivity of the fin is 401 W/(m.K).
- The maximum allowable chip temperature T_c is 80 °C.
- The heat transfer coefficient on the dielectric liquid side h_o is 975 W/m^2.K
- The heat transfer coefficient on the ambient air side h_i is 40 W/m^2.K
- The free stream temperature of the fluids is 22 °C.
- The thermal contact between the chip and the board is perfect.
- The board thickness and thermal conductivity are 5.5 mm and 1.2 W/(m.K), respectively.

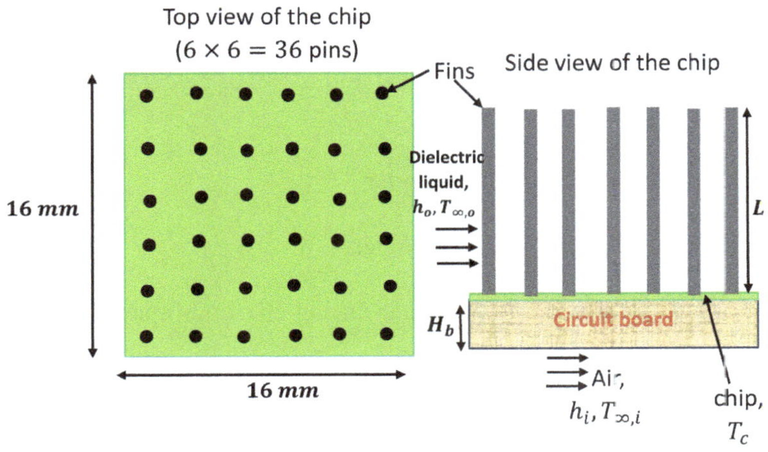

Estimate: (a) the effectiveness of a fin, (b) the efficiency of a fin, (c) the overall efficiency of the array of fins, (d) the rate of heat removal from the chip when the fins are not used, and (e) the rate of heat removal from the chip when the pins are in place. In your calculations, you can assume that the chip is at a maximum allowable temperature of 80 °C.

Problem 10

Consider a pin fin system consisting of a base of metal block, square in shape, 75 mm by 75 mm, at a constant temperature of 80 °C. An array of 13 by 13 metal pin fins (pin diameter 2.2 mm and pin length 22 mm) is mounted on the base of the metal block. The thermal conductivity of the metal (base and fins) is 195 W/m.K. The system is exposed to air at 22 °C with a convective heat transfer coefficient of 15 $W/m^2.K$. Calculate the following:

a) effectiveness of a fin

b) efficiency of a fin

c) overall efficiency of the array of fins

d) the total rate of heat loss by the system

e) the fraction of the total rate of heat loss by the fins alone

Problem 11

Re-do example 7 (section 11.7.7) with the following changes:

Worst anticipated conditions during the winter are: ground surface temperature of -35^oC for full 45 days. You can assume that the ground is at a uniform temperature of 10^oC before the cold spell. The properties of the soil are the same as indicated in example 7.

Problem 12

A large aluminium plate, 4.5 cm in thickness, is initially at a uniform temperature of 220 °C. The plate is suddenly subjected to a convection environment of 75 °C with a heat transfer coefficient of 500 $W/m^2.K$. Estimate the following: a) the temperature of the plate after 45 seconds of exposure to the convection environment, b) the heat lost by the plate per unit lateral area after 45 seconds of exposure to the convection environment. The properties of aluminium are as follows: $k = 237\ W/m.K$ and $\alpha = 97\, mm^2/s$.

Problem 13

The manufacturing process of metal components involves quenching of a hot component in water. A cylindrical metal component of 0.18 m diameter and 1.9 m length is initially at a uniform temperature of 420 °C. It is suddenly quenched in water at 45 °C. The heat transfer coefficient is 250 $W/m^2.K$. The properties of metal are as follows: $k = 38\ W/m.K$ and $\alpha = 10^{-5}\, m^2/s$. Estimate the following after 25 min of quenching:

a) The temperature at the centre of the cylinder

b) The temperature at the surface of the cylinder

c) The amount of heat transferred to the water during 25 min of quenching

Problem 14

Consider a 25 m long, 12 cm diameter pipe carrying hot water. The pipe is buried in the ground 1 m below the ground surface. The ground surface temperature is 15 °C and the outer surface temperature of the pipe is 78°C. Estimate the rate of heat loss from the pipe. The thermal conductivity of the soil ia about 1 W/m.K.

FURTHER READING

1. T.L. Bergman, A.S. Levine, F.P. Incropera, and D.P. Dewitt. *Fundamentals of Heat and Mass Transfer*. 7th Edition, New York, Wiley & Sons, 2011.
2. R.B. Bird, W.E. Stewart, and E.N. Lightfoot. 2007. *Transport Phenomena*. 2nd Edition, Wiley & Sons, New York, 2007.
3. J.R. Welty, C.E. Wicks, R.E. Wilson, and G.L. Rorrer. Fundamentals of Momentum, Heat, and Mass Transfer. 5th Edition, Wiley & Sons, New York, 2008.
4. F. Keith, M.S. Bohn. *Principles of Heat Transfer*. 5th Edition, West Publishing Co., St. Paul, MN, USA, 1993.
5. A. Bejan. *Heat Transfer*. Wiley & Sons, New York, 1993.
6. J.P. Holman. *Heat Transfer*, 10th Edition, McGraw Hill, New York, 2009.
7. Y.A. Cengel. *Heat and Mass Transfer*, 3rd Edition, McGraw Hill, New York, 2007.
8. L.C. Thomas. *Heat Transfer*. Prentice Hall, New Jersey, 1992.
9. W.S. Janna. *Engineering Heat Transfer*. PWS Publishers, Boston, 1986.
10. C.J. Geankoplis, A.A. Hersel, D.H. Lepek. *Transport Processes and Separation Process Principles*. 5th Edition, Prentice Hall, New Jersey, 2018.
11. W.L. McCabe, J.C. Smith, P. Harriott. *Unit Operations of Chemical Engineering*. 6th Edition, McGraw Hill, New York, 2001
12. R. Pal. *Electromagnetic, Mechanical, and Transport Properties of Composite Materials*. CRC Press, Boca Raton, Florida, 2015.

Chapter 12

Heat Transfer in a Moving Fluid (Convective Heat Transfer)

The transfer of heat from a high temperature region to a low temperature region of a fluid in the presence of bulk macroscopic motion of the fluid is referred to as "convective heat transfer". Heat transfer between a solid boundary and a fluid flowing adjacent to the solid boundary when there exists a temperature difference between the fluid and the solid surface is also referred to as "convective heat transfer". Thus, convective heat transfer operations involve bulk fluid motion as well as temperature variations within the fluid and between the fluid and contact surface.

Convection can be classified into two broad groups: free or natural convection and forced convection. In free or natural convection, fluid motion is induced by buoyancy forces which arise from density differences caused by temperature variations in the fluid. There is no external mechanical means (pump, fan, etc.) causing the fluid motion. When the fluid motion is caused by some external means such as a pump, fan, blower, or compressor, the convection is called forced convection.

In convective heat transfer process, energy is generally transported from one region of fluid to another by two mechanisms acting simultaneously: a) transport of energy by random molecular motion (diffusion) as in conduction, and b) transport of energy by bulk or macroscopic motion of the fluid (advection). Thus, convective heat transfer involves both diffusion and advection mechanisms of energy transfer.

12.1 GOVERNING EQUATIONS OF FLUID FLOW AND CONVECTIVE HEAT TRANSFER

The basic principles that govern the flow of fluid and convective heat transfer are a) conservation of mass, b) conservation of linear momentum (Newton's second law of motion), and c) conservation of energy (first law of thermodynamics). The applications of these principles to fluid flow generate the fundamental differential equations, namely the continuity equation, the equation of motion, and the energy equations. These equations allow us to determine the flow field (velocity distribution) and temperature field.

The continuity equation and the equations of motion (stress equation of motion and Navier-Stokes equation) were developed in Chapter 4. They are as follows:

$$\frac{\partial \rho}{\partial t} + \nabla \cdot (\rho \vec{V}) = 0 \tag{12.1}$$

$$\rho \frac{D\vec{V}}{Dt} = \rho \left(\frac{\partial \vec{V}}{\partial t} + \vec{V} \cdot \nabla \vec{V} \right) = \rho \vec{g} + \nabla \cdot \bar{\bar{\pi}} \tag{12.2}$$

$$\rho \frac{D\vec{V}}{Dt} = \rho \left(\frac{\partial \vec{V}}{\partial t} + \vec{V} \cdot \nabla \vec{V} \right) = \rho \vec{g} - \nabla P + \eta \nabla^2 \vec{V} \tag{12.3}$$

Equation (12.1) is the continuity equation, Equation (12.2) is the stress equation of motion, and Equation (12.3) is the Navier-Stokes equation. The Navier-Stokes equation together with the continuity equation describe the flow of Newtonian fluids of constant density and viscosity.

The flow of isothermal (constant temperature) fluids can be described adequately using the Navier-Stokes equation together with the continuity equation. However, when the flow is non-isothermal, we need another equation to determine the temperature distribution in the fluid. This equation comes from the energy balance on the fluid. In what follows, the total energy equation is first developed using the first law of thermodynamics. From the total energy equation, the thermal energy equation is extracted by subtracting the mechanical energy equation. The thermal energy equation when expressed in terms of temperature is needed to solve for the temperature distribution in the fluid.

12.2 TOTAL ENERGY EQUATION

The principle of conservation of energy is equivalent to the first law of thermodynamics. It states that the time rate of change of total energy of a material system (mass of fixed identity) is equal to the rate of heat addition to the system minus the rate of work done by the system on the surroundings. Thus,

$$\frac{D}{Dt}\iiint_{system}\rho e dV = \dot{Q} - \dot{W} \tag{12.4}$$

where e is the total energy (sum of internal, kinetic, and potential energies) per unit mass, $\dot{Q}$ is rate of heat input to the system, and $\dot{W}$ is rate of work done by the system on its surroundings.

The rate of heat input to the system can be expressed in terms of the heat flux vector $\vec{q}''$ as follows:

$$\dot{Q} = -\iint_S(\hat{n}\cdot\vec{q}'')dS \tag{12.5}$$

The rate of work done by the fluid system on its surroundings can be expressed in terms of the stress vector $\vec{\pi}_{(\hat{n})}$ as follows:

$$\dot{W} = -\iint_S(\vec{\pi}_{(\hat{n})}\cdot\vec{V})dS \tag{12.6}$$

From Equations (12.4) - (12.6), it follows that:

$$\frac{D}{Dt}\iiint_{system}\rho e dV = -\iint_S(\hat{n}\cdot\vec{q}'')dS + \iint_S(\vec{\pi}_{(\hat{n})}\cdot\vec{V})dS \tag{12.7}$$

Using the Reynolds transport theorem, the substantial time derivative in Equation (12.7) can be written as:

$$\frac{D}{Dt}\iiint_{system}\rho e dV = \iiint_{CV}\frac{\partial(\rho e)}{\partial t}dV + \iint_{CS}(\hat{n}\cdot\rho e\vec{V})dS \tag{12.8}$$

Applying the Gauss divergence theorem to the surface integral on the right-hand side of Equation (12.8), we get:

$$\frac{D}{Dt}\iiint_{system}\rho e dV = \iiint_{CV}\frac{\partial(\rho e)}{\partial t}dV + \iiint_{CV}(\nabla\cdot\rho e\vec{V})dV \tag{12.9}$$

Substituting the expression for substantial derivative from Equation (12.9) into Equation (12.7) gives:

$$\iiint_{CV}\frac{\partial(\rho e)}{\partial t}dV + \iiint_{CV}(\nabla\cdot\rho e\vec{V})dV = -\iint_S(\hat{n}\cdot\vec{q}'')dS + \iint_S(\vec{\pi}_{(\hat{n})}\cdot\vec{V})dS \tag{12.10}$$

The stress vector $\vec{\pi}_{(\hat{n})}$ can be expressed in terms of the stress tensor as:

$$\vec{\pi}_{(\hat{n})} = \hat{n}\cdot\bar{\bar{\pi}} \tag{12.11}$$

Consequently,

$$\iiint_{CV} \frac{\partial(\rho e)}{\partial t} dV + \iiint_{CV} (\nabla \cdot \rho e\vec{V}) dV = -\iint_S (\hat{n} \cdot \vec{q}'') dS + \iint_S (\hat{n} \cdot \bar{\bar{\pi}} \cdot \vec{V}) dS \quad (12.12)$$

The surface integrals on the right-hand side of Equation (12.12) can be converted into volume integrals using the Gauss divergence theorem. Thus:

$$\iiint_{CV} \frac{\partial(\rho e)}{\partial t} dV + \iiint_{CV} (\nabla \cdot \rho e\vec{V}) dV = -\iiint_{CV} \nabla \cdot \vec{q}'' \, dV + \iiint_{CV} (\nabla \cdot \bar{\bar{\pi}} \cdot \vec{V}) dV \quad (12.13)$$

Thus, energy balance can be written as:

$$\iiint_{CV} \left[\frac{\partial(\rho e)}{\partial t} + \nabla \cdot (\rho e\vec{V}) + \nabla \cdot \vec{q}'' - \nabla \cdot \bar{\bar{\pi}} \cdot \vec{V}\right] dV = 0 \quad (12.14)$$

As this equation holds for any arbitrary volume, the integrand must be zero:

$$\frac{\partial(\rho e)}{\partial t} + \nabla \cdot (\rho e\vec{V}) + \nabla \cdot \vec{q}'' - \nabla \cdot \bar{\bar{\pi}} \cdot \vec{V} = 0 \quad (12.15)$$

Using the continuity equation, it can be readily shown that:

$$\rho \frac{De}{Dt} = \frac{\partial(\rho e)}{\partial t} + \nabla \cdot (\rho e\vec{V}) \quad (12.16)$$

Thus, energy balance becomes:

$$\rho \frac{De}{Dt} = -\nabla \cdot \vec{q}'' + \nabla \cdot \bar{\bar{\pi}} \cdot \vec{V} \quad (12.17)$$

This is the general form of the total energy equation. Note that e in Equation (12.17) is $u + V^2/2 + \varphi$, where u is internal energy per unit mass, $V^2/2$ is kinetic energy per unit mass, and φ is potential energy per unit mass.

12.3 THERMAL ENERGY EQUATION

In solving the convective heat transfer problems, the total energy equation is not directly useful. It is the thermal energy (internal energy) equation that is required. The thermal energy equation is obtained from the total energy equation by subtracting the mechanical energy equation. The mechanical energy equation is obtained readily by taking dot product of the local velocity vector $\vec{V}$ with the stress equation of motion, Equation (12.2), that is:

$$\vec{V} \cdot \rho \frac{D\vec{V}}{Dt} = \vec{V} \cdot \rho\vec{g} + \vec{V} \cdot \nabla \cdot \bar{\bar{\pi}} \quad (12.18)$$

Substituting

$$\bar{\bar{\pi}} = -P\bar{\bar{\delta}} + \bar{\bar{\tau}} \quad (12.19)$$

and simplifying, the mechanical energy balance, Equation (12.18), becomes:

$$\rho \frac{D}{Dt}\left(\frac{V^2}{2}\right) = -\vec{V} \cdot \nabla P + \rho\vec{V} \cdot \vec{g} + \vec{V} \cdot (\nabla \cdot \bar{\bar{\tau}}) \quad (12.20)$$

As $\vec{g} = -\nabla\varphi$ where φ is potential energy per unit mass, Equation (12.20) can be re-written as:

$$\rho \frac{D}{Dt}\left(\frac{V^2}{2} + \varphi\right) = -\vec{V} \cdot \nabla P + \vec{V} \cdot (\nabla \cdot \bar{\bar{\tau}}) \tag{12.21}$$

Upon subtracting Equation (12.21) from the total energy equation, Equation (12.17), the thermal energy equation is obtained as follows:

$$\rho \frac{Du}{Dt} = -\nabla \cdot \vec{q}'' - P(\nabla \cdot \vec{V}) + \bar{\bar{\tau}} : \nabla \vec{V} \tag{12.22}$$

The left-hand side term of the thermal energy equation, Equation (12.22), represents the rate of internal energy increase per unit volume of the fluid. The first term on the right-hand side of Equation (12.22) represents the rate at which heat is being added to the fluid per unit volume through conduction, the second term represents the reversible rate of internal energy increase per unit volume by compression, and the last term represents the irreversible rate of internal energy increase per unit volume by viscous dissipation.

It can be readily shown that the left-hand side term of Equation (12.22) is equivalent to:

$$\rho \frac{Du}{Dt} = \rho c_v \frac{DT}{Dt} + \left[-P + T\left(\frac{\partial P}{\partial T}\right)_v\right](\nabla \cdot \vec{V}) \tag{12.23}$$

The derivation of Equation (12.23) involves the application of thermodynamic relations. Combining Equations (12.22) and (12.23) yields:

$$\rho c_v \frac{DT}{Dt} = -\nabla \cdot \vec{q}'' - T\left(\frac{\partial P}{\partial T}\right)_v (\nabla \cdot \vec{V}) + \bar{\bar{\tau}} : \nabla \vec{V} \tag{12.24}$$

Using Fourier's Law of heat conduction, Equation (12.24) modifies to

$$\rho c_v \frac{DT}{Dt} = \nabla \cdot (k \nabla T) - T\left(\frac{\partial P}{\partial T}\right)_v (\nabla \cdot \vec{V}) + \bar{\bar{\tau}} : \nabla \vec{V} \tag{12.25}$$

For incompressible fluid of constant thermal conductivity, Equation (12.25) simplifies to:

$$\rho c_p \frac{DT}{Dt} = \rho c_p \left(\frac{\partial T}{\partial t} + \vec{V} \cdot \nabla T\right) = k \nabla^2 T + \bar{\bar{\tau}} : \nabla \vec{V} \tag{12.26}$$

Note that $c_p = c_v$ for incompressible fluids. The viscous dissipation term $\bar{\bar{\tau}} : \nabla \vec{V}$ is a function of fluid viscosity and strain rates and is always positive.

The thermal energy equation, Equation (12.26), is given in Table 12.1 for different coordinate systems. In the special case of stationary fluid, the equation reduces to:

$$\rho c_p \frac{\partial T}{\partial t} = k \nabla^2 T \tag{12.27}$$

In summary, we have at our disposal five equations governing the flow and heat transfer in fluids of constant properties: Continuity equation (Equation (12.1)), three components of Navier-Stokes equation (Equation (12.3)) and thermal energy equation (Equation (12.26)). There are five unknowns, namely: pressure, three velocity components, and temperature. Thus, in principle, one can obtain pressure distribution, velocity distribution, and temperature distribution in non-isothermal fluid flow by solving the five equations subject to the given boundary and initial conditions.

Table 12.1 Thermal Energy Equation for Newtonian Fluids of Constant Properties

Co-ordinate System	Thermal Energy Equation
Cartesian	$\rho c_p\left(\frac{\partial T}{\partial t}+V_x\frac{\partial T}{\partial x}+V_y\frac{\partial T}{\partial y}+V_z\frac{\partial T}{\partial z}\right)=k\left[\frac{\partial^2 T}{\partial x^2}+\frac{\partial^2 T}{\partial y^2}+\frac{\partial^2 T}{\partial z^2}\right]+\bar{\bar{\tau}}:\nabla\vec{V}$ where $\bar{\bar{\tau}}:\nabla\vec{V}=2\eta\left\{\left(\frac{\partial V_x}{\partial x}\right)^2+\left(\frac{\partial V_y}{\partial y}\right)^2+\left(\frac{\partial V_z}{\partial z}\right)^2\right\}$ $+\eta\left\{\left(\frac{\partial V_x}{\partial y}+\frac{\partial V_y}{\partial x}\right)^2+\left(\frac{\partial V_x}{\partial z}+\frac{\partial V_z}{\partial x}\right)^2+\left(\frac{\partial V_y}{\partial z}+\frac{\partial V_z}{\partial y}\right)^2\right\}$
Cylindrical	$\rho c_p\left(\frac{\partial T}{\partial t}+V_r\frac{\partial T}{\partial r}+\frac{V_\theta}{r}\frac{\partial T}{\partial y}+V_z\frac{\partial T}{\partial z}\right)=k\left[\frac{1}{r}\frac{\partial}{\partial r}\left(r\frac{\partial T}{\partial r}\right)+\frac{1}{r^2}\frac{\partial^2 T}{\partial \theta^2}+\frac{\partial^2 T}{\partial z^2}\right]+\bar{\bar{\tau}}$ $:\nabla\vec{V}$ where $\bar{\bar{\tau}}:\nabla\vec{V}=2\eta\left\{\left(\frac{\partial V_r}{\partial r}\right)^2+\left[\frac{1}{r}\left(\frac{\partial V_\theta}{\partial \theta}\right)+V_r\right]^2+\left(\frac{\partial V_z}{\partial z}\right)^2\right\}$ $+\eta\left\{\left(\frac{\partial V_\theta}{\partial z}+\frac{1}{r}\frac{\partial V_z}{\partial \theta}\right)^2+\left(\frac{\partial V_z}{\partial r}+\frac{\partial V_r}{\partial z}\right)^2\right.$ $\left.+\left[\frac{1}{r}\frac{\partial V_r}{\partial \theta}+r\frac{\partial}{\partial r}\left(\frac{V_\theta}{r}\right)\right]^2\right\}$
Spherical	$\rho c_p\left(\frac{\partial T}{\partial t}+V_r\frac{\partial T}{\partial r}+\frac{V_\theta}{r}\frac{\partial T}{\partial y}+\frac{V_\varphi}{rsin\theta}\frac{\partial T}{\partial \varphi}\right)$ $=k\left[\frac{1}{r^2}\frac{\partial}{\partial r}\left(r^2\frac{\partial T}{\partial r}\right)+\frac{1}{r^2sin\theta}\frac{\partial}{\partial \theta}\left(sin\theta\frac{\partial T}{\partial \theta}\right)\right.$ $\left.+\frac{1}{r^2sin^2\theta}\frac{\partial^2 T}{\partial \varphi^2}\right]+\bar{\bar{\tau}}:\nabla\vec{V}$ where $\bar{\bar{\tau}}:\nabla\vec{V}=2\eta\left\{\left(\frac{\partial V_r}{\partial r}\right)^2+\left(\frac{1}{r}\frac{\partial V_\theta}{\partial \theta}+\frac{V_r}{r}\right)^2+\left(\frac{1}{rsin\vartheta}\frac{\partial V_\varphi}{\partial \varphi}+\frac{V_r}{r}+\frac{V_\theta cot\theta}{r}\right)^2\right\}$ $+\eta\left\{\left[r\frac{\partial}{\partial r}\left(\frac{V_\theta}{r}\right)+\frac{1}{r}\frac{\partial V_r}{\partial \theta}\right]^2+\left[\frac{1}{rsin\theta}\frac{\partial V_r}{\partial \varphi}+r\frac{\partial}{\partial r}\left(\frac{V_\varphi}{r}\right)\right]^2\right.$ $\left.+\left[\frac{sin\theta}{r}\frac{\partial}{\partial \theta}\left(\frac{V_\varphi}{sin\theta}\right)+\frac{1}{rsin\theta}\frac{\partial V_\theta}{\partial \varphi}\right]^2\right\}$

12.4 EXAMPLES

12.4.1 Flow Between Large Parallel Plates with Viscous Dissipation

Problem

Consider flow between two large parallel plates. Fluid is placed between two plates separated by a distance of δ. The upper plate at temperature T_2 moves at velocity V_o while the lower plate at temperature T_1 is stationary (see Figure 12.1). Determine the temperature distribution in the fluid and the heat flux at the surface of the lower plate. You can assume steady laminar flow of a Newtonian fluid of constant properties. There is no pressure gradient applied on the fluid.

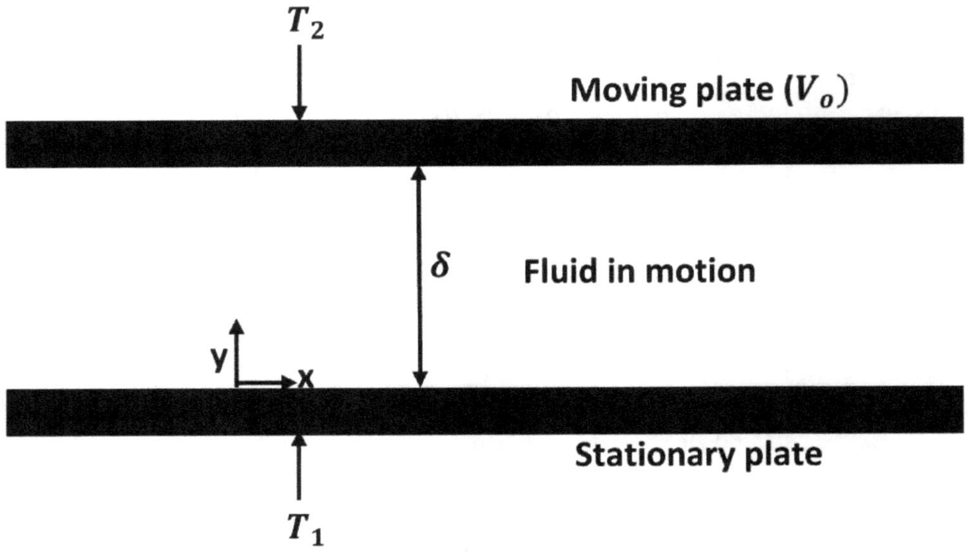

Figure 12.1 Flow between large parallel plates with viscous dissipation.

Solution

Velocity Distribution:

As the flow is in x -direction, only V_x is non-zero. The continuity equation gives:

$$\frac{\partial V_x}{\partial x} = 0 \tag{12.28}$$

The x-direction Navier-Stokes equation in Cartesian coordinates gives:

$$0 = \eta\left(\frac{d^2 V_x}{dy^2}\right) \tag{12.29}$$

Upon integration, Equation (12.29) yields:

$$V_x = C_1 y + C_2 \tag{12.30}$$

The integration constants C_1 and C_2 can be determined from the following boundary conditions:

B.C. 1: at $y = 0,\ V_x = 0$ (no-slip boundary condition) (12.31)

B.C. 2: at $y = \delta,\ V_x = V_o$ (no-slip boundary condition) (12.32)

The solution is:

$$C_1 = \frac{V_o}{\delta},\ C_2 = 0 \tag{12.33}$$

$$V_x = \left(\frac{V_o}{\delta}\right) y \tag{12.34}$$

Temperature Distribution:

The thermal energy equation gives:

$$k\frac{d^2T}{dy^2} + \eta\left(\frac{dV_x}{dy}\right)^2 = 0 \quad (12.35)$$

From Equations (12.34) and (12.35), it follows that:

$$k\frac{d^2T}{dy^2} + \eta\left(\frac{V_o}{\delta}\right)^2 = 0 \quad (12.36)$$

Upon integration, Equation (12.36) gives:

$$\frac{dT}{dy} = -\frac{\eta}{k}\left(\frac{V_o}{\delta}\right)^2 y + C_1 \quad (12.37)$$

$$T = -\frac{\eta}{k}\left(\frac{V_o}{\delta}\right)^2 \frac{y^2}{2} + C_1 y + C_2 \quad (12.38)$$

The integration constants C_1 and C_2 can be determined from the following boundary conditions:

B.C. 1: at $y = 0,\ T = T_1$ (12.39)

B.C. 2: at $y = \delta,\ T = T_2$ (12.40)

The solution is:

$$C_1 = \frac{T_2 - T_1}{\delta} + \frac{\eta}{k}\left(\frac{V_o^2}{2\delta}\right) \qquad C_2 = T_1 \quad (12.41)$$

$$T - T_1 = \left(\frac{T_2 - T_1}{\delta}\right) y + \left(\frac{\eta V_o^2}{2k}\right)\left[\frac{y}{\delta} - \left(\frac{y}{\delta}\right)^2\right] \quad (12.42)$$

Heat Flux:

From Fourier's law of heat conduction, the heat flux is:

$$q_y'' = -k\left(\frac{dT}{dy}\right) \quad (12.43)$$

From Equations (12.42) and (12.43), it follows that:

$$q_y'' = -k\left(\frac{T_2 - T_1}{\delta}\right) - \left(\frac{\eta V_o^2}{2\delta}\right)\left(1 - 2\frac{y}{\delta}\right) \quad (12.44)$$

Heat flux at lower plate is:

$$q_y''\Big|_{y=0} = -k\left(\frac{T_2 - T_1}{\delta}\right) - \left(\frac{\eta V_o^2}{2\delta}\right) \quad (12.45)$$

Comments:

Note that when the temperatures of the two plates are the same, that is,

$$T_2 = T_1 = T_o \quad (12.46)$$

the temperature in the fluid is still not uniform and is given as:

$$T - T_o = \left(\frac{\eta V_o^2}{2k}\right)\left[\frac{y}{\delta} - \left(\frac{y}{\delta}\right)^2\right] \tag{12.47}$$

Likewise, the heat flux is also not zero:

$$q_y'' = -\left(\frac{\eta V_o^2}{2\delta}\right)\left(1 - 2\frac{y}{\delta}\right) \tag{12.48}$$

The temperature variation and non-zero heat flux are due to frictional heating in the fluid. Only if the fluid viscosity is zero, that is, the flow is inviscid frictionless, the temperature is uniform in the fluid $T = T_o$ and the heat flux is zero.

12.4.2 Flow in a Pipe with Viscous Dissipation

Problem

Consider steady laminar flow of a Newtonian fluid of constant properties in a pipe of radius R. The wall of the pipe is maintained at a constant temperature of T_w (see Figure 12.2). Neglecting axial heat conduction, determine the temperature distribution in the fluid and the heat flux at the pipe wall.

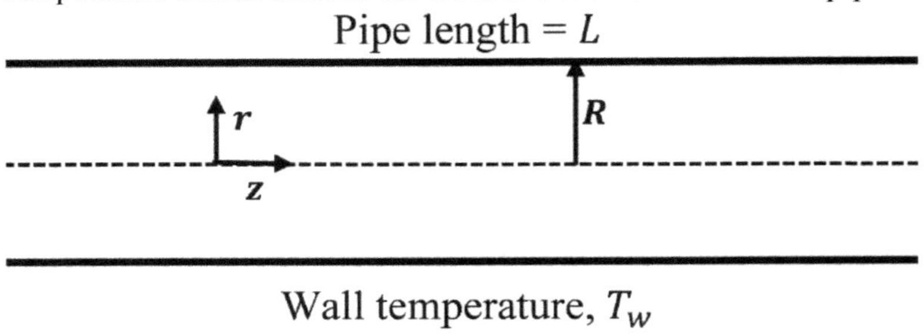

Figure 12.2 Flow in a pipe with viscous dissipation.

Solution

Velocity Distribution:

As the flow is in z-direction, only V_z is non-zero. From continuity equation:

$$\frac{\partial V_z}{\partial z} = 0 \tag{12.49}$$

The z -component of the Navier-Stokes equation in cylindrical coordinates gives:

$$-\frac{\partial P}{\partial z} + \rho g_z + \frac{\eta}{r}\frac{\partial}{\partial r}\left(r\frac{\partial V_z}{\partial r}\right) = 0 \tag{12.50}$$

P is a function of z only, V_z is a function of r only, and $g_z = 0$. Thus, Equation (12.50) can be re-written as:

$$\frac{d}{dr}\left(r\frac{dV_z}{dr}\right) = \left(\frac{dP}{dz}\right)\frac{r}{\eta} \tag{12.51}$$

Upon integration and using the boundary conditions, Equation (12.51 gives:

$$V_z(r) = V_{max}\left[1 - \left(\frac{r}{R}\right)^2\right] \tag{12.52}$$

where

$$V_{max} = \frac{R^2}{4\eta}\left(-\frac{dP}{dz}\right) \tag{12.53}$$

Temperature Distribution:

The thermal energy equation gives:

$$\frac{k}{r}\frac{d}{dr}\left(r\frac{dT}{dr}\right) + \eta\left(\frac{dV_z}{dr}\right)^2 = 0 \tag{12.54}$$

From Equations (12.52) and (12.54), it follows that:

$$\frac{k}{r}\frac{d}{dr}\left(r\frac{dT}{dr}\right) + \eta\left(-\frac{2r}{R^2}V_{max}\right)^2 = 0 \tag{12.55}$$

Upon rearrangement, Equation (12.55) becomes:

$$\frac{d}{dr}\left(r\frac{dT}{dr}\right) = \left(-\frac{4\eta V_{max}^2}{kR^4}\right)r^3 \tag{12.56}$$

Upon integration, Equation (12.56) gives:

$$\frac{dT}{dr} = \left(-\frac{\eta V_{max}^2}{kR^4}\right)r^3 + \frac{C_1}{r} \tag{12.57}$$

We must take $C_1 = 0$ as otherwise, the temperature gradient diverges at $r = 0$. Taking $C_1 = 0$ and further integration, Equation (12.57) gives:

$$T = \left(-\frac{\eta V_{max}^2}{4kR^4}\right)r^4 + C_2 \tag{12.58}$$

The integration constant C_2 can be determined from the following boundary condition:

$$\text{B.C.:}\quad \text{at } r = R,\ T = T_w \tag{12.59}$$

The solution is:

$$C_2 = T_w + \left(\frac{\eta V_{max}^2}{4k}\right) \tag{12.60}$$

$$T - T_w = \left(\frac{\eta V_{max}^2}{4k}\right)\left[1 - \left(\frac{r}{R}\right)^4\right] \tag{12.61}$$

Heat Flux:

From Fourier's law of heat conduction, the heat flux is:

$$q_r'' = -k\left(\frac{dT}{dr}\right) \tag{12.62}$$

From Equations (12.61) and (12.62), it follows that:

$$q_r'' = \left(\frac{\eta V_{max}^2}{R^4}\right) r^3 \tag{12.63}$$

Heat flux at the pipe wall is:

$$q_r''|_{r=R} = \frac{\eta V_{max}^2}{R} \tag{12.64}$$

Note that heat flux is zero when the flow is frictionless ($\eta = 0$).

12.4.3 Flow of a Falling Film with Viscous Dissipation

Problem

Consider the steady laminar flow of a Newtonian liquid film flowing down a flat inclined plate due to gravity (see Figure 12.3). The plate is inclined at an angle α with the horizontal. The length and width of the plate are L and W, respectively. The thickness of the liquid film is δ. The free surface of the liquid is at temperature T_o. The plate temperature is maintained constant at T_w. Determine the velocity and temperature distributions in the fluid and the heat flux at the plate surface.

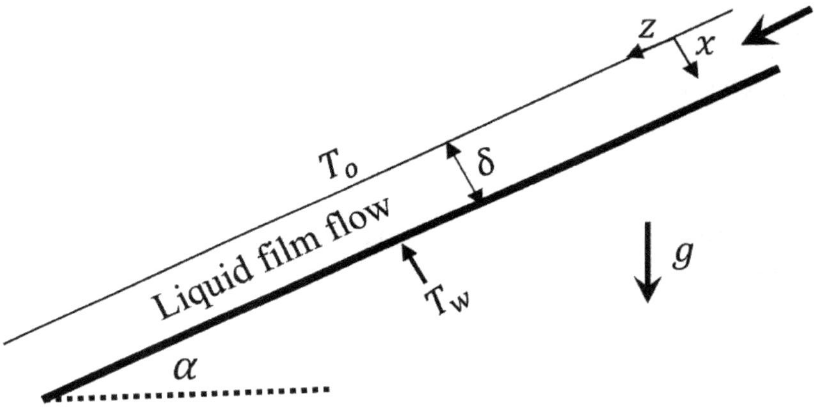

Figure 12.3 Flow of a falling film with viscous dissipation.

Solution

Velocity Distribution:

As the flow is in z-direction, only V_z is non-zero. From continuity equation:

$$\frac{\partial V_z}{\partial z} = 0 \tag{12.65}$$

The z -component of the Navier-Stokes equation in Cartesian coordinates gives:

$$\eta \frac{d^2 V_z}{dx^2} = -\rho g_z \tag{12.66}$$

As $g_z = g\, sin\, \alpha$, Equation (12.66) becomes:

$$\eta \frac{d^2 V_z}{dx^2} = -\rho g \sin \alpha \tag{12.67}$$

Upon integration:

$$V_z = -\left(\frac{\rho g \sin \alpha}{\eta}\right)\frac{x^2}{2} + C_1 x + C_2 \tag{12.68}$$

The integration constants C_1 and C_2 can be determined from the following boundary conditions:

B.C. 1: at $x = 0, \tau_{xz} = 0, \frac{dV_z}{dx} = 0$ (shear stress at liquid surface is zero) (12.69)

B.C. 2: $at\ x = \delta, V_z = 0$ (no-slip boundary condition) (12.70)

The solution is:

$$C_1 = 0\,,\quad C_2 = \left(\frac{\rho g \sin \alpha}{\eta}\right)\frac{\delta^2}{2} \tag{12.71}$$

$$V_z(x) = V_{max}\left[1 - \left(\frac{x}{\delta}\right)^2\right] \tag{12.72}$$

where

$$V_{max} = \frac{\rho g (\sin \alpha)\delta^2}{2\eta} \tag{12.73}$$

Temperature Distribution:

The thermal energy equation gives:

$$0 = k\left[\frac{d^2 T}{dx^2}\right] + \eta\left(\frac{dV_z}{dx}\right)^2 \tag{12.74}$$

From velocity distribution, Equation (12.72), upon differentiation we get:

$$\frac{dV_z}{dx} = -\frac{2V_{max}x}{\delta^2} \tag{12.75}$$

The thermal energy equation becomes:

$$\frac{d^2 T}{dx^2} = -\frac{4\eta V_{max}^2 x^2}{k\delta^4} \tag{12.76}$$

$$\frac{dT}{dx} = -\frac{4\eta V_{max}^2 x^3}{3k\delta^4} + C_1 \tag{12.77}$$

$$T = -\frac{\eta V_{max}^2 x^4}{3k\delta^4} + C_1 x + C_2 \tag{12.78}$$

The integration constants C_1 and C_2 can be determined from the following boundary conditions:

B.C. 1: at $x = 0, T = T_o$ (12.79)

$$\text{B.C. 2:} \quad at\ x = \delta, \mathrm{T} = T_w \tag{12.80}$$

Thus, the temperature distribution is:

$$T - T_o = (T_w - T_o)\left(\frac{x}{\delta}\right) + \frac{\eta V_{max}^2}{3k}\left[\left(\frac{x}{\delta}\right) - \left(\frac{x}{\delta}\right)^4\right] \tag{12.81}$$

Heat Flux:

The heat flux at any point in the fluid can be obtained as follows:

$$q_x'' = -k\frac{dT}{dx} = -k\left(\frac{T_w - T_o}{\delta}\right) - \frac{\eta V_{max}^2}{3}\left[\left(\frac{1}{\delta}\right) - 4\frac{x^3}{\delta^4}\right] \tag{12.82}$$

Hence the heat flux at the wall is:

$$q_x''|_{x=\delta} = -k\left.\frac{dT}{dx}\right|_{x=\delta} = -k\left(\frac{T_w - T_o}{\delta}\right) + \frac{\eta}{\delta}V_{max}^2 \tag{12.83}$$

Note that for frictionless flow ($\eta = 0$), the heat flux present in the fluid is just the conductive heat flux corresponding to linear temperature distribution.

12.4.4 Flow of a Falling Film with Temperature Variation in Two Directions

Problem

As an example, consider the flow of a liquid film down a flat inclined wall maintained at a uniform surface temperature of T_w (see Figure 12.4). At the entrance, the liquid temperature is T_o. Assuming $T_o < T_w$, heat transfer takes place from the wall to the fluid. The length and width of the wall are L and W, respectively. The thickness of the liquid film is δ. We are interested in determining the temperature distribution in the fluid and the rate of heat transfer from the wall to the fluid. Assume that the flow is steady laminar, and the fluid properties are constant. Also, the temperature changes are restricted to the region close to the wall due to short contact time between the fluid and the wall (L/V_{max} is small).

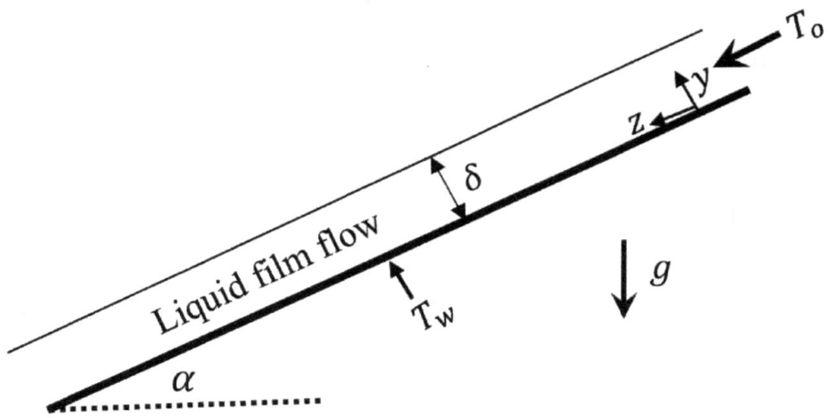

Figure 12.4 Flow of a falling film with temperature variation in two directions.

Solution

Velocity Distribution:

The velocity distribution in the film is as follows:

$$V_z(y) = V_{max}\left[1-\left(1-\frac{y}{\delta}\right)^2\right] = V_{max}\left[2\left(\frac{y}{\delta}\right)-\left(\frac{y}{\delta}\right)^2\right] \quad \text{where} \quad V_{max} = \frac{\rho g(\sin\alpha)\delta^2}{2\eta} \tag{12.84}$$

Equation (12.84) follows from Equation (12.72) by substituting $x = \delta - y$ in Equation (12.72). Close to the wall, $y/\delta \gg (y/\delta)^2$ and therefore, the velocity distribution is approximately linear in the wall region:

$$V_z = 2V_{max}\left(\frac{y}{\delta}\right) = \left(\frac{\rho g\delta \sin\alpha}{\eta}\right)y \tag{12.85}$$

Temperature Distribution:

Under steady state condition, the temperature distribution in the film is governed by the following thermal energy equation:

$$\rho C_p V_z \frac{\partial T}{\partial z} = k\frac{\partial^2 T}{\partial y^2} + \eta\left(\frac{\partial V_x}{\partial y}\right)^2 \tag{12.86}$$

Neglecting viscous heating and assuming short contact time, Equation (12.86) simplifies to:

$$y\frac{\partial T}{\partial z} = \beta\frac{\partial^2 T}{\partial y^2} \quad \text{where} \quad \beta = \eta k/(\rho^2 C_p g\delta \sin\alpha) \tag{12.87}$$

Equation (12.87) needs to be solved for temperature distribution subject to the following boundary conditions:

B.C.1: $z = 0,\ T = T_o$ the liquid enters at a uniform temperature T_o (12.88)

B.C. 2: $y = 0,\ T = T_w$ the temperature at the wall is T_w (12.89)

B.C. 3: $y \to \infty,\ T = T_o$ as heat penetrates only a short distance into the film, $T = T_o$ for large y (12.90)

Equation (12.87) can be solved by using the combination of variables. Let ε be a new variable defined as:

$$\varepsilon = y/(9\beta z)^{1/3} \tag{12.91}$$

The partial differential equation in Equation (12.87) can be reduced to the following ordinary differential equation by substitution of the new variable ε:

$$\left(\frac{d^2\Upsilon}{d\varepsilon^2}\right) + 3\varepsilon^2\frac{d\Upsilon}{d\varepsilon} = 0 \quad \text{where} \quad \Upsilon = \frac{T-T_o}{T_s-T_o} \tag{12.92}$$

The new boundary conditions become:

$$\text{B.C.1: } \varepsilon = 0,\ \ \Upsilon = 1 \tag{12.93}$$

$$\text{B.C.2: } \varepsilon = \infty,\ \ \Upsilon = 0 \tag{12.94}$$

Note that the previous boundary conditions B.C. 1, Equation (12.88), and B.C. 3, Equation (12.90), now reduce to the same new boundary condition, that is, Equation (12.94).

The following solution is obtained for Equation (12.92) subject to the given boundary conditions:

$$\Upsilon = \frac{T-T_o}{T_s-T_o} = \frac{\int_\varepsilon^\infty e^{-\lambda^3} d\lambda}{\int_0^\infty e^{-\lambda^3} d\lambda} = \frac{\int_\varepsilon^\infty e^{-\lambda^3} d\lambda}{\Gamma(4/3)} \tag{12.95}$$

where Γ is the gamma function.

Heat Flux:

The local heat flux at the solid-liquid interface ($y = 0$) can now be obtained as follows:

$$q''_y\Big|_{y=0} = -k\frac{\partial T}{\partial y}\Big|_{y=0} = \frac{k}{\Gamma(4/3)}(9\beta z)^{-1/3}(T_s - T_o) = \frac{k}{\Gamma(4/3)}\left(\frac{\rho^2 C_p g\delta \sin\alpha}{9k\eta z}\right)^{1/3}(T_s - T_o) \tag{12.96}$$

Note that:

$$\left(\frac{\partial T}{\partial y}\right)_{y=0} = \frac{(T_s-T_o)}{\Gamma(4/3)}\left(\frac{\partial \varepsilon}{\partial y}\right)_{y=0}\left[\frac{d}{d\varepsilon}\int_\varepsilon^\infty e^{-\lambda^3} d\lambda\right]_{\varepsilon=0} \tag{12.97}$$

The square bracketed term on the right-hand side of Equation (12.97) is equal to -1 and therefore,

$$\left(\frac{\partial T}{\partial y}\right)_{y=0} = -\frac{(T_s-T_o)}{\Gamma(4/3)}\left(\frac{\partial \varepsilon}{\partial y}\right)_{y=0} = -\frac{(T_s-T_o)}{\Gamma(4/3)}(9\beta z)^{-1/3} \tag{12.98}$$

The total heat transfer rate from the wall to the liquid can be determined as follows:

$$q = \int_0^L q''_y\Big|_{y=0} W dz = \frac{3}{2}\frac{WL}{\Gamma(4/3)}\left(\frac{\rho^2 C_p g\delta \sin\alpha}{9k\eta L}\right)^{1/3} k(T_s - T_o) \tag{12.99}$$

12.4.5 Tangential Flow in an Annulus of Two Co-axial Cylinders with Viscous Dissipation

Problem
A co-axial cylinder device consists of two concentric cylinders with the annular space between the cylinders filled with a Newtonian liquid of constant properties. The outer cylinder is rotated at an angular velocity of ω while the inner cylinder is held stationary. The temperature of the inner cylinder is maintained at T_i and the outer cylinder is maintained at T_o (see Figure 12.5). determine the temperature distribution and heat flux in the fluid.

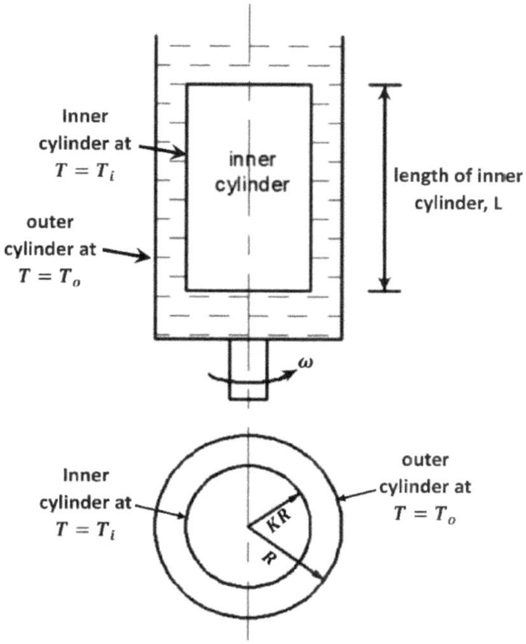

Figure 12.5 Tangential flow in an annulus of two co-axial cylinders with viscous dissipation.

Solution:

Velocity Distribution:

As the flow is in θ-direction, only V_θ is non-zero. From continuity equation:

$$\frac{\partial V_\theta}{\partial \theta} = 0 \tag{12.100}$$

Thus, V_θ is a function of r only. The θ-direction Navier-Stokes equation in cylindrical coordinates gives:

$$\frac{d}{dr}\left(\frac{1}{r}\frac{d}{dr}(rV_\theta)\right) = 0 \tag{12.101}$$

Upon integration, we get:

$$V_\theta = \frac{C_1 r}{2} + \frac{C_2}{r} \tag{12.102}$$

The integration constants C_1 and C_2 can be determined from the following boundary conditions:

B.C. 1: at $r = R, V_\theta = \omega R$ (no-slip boundary condition) (12.103)

B.C. 2: at r $=$ KR, $V_\theta = 0$ (no-slip boundary condition) (12.104)

The solution is:

$$C_1 = \frac{2\omega}{K^2-1}, \quad C_2 = -\frac{K^2R^2\omega}{K^2-1} \tag{12.105}$$

$$V_\theta = \frac{\omega RK}{K^2-1}\left[\frac{KR}{r} - \frac{r}{KR}\right] \tag{12.106}$$

Temperature Distribution:

The thermal energy equation gives:

$$\frac{k}{r}\frac{d}{dr}\left(r\frac{dT}{dr}\right)+\eta\left[r\frac{d}{dr}\left(\frac{V_\theta}{r}\right)\right]^2=0 \tag{12.107}$$

From Equation (12.106):

$$\frac{d}{dr}\left(\frac{V_\theta}{r}\right)=\frac{2\omega R^2K^2}{1-K^2}\left(\frac{1}{r^3}\right) \tag{12.108}$$

From Equations (12.107) and (12.108), we get:

$$\frac{k}{r}\frac{d}{dr}\left(r\frac{dT}{dr}\right)+\eta\left[\frac{2\omega R^2K^2}{1-K^2}\left(\frac{1}{r^2}\right)\right]^2=0 \tag{12.109}$$

Note that k is thermal conductivity whereas K is a parameter specifying the radius of the inner cylinder (see Figure 12.4).

For the sake of convenience, we re-write Equation (12.109) in terms of the dimensionless variables defined below:

$$\eta=\frac{r}{R} \tag{12.110}$$

$$\Upsilon=\frac{T-T_i}{T_o-T_i} \tag{12.111}$$

$$N=\frac{\eta(\omega R)^2}{k(T_o-T_i)}\left[\frac{K^4}{(1-K^2)^2}\right] \tag{12.112}$$

Thus, dimensionless form of Equation (12.109) is as follows:

$$\frac{1}{\eta}\frac{d}{d\eta}\left(\eta\frac{d\Upsilon}{d\eta}\right)=-\frac{4N}{\eta^4} \tag{12.113}$$

Upon integration, Equation (12.113) gives:

$$\frac{d\Upsilon}{d\eta}=\frac{2N}{\eta^3}+\frac{C_1}{\eta} \tag{12.114}$$

$$\Upsilon=-\frac{N}{\eta^2}+C_1\ln(\eta)+C_2 \tag{12.115}$$

The integration constants C_1 and C_2 can be determined from the following boundary conditions:

$$\text{B.C. 1:}\quad \text{at } \eta=K, \Upsilon=0 \tag{12.116}$$

$$\text{B.C. 2:}\quad at\ \eta=1, \Upsilon=1 \tag{12.117}$$

The solution is:

$$C_1=\left[\frac{N}{K^2}-(1+N)\right]\left(\frac{1}{lnK}\right),\quad C_2=1+N \tag{12.118}$$

$$\Upsilon = \left[-\frac{N}{\eta^2} + (1+N)\right] + \left[\frac{N}{K^2} - (1+N)\right]\left[\frac{\ln(\eta)}{\ln(K)}\right] \tag{12.119}$$

Heat Flux:

The heat flux at any point in the fluid can be obtained as follows:

$$q_r'' = -k\frac{dT}{dr} \tag{12.120}$$

From Equations (12.119) and (12.120):

$$q_r'' = -k\frac{dT}{dr} = -k\left(\frac{T_o - T_i}{R}\right)\frac{d\Upsilon}{d\eta} \tag{12.121}$$

Upon differentiation of Equation (12.119), we get:

$$\frac{d\Upsilon}{d\eta} = \left[\frac{a(1+\eta^2)}{\eta^2} + \frac{a}{K^2 lnK}(1 + lnK) - \frac{1}{lnK}\left(\frac{1}{\eta} + a(1 + lnK)\right)\right] \tag{12.122}$$

where $a = N/\eta$. From Equations (12.121) and (12.122), the heat flux at any point in the fluid is given as:

$$q_r'' = -k\frac{dT}{dr} = -k\left(\frac{T_o - T_i}{R}\right)\left[2a + \frac{a}{K^2 lnK} - \frac{1}{lnK}(1 + a)\right] \tag{12.123}$$

NOTATION

CS	control surface
CV	control volume
c_p	constant pressure specific heat capacity
c_v	constant volume specific heat capacity
e	total energy per unit mass
g	magnitude of acceleration due to gravity
$\vec{g}$	acceleration due to gravity
g_z	component of gravity in z-direction
k	thermal conductivity
L	length
$\hat{n}$	unit normal vector
P	pressure
$\dot{Q}$	rate of heat input
$\vec{q}''$	heat flux vector
q_r''	heat flux in radial direction
q_x''	heat flux in x-direction
q_y''	heat flux in y-direction
R	radius
r	radial coordinate or distance in radial direction
S	surface area
T	temperature
T_w	wall temperature
u	specific internal energy
V	magnitude of velocity vector or volume

V_{max} maximum velocity
$\vec{V}$ velocity vector
V_x, V_y, V_z velocity components in cartesian coordinates
V_r, V_θ, V_z velocity components in cylindrical coordinates
V_r, V_θ, V_φ velocity components in spherical coordinates
$\dot{W}$ rate of work output
x, y, z space coordinates in cartesian coordinates

Greek Letters

∇ mathematical del operator
$\bar{\bar{\delta}}$ unit tensor
η viscosity of fluid
θ space coordinate in cylindrical and spherical coordinates
$\vec{\pi}_{(\hat{n})}$ total stress vector
$\bar{\bar{\pi}}$ total stress tensor
ρ density
$\bar{\bar{\tau}}$ viscous stress tensor
υ specific volume
φ space coordinate in spherical coordinates
ω angular velocity

PRACTICE PROBLEMS

Problem 1

Consider the flow of a Newtonian fluid between two large parallel plates separated by a distance δ as shown in the figure below. The bottom plate is stationary and is insulated from heat transfer. The top plate is moving at a constant velocity V_o and is maintained at a constant temperature of T_o. No pressure gradient is applied to the fluid. Determine the following:

a) Velocity and temperature distributions in the fluid

b) The temperature of the bottom stationary plate

c) Heat flux distribution in the fluid

d) Haet flux at the top moving plate

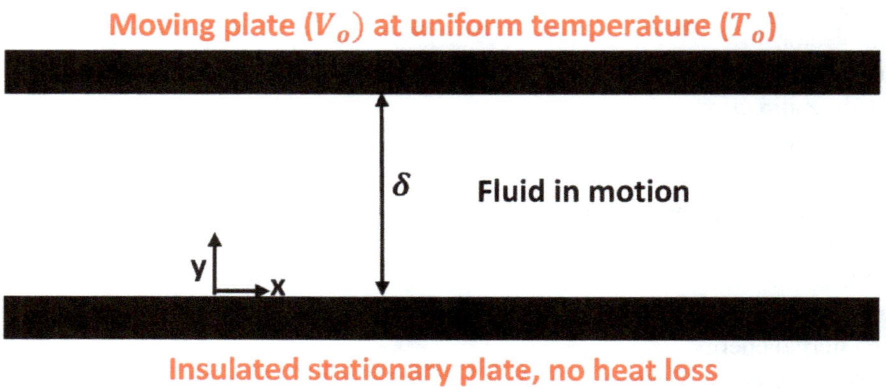

Problem 2

Consider flow between two large parallel plates. Fluid is placed between two isothermal plates separated by a distance of $2B$. Both plates are stationary and are maintained at a constant temperature of T_o (see figure). The motion of fluid is generated by applying a constant pressure gradient $(dP/dx < 0)$. Determine the temperature and heat flux distributions in the fluid. What is the heat flux at the surface of the plates. You can assume steady laminar flow of a Newtonian fluid of constant properties.

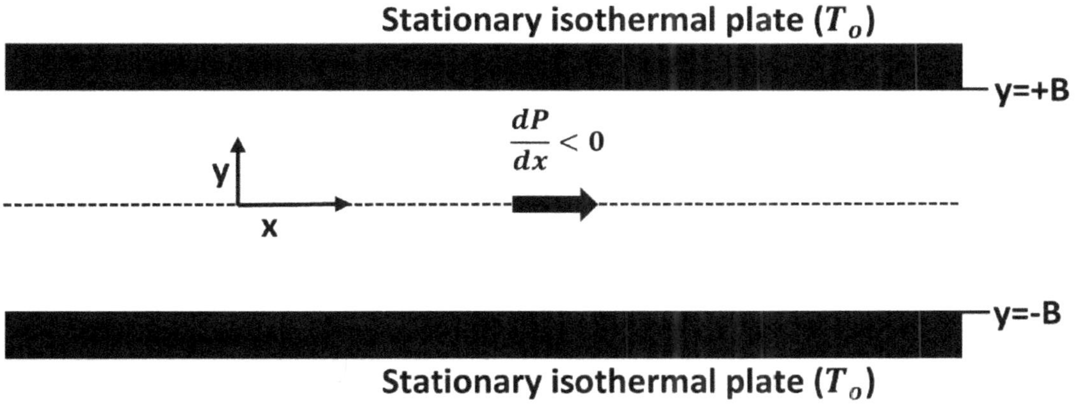

Problem 3

Consider the steady laminar flow of a Newtonian fluid between two infinitely long horizontal concentric cylinders. The fluid motion is caused by the longitudinal motion of the inner cylinder at a constant velocity of V_o (see figure). The outer cylinder is stationary. The temperatures of the inner and outer cylinders are maintained constant at T_o. There is no pressure-gradient applied. Determine the following:

i) temperature distribution in the fluid
ii) heat flux distribution in the fluid
iii) Heat fluxes at the surfaces of inner and outer cylinders

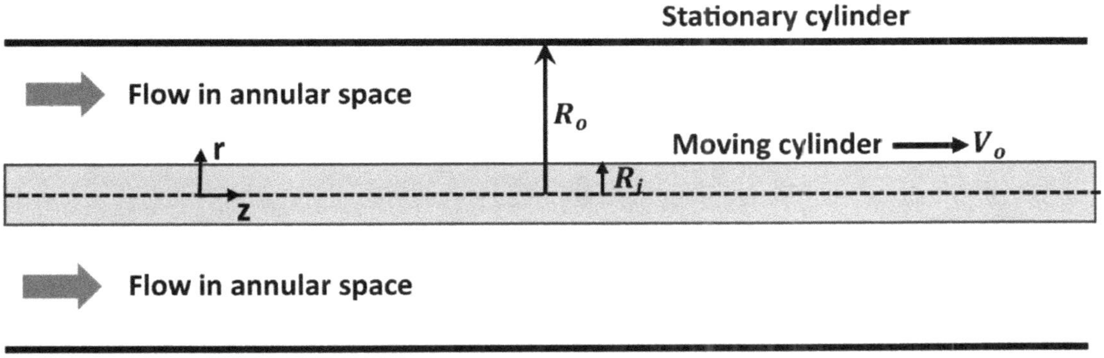

FURTHER READING

1. R.B. Bird, W.E. Stewart, and E.N. Lightfoot. 2007. *Transport Phenomena.* 2nd Edition, Wiley & Sons, New York, 2007.
2. T.L. Bergman, A.S. Levine, F.P. Incropera, and D.P. Dewitt. *Fundamentals of Heat and Mass Transfer.* 7th Edition, New York, Wiley & Sons, 2011.
3. A. Bejan. *Heat Transfer.* Wiley & Sons, New York, 1993.
4. L.C. Thomas. *Heat Transfer.* Prentice Hall, New Jersey, 1992.

Chapter 13

Heat Transfer Coefficient: Correlations and Applications

In Chapter 12, analytical solutions of temperature and heat flux distributions in fluid were obtained for many problems involving steady laminar flow of Newtonian fluids of constant properties. However, it is not always possible to obtain an analytical solution to the microscopic equations of motion and thermal energy even in the case of laminar flows when the geometry is complicated. Furthermore, turbulent flows are completely out of reach as it is not at all possible to obtain any analytical solutions of turbulent flows although the microscopic equations of motion and thermal energy equation are still valid. Fortunately, in many practical applications involving convective heat transfer, detailed temperature and heat flux distributions in the fluid are not required. Often, the quantity of practical interest is the wall heat flux, that is, the heat flux at the wall on the flow geometry. The heat transfer rate from the solid boundary (wall) to the fluid and vice versa can be calculated from the knowledge of heat flux at the solid boundary and the heat transfer area.

Two types of flow situations are encountered in practical applications of convective heat transfer: external flows (see Figure 13.1) and internal flows (see Figure 13.2). There is enough experimental evidence to suggest that the wall heat flux q''_w is directly proportional to the characteristic temperature difference between the solid boundary and the fluid moving adjacent to the solid boundary, that is,

$$q''_w \propto (T_w - T_c) \tag{13.1}$$

where T_w is the wall (solid boundary) temperature and T_c is the characteristic temperature of the fluid.

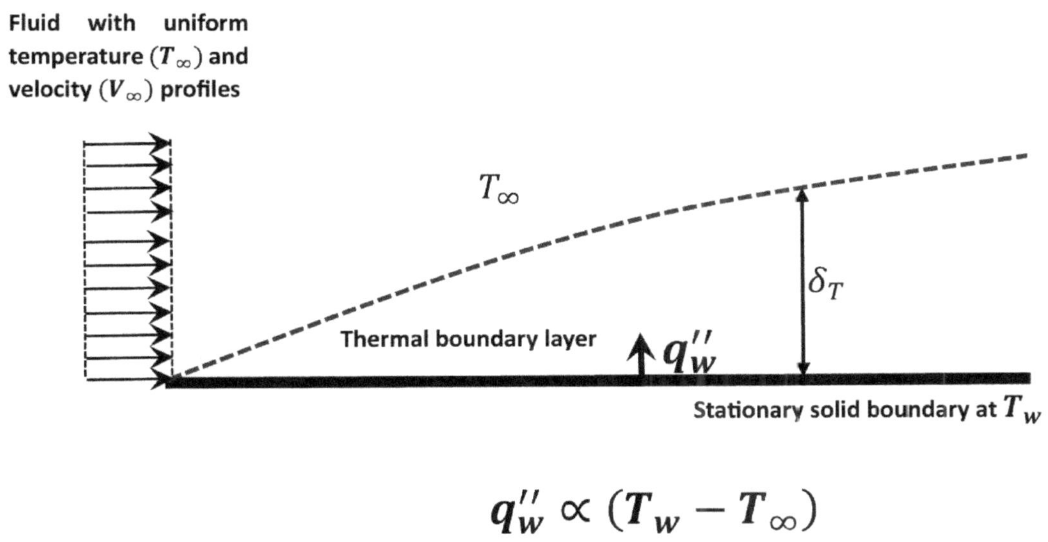

Figure 13.1 Convective heat transfer in external flow (cold fluid flow over a hot plate).

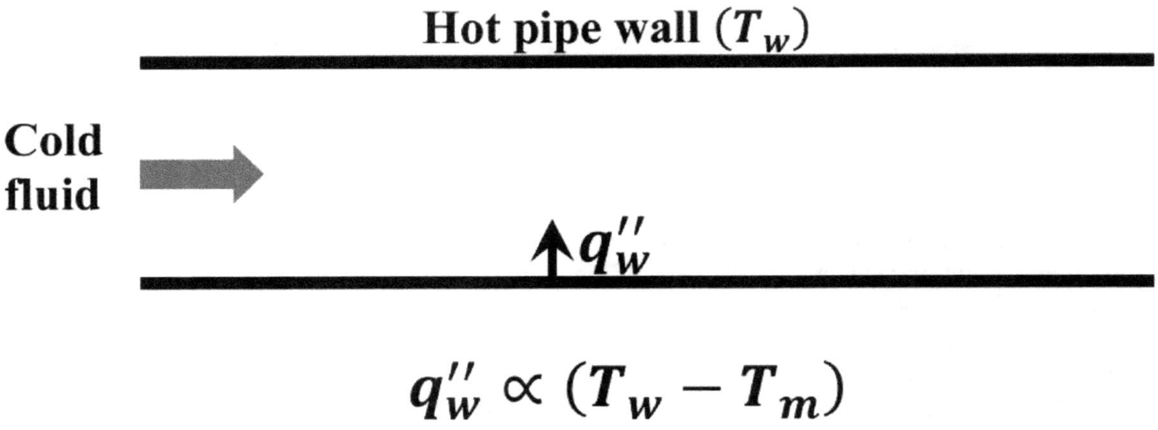

Figure 13.2 Convective heat transfer in internal flow (cold fluid flow in a hot pipe).

For external flows:

$$T_c = T_\infty \tag{13.2}$$

where T_∞ is the free-stream temperature of the fluid. For internal flows:

$$T_c = T_m \tag{13.3}$$

where T_m is the mean temperature over the cross-section of the pipe at any given axial location.

The heat transfer coefficient is defined as the proportionality factor between q''_w and $(T_w - T_c)$. Thus, Equation (13.1) can be re-written as:

$$q''_w = h(T_w - T_c) \tag{13.4}$$

Equation (13.4) is often referred to as Newton's law of cooling.

For external flows,

$$q''_w = h(T_w - T_\infty) \tag{13.5}$$

For internal flows,

$$q''_w = h(T_w - T_m) \tag{13.6}$$

The heat flux at the wall and hence the heat transfer rate between the solid boundary and the fluid can be calculated from Equations (13.4) - (13.6) provided that the heat transfer coefficient h for the given situation is known or could be estimated from correlations.

Note that the heat transfer coefficient is linked to temperature gradient in the fluid at the solid boundary through the following relationship:

$$h = \frac{-k\left(\frac{\partial T}{\partial y}\right)_{y=0}}{(T_w - T_c)} \tag{13.7}$$

where k is the thermal conductivity of the fluid and $(\partial T/\partial y)_{y=0}$ is the temperature gradient in the fluid at the solid boundary. Equation (13.7) follows from the equality of Fourier's law of heat conduction and Newton's law of cooling at the solid boundary.

13.1 HEAT TRANFER COEFFICIENT IN EXTERNAL FLOWS

External flow refers to the flow of unbounded fluid over a solid surface. Examples include flow over a flat plate and flow over a solid sphere or solid cylinder. In many practical applications, the solid surface and the fluid have different temperatures and one is interested in the calculation of the rate of heat transfer between the solid surface and the fluid moving over the solid surface. The total rate of heat transfer q can be expressed as:

$$q = \int_A q'' dA \tag{13.8}$$

where q'' is the local heat flux at any point on the surface of the solid, normal to the surface, and A is the surface area of the solid object over which heat transfer is taking place.

At a solid/fluid interface, the local heat flux normal to the solid surface is related to the difference between the local solid surface temperature T_w and the free stream temperature of the fluid T_∞ as given by the Newton's law of cooling, that is, Equation (13.5). Thus, the heat transfer rate can be expressed as:

$$q = \int_A h(T_w - T_\infty) dA \tag{13.9}$$

For a given geometry (shape of the solid boundary) and temperature difference $(T_w - T_\infty)$, q can be calculated from Equation (13.9) provided that h is known. Note that h is by no means constant. It varies from point to point on the surface of the solid boundary. It is a complicated function of fluid properties, fluid motion, and geometry.

13.1.1 Flow Over a Flat Plate

The heat transfer coefficient is usually expressed in a dimensionless form as Nusselt number defined below:

$$Nu_x = \frac{hx}{k} \tag{13.10}$$

where Nu_x is the local Nusselt number, x is the distance from the leading edge of the solid boundary (plate in the present case). Dimensional analysis indicates that Nusselt number is a function of Reynolds number and Prandtl number:

$$Nu_x = f(Re_x, Pr) \tag{13.11}$$

where Re_x is the local Reynolds number defined as $\rho V_\infty x/\eta$ and Pr is the Prandtl number defined as:

$$Pr = \frac{\nu}{\alpha} = \frac{c_p \eta}{k} \tag{13.12}$$

The following relations are valid for local heat transfer in laminar and turbulent boundary layers over a smooth plate:

$$Nu_x = \frac{hx}{k} = 0.332(Re_x)^{1/2}\left(Pr^{1/3}\right) \tag{13.13}$$

$$Nu_x = \frac{hx}{k} = 0.0296(Re_x)^{4/5}\left(Pr^{1/3}\right) \tag{13.14}$$

Equation (13.13) is valid for laminar boundary layer and Equation (13.14) is valid for turbulent boundary layer. The temperature of the plate is assumed to be constant and $Pr \gg 1$. Note that $Pr \gg 1$ is for non-metallic liquids. Equation (13.13) is an exact theoretical relationship obtained from the solution of boundary layer equations. The solution is given in Appendix A. Equation (13.14) is an empirical relation based on Chilton-Colburn analogy between momentum and heat transfer (see Appendix A). The Chilton-Colburn analogy between momentum and heat transfer links friction factor with heat transfer coefficient and is given as:

$$St_x\, Pr^{2/3} = \frac{f}{2} \tag{13.15}$$

where St_x is called Stanton number, defined as $Nu_x/(Re_x\, Pr)$. Thus, the heat transfer coefficient can be obtained if only the friction factor is known.

As already noted, the heat transfer coefficient equations, Equations (13.13) and (13.14), assume that the solid surface (flat plate) is isothermal. Heat transfer coefficient equations have also been developed for constant wall-heat flux problems. Constant wall heat flux problems are equally important from a practical point of view. The results for $Pr \gg 1$ are given here without going into the derivations (see Appendix A):

$$Nu_x = \frac{hx}{k} = 0.453(Re_x)^{1/2}\left(Pr^{1/3}\right) \tag{13.16}$$

$$Nu_x = \frac{hx}{k} = 0.0308(Re_x)^{4/5}\left(Pr^{1/3}\right) \tag{13.17}$$

Figure 13.3 shows the plots of local heat transfer coefficient h versus position x on the plate from the leading edge of the plate. The plots are generated from Equations (13.13) and (13.14) under the conditions: $V_\infty = 0.1\, m/s\, , Pr = 8, k = 0.59\, W/m.K\, , \rho = 1000\, kg/m^3\, , \eta = 0.001\, Pa.s$. The heat transfer coefficient decreases with the increase in distance from the leading edge of the plate. However, the decrease is steeper when the flow is laminar in the boundary layer. Upon transition from laminar to turbulent flow in the boundary layer, a sharp jump occurs in the heat transfer coefficient.

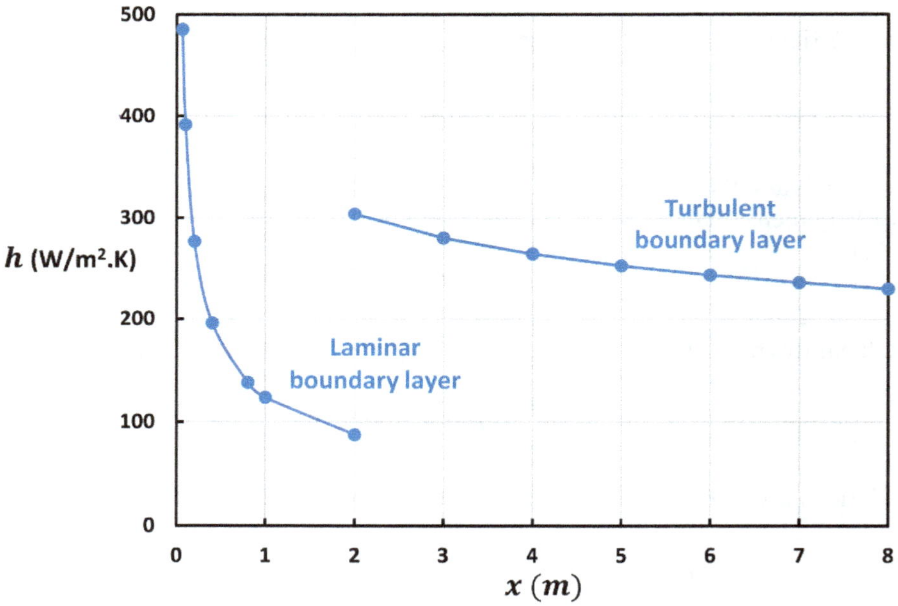

Figure 13.3 Variation of local heat transfer coefficient h with position x on the plate from the leading edge of the plate.

Average Heat Transfer Coefficient:

The average heat transfer coefficient for convective heat transfer from a flat plate is defined as:

$$\bar{h} = \frac{1}{L}\int_0^L h dx \tag{13.18}$$

.

For laminar flow over a flat plate, it can be readily shown that:

$$\bar{h} = 2h|_{x=L} \tag{13.19}$$

$$N\bar{u} = \frac{\bar{h}L}{k} = 2Nu|_{x=L} \tag{13.20}$$

Thus, the average heat transfer coefficient and the average Nusselt number are twice the local values at the end of the plate. This result holds for both isothermal-surface and constant-flux conditions over the full range of Pr. Note that the local Nusselt number dependence on the local Reynolds number is the same in these situations, that is,

$$Nu_x \; \alpha \; Re_x^{1/2} \tag{13.21}$$

The total heat transfer rate q from a given surface can be obtained as follows:

$$q = \bar{q}_w'' A = \bar{h}A(T_w - T_\infty) \tag{13.22}$$

where $\bar{q}_w''$ is the average heat flux at the surface and A is the heat transfer area. Equation (13.22) is appropriate only for isothermal-surface condition.

For constant-flux condition, the total heat transfer rate is simply the product of the uniform heat flux q_o'' and the heat transfer area: $q = q_o'' A$. The surface temperature for constant wall-flux condition is not constant. It increases with distance along the plate as:

$$T_w(x) = T_\infty + \frac{q_o''}{h} \tag{13.23}$$

The average temperature difference between the surface and fluid stream can be determined as follows:

$$(T_w - T_\infty)_{avg} = \frac{1}{L}\int_0^L (T_w - T_\infty)\, dx = \frac{q_o''}{kL}\int_0^L \frac{x}{Nu_x} dx \tag{13.24}$$

In practical problems involving constant-flux condition, prediction of the surface temperature is important to address concerns such as overheating and meltdown or burnout.

Flow is laminar in the boundary layer only in the initial portion (leading edge) of the surface (plate). At some point downstream of the leading edge, the boundary layer becomes turbulent. If the transition from laminar to turbulent boundary layer occurs sharply at a single critical point (x_c), one can calculate the average heat transfer coefficient as follows:

$$\bar{h} = \frac{1}{L}\int_0^L h dx = \frac{1}{L}\left[\int_0^{x_c} h_{laminar} dx + \int_{x_c}^L h_{turbulent} dx\right] \tag{13.25}$$

As an example, consider mixed boundary layer (initial portion laminar and rear portion turbulent) over an isothermal flat plate. Using Equation (13.13) for laminar region and Equation (13.14) for turbulent region, the average heat transfer coefficient and hence average Nusselt number can be determined from Equation (13.25) as:

$$N\bar{u} = \frac{\bar{h}L}{k} = \left[0.037\, Re_L^{4/5} - \left(0.037\, Re_{x_c}^{4/5} - 0.664 Re_{x_c}^{1/2}\right)\right] Pr^{1/3} \tag{13.26}$$

where Re_L is the Reynolds number at the end of the plate and Re_{x_c} is the critical Reynolds number where the boundary layer undergoes a sharp change in flow regime form laminar to turbulent.

13.1.2 Flow Across a Cylinder

In the preceding section we covered convective heat transfer from a flat plate. Another commonly encountered geometry in convective heat transfer applications is a cylinder.

Consider a long cylindrical object oriented with its axis normal to a fluid stream with free stream velocity of V_∞ and free stream temperature of T_∞ (Figure 13.4). Let the temperature of the surface of the cylinder be uniform (=T_W). The local heat transfer coefficient now varies with the angle θ (measured from the forward stagnation point) at any given Reynolds number ($Re_D = \rho D V_\infty / \eta$). At low to moderate Reynolds numbers ($\leq 10^5$), the heat transfer coefficient (and hence Nusselt number) decreases initially with the increase in θ due to the growth of the laminar boundary layer. However, a minimum in heat transfer coefficient is reached at a θ value of about 80°. Above θ of 80°, separation of the boundary layer occurs and due to mixing action of vortex/eddy formation, h begins to increase. At high Reynolds numbers (>10^5), a more complicated behavior is observed. Now the variation of h with θ is characterized by two minima. The first minimum occurs at a θ value of about 80° and the second minimum occurs at θ value of about 140°. The initial decrease in h with θ is again due to the growth of the laminar boundary layer. A sharp increase in h is observed for θ between 80° and 100° due to flow transition from laminar to turbulent regime in the boundary layer. With further increase in angle θ, the heat transfer coefficient declines due to the growth of the turbulent boundary layer. At a θ value of about 140°, separation of the boundary layer occurs and due to mixing action caused by boundary layer separation in the wake region, the heat transfer coefficient begins to increase.

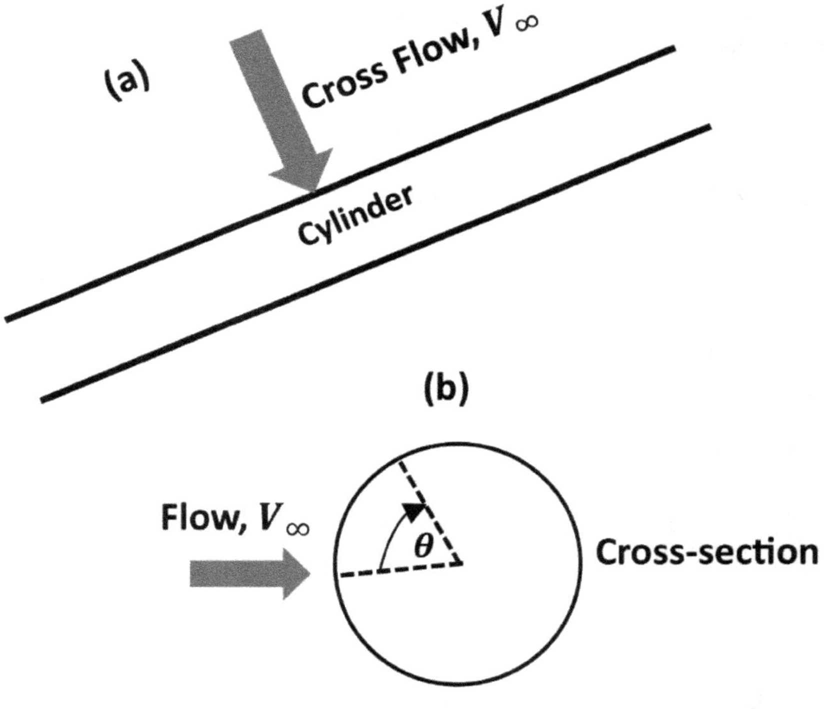

Figure 13.4 Flow across a cylinder.

From a practical point of view, it is the average heat transfer coefficient that is important. The average heat transfer coefficient for a single cylinder in crossflow is defined as:

$$\bar{h} = \frac{1}{2\pi}\int_0^{2\pi} h d\theta \qquad (13.27)$$

There are several empirical equations available in the literature to estimate the average Nusselt number. For example, Churchill and Bernstein proposed the following correlation:

$$N\bar{u}_D = \frac{\bar{h}D}{k} = 0.3 + \frac{0.62\,Re_D^{1/2}\,Pr^{1/3}}{\left[1+(0.4/Pr)^{2/3}\right]^{1/4}}\left[1+\left(\frac{Re_D}{282000}\right)^{5/8}\right]^{4/5} \qquad (13.28)$$

This equation is valid for $Re_D Pr \geq 0.2$.

13.1.3 Flow Across a Bundle of Tubes

We will now consider flow across a bundle of tubes (cylinders). Heat transfer to or from a bank of tubes in cross flow (fluid flowing at right angles or normal to the tubes) is important in the design and analysis of heat exchangers.

From a practical point of view, we need to know the average heat transfer coefficient for the entire tube bank. Equations available for the calculation of average heat transfer coefficients in flow over tube banks are based entirely on experimental data because the flow pattern is too complex to be treated analytically. The average heat transfer coefficient depends on the tube layout, in addition to flow properties and fluid motion. Two types of tube layouts are in common use: in-line (aligned) arrangement and staggered arrangement. In aligned arrangement, the next row of tubes is placed exactly behind the previous row of tubes and the centers of the cross sections of the tubes form rectangles (see Figure 13.5). In staggered arrangement, the centers of the cross sections of the tubes form isosceles triangles; any tube in the next row of tubes falls in between the two neighboring tubes of the previous row of tubes (see Figure 13.6).

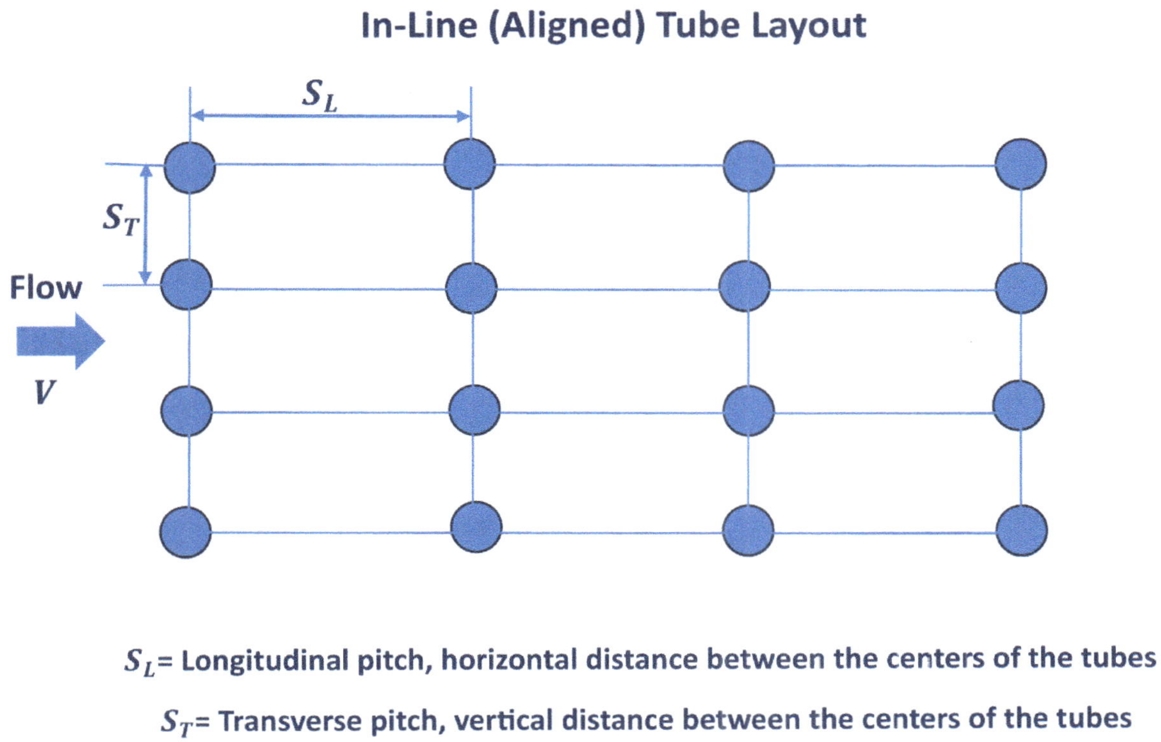

Figure 13.5 Flow across a bundle of tubes with aligned tube layout.

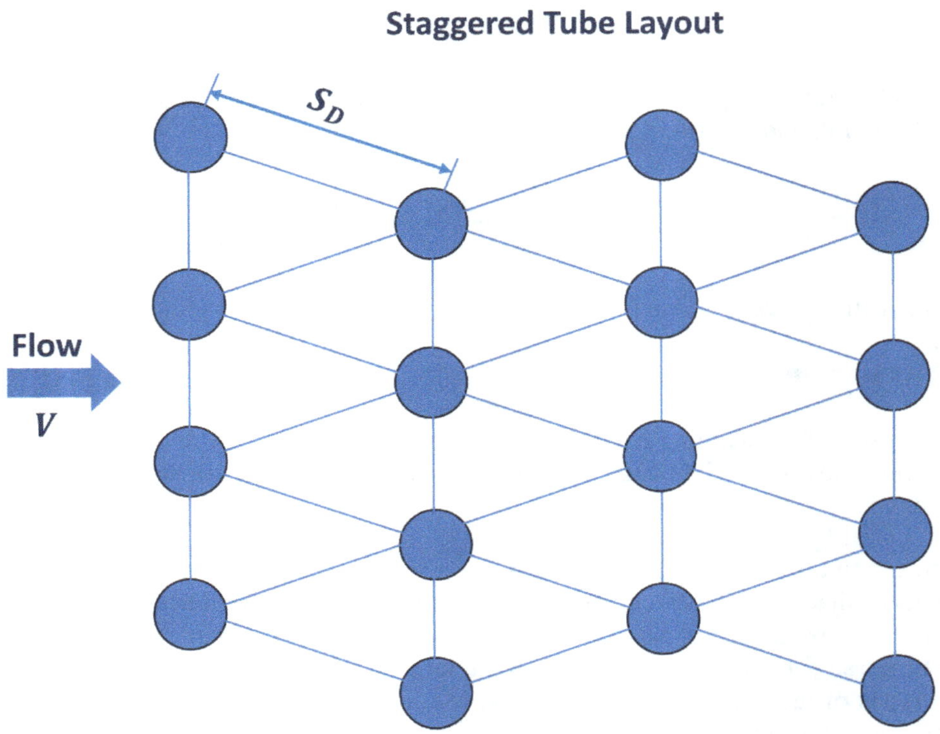

$$S_D = \sqrt{S_L^2 + \left(\frac{S_T}{2}\right)^2} \quad \textbf{(diagonal pitch)}$$

Figure 13.6 Flow across a bundle of tubes with staggered tube layout.

Zukauskas proposed the following correlation:

$$N\bar{u}_D = \frac{\bar{h}D}{k} = C_1 Re_{D,max}^m Pr^{0.36} \left(\frac{Pr}{Pr_w}\right)^{1/4} \tag{13.29}$$

The constants C_1 and m are different for different tube layouts. They also vary from one $Re_{D,max}$ range to another.

The values of the constants C_1 and m for heat transfer in tube banks of 20 or more rows are listed in Table 13.1 where S_T is transverse pitch and S_L is longitudinal pitch of the tube layout.

Table 13.1 The values of C_1 and m for tube banks of 20 or more rows

Configuration	$Re_{D,max}$ range	C_1	m	Comments
In-Line	10-100	0.8	0.4	Laminar regime
In-Line	100-10^3	treat as a single isolated tube (cylinder)		Laminar regime
In-Line	10^3-$2\text{x}10^5$	0.27	0.63	Transition regime
In-Line	> $2\text{x}10^5$	0.021	0.84	Turbulent regime
Staggered	10-100	0.9	0.4	Laminar regime
Staggered	100-10^3	treat as a single isolated tube (cylinder)		Laminar regime
Staggered	10^3-$2\text{x}10^5$	$0.35(S_T/S_L)^{1/5}$ (when $S_T/S_L < 2$)	0.6	Transition regime
Staggered	10^3-$2\text{x}10^5$	0.4 (when $S_T/S_L > 2$)	0.6	Transition regime
Staggered	> $2\text{x}10^5$	0.022	0.84	Turbulent regime

The Reynolds number $Re_{D,max}$ is based on the maximum fluid velocity occurring within the tube bank (that is, velocity through the minimum flow area) and is defined as:

$$Re_{D,max} = \frac{\rho V_{max} D}{\eta} \tag{13.30}$$

where D is the tube outside diameter.

For flow across aligned tube configuration, the maximum flow velocity will occur through the minimum frontal area and is given as:

$$V_{max} = VS_T/(S_T - D) \tag{13.31}$$

where S_T , the transverse pitch, is the vertical distance between tube centers, and V is fluid velocity at the entrance of the tube bank.

For flow across staggered tube configuration, the maximum flow velocity is given by the above equation provided that:

$$2 \times \left[\sqrt{(S_T/2)^2 + S_L^2} - D\right] \geq S_T - D \tag{13.32}$$

where S_L is longitudinal pitch (horizontal distance between tube centers). When the following condition is met,

$$2 \times \left[\sqrt{(S_T/2)^2 + S_L^2} - D\right] < S_T - D \tag{13.33}$$

the maximum velocity is given by:

$$V_{max} = \frac{(S_T/2)V}{S_D - D} \tag{13.34}$$

where S_D is diagonal pitch defined as:

$$S_D = \sqrt{(S_T/2)^2 + S_L^2} \tag{13.35}$$

13.1.4 Natural Convection Over a Vertical Plate

In the preceding sections we have considered only forced convection where the fluid motion was induced by a fan or some mechanical means. Now we will consider free or natural convection where fluid motion is caused by buoyancy forces acting on the fluid in which density gradients are present.

Consider a vertical hot plate immersed in a quiescent fluid. The fluid adjacent to the plate gets heated up due to conduction. The hot fluid in the vicinity of the plate, being less dense than the fluid further away from the plate, rises vertically due to buoyancy force. Buoyancy forces arise when there exist density gradients in gravitational field. As the hot fluid rises, it also entrains fluid from the quiescent region. At steady state, a boundary layer is established on the surface of the plate as shown in Figure 13.7. Over the initial portion of the solid surface, the boundary layer flow is laminar. However, after some distance from the leading edge of the surface, the flow may become turbulent.

The velocity distribution in the boundary layer of free convection is quite different from forced convection. The velocity of the fluid is not only zero at the solid surface (no-slip boundary condition), but also zero outside the boundary layer, that is, $V_\infty = 0$. Because $V_\infty = 0$ in free convection, the Reynolds number $\rho V_\infty x/\eta$ is also zero and is no longer suitable to describe the flow. Instead, a new dimensionless group called Grashof number governs the flow. The local Grashof number (Gr_x) is defined as:

$$Gr_x = \frac{g\beta(T_w - T_\infty)x^3}{\nu^2} \tag{13.36}$$

where β is the volumetric coefficient of thermal expansion defined as:

$$\beta = \frac{1}{V}\left(\frac{\partial V}{\partial T}\right)_P = -\frac{1}{\rho}\left(\frac{\partial \rho}{\partial T}\right)_P \tag{13.37}$$

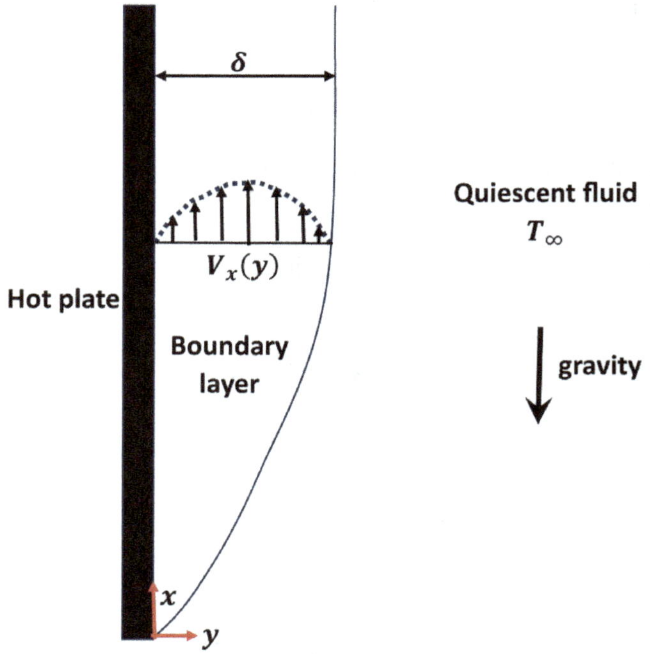

Figure 13.7 Development of natural convection boundary layer on a vertical plate.

The Grashof number can be interpreted as the ratio of buoyancy force to viscous force. It plays the same role in free convection as the Reynolds number plays in forced convection. For example, the transition from laminar to turbulent flow in free convection boundary layer is governed by Grashof number.

The Nusselt number in free convection is a function of Grashof number and Prandtl number. The relationship is of the form:

$$Nu_x = f(Gr_x, Pr) = C Gr_x^m Pr^n \tag{13.38}$$

where C, m, and n are constants. Such correlations could also be expressed in the form:

$$Nu_x = C(Gr_x\, Pr)^a\, Pr^b = C Ra_x^a Pr^b \tag{13.39}$$

where the product of Grashof number and Prandtl number is called Rayleigh number (Ra_x), that is,

$$Ra_x = GrPr \tag{13.40}$$

For laminar free convection on an isothermal vertical plate, Nu_x is given as:

$$Nu_x = g(Pr)\left[\frac{Gr_x}{4}\right]^{1/4} \tag{13.41}$$

where

$$g(Pr) = \frac{0.75 Pr^{1/2}}{\left(0.609 + 1.221\, Pr^{1/2} + 1.238 Pr\right)^{1/4}} \tag{13.42}$$

Equation (13.41) is valid over the full range of Pr. In the limits of $Pr \gg 1$ and $Pr \ll 1$, this equation gives:

$$Nu_x = 0.503[Ra_x]^{1/4} \qquad Pr \gg 1 \tag{13.43}$$

$$Nu_x = 0.60[Ra_x Pr]^{1/4} \qquad Pr \ll 1 \tag{13.44}$$

For laminar free convection from vertical surface under constant-flux condition, the numerical solution of the boundary layer equations leads to the following expression:

$$Nu_x = 0.616\left[\frac{Pr}{Pr + 0.8}\right][Ra_x^*]^{1/5} \tag{13.45}$$

where Ra_x^* is the modified Rayleigh number, defined as:

$$Ra_x^* = \frac{g\beta q_w'' x^4}{\alpha \nu k} \tag{13.46}$$

The flow in the boundary layer of a vertical isothermal plate is laminar when Rayleigh number (Ra_x) $\leq 10^9$. At higher Rayleigh numbers ($> 10^9$), the boundary layer becomes turbulent. Turbulent flows are difficult to handle analytically. Making use of the Chilton-Colburn analogy, it can be shown that for turbulent free convection from vertical isothermal plate, the local Nusselt number is given as:

$$Nu_x = \frac{hx}{k} = 0.039(Ra_x)^{2/5}(Pr)^{-1/5} \tag{13.47}$$

This equation is valid for $Pr \gg 1$. For turbulent flow under constant flux condition ($Ra_x^* \geq 10^{13}$) the following equation is applicable:

$$Nu_x = \frac{hx}{k} = 0.568(Ra_x^*)^{0.22} \tag{13.48}$$

From a practical point of view, it is the average heat transfer coefficient or average Nusselt number that is important. In the special case of laminar free convection from a vertical isothermal plate, the average heat transfer coefficient and average Nusselt number are given as:

$$\bar{h} = \frac{4}{3} h\Big|_{x=L} \tag{13.49}$$

$$N\bar{u} = \frac{\bar{h}L}{k} = \frac{4}{3} Nu_x\Big|_{x=L} \tag{13.50}$$

Empirical correlations are available in the literature for the average Nusselt number covering the entire range of Rayleigh numbers. For example, Churchill and Chu proposed the following correlations:

$$N\bar{u} = \frac{\bar{h}L}{k} = \left\{0.825 + \frac{0.387 Ra_L^{1/6}}{\left[1+(0.492/Pr)^{9/16}\right]^{8/27}}\right\}^2 \quad \text{(isothermal plate)} \tag{13.51}$$

$$N\bar{u} = \frac{\bar{h}L}{k} = \left\{0.825 + \frac{0.387 Ra_L^{1/6}}{\left[1+(0.437/Pr)^{9/16}\right]^{8/27}}\right\}^2 \quad \text{(constant heat flux)} \tag{13.52}$$

These equations are valid regardless of the flow regime. All properties are to be evaluated at the film temperature.

13.2 HEAT TRANFER COEFFICIENT IN INTERNAL FLOWS

13.2.1 Laminar Flow

For hydrodynamically and thermally fully developed laminar flow in a pipe subjected to constant wall temperature:

$$Nu = \frac{hD}{k} = 3.66 \tag{13.53}$$

For hydrodynamically and thermally fully developed laminar flow in a cylindrical pipe subjected to constant wall heat flux:

$$Nu = \frac{hD}{k} = \frac{48}{11} = 4.36 \tag{13.54}$$

Equations (13.55) and (13.56) are the exact theoretical results. They are derived in Appendix C. When the flow is hydrodynamically fully developed laminar but thermally developing:

$$Nu_z = \frac{hD}{k} = f(Gz_z) \tag{13.55}$$

where h is the local heat transfer coefficient, z is the direction of flow in the pipe, and Gz_z is the local Graetz number, defined as:

$$Gz_z = \frac{\pi}{4}\left(\frac{D}{z}\right) RePr \tag{13.56}$$

Note that the product $RePr$ is also referred to as Peclet number.

The analytical solutions of Nu_z for thermally developing but hydrodynamically fully developed flows are available in the literature in the form of Equation (13.55). The local Nusselt number Nu_z decreases with the increase in Gz_z^{-1} (that is, increase in distance z from entrance) and reaches a constant value of either 3.66 (constant wall temperature case) or 4.36 (constant wall flux case) as predicted by Equations (13.53) and (13.54) for fully developed flows.

To calculate the total heat transfer rate from the pipe wall to the fluid, the value of the average heat transfer coefficient or average Nusselt number is required, especially for the constant wall temperature case (in the constant wall flux case, $q = \pi D L q''_w$ where q''_w is the imposed constant flux). The following Hausen correlation is applicable for the average Nusselt number in laminar flow through pipes subjected to uniform surface temperature:

$$N\bar{u} = \frac{\bar{h}D}{k} = 3.66 + \frac{0.0668(D/L)(RePr)}{1+0.04[(D/L)(RePr)]^{2/3}} \tag{13.57}$$

Equation (13.57) is valid for thermal entry length problem (hydrodynamically developed but thermally developing laminar flow). It gives the correct behavior in the limit of fully developed flow (thermally and hydrodynamically), that is, $Nu = 3.66$ when $(L/D)/(RePr)$ is large.

The combined entry length problem where the flow is hydrodynamically and thermally developing is much more difficult to solve as velocity and temperature both depend on radial as well as axial position. For the calculation of the average Nusselt number in the case of combined entry length problem dealing with laminar flow through a pipe subjected to constant wall temperature, a simple correlation proposed by Sieder and Tate is recommended:

$$N\bar{u} = 1.86(RePr)^{1/3}\left(\frac{D}{L}\right)^{1/3}\left(\frac{\eta}{\eta_w}\right)^{0.14} \tag{13.58}$$

where all properties (except η_w) are to be evaluated at the arithmetic average of the inlet and outlet temperatures and η_w is the viscosity at the wall temperature.

13.2.2 Turbulent Flow

The critical Reynolds number for pipe flow is 2100. At higher values of Re, the flow becomes turbulent. The entry length (thermal or hydrodynamic) for turbulent flow in pipes is usually short and therefore, it is reasonable to assume that the flow is fully developed. Due to the complex nature of turbulent flows, they are often handled in an empirical or semi-empirical manner. For example, the heat transfer coefficient in turbulent flow can be determined from Chilton-Colburn analogy, Equation (13.15), provided that the friction factor relationship is known. For smooth pipes, the following empirical equation is often used to describe friction factor in turbulent flow over the Re range of $2\text{x}10^4 < Re < 10^6$:

$$f = \frac{0.046}{Re^{1/5}} \tag{13.59}$$

Substituting Equation (13.59) in Chilton-Colburn analogy, Equation (13.15), one obtains:

$$Nu = \frac{hD}{k} = 0.023\,Re^{4/5}\,Pr^{1/3} \tag{13.60}$$

This equation, usually referred to as Colburn equation, is valid for $Pr > 0.5$ and $2\text{x}10^4 < Re < 10^6$. A slightly modified version of Colburn equation is the Dittus-Boelter equation given as:

$$Nu = \frac{hD}{k} = 0.023\,Re^{4/5}\,Pr^{n} \tag{13.61}$$

where $n = 0.3$ for cooling ($T_w < T_m$) and 0.4 for heating ($T_w > T_m$). This Dittus-Boelter equation is quite popular.

For fluids whose viscosity is quite sensitive to temperature, the equation proposed by Sieder and Tate is recommended:

$$Nu = \frac{hD}{k} = 0.027\, Re^{4/5}\, Pr^{1/3} \left(\frac{\eta}{\eta_w}\right)^{0.14} \tag{13.62}$$

All properties (except η_w) are evaluated at the arithmetic average of the inlet and outlet temperatures. The fluid viscosity η_w is evaluated at the wall temperature.

The foregoing correlations (Equations (13.60) - (13.62)) can be applied for both constant wall flux and constant wall temperature conditions when $Pr > 0.5$ and $Re > 10^4$. In the case of liquid metals where $Pr << 1.0$, the following correlations are recommended for fully developed turbulent flow in smooth tubes:

$$Nu = \frac{hD}{k} = 4.80 + 0.0156\, Re^{0.85}\, Pr^{0.93} \qquad (T_w \text{ constant}) \tag{13.63}$$

$$Nu = \frac{hD}{k} = 6.30 + 0.0167\, Re^{0.85}\, Pr^{0.93} \qquad (q''_w \text{ constant}) \tag{13.64}$$

13.3 HEAT EXCHANGERS

Heat exchangers are devices used to transfer heat from one fluid stream to another. There are a variety of heat exchangers available but the most used heat exchangers in chemical industry are: double-pipe heat exchangers and shell-and-tube heat exchangers.

13.3.1 Double Pipe Heat Exchangers

Double pipe heat exchangers consist of two concentric pipes. One stream flows in the inner tube and the other stream flows in the annular space, as shown in Figures 13.8 and 13.9. Figure 13.8 shows the counter-current flow arrangement where the two fluid streams flow in opposite directions. Figure 13.9 shows the co-current or parallel flow arrangement where the two fluid streams flow in the same directions. The temperature distributions for counter-current and co-current flow arrangements are shown schematically in Figures 13.10 and 13.11, respectively. Note that in co-current flow, the hot fluid outlet temperature $T_{h,o}$ is always higher than the cold fluid outlet temperature $T_{c,o}$, that is, we cannot cool the hot fluid to a temperature below the cold fluid outlet temperature. However, in counter-current flow, $T_{h,o}$ can be lower than $T_{c,o}$, that is, the hot fluid can be cooled to a temperature lower than the cold fluid outlet temperature. Furthermore, counter-current heat exchanger is a more effective heat exchanger as compared with the co-current heat exchanger. For the same inlet and outlet temperatures, the average driving force for heat transfer, that is, the log-mean temperature difference (LMTD) between the two fluid streams for counter-current flow exceeds that for co-current flow. Hence the heat transfer area required to affect a prescribed heat transfer rate is smaller for the counter-current arrangement. Therefore, in most applications, counter-current flow arrangement is used. However, co-current arrangement is also desired in some applications. When a high initial rate of heating is desired, it is preferred to use co-current flow. The temperature difference between the two fluid streams, that is, the driving force for heat transfer, is large initially in co-current flow (see Figure 13.11). When one of the fluids involved is very viscous, co-current flow provides rapid initial heating of the fluid resulting in a large decrease in viscosity and hence a significant reduction in pumping requirements.

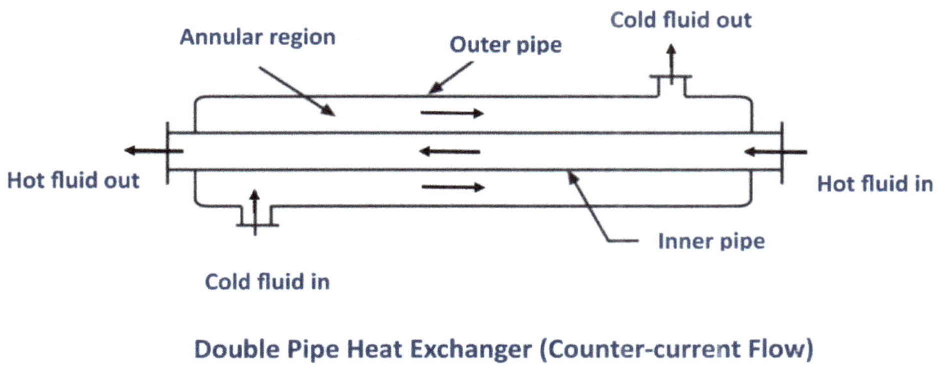

Figure 13.8 Double pipe heat exchanger with counter-current flow.

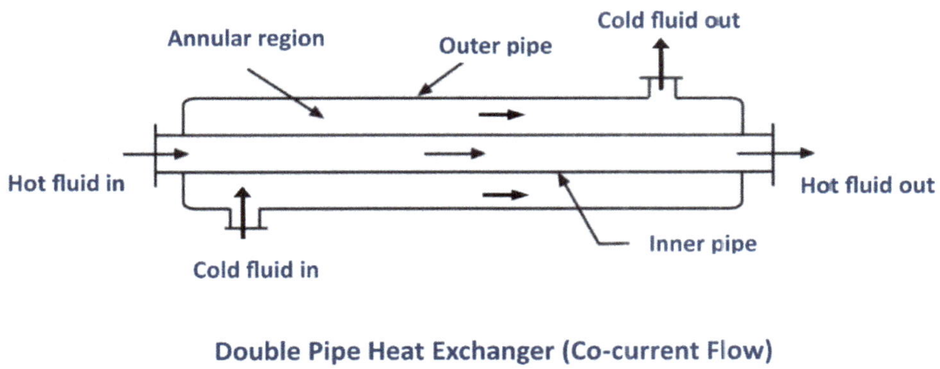

Figure 13.9 Double pipe heat exchanger with co-current flow.

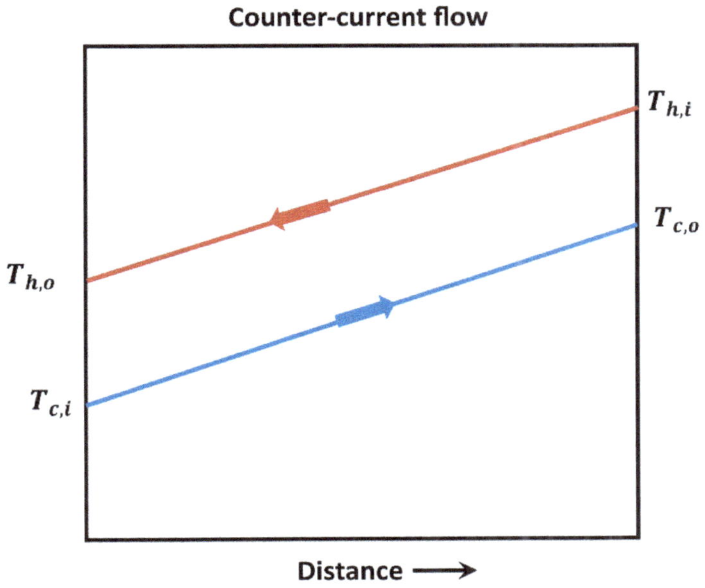

Figure 13.10 Temperature distribution in a double pipe heat exchanger with counter-current flow.

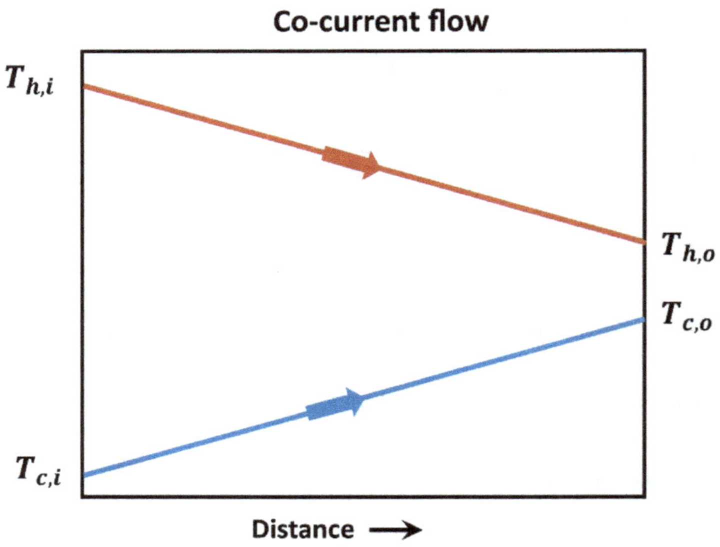

Figure 13.11 Temperature distribution in a double pipe heat exchanger with co-current flow.

Rate of Heat Transfer in Double Pipe Heat Exchanger

Energy balances over hot and cold fluid streams can be written as:

$$q = \dot{m}_h c_{p,h}(T_{h,i} - T_{h,o}) \tag{13.65}$$

$$q = \dot{m}_c c_{p,c}(T_{c,o} - T_{c,i}) \tag{13.66}$$

where q is the rate of heat transfer in the exchanger, $\dot{m}_h$ and $\dot{m}_c$ are mass flow rates of hot and cold streams, $T_{h,i}$ and $T_{h,o}$ are hot fluid inlet and outlet temperatures, $T_{c,i}$ and $T_{c,o}$ are cold fluid inlet and outlet temperatures.

The heat transfer rate in the heat exchanger is also given as:

$$q = \frac{\text{log-mean temperature difference}}{\text{total resistance}} = \frac{LMTD}{R_t} \tag{13.67}$$

There are three thermal resistances in series in the heat exchanger (see Figure 13.12): convective resistance on the inside of the smaller diameter pipe, conductive resistance of smaller diameter pipe wall, and convective resistance on the outside of the smaller diameter pipe. Thus,

$$R_t = \frac{1}{h_i A_i} + \frac{ln\left(\frac{r_o}{r_i}\right)}{2\pi k_w L} + \frac{1}{h_o A_o} \tag{13.68}$$

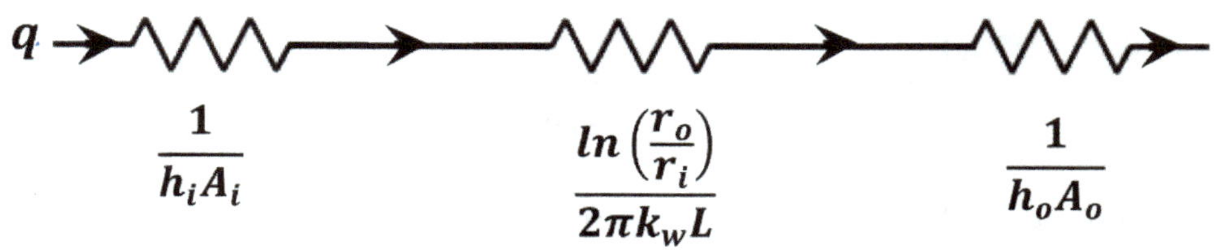

Figure 13.12 Electric analog of double pipe heat exchanger.

The total resistance R_t could also be expressed in terms of the overall heat transfer coefficient U_o based on the outside area of the smaller diameter pipe as:

$$\frac{1}{R_t} = U_o A_0 = \frac{1}{\frac{1}{h_i A_i} + \frac{ln\left(\frac{r_O}{r_i}\right)}{2\pi k_W L} + \frac{1}{h_O A_O}} \tag{13.69}$$

Consequently, Equation (13.67) can be re-written as:

$$q = U_o A_0 LMTD \tag{13.70}$$

where $A_0 = 2\pi r_o L$ and U_o is given as:

$$U_o = \frac{1}{\frac{A_0}{h_i A_i} + \frac{A_0 ln\left(\frac{r_O}{r_i}\right)}{2\pi k_W L} + \frac{1}{h_O}} = \frac{1}{\left(\frac{r_O}{r_i}\right)\left(\frac{1}{h_i}\right) + \left(\frac{r_O}{k_W}\right) ln\left(\frac{r_O}{r_i}\right) + \frac{1}{h_O}} \tag{13.71}$$

Note that r_i and r_o are inside and outside radii of the smaller diameter pipe, h_i and h_o are the inside and outside average convective heat transfer coefficients of the smaller diameter pipe, and k_w is the wall thermal conductivity of the smaller diameter pipe.

Calculation of LMTD for Double Pipe Heat Exchanger

For the calculation of LMTD for counter-current double pipe heat exchanger, consider Figure 13.13. The LMTD for this arrangement is given as:

$$LMTD = \frac{(T_{h,o} - T_{c,i}) - (T_{h,i} - T_{c,o})}{ln\left(\frac{T_{h,o} - T_{c,i}}{T_{h,i} - T_{c,o}}\right)} \tag{13.72}$$

Figure 13.13 Estimation of LMTD for counter-current flow.

For the calculation of LMTD for co-current double pipe heat exchanger, consider Figure 13.14. The LMTD for this arrangement is given as:

$$LMTD = \frac{(T_{h,o} - T_{c,o}) - (T_{h,i} - T_{c,i})}{ln\left(\frac{T_{h,o} - T_{c,o}}{T_{h,i} - T_{c,i}}\right)} \tag{13.73}$$

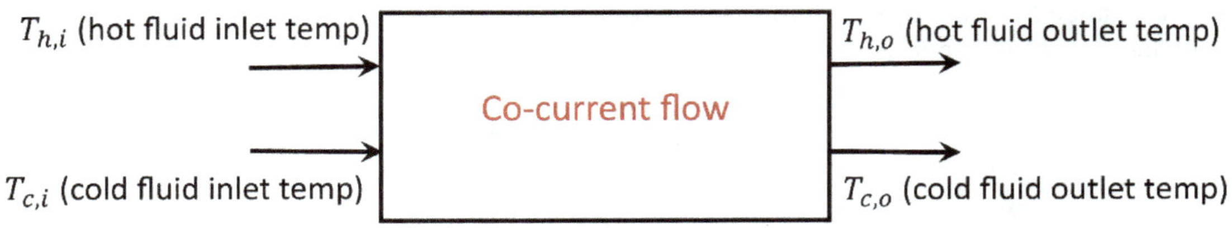

Figure 13.14 Estimation of LMTD for co-current flow.

13.3.2 Shell and Tube Heat Exchangers

Shell and tube heat exchangers are the most widely used heat exchangers in industry. They are used when large quantities of fluid must be heated or cooled which requires large heat transfer areas. Shell and tube heat exchangers are specified as $n - m$ heat exchangers where n is the number of shell passes and m is the number of tube passes. The shell passes indicate the number of times the shell side fluid passes through the heat exchanger whereas the tube passes indicate the number of times the tube side fluid passes through the heat exchanger. The simplest type of shell and tube heat exchanger is $1 - 1$, that is, one shell pass and one tube pass, as shown in Figure 13.15. Figure 13.16 shows a $1 - 2$ heat exchanger with one shell pass and two tube passes. Note that baffles are installed on the shell side to increase the heat transfer coefficient of the shell side fluid by inducing turbulence and cross flow.

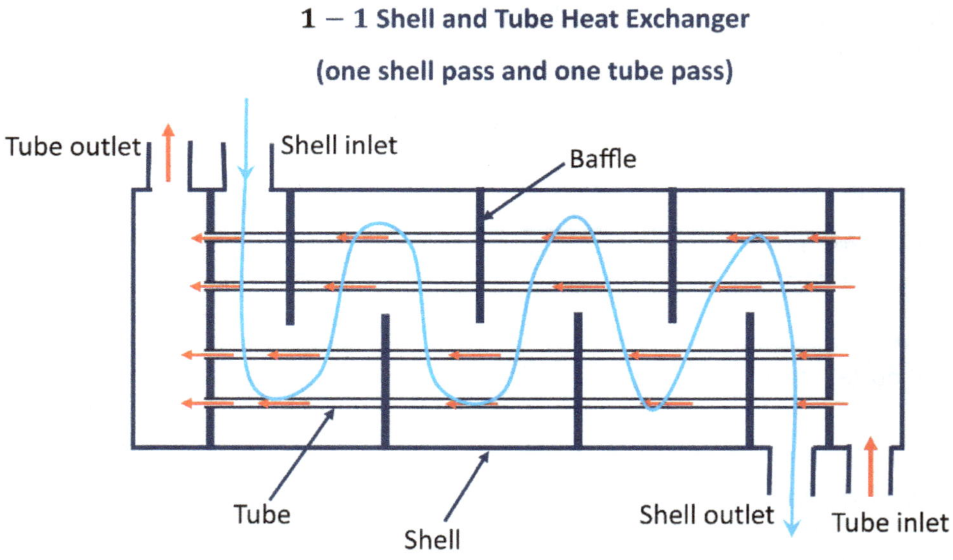

Figure 13.15 Shell and tube heat exchanger with one shell pass and one tube pass.

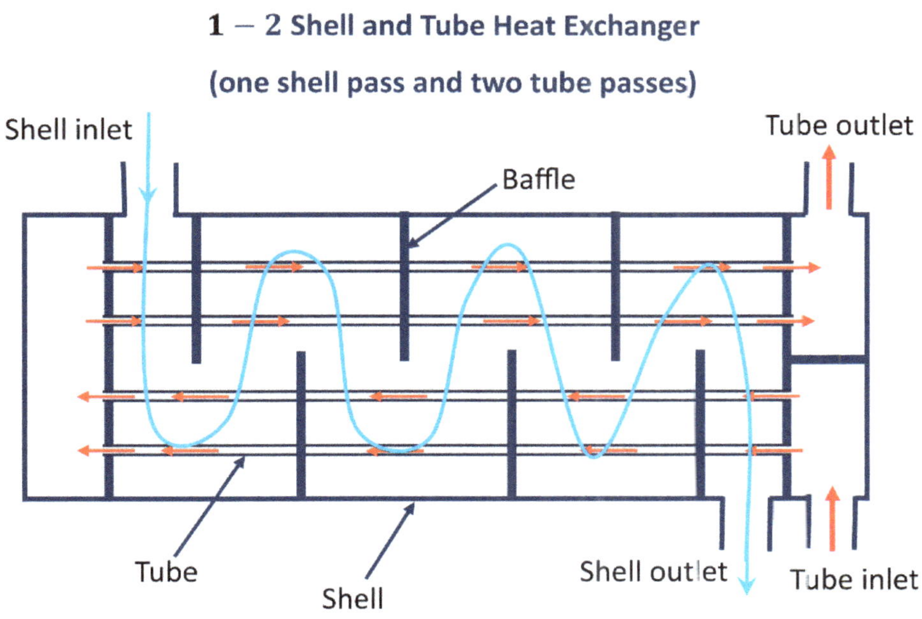

Figure 13.16 Shell and tube heat exchanger with one shell pass and two tube passes.

Rate of Heat Transfer in Shell and Tube Heat Exchanger

The energy balances equations of double pipe heat exchanger, that is, Equations (13.65) and (13.66) are still applicable. They are re-written as:

$$q = \dot{m}_s c_{p,s} |(T_i - T_o)| \tag{13.74}$$

$$q = \dot{m}_t c_{p,t} |(t_i - t_o)| \tag{13.75}$$

where q is the rate of heat transfer in the exchanger, $\dot{m}_s$ and $\dot{m}_t$ are mass flow rates of shell side and tube side flow streams, T_i and T_o are the inlet and outlet temperatures of shell side fluid, t_i and t_o are the inlet and outlet temperatures of tube side fluid. Note that $|(T_i - T_o)|$ and $|(t_i - t_o)|$ are absolute (positive) values of the temperature differences.

Equation (13.70) is also applicable. It is re-written as:

$$q = U_o A_0 (LMTD)_{ST} \tag{13.76}$$

where A_0 is the total outside area of all the tubes, $(LMTD)_{ST}$ is the log-mean temperature difference between the shell and tube side fluids of the shell and tube heat exchanger, and U_o is the overall heat transfer coefficient based on the outside area of the tubes.

The overall heat transfer coefficient U_o is given as:

$$U_o = \frac{1}{\left(\frac{r_o}{r_i}\right)\left(\frac{1}{h_i}\right) + \left(\frac{r_o}{k_w}\right) ln\left(\frac{r_o}{r_i}\right) + \frac{1}{h_o}} \tag{13.77}$$

For thin tubes, $r_o \cong r_i$. Thus,

$$U_o = \frac{h_i h_o}{h_i + h_o} \tag{13.78}$$

For the inside heat transfer coefficient h_i, equations discussed in section 13.2 can be used. For the outside heat transfer coefficient h_o, the following Kern equation can be used:

$$h_o D_H/k = 0.36\left(\frac{G_s D_H}{\eta}\right)^{0.55}\left(\frac{C_p \eta}{k}\right)^{1/3}\left(\frac{\eta}{\eta_w}\right)^{0.14} \tag{13.79}$$

where D_H is the hydraulic diameter and G_s is the shell-side mass velocity (kg/m^2.s).

The hydraulic diameter, defined below, depends on the tube layout on the shell-side:

$$D_H = 4\left(\frac{\text{flow−area}}{\text{wetted−perimeter}}\right) \tag{13.80}$$

For tube layout with square pitch (see Figure 13.17):

$$D_H = \frac{4(p^2 - \pi d_o^2/4)}{\pi d_o} \tag{13.81}$$

For tube layout with equilateral triangular pitch (see Figure 13.17):

$$D_H = \frac{4\{(p/2)(0.87p) - (\pi d_o^2/8)\}}{(\pi d_o/2)} \tag{13.82}$$

where p is the tube pitch defined as the distance between the centers of the neighboring tubes and d_o is the outside diameter of the tubes.

The mass velocity on the shell side (G_s) is given as the ratio of shell-side fluid mass flow rate ($\dot{m}_s$) to the crossflow area of shell (a_s). The crossflow area depends on baffle spacing (b), clearance between two adjacent tubes (c), shell diameter (D_s), and tube pitch (p). The mass velocity and crossflow area can be expressed as:

$$G_s = \frac{\dot{m}_s}{a_s} \tag{13.83}$$

$$a_s = \frac{c b D_s}{p} \tag{13.84}$$

Note that $c = p - d_o$.

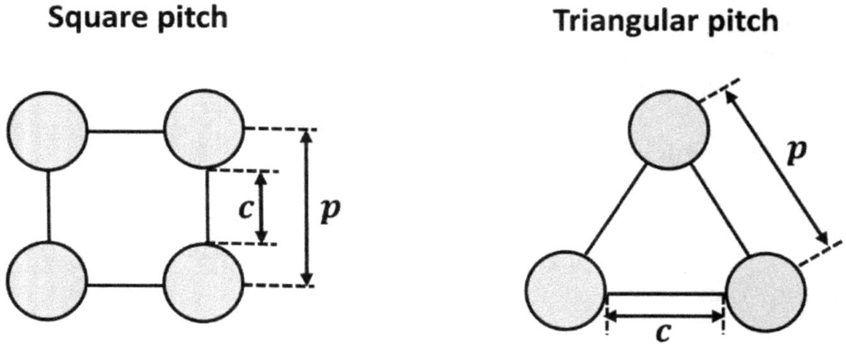

Figure 13.17 Tube layouts with square and triangular pitch.

Calculation of LMTD for Shell and Tube Heat Exchanger

In shell and tube heat exchangers, the flow pattern is neither pure co-current nor pure counter-current (see Figures 13.15 and 13.16). They have a mixed flow pattern, that is, some portions of the heat exchanger have co-current flow, some have counter-current flow, and some have crossflow. Therefore, the LMTD expressions for pure co-current or counter-current flows are not directly applicable.

In a shell and tube heat exchanger, the $(LMTD)_{ST}$ is calculated by using a correction factor (F) applied to the LMTD for pure counter-current flow. Thus:

$$(LMTD)_{ST} = F(LMTD)_{\text{counter-current flow}} \tag{13.85}$$

The LMTD correction factor F is a function of the shell side and tube side fluid temperatures and the number of tube and shell passes. It is expressed as a function of two dimensionless temperature ratios:

$$R = \frac{(T_i - T_o)}{(t_o - t_i)} \quad ; \quad P = \frac{(t_o - t_i)}{(T_i - t_i)} \tag{13.86}$$

where T is shell side fluid temperature and t is tube side fluid temperature.

For $1 - 2$ (or one shell pass and any multiple of 2 as tube passes) shell and tube heat exchanger, the correction factor is given by:

$$F = \left[\frac{\sqrt{R^2+1}}{R-1}\right]\left[\frac{ln\left(\frac{1-P}{1-RP}\right)}{ln\left(\frac{2-P\left(R+1-\sqrt{R^2+1}\right)}{2-P\left(R+1+\sqrt{R^2+1}\right)}\right)}\right] \tag{13.87}$$

The corrections factors are commonly expressed in the form of charts. Figure 13.18 shows the correction factor chart for a shell and tube heat exchanger with 1 shell pass and any multiple of 2 tube passes. Figure 13.19 shows the correction factor chart for a shell and tube heat exchanger with 2 shell passes and any multiple of 4 tube passes.

Note that F curves drop off rapidly below about 0.75. If the heat exchanger design indicates an F value of less than 0.75, the exchanger probably needs to be redesigned.

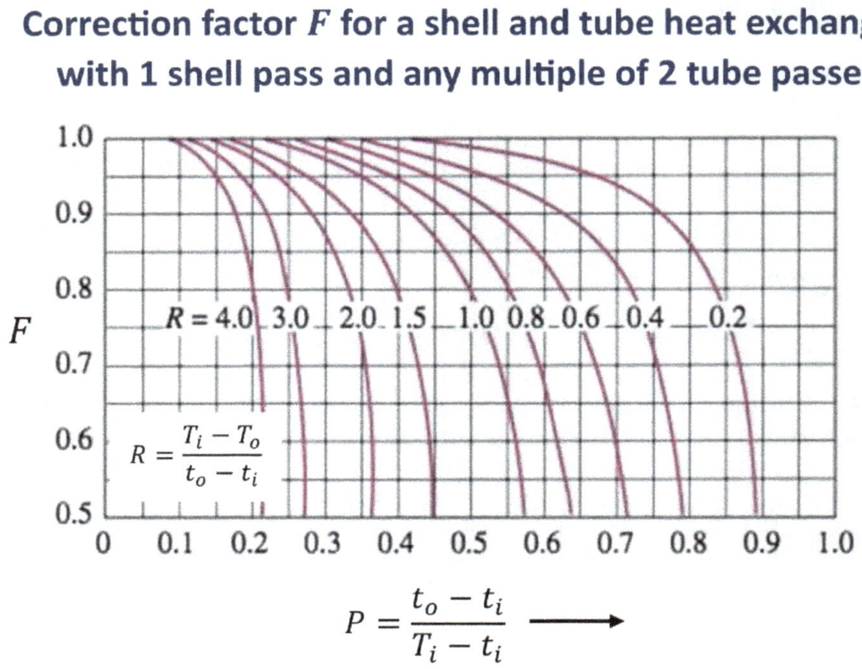

Figure 13.18 Correction factor for a heat exchanger with one shell pass and any multiple of two tube passes.

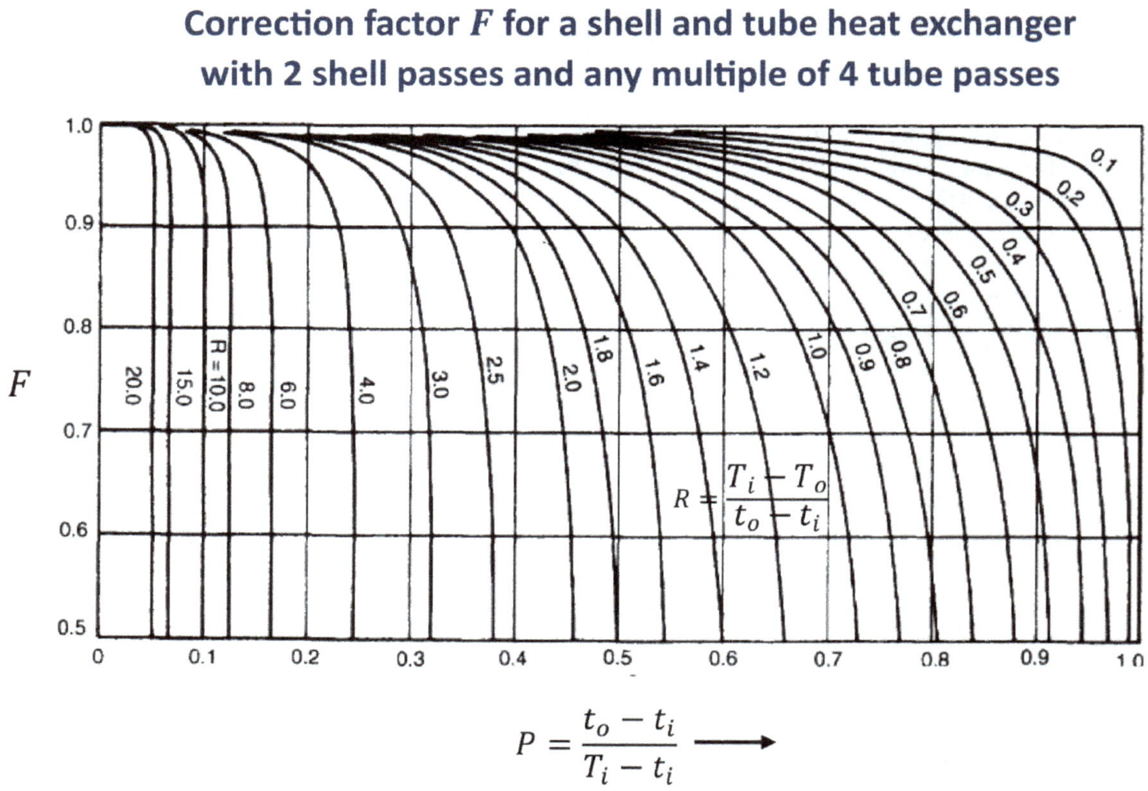

Figure 13.19 Correction factor for a heat exchanger with two shell passes and any multiple of four tube passes.

13.4 EXAMPLES

13.4.1 Example 1

Consider the flow of water at 4.45 °C over a heated metal plate maintained at 37.8 °C. The free stream velocity of water is 0.366 m/s. The length and width of the plate are 0.61 m (in the flow direction) and 0.46 m, respectively. Estimate the following:

a) The heat transfer rate from plate to water over the first-half (0.305 m) of the plate

b) The heat transfer rate from plate to water over the full length of the plate

You can use the following properties of water: $\nu = 1.006 \times 10^{-6}$ m²/s, $k = 0.597\ W/m.K$, and $Pr = 7.02$. You can assume $Re_{x_c} = 5 \times 10^5$.

Solution

Refer to the figure below.

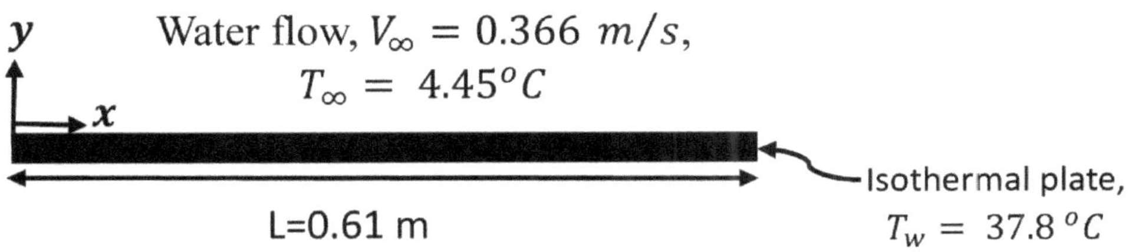

a) The heat transfer rate from plate to water over the first-half (0.305 m) of the plate

$$Re_x|_{x=0.305\ m} = \frac{\rho V_\infty x}{\eta} = \frac{V_\infty x}{\nu} = \frac{0.366 \times 0.305}{1.006 \times 10^{-6}} = 1.11 \times 10^5$$

Since $Re_x < 5 \times 10^5$, the flow is laminar over the first-half of the plate. Thus,

$$\bar{h} = 2h|_{x=0.305\ m}$$

$$Nu_x = \frac{hx}{k} = 0.332(Re_x)^{1/2}(Pr^{1/3}) = 0.332(1.11 \times 10^5)^{1/2}(7.02^{1/3}) = 211.793$$

$$h|_{x=0.305\ m} = \left(\frac{k}{x}\right) \times 211.793 = \left(\frac{0.597}{0.305}\right) \times 211.793 = 414.559\quad W/m^2.K$$

$$\bar{h} = 2h|_{x=0.305\ m} = 2 \times 414.559 = 829.118\ \ W/m^2.K$$

$$q = \bar{h}A(T_w - T_\infty) = 829.118 \times (0.305 \times 0.46)(37.8 - 4.45) = 3879.45\ W$$

b) The heat transfer rate from plate to water over the full length of the plate

$$Re_x|_{x=0.61\ m} = \frac{\rho V_\infty x}{\eta} = \frac{V_\infty x}{\nu} = \frac{0.366 \times 0.61}{1.006 \times 10^{-6}} = 2.22 \times 10^5$$

Since $Re_x < 5 \times 10^5$, the flow is laminar over the entire plate. Thus,

$$\bar{h} = 2h|_{x=0.61\ m}$$

$$Nu_x = \frac{hx}{k} = 0.332(Re_x)^{1/2}(Pr^{1/3}) = 0.332(2.22 \times 10^5)^{1/2}(7.02^{1/3}) = 299.52$$

$$h|_{x=0.61\ m} = \left(\frac{k}{x}\right) \times 299.52 = \left(\frac{0.597}{0.61}\right) \times 299.52 = 293.137 \quad W/m^2.K$$

$$\bar{h} = 2h|_{x=61\ m} = 2 \times 293.137 = 586.274 \quad W/m^2.K$$

$$q = \bar{h}A(T_w - T_\infty) = 586.274 \times (0.61 \times 0.46)(37.8 - 4.45) = 5486.32\ W$$

Thus,

$$\frac{q_{half-plate}}{q_{full-plate}} = \frac{3879.45}{5486.32} = 0.7071$$

Interestingly, the rate of heat transfer from the first one-half of the plate is 70.71% of the rate of heat transfer from the entire plate. The reason for this is the decrease in the local heat transfer coefficient with the increase in distance x from the leading edge of the plate (see laminar curve of Figure 13.3).

13.4.2 Example 2

Twenty electric strip heaters, each 50 mm wide and 1 m long, are mounted on a square plate $1m \times 1m$ in size, as shown in the figure below. Each strip heater is independently controlled. The temperature of the plate is maintained at 230 °C. Air flows over the plate with free-stream velocity and temperature of 60 m/s and 25 °C, respectively. Which strip heater requires the maximum electric power and what is the value of the power required? The properties of air are: $\nu = 2.641 \times 10^{-5}$ m²/s, $k = 0.0338\ W/m.K$, and $Pr = 0.69$. You can assume $Re_{x_c} = 5 \times 10^5$.

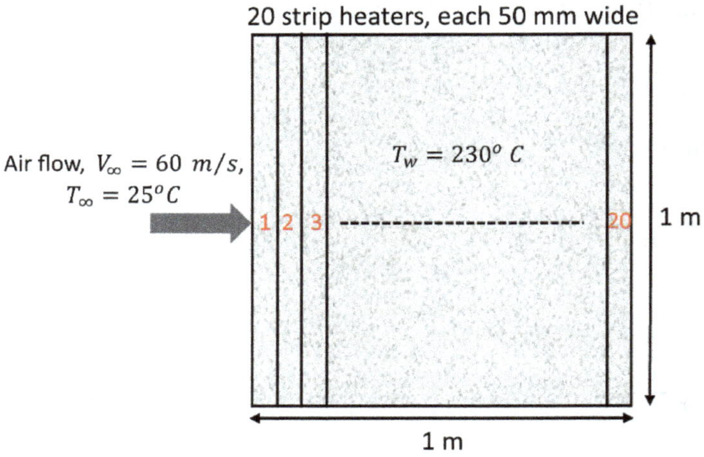

Solution

Point of transition from laminar to turbulent flow:

$$Re_{x_c} = \frac{V_\infty x_c}{\nu} = 5 \times 10^5$$

$$x_c = 5 \times 10^5 \left(\frac{\nu}{V_\infty}\right) = 5 \times 10^5 \left(\frac{2.641 \times 10^{-5}}{60}\right) = 0.22\, m$$

The transition to turbulence occurs on the 5th electrical strip heater. Referring to the heat transfer coefficient plot of Figure 13.3, the heat transfer coefficient in the laminar region is the highest at the beginning, that is, the first electrical strip heater. There also occurs a jump in the heat transfer coefficient on the 5th electrical strip heater. Depending on the point of transition, the average heat transfer coefficient on the 6th strip heater may be higher than that on the 5th heater, Thus, we need to compare three strip heaters: first, fifth, and sixth hater and see which one gives the highest rate of heat transfer.

Heat transfer rate from the first heater, q_1:

$$q_1 = \bar{h}_1 A_1 (T_w - T_\infty) = \bar{h}_1 (0.05 \times 1)(230 - 25)$$

As the flow is laminar over the first heater,

$$\bar{h}_1 = 2h|_{x=05\, m}$$

$$Re_x|_{x=0.05\, m} = \frac{V_\infty x}{\nu} = \frac{60 \times 0.05}{2.641 \times 10^{-5}} = 1.136 \times 10^5$$

$$Nu_x = \frac{hx}{k} = 0.332 (Re_x)^{1/2} (Pr^{1/3}) = 0.332 (1.136 \times 10^5)^{1/2} (0.69^{1/3}) = 98.88$$

$$\bar{h}_1 = 2h|_{x=05\, m} = 2\left(\frac{0.0338}{0.05}\right) \times 98.88 = 133.686 \;\; W/m^2 . K$$

Thus,

$$q_1 = \bar{h}_1 A_1 (T_w - T_\infty) = \bar{h}_1 (0.05 \times 1)(230 - 25) = 133.686\, (0.05 \times 1)(230 - 25) = 1370.28\, W$$

Heat transfer rate from the fifth heater, q_5:

$$q_5 = q_{1-5} - q_{1-4} = \bar{h}_{1-5} (0.25 \times 1)(230 - 25) - \bar{h}_{1-4} (0.20 \times 1)(230 - 25)$$
$$= \left(0.25 \bar{h}_{1-5} - 0.20 \bar{h}_{1-4}\right)(230 - 25)$$

Estimation of $\bar{h}_{1-5}$

As we have a mixed boundary layer so we use mixed boundary layer equation,

$$N\bar{u} = \frac{\bar{h}L}{k} = \left[0.037\, Re_L^{4/5} - \left(0.037\, Re_{x_c}^{4/5} - 0.664 Re_{x_c}^{1/2}\right)\right] Pr^{1/3}$$

Using $Re_{x_c} = 5 \times 10^5$, we get:

$$N\bar{u} = \frac{\bar{h}L}{k} = \left[0.037\, Re_L^{4/5} - 871.3\right] Pr^{1/3}$$

$L = 0.25\ m$ and $Re_L = \frac{V_\infty L}{\nu} = \frac{60 \times 0.25}{2.641 \times 10^{-5}} = 5.68 \times 10^5$

Thus,

$$\bar{h}_{1-5} = \left(\frac{0.0338}{0.25}\right)\left[0.037\,(5.68 \times 10^5)^{4/5} - 871.3\right](0.69)^{1/3} = 73.3\ W/m^2.K$$

Estimation of $\bar{h}_{1-4}$

We have only a laminar boundary layer over the first four heaters. Thus,

$$\bar{h}_{1-4} = 2h|_{x=20\ m}$$

$$Re_x|_{x=0.20\ m} = \frac{V_\infty x}{\nu} = \frac{60 \times 0.20}{2.641 \times 10^{-5}} = 4.544 \times 10^5$$

$$Nu_x = \frac{hx}{k} = 0.332(Re_x)^{1/2}(Pr^{1/3}) = 0.332(4.544 \times 10^5)^{1/2}(0.69^{1/3}) = 197.761$$

$$\bar{h}_{1-4} = 2h|_{x=20\ m} = 2\left(\frac{0.0338}{0.20}\right) \times 197.761 = 66.843\ \ W/m^2.K$$

Thus,

$$q_5 = (0.25\bar{h}_{1-5} - 0.20\bar{h}_{1-4})(230 - 25) = (0.25 \times 73.3 - 0.20 \times 66.843)(230 - 25) = 1016.06\ W$$

Note that $q_1 > q_5$. We still need to check strip heater number 6.

Heat transfer rate from the sixth heater, q_6*:*

$$q_6 = q_{1-6} - q_{1-5} = \bar{h}_{1-6}(0.30 \times 1)(230 - 25) - \bar{h}_{1-5}(0.25 \times 1)(230 - 25)$$
$$= (0.30\bar{h}_{1-6} - 0.25\bar{h}_{1-5})(230 - 25)$$

Estimation of $\bar{h}_{1-6}$

As we have a mixed boundary layer so we use mixed boundary layer equation,

$$N\bar{u} = \frac{\bar{h}L}{k} = \left[0.037\, Re_L^{4/5} - 871.3\right] Pr^{1/3}$$

$L = 0.30\ m$ and $Re_L = \frac{V_\infty L}{\nu} = \frac{60 \times 0.30}{2.641 \times 10^{-5}} = 6.816 \times 10^5$

Thus,

$$\bar{h}_{1-6} = \left(\frac{0.0338}{0.30}\right)\left[0.037\,(6.816 \times 10^5)^{4/5} - 871.3\right](0.69)^{1/3} = 84.29\ W/m^2.K$$

Thus,

$$q_6 = (0.30\bar{h}_{1-6} - 0.25\bar{h}_{1-5})(230 - 25) = (0.30 \times 84.29 - 0.25 \times 73.3)(230 - 25) = 1427.21\ W$$

Comparison of heat loss from different heaters:

Strip heater number	Rate of heat loss, q
first	$1370.28\ W$
fifth	$1016.06\ W$
sixth	$1427.21\ W$

Clearly, the rate of heat loss is maximum from the 6th strip heater. Thus, maximum electric power is needed for the 6th electrical strip heater to maintain constant temperature of the plate.

13.4.3 Example 3

Air at 25 °C cross flows over a pipe 25 mm in diameter at a free stream velocity of 15 m/s (see figure below). The surface temperature of the pipe is maintained constant at 100 °C. Determine the rate of heat loss from the pipe per unit length. The properties of air are: $\nu = 1.931 \times 10^{-5}$ m²/s, $k = 0.0288\ W/m.K$, and $Pr = 0.702$.

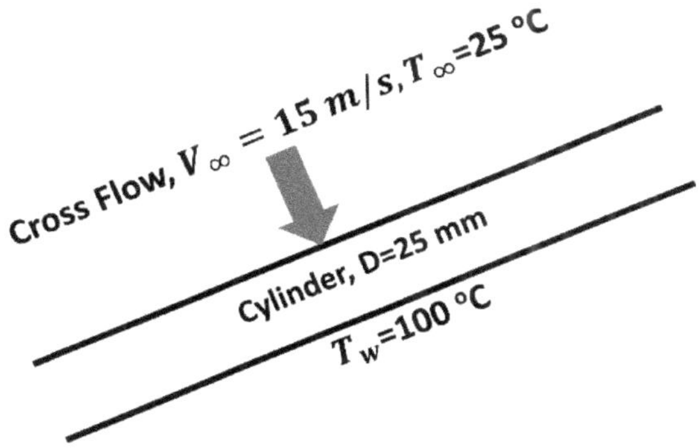

Solution

The Reynolds number Re_D is:

$$Re_D = \frac{DV_\infty}{\nu} = \frac{(25 \times 10^{-3}) \times 15}{1.931 \times 10^{-5}} = 1.942 \times 10^4$$

Use Churchill and Bernstein equation:

$$N\bar{u}_D = \frac{\bar{h}D}{k} = 0.3 + \frac{0.62\, Re_D^{\frac{1}{2}}\, Pr^{\frac{1}{3}}}{\left[1+(0.4/Pr)^{\frac{2}{3}}\right]^{1/4}}\left[1 + \left(\frac{Re_D}{282000}\right)^{\frac{5}{8}}\right]^{\frac{4}{5}}$$

$$= 0.30 + \frac{0.62\,(1.942\times10^4)^{1/2}(0.702)^{1/3}}{[1+(0.4/0.702)^{2/3}]^{1/4}}\left[1 + \left(\frac{1.942\times10^4}{282000}\right)^{5/8}\right]^{4/5} = 77.62$$

$$\bar{h} = N\bar{u}_D\left(\frac{k}{D}\right) = 77.62 \times \left(\frac{0.0288}{25 \times 10^{-3}}\right) = 89.42\ W/m^2.K$$

$$\frac{q}{L} = \bar{h}(\pi D)(T_w - T_\infty) = 89.42 \times \pi \times (25 \times 10^{-3})(100 - 25) = 526.73 \;\; W/m$$

13.4.4 Example 4

Consider a tube bank consisting of 20 rows of tubes with each row containing 50 tubes. The tube layout is in-line (aligned) type. The tube diameter is 10 mm and $S_T = S_L = 20$ mm. Cold water flows through the tubes and hot flue gas at 427 ^{0}C cross flows over the tube bank at a velocity of 5 m/s. The outer surface temperature of the tubes is maintained at 27 ^{0}C. Determine the total rate of heat transfer per unit length of the tubes in the bank.

Properties of the gas: $\nu = 68.1 \times 10^{-6}$ m^2/s, $k = 0.0524$ W/mK, $Pr = 0.695$, $c_p = 1075$ J/kgK, $\rho = 0.498$ kg/m^3 , $Pr_w = 0.707$ at 300 K

Solution

Refer to the figure given below.

There are 20 rows of tubes with 50 tubes in each row. Thus, $N = 20 \times 50 = 1000$ and $N_T = 50$. Use the following to predict the average heat transfer coefficient:

$$N\bar{u}_D = \frac{\bar{h}D}{k} = C_1 Re_{D,max}^m Pr^{0.36} \left(\frac{Pr}{Pr_w}\right)^{1/4}$$

$$V_{max} = VS_T/(S_T - D) = 5 \times 20/(20 - 10) = 10 \; m/s$$

$$Re_{D,max} = \frac{V_{max} D}{\nu} = \frac{10 \times 10 \times 10^{-3}}{68.1 \times 10^{-6}} = 1468.43$$

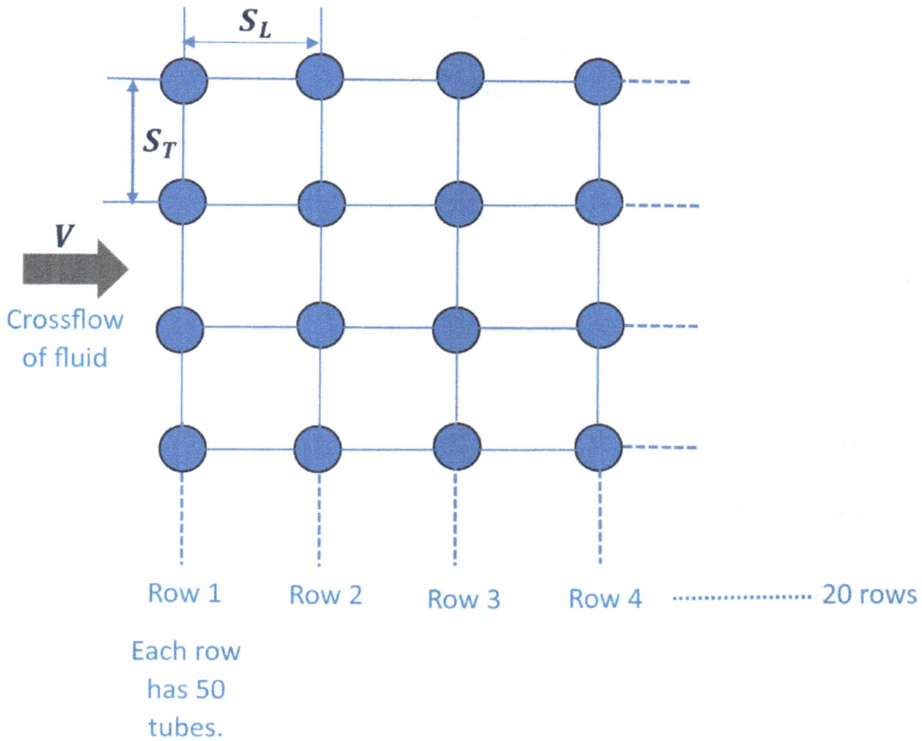

Zukauskas equation parameters from Table 13.1 are: $C_1 = 0.27$ and $m = 0.63$. Thus,

$$N\bar{u}_D = \frac{\bar{h}D}{k} = C_1 Re_{D,max}^m Pr^{0.36}\left(\frac{Pr}{Pr_w}\right)^{1/4} = 0.27 \times (1468.43)^{0.63} 0.695^{0.36}\left(\frac{0.595}{0.707}\right)^{0.25} = 23.316$$

$$\bar{h} = N\bar{u}_D\left(\frac{k}{D}\right) = 23.316\left(\frac{0.0524}{10\times10^{-3}}\right) = 122.177\ W/m^2.K$$

Estimation of outlet temperature of fluid (flue gas):

$$T_o = T_w - (T_w - T_i)exp\left[-\frac{\pi DN\bar{h}}{\rho VN_T S_T C_p}\right] = 27 - (27 - 427)exp\left[-\frac{\pi\times10\times10^{-3}\times1000\times122.177}{0.498\times5\times50\times20\times10^{-3}\times1075}\right] = 122.347^oC$$

Calculation of q

$$q = \bar{h}N(\pi DL)LMTD$$

$$LMTD = \frac{(T_o - T_w) - (T_i - T_w)}{ln\left(\frac{T_o - T_w}{T_i - T_w}\right)} = \frac{(122.347 - 27) - (427 - 27)}{ln\left(\frac{122.347 - 27}{427 - 27}\right)} = 218.956^oC$$

$$\frac{q}{L} = \bar{h}N(\pi D)LMTD = 122.177 \times 1000 \times (\pi \times 10 \times 10^{-3}) \times 218.956 = 8.404 \times 10^5\ W/m$$

13.4.5 Example 5

A fireplace installed in a room has a square glass fire screen, 1 m by 1 m (see figure below). When the fireplace is on, the temperature of the glass screen reaches 215 °C. The room temperature is 20 °C. Calculate the natural convection heat transfer rate from the fireplace to the room air. The properties of air are: $\nu = 26.4 \times 10^{-6}$ m²/s, $k = 33.8 \times 10^{-3}$ W/m.K, $\beta = 0.0025\ K^{-1}$, $Pr = 0.69$, $\alpha = 38.3 \times 10^{-6}\ m^2/s$.

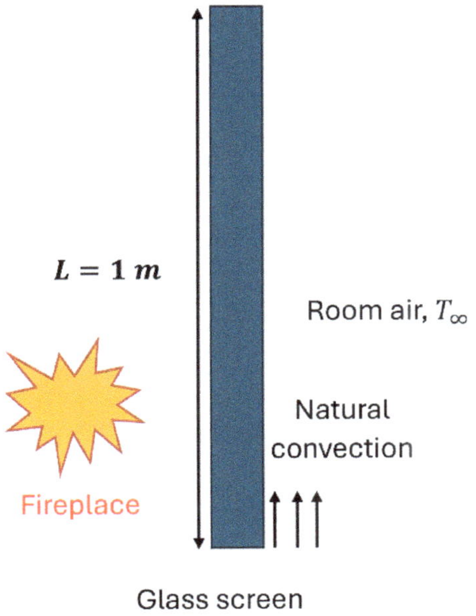

Solution

$$Ra_x = GrPr = \frac{g\beta(T_w - T_\infty)x^3}{\alpha\nu}$$

$$Ra_L = \frac{g\beta(T_w - T_\infty)L^3}{\alpha\nu} = \frac{9.81 \times 0.0025 \times (215 - 20) \times 1^3}{38.3 \times 10^{-6} \times 26.4 \times 10 - 6} = 4.73 \times 10^9$$

Since $Ra_L > 10^9$, there occurs turbulence in the boundary layer at the end (top) portion of the screen. Thus, the boundary layer formed on the screen is a mixed boundary layer (laminar at the leading edge and turbulent at the trailing edge of the screen).

We can use Churchill and Chu equation:

$$N\bar{u} = \frac{\bar{h}L}{k} = \left\{0.825 + \frac{0.387Ra_L^{1/6}}{\left[1+(0.492/Pr)^{9/16}\right]^{8/27}}\right\}^2 = \left\{0.825 + \frac{0.387\left(4.73\times10^9\right)^{1/6}}{\left[1+(0.492/0.69)^{9/16}\right]^{8/27}}\right\}^2 = 198.48$$

$$\bar{h} = N\bar{u}\left(\frac{k}{L}\right) = 198.48 \times \frac{33.8 \times 10^{-3}}{1} = 6.709\ W/m^2.K$$

$$q = \bar{h}A(T_w - T_\infty) = 6.709 \times (1 \times 1) \times (215 - 20) = 1308.26\ W$$

13.4.6 Example 6

Consider a double pipe heat exchanger with the following dimensions: The ID (inside diameter) and OD (outside diameter) of the inner pipe are: 5.5 cm and 6.25 cm, respectively. The ID of the outer pipe is 8.2 cm. A cold gas is flowing through the annular region of the heat exchanger and a hot liquid is flowing through the inner pipe. The convective heat transfer coefficients are 100 on the gas side and 600 on the water side in units of W/m^2°C. Estimate the overall heat transfer coefficient U_o based on the outside area of the inner pipe. The thermal conductivity of the inner pipe is 39 W/m°C.

Solution

The overall heat transfer coefficient U_o is given as:

$$U_o = \frac{1}{\left(\frac{r_o}{r_i}\right)\left(\frac{1}{h_i}\right) + \left(\frac{r_o}{k_w}\right)ln\left(\frac{r_o}{r_i}\right) + \frac{1}{h_o}}$$

Given that: $r_o = 3.125\ cm$, $r_i = 2.75\ cm$, $h_i = 600$ W/m^2°C, $h_o = 100$ W/m^2°C, $k_w = 39$ W/m°C. Thus,

$$U_o = \frac{1}{\left(\frac{r_o}{r_i}\right)\left(\frac{1}{h_i}\right)+\left(\frac{r_o}{k_w}\right)ln\left(\frac{r_o}{r_i}\right)+\frac{1}{h_o}} = \frac{1}{\left(\frac{3.125}{2.75}\right)\left(\frac{1}{600}\right)+\left(\frac{3.125\times10^{-2}}{39}\right)ln\left(\frac{3.125}{2.75}\right)+\frac{1}{100}} = 83.36\ \text{W/m}^2\text{°C}$$

13.4.7 Example 7

A liquid available at 77 °C is to be cooled to 45 °C. It is supplied at a mass flow rate of 3600 kg/hr. The cooling water is available at 13 °C and a mass flow rate of 2250 kg/hr. Determine the required heat exchanger area for the counter-current double pipe heat exchanger.

Given that:

U = 285 W/m^2.K
c_p, H_2O = 4187 J/kg.K
c_p, liquid = 1758 J/kg.K

Solution

To determine the unknown temperature $T_{c,o}$ (see figure above), we do energy balance on hot and cold fluids as:

$$q = \dot{m}_h c_{p,h}(T_{h,i} - T_{h,o}) = 3600 \times 1758 \times (77 - 45) = 2.0252 \times 10^8\ J/h$$

$$q = \dot{m}_c c_{p,c}(T_{c,o} - T_{c,i}) = 2250 \times 4187 \times (T_{c,o} - 13)$$

From the above equations,

$$T_{c,o} = 13 + \frac{2.0252\times10^8}{2250\times4187} = 34.5\ °\text{C}$$

$$LMTD = \frac{(T_{h,o}-T_{c,i})-(T_{h,i}-T_{c,o})}{ln\left(\frac{T_{h,o}-T_{c,i}}{T_{h,i}-T_{c,o}}\right)} = \frac{(45-13)-(77-34.5)}{ln\left(\frac{45-13}{77-34.5}\right)} = 37\ °\text{C}$$

$$q = UA(LMTD)$$

$$A = \frac{q}{U(LMTD)} = \frac{2.0252 \times 10^8/3600}{285 \times 37} = 5.335\ m^2$$

13.4.8 Example 8

A shell-and-tube heat exchanger with one shell pass and two tube passes, that is, 1-2 shell and tube heat exchanger, is being used to heat water on the shell side from 21.1 to 54.4°C at a mass flow rate of 2.52 kg/s. Hot water under pressure enters the tubes of the heat exchanger at 115.6°C and leaves the exchanger at 48.9°C. The total outside surface area of the tubes of the exchanger is A_o = 8.35 m^2. The heat capacity of water is 4187 J/kg.K.

a) Determine the actual LMTD of the exchanger.

b) What is the overall heat-transfer coefficient U_o of the exchanger?

c) If a 2-4 shell-and-tube heat exchanger is used instead of a 1-2 heat exchanger for the same temperatures, what would be the actual LMTD?

Solution

a) Determine the actual LMTD of the exchanger.

$$(LMTD)_{actual} = F(LMTD)_{counter-current}$$

Refer to the following figure for $(LMTD)_{counter-current}$ calculation:

$$(LMTD)_{counter-current} = \frac{(T_{h,o}-T_{c,i})-(T_{h,i}-T_{c,o})}{ln\left(\frac{T_{h,o}-T_{c,i}}{T_{h,i}-T_{c,o}}\right)} = \frac{(48.9-21.1)-(115.6-54.4)}{ln\left(\frac{48.9-21.1}{115.6-54.4}\right)} = 42.3261\ ^{\circ}\mathrm{C}$$

Correction factor, F:

$$R = \frac{(T_i-T_o)}{(t_o-t_i)} = \frac{21.1-54.4}{48.9-115.6} = 0.4992$$

$$P = \frac{(t_o-t_i)}{(T_i-t_i)} = \frac{48.9-115.6}{21.1-115.6} = 0.7058$$

From Figure 13.18, $F = 0.75$

Thus,

$$(LMTD)_{actual} = F(LMTD)_{counter-current} = 0.75 \times 42.3261 = 31.745\ ^{\circ}\mathrm{C}.$$

b) What is the overall heat-transfer coefficient U_o of the exchanger?

$$q = U_o A_o (LMTD)_{actual}$$

$$U_o = \frac{q}{A_o (LMTD)_{actual}}$$

From energy balance on the cold fluid side,

$$q = \dot{m}_c c_{p,c}(T_{c,o} - T_{c,i}) = 2.52 \times 4187 \times (54.4 - 21.1) = 351356.29\ W$$

$$U_o = \frac{q}{A_o (LMTD)_{actual}} = \frac{351356.29}{8.35 \times 31.745} = 1325.52\ W/m^2.K$$

c) If a 2-4 shell-and-tube heat exchanger is used instead of a 1-2 heat exchanger for the same temperatures, what would be the actual LMTD?

As temperatures are the same, $(LMTD)_{counter-current}$ remains the same, that is, 42.3261 °C. However, the $(LMTD)_{actual}$ will not be the same as the correction factor F is not the same. For $R = 0.4992$ and $P = 0.7058$, Figure 13.19 gives $F = 0.95$. Thus,

$$(LMTD)_{actual} = F(LMTD)_{counter-current} = 0.95 \times 42.3261 = 40.21\ \text{°C}.$$

NOTATION

A	heat transfer area
A_i	inside surface area of a pipe
A_o	outside surface area of a pipe or total outside area of a bundle of tubes
a_s	crossflow area of shell of a shell and tube heat exchanger
b	spacing of baffles in a shell and tube heat exchanger
c	clearance between two adjacent tubes of a shell and tube heat exchanger
c_p	constant pressure specific heat capacity
$c_{p,c}$	constant pressure specific heat capacity of a cold fluid
$c_{p,h}$	constant pressure specific heat capacity of a hot fluid
$c_{p,s}$	constant pressure specific heat capacity of a shell side fluid
$c_{p,t}$	constant pressure specific heat capacity of a tube side fluid
D	diameter
D_H	hydraulic diameter
D_s	diameter of the shell of a shell and tube heat exchanger
d_o	outside diameter of the tubes of a shell and tube heat exchanger
f	friction factor
F	LMTD correction factor for a shell and tube heat exchanger
g	acceleration due to gravity
$g(Pr)$	function of Pr, defined in Equation (13.42)
G_s	mass velocity on the shell side of a shell and tube heat exchanger
Gr_x	local Grashof number
Gz_z	local Graetz number
h	heat transfer coefficient
h_i	convective heat transfer coefficient for flow inside the pipe
h_o	convective heat transfer coefficient for flow outside the pipe
$\bar{h}$	average heat transfer coefficient
k	thermal conductivity
k_w	thermal conductivity of the pipe wall
L	length
$LMTD$	log-mean temperature difference
$LMTD_{ST}$	log-mean temperature difference for a shell and tube heat exchanger
m	constants in Equations (13.29) and (13.38)

$\dot{m}$	mass flow rate
$\dot{m}_c$	mass flow rate of cold fluid
$\dot{m}_h$	mass flow rate of hot fluid
$\dot{m}_s$	mass flow rate of shell side fluid of a shell and tube heat exchanger
$\dot{m}_t$	total mass flow rate through the tubes of the shell and tube heat exchanger
Nu_x	local Nusselt number
$N\bar{u}$	average Nusselt number
P	pressure
p	tube pitch
Pr	Prandtl number
Pr_w	Prandtl number corresponding to wall conditions
q	heat transfer rate
q_o''	uniform wall heat flux
q_w''	wall heat flux
$\bar{q}_w''$	average wall heat flux
r_i	inside radius of a pipe
r_o	outside radius of a pipe
R	dimensionless temperature defined in Equation (13.86)
R_t	total thermal resistance
Ra_x	local Rayleigh number
Ra_x^*	modified local Rayleigh number, Equation (13.46)
Ra_L	local Rayleigh number at the end of the plate at $x = L$
Re	Reynolds number
Re_D	Reynolds number based on diameter
Re_L	local Reynolds number at the end of the plate at $x = L$
Re_x	local Reynolds number
Re_{x_c}	critical Reynolds number
S_D	diagonal pitch
S_L	longitudinal pitch
S_T	transverse pitch
St_x	local Stanton number
t_i	inlet temperature of tube side fluid
t_o	outlet temperature of tube side fluid
T	temperature
T_c	temperature of cold fluid
T_h	temperature of hot fluid
T_i	inlet temperature of shell side fluid
T_m	mean temperature inside the pipe
T_o	inlet temperature of shell side fluid
T_w	wall temperature
T_∞	free stream temperature of fluid
U_o	overall heat transfer coefficient based on outside area
V	fluid velocity or volume
V_{max}	maximum fluid velocity
V_∞	free stream velocity of fluid
x	space coordinate or distance in x-direction
x_c	critical distance where transition occurs from laminar to turbulent flow
y	space coordinate or distance in y-direction

Greek Letters

α	thermal diffusivity
β	volumetric coefficient of thermal expansion
δ_T	thermal boundary layer thickness
η	viscosity of fluid
η_w	viscosity of fluid at wall temperature
θ	angle
ν	kinematic viscosity of fluid
ρ	density

PRACTICE PROBLEMS

Problem 1

Consider a square flat plate, 1 m by 1m, maintained at 58 °C. Air at 22 °C blows over the plate at a free stream velocity of 37 m/s. Estimate the rate of heat loss from the plate. The properties of air are: $\nu = 1.7 \times 10^{-5}$ m²/s, $k = 2.8 \times 10^{-2}$ W/m.K, $Pr = 0.69$. You can assume $Re_{x_c} = 5 \times 10^5$.

Problem 2

Consider a very thin metal wire of diameter 0.04 mm carrying an electric current. The surface temperature of the wire is maintained at 52 °C. An air stream at 22 °C flows over the wire at a velocity of 48 m/s in a direction perpendicular to the wire (that is, the flow is cross flow). Estimate the rate heat loss from the wire per unit length. The properties of air are: $\nu = 1.7 \times 10^{-5}$ m²/s, $k = 2.8 \times 10^{-2}$ W/m.K, $Pr = 0.70$.

Problem 3

Consider a tube bank consisting of 21 rows of tubes with each row containing 21 tubes. Each tube is 1.2 m long and has an outside diameter is 12 mm. The tube layout is in-line (aligned) type with $S_T = S_L = 18$ mm. Steam is condensing inside the tubes at 100 °C. Air at 23 ^{0}C flows at 4.9 m/s in cross flow over the tubes. Determine the total rate of heat transfer to the air.

Properties of the air are: $\nu = 16 \times 10^{-6}$ m²/s, $k = 0.025$ W/mK, $Pr = 0.71$, $C_p = 1010$ J/kgK, $\rho = 1.18$ kg/m³, $Pr_w = 0.698$ at 300 K

Problem 4

A vertical square plate 5 m high and 5 m wide is maintained at 62 °C. It is exposed to air at 12 °C. Calculate the natural convection heat transfer rate from the plate to air. The properties of air are: $\nu = 17 \times 10^{-6}$ m²/s, $k = 27 \times 10^{-3}$ W/m.K, $\beta = 0.003\ K^{-1}$, $Pr = 0.69$.

Problem 5

Consider the flow of water at 93 °C in a 1.5 cm diameter tube at an average velocity of 0.015 m/s. The length of the tube is 1 m. The wall temperature of the tube is maintained at 38 °C. Estimate the average heat transfer coefficient in the pipe. You can assume that the flow is hydrodynamically fully developed.

The properties of water are: $\eta = 0.305\ mPa.s$, $k = 0.702\ W/m.K$, $\rho = 963\ kg/m^3$, $Pr = 1.85$, $\eta_w = 0.686\ mPa.s$

Note: You can use more than one valid correlation to estimate and compare the heat transfer coefficient values.

Problem 6

Air at 22 °C flows through a 5 cm diameter tube maintained at a wall temperature of 105 °C. The average velocity of air in the tube is 33 m/s. Estimate the average heat transfer coefficient for this internal flow. The properties of air are: $\nu = 18.5 \times 10^{-6}$ m^2/s, $k = 3 \times 10^{-2}$ W/m.K, $Pr = 0.70$.

Problem 7

A counter-current double pipe heat exchanger is to be designed to heat liquid A from 23 °C to 82 °C using a hot liquid B on the annular side. Liquid B is available at 157 °C and is to be discharged at 137 °C. The inner tube (thin walled) of the heat exchanger has a diameter of 18 mm. The overall heat transfer coefficient is 495 W/m^2.K. What would be the length of the heat exchanger if the total rate of heat transfer is 2995 W.

Problem 8

A liquid at 24 °C enters the inner thin-walled tube of a double pipe heat exchanger at a mass flow rate of 0.22 kg/s. The diameter of the tube is 24 mm. Water is used on the annular side of the heat exchanger. Water enters the heat exchanger at 4.5 °C with a flow rate of 0.5 kg/s. The liquid leaves the heat exchanger at 13.5 °C. The diameter of the outer annulus is 47 mm.

a) Estimate the heat transfer coefficient for flow in the inner tube.

b) Using the concept of hydraulic diameter, estimate the heat transfer coefficient in the annular region.

c) How long should be the heat exchanger to achieve the desired heat transfer.

Given that the properties are a) liquid: $C_p = 1040\ J/kg.K,\ \eta = 0.765\ mPa.s,\ k = 0.058\ W/m.K,\ Pr = 14$, b) water: $C_p = 4200\ J/kg.K,\ \eta = 1.3\ mPa.s,\ \ k = 0.58\ W/m.K,\ \ Pr = 9.5$.

Problem 9

Consider a shell and tube heat exchanger with shell diameter (ID) of 35 inch. The tubes of the heat exchanger are 0.75 inch in diameter (OD), 12 ft long, and are laid out on a 1-inch square pitch. Standard baffles are utilized with a spacing of 12 inches. A liquid is flowing on the shell side of the heat exchanger at a mass flow rate of 12 kg/s. Estimate the heat transfer coefficient on the shell side of the heat exchanger using the Kern equation. The properties of the liquid are: $\eta = 0.70\ mPa.s,\ \eta_w = 0.38\ mPa.s,\ C_p = 1717\ J/kg.K\ ,\ k = 0.159\ W/m.K.$

Problem 10

A shell-and-tube heat exchanger consisting of one shell pass and two tube passes is to be designed to heat liquid A on the shell side from 35 to 120 °C at a mass flow rate of 10,000 kg/h. A hot liquid B enters the exchanger on the tube side at 300 °C and leaves at 147 °C. The overall heat transfer coefficient of the exchanger is 1500 W/m^2.K. Determine the following: a) the mass flow rate of liquid B, and b) the required heat exchanger area. The heat capacities of the liquids are: $C_{p,A} = 4195$ J/kg.K and $C_{p,B} = 4660$ J/kg.K

FURTHER READING

1. T.L. Bergman, A.S. Levine, F.P. Incropera, and D.P. Dewitt. *Fundamentals of Heat and Mass Transfer*. 7th Edition, New York, Wiley & Sons, 2011.
2. R.B. Bird, W.E. Stewart, and E.N. Lightfoot. 2007. *Transport Phenomena*. 2nd Edition, Wiley & Sons, New York, 2007.

3. J.R. Welty, C.E. Wicks, R.E. Wilson, and G.L. Rorrer. *Fundamentals of Momentum, Heat, and Mass Transfer*. 5th Edition, Wiley & Sons, New York, 2008.
4. F. Keith, M.S. Bohn. *Principles of Heat Transfer*. 5th Edition, West Publishing Co., St. Paul, MN, USA, 1993.
5. A. Bejan. *Heat Transfer*. Wiley & Sons, New York, 1993.
6. J.P. Holman. *Heat Transfer*, 10th Edition, McGraw Hill, New York, 2009.
7. Y.A. Cengel. *Heat and Mass Transfer*, 3rd Edition, McGraw Hill, New York, 2007.
8. L.C. Thomas. *Heat Transfer*. Prentice Hall, New Jersey, 1992.
9. W.S. Janna. *Engineering Heat Transfer*. PWS Publishers, Boston, 1986.
10. C.J. Geankoplis, A.A. Hersel, D.H. Lepek. *Transport Processes and Separation Process Principles*. 5th Edition, Prentice Hall, New Jersey, 2018.
11. W.L. McCabe, J.C. Smith, P. Harriott. *Unit Operations of Chemical Engineering*. 6th Edition, McGraw Hill, New York, 2001

Chapter 14

Heat Transfer with Phase Change

Boilers and condensers are used extensively in industrial applications including power plants and refrigeration processes. Boiling and condensation involve heat transfer with phase change. The unique feature of boiling and condensation heat transfer is that the values of the heat transfer coefficients involved are usually very high. A good understanding of heat transfer with phase change is necessary in the design and operation of boilers and condensers.

14.1 BOILING

Boiling heat transfer occurs when a liquid is exposed to a hot (solid) surface with temperature sufficiently greater than the saturation temperature T_{sat}. The heating surface may be a plate, tube, or a rod heated electrically or otherwise. The process of boiling heat transfer is characterized by the formation of vapor bubbles at the solid-fluid interface, which grow and subsequently detach from the surface. The bubbles rise through the mass of liquid and enter the vapor space above the liquid surface. The heat flux at the solid surface is given by the following form of Newton's law of cooling:

$$q_w'' = h(T_w - T_{sat}) = h\Delta T_e \tag{14.1}$$

where ΔT_e is called excess temperature.

14.1.1 Two Types of Boiling

Two types of boiling are found in practice: Pool boiling and Flow boiling (or forced convection boiling). In pool boiling there occurs no bulk fluid flow (the bulk liquid is quiescent and no external means such as a pump are used to induce fluid flow). In flow boiling, there occurs bulk fluid flow. The fluid motion is induced by some external means such as a pump.

Pool and Flow boiling processes are further classified according to whether the bulk liquid temperature is equal to or less than the saturation temperature. If the bulk temperature of the liquid is below T_{sat}, the process is referred to as subcooled boiling. Otherwise, the process is known as saturated boiling. In the case of subcooled boiling, the bubbles condense before reaching the liquid surface. An example of pool boiling is boiling of a pan of water (see Figure 14.1) and an example of flow boiling is forced convection boiling in a heated tube (see Figure 14.2).

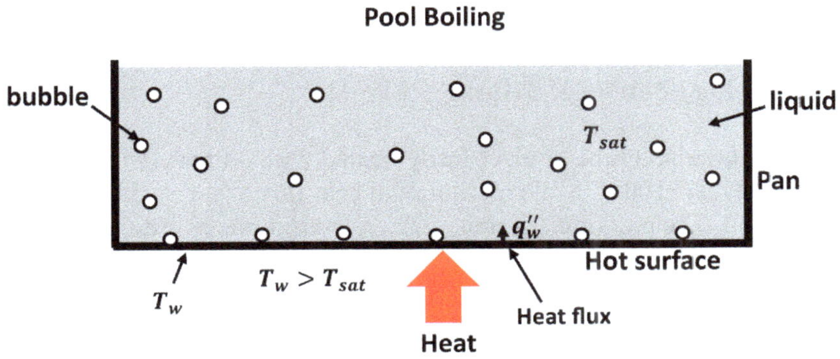

Figure 14.1 Saturated pool boiling.

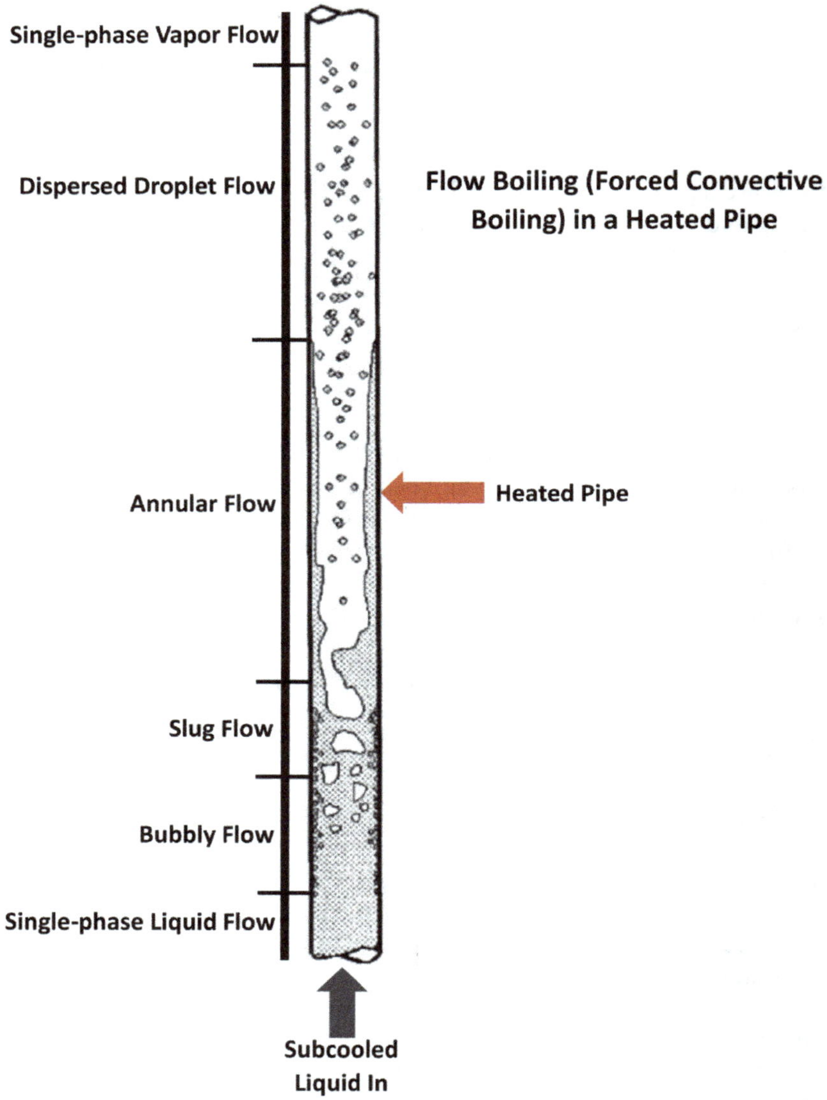

Figure 14.2 Flow boiling in a heated pipe with constant wall heat flux.

Figure 14.2 shows heating with constant surface heat flux. Forced convection boiling is a complicated process due to various flow patterns shown in Figure 14.2. It is still not fully understood. Therefore, we will focus mainly on pool boiling of saturated liquid.

14.1.2 Boiling Curve for Pool Boiling of Saturated Liquid

Consider a small heated horizontal tube or electrically heated wire immersed in a vessel containing water at saturation temperature (T_{sat}=100 ^{0}C). We measure the heat flux at the surface q_w'' and the excess temperature ΔT_e ($= T_w - T_{sat}$). The plot of q_w'' versus ΔT_e on logarithmic coordinates gives boiling curve for pool boiling. For electrically heated wire, the heat flux is determined from the power input. The power input or the rate at which energy is generated by passing a current I through a wire is:

$$P = I^2R = EI \tag{14.2}$$

where P is power input, R is electric resistance, E is the voltage drop. By measuring the potential drop (E) and current flow (I), P can be calculated. The heat flux is then P/A_s where A_s is the surface area of the wire (πDL). The wire temperature can be determined from knowledge of the resistance and the way in which the electrical resistance varies with temperature. However, the power-controlled method of heating, where ΔT_e is a dependent variable, has the disadvantage of exhibiting hysteresis effects on the boiling curve and it is not possible to obtain the full boiling curve. Thus, it is preferred to use a heating method permitting the independent control of ΔT_e to trace the full boiling curve. For example, by condensing steam inside a tube at different pressures, one can control the surface temperature, hence ΔT_e, of the tube and obtain boiling curve of a low boiling point organic fluid present at the outer surface of the tube.

Figure 14.3 shows the typical pool boiling curve. The heat flux is plotted as a function of excess temperature ΔT_e. As T_{sat} is constant, an increase in ΔT_e means an increase is the surface temperature. As ΔT_e is increased, one would normally expect a proportional (linear) increase in the rate of heat transfer (hence heat flux). However, the heat transfer coefficient is not constant in the present situation. Thus, the heat flux given by Equation (14.1) reflects the combined effects of h and ΔT_e variations.

The boiling curve exhibits four distinct regimes (see Figure 14.3): Free or natural convection, nucleate boiling, transition boiling, and film boiling.

Free or Natural Convection Regime

When the excess temperature ΔT_e is small (less than about 5 ^{0}C), free convection boiling occurs. The heat is transferred to liquid near the heating surface by natural convection. The convection currents circulate the superheated liquid and evaporation takes place at the free surface of the liquid. No boiling per se occurs in the natural convection regime (the very few bubbles which are formed on the surface of the heater, rise to the surface of the liquid, and are disengaged into the vapor space; but they are too few to disturb to any significant extent the normal currents of free convection). In the natural convection regime, the boiling curve is a straight line (h is proportional to $(\Delta T_e)^{0.25}$) with a slope of 1.25. That is,

$$q_w'' = a\Delta T_e^{\,1.25} \tag{14.3}$$

This variation is the same as that observed for natural convection in single-phase liquids.

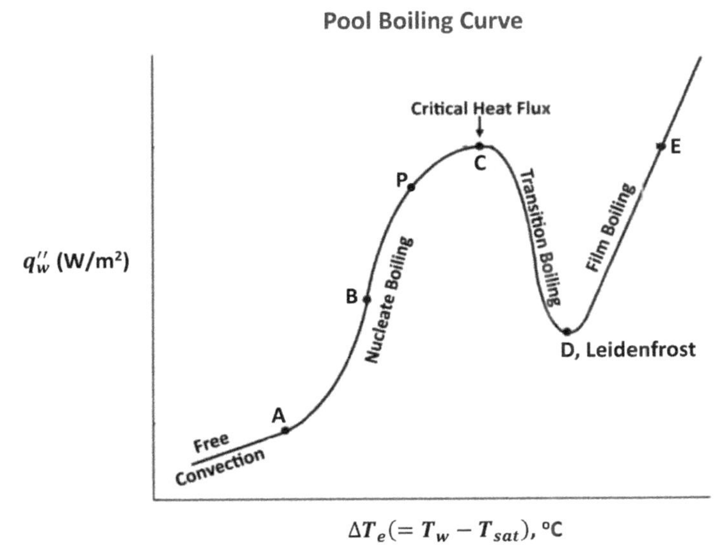

Figure 14.3 Pool boiling curve.

Nucleate Boiling Regime

At larger excess temperatures, ΔT_e lying between 5 and 30 ^{0}C, nucleate boiling occurs. In nucleate boiling regime, bubbles form at various preferential (nucleation) sites on the heating surface. The bubbles grow and separate from the heated surface and rise toward the surface of the pool. The rate of bubble production is large enough to induce considerable fluid mixing near the heating surface. The increase in the velocity of the circulation currents in the mass of liquid result in a substantial increase in the heat transfer coefficient. Now the heat transfer coefficient is much larger than that in unperturbed natural convection. As ΔT_e is increased, the rate of bubble production increases and the coefficient increases rapidly.

The nucleate boiling regime can be subdivided into two different regimes depending upon the behavior of bubbles: Isolated bubbles nucleate boiling, and jets/columns/slugs nucleate boiling. In the isolated bubble regime corresponding to the region between ONB (onset of nucleate boiling) indicated by point A and point B (see Figure 14.3), bubbles formed remain separate. There is little interaction (collision, aggregation and coalescence) between different bubbles. Most of the heat is transferred from the heated surface to the surrounding liquid by a bubble-liquid exchange action. As bubbles rise, they push hot liquid from the vicinity of the heated surface into the colder bulk of the liquid. The colder liquid from the bulk rushes towards the heated surface to fill the void. It is this intermittent entrainment of cold liquid onto the hot surface which is mainly responsible for the increase in heat transfer. Note that in this regime most of the heat exchange is through direct transfer from the heated surface to liquid in motion at the surface. The bubbles occupy but a small portion of the heated surface at any time, and most of the surface is in direct contact with the liquid. Some contribution to heat transfer comes from the formation and rise of bubbles. However, the energy carried away by vapor bubbles is not the main mechanism of heat transfer in nucleate boiling.

In the jets/columns/slugs regime corresponding to the region between points B and C (see Figure 14.3), the number of active nucleation sites becomes so large that the neighboring bubbles coalesce and escape from the heated surface in the form of jets and columns, which subsequently merge into slugs of vapor. The large heat flux in this region is due to the combined effects of liquid entrainment and evaporation (energy carried away by bubbles). Whereas the effect of evaporation increase with ΔT_e , the contribution of the liquid entrainment reaches a maximum at point P. Beyond this point the inflow of cooler liquid toward the heated surface is inhibited by the escaping bubbles with the result that the slope of heat flux versus excess temperature decreases until the point C is reached at which the surface is no longer continually wetted by the liquid and the heat flux is maximum. This maximum heat flux is commonly referred to as CHF (critical heat flux). Note that beyond point P, the heat transfer coefficient decreases with the increase in ΔT_e.

At CHF, considerable vapor is formed making it difficult for liquid to continuously wet the surface. Finally note that experimental evidence indicates that the heat flux is approximately proportional to ΔT_e^3 throughout most of the nucleate boiling regime (h is approximately proportional to ΔT_e^2; actually, the power of ΔT_e varies from 2 to 3).

Transition Boiling Regime

The region between points C and D (corresponding to ΔT_e between 30 to 120 ^{0}C) is called transition boiling, unstable film boiling, or partial film boiling. The bubble formation is now so rapid that a vapor film or blanket begins to form on the heated surface. The vapor layer is highly unstable from which miniature explosions send jets of vapor away from the heating element into the bulk of the liquid. At any point on the surface, conditions may oscillate between film and nucleate boiling, that is, nucleate and film boiling occur alternatively at a given location on the heater surface). The film is intermittently built up and partially destroyed. However, the fraction of the total surface covered by the vapor film increases with the increase in ΔT_e. The heat transfer coefficient and heat flux both decrease with increasing ΔT_e in the transition boiling regime. Note that vapor has a low thermal conductivity and is a poor conductor of heat as compared with liquid.

Film Boiling Regime

A continuous stable vapor film covers the heating surface in the film boiling regime, that is, the surface is completely covered by a vapor blanket, corresponding to excess temperatures larger than 120 ^{0}C (region beyond point D, see Figure 14.3). The location D at which the heat flux is minimum is known as the Leidenfrost point. In the film boiling region, the heat transfer from the heated surface to the liquid occurs by conduction and radiation through the vapor. The radiation heat transfer becomes more important as ΔT_e is increased. Because vapor has a much lower thermal conductivity than the liquid, very large temp differences occur across the vapor film. With the increase in ΔT_e in the film boiling regime, a point may be reached where the surface temperature of the heating element may reach the melting point of the material and burnout occurs.

Film boiling is not usually desired in commercial applications because the heat transfer rate is low for such a large excess temperature ΔT_e . It is desirable to operate many engineering devices in the nucleate boiling regime because of high heat transfer rates and convection coefficients at modest values of excess temperature.

14.1.3 Pool Boiling Heat Transfer Correlations

The key dimensionless groups are:

Average Nusselt number, $N\bar{u}_L (= \bar{h}L/k$ where L is the appropriate length scale), ratio of convective to conductive heat transfer.

Grashof number, Gr_L, defined as $Gr_L = \frac{\rho g(\rho_l - \rho_g)L^3}{\eta^2}$, ratio of the buoyant to viscous forces.

Jacob number, Ja, defined as $Ja = c_p \Delta T_e / \lambda$, ratio of sensible heat to the latent heat of vaporization.

Prandtl number, Pr, defined as $\eta c_p / k$, ratio of momentum diffusivity to thermal diffusivity.

Bond number, Bo_L, defined as $Bo_L = \frac{g(\rho_l - \rho_g)L^2}{\sigma}$, ratio of gravitational to capillary forces.

Rayleigh number, Ra_L, defined as $Ra_L = \frac{g\beta\Delta T_e L^3}{\nu\alpha}$, ratio of buoyancy force to thermal and momentum diffusive forces.

Thus, one can write the pool boiling correlation in the following general form:

$$N\bar{u}_L = f(Gr_L, Ja, Pr, Bo_L, Ra_L) \tag{14.4}$$

Free or Natural Convection Regime

In the natural convection regime, assuming the heating element is a long isothermal cylinder in horizontal position, the following Churchill and Chu correlation is applicable:

$$N\bar{u}_D = \frac{\bar{h}D}{k_l} = \left[0.60 + \frac{0.387 Ra_D^{1/6}}{\{1 + (0.559/Pr)^{9/16}\}^{8/27}}\right]^2 \tag{14.5}$$

where $\bar{h}$ is the average heat transfer coefficient and Ra_D is given as:

$$Ra_D = \frac{g\beta\Delta T_e D^3}{\nu\alpha} = \left(\frac{g\beta\Delta T_e D^3}{\nu^2}\right) Pr \tag{14.6}$$

In Equation (14.6), β is the volumetric coefficient of thermal expansion, ν is the kinematic viscosity, and α is the thermal diffusivity of the liquid. Note that Churchill and Chu correlation is of the form:

$$N\bar{u}_D = \bar{h}D/k_l = f(Pr, Ra_D) \tag{14.7}$$

In using the Churchill and Chu correlation, the fluid (liquid) properties are to be evaluated at the film temperature, that is, $(T_w + T_{sat})/2$. Also note that:

$$q''_w = \bar{h}\Delta T_e = \left(\frac{k_l}{D} N\bar{u}_D\right)\Delta T_e \tag{14.8}$$

Nucleate Boiling Regime

The most widely used correlation for nucleate boiling is the Rohsenow equation:

$$q''_w = \eta_l \lambda \left[\frac{g(\rho_l - \rho_g)}{\sigma}\right]^{1/2} \left(\frac{c_{p,l}\Delta T_e}{C_{s,l}\lambda Pr_l^n}\right)^3 \tag{14.9}$$

where λ is latent heat of vaporization, subscript l refers to liquid, and subscript g refers to gas/vapor. The coefficient $C_{s,l}$ and the exponent n depend on the solid-liquid combination. They are listed in Table 14.1 for various solid-liquid combinations. For example, $C_{s,l}$ is 0.013 for copper-water interface. n is generally equal to 1 for water and 1.7 for other liquids. Equation (14.9) is valid for tubes, wires, and plates in horizontal position. In using Equation (14.9), the properties are to be evaluated at T_{sat}.

Note that Rohsenow equation applies only for clean surfaces. It can also be written in dimensionless form as:

$$N\bar{u}_L = \left[Ja_l^2 Bo_L^{1/2} Pr_l^{1-3n}\right]/C_{s,l} \tag{14.10}$$

It should be emphasized that this correlation for nucleate pool boiling applies only to that part of the boiling curve for which $\Delta T_e < \Delta T_{e,max}$ ($\Delta T_{e,max}$ is ΔT_e corresponding to CHF)

Table 14.1 $C_{s,l}$ and n values for various solid-liquid combinations

Surface-Liquid Combination	$C_{s,l}$	n
Water-copper (scored)	0.0068	1
Water-copper (polished)	0.0128	1
Water-stainless steel (chemically etched)	0.0133	1
Water-stainless steel (mechanically polished)	0.0132	1
Water-stainless steel (ground and polished)	0.0080	1
Water-brass	0.0060	1
Water-nickel	0.006	1
Water-platinum	0.0130	1
n-Pentane-copper (polished)	0.0154	1.7
n-Pentane-copper (lapped)	0.0049	1.7
Benzene-chromium	0.0101	1.7
Ethyl alcohol-chromium	0.0027	1.7

Critical Heat Flux (CHF)

The following equation developed by Zuber can be used to calculate the peak flux:

$$q''_{w,max} = C\rho_g\lambda\left[\frac{g\sigma(\rho_l-\rho_g)}{\rho_g^2}\right]^{1/4}\left[1+\frac{\rho_g}{\rho_l}\right]^{1/2} \cong C\rho_g\lambda\left[\frac{g\sigma(\rho_l-\rho_g)}{\rho_g^2}\right]^{1/4} \tag{14.11}$$

The constant C in Equation (14.11) is weakly dependent on the geometry. For large horizontal plates, $C = 0.149$. For large horizontal cylinders and for spheres, $C = \pi/24 = 0.131$. In using Equation (14.11), the properties are to be evaluated at T_{sat}.

In dimensionless form, Equation (14.11) can be written as:

$$N\bar{u}_L(Ja_l) = q''_{w,max}Lc_{p,l}/(\lambda k_l) = C_1 Gr_{L,l}^{1/2}Pr_l/Bo_L^{1/4} \tag{14.12}$$

The CHF depends strongly on pressure, mainly through the pressure dependence of surface tension and the heat of vaporization.

Minimum Heat Flux

Zuber proposed the following equation for the minimum heat flux associated with the Leidenfrost point:

$$q''_{w,min} = 0.09\rho_g\lambda\left[\frac{g\sigma(\rho_l-\rho_g)}{(\rho_l+\rho_g)^2}\right]^{1/4} \tag{14.13}$$

Equation (14.13) is valid for large horizontal plates. In using Equation (14.13), the properties are to be evaluated at T_{sat}.

In dimensionless form, the Zuber equation can be written as:

$$N\bar{u}_L(Ja_l) = q''_{w,min}Lc_{p,l}/(\lambda k_l) = C_2 Gr_{L,l}^{1/2}\,Pr_l/Bo_L^{1/4} \tag{14.14}$$

Film Pool Boiling

Correlations for film boiling on the surface of a horizontal tube or sphere of diameter D have been developed that take the form:

$$N\bar{u}_D = \frac{\bar{h}_{conv}D}{k_g} = C\left[\frac{g(\rho_l-\rho_g)\lambda' D^3}{\nu_g k_g(T_w-T_{sat})}\right]^{1/4} \tag{14.15}$$

$$\lambda' = \lambda + 0.40c_{p,g}(T_w - T_{sat}) \tag{14.16}$$

where C=0.62 for the tube and C=0.67 for the sphere. The vapor properties are evaluated at the film temperature of $(T_w + T_{sat})/2$. This equation could also be written as:

$$\bar{h}_{conv} = C\left[\frac{g(\rho_l-\rho_g)\lambda' k_g^3}{\nu_g D(T_w-T_{sat})}\right]^{1/4} \tag{14.17}$$

In dimensionless form, the equation can be written as:

$$N\bar{u}_D = \frac{\bar{h}_{conv}D}{k_g} = C\left[Gr_{D,g}\,Pr_g\left(\frac{1}{Ja_g}+0.4\right)\right]^{1/4} \tag{14.18}$$

At elevated temperatures (T_w > 300 ^{0}C), the radiation heat transfer across the vapor film becomes significant. Bromley proposed the following correlation to account for radiation:

$$\bar{h}^{4/3} = \bar{h}_{conv}^{4/3} + \bar{h}_{rad}\bar{h}^{1/3} \tag{14.19}$$

If $\bar{h}_{rad} < \bar{h}_{conv}$, a simpler form may be used:

$$\bar{h} = \bar{h}_{conv} + (3/4)\bar{h}_{rad} \tag{14.20}$$

The radiation heat transfer coefficient is given as:

$$\bar{h}_{rad} = \frac{\varepsilon\sigma_{SB}(T_w^4 - T_{sat}^4)}{T_w - T_{sat}} \tag{14.21}$$

where ε is the emissivity of the solid surface and σ_{SB} is the Stefan-Boltzmann constant.

14.1.4 Enhancement of Nucleate Pool Boiling

As noted earlier, it is desirable to operate many engineering devices in the nucleate boiling regime because of high heat transfer rates and convection coefficients at modest values of excess temperature. The effectiveness of nucleate pool boiling depends primarily on the ease with which bubbles form and free themselves from the heating surface. To increase the number of nucleation sites and hence rate of bubble generation, the use of rough surfaces and extended surfaces (fins) is recommended. The other factor (other than the number of nucleation sites) affecting nucleate pool boiling is the ease or difficulty with which the bubble detaches from the surface once it is formed. The important factor in controlling the rate of bubble detachment is the contact angle between the bubble and heating surface. The contact angle should be small for the bubble to pinch off easily. Note that the coefficient $C_{s,l}$ in Rohsenow equation (Equation (14.9)) depends on the surface roughness and angle of contact between the bubble and heating surface. The contact angle is a measure of the wettability of a surface with a particular fluid. It decreases with greater wetting. For example, a totally wetted surface, wetted by liquid phase, has the smallest area covered by vapor and consequently represents the most favorable condition for efficient heat transfer. It should further be noted that the contact angle is related to interfacial tension between liquid and heated surface. If interfacial tension is large, the contact angle between the liquid and surface is large and the bubble tends to spread along the surface and blanket the heat transfer area.

14.2 CONDENSATION

14.1.1 Film-wise and Dropwise Condensation

When a saturated vapor meets a solid surface whose temperature (T_w) is below the saturation temperature (T_{sat}), condensation occurs. The condensate that is formed flows down the surface under the influence of gravity. If the liquid wets the surface, the condensate forms a film on the surface (see Figure 14.4). This type of condensation is called film-wise condensation. However, if the liquid does not wet the surface, droplets (instead of a liquid film) are formed on the surface. The droplets grow and coalesce with neighboring droplets. Eventually when they become large enough to overcome interfacial tension force, they roll off the surface under the influence of gravity. This type of condensation is called dropwise condensation (see Figure 14.4).

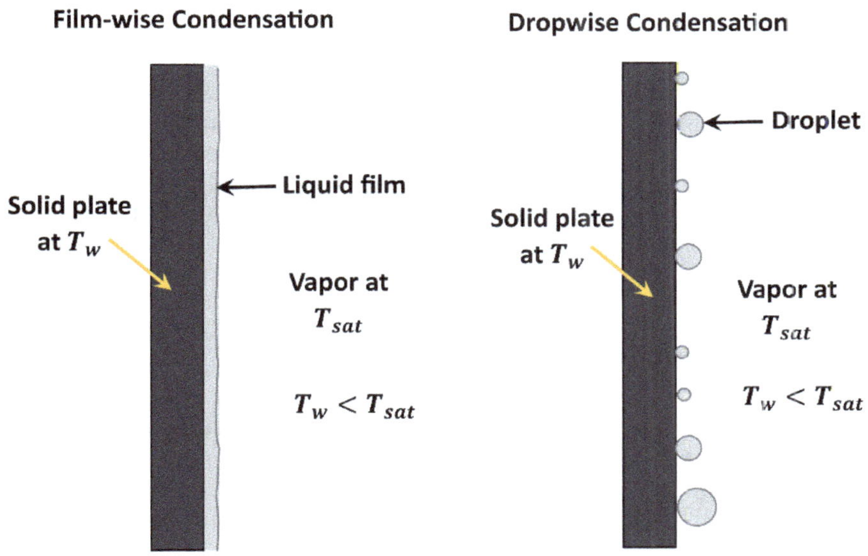

Figure 14.4 Film-wise and dropwise condensation.

The rate of heat transfer during the two types of condensation is quite different. Figure 14.5 compares heat flux versus $(T_{sat} - T_w)$ plots for the two types of condensation. For the same $(T_{sat} - T_w)$, the heat flux is much higher in the case of dropwise condensation. The average heat transfer coefficient in dropwise condensation may be 5 to 10 times larger than the film-wise coefficient. Thus, dropwise condensation is more effective and desirable. In film-wise condensation, the existence of the condensate film on the solid surface increases the resistance to heat transfer.

However, it is difficult to maintain dropwise condensation in practice. Dropwise condensation may occur initially on a solid surface, but eventually film-wise condensation takes over. Most surfaces become wetting when exposed to a condensing vapor over an extended period even though they are non-wetting initially. Therefore, the usual practice is to design heat transfer equipment under the assumption that condensation is of film type. Note that although the heat transfer coefficient in film-wise condensation is lower than that in dropwise condensation, it is still high as compared with single-phase heat transfer.

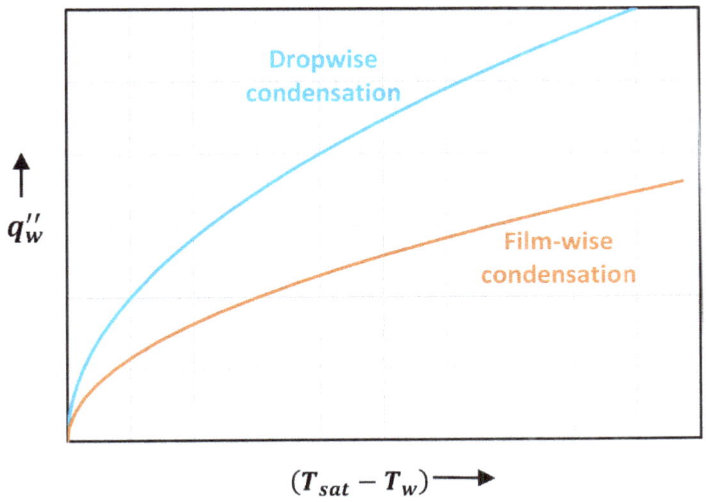

Figure 14.5 Comparison of heat flux in film-wise and dropwise condensation.

14.2.2 Analysis of Film-wise Condensation

Consider film-wise condensation on a vertical isothermal plate, as shown in Figure 14.6. A pure vapor at saturation temperature T_{sat} is condensing on a plate at constant temperature T_w ($T_w < T_{sat}$). The film thickness, denoted by δ, increases downstream the plate as more and more vapor condenses at the liquid-vapor interface. Thus, both $\delta(x)$ and mass flow rate of liquid $\dot{m}(x)$ increase in the x-direction. We are interested in local and average heat transfer coefficients over the plate. We will assume:

- The flow is steady laminar in the condensate film.
- The fluid properties are constant.
- Vapor is at a uniform temperature of T_{sat}.
- Shear stress at the liquid-vapor interface is negligible.
- Momentum and energy transfer by advection in the condensate film are negligible due to low velocities in the liquid film. Thus, we can neglect certain terms in the Navier-Stokes equation and heat transfer across the film occurs only by conduction.

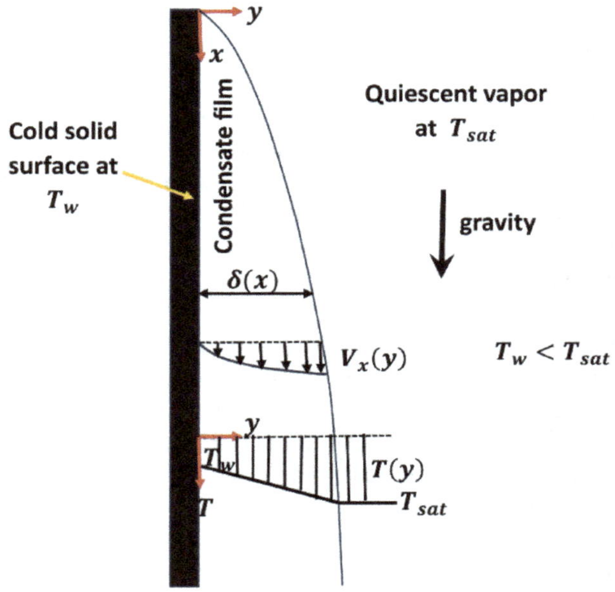

Figure 14.6 Film-wise condensation on a vertical isothermal plate.

Energy Balance

Consider a differential control volume (CV) shown in Figure 14.7. From energy balance, the rate of heat transfer into the control volume is equal to the rate of energy released due to condensation at the interface. Thus,

$$dq = \lambda d\dot{m} = q''_w(Wdx) \tag{14.22}$$

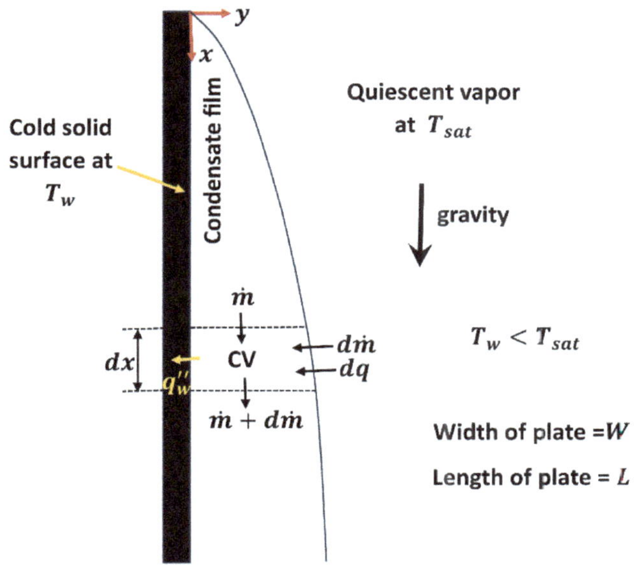

Figure 14.7 Energy balance over a differential control volume.

where λ is the latent heat of vaporization and W is the width of the plate. Equation (14.22) can be re-written as:

$$\frac{d\dot{m}}{dx} = \frac{q''_w W}{\lambda} \tag{14.23}$$

Heat is transferred through the condensate film solely by conduction (neglecting any thermal advection effects). Assuming that the temperature gradient in the condensate film is linear (see Figure 14.6), Fourier's law of heat conduction can be used to express the wall heat flux q''_w as:

$$q''_w = \frac{k_l(T_{sat}-T_w)}{\delta} \tag{14.24}$$

From Equations (14.23) and (14.24), we get:

$$\frac{d\dot{m}}{dx} = \frac{k_l(T_{sat}-T_w)}{\delta}\left(\frac{W}{\lambda}\right) \tag{14.25}$$

Velocity Distribution

The x -component of the Navier-Stokes equation in Cartesian coordinates gives:

$$\eta_l \frac{d^2 V_x}{dy^2} = -\rho g_x + \frac{\partial P}{\partial x} \tag{14.26}$$

Note that we have neglected the following terms in the Navier-Stokes equation as V_x is small:

$$\frac{\partial^2 V_x}{\partial x^2} \cong 0 \ ; V_x \frac{\partial V_x}{\partial x} \cong 0 \tag{14.27}$$

The pressure gradient in the x direction can be expressed as:

$$\frac{\partial P}{\partial x} = \rho_g g \tag{14.28}$$

Thus, Equation (14.26) becomes:

$$\eta_l \frac{d^2 V_x}{dy^2} = -g(\rho_l - \rho_g) \tag{14.29}$$

Upon integration:

$$V_x = -\frac{g(\rho_l - \rho_g)y^2}{2\eta_l} + C_1 y + C_2 \tag{14.30}$$

The integration constants C_1 and C_2 can be determined from the following boundary conditions:

B.C. 1: at $y = 0$, $V_x = 0$ (no-slip boundary condition) (14.31)

B.C. 2: at $y = \delta$, $\tau_{yx} = 0, \frac{dV_x}{dy} = 0$ (shear stress at liquid surface is zero) (14.32)

The solution is:

$$C_1 = \frac{g(\rho_l - \rho_g)\delta}{\eta_l}, \quad C_2 = 0 \tag{14.33}$$

$$V_x(y) = \frac{g(\rho_l - \rho_g)\delta^2}{\eta_l}\left[\frac{y}{\delta} - \frac{1}{2}\left(\frac{y}{\delta}\right)^2\right] \tag{14.34}$$

Condensate Mass Flow Rate

The condensate mass flow rate per unit width $\Gamma(x)$ can be obtained as follows:

$$\Gamma(x) = \frac{\dot{m}(x)}{W} = \rho_l \int_0^\delta V_x dy = \frac{g\rho_l(\rho_l - \rho_g)\delta^2}{\eta_l}\int_0^\delta \left[\frac{y}{\delta} - \frac{1}{2}\left(\frac{y}{\delta}\right)^2\right] dy \tag{14.35}$$

Upon integration, we get:

$$\Gamma(x) = \frac{g\rho_l(\rho_l - \rho_g)\delta^3}{3\eta_l} \tag{14.36}$$

The change in the condensate mass flow rate per unit width with the distance x is:

$$\frac{d\Gamma}{dx} = \frac{g\rho_l(\rho_l - \rho_g)\delta^2}{\eta_l}\left(\frac{d\delta}{dx}\right) \tag{14.37}$$

Condensate Film Thickness

From Equations (14.25) and (14.37), it follows that:

$$\frac{d\delta}{dx} = \frac{k_l \eta_l (T_{sat} - T_w)}{g\rho_l(\rho_l - \rho_g)\delta^3 \lambda} \tag{14.38}$$

Upon integration with the limit at $x = 0$, $\delta = 0$, we obtain:

$$\delta = \left[\frac{4k_l \eta_l (T_{sat} - T_w)}{g\rho_l(\rho_l - \rho_g)\lambda}\right]^{1/4} (x)^{1/4} \tag{14.39}$$

Local Heat Transfer Coefficient

The heat flux at the wall q''_w is:

$$q''_w = -k_l \left(\frac{dT}{dy}\right)_{y=0} = h(T_w - T_{sat}) \tag{14.40}$$

The temperature gradient is:

$$\left(\frac{dT}{dy}\right)_{y=0} = \frac{T_{sat}-T_w}{\delta} \tag{14.41}$$

From Equations (14.40) and (14.41), we obtain:

$$h = \frac{k_l}{\delta} \tag{14.42}$$

Substituting the expression for δ from Equation (14.39) into Equation (14.42), the following result is obtained:

$$h = \left[\frac{g\rho_l(\rho_l-\rho_g)\lambda k_l^3}{4\eta_l(T_{sat}-T_w)}\right]^{1/4} (x)^{-1/4} \tag{14.43}$$

Note that the heat transfer coefficient decreases with the increase in x as the film thickness and associated heat transfer resistance increases with the increase in x.

Average Heat Transfer Coefficient

The average heat transfer coefficient for the whole plate is:

$$\bar{h} = \frac{1}{L}\int_0^L h dx \tag{14.44}$$

From Equations (14.43) and (14.44), we get for laminar condensate film:

$$\bar{h} = \frac{4}{3} h|_{x=L} \tag{14.45}$$

$$\bar{h} = 0.943 \left[\frac{g\rho_l(\rho_l-\rho_g)\lambda k_l^3}{\eta_l(T_{sat}-T_w)L}\right]^{1/4} \tag{14.46}$$

The average Nusselt number is:

$$N\bar{u} = \frac{\bar{h}L}{k_l} = 0.943 \left[\frac{g\rho_l(\rho_l-\rho_g)\lambda L^3}{\eta_l(T_{sat}-T_w)k_l}\right]^{1/4} \tag{14.47}$$

In using Equation (14.46) or (14.47), all liquid properties should be evaluated at the film temperature, that is, $(T_{sat} + T_w)/2$. The vapor density ρ_g and latent heat of vaporization λ should be evaluated at T_{sat}.

We can re-write Equation (14.47) in dimensionless form as:

$$N\bar{u} = 1.468 Re_\delta^{-1/3} Gr_{L,l}^{1/3} \tag{14.48}$$

where the film Reynolds number is given as:

$$Re_\delta = \frac{4\Gamma}{\eta_l} = \frac{4g\rho_l(\rho_l-\rho_g)\delta^3}{3\eta_l^2} \tag{14.49}$$

and the Grashof number $Gr_{L,l}$ is given as:

$$Gr_{L,l} = \frac{g\rho_l(\rho_l-\rho_g)L^3}{\eta_l^2} \tag{14.50}$$

The film Reynolds number Re_δ in Equation (1.468) is at the end of the plate, that is, the film thickness δ is evaluated at $x = L$.

Effect of Turbulence

The critical Reynolds number for the condensate film is 1800:

$$(Re_\delta)_{\text{critical}} = 1800 \tag{14.51}$$

When $Re_\delta \leq 30$ at the end of the plate, the film undergoes wave-free laminar flow. When $30 \leq Re_\delta \leq 1800$ at the end of the plate, ripples or waves are formed on the condensate film. When $Re_\delta > 1800$ at the end of the plate, the film becomes turbulent.

Equation (14.48) is strictly valid for $Re_\delta \leq 30$, that is, when the flow at the end of the plate is wave-free laminar, the average Nusselt number associated with condensation over the entire plate is given by Equation (14.48).

When the flow at the end of the plate is wavy-laminar ($30 \leq Re_\delta \leq 1800$), the average Nusselt number associated with condensation over the entire plate can be estimated from the following Kutateladze equation:

$$N\bar{u} = \left[\frac{Re_\delta}{1.08Re_\delta^{1.22}-5.2}\right] Gr_{L,l}^{1/3} \tag{14.52}$$

When the flow at the end of the plate is turbulent ($Re_\delta > 1800$), the average Nusselt number associated with condensation over the entire plate can be estimated by the following Labuntsov equation:

$$N\bar{u} = \left[\frac{Re_\delta}{8750+58\,Pr_l^{-0.5}(Re_\delta^{0.75}-253)}\right] Gr_{L,l}^{1/3} \tag{14.53}$$

Note that although the forgoing analysis of film-wise condensation is made specifically for a vertical flat plate, the analysis is also valid for inside or outside surface of a vertical tube provided that the tube is large in diameter compared with the film thickness.

14.3 EXAMPLES

14.3.1 Example 1

A stainless steel (mechanically polished) pan is used to boil water. The water level in the pan is 3 cm above the surface of the pan. The surface temperature of the pan (in contact with the water) is maintained at 120 °C by placing the pan on an electric stove. The area of the pan is 1.56×10^{-2} m^2.

a) Estimate the power needed to boil the water,

b) What is the evaporation rate of water?

c) What is critical heat flux?

The properties of water at $T_{sat} = 100$ °C are as follows:
$\rho_l = 958$ kg/m^3, $\rho_g = 0.596$ kg/m^3 , $c_{p,l} = 4217$ J/kg.K , λ=2257 kJ/kg, $\nu_l = 2.91 \times 10^{-7}$ m^2/s, $Pr_l = 1.76$, $\sigma = 0.0589\ N/m$.

Solution

Refer to the figure below.

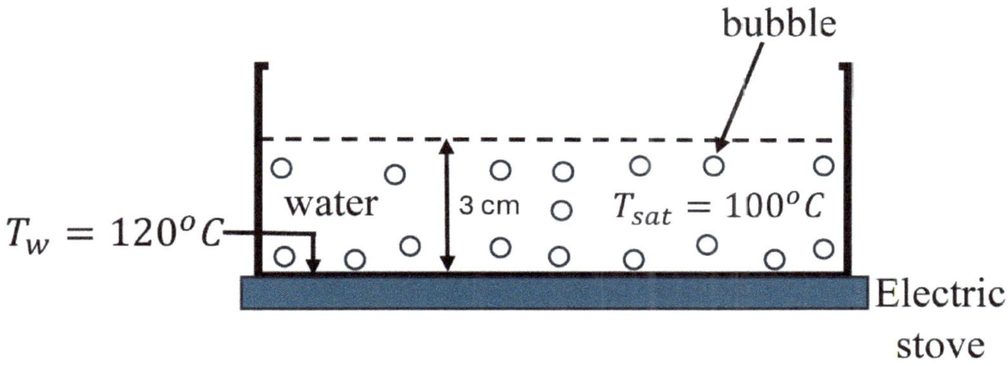

a) Estimate the power needed to boil the water.

As the excess temperature ΔT_e ($= T_w - T_{sat}$) is 20 °C, we expect nucleate boiling regime. Use Rohsenow equation to calculate the wall heat flux:

$$q''_w = \eta_l \lambda \left[\frac{g(\rho_l-\rho_g)}{\sigma}\right]^{1/2} \left(\frac{c_{p,l}\Delta T_e}{C_{s,l}\lambda Pr_l^n}\right)^3 = (958 \times 2.91 \times 10^{-7})(2257 \times 10^3) \left[\frac{9.81\times(958-0.596)}{0.0589}\right]^{1/2} \left[\frac{4217\times 20}{0.0132\times 2257\times 10^3 (1.76)^1}\right]^3 = 1.0456 \times 10^6\ W/m^2$$

Note that $C_{s,l} = 0.0132$ and $n = 1$ (see Table 14.1).

Power needed = $q''_w A = 1.0456 \times 10^6 \times 1.56 \times 10^{-2} = 1.6311 \times 10^4 = 16.311\ kW$

b) What is the evaporation rate of water?

$$\dot{m}_{evaporation} = \frac{q}{\lambda} = \frac{1.6311 \times 10^4}{2257 \times 10^3} = 7.23 \times 10^{-3}\ \ kg/s$$

c) What is critical heat flux?

Use Zuber equation:

$$q''_{w,max} = C\rho_g \lambda \left[\frac{g\sigma(\rho_l-\rho_g)}{\rho_g^2}\right]^{1/4} = 0.149 \times 0.596 \times 2257 \times 10^3 \left[\frac{9.81\times 0.0589\times(958-0.596)}{(0.596)^2}\right]^{1/4} = 1.262 \times 10^6\ \ W/m^2$$

Note that for large horizontal plates, $C = 0.149$.

14.3.2 Example 2

A 0.25 m long, 5 cm in diameter, aluminium cylinder is immersed horizontally in a pool of saturated liquid nitrogen at -196 °C. The cylinder temperature is maintained at 500 °C. Calculate the heat transfer rate from the cylinder.

Note that ΔT_e $(= T_w - T_{sat})$ is 696 °C. At this high ΔT_e, we expect a film boiling regime.

The nitrogen gas properties at the film temperature of $(500-196)/2 = 152^oC$ are: $k_g = 0.0349\ W/m.K$, $\eta_g = 23\times10^{-6}\ Pa.s$, ρ_g= 0.80 kg/m^3, $c_{p,g}$=1048 J/kg.K.

At saturation temperature of -196 °C, ρ_l= 800 kg/m^3 and λ = 201x10^3 J/kg.

Solution

As the boiling regime is film-type boiling regime, we can use the Bromley equation to estimate the average Nusselt number, neglecting any radiation effects:

$$N\bar{u}_D = \frac{\bar{h}_{conv}D}{k_g} = 0.62\left[\frac{g(\rho_l-\rho_g)\lambda' D^3}{\nu_g k_g(T_w-T_{sat})}\right]^{1/4}$$

$$\lambda' = \lambda + 0.40c_{p,g}(T_w - T_{sat})$$

$$\lambda' = \lambda + 0.40c_{p,g}(T_w - T_{sat}) = 201\times10^3 + 0.40\times1048\times696 = 492763.2\ J/kg$$

$$N\bar{u}_D = \frac{\bar{h}_{conv}D}{k_g} = 0.62\left[\frac{9.81(800-0.80)\times492763.2\times(5\times10^{-2})^3}{\left(\frac{23\times10^{-6}}{0.80}\right)\times0.0349\times(500+196)}\right]^{1/4} = 100.54$$

$$\bar{h}_{conv} = 100.54\times\frac{k_g}{D} = 100.54\times\frac{0.0349}{0.05} = 70.177\ W/m^2.K$$

$$q = \bar{h}_{conv}A\Delta T_e = 70.177\ \times\pi\times(0.05)\times0.25\times696 = 1918.07\ W$$

14.3.3 Example 3

Consider a vertical plate that is 2 ft tall and 3 ft wide. The plate is maintained at 325 °F. The plate is immersed in a saturated steam at T_{sat} = 327.81°F and film wise condensation occurs on the plate. Plot film thickness and local convection coefficient as a function of height. Also, calculate the heat-transfer rate and the mass flow rate of steam condensation.

The properties of water are $\rho = 56.7216\ lb_m/ft^3$, k =0.393 Btu/hr.ft.°R, $\nu = 0.204\times10^{-5}\ ft^2/s$. The steam density is $\rho_g = 0.202\ \ lb_m/ft^3$ and the latent heat of vaporization is λ = 888.8 Btu/lb$_m$

Solution

Refer to the figure below.

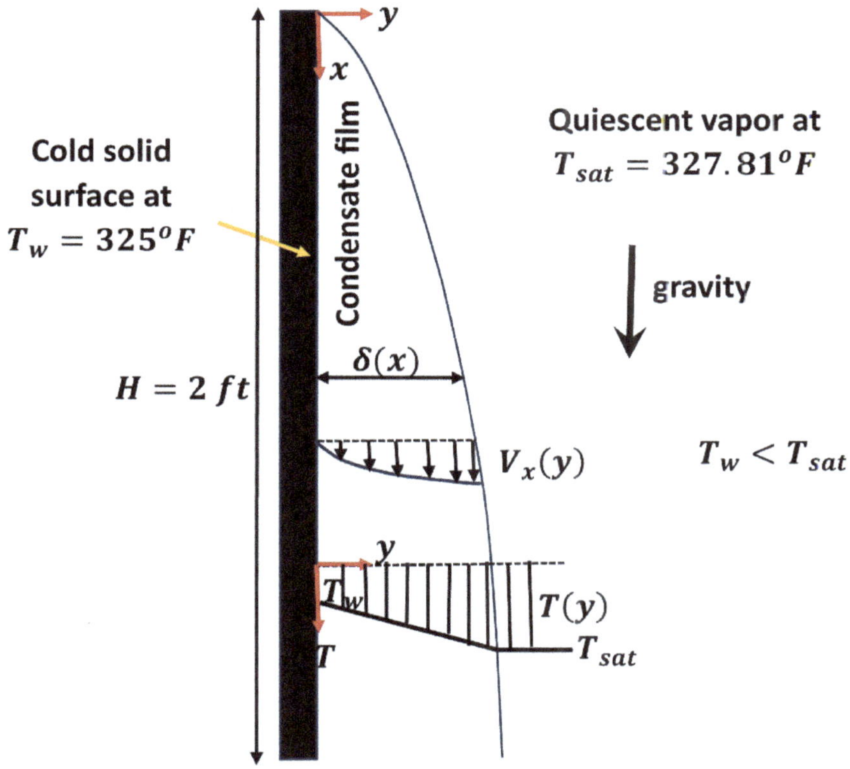

We assume that the flow of the condensate film is laminar. This assumption can be verified at a later stage.

Film thickness (δ)

$$\delta = \left[\frac{4k_l\nu_l(T_{sat}-T_w)}{g(\rho_l-\rho_g)\lambda}\right]^{1/4}(x)^{1/4} = \left[\frac{4\times(0.393/3600)(0.204\times10^{-5})(327.81-325)}{32.2\times(56.7216-0.202)\times888.8}\right]^{1/4}(x)^{1/4} = (1.983\times10^{-4})(x)^{1/4}\ \text{ft}$$

Local heat transfer coefficient (h)

$$h = \frac{k_l}{\delta} = \left(\frac{0.393}{1.983\times10^{-4}}\right)(x)^{-1/4} \quad \text{Btu/hr.ft}^2.^{\circ}\text{R}$$

The film thickness and local heat transfer coefficient are calculated for different x values. They are summarized below. As expected, the film thickness increases with the increase in x and consequently, the local heat transfer coefficient decreases with the increase in x.

x, ft	$\delta \times 10^4, ft$	h, Btu/hr.ft^2.°R
0	0	-
0.2	1.33	2963.6
0.4	1.58	2492.0
0.6	1.75	2251.8
0.8	1.88	2095.5
1.0	1.98	1981.8
1.2	2.08	1893.5

1.4	2.16	1822.0
1.6	2.23	1762.1
1.8	2.30	1711.0
2.0	2.36	1666.5

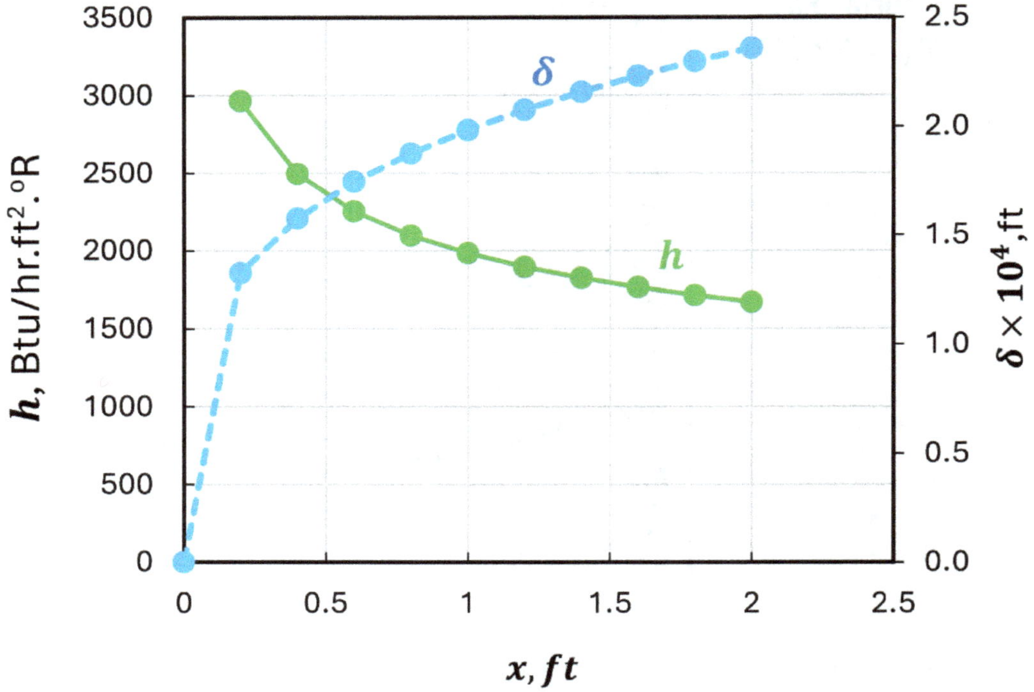

Heat transfer rate (q)

$$q = \bar{h}A(T_{sat} - T_w)$$

$$\bar{h} = \frac{4}{3}h|_{x=L}$$

From the table given above, $h|_{x=L} = 1666.5$ Btu/hr.ft^2.°R. Thus,

$$\bar{h} = \frac{4}{3}h|_{x=L} = \frac{4}{3} \times 1666.5 = 2222 \text{ Btu/hr.ft}^2.^\circ\text{R}.$$

$$q = \bar{h}A(T_{sat} - T_w) = 2222 \times (2 \times 3) \times (327.81 - 325) = 3.746 \times 10^4 \text{ Btu/hr}$$

Mass flow rate of steam condensation ($\dot{m}$)

$$\dot{m}\lambda = q = 3.746 \times 10^4 \text{ Btu/hr}$$

$$\dot{m} = \frac{3.746 \times 10^4}{888.8} = 42.15\ lb_m/hr$$

Now check the assumption of Laminar flow in the condensate film

$$Re_\delta = \frac{4\,\dot{m}/W}{\eta_l} = \frac{4\,(42.15/3600)/3}{0.204 \times 10^{-5} \times 56.7216} = 134.9$$

Since $Re_\delta < 1800$, we have laminar flow in the condensate film. However, the flow is wavy-laminar. For wavy laminar, a more accurate equation for the average Nusselt number and hence average heat transfer coefficient is the following Kutateladze equation:

$$N\bar{u} = \left[\frac{Re_\delta}{1.08Re_\delta^{1.22} - 5.2}\right] Gr_{L,l}^{1/3}$$

$$Gr_{L,l} = \frac{g\rho_l(\rho_l - \rho_g)L^3}{\eta_l^2} = \frac{32.2 \times 56.7216(56.7216 - 0.202\,)}{(0.204 \times 10^{-5} \times 56.7216)^2} = 6.168 \times 10^{13}$$

$$N\bar{u} = \left[\frac{134.9}{1.08(134.9)^{1.22} - 5.2}\right](6.168 \times 10^{13})^{1/3} = 12588.89$$

$$\bar{h} = N\bar{u}\left(\frac{k_l}{L}\right) = 12588.89\left(\frac{0.393}{2}\right) = 2473.716 \text{ Btu/hr.ft}^2.^\circ\text{R.}$$

This value of the average heat transfer coefficient is somewhat larger than the value of 2222 Btu/hr.ft^2.°R obtained earlier assuming wave-free laminar film. However, the values are close to each other.

NOTATION

a	constant in Equation (14.3)
A_s	surface area of the wire
Bo_L	Bond number
C	constant in Equations (14.11), (14.15), (14.17), and (14.18)
C_1	constant in Equation (14.12)
C_2	constant in Equation (14.14)
CHF	critical heat flux
c_p	constant pressure specific heat capacity
$c_{p,l}$	constant pressure specific heat capacity of liquid
$C_{s,l}$	constant in Equations (14.9) and (14.10), also see Table 14.1
D	diameter
E	voltage drop
g	acceleration due to gravity
g_x	acceleration due to gravity in x-direction
$Gr_{D,g}$	Grashof number of gas (vapor) based on diameter (cylinder or sphere)_
Gr_L	Grashof number based on length of plate
$Gr_{L,l}$	Grashof number of liquid based on length of plate
h	heat transfer coefficient
$\bar{h}$	average heat transfer coefficient
$\bar{h}_{conv}$	average convection coefficient in film boiling
$\bar{h}_{rad}$	average radiation coefficient in film boiling
I	electric current
Ja	Jacob number
Ja_g	Jacob number for gas (vapor)
Ja_l	Jacob number for liquid
k	thermal conductivity

k_g thermal conductivity of gas (vapor)
k_l thermal conductivity of liquid
L length
$\dot{m}$ mass flow rate
n constant in Equations (14.9) and (14.10), also see Table 14.1
$N\bar{u}$ average Nusselt number
$N\bar{u}_L$ average Nusselt number for a plate
P power input
Pr Prandtl number
Pr_l Prandtl number for liquid
Pr_g Prandtl number for gas (vapor)
q heat transfer rate
q''_w heat flux at the wall
$q''_{w,max}$ maximum heat flux, same as CHF
$q''_{w,min}$ minimum heat flux corresponding to Leidenfrost point
R electric resistance
Ra_D Rayleigh number for a long isothermal cylinder
Ra_L Rayleigh number for a plate of length L
Re_δ film Reynolds number
$(Re_\delta)_{critical}$ critical film Reynolds number

T_{sat} saturation temperature
T_w wall temperature
V_x velocity component in x-direction
W width of a plate
x space coordinate or distance in the x-direction
y space coordinate or distance in the y-direction

Greek Letters

α thermal diffusivity
β volumetric coefficient of thermal expansion
Γ condensate mass flow rate per unit width
δ condensate film thickness
ΔT_e excess temperature, defined in Equation (14.1)
$\Delta T_{e,max}$ ΔT_e corresponding to CHF
ε emissivity of the solid surface
η viscosity of fluid
η_l viscosity of liquid
λ latent heat of vaporization
λ' modified latent heat of vaporization, given in Equation (14.16)
ν kinematic viscosity
ν_g kinematic viscosity of gas (vapor)
ρ density
ρ_g density of gas (vapor)
ρ_l density of liquid
σ_{SB} Stefan-Boltzmann constant
τ_{yx} shear stress

PRACTICE PROBLEMS

Problem 1

A 1.5 mm diameter platinum wire is held horizontally in a pool of saturated water at 100 °C and 1.013 bar. The wire is heated electrically by passing current through the wire. Estimate the heat flux at the wire surface under the following conditions: a) $\Delta T_e (= T_w - T_{sat}) = 1^oC$, b) $\Delta T_e = 15^oC$.

Given that:

At a film temperature of 100.5°C, $\beta = 7.55x10^{-4}K^{-1}$, $\nu = 0.294x10^{-6}\ m^2/s$, $k = 0.683\ W/m.K$, $Pr_l = 1.74$.

At a saturation temperature of 100 °C , $C_{p,l}$ = 4220 J/kg.K, $\eta_l = 282.4x10^{-6}$ Pa.s, $\sigma = 589x10^{-4}$ N/m, ρ_l = 958.4 kg/m^3 , ρ_g= 0.598 kg/m^3 , $Pr_l = 1.75$, λ = 2257 x 10^3 J/kg.K.

Problem 2

A heating element, cylindrical with a diameter of 1.8 cm and length 37 cm, is used to boil saturated water at atmospheric pressure (T_{sat} = 100°C). The element is placed horizontally in a pool of saturated water. The surface of the cylindrical heating element is plated with nickel.

a) Estimate the heat flux q''_w and total heat transfer rate from the element when the surface temperature of the element is $T_w = 109^oC$.

b) Estimate the critical heat flux $q''_{w.max}$

The properties of saturated water at 100 °C are as follows:

$C_{p,l}$= 4216 J/kg.K, $\eta_l = 283x10^{-6}$Pa.s, $\sigma = 590x10^{-4}$ N/m, ρ_l= 958 kg/m^3 , ρ_g = 0.6 kg/m^3 , Pr_l = 1.78, λ= 2257 x 10^3 J/kg.K

Problem 3

Consider a vertical tube over which saturated steam at 110 ^{0}C is condensing. The tube dimensions are: 21 mm OD and 22 cm long. The surface temperature of the tube is maintained at 109 ^{0}C.

Calculate:

(a) the local heat transfer coefficient at the end of the tube

(b) the thickness of the condensate film at the end of the tube

(c) the average heat transfer coefficient

(d) the total heat transfer rate

(e) the rate of condensation of steam

(f) film Reynolds number at the end of the tube

Given that the properties of condensate (water) at the film temperature are:

$\rho = 951.4kg/m^3$; $k = 0.685W/m.K$; $\eta = 260.1x10^{-6}Pa.s$

The latent heat of vaporization at saturation temperature is $\lambda = 2230x10^3\ J/kg$

Problem 4

A plane vertical wall of height 2.5 m and width 1 m is maintained at a temperature of 60 °C. It faces a space filled with saturated steam at $T_{sat} = 100^oC$. Estimate the following:

a) the average heat transfer coefficient over the wall,

b) the mass flow rate of steam condensation on the wall.

The properties of condensate (water) at the film temperature of 80 °C are:

$\rho = 972kg/m^3$; $k = 0.67W/m.K$; $\eta = 355.75x10^{-6}Pa.s$

The latent heat of vaporization is $\lambda = 2257x10^3 J/kg$ and g = 9.81 m/s^2

FURTHER READING

1. T.L. Bergman, A.S. Levine, F.P. Incropera, and D.P. Dewitt. *Fundamentals of Heat and Mass Transfer*. 7th Edition, New York, Wiley & Sons, 2011.
2. R.B. Bird, W.E. Stewart, and E.N. Lightfoot. 2007. *Transport Phenomena*. 2nd Edition, Wiley & Sons, New York, 2007.
3. J.R. Welty, C.E. Wicks, R.E. Wilson, and G.L. Rorrer. *Fundamentals of Momentum, Heat, and Mass Transfer*. 5th Edition, Wiley & Sons, New York, 2008.
4. F. Keith, M.S. Bohn. *Principles of Heat Transfer*. 5th Edition, West Publishing Co., St. Paul, MN, USA, 1993.
5. A. Bejan. *Heat Transfer*. Wiley & Sons, New York, 1993.
6. J.P. Holman. *Heat Transfer*, 10th Edition, McGraw Hill, New York, 2009.
7. Y.A. Cengel. *Heat and Mass Transfer*, 3rd Edition, McGraw Hill, New York, 2007.
8. L.C. Thomas. *Heat Transfer*. Prentice Hall, New Jersey, 1992.
9. W.S. Janna. *Engineering Heat Transfer*. PWS Publishers, Boston, 1986.
10. C.J. Geankoplis, A.A. Hersel, D.H. Lepek. *Transport Processes and Separation Process Principles*. 5th Edition, Prentice Hall, New Jersey, 2018.
11. W.L. McCabe, J.C. Smith, P. Harriott. *Unit Operations of Chemical Engineering*. 6th Edition, McGraw Hill, New York, 2001

Chapter 15

Radiative Heat Transfer

When two bodies or surfaces at different temperatures are placed in view of each other and are separated by vacuum, the sole mechanism of heat transfer between them is the exchange of thermal radiation.

15.1 NATURE OF THERMAL RADIATION

Thermal radiation is a stream of electromagnetic radiation emitted by a body because of its finite absolute temperature. All forms of matter at finite absolute temperature emit thermal radiation because of atomic and molecular scale motions, such as translational, vibrational, and rotational movement of molecules, vibration of atoms, and certain electronic transitions.

Thermal radiation is a part of the electromagnetic spectrum confined to wavelengths in the range 0.1 to 1000 μm (see Figure 15.1). It includes a portion of ultraviolet region (0.1 to 0.38 μm), all the visible (0.38 to 0.76 μm) and infrared (0.76 to 1000 μm) regions.

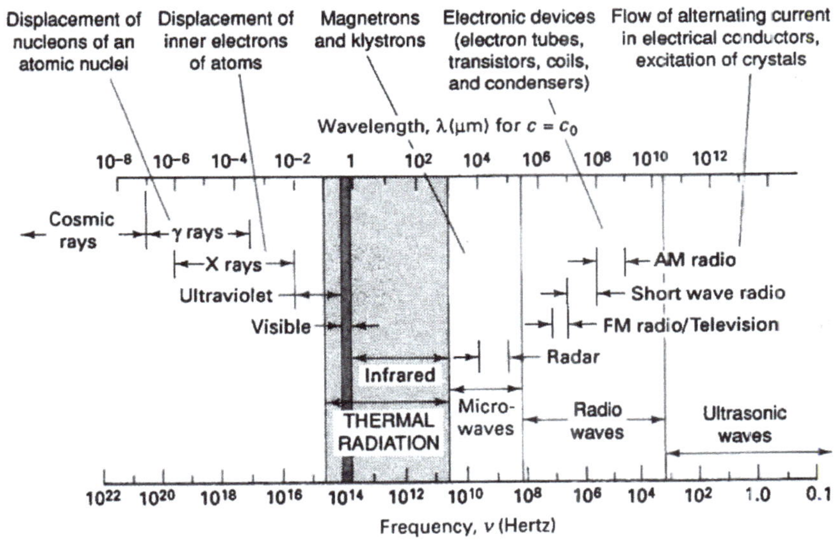

Figure 15.1 Electromagnetic spectrum.

The wavelength (λ) of thermal radiation is related to frequency (ν) and speed of light (c) as:

$$\lambda = c/\nu \tag{15.1}$$

While the propagation velocity and wavelength of electromagnetic waves are affected by the medium in which they are travelling, the frequency is independent of the medium and depends only on the radiating source. The speed of electromagnetic waves in vacuum is $c_0 = 2.998 \times 10^8$ m/s.

Interestingly electromagnetic radiation including thermal radiation has both a particulate and wavelike nature. Thus, thermal radiation can be thought of as a stream of discrete packets of energy called photons or quanta. Each photon or quantum has energy of

$$e = h\nu \tag{15.2}$$

where h is Planck's constant (=6.625 x 10^{-34} J.s)

The thermal radiation emitted by a surface at a finite absolute temperature is not monochromatic (single wavelength). It covers a range of wavelengths and is emitted in all directions. The intensity of the radiation varies with the wavelength and with the direction of the emission. The variation of intensity with λ is called spectral distribution and the variation of intensity with direction is called directional distribution.

15.1.1 Emissive power, Irradiation, and Radiosity

The total emissive power E (W/m^2) of the radiation emitted by a surface is defined as:

$$E = \int_0^\infty E_\lambda d\lambda \tag{15.3}$$

where E_λ (W/m^3) is the spectral emissive power of radiation. It is the rate at which radiation energy is emitted by a surface per unit wavelength interval about λ in all directions per unit surface area. As a hemisphere placed above the emitting surface intercepts all the radiation rays emitted by the surface, E_λ is called spectral hemispherical emissive power of radiation and E is called total hemispherical emissive power.

The total irradiation G (W/m^2) is defined as:

$$G = \int_0^\infty G_\lambda d\lambda \tag{15.4}$$

where G_λ (W/m^3) is the spectral irradiation. It is the rate at which radiation energy is incident on a surface per unit wavelength interval about λ from all directions per unit surface area.

The total radiosity J (W/m^2) is defined as:

$$J = \int_0^\infty J_\lambda d\lambda \tag{15.5}$$

where J_λ (W/m^3) is the spectral radiosity. It is the rate at which radiation energy leaves a surface per unit wavelength interval about λ from all directions per unit surface area. It includes the reflected portion of the irradiation as well as direct emission.

15.1.2 Absorptivity, Reflectivity, and Transmissivity

When radiation falls on the surface, some of it is absorbed, some of it is reflected, and the remaining part is transmitted. Let the total incident radiation flux on a surface from all directions and over all wave lengths be G.

The total absorptivity (α) is the fraction of G absorbed by the surface, that is,

$$\alpha = \frac{G_{abs}}{G} \tag{15.6}$$

where G_{abs} is the absorbed portion of incident radiation G.

The total reflectivity (ρ) is the fraction of G reflected by the surface, that is,

$$\rho = \frac{G_{ref}}{G} \tag{15.7}$$

where G_{ref} is the reflected portion of incident radiation G.

The total transmissivity (τ) is the fraction of G transmitted through the surface, that is,

$$\tau = \frac{G_{trans}}{G} \tag{15.8}$$

where G_{trans} is the transmitted portion of incident radiation G.

Using energy balance:

$$G_{abs} + G_{ref} + G_{trans} = G \tag{15.9}$$

Or,

$$\alpha + \rho + \tau = 1 \tag{15.10}$$

In the case of opaque surfaces, $\tau = 0$ and therefore,

$$\alpha + \rho = 1 \tag{15.11}$$

15.1.3 Black Bodies

A blackbody is one which absorbs all incident radiation, regardless of wavelength and direction. Blackbodies are not only perfect absorbers of radiation, but they are also the best emitters of radiation. At any given temperature, radiation emitted from a blackbody has the maximum possible value of spectral emissive power. No surface can emit radiation with spectral emissive power higher than that of a black surface when comparison is made at the same temperature. Furthermore, blackbodies are diffuse emitters of radiation, that is, the spectral emissive power is independent of direction and depends only on wavelength and temperature.

It should be noted that a surface that appears black to the eye absorbs radiation in the narrow band of visible portion of electromagnetic spectrum (approximately 0.38 to 0.76 μm) whereas a blackbody absorbs all incident radiation of all wavelengths. Thus, a surface that appears black to the eye may not be a black surface from thermal radiation point of view. Likewise, a surface that is almost black from thermal radiation point of view may not appear black to the eye. For example, snow appears white to the eye, but it is essentially black with respect to long-wavelength (infra-red) thermal radiation.

The spectral emissive power emitted by a blackbody is given by Planck's radiation law:

$$E_{\lambda,b}(\lambda, T) = \frac{2\pi h c_o^2}{\lambda^5[exp(hc_o/\lambda kT)-1]} \tag{15.12}$$

where h is Planck's constant (=$6.6256\text{x}10^{-34}$ J.s), k is Boltzmann constant (=$1.3805\text{x}10^{-23}$ J/K), c_o is the speed of light in vacuum (=$2.998\text{x}10^{8}$ m/s) and T is the absolute temperature. The Planck radiation law indicates that: a) the spectral emissive power of a blackbody is a continuous function of wavelength (λ) at any given temperature, b) the spectral emissive power increases with the increase in temperature at any given wavelength, and c) the wavelength at which the spectral emissive power has its maximum value decreases with the increase in temperature. Figure 15.2 shows the plots of $E_{\lambda,b}$ at different temperatures generated from the Planck's radiation law.

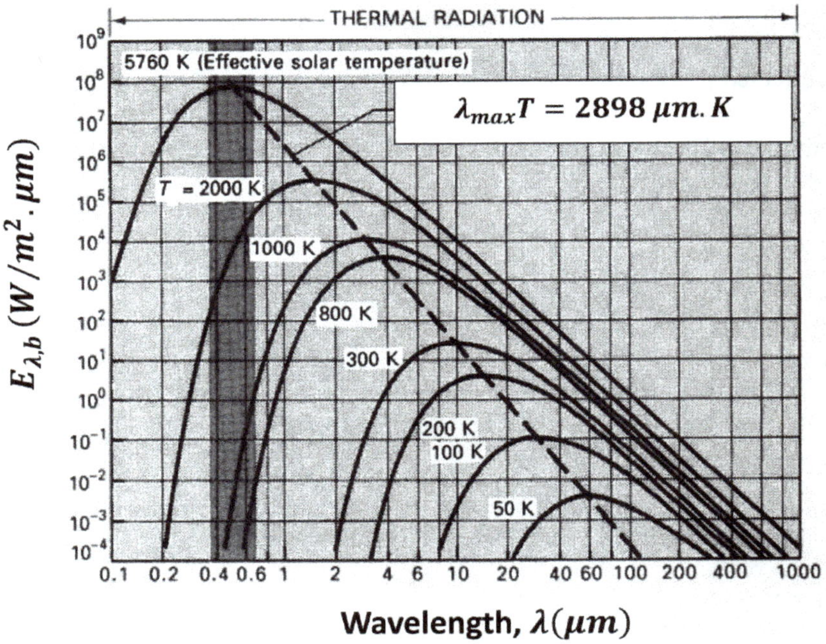

Figure 15.2 Monochromatic emissive power of a black body at different temperatures.

The wavelength at which the spectral emissive power is maximum can be obtained from the Planck law by setting $dE_\lambda/d\lambda = 0$ (see Figure 15.2). Thus:

$$\lambda_{max}T = 2.898 \times 10^{-3}\ mK \tag{15.13}$$

where λ_{max} is the value of λ where $E_{\lambda,b}$ is maximum. This equation is referred to as Wien's displacement law.

The total emissive power of a blackbody is given by the Stefan-Boltzmann law:

$$E_b = \sigma T^4 \tag{15.14}$$

where σ is the Stefan-Boltzmann constant ($= 5.67 \times 10^{-8}\ W/m^2.K^4$). The Stefan-Boltzmann law follows from the Planck distribution, Equation (15.12):

$$E_b = \int_0^\infty \frac{2\pi h c_0^2 d\lambda}{\lambda^5[exp(hc_o/\lambda kT)-1]} = \sigma T^4 \tag{15.15}$$

In practice, it is often required to estimate the band emission from a radiating body. Band emission is a fraction of the total emission that is in a certain wavelength interval (band) at a given temperature. One can easily calculate band emission from a blackbody using the Planck distribution and Stefan-Boltzmann law. Thus, the band emission $F_{(\lambda_1 \to \lambda_2)}$ over the wavelength band of $(\lambda_1 \to \lambda_2)$ is:

$$F_{(\lambda_1 \to \lambda_2)} = F_{(0 \to \lambda_2)} - F_{(0 \to \lambda_1)} = \frac{\int_0^{\lambda_2} E_{\lambda,b} d\lambda - \int_0^{\lambda_1} E_{\lambda,b} d\lambda}{\sigma T^4} \tag{15.16}$$

15.1.4 Real Bodies

Real surfaces are not perfect absorbers of incident radiation. They absorb only a fraction of the incident radiation. Some of the incident radiation is reflected and some of it is transmitted. Real surfaces also emit less radiation energy at any given temperature as compared with black surfaces at the same temperature.

The total emissivity (ε) of a real surface is defined as the ratio of total emissive power of a real surface (E) to total emissive power of a black surface at the same temperature, that is:

$$\varepsilon = \frac{E}{E_b} \tag{15.17}$$

Note that the emissivity of a real surface can be defined in different ways. For example, the spectral or monochromatic emissivity ε_λ is defined as the ratio of spectral emissive power of a real surface to spectral power of a black surface at the same temperature:

$$\varepsilon_\lambda = \frac{E_\lambda}{E_{\lambda,b}} \tag{15.18}$$

The total emissivity is related to spectral emissivity as follows:

$$\varepsilon = \frac{E}{E_b} = \frac{\int_0^\infty \varepsilon_\lambda E_{\lambda,b} d\lambda}{\sigma T^4} \tag{15.19}$$

15.1.5 Some More Definitions

Diffuse Emitter: A diffuse emitter is one whose emissivity is independent of the direction, that is, the radiation is emitted isotropically (equally in all directions).

Diffuse Reflector: A diffuse reflector is one which reflects incident radiation equally in all directions. There is no directional dependence of reflected radiation. Rough surfaces give diffuse reflection.

Diffuse Surface: A surface which reflects and emits radiation in a diffuse manner.

Gray Surface: In general, the radiation properties of the material (emissivity, reflectivity, absorptivity, transmissivity) are functions of wavelength of the radiation. Gray surface is one whose radiation properties are independent of the wavelength.

Diffuse Gray Surface: A surface whose radiation properties are independent of both direction and wavelength (diffuse implies directional independence and gray implies wavelength independence)

15.1.6 Kirchhoff's Radiation Law

Kirchhoff's radiation law states that the monochromatic directional emissivity ($\varepsilon_{\lambda,\theta}$) of a surface is equal to the monochromatic directional absorptivity ($\alpha_{\lambda,\theta}$) of the surface when comparison is made at the same temperature. Thus:

$$\varepsilon_{\lambda,\theta} = \alpha_{\lambda,\theta} \tag{15.20}$$

Note that, in general, $\varepsilon_{\lambda,\theta}$ and $\alpha_{\lambda,\theta}$ are both functions of direction and wavelength. This form of Kirchhoff's radiation law is very general as $\varepsilon_{\lambda,\theta}$ and $\alpha_{\lambda,\theta}$ are inherent surface properties.

Another form of Kirchhoff's law that is less general is as follows:

$$\varepsilon = \alpha \tag{15.21}$$

where ε and α are total hemispherical emissivity and absorptivity of the surface. This form of Kirchhoff's law is valid under the condition that the surface is in thermal equilibrium with its surroundings.

For diffuse gray surface, the emissivity and absorptivity are equal and are independent of direction and wavelength. This is true whether the surface is in thermal equilibrium or not with its surroundings.

15.2 RADIATIVE HEAT TRANSFER BETWEEN SURFACES

Thus far, thermal radiation and thermal radiation properties of bodies have been described. In what follows, radiative heat exchange among the surfaces is described.

The exchange of radiation energy between two surfaces depends on their view factors. The view factor F_{ij} is the fraction of radiant energy leaving surface i that is intercepted by surface j, that is:

$$F_{ij} = \frac{q_{i\to j}}{A_i J_i} \tag{15.22}$$

where $q_{i\to j}$ is radiation energy leaving surface i and being intercepted by surface j, J_i is the radiosity of surface i, and A_i is area of the surface through which radiation is being emitted.

Likewise, the view factor F_{ji} is the fraction of radiant energy leaving surface j that is intercepted by surface i , that is:

$$F_{ji} = \frac{q_{j\to i}}{A_j J_j} \tag{15.23}$$

where $q_{j\to i}$ is radiation energy leaving surface j and being intercepted by surface i, J_j is the radiosity of surface j, and A_j is area of the surface through which radiation is being emitted.

The total radiation leaving from any surface is the sum of emitted radiation and reflected radiation. The radiosity J_i of a surface is the sum of emitted radiation (E_i) and reflected irradiation ($\rho_i G_i$):

$$J_i = E_i + \rho_i G_i \tag{15.24}$$

The net rate of radiative heat transfer from surface i to surface j is given by:

$$q_{ij} = q_{i\to j} - q_{j\to i} = A_i J_i F_{ij} - A_j J_j F_{ji} \tag{15.25}$$

Also note that:

$$q_{ji} = -q_{ij} = q_{j\to i} - q_{i\to j} = A_j J_j F_{ji} - A_i J_i F_{ij} \tag{15.26}$$

The view factors F_{ij} and F_{ji} are purely geometric quantities. They depend only on size, orientation, and relative positions of surfaces i and j.

It can be readily shown that:

$$A_i F_{ij} = A_j F_{ji} \tag{15.27}$$

This equation is called the reciprocity relation for view factors and is applicable to any pair of radiating surfaces if they emit and reflect radiation diffusively.

Most radiative heat transfer problems of practical interest involve enclosed spaces (enclosures). As the total radiation leaving any surface i of an enclosure must be intercepted by the surfaces of the enclosures (including surface i itself), energy balance gives:

$$\sum_{j=1}^{N} F_{ij} = \sum_{j=1}^{N} \frac{q_{i\to j}}{A_i J_i} = 1 \tag{15.28}$$

where N is the number of surfaces forming the enclosure. This equation is referred to as the summation rule. The summation rule can be applied to each surface present in the enclosure. If there are N surfaces present in an enclosure, N^2 view factors are required. For example, in a two- surface enclosure the four view factors required are: F_{11}, F_{22}, F_{12} and F_{21}. As the summation rule is applicable to each surface of the enclosure, it leads to N equations. The reciprocity relation gives additional $N(N-1)/2$ equations. Thus, the number of view factors that must be obtained from geometric considerations is $N(N-1)/2$, as shown below:

$$\text{Unknown View Factors} = \text{Total Number of View Factors} - \text{Number of Equations} \tag{15.29}$$

$$\text{Unknown View Factors} = N^2 - [N + N(N-1)/2] = N(N-1)/2 \tag{15.30}$$

15.2.1 Surface Resistance

Consider now a single surface i (not necessarily in an enclosure) which is receiving radiation from other surfaces and is emitting radiation. Assume that the surface is opaque, diffuse, and gray (radiation properties are independent of λ and direction and $\alpha = \varepsilon = 1 - \rho$). The surface is irradiated with incident radiation energy rate of $G_i A_i$. The surface is emitting and reflecting radiation energy at a rate of $J_i A_i$. Thus, the net rate of radiation energy leaving surface i is:

$$q_i = A_i (J_i - G_i) \tag{15.31}$$

The radiosity J_i of the surface is the sum of emitted radiation (E_i) and reflected irradiation ($\rho_i G_i$), given by Equation (15.24). Since $E_i = \varepsilon_i E_{b,i}$ and $\rho_i = 1 - \varepsilon_i$, Equation (15.24) gives:

$$G_i = \frac{J_i - \varepsilon_i E_{b,i}}{1 - \varepsilon_i} \tag{15.32}$$

Substitution of G_i from Equation (15.32) into Equation (15.31) gives:

$$q_i = \frac{E_{b,i} - J_i}{\left[\frac{1-\varepsilon_i}{\varepsilon_i A_i}\right]} \tag{15.33}$$

Equation (15.33) can be interpreted in terms of electrical analog as follows: q_i, the net radiative heat transfer rate from a surface, is the "current" through the surface, $E_{b,i} - J_i$, the difference in the radiosities of a black surface and actual surface, is the "driving potential" for the current, and $(1 - \varepsilon_i)/\varepsilon_i A_i$ is the "resistance" to current flow through the surface (see Figure 15.3).

Electrical Analog of a Radiating Surface

$$R = \left(\frac{1-\varepsilon}{\varepsilon A}\right)$$

$$q = \frac{E_b - J}{\left(\frac{1-\varepsilon}{\varepsilon A}\right)}$$

Figure 15.3 Surface resistance of a radiating surface.

When the radiosity of the actual surface is less than the radiosity of the surface if it were black, the driving potential for net radiative heat transfer is positive and there occurs a net radiation heat transfer from the surface. When the radiosity of the actual surface is more than the radiosity of the surface if it were black, there occurs a net radiation heat transfer to the surface. The larger the difference in radiosities, greater is the net radiative heat transfer rate. The resistance term $(1-\varepsilon_i)/\varepsilon_i A_i$ is also called "surface resistance". If the actual surface is a black surface, the surface resistance is zero and $E_{b,i} = J_i$. In this case, $q_i = A_i(E_{b,i} - G_i)$.

Consider another example where a real surface at constant temperature is just emitting radiation. No radiation is falling on the surface and hence there is no reflected radiation. The surface has an emissivity of ε_i. In this case, the radiosity J_i is just $\varepsilon_i E_{b,i}$ as there is no reflected radiation present. Thus, the net rate of radiative heat transfer from the surface is given as:

$$q_i = \frac{E_{b,i} - \varepsilon_i E_{b,i}}{\left[\frac{1-\varepsilon_i}{\varepsilon_i A_i}\right]} = \varepsilon_i E_{b,i} A_i \tag{15.34}$$

15.2.2 Space or View Resistance

Consider now radiation exchange between two surfaces. As noted earlier, the net rate of radiative heat transfer from surface i to surface j is given by:

$$q_{ij} = q_{i\to j} - q_{j\to i} = A_i J_i F_{ij} - A_j J_j F_{ji} \tag{15.35}$$

Using the reciprocity relation and re-arranging Equation (15.35), the following result is obtained:

$$q_{ij} = \frac{J_i - J_j}{\left(\frac{1}{A_i F_{ij}}\right)} \tag{15.36}$$

Thus, the driving potential for net radiative heat transfer between two surfaces is the difference in the radiosities of the two surfaces, and the resistance to radiative heat transfer between two surfaces is $1/A_i F_{ij}$. The resistance $1/A_i F_{ij}$ to radiative heat transfer is referred to as space or view resistance (see Figure 15.4)

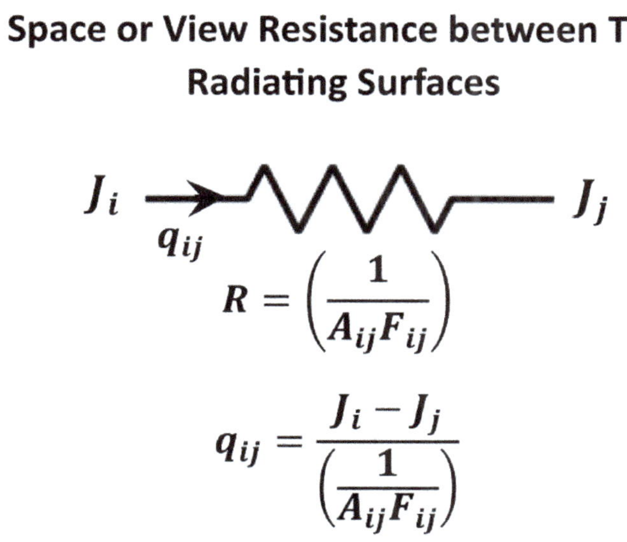

Figure 15.4 Space or view resistance between two radiating surfaces.

If surface i is in an enclosure, it will exchange radiation with each of the N surfaces present in the enclosure. The net rate of radiative heat transfer from surface i in an enclosure is equal to the sum of the net rates of radiative heat transfer from surface i to N surfaces:

$$q_i = \frac{E_{b,i}-J_i}{\left[\frac{(1-\varepsilon_i)}{\varepsilon_i A_i}\right]} = \sum_{j=1}^{N} q_{ij} = \sum_{j=1}^{N} \frac{J_i-J_j}{\left(\frac{1}{A_i F_{ij}}\right)} \tag{15.37}$$

Figure 15.5 shows the electric analog of the net rate of radiative heat transfer from surface 1 to three other surfaces (numbered 2 to 4) present in an enclosure.

It should be noted that in radiative heat transfer between surfaces, two different types of resistances are encountered: the surface resistance given by $(1-\varepsilon_i)/\varepsilon_i A_i$ and space or view resistance given as $1/A_i F_{ij}$. If surface i is a black surface and is exchanging radiation with surface j, the surface resistance is zero but the space resistance between surfaces i and j is not zero.

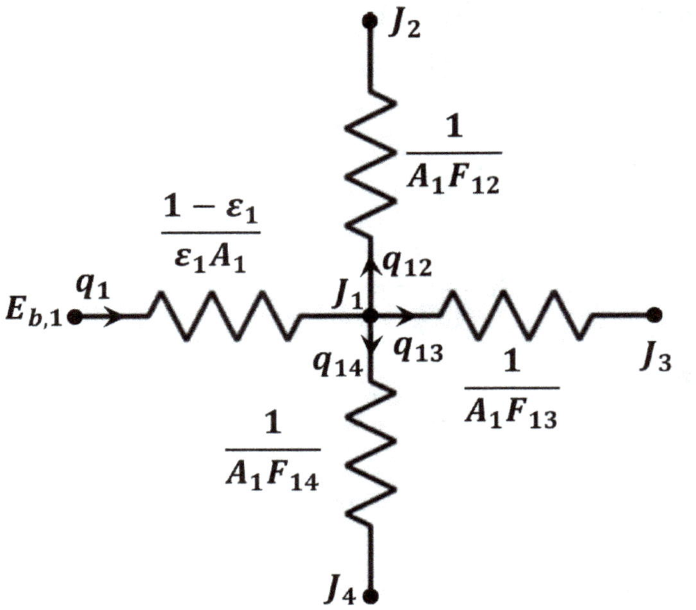

$$q_1 = \frac{(E_{b,1} - J_1)}{\left(\frac{1-\varepsilon_1}{\varepsilon_1 A_1}\right)} = q_{12} + q_{13} + q_{14} = \frac{(J_1 - J_2)}{\frac{1}{A_1F_{12}}} + \frac{(J_1 - J_3)}{\frac{1}{A_1F_{13}}} + \frac{(J_1 - J_4)}{\frac{1}{A_1F_{14}}}$$

Figure 15.5 Net radiative heat transfer from surface 1 to surfaces 2,3, and 4 present in an enclosure.

15.2.3 Radiation Heat Exchange Between Two Real Surfaces

Open systems:

Consider two real surfaces maintained at constant temperatures T_1 and T_1. Assume that the surfaces are diffuse and gray and their emissivities are ε_1 and ε_2. The surfaces are open in that they do not form an enclosure. The surfaces may be exchanging thermal radiation with other surfaces present in the surroundings. The net rate of radiative heat transfer from surface 1 to surface 2 is as follows:

$$q_{12} = \frac{J_1 - J_2}{\left(\frac{1}{A_1F_{12}}\right)} \tag{15.38}$$

It is important to realize that q_{12} is not equal to q_1, the net rate of radiative heat transfer from surface 1 to all the surrounding surfaces (including surface 2). If the surfaces 1 and 2 are black surfaces, the above expression simplifies to:

$$q_{12} = \frac{E_{b,1} - E_{b,2}}{\left(\frac{1}{A_1F_{12}}\right)} = \frac{\sigma(T_1^4 - T_2^4)}{\left(\frac{1}{A_1F_{12}}\right)} \tag{15.39}$$

Enclosures:

Consider two real surfaces which together form an enclosure such that the two surfaces exchange radiation only with each other. As there are only two surfaces exchanging radiation with each other, it follows that:

$$q_1 = -q_2 = q_{12} = -q_{21} \tag{15.40}$$

Now there are three radiative resistances in series: surface resistance corresponding to surface 1, space resistance between the two surfaces, and surface resistance corresponding to surface 2. Thus:

$$R_{total} = \frac{1-\varepsilon_1}{\varepsilon_1 A_1} + \frac{1}{A_1 F_{12}} + \frac{1-\varepsilon_2}{\varepsilon_2 A_2} \tag{15.41}$$

The overall driving potential for radiative heat transfer from surface 1 to surface 2 is $E_{b,1} - E_{b,2}$. Thus:

$$q_1 = \frac{E_{b,1} - E_{b,2}}{\frac{1-\varepsilon_1}{\varepsilon_1 A_1} + \frac{1}{A_1 F_{12}} + \frac{1-\varepsilon_2}{\varepsilon_2 A_2}} = \frac{\sigma\left(T_1^4 - T_2^4\right)}{\frac{1-\varepsilon_1}{\varepsilon_1 A_1} + \frac{1}{A_1 F_{12}} + \frac{1-\varepsilon_2}{\varepsilon_2 A_2}} \tag{15.42}$$

Figure 15.6 shows the corresponding electric analog for two surfaces in an enclosure.

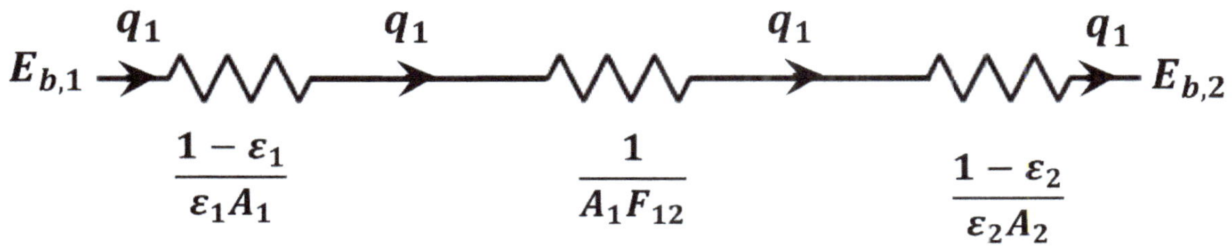

Figure 15.6 Electric analog of two radiating surfaces in an enclosure.

If the two surfaces are infinitely large parallel surfaces maintained at temperatures T_1 and T_1, $A_1 = A_2 = A$, $F_{12} = 1$, and Equation (15.42) simplifies to:

$$q_1 = \frac{\sigma A\left(T_1^4 - T_2^4\right)}{\frac{1-\varepsilon_1}{\varepsilon_1} + 1 + \frac{1-\varepsilon_2}{\varepsilon_2}} = \frac{\sigma A\left(T_1^4 - T_2^4\right)}{\frac{1}{\varepsilon_1} + \frac{1}{\varepsilon_2} - 1} \tag{15.43}$$

For systems such as two long concentric cylinders or two concentric spheres, $F_{12} = 1$. In this case, the net rate of radiative heat transfer from surface 1 to surface 2 as given by Equation (15.42) is as follows:

$$q_1 = \frac{\sigma A_1\left(T_1^4 - T_2^4\right)}{\frac{1}{\varepsilon_1} + \frac{A_1}{A_2}\left(\frac{1}{\varepsilon_2} - 1\right)} \tag{15.44}$$

In the special case of a small convex body enclosed in a large enclosure, A_1/A_2 is approximately zero. Thus:

$$q_1 = \sigma A_1 \varepsilon_1 \left(T_1^4 - T_2^4\right) \tag{15.45}$$

This same result is obtained if a small convex body is enclosed in a black enclosure. Consequently, a large enclosure can be approximately treated as a black surface.

15.3 RADIATION SHIELDING

In many practical situations, it is desired to reduce radiative heat transfer between two surfaces. This can be achieved by inserting a thin sheet of high reflectivity and low emissivity material between the two surfaces. Such thin sheets of highly reflective material are called "radiation shields". The radiation shield adds another resistance in the path of heat flow so that the overall rate of radiative heat transfer is reduced.

Insertion of any surface (regardless of its surface radiative properties) that intercepts the radiation path always reduces the net radiative heat transfer rate. However, the higher the reflectivity (that is, smaller the emissivity) of the shield, greater is the resistance associated with the shield.

Consider two infinitely large parallel surfaces maintained at temperatures T_1 and T_1 exchanging thermal radiation with each other. In the absence of any radiation shield, there are three radiative resistances in series (two surface resistances and one space resistance) and the net rate of radiative heat transfer from surface 1 to surface 2 is as follows:

$$q_1 = \frac{\sigma A\left(T_1^4 - T_2^4\right)}{\frac{1}{\varepsilon_1} + \frac{1}{\varepsilon_2} - 1} \tag{15.46}$$

When a radiation shield (large parallel sheet) is inserted between the two parallel surfaces, two additional surface resistances (corresponding to two sides of the shield) come into play (see Figure 15.7). The net rate of radiative heat transfer from surface 1 to surface 2 now becomes:

$$q_1 = \frac{\sigma\left(T_1^4 - T_2^4\right)}{\frac{1-\varepsilon_1}{\varepsilon_1 A_1} + \frac{1}{A_1 F_{13}} + \frac{1-\varepsilon_{3(1)}}{\varepsilon_{3(1)} A_3} + \frac{1-\varepsilon_{3(2)}}{\varepsilon_{3(2)} A_3} + \frac{1}{A_3 F_{32}} + \frac{1-\varepsilon_2}{\varepsilon_2 A_2}} \tag{15.47}$$

where $\varepsilon_{3(1)}$ is emissivity of radiation shield facing surface 1, $\varepsilon_{3(2)}$ is emissivity of radiation shield facing surface 2, and A_3 is surface area of the shield (same on both sides). Figure 15.8 shows the electric analog of net radiative heat transfer between two surfaces with radiation shield inserted between the surfaces.

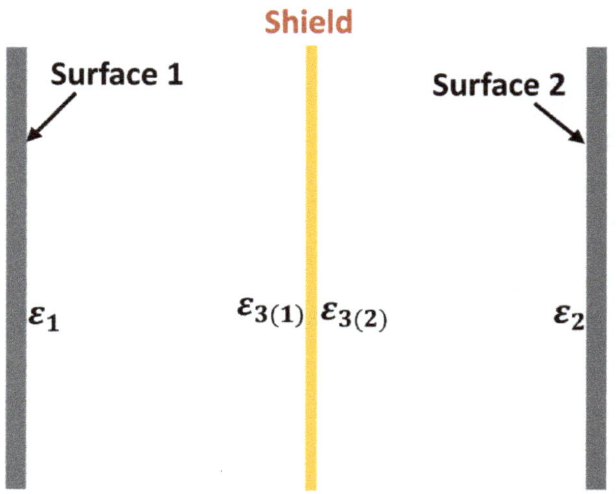

Figure 15.7 Thermal radiation exchange between large parallel plates with radiation shield.

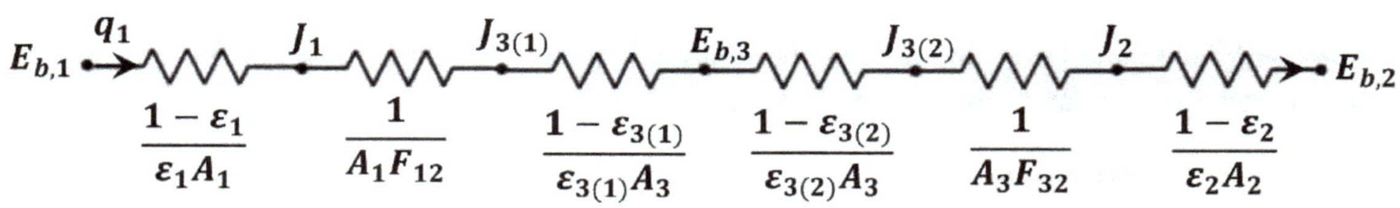

Figure 15.8 Electric analog of net radiative heat transfer between two surfaces with radiation shield inserted between the surfaces.

Assuming $A_1 = A_2 = A_3 = A$ and $\varepsilon_{3(1)} = \varepsilon_{3(2)} = \varepsilon_s$, this expression simplifies to:

$$q_1 = \frac{\sigma A(T_1^4 - T_2^4)}{\frac{1}{\varepsilon_1} + \frac{1}{\varepsilon_2} + 2\left(\frac{1}{\varepsilon_s} - 1\right)} \tag{15.48}$$

Note that $F_{13} = 1$ and $F_{32} = 1$. In the special case of emissivities of all surfaces being equal ($\varepsilon_1 = \varepsilon_2 = \varepsilon_3 = \varepsilon$), the net rates of radiative heat transfer from surface 1 to surface 2 with and without shield are given as:

$$q_1 = \frac{\sigma A(T_1^4 - T_2^4)}{2\left(\frac{2}{\varepsilon} - 1\right)} \text{ (with shield)} \tag{15.49}$$

$$q_1 = \frac{\sigma A(T_1^4 - T_2^4)}{\left(\frac{2}{\varepsilon} - 1\right)} \text{ (without shield)} \tag{15.50}$$

Thus, the insertion of a radiation shield reduces the net heat transfer rate to one-half the rate that prevails when the shield is not present.

15.4 RADIATIVE HEAT TRANSFER COEFFICIENT

Thus far, radiative heat transfer is treated as an isolated phenomenon. In many practical problems, radiation and convection occur simultaneously. In such situations it is convenient to express the net radiative heat transfer in terms of Newton's law of cooling. Assuming a two-surface enclosure, the net radiative heat transfer rate from surface 1 to surface 2 can be expressed as:

$$q_1 = h_r(T_1 - T_2)A_1 \tag{15.51}$$

where h_r is the radiative heat transfer coefficient, given as:

$$h_r = \frac{\sigma(T_1 + T_2)(T_1^2 + T_2^2)}{\frac{1-\varepsilon_1}{\varepsilon_1} + \frac{1}{F_{12}} + \left(\frac{A_1}{A_2}\right)\frac{1-\varepsilon_2}{\varepsilon_2}} \tag{15.52}$$

This expression for h_r follows from Equations (15.42) and (15.52). For a small body in a large enclosure ($F_{12} = 1$ and $A_1/A_2 = 0$), this expression reduces to:

$$h_r = \sigma\varepsilon_1(T_1 + T_2)(T_1^2 + T_2^2) \tag{15.53}$$

Note that the radiation heat transfer coefficient h_r is a strong function of temperature whereas convective heat transfer coefficient depends only weakly on temperature (through temperature dependence of properties).

When radiation and convection occur in parallel, the total rate of heat transfer can be expressed as:

$$q_{total} = q_{radiation} + q_{convection} = \frac{(T_1 - T_2)}{\left(\frac{1}{h_r A_1}\right)} + \frac{(T_1 - T_2)}{\left(\frac{1}{h_c A_1}\right)} \tag{15.54}$$

where h_c is the convective heat transfer coefficient.

15.5 RADIATION PROPERTIES OF MATERIALS

The thermal emissivity of a material is the key property involved in radiative heat transfer calculations. Some applications require that the thermal emissivity of the material be as low as possible whereas other applications require a high thermal emissivity. It is well known that metals have low values of thermal emissivity whereas polymers exhibit high values of thermal emissivity due to the presence of certain functional groups. Therefore, by mixing these two materials (metals and polymers) in different proportions, one can generate materials of any desired thermal emissivity. It is due to this reason that composites of metal filler particles and polymer matrix are used widely as coating materials of designed thermal emissivity.

Assuming transmittivity to be negligible, the absorptivity and reflectivity are related to each other as:

$$\alpha + \rho = 1 \tag{15.55}$$

From Kirchhoff's Law:

$$\varepsilon = \alpha \tag{15.56}$$

From Equations (15.55) and (15.56):

$$\varepsilon = 1 - \rho \tag{15.57}$$

Thus, one can calculate emissivity once reflectivity of a material is known. According to Equation (15.57), a high reflectivity means low emissivity.

The reflectivity of a material is related to refractive index via the Fresnel equations. For normal incidence of radiation on the surface, the Fresnel equation is given as follows:

$$\rho = \left[\frac{n-1}{n+1}\right]^2 \tag{15.58}$$

where n is the refractive index of the material which is related to dielectric constant and magnetic permeability as follows:

$$n = \sqrt{\in \mu} \tag{15.59}$$

Here $\in$ is the dielectric constant (relative permittivity) and μ is the relative magnetic permeability.

For non-magnetic materials, $\mu = 1$ and therefore,

$$n = \sqrt{\in} \tag{15.60}$$

From Equations (15.58) and (15.60), we get:

$$\rho = \frac{(\sqrt{\in}-1)^2}{(\sqrt{\in}-1)^2} \tag{15.61}$$

Thus, the emissivity of the material is related to dielectric constant as follows:

$$\varepsilon = 1 - \frac{\left(\sqrt{\epsilon}-1\right)^2}{\left(\sqrt{\epsilon}-1\right)^2} \tag{15.62}$$

It should be noted that the dielectric constant is a bulk property whereas the emissivity is a surface property.

15.6 EXAMPLES

15.6.1 Example 1

Two large parallel surfaces facing each other are exchanging thermal radiation between themselves, as shown below. The top surface is at a temperature of $T_1 = 1100\ K$. The bottom surface is at $T_2 = 520\ K$. The top surface is a black surface with $\varepsilon_1 = 1$. The bottom surface is a diffuse gray surface with $\varepsilon_2 = 0.75$.

a) Determine the irradiation and radiosity of the top surface.

b) What is the radiosity of the bottom surface.

c) What is the net radiative heat transfer between the surface per unit area of the surface.

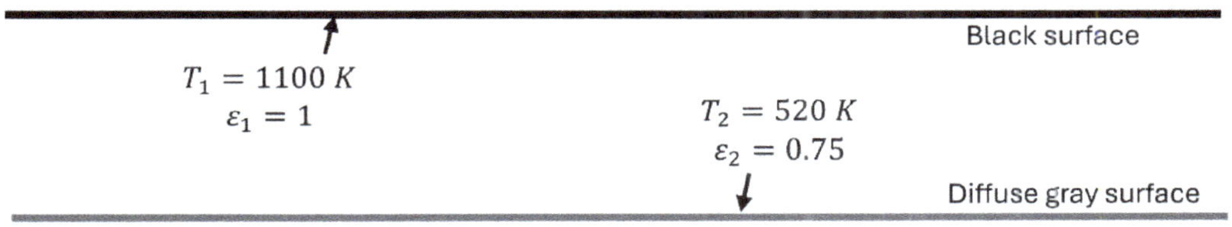

Solution:

Refer to the figure below.

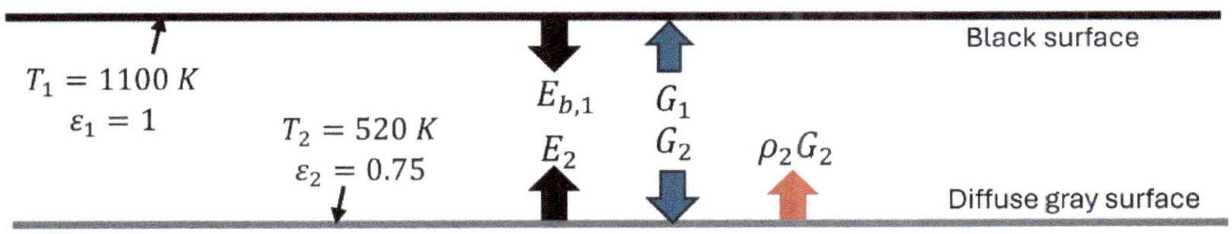

a) Determine the irradiation and radiosity of the top surface.

$$J_1 = E_{b,1} = \sigma T_1^4 = (5.67 \times 10^{-8})(1100)^4 = 8.30 \times 10^4\ W/m^2$$

$$G_1 = E_2 + \rho_2 G_2 = E_2 + \rho_2 E_{b,1} = \varepsilon_2 E_{b,2} + (1-\varepsilon_2)E_{b,1}$$
$$= \sigma[\varepsilon_2 T_2^4 - (1-\varepsilon_2)T_1^4] = (5.67 \times 10^{-8})[0.75(520^4) - (1-0.75)1100^4] = 2.386 \times 10^4\ W/m^2$$

b) What is the radiosity of the bottom surface.

$$J_2 = G_1 = 2.386 \times 10^4\ W/m^2$$

c) What is the net radiative heat transfer between the surface per unit area of the surface.

$$\frac{q_{12}}{A} = J_1 - J_2 = (8.30 - 2.386) \times 10^4 \; W/m^2$$

15.6.2 Example 2

We can model radiation from a light bulb as a black body radiation. Estimate the percentage of total energy emitted by the bulb that lies in the visible range. You can take the bulb temperature as 2800 K.

Solution:

This problem deals with band emission. Band emission is a fraction of the total emission that is in a certain wavelength interval (band) at a given temperature. The band emission $F_{(\lambda_1 \to \lambda_2)}$ over the wavelength band of $(\lambda_1 \to \lambda_2)$ is given as:

$$F_{(\lambda_1 \to \lambda_2)} = F_{(0 \to \lambda_2)} - F_{(0 \to \lambda_1)} = \frac{\int_0^{\lambda_2} E_{\lambda,b} d\lambda - \int_0^{\lambda_1} E_{\lambda,b} d\lambda}{\sigma T^4}$$

The wavelength interval (band) for the visible region is $0.38 \; \mu m \leq \lambda \leq 0.76 \; \mu m$. Thus, we need to evaluate:

$$F_{(\lambda_1 \to \lambda_2)} = \frac{\int_0^{\lambda_2} E_{\lambda,b} d\lambda - \int_0^{\lambda_1} E_{\lambda,b} d\lambda}{\sigma T^4}$$

The integrals in the above equation are more conveniently evaluated if the variable λ is changed to λT as shown below.

$$F_{(\lambda_1 \to \lambda_2)} = \int_0^{\lambda_2 T} \frac{E_{\lambda,b} d(\lambda T)}{\sigma T^5} - \int_0^{\lambda_1 T} \frac{E_{\lambda,b} d(\lambda T)}{\sigma T^5}$$

The values of the integrals on the right-hand side are available in the literature for different values of λT. They are given in a table form in Appendix D for a range λT values. In the present situation,

$$\lambda_1 T = 0.38 \times 2800 = 1064 \; \mu m. K$$

$$\lambda_2 T = 0.76 \times 2800 = 2128 \; \mu m. K$$

From tables given in the Appendix D, we get:

$$\int_0^{\lambda_1 T} \frac{E_{\lambda,b} d(\lambda T)}{\sigma T^5} = 0.0009$$

$$\int_0^{\lambda_2 T} \frac{E_{\lambda,b} d(\lambda T)}{\sigma T^5} = 0.0886$$

Consequently,

$$F_{(\lambda_1 \to \lambda_2)} = 0.0886 - 0.0009 = 0.0877$$

Thus, the percentage of total energy emitted by the bulb that lies in the visible range is 8.77.

15.6.3 Example 3

Use the Wein displacement law to estimate the surface temperature of the sun. The maximum intensity of radiation from the sun is at a wavelength of 0.50 μm based on the experimental measurements.

Solution:

The Wein displacement is:

$$\lambda_{max} T = 2.898 \times 10^{-3}\ mK$$

where λ_{max} is the value of λ where $E_{\lambda,b}$ is maximum. Since $\lambda_{max} = 0.50\ \mu$m for the sun, we get:

$$T = \frac{2.898 \times 10^{-3}}{\lambda_{max}} = \frac{2.898 \times 10^{-3}}{0.50 \times 10^{-6}} = 5796\ K$$

Thus, the surface temperature of the sun is 5796 K.

15.6.4 Example 4

The human skin does not absorb all the incident solar radiation. It is interesting to know what fraction of the incident solar radiation is absorbed by the human skin. It is known that the skin absorbs only 50 percent of the incident radiation with wavelengths in the range of $\lambda_1 = 0.52$ μm and $\lambda_2 = 1.55$ μm. Outside this wavelength range, the radiation is fully absorbed. Thus, radiation in the wavelength ranges of $0 \leq \lambda \leq 0.52$ μm and $1.55\ \mu m \leq \lambda \leq \infty$ μm is fully absorbed.

Estimate the fraction of the incident solar radiation that is absorbed by the human skin. You can model solar radiation as a black body radiation at a temperature T = 5800 K.

Solution:

Fraction of incident solar radiation absorbed by human skin $= 1 - 0.5 \times F_{(\lambda_1 \to \lambda_2)}$ where

$$F_{(\lambda_1 \to \lambda_2)} = \int_0^{\lambda_2 T} \frac{E_{\lambda,b} d(\lambda T)}{\sigma T^5} - \int_0^{\lambda_1 T} \frac{E_{\lambda,b} d(\lambda T)}{\sigma T^5}$$

$\lambda_1 = 0.52$ μm, $T = 5800\ K$, $\lambda_1 T = 3016$ μm.K

$\lambda_2 = 1.55$ μm, $T = 5800\ K$, $\lambda_2 T = 8990$ μm.K

From tables in Appendix D

$$\int_0^{\lambda_1 T} \frac{E_{\lambda,b} d(\lambda T)}{\sigma T^5} = 0.2768$$

$$\int_0^{\lambda_2 T} \frac{E_{\lambda,b} d(\lambda T)}{\sigma T^5} = 0.8897$$

$$F_{(\lambda_1 \to \lambda_2)} = \int_0^{\lambda_2 T} \frac{E_{\lambda,b} d(\lambda T)}{\sigma T^5} - \int_0^{\lambda_1 T} \frac{E_{\lambda,b} d(\lambda T)}{\sigma T^5} = 0.8897 - 0.2768 = 0.6129$$

Thus, the fraction of incident solar radiation absorbed by human skin $= 1 - 0.5 \times F_{(\lambda_1 \to \lambda_2)} = 1 - 0.5 \times 0.6129 = 0.6936$

15.6.5 Example 5

The surface temperature of a horizontal 4-inch OD (outer diameter) pipe carrying steam is 125 °C. To minimize the radiative heat loss from the pipe, a cylindrical radiation shield is installed around the pipe, see figure. The gap between the tube and the shield is 12.5 mm. At steady state, the surface temperature of the shield reaches 37 °C. Assuming that the pipe and the shield are diffuse, gray surfaces with emissivities of 0.82 and 0.12, respectively, estimate the net radiative rate of heat loss from the pipe per unit length? Show the electrical analog of this thermal problem.

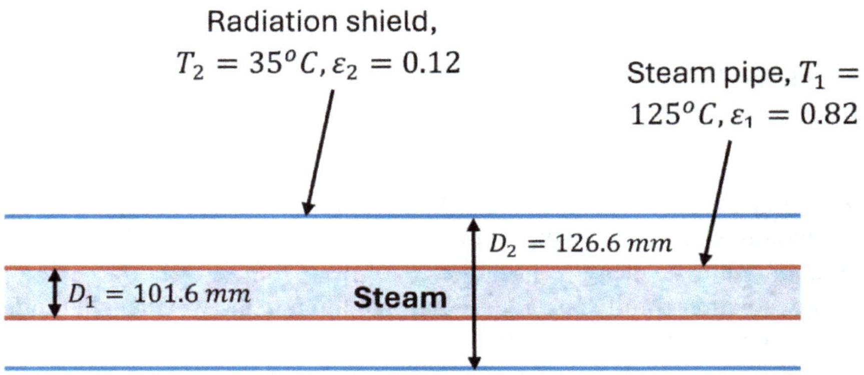

Solution:

This involves a two-surface enclosure. The electric analog of the heat transfer problem is as follows:

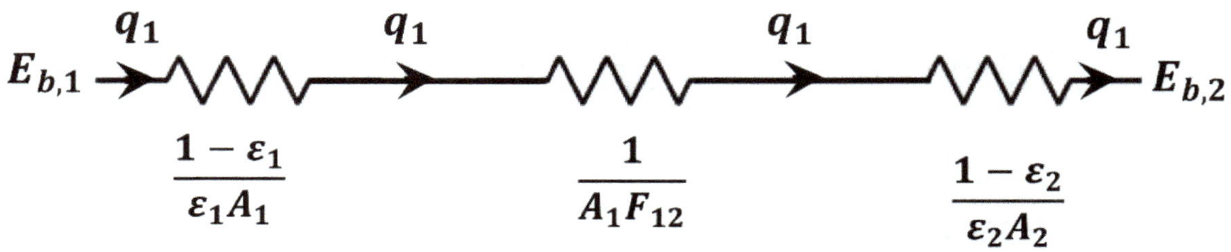

Thus,

$$q_1 = q_{12} = \frac{E_{b,1} - E_{b,2}}{\frac{1-\varepsilon_1}{\varepsilon_1 A_1} + \frac{1}{A_1 F_{12}} + \frac{1-\varepsilon_2}{\varepsilon_2 A_2}} = \frac{\sigma(T_1^4 - T_2^4)}{\frac{1-\varepsilon_1}{\varepsilon_1 A_1} + \frac{1}{A_1 F_{12}} + \frac{1-\varepsilon_2}{\varepsilon_2 A_2}}$$

For the given system of two long concentric cylinders, $F_{12} = 1$. Consequently, the net rate of radiative heat transfer from surface 1 to surface 2 is as follows:

$$q_1 = \frac{\sigma A_1 (T_1^4 - T_2^4)}{\frac{1}{\varepsilon_1} + \frac{A_1}{A_2}\left(\frac{1}{\varepsilon_2} - 1\right)}$$

$$\frac{q_1}{L} = \frac{\sigma(\pi D_1)(T_1^4 - T_2^4)}{\frac{1}{\varepsilon_1} + \frac{D_1}{D_2}\left(\frac{1}{\varepsilon_2} - 1\right)} = \frac{5.67\times10^{-8}(\pi\times101.6\times10^{-3})(398.15^4 - 308.15^4)}{\frac{1}{0.82} + \frac{101.6}{126.6}\left(\frac{1}{0.12} - 1\right)} = \frac{291.61}{7.105} = 41.04 \text{ W/m}$$

15.6.6 Example 6

A large black surface at -18 °C exchanges thermal radiation with a parallel surface (diffuse, gray) of emissivity of 0.49 at 820 °C. If a radiation shield is placed between the two surfaces (see figure below) determine the percent reduction in the net radiative heat transfer between the surfaces. Also determine the temperature of the radiation shield. You can treat the radiation shield as a diffuse gray body with emissivity on both sides as 0.075.

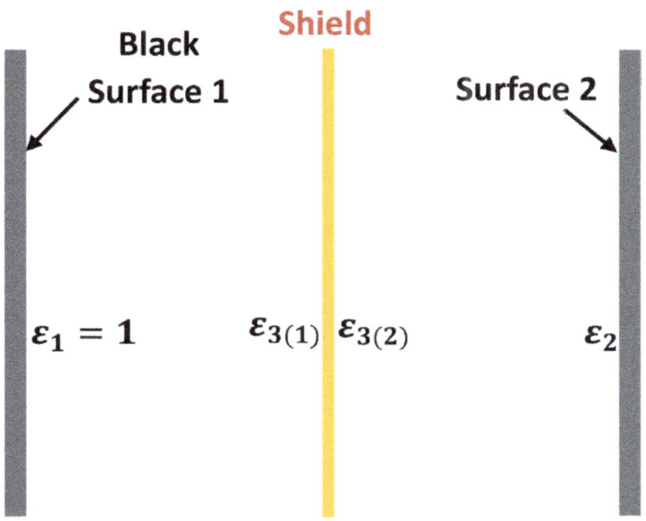

Solution:

No radiation shield present

The electric analog of the problem when no radiation shield is present involves only two resistors in series, as shown below.

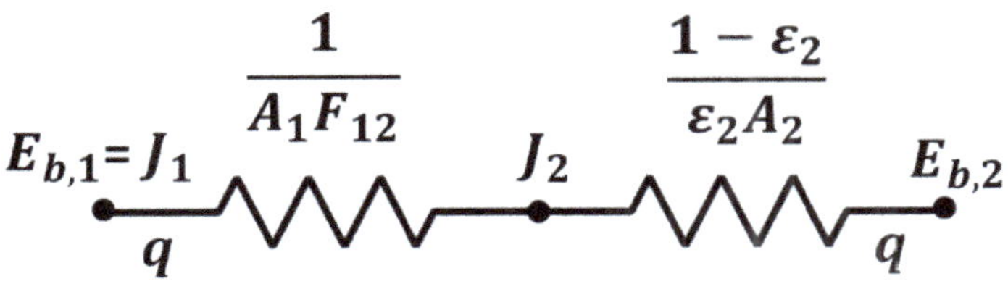

$$q = \frac{E_{b,1} - E_{b,2}}{\frac{1}{A_1F_{12}} + \frac{1-\varepsilon_2}{\varepsilon_2 A_2}} = \frac{\sigma A_1(T_1^4 - T_2^4)}{1 + \left(\frac{A_1}{A_2}\right)\frac{1-\varepsilon_2}{\varepsilon_2}}$$

As $A_1 = A_2 = A$,

$$\frac{q}{A}=\frac{\sigma(T_1^4-T_2^4)}{1+\dfrac{1-\varepsilon_2}{\varepsilon_2}}=\frac{(5.67\times10^{-8})(255.15^4-1093.15^4)}{1+\dfrac{1-0.49}{0.49}}=-3.956\times10^4\ W/m^2$$

Note that q is negative as it flows from surface 2 to surface 1.

Radiation shield present

The electric analog of the problem when radiation shield is present involves five resistors in series, as shown below.

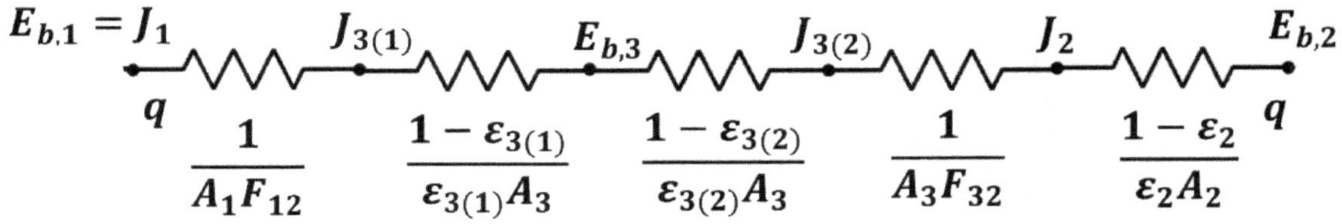

The net rate of radiative heat transfer from surface 1 to surface 2 now becomes:

$$q=\frac{\sigma(T_1^4-T_2^4)}{\dfrac{1}{A_1F_{13}}+\dfrac{1-\varepsilon_{3(1)}}{\varepsilon_{3(1)}A_3}+\dfrac{1-\varepsilon_{3(2)}}{\varepsilon_{3(2)}A_3}+\dfrac{1}{A_3F_{32}}+\dfrac{1-\varepsilon_2}{\varepsilon_2A_2}}$$

Given that: $A_1=A_3=A_2=A$, $\varepsilon_{3(1)}=\varepsilon_{3(2)}=0.075$, $\varepsilon_2=0.49$. Also, $F_{13}=F_{32}=1$. Thus,

$$\frac{q}{A}=\frac{\sigma(T_1^4-T_2^4)}{1+\dfrac{1-\varepsilon_{3(1)}}{\varepsilon_{3(1)}}+\dfrac{1-\varepsilon_{3(2)}}{\varepsilon_{3(2)}}+1+\dfrac{1-\varepsilon_2}{\varepsilon_2}}=\frac{(5.67\times10^{-8})(255.15^4-1093.15^4)}{2+2\left(\dfrac{1-0.075}{0.075}\right)+\left(\dfrac{1-0.49}{0.49}\right)}$$
$$=-2.9135\times10^3\ W/m^2$$

Percent reduction in net radiative heat transfer between the surfaces

$$=\left(\frac{3.956\times10^4-2.9135\times10^3}{3.956\times10^4}\right)\times100=92.64\%$$

Temperature of the radiation shield

From the electric analog, we can express:

$$\frac{q}{A}=\frac{E_{b,1}-E_{b,3}}{\dfrac{1}{F_{12}}+\dfrac{1-\varepsilon_{3(1)}}{\varepsilon_{3(1)}}}=\frac{E_{b,1}-E_{b,3}}{1+\left(\dfrac{1-0.075}{0.075}\right)}=-2.9135\times10^3\ W/m^2$$

Thus,

$$E_{b,1}-E_{b,3}=-2.9135\times10^3\times13.3333=-3.8847\times10^4\quad W/m^2$$

$$E_{b,3}=E_{b,1}+3.8847\times10^4=(5.67\times10^{-8})(255.15^4)+3.8847\times10^4=39087.306\ W/m^2$$

$$E_{b,3} = \sigma T_s^4 = 39087.306\ W/m^2$$

Hence, the temperature of the radiation shield T_s is:

$$T_s = 911.2\ K = 638.05^oC$$

15.6.7 Example 7

A horizontal pipe carrying a hot fluid is losing heat by radiation and natural convection. The outside diameter of the pipe is 5.5 inches, and the outer surface temperature is 100 °F. The pipe is in a large enclosure with surrounding walls and air at 80 °F. The emissivity of the pipe surface (diffuse, gray) is 0.93 and σ = 0.1714 x 10^{-8} Btu/(h.ft^2. °R^4).

Estimate:

i) the radiation heat transfer coefficient in Btu/(h.ft^2 °R)

ii) the average heat transfer coefficient for natural convection in Btu/(h.ft^2 °R)

iii) the total rate of heat loss from a unit length of a pipe in Btu/(h.ft)

The properties of air to be used for the calculation of natural convection heat transfer coefficient are given as: Pr = 0.73; Ra_d = 3.4x10^6; k = 0.0152 Btu/(h.ft.°R)

Solution:

i) the radiation heat transfer coefficient in Btu/(h.ft^2 °R)

For a small body (pipe) in a large enclosure ($F_{12} = 1$ and $A_1/A_2 = 0$), the radiative heat transfer coefficient is:

$$h_r = \sigma\varepsilon_1(T_1 + T_2)(T_1^2 + T_2^2)$$

where $\sigma = 0.1714 \times 10^{-8}$ Btu/(h.ft^2.°R^4), $\varepsilon_1 = 0.93$, $T_1 = 100 + 460 = 560$ °R, $T_2 = 80 + 460 =$ 540 °R. Thus,

$$h_r = (0.1714 \times 10^{-8}) \times 0.93 \times (560 + 540)(560^2 + 540^2) = 1.06117 \text{ Btu/(h.ft}^2\text{ °R)}$$

ii) the average heat transfer coefficient for natural convection in Btu/(h.ft^2 °R)

Use the following Churchill and Chu equation,

$$N\bar{u}_D = \frac{\bar{h}D}{k} = \left\{0.60 + \frac{0.387Ra_D^{1/6}}{[1+(0.559/Pr)^{9/16}]^{8/27}}\right\}^2 = \left\{0.60 + \frac{0.387(3.4\times10^6)^{1/6}}{[1+(0.559/0.73)^{9/16}]^{8/27}}\right\}^2 = 20.6857$$

$$\bar{h} = N\bar{u}_D\left(\frac{k}{D}\right) = 20.6857\left(\frac{0.0152}{5.5/12}\right) = 0.686 \text{ Btu/(h.ft}^2\text{ °R)}$$

iii) the total rate of heat loss from a unit length of a pipe in Btu/(h.ft)

$$q_{total} = q_{radiation} + q_{convection} = \frac{(T_1 - T_2)}{\left(\frac{1}{h_r A_1}\right)} + \frac{(T_1 - T_2)}{\left(\frac{1}{h_c A_1}\right)} = (h_r + h_c)A_1(T_1 - T_2)$$

$$\frac{q_{total}}{L} = (h_r + h_c)(\pi D)(T_1 - T_2) = (1.06117 + 0.686) \times \pi \times \left(\frac{5.5}{12}\right)(100 - 80) = 50.315 \text{ Btu/h}$$

15.6.8 Example 8

Two square parallel surfaces 1.0 by 1.0 m are spaced 1 m apart. One surface (surface 1) is maintained at 1100 °C and the other surface (surface 2) is maintained at 550 °C. The emissivities of the surfaces (diffuse, gray) are: $\varepsilon_1 = 0.25$ and $\varepsilon_2 = 0.55$. The surfaces are situated in a large room whose walls are at 25 °C. The surfaces facing each other are exchanging thermal radiation between themselves and with the room walls.

Estimate:

i) the rate of heat loss from surface 1

ii) the rate of heat loss from surface 2

iii) the rate of heat gain by the room

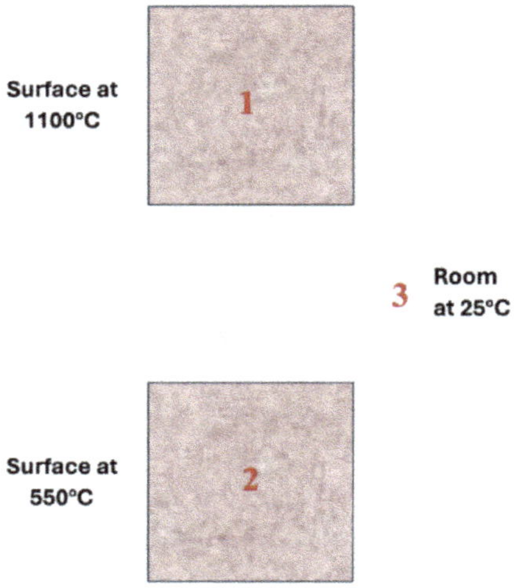

Solution:

Refer to the electric analog of the problem given below.

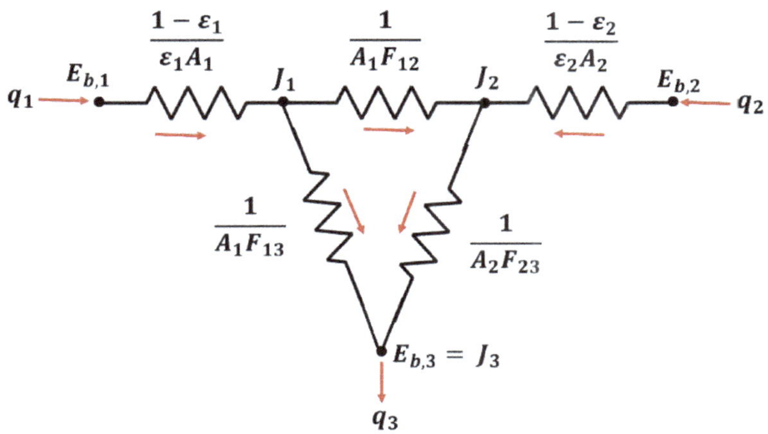

Given that:
$T_1 = 1100^oC + 273.15 = 1373.15\ K$; $T_2 = 550^oC + 273.15 = 823.15\ K$; $T_3 = 25^oC + 273.15 = 298.15\ K$

$A_1 = A_2 = 1\ m^2$; $\varepsilon_1 = 0.25, \varepsilon_2 = 0.55$

Surface 3 (large room) has a large area. Thus, its surface resistance is negligible. As $\frac{1-\varepsilon_3}{\varepsilon_3 A_3} \to 0$, $E_{b,3} = J_3$ as shown in the electric analog.

Shape factors:

From geometric considerations, $F_{12} \cong 0.20$, $F_{11} = 0$, $F_{22} = 0$. Using summation rules for enclosures:

$$F_{11} + F_{12} + F_{13} = 0 \ ; \quad F_{13} = 1 - F_{12} = 1 - 0.2 = 0.80$$

$$F_{21} + F_{22} + F_{23} = 0 \ ; \quad F_{23} = 1 - F_{21} = 1 - 0.2 = 0.80$$

Note that reciprocity relation gives: $F_{12} = F_{21}$

Resistances:

$$\frac{1-\varepsilon_1}{\varepsilon_1 A_1} = \frac{1-0.25}{0.25 \times 1} = 3$$

$$\frac{1}{A_1 F_{12}} = \frac{1}{1 \times 0.20} = 5$$

$$\frac{1-\varepsilon_2}{\varepsilon_2 A_2} = \frac{1-0.55}{0.55 \times 1} = 0.8182$$

$$\frac{1}{A_1 F_{13}} = \frac{1}{1 \times 0.80} = 1.25$$

$$\frac{1}{A_2F_{23}} = \frac{1}{1 \times 0.80} = 1.25$$

Calculate the unknown radiosities J_1 and J_2:

To calculate the unknown radiosities J_1 and J_2, we use the fact that the sum of heat currents entering a node is zero.

At node J_1 (see figure),

$$\frac{E_{b,1} - J_1}{\left(\frac{1-\varepsilon_1}{\varepsilon_1 A_1}\right)} + \frac{J_2 - J_1}{\left(\frac{1}{A_1F_{12}}\right)} + \frac{E_{b,3} - J_1}{\left(\frac{1}{A_1F_{13}}\right)} = 0$$

At node J_2 (see figure),

$$\frac{J_1 - J_2}{\left(\frac{1}{A_1F_{12}}\right)} + \frac{E_{b,2} - J_2}{\left(\frac{1-\varepsilon_2}{\varepsilon_2 A_2}\right)} + \frac{E_{b,3} - J_2}{\left(\frac{1}{A_2F_{23}}\right)} = 0$$

$$E_{b,1} = \sigma T_1^4 = (5.67 \times 10^{-8})(1373.15)^4 = 2.016 \times 10^5 \quad W/m^2$$

$$E_{b,2} = \sigma T_2^4 = (5.67 \times 10^{-8})(823.15)^4 = 2.603 \times 10^4 \quad W/m^2$$

$$E_{b,3} = \sigma T_3^4 = (5.67 \times 10^{-8})(298.15)^4 = 4.48 \times 10^2 \quad W/m^2$$

Thus, the node equations become:

$$\frac{2.016 \times 10^5 - J_1}{(3)} + \frac{J_2 - J_1}{(5)} + \frac{4.48 \times 10^2 - J_1}{(1.25)} = 0$$

$$\frac{J_1 - J_2}{(5)} + \frac{2.603 \times 10^4 - J_2}{(0.8182)} + \frac{4.48 \times 10^2 - J_2}{(1.25)} = 0$$

Upon simplification, the node equations reduce to:

$$1266.72 - 25J_1 + 3.75J_2 = 0$$

$$1.02275J_1 - 11.36375J_2 + 164.52027 = 0$$

In the above equations, the units of radiosities have been changed from W/m^2 to kW/m^2. The equations can be solved easily for the unknowns J_1 and J_2. The solution is:

$$J_1 = 53.5636 \quad kW/m^2 \;\; ; \;\; J_2 = 19.2984 \quad kW/m^2$$

i) the rate of heat loss from surface 1

$$q_1 = \frac{E_{b,1} - J_1}{\left(\frac{1-\varepsilon_1}{\varepsilon_1 A_1}\right)} = \frac{201.6 - 53.5636}{3} = 49.3455\ kW$$

ii) the rate of heat loss from surface 2

$$q_2 = \frac{E_{b,2} - J_2}{\left(\frac{1-\varepsilon_2}{\varepsilon_2 A_2}\right)} = \frac{26.03 - 19.2984}{0.8182} = 8.2273\ kW$$

iii) the rate of heat gain by the room

$$q_3 = \frac{J_1 - E_{b,3}}{\left(\frac{1}{A_1 F_{13}}\right)} + \frac{J_2 - E_{b,3}}{\left(\frac{1}{A_2 F_{23}}\right)} = \frac{53.5636 - 0.448}{1.25} + \frac{19.2984 - 0.448}{1.25} = 42.49 + 15.08 = 57.57\ kW$$

Note that as expected, $q_3 = q_1 + q_2 = 49.3455 + 8.2273 = 57.57\ kW$

15.6.9 Example 9

Consider a long duct of an equilateral triangular shaped cross-section with 1.2 m side. One surface of the duct is heated and maintained at 1225 K. One surface is insulated, and the third surface is maintained at 525 K. The heated and insulated surfaces have an emissivity of 0.75 and the third surface has an emissivity of 0.39.

i) At steady state, what is the rate of heat supplied to the heated side of the duct per unit length?

ii) What are the radiosities J_1 and J_3?

iii) What is the temperature T_2 of the insulated surface?

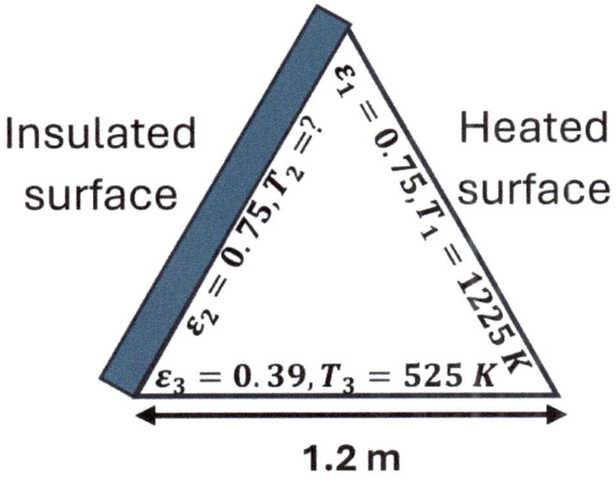

Solution:

Refer to the electric analog of the problem given below.

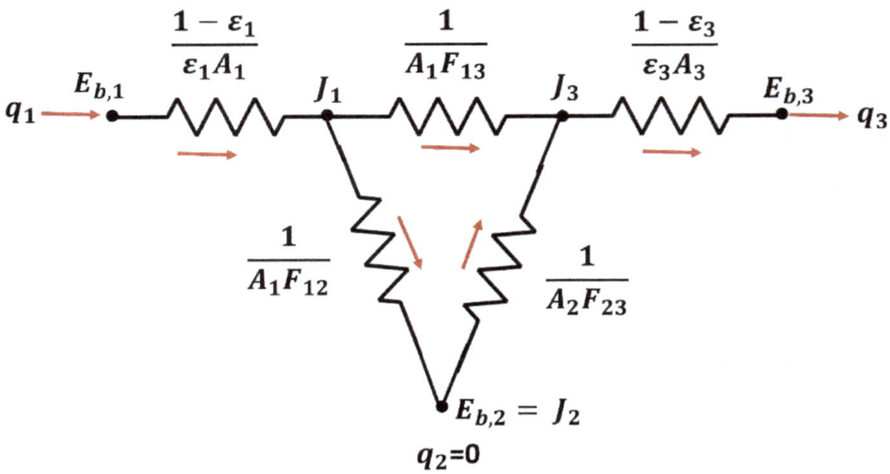

Using series-parallel type of approach for resistors, we can re-draw the above thermal circuit as follows:

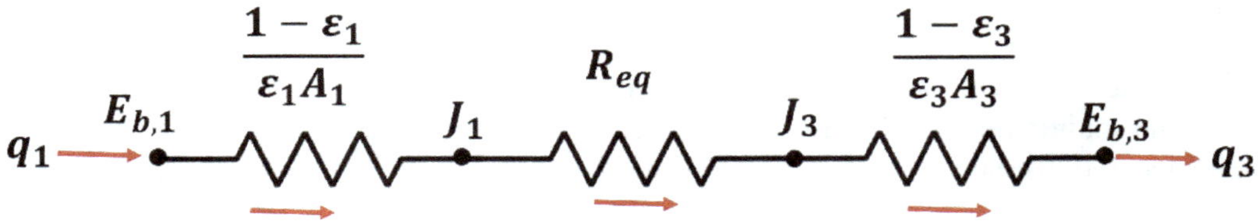

where the equivalent resistance R_{eq} is given as:

$$R_{eq} = \frac{1}{A_1F_{13} + \left(\frac{1}{A_1F_{12}} + \frac{1}{A_2F_{23}}\right)^{-1}}$$

Thus,

$$q_1 = q_3 = \frac{E_{b,1} - E_{b,3}}{\left(\frac{1-\varepsilon_1}{\varepsilon_1 A_1}\right) + \frac{1}{A_1F_{13} + \left(\frac{1}{A_1F_{12}} + \frac{1}{A_2F_{23}}\right)^{-1}} + \left(\frac{1-\varepsilon_3}{\varepsilon_3 A_3}\right)}$$

Given that:

$T_1 = 1225\ K\ ;\ T_2 = ?\ ;\ T_3 = 525\ K$

$A_1 = A_2 = A_3 = 1.2 \times 1 = 1.2\ m^2\ ;\ \varepsilon_1 = 0.75,\ \varepsilon_2 = 0.75,\ \varepsilon_3 = 0.39$

$E_{b,1} = \sigma T_1^4 = (5.67 \times 10^{-8})(1225)^4 = 1.2768 \times 10^5\ \ W/m^2$

$E_{b,3} = \sigma T_3^4 = (5.67 \times 10^{-8})(525)^4 = 4.307 \times 10^3\ \ W/m^2$

Shape factors:

From geometric considerations,

$$F_{12} = F_{13} = F_{23} = 0.5$$

Resistances:

$$\frac{1-\varepsilon_1}{\varepsilon_1 A_1} = \frac{1-0.75}{0.75 \times 1.2} = 0.2778$$

$$\frac{1}{A_1 F_{12}} = \frac{1}{1.2 \times 0.50} = 1.667$$

$$\frac{1}{A_1 F_{13}} = \frac{1}{1.2 \times 0.50} = 1.667$$

$$\frac{1}{A_2 F_{23}} = \frac{1}{1.2 \times 0.50} = 1.667$$

$$\frac{1-\varepsilon_3}{\varepsilon_3 A_3} = \frac{1-0.39}{0.39 \times 1.2} = 1.3034$$

$$R_{eq} = \frac{1}{A_1 F_{13} + \left(\frac{1}{A_1 F_{12}} + \frac{1}{A_2 F_{23}}\right)^{-1}} = \frac{1}{\frac{1}{1.667} + (1.667 + 1.667)^{-1}} = 1.1113$$

i) At steady state, what is the rate of heat supplied to the heated side of the duct per unit length?

Calculate the unknown q_1 and q_3:

$$q_1 = q_3 = \frac{E_{b,1} - E_{b,3}}{\left(\frac{1-\varepsilon_1}{\varepsilon_1 A_1}\right) + \frac{1}{A_1 F_{13} + \left(\frac{1}{A_1 F_{12}} + \frac{1}{A_2 F_{23}}\right)^{-1}} + \left(\frac{1-\varepsilon_3}{\varepsilon_3 A_3}\right)} = \frac{1.2768 \times 10^5 - 4.307 \times 10^3}{0.2778 + 1.1113 + 1.3034} = \frac{123373}{2.6925} = 45820.98\ W$$

ii) What are the radiosities J_1 and J_3?

$$q_1 = \frac{E_{b,1} - J_1}{\left(\frac{1-\varepsilon_1}{\varepsilon_1 A_1}\right)}$$

$$J_1 = E_{b,1} - q_1\left(\frac{1-\varepsilon_1}{\varepsilon_1 A_1}\right) = 1.2768 \times 10^5 - 45820.98 \times 0.2778 = 1.1495 \times 10^5\ W/m^2$$

$$q_3 = \frac{J_3 - E_{b,3}}{\left(\frac{1-\varepsilon_3}{\varepsilon_3 A_3}\right)}$$

$$J_3 = E_{b,3} + q_3\left(\frac{1-\varepsilon_3}{\varepsilon_3 A_3}\right) = 4.307 \times 10^3 + 45820.98 \times 1.3034 = 6.403 \times 10^4\ W/m^2$$

iii) What is the temperature T_2 of the insulated surface?

Consider node J_2 (see figure),

$$\frac{J_1 - E_{b,2}}{\left(\frac{1}{A_1 F_{12}}\right)} + \frac{J_3 - E_{b,2}}{\left(\frac{1}{A_2 F_{23}}\right)} = 0$$

$$\frac{1.1495 \times 10^5 - E_{b,2}}{1.667} + \frac{6.403 \times 10^4 - E_{b,2}}{1.667} = 0$$

$$E_{b,2} = 89490\ W/m^2$$

$$E_{b,2} = \sigma T_2^4 = 89490\ W/m^2$$

Thus,

$$T_2 = \left(\frac{89490}{5.67 \times 10^{-8}}\right)^{1/4} = 1120.851\ K$$

NOTATION

A_i	radiation area of the surface i
c	speed of electromagnetic radiation
c_o	speed of electromagnetic radiation in vacuum
e	energy of photon
E	emissive power of a surface
E_b	emissive power of a black surface
E_λ	monochromatic emissive power of a surface
$E_{\lambda,b}$	monochromatic emissive power of black surface
F_{ij}	view factor, fraction of radiant energy leaving surface i that is intercepted by surface j
$F_{(\lambda_1 \to \lambda_2)}$	band emission over the wavelength band of $(\lambda_1 \to \lambda_2)$
G	total incident radiation flux on a surface from all directions and over all wave lengths
G_λ	spectral irradiation. It is the rate at which radiation energy is incident on a surface per unit wavelength interval about λ from all directions per unit surface area.
G_{abs}	portion of incident radiation that is absorbed
G_{ref}	portion of incident radiation that is reflected
G_{trans}	portion of incident radiation that is transmitted
h	Planck's constant
h_c	convective heat transfer coefficient
h_r	radiative heat transfer coefficient
J	total radiosity
J_i	total radiosity of surface i
J_λ	spectral radiosity. It is the rate at which radiation energy leaves a surface per unit wavelength interval about λ from all directions per unit surface area
k	Boltzmann constant
n	refractive index of the material
N	number of surfaces forming the enclosure
q_i	net rate of radiative heat transfer from surface i
q_{ij}	net rate of radiative heat transfer from surface i to surface j
$q_{i \to j}$	radiation energy leaving surface i and being intercepted by surface j

$q_{radiation}$ net rate of radiative heat transfer
$q_{convection}$ rate of convective heat transfer
q_{total} total rate of heat transfer
R_{total} total resistance to heat transfer
T absolute temperature

Greek Letters

α total hemispherical absorptivity of the surface
$\alpha_{\lambda,\theta}$ monochromatic directional absorptivity of the surface
ε total hemispherical emissivity of the surface
$\varepsilon_{\lambda,\theta}$ monochromatic directional emissivity of a surface
λ wavelength
ν frequency of radiation
μ relative magnetic permeability
ρ total reflectivity
σ Stefan-Boltzmann constant
τ total transmittivity

PRACTICE PROBLEMS

Problem 1

The sun could be modelled as a blackbody radiator with a surface temperature of approximately 5800 K. Assuming sun to be a black surface, estimate the fraction of total energy emitted by the sun that is within the visible region $38\ \mu m \leq \lambda \leq 76\ \mu m$.

Problem 2

Derive the Wein displacement law using Planck's distribution of spectral emissive power emitted by a blackbody, that is, Equation (15.12).

Problem 3

Derive the Stefan-Boltzmann using Planck's distribution of spectral emissive power emitted by a blackbody, that is, Equation (15.12).

Problem 4

The surface of a spacecraft is maintained at 4.5 °C. The surface can be treated as a gray diffuse body with an emissivity of 0.09. The spacecraft is surrounded by space which can be treated as a black body with absolute zero temperature. Calculate the rate at which heat is lost by the spacecraft to the surroundings due to radiation per unit surface area of the craft.

Problem 5

Consider two very large parallel surfaces maintained at temperatures of 1100 K and 525 K. The emissivity of the surface with a temperature of 1100 K is 0.25 whereas the emissivity of the surface with a temperature of 525 K is 0.75. What is the net rate of radiation heat transfer between the surfaces. What is the percentage reduction in the radiation heat transfer between the surfaces when a radiation shield with the same emissivity of 0.045 on both sides is placed between them?

Problem 6

Consider two concentric spherical surfaces of diameters D_1=0.75 m and D_2=1.25 m maintained at surface temperatures of T_1 = 398 K and T_2 = 296 K. They are separated by a vacuum.

i) What is the net rate of radiation heat transfer between the surfaces if they are treated as black surfaces?

ii) What is the net rate of radiation heat transfer between the surfaces if the surfaces are diffuse and gray with emissivities of $\varepsilon_1 = 0.49$ and ε_2=0.045?

iii) How would the net rate of radiation heat transfer change if D_2 is increased from 1.25 m to 23 m keeping $\varepsilon_1 = 0.49$ and ε_2=0.045 and D_1=0.75m? In this case, can we treat the outer spherical surface as black surface?

Problem 7

Consider two square surfaces 75 cm by 75 cm that are placed perpendicular to each other as shown in the figure below. Surface 1 is maintained at $T_1 = 1100\ K$ and has an emissivity of $\varepsilon_1 = 0.58$. The surface 2 is insulated. The two surfaces are placed in a large room with walls maintained at $T_3 = 298\ K$. The surfaces are exchanging radiation between themselves and with the room walls.

i) What is the net rate of radiation heat loss from surface 1?

ii) Estimate the temperature of the insulated surface?

Note that this is a 3-surface enclosure problem. From geometrical considerations, the shape factor is $F_{12} \cong 0.20$. Also note that $F_{11} = 0$ and $F_{22} = 0$.

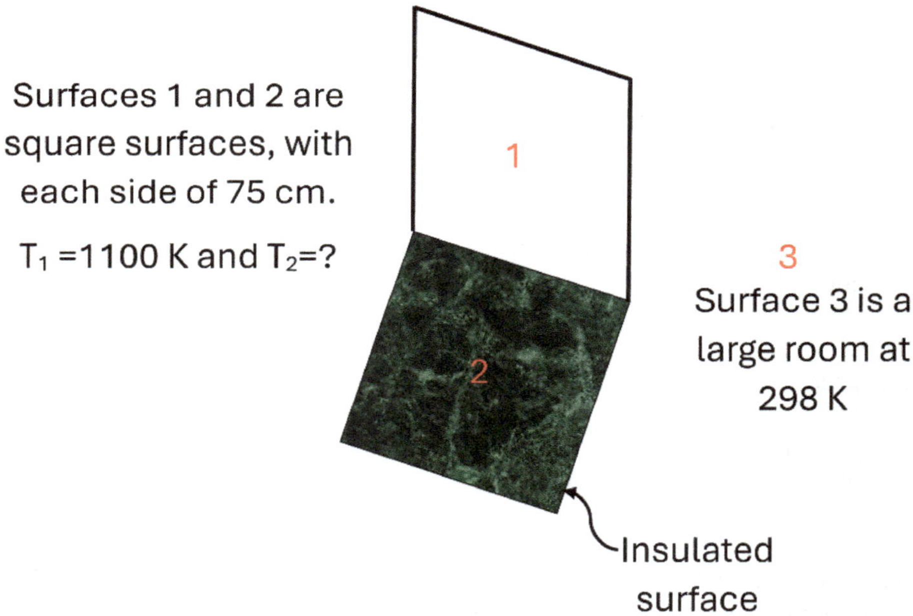

Problem 8

For heat transfer purposes, a human body can be modelled as a vertical cylinder 1.8 m tall and o.3 m in diameter. The surface temperature of the cylinder is 25 °C, the natural convective heat transfer coefficient is 15 W/m^2.K, the surface emissivity of the cylinder is 0.88. The cylinder is situated in a large room where the air temperature is 18 °C and the wall temperature is 7 °C. Estimate the rate of heat loss from the cylinder considering both natural convection and radiative heat transfer. What happens when the wall temperature is increased from 7 °C to 25 °C?

Problem 9

Consider a long rod, 5 cm in diameter, installed in a vacuum furnace. The rod is electrically heated to maintain a surface temperature of 1200 K. The walls of the furnace are at a temperature of 750 K. The rod is a gray diffuse body with an emissivity of 0.90.

i) Estimate the radiation heat transfer coefficient for the rod

ii) What is the net rate of radiative heat transfer from the rod per unit length.

Problem 10

Consider a steam pipe passing through a large enclosure in which the air and walls are at 22 °C. The outside diameter of the pipe is 76 mm. There is no insulation on the pipe and its surface temperature is 179 °C. The emissivity of the pipe surface is 0.75. In addition to radiative heat loss, the pipe also loses heat by natural convection to air at a free stream temperature of 22 °C. The natural convection heat transfer coefficient is approximately 14.5 $W/m^2.K$.

Estimate the rate of heat loss from the pipe per unit length?

FURTHER READING

1. T.L. Bergman, A.S. Levine, F.P. Incropera, and D.P. Dewitt. *Fundamentals of Heat and Mass Transfer*. 7th Edition, New York, Wiley & Sons, 2011.
2. R.B. Bird, W.E. Stewart, and E.N. Lightfoot. 2007. *Transport Phenomena*. 2nd Edition, Wiley & Sons, New York, 2007.
3. J.R. Welty, C.E. Wicks, R.E. Wilson, and G.L. Rorrer. *Fundamentals of Momentum, Heat, and Mass Transfer*. 5th Edition, Wiley & Sons, New York, 2008.
4. F. Keith, M.S. Bohn. *Principles of Heat Transfer*. 5th Edition, West Publishing Co., St. Paul, MN, USA, 1993.
5. A. Bejan. *Heat Transfer*. Wiley & Sons, New York, 1993.
6. J.P. Holman. *Heat Transfer*, 10th Edition, McGraw Hill, New York, 2009.
7. Y.A. Cengel. *Heat and Mass Transfer*, 3rd Edition, McGraw Hill, New York, 2007.
8. L.C. Thomas. *Heat Transfer*. Prentice Hall, New Jersey, 1992.
9. W.S. Janna. *Engineering Heat Transfer*. PWS Publishers, Boston, 1986.
10. C.J. Geankoplis, A.A. Hersel, D.H. Lepek. *Transport Processes and Separation Process Principles*. 5th Edition, Prentice Hall, New Jersey, 2018.
11. W.L. McCabe, J.C. Smith, P. Harriott. *Unit Operations of Chemical Engineering*. 6th Edition, McGraw Hill, New York, 2001.
12. D.R.Gaskell. *An Introduction to Transport Phenomena in Materials Engineering*. Macmillan Publishing Company, New York, 1992.
13. R. Pal. *Electromagnetic, Mechanical, and Transport Properties of Composite Materials*. CRC Press, Boca Raton, Florida, 2015.

Part 4

Mass Transport

Chapter 16

Mass Transport in a Stationary Medium

When different parts of a material medium have a different concentration of some species A, that is, concentration gradient of species A exists in the medium, species A moves (diffuses) spontaneously from a region at high concentration to a region at low concentration. The driving force for this molecular mass transport of species A in the direction of decreasing concentration of A is the concentration gradient in A. Larger the concentration gradient in A, faster is the rate of diffusion of A. Figure 16.1 shows the diffusion process schematically. The species A is diffusing through a stationary matrix of species B.

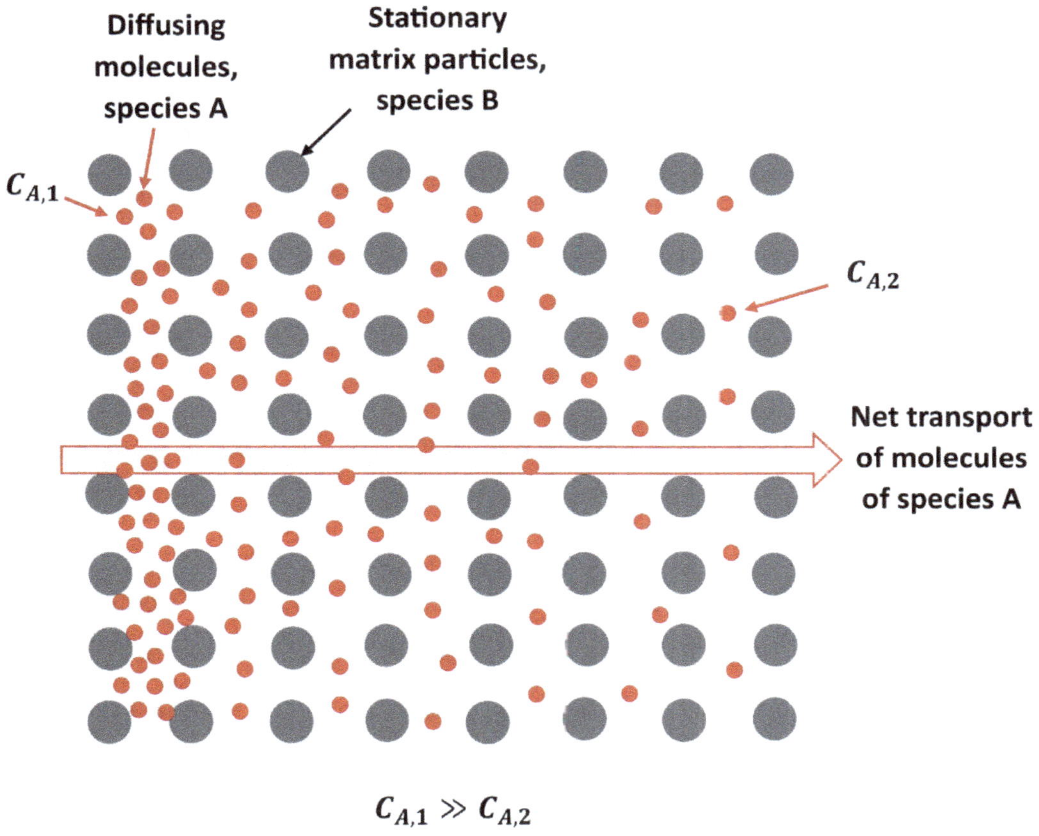

Figure 16.1 Diffusion of species A in a stationary medium B.

Figure 16.2 shows how a single molecule of species A diffuses from left to right of the stationary material medium B through random motion and collision with particles of species B. It should be noted that the molecules of A are present everywhere in the matrix and they move randomly in both forward and backward directions, that is, from left to right and from right to left. However, the number of molecules of species A moving in the forward direction (left to right) is much larger than the number of molecules of A moving in the opposite direction as the concentration of molecules of A is much higher on the left side of the medium. Consequently, there occurs a net transport of species A in the forward direction (left to right) due to concentration gradient, as illustrated in Figure 16.1. Thus, the net movement of the molecules of species A through the material medium B, under the influence of concentration gradient in A, is referred to as molecular mass transport or simply diffusion. Note that when the concentration gradient is absent, that is, the concentration of species A is uniform in the medium, the number of molecules of species A moving from left to right is balanced by the number of molecules of species A moving from right to left. Consequently, there is no net transport of molecules of species A or diffusion of A through the medium.

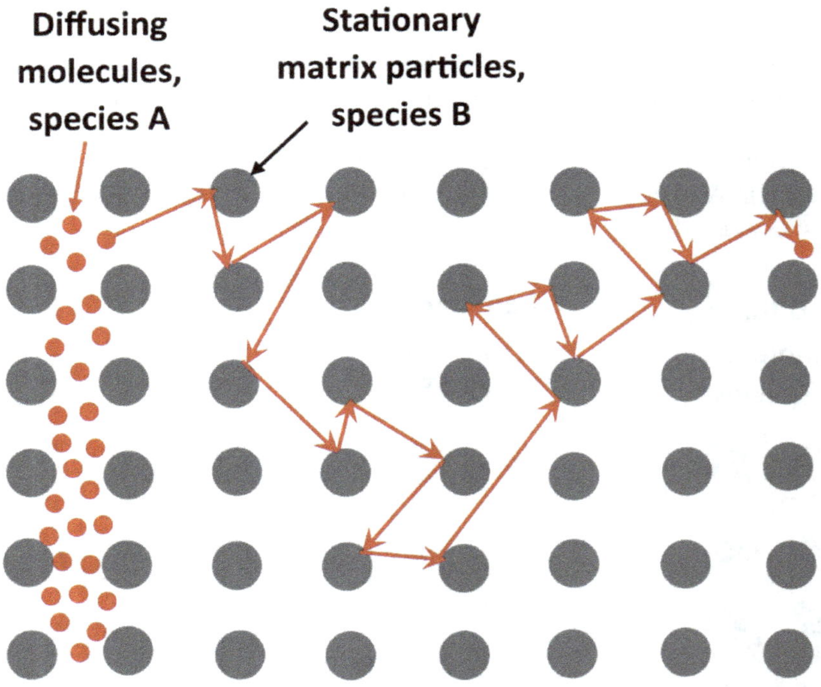

Figure 16.2 Diffusion of a molecule of species A in a stationary medium B.

16.1 FICK'S LAW OF DIFFUSION

Consider a binary mixture of species A and B. Let the molar concentration of species A, C_A, be non-uniform in the mixture and let the mixture be at rest without any macroscopic (bulk) motion, that is, the mixture velocity $\vec{V} = 0$. According to the Fick's law of diffusion, the presence of a concentration gradient dC_A/dx in the system gives rise to a diffusional molar flux J_A (moles/m^2.s) whose magnitude is proportional to the concentration gradient, that is:

$$J_A = -D_{AB}\frac{dC_A}{dx} \tag{16.1}$$

where D_{AB} (m^2/s) is the diffusion coefficient (diffusivity) of species A in B. Equation (16.1) is a one-dimensional form of the Fick's law of diffusion. A more general vector-form of the Fick's law is as follows:

$$\vec{J}_A = -D_{AB}\nabla C_A \tag{16.2}$$

where $\vec{J}_A$ is the molar flux of species A (moles/m^2.s) defined as the rate of moles of species A diffusing per unit area perpendicular to the direction of diffusion, and ∇C_A is the molar concentration gradient of species A in the medium given as:

$$\nabla C_A = \frac{\partial C_A}{\partial x}\hat{\imath} + \frac{\partial C_A}{\partial y}\hat{\jmath} + \frac{\partial C_A}{\partial z}\hat{k} \tag{16.3}$$

Fick's law of diffusion, Equation (16.2), is given in different co-ordinate systems in Table 16.1.

Table 16.1 Fick's law of diffusion in different co-ordinate systems

Co-ordinate system	Fick's law for a given co-ordinate system
Cartesian	$\vec{J}_A = J_{A,x}\hat{\imath} + J_{A,y}\hat{\jmath} + J_{A,z}\hat{k} = -D_{AB}\left[\left(\frac{\partial C_A}{\partial x}\right)\hat{\imath} + \left(\frac{\partial C_A}{\partial y}\right)\hat{\jmath} + \left(\frac{\partial C_A}{\partial z}\right)\hat{k}\right]$
Cylindrical	$\vec{J}_A = J_{A,r}\hat{r} + J_{A,\theta}\hat{\theta} + J_{A,z}\hat{z} = -D_{AB}\left[\left(\frac{\partial C_A}{\partial r}\right)\hat{r} + \frac{1}{r}\left(\frac{\partial C_A}{\partial \theta}\right)\hat{\theta} + \left(\frac{\partial C_A}{\partial z}\right)\hat{z}\right]$
Spherical	$\vec{J}_A = J_{A,r}\hat{r} + J_{A,\theta}\hat{\theta} + J_{A,\varphi}\hat{\varphi} = -D_{AB}\left[\left(\frac{\partial C_A}{\partial r}\right)\hat{r} + \frac{1}{r}\left(\frac{\partial C_A}{\partial \theta}\right)\hat{\theta} + \frac{1}{r sin\theta}\left(\frac{\partial C_A}{\partial \varphi}\right)\hat{\varphi}\right]$

The Fick's law of diffusion could also be expressed on a mass basis (instead of molar basis) as follows:

$$\vec{j}_A = -D_{AB}\nabla\rho_A \tag{16.4}$$

where $\vec{j}_A$ is the mass flux of species A (kg/m^2.s) and ρ_A is the mass concentration (kg/m^3).

The diffusion coefficient D_{AB} has the units of m^2/s. It is a measure of the ability of species A to diffuse through species B. Higher the D_{AB}, faster is the rate of diffusion of A in B. D_{AB} depends on temperature, pressure, composition of mixture, chemical nature of species A and B, size of the diffusing molecules of species A, and the spacing between the molecules. D_{AB} increases with temperature and decreases with pressure. The larger the molecular spacing, the higher the diffusivity. Thus, diffusion rates are typically much higher in gases than they are in liquids and much higher in liquids than in solids. In gases, the D_{AB} is typically in the range of $5 \times 10^{-6} - 10^{-5}\, m^2/s$ whereas in liquids it is in the range of $5 \times 10^{-10} - 10^{-9}\, m^2/s$ and in solids it is in the range of $10^{-14} - 10^{-10}\, m^2/s$.

16.2 DIFFUSIVITY

16.2.1 Gases

Due to the complicated nature of diffusion, the diffusion coefficient or diffusivity is usually determined experimentally. However, the diffusivity of a binary gas mixture at ordinary pressures can be predicted reasonably well using the kinetic theory of gases. The following Chapman-Enskog equation for diffusivity is based on the kinetic theory of gases:

$$D_{AB} = 0.0018583\frac{\sqrt{T^3\left(\frac{1}{M_A}+\frac{1}{M_B}\right)}}{p\sigma_{AB}^2\Omega_{AB}} \tag{16.5}$$

where D_{AB} is in cm²/s, T is in K, p is in atm, σ_{AB} is the average collision diameter in A^o, and Ω_{AB} is dimensionless collision integral based on the Lennard-Jones potential. The average collision diameter σ_{AB} is given as:

$$\sigma_{AB} = \frac{\sigma_A + \sigma_B}{2} \tag{16.6}$$

The collision integral Ω_{AB} is a function of dimensionless temperature defined as:

$$T^* = T/(\varepsilon_{AB}/k_B) \tag{16.7}$$

where ε_{AB} is the characteristic energy of molecular interaction for the mixture and k_B is the Boltzmann constant. ε_{AB}/k_B is given as:

$$\varepsilon_{AB}/k_B = \sqrt{(\varepsilon_A/k_B)(\varepsilon_B/k_B)} \tag{16.8}$$

where ε_A and ε_B are the characteristic molecular interaction energies of pure A and B. The collision integral Ω_{AB} can be calculated from the following equation:

$$\Omega_{AB} = 1.06036(T^*)^{-0.1561} + 0.193exp(-0.47635T^*) + 1.03587exp(-1.52996T^*) + 1.76474exp(-3.894117T^*) \tag{16.9}$$

Table 16.2 gives the relevant parameters for some gases needed to estimate the diffusivity from Equation (16.5).

Table 16.2 Collision diameters and characteristic molecular interaction energies of common gases

Gas	Chemical formula	σ (angstroms)	ε/k_B (K)	M
Methane	CH_4	3.76	149	16
Carbon monoxide	CO	3.69	91.7	28
Carbon dioxide	CO_2	3.94	195	44
Ethane	C_2H_6	4.44	216	30
Propane	C_3H_8	5.12	237	44
n-Butane	C_4H_{10}	4.69	531	58
Hydrogen	H_2	2.83	59.7	2
Ammonia	NH_3	2.9	558	17
Nitrogen	N_2	3.8	71.4	28
Oxygen	O_2	3.47	107	32
Helium	He	2.58	10.2	4

16.2.2 Liquid Solutions

The Nernst-Einstein equation relates the diffusivity of a solute molecule (species A) through a stationary liquid medium (solvent species B) as follows:

$$\frac{D^o_{AB}}{k_B T} = \frac{u_A}{F_A} \tag{16.10}$$

where D^o_{AB} is the diffusivity of a solute molecule, T is the absolute temperature, u_A/F_A is the "mobility" of a molecule, u_A is the velocity of a molecule, and F_A is the force acting on the molecule. Assuming the molecule to be rigid and spherical of R_A, the force F_A is given by the Stokes law:

$$F_A = 6\pi\eta_B R_A u_A \tag{16.11}$$

where η_B is the solvent viscosity. From Equations (16.10) and (16.11), we get:

$$\frac{D^o_{AB}}{k_B T} = \frac{1}{6\pi\eta_B R_A} \tag{16.12}$$

Thus,

$$D^o_{AB} = \frac{k_B T}{6\pi\eta_B R_A} \tag{16.13}$$

This is the Stokes-Einstein equation. It could be used to estimate the diffusivity D^o_{AB} of large spherical molecules in solvents of low molecular weight in an infinitely dilute solution.

The Stokes-Einstein equation can be re-written in the following form:

$$\frac{D^o_{AB}\eta_B}{k_B T} = \frac{1}{6\pi\left(\frac{3V_A}{4\pi N}\right)^{1/3}} = f(V_A) \tag{16.14}$$

where $f(V_A)$ is a function of molar volume V_A of the diffusing solute and N is the Avogadro number.

Several semi-empirical equations of the form of Equation (16.14) have been proposed in the literature to estimate the diffusivity of solute molecules in an infinitely dilute solution. Wilke and Chang correlation is quite popular among them. The Wilke-Chang correlation is given as:

$$\frac{D^o_{AB}\eta}{T} = \frac{7.4\times10^{-8}}{V_A^{0.6}}(\psi_B M_B)^{1/2} \tag{16.15}$$

where D^o_{AB} is the diffusivity of solute molecules in an infinitely dilute solution in units of cm^2/s, η is solution viscosity in centipoises, T is the absolute temperature in K, V_A is molar volume of solute A at normal boiling point in units of $cm^3/gmol$, M_B is the molecular weight of solvent B, and ψ_B is the "association parameter" for the solvent. The recommended values of ψ_B for some common solvents are as follows: 2.26 for water; 1.9 for methanol; 1.5 for ethanol; and 1.0 for benzene, ether, heptane, and other unassociated solvents. The Wilke-Chang correlation generally gives good predictions of D^o_{AB}, usually within $\pm10\%$.

16.3 MATERIAL BALANCE

Fick's law of diffusion can be used to calculate the molar or mass transfer rate of diffusing species provided that the concentration distribution of the species in the medium is known. To obtain the concentration distribution, material balance of species A is required. Consider the one-dimensional diffusion of species A in the x-direction as shown in Figure 16.3. The material or mole balance over the control volume in the absence of any chemical reaction gives:

$$J_{A,x}\big|_x A - J_{A,x}\big|_{x+\Delta x} A = (A\Delta x)\frac{\partial C_A}{\partial t} \tag{16.16}$$

Upon re-arrangement, Equation (16.6) becomes:

$$-\left[\frac{J_{A,x}\big|_{x+\Delta x} - J_{A,x}\big|_x}{\Delta x}\right] = \frac{\partial C_A}{\partial t} \tag{16.17}$$

Taking the limit $\Delta x \rightarrow 0$, we get:

$$-\frac{\partial J_{A,x}}{\partial x} = \frac{\partial C_A}{\partial t} \tag{16.18}$$

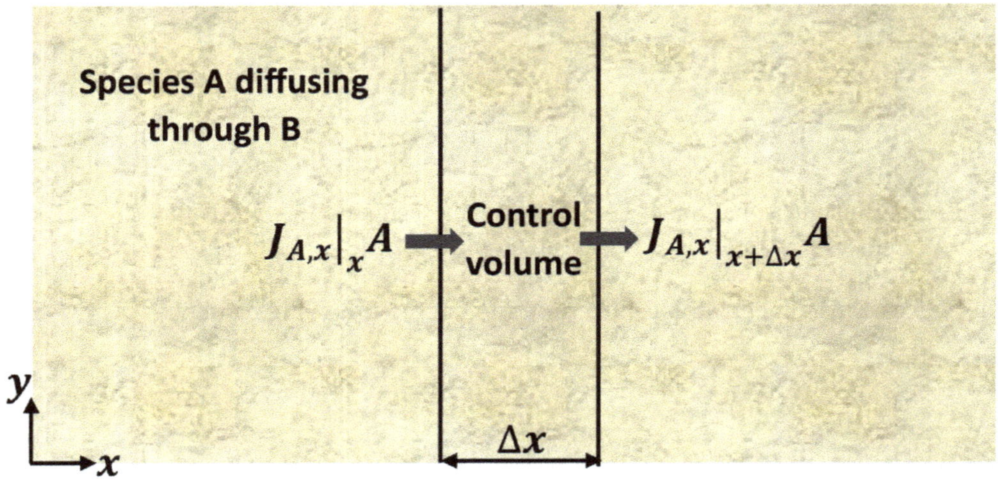

Figure 16.3 Material balance over differential control volume.

Equation (16.18) can now be generalized to a three-dimensional form as:

$$-\frac{\partial J_{A,x}}{\partial x} - \frac{\partial J_{A,y}}{\partial y} - \frac{\partial J_{A,z}}{\partial z} = \frac{\partial C_A}{\partial t} \tag{16.19}$$

Thus,

$$-\nabla \cdot \vec{J}_A = \frac{\partial C_A}{\partial t} \tag{16.20}$$

From Equations (16.2) and (16.20), we get:

$$\nabla \cdot (D_{AB} \nabla C_A) = \frac{\partial C_A}{\partial t} \tag{16.21}$$

Assuming D_{AB} is constant, Equation (16.21) gives:

$$\frac{\partial C_A}{\partial t} = D_{AB} \nabla^2 C_A \tag{16.22}$$

Equation (16.22) is the mass diffusion equation, also referred to as Fick's second law of diffusion. Equation (16.22) needs to be solved subject to the given initial condition and boundary conditions to obtain the concentration distribution of species A in the medium. Equation (16.22) is given in different co-ordinate systems in Table 16.3.

Table 16.3 Mass diffusion equation in different co-ordinate systems

Co-ordinate system	Mass diffusion equation for a given co-ordinate system
Cartesian	$\frac{\partial C_A}{\partial t} = D_{AB}\left[\frac{\partial^2 C_A}{\partial x^2} + \frac{\partial^2 C_A}{\partial y^2} + \frac{\partial^2 C_A}{\partial z^2}\right]$
Cylindrical	$\frac{\partial C_A}{\partial t} = D_{AB}\left[\frac{1}{r}\frac{\partial}{\partial r}\left(r\frac{\partial C_A}{\partial r}\right) + \frac{1}{r^2}\frac{\partial^2 C_A}{\partial \theta^2} + \frac{\partial^2 C_A}{\partial z^2}\right]$
Spherical	$\frac{\partial C_A}{\partial t} = D_{AB}\left[\frac{1}{r^2}\frac{\partial}{\partial r}\left(r^2\frac{\partial C_A}{\partial r}\right) + \frac{1}{r^2 \sin\theta}\frac{\partial}{\partial \theta}\left(\sin\theta\frac{\partial C_A}{\partial \theta}\right) + \frac{1}{r^2 \sin^2\theta}\frac{\partial^2 C_A}{\partial \varphi^2}\right]$

Note that the mass diffusion equation could also be expressed on a mass basis (instead of molar basis) as follows:

$$\frac{\partial \rho_A}{\partial t} = D_{AB}\nabla^2 \rho_A \tag{16.23}$$

16.4 STEADY STATE ONE-DIMENSONAL DIFFUSION

16.4.1 Cartesian Co-ordinates

Consider a plane solid wall of material B through which species A (say gaseous) is diffusing. Let the molar concentration of species A be $C_{A,1}$ within the material B at $x = x_1$ and let the molar concentration of species A be $C_{A,2}$ within the material B at $x = x_2$ (see Figure 16.4). At steady state, the concentration distribution of A within the material B is given by the mass diffusion equation, Equation (16.22), as:

$$\frac{d^2 C_A}{dx^2} = 0 \tag{16.24}$$

$$C_A = C_{A,1} + \frac{C_{A,2} - C_{A,1}}{x_2 - x_1}(x - x_1) \tag{16.25}$$

Thus, the concentration distribution is linear.

The molar flux of species A is given as:

$$J_{A,x} = -D_{AB}\frac{dC_A}{dx} = -\frac{D_{AB}(C_{A,2} - C_{A,1})}{x_2 - x_1} \tag{16.26}$$

The molar flux of A is constant throughout the medium.

The resistance to diffusional mass transfer can be defined as:

$$R_m = \frac{\Delta C_A}{\dot{M}_A} \tag{16.27}$$

where $\Delta C_A = C_{A,1} - C_{A,2}$, and $\dot{M}_A$ is the mass transfer rate of species A in units of moles/time. Note that $\dot{M}_A = J_A A$, where A is the area of mass transfer (area normal to the direction of mass transfer).
From Equations (16.26) and (16.27), the resistance to diffusional mass transfer for plane objects is:

$$R_m = \frac{L}{D_{AB} A} \tag{16.28}$$

where L is the thickness of the planar object.

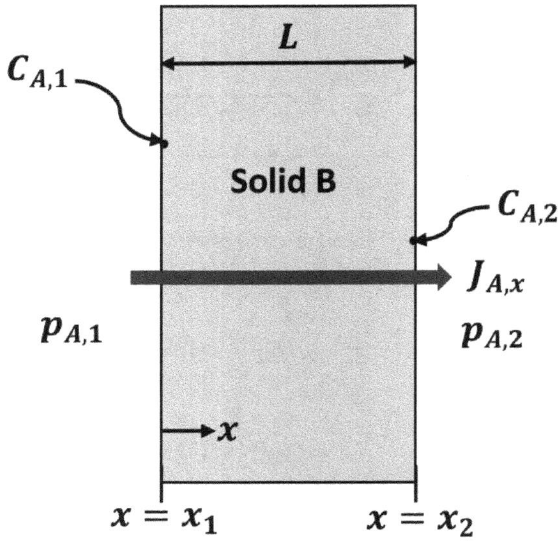

Figure 16.4 Diffusion through a plane wall.

In practical problems involving diffusion of gaseous species A through a stationary solid material B, the concentration of species A in the solid at the gas/solid interface is not known. The concentration of gaseous species A in the solid at the gas/solid interface can be determined from the following equilibrium relationship:

$$C_{A,\text{solid-side}} = S p_{A,\text{gas-side}} \tag{16.29}$$

where S is the solubility of A in the solid material B, units of mol/(m^3.Pa), and p_A is the partial pressure of species A in the gas. Thus, the molar flux of species A through material B can be expressed as:

$$J_{A,x} = \frac{p_{A,1} - p_{A,2}}{\left(\frac{L}{S D_{AB}}\right)} \tag{16.30}$$

Or,

$$\dot{M}_A = J_{A,x}A = \frac{A(p_{A,1}-p_{A,2})}{\left(\frac{L}{SD_{AB}}\right)} = \frac{\Delta p_A}{\left(\frac{L}{SAD_{AB}}\right)} \tag{16.31}$$

where $\Delta p_A = (p_{A,1} - p_{A,2})$. Equation (16.31) indicates that the resistance to diffusional mass transfer is $L/(SAD_{AB})$ if the driving force is taken to be difference in partial pressures. The product of solubility (S) and diffusion coefficient (D_{AB}) is referred to as permeability or permeability coefficient (P), that is:

$$P = SD_{AB} \tag{16.32}$$

The permeability is the flux per unit pressure-gradient, often expressed in units of Barrers (1 Barrer = 10^{-10} cm^3 (STP)/[(cm^2.s)(cmHg/cm)] where cm^3 (STP) is standard cubic centimeter). It reflects the combination of solubility and diffusion characteristics of species A in material B. The resistance to diffusional mass transfer can now be expressed as $L/(PA)$ provided that the driving force for mass transfer is taken to be the difference in partial pressures of species A on the two sides of material B. Note that in SI system of units, 1 Barrer is 3.35×10^{-16} mol.m/m^2.s.Pa.

16.4.2 Cylindrical Co-ordinates

Consider a solid cylindrical shell of material B with inner and outer radii r_1 and r_2 , respectively. Gaseous species A is diffusing through the shell in the radial direction. Let $C_{A,1}$ be the molar concentration of species A inside the cylindrical solid at $r = r_1$ and let $C_{A,2}$ be the molar concentration of A inside the cylindrical solid at $r = r_2$ (see Figure 16.5).

At steady state, the concentration distribution of A within B is given by the mass diffusion equation, Equation (16.22), as:

$$\frac{d}{dr}\left(r\frac{dC_A}{dr}\right) = 0 \tag{16.33}$$

Upon integration subject to the given boundary conditions, we obtain:

$$C_A = C_{A,1} + \frac{C_{A,2}-C_{A,1}}{ln\left(\frac{r_2}{r_1}\right)} ln\left(\frac{r}{r_1}\right) \tag{16.34}$$

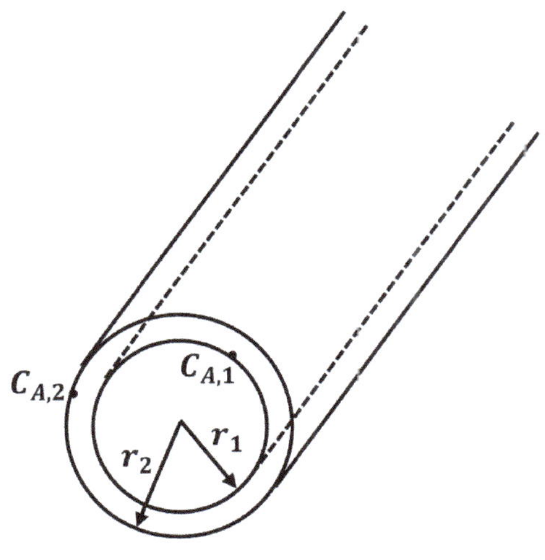

Figure 16.5 Diffusion through a cylindrical shell.

The molar flux of species A is given as:

$$J_{A,r} = -D_{AB}\frac{dC_A}{dr} = -\frac{D_{AB}(C_{A,2}-C_{A,1})}{ln\left(\frac{r_2}{r_1}\right)}\left(\frac{1}{r}\right) \qquad (16.35)$$

As expected, the molar flux is not constant. It decreases with the increase in radial distance due to the increase in the mass transfer area. Note that the molar flow rate of species A ($\dot{M}_A$) is constant:

$$\dot{M}_A = J_{A,r}(2\pi rL) = -\frac{D_{AB}(C_{A,2}-C_{A,1})}{ln\left(\frac{r_2}{r_1}\right)}(2\pi L) \qquad (16.36)$$

where L is the length of the cylindrical shell. The resistance to diffusional mass transfer in cylindrical geometry is:

$$R_m = \frac{\Delta C_A}{\dot{M}_A} = \frac{ln\left(\frac{r_2}{r_1}\right)}{2\pi D_{AB}L} \qquad (16.37)$$

where $\Delta C_A = C_{A,1} - C_{A,2}$. Using the equilibrium relationship Equation (16.29), the molar flux can also be expressed as:

$$J_{A,r} = \frac{(p_{A,1}-p_{A,2})}{\left[\frac{ln(r_2/r_1)}{D_{AB}S}\right]}\left(\frac{1}{r}\right) \qquad (16.38)$$

or,

$$\dot{M}_A = \frac{(p_{A,1}-p_{A,2})}{\left(\frac{ln(r_2/r_1)}{2\pi SLD_{AB}}\right)} = \frac{(p_{A,1}-p_{A,2})}{\left(\frac{ln(r_2/r_1)}{2\pi PL}\right)} \qquad (16.39)$$

where $p_{A,1}$ is the partial pressure of species A in the gas phase at gas/liquid interface (at $r = r_1$), $p_{A,2}$ is the partial pressure of species A in the gas phase at gas/solid interface (at $r = r_2$), and P is permeability.

16.4.3 Spherical Co-ordinates

For gaseous species A diffusing through a solid spherical shell of material B, the mass diffusion equation, Equation (16.2) reduces to:

$$\frac{d}{dr}\left(r^2\frac{dC_A}{dr}\right) = 0 \qquad (16.40)$$

Upon integration,

$$C_A = C_{A,1} + \frac{C_{A,2}-C_{A,1}}{\left(\frac{1}{r_1}-\frac{1}{r_2}\right)}\left(\frac{1}{r}-\frac{1}{r_1}\right) \qquad (16.41)$$

where $C_{A,1}$ is the molar concentration of A within B at $r = r_1$ and $C_{A,2}$ is the molar concentration of A within B at $r = r_2$ (see Figure 16.6).

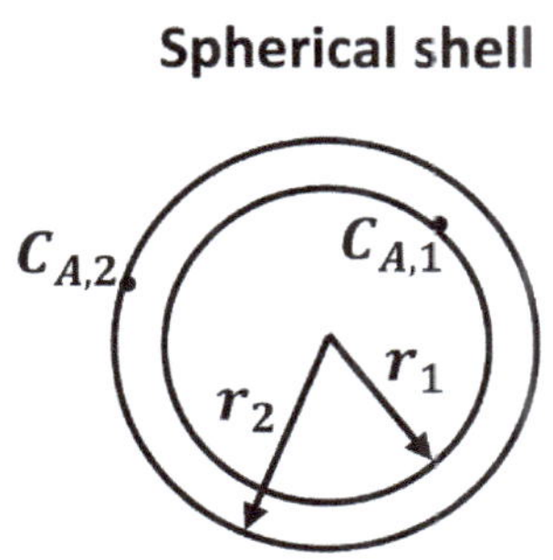

Figure 16.6 Diffusion through a spherical shell.

The molar flux of species A is:

$$J_{A,r} = -D_{AB}\frac{dC_A}{dr} = -\frac{D_{AB}(C_{A,2}-C_{A,1})}{\left(\frac{1}{r_1}-\frac{1}{r_2}\right)}\left(\frac{1}{r^2}\right) \tag{16.42}$$

As expected, the molar flux is not constant. It decreases with the increase in radial distance due to the increase in the mass transfer area. Note that the molar flow rate of species A ($\dot{M}_A$) is constant:

$$\dot{M}_A = J_{A,r}(4\pi r^2) = -\frac{4\pi D_{AB}(C_{A,2}-C_{A,1})}{\left(\frac{1}{r_1}-\frac{1}{r_2}\right)} \tag{16.43}$$

The resistance to diffusional mass transfer in a spherical geometry is:

$$R_m = \frac{\Delta C_A}{\dot{M}_A} = \frac{r_2-r_1}{4\pi D_{AB}r_1r_2} \tag{16.44}$$

where $\Delta C_A = C_{A,1} - C_{A,2}$. The molar flux of species A in terms of partial pressures of A can be obtained using Equation (16.29). Thus,

$$J_{A,r} = \frac{(p_{A,1}-p_{A,2})}{\frac{r_2-r_1}{SD_{AB}r_1r_2}}\left(\frac{1}{r^2}\right) \tag{16.45}$$

Or,

$$\dot{M}_A = \frac{(p_{A,1}-p_{A,2})}{\frac{r_2-r_1}{4\pi P r_1 r_2}} \tag{16.46}$$

where P is the permeability of A in B, given as SD_{AB}.

16.4.4 Summary of Mass Transfer Resistances in One-Dimensional Diffusion

Table 16.4 summarizes the diffusion mass transfer resistance formulae for different geometries.

Table 16.4 Diffusion mass transfer resistance for different geometries

Geometry	R_m (ΔC_A as the driving force)	R_m (Δp_A as the driving force)
Plane wall	$L/(D_{AB}A)$	$L/(PA)$
Cylindrical shell	$ln(r_2/r_1)/(2\pi D_{AB}L)$	$ln(r_2/r_1)/(2\pi PL)$
Spherical shell	$(r_2-r_1)/(4\pi D_{AB}r_1r_2)$	$(r_2-r_1)/(4\pi P r_1 r_2)$

The general formula for mass transfer resistance in one-dimensional diffusion is as follows:

$$R_m = \frac{\text{thickness}}{\text{Area}\times D_{AB}} \tag{16.47}$$

where thickness and area are defined in Table 16.5 for various geometries.

Table 16.5 Relevant thicknesses and areas in diffusion mass transfer through different geometries

Geometry	Thickness	Area
Plane wall	L	A
Cylindrical shell	$(r_2 - r_1)$	$A_{log-mean} = (A_2 - A_1)/ln(A_2/A_1)$
Spherical shell	$(r_2 - r_1)$	$A_{geometric-mean} = \sqrt{A_1 A_2}$

16.5 STEADY STATE ONE-DIMENSIONAL DIFFUSION WITH CHEMICAL REACTION

Many processes of practical interest involve the simultaneous diffusion of a chemical species and the consumption or generation of the same species through a chemical reaction. In such situations, the mass diffusion equation, Equation (16.22), is modified as follows:

$$\frac{\partial C_A}{\partial t} = D_{AB}\nabla^2 C_A + R_A \tag{16.48}$$

where R_A is the rate of production of species A per unit volume of the medium.

As an example of the application of Equation (16.48), consider a plane layer of organic matter of thickness L. The oxygen is diffusing through organic matter. The oxygen is also being consumed in the organic matter by biochemical reactions (say by bacteria that is uniformly distributed in the organic matter). The consumption of oxygen depends on the local concentration of oxygen and can be expressed as $R_A = -kC_A$ (mol/s.m^3). Let $C_{A,o}$ and $C_{A,L}$ be the molar concentrations of oxygen in the organic matter at $x = 0$ and $x = L$, respectively. Assuming that $C_{A,o}$ and $C_{A,L}$ are constant (independent of time), we can determine the concentration distribution of oxygen in the organic matter by solving Equation (16.48). Thus,

$$D_{AB}\frac{d^2C_A}{dx^2} - kC_A = 0 \tag{16.49}$$

Equation (16.49) needs to be solved subject to the boundary conditions: at $x = 0$, $C_A = C_{A,o}$ and at $x = L$, $C_A = C_{A,L}$. The solution is as follows:

$$C_A = \frac{C_{A,0}\left(e^{-(\sqrt{k/D_{AB}})(L-x)} - e^{(\sqrt{k/D_{AB}})(L-x)}\right) + C_{A,L}\left(e^{-(\sqrt{k/D_{AB}})x} - e^{(\sqrt{k/D_{AB}})x}\right)}{\left(e^{-(\sqrt{k/D_{AB}})L} - e^{(\sqrt{k/D_{AB}})L}\right)} \tag{16.50}$$

16.6 TRANSIENT DIFFUSION

In steady state analysis of diffusion considered in the preceding sections, the concentration and flux are independent of time. Now transient mass diffusion is considered where the concentration and flux vary with both location and time.

Consider a semi-infinite body of material B. This body is exposed to an environment consisting of species A. Species A diffuses into the semi-infinite body. We are interested in concentration distribution of species A in the body as a function of time, given that:

$$C_A(x, t = 0) = C_{A,i} \tag{16.51}$$

$$C_A(x = 0, t) = C_{A,s} \tag{16.52}$$

$$C_A(x \rightarrow \infty, t) = C_{A,i} \tag{16.53}$$

Equation (16.51) represents the initial condition that the mass concentration of species A in the body is uniform at $t = 0$, Equation (16.52) is the surface boundary condition that the concentration of species A within the body at the surface is constant, and Equation (16.53) represents the second boundary condition that far away from the surface, the concentration of species A in the body is the same as initial concentration.

The equation governing the concentration distribution is Equation (16.22) if D_{AB} is constant. Thus:

$$\frac{\partial C_A}{\partial t} = D_{AB}\frac{\partial^2 C_A}{\partial x^2} \tag{16.54}$$

Equation (16.54) needs to be solved subject to the initial and boundary conditions specified in Equations (16.51) - (16.53). The solution is as follows:

$$\frac{C_A - C_{A,i}}{C_{A,s} - C_{A,i}} = 1 - erf\left[\frac{x}{2\sqrt{D_{AB}t}}\right] \tag{16.55}$$

where "erf" refers to error function. If $C_{A,i} = 0$, Equation (16.55) reduces to:

$$\frac{C_A}{C_{A,s}} = 1 - erf\left[\frac{x}{2\sqrt{D_{AB}t}}\right] \tag{16.56}$$

These results can be applied to several problems of practical interest.

16.7 GENERAL FORM OF FICK'S LAW OF DIFFUSION

The Fick's law of diffusion expressed in the form of Equation (16.2) is valid under the restriction that the total molar concentration of the mixture (species A and B) is constant. A more general form of the Fick's law of diffusion is as follows:

$$\vec{J}_A = -CD_{AB}\nabla x_A \tag{16.57}$$

where the molar flux $\vec{J}_A$ is the rate of transfer of species A in moles per unit area perpendicular to the direction of transfer, C is the total molar concentration of the mixture ($C = C_A + C_B$), and x_A is the mole fraction of species A given as C_A/C. When the total concentration of the mixture is constant ($C =$ constant), Equation (16.57) reduces to Equation (16.2).

The general form of the Fick's law of diffusion on mass basis is as follows:

$$\vec{j}_A = -\rho D_{AB}\nabla\omega_A \tag{16.58}$$

where ρ is the total mass concentration of the mixture and ω_A is the mass fraction of species A.

Thus, the Fick's law of diffusion expressed in the form of Equation (16.2) is valid under the conditions:

- Diffusion is isothermal and isobaric if the species A and B are gaseous species, and the gas mixture is an ideal gas mixture. Note that for ideal gas mixtures, $C = p/RT$. Therefore, for isothermal and isobaric gas phase diffusion, C =constant.
- Species B is stagnant (non-diffusing) and species A is present in dilute amounts such that: $C = C_A + C_B \cong C_B$ =constant.
- The mixture is a binary mixture and there is no bulk motion involved.

16.8 EXAMPLES

16.8.1 Example 1

Using the Chapman-Enskog equation, estimate D_{AB} for the binary gaseous system of carbon monoxide (species A) and carbon dioxide (species B) at T=25 °C and P=1 atm.

Solution

From Table 16.2,

$$M_A = 28,\ \sigma_A = 3.69\ A^o,\ \ \varepsilon_A/k_B = 91.7\ K$$

$$M_B = 44,\ \sigma_B = 3.94\ A^o,\ \ \varepsilon_B/k_B = 195\ K$$

$$\sigma_{AB} = \frac{\sigma_A + \sigma_B}{2} = \frac{3.69 + 3.94}{2} = 3.815\ A^o$$

$$\varepsilon_{AB}/k_B = \sqrt{(\varepsilon_A/k_B)(\varepsilon_B/k_B)} = \sqrt{91.7 \times 195} = 133.722\ K$$

$$T^* = T/(\varepsilon_{AB}/k_B) = 298.15/133.722 = 2.23$$

From collision integral equation, Equation (16.9), we get $\Omega_{AB} = 1.036763$. Thus,

$$D_{AB} = 0.0018583 \frac{\sqrt{T^3\left(\frac{1}{M_A} + \frac{1}{M_B}\right)}}{p\sigma_{AB}^2\Omega_{AB}} = 0.0018583 \frac{\sqrt{298.15^3\left(\frac{1}{28} + \frac{1}{44}\right)}}{1 \times 3.815^2 \times 1.036763} = 0.15327\ cm^2/s$$

16.8.2 Example 2

Estimate D_{AB} for a binary liquid solution of species A and species B at T=20 °C. The species A is present in small amounts, that is, the solution is very dilute in A. Given that: $\eta = 1.18\ cP$, $V_A = 40.5\ cm^3/gmol$, $\psi_B = 1.5$, and $M_B = 46$.

Solution

We can use the Wilke-Chang correlation:

$$\frac{D_{AB}^{o}\eta}{T} = \frac{7.4 \times 10^{-8}}{V_A^{0.6}}(\psi_B M_B)^{1/2}$$

where D_{AB}^{o} is the diffusivity of solute molecules in an infinitely dilute solution in units of cm^2/s, η is solution viscosity in centipoises, T is the absolute temperature in K, V_A is molar volume of solute A at normal boiling point in units of $cm^3/gmol$, M_B is the molecular weight of solvent B, and ψ_B is the "association parameter" for the solvent.

$$D_{AB}^{o} = \left(\frac{T}{\eta}\right)\left(\frac{7.4 \times 10^{-8}}{V_A^{0.6}}\right)(\psi_B M_B)^{1/2} = \left(\frac{293.15}{1.18}\right)\left(\frac{7.4 \times 10^{-8}}{40.5^{0.6}}\right)(1.5 \times 46)^{1/2} = 1.66 \times 10^{-5}\ cm^2/s$$

16.8.3 Example 3

Oxygen gas is diffusing through a rubber membrane at 24 °C. The membrane is 0.6 mm thick. The oxygen is maintained at a pressure of 1.95 bars on one side and 1.05 bar on the other side. The solubility of O_2 gas in rubber is 3.12×10^{-3} $kmol/m^3.bar$ and the diffusivity of O_2 in rubber is 0.21×10^{-9} m^2/s.

i) Calculate the molar diffusion flux of O_2 through the membrane

ii) Determine the molar concentrations of O_2 outside the membrane on both sides of the membrane.

iii) Plot concentration distribution across the membrane

Solution

Refer to the figure below.

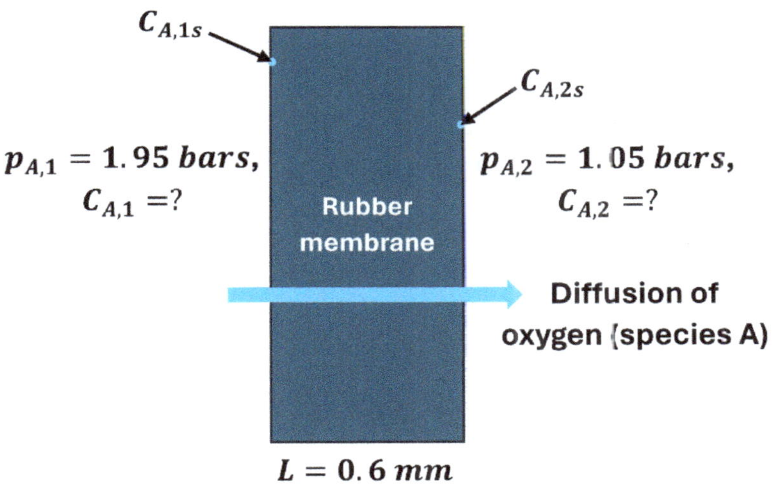

i) Calculate the molar diffusion flux of O_2 through the membrane

$$C_{A,1s} = Sp_{A,1} = 3.12 \times 10^{-3} \times 1.95 = 6.084 \times 10^{-3}\ kmol/m^3$$

$$C_{A,2s} = Sp_{A,2} = 3.12 \times 10^{-3} \times 1.05 = 3.276 \times 10^{-3}\ kmol/m^3$$

$J_{A,x} = -D_{AB}\frac{dC_A}{dx} = -\frac{D_{AB}(C_{A,2s}-C_{A,1s})}{L} = -0.21 \times 10^{-9}\left(\frac{3.276\times10^{-3}-6.084\times10^{-3}}{0.6\times10^{-3}}\right) = 9.828 \times 10^{-10}$ kmol/m^2.s

ii) Determine the molar concentrations of O_2 outside the membrane on both sides of the membrane.

Assume that ideal gas law is valid,

$$C_{A,1} = \frac{p_{A,1}}{RT} = \frac{1.95 \times 10^5}{8.314 \times 297.15} = 78.93\ moles/m^3$$

$$C_{A,2} = \frac{p_{A,2}}{RT} = \frac{1.05 \times 10^5}{8.314 \times 297.15} = 42.50\ moles/m^3$$

iii) Plot concentration distribution across the membrane

The plot of concentration distribution across the membrane is shown below. Note that there occur jumps in concentration at the interfaces. Interestingly, mass transport of species A occurs from left to right side of the membrane despite $C_{A,2s} \ll C_{A,2}$.

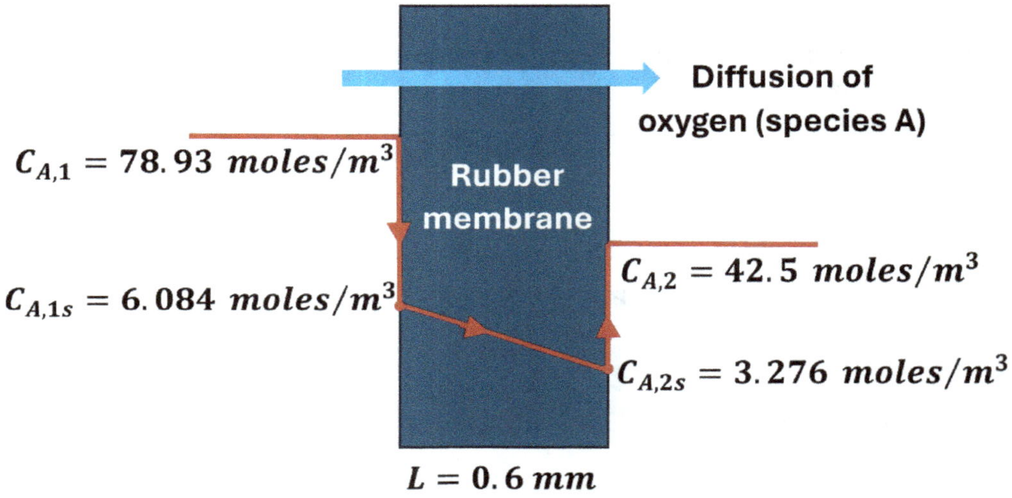

16.8.4 Example 4

Convert 1 Barrer from non-SI unit of cm^3 (STP)/[(cm^2.s)(cmHg/cm)] to SI unit of moles.m/[(m^2.s)(Pa)]

Solution

$1\ Barrer = 10^{-10}\ cm^3(STP).cm/(cm^2.s)(cm.Hg)$

Convert $cm^3(STP)$ to $moles$ using standard temperature (0°C) and pressure (1 atm):

$1\ cm^3(STP) = \frac{PV}{RT} = \frac{101.325\times10^3\times1\times10^{-6}}{8.314\times273.15} = 44.62 \times 10^{-6}\ moles$

Thus,

$$1\ \frac{cm^3(STP).cm}{(cm^2.s)(cm.Hg)} = \frac{44.62 \times 10^{-6} moles \times 10^{-2} m}{10^{-4} m^2.s \times 1333.2239\ Pa} = 3.35 \times 10^{-6} \frac{moles.m}{(m^2.s)(Pa)}$$

$$1\ Barrer = 10^{-10}\ cm^3(STP).cm/(cm^2.s)(cm.Hg) = 3.35 \times 10^{-16} \frac{moles.m}{(m^2.s)(Pa)}$$

16.8.5 Example 5

Consider a 12-liter vessel containing CO_2 at 23 °C and 4.8 bars (see figure). The vessel is connected to a rubber plug that is 28 mm thick and has a surface area of 325 mm^2. The concentration of CO_2 in the rubber plug at the outer surface is negligible.

i) Determine the permeability of the membrane in unite of Barrers.

ii) Determine the rate of mass loss of CO_2 from the vessel.

The solubility of CO_2 gas in rubber is 40.15 x 10^{-3} kmol/m^3.bar and the diffusivity of CO_2 in rubber is 0.11x10^{-9} m^2/s. Molecular weight of CO_2 is 44.01 g/mol.

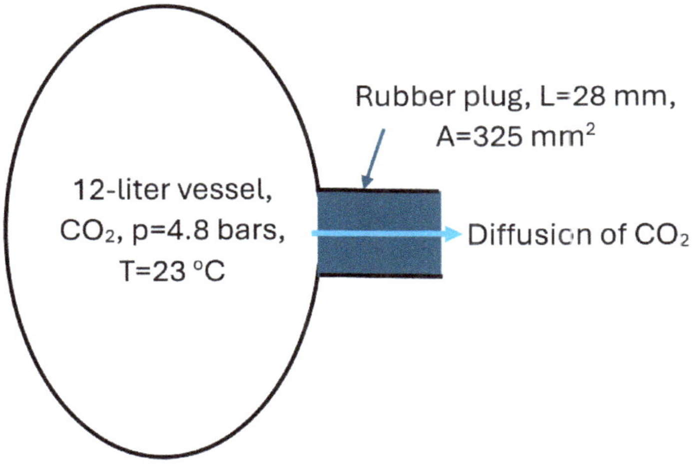

Solution

i) Determine the permeability of the membrane in unite of Barrers.

$$P = SD_{AB}$$

$$S = 40.15 \times 10^{-3} \frac{kmol}{m^3.bar} = \frac{40.15 \times 10^{-3} \times 10^3 moles}{10^5\ m^3.Pa} = 40.15 \times 10^{-5} \frac{moles}{m^3.Pa}$$

$$D_{AB} = 0.11 \times 10^{-9}\ m^2/s$$

$$P = SD_{AB} = 40.15 \times 10^{-5} \times 0.11 \times 10^{-9} = 4.4165 \times 10^{-14} \frac{moles.m}{m^2.s.Pa}$$

Thus,

$$P = SD_{AB} = \frac{4.4165 \times 10^{-14}}{3.35 \times 10^{-16}} = 131.836\ Barrers$$

ii) Determine the rate of mass loss of CO_2 from the vessel.

$$J_{A,x} = \frac{p_{A,1} - p_{A,2}}{\left(\frac{L}{SD_{AB}}\right)} = P\left(\frac{p_{A,1} - p_{A,2}}{L}\right) = 4.4165 \times 10^{-14} \times \frac{(4.8 \times 10^5 - 0)}{28 \times 10^{-3}} = 7.57 \times 10^{-7}\ \frac{moles}{m^2.s}$$

$$\dot{M}_A = J_{A,x}A = 7.57 \times 10^{-7} \times 325 \times 10^{-6} = 2.46 \times 10^{-10}\ \frac{moles}{s}$$

$$\dot{m}_A = \dot{M}_A M = 2.46 \times 10^{-10} \times 44.01 = 1.083 \times 10^{-8}\ \frac{g}{s}$$

16.8.6 Example 6

Consider a glass cylinder of inside diameter 110 mm, thickness 15 mm, and length 1 m. Helium gas is stored in the cylinder at 23 °C and 3.9 bars. Estimate the rate of mass loss of helium gas from the container. The concentration of helium in glass at the outer surface is negligible.

The solubility of He gas in glass is 0.45×10^{-3} kmol/m^3.bar and the diffusivity of He in glass is 0.40×10^{-13} m^2/s. Molecular weight of He is 4 g/mol.

Solution

Refer to the figure below.

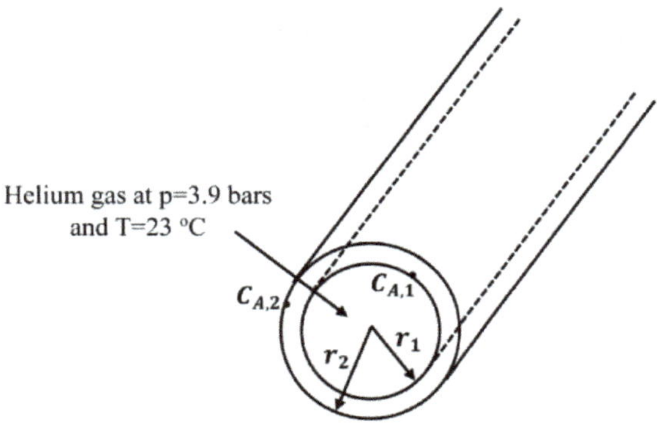

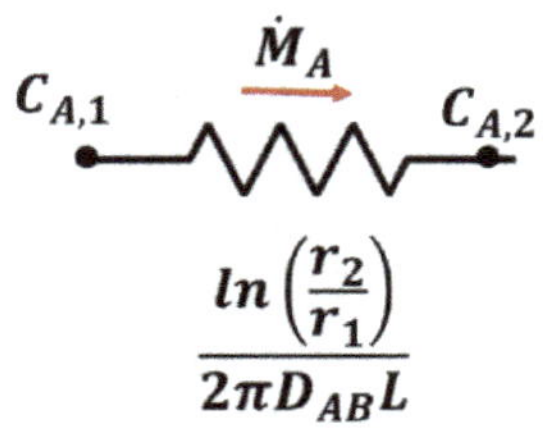

$$C_{A,1} = pS = 3.9 \times 0.45 \times 10^{-3} = 1.755 \times 10^{-3}\ \frac{kmol}{m^3}$$

$$C_{A,2} = 0$$

$r_2 = 70\ mm\ ,\ r_1 = 55\ mm$

$$R = \frac{ln(r_2/r_1)}{2\pi D_{AB}L} = \frac{ln(70/55)}{2\pi \times 0.40 \times 10^{-13} \times 1} = 9.596 \times 10^{11}\ \ s/m^3$$

$$\dot{M}_A = \frac{(C_{A,1} - C_{A,2})}{\frac{ln(r_2/r_1)}{2\pi D_{AB}L}} = \frac{1.755 \times 10^{-3} - 0}{9.596 \times 10^{11}} = 1.829 \times 10^{-15}\ \frac{kmol}{s}$$

$$\dot{m}_A = \dot{M}_A M = 1.829 \times 10^{-15} \times 4 = 7.316 \times 10^{-15}\ \text{kg/s}$$

16.8.7 Example 7

A spherical container made of tempered glass contains helium gas at 23 °C and 3.9 bars. The inside diameter of the container is 196 mm and its thickness is 15 mm. The concentration of helium in the glass at the outer surface is negligible.

The solubility of He gas in Pyrex is 0.45 x 10^{-3} kmol/m^3.bar and the diffusivity of He in Pyrex is 0.40 x 10^{-13} m^2/s. Molecular weight of He is 4 g/mol.

Estimate the rate of mass loss of helium gas from the container.

Solution

Refer to the figure below.

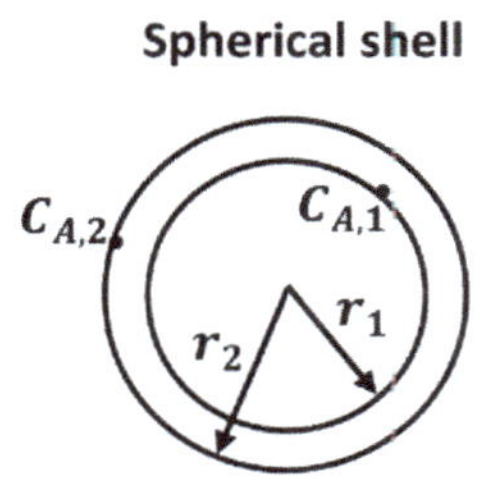

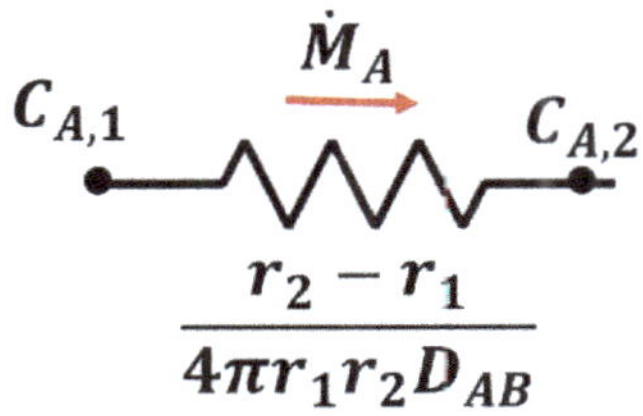

$$C_{A,1} = pS = 3.9 \times 0.45 \times 10^{-3} = 1.755 \times 10^{-3}\ \frac{kmol}{m^3}$$

$C_{A,2} = 0$
$r_2 = 113\ mm\ ,\ r_1 = 98\ mm$

$$R = (r_2 - r_1)/(4\pi D_{AB} r_1 r_2) = (113 - 98) \times 10^{-3}/(4\pi \times 113 \times 98 \times 10^{-6} \times 0.40 \times 10^{-13}) = 2.695 \times 10^{12}\ \ s/m^3$$

$$\dot{M}_A = \frac{(C_{A,1} - C_{A,2})}{R} = \frac{1.755 \times 10^{-3} - 0}{2.695 \times 10^{12}} = 0.6512 \times 10^{-15}\ \frac{kmol}{s}$$

$$\dot{m}_A = \dot{M}_A M = 0.6512 \times 10^{-15} \times 4 = 2.6048 \times 10^{-15}\ \text{kg/s}$$

16.8.8 Example 8

Hydrogen is diffusing through a composite wall consisting of two layers of metals, nickel (Ni) and palladium (Pd), in intimate contact with each other (see figure below). Each layer is 1 mm thick. On the nickel side, the hydrogen is at 1 atm and on the palladium side, the hydrogen is at 0.1 atm. The temperature of the system is 400 °C. At the interface between H_2 gas and metal (Ni or Pd), the H_2 gas dissolves in metal in atomic form according to the following law:

$$\rho_H = K\sqrt{p_{H_2}}$$

where ρ_H is concentration of dissolved atomic hydrogen in kg/m^3, $K_{Ni} = 3.73\text{x}10^{-2}$ and $K_{Pd} = 1.11$ in the units of (kg/m^3)/(atm)$^{1/2}$. Given that $D_{H\text{-}Ni}$=4.9x10^{-10} m^2/s and $D_{H\text{-}Pd}$=6.4x10^{-9} m^2/s, and the concentration of atomic hydrogen on the Nickel side of the Ni/Pd interface is 1.19x10^{-2} kg/m^3, do the following:

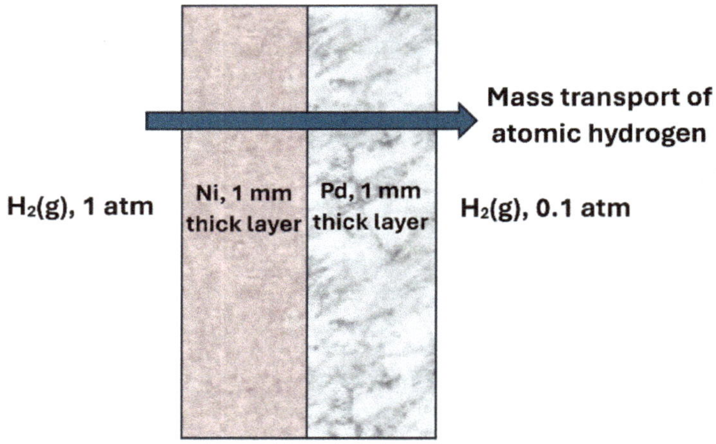

i) Calculate the mass flux of atomic hydrogen in the composite wall

ii) Calculate the concentration of atomic hydrogen on the Pd side of Ni/Pd interface

iii) Plot the concentration of atomic hydrogen in the composite wall

Solution

We first consider the nickel layer (see figure below)

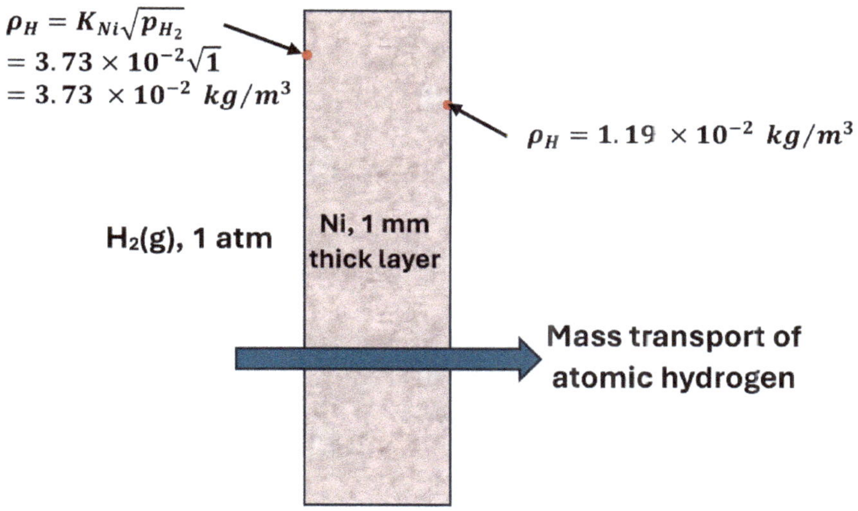

The mass flux of atomic hydrogen through the nickel layer is:

$$j_H = -D_{H-Ni}\frac{d\rho_H}{dx} = -\frac{D_{AB}(\rho_{H,2} - \rho_{H,1})}{L}$$

where j_H is the mass flux of atomic hydrogen, $D_{H-Ni}$is diffusivity of atomic hydrogen in nickel, and ρ_H is mass concentration of atomic hydrogen in the medium. From the above equation:

$$j_H = -\frac{D_{H-Ni}(\rho_{H,2}-\rho_{H,1})}{L} = -\frac{4.9\times10^{-10}(1.19\times10^{-2}-3.73\times10^{-2})}{1\times10^{-3}} = 1.2446 \times 10^{-8}\ kg/m^2.s$$

i) Calculate the mass flux of atomic hydrogen in the composite wall

At steady state, the mass flux of atomic hydrogen through the nickel and palladium layers is the same. Thus, the mass flux of atomic hydrogen in the composite wall is $1.2446 \times 10^{-8}\ kg/m^2.s$

ii) Calculate the concentration of atomic hydrogen on the Pd side of Ni/Pd interface

We now consider the palladium layer (see figure below). At steady state,

$$j_H(Ni) = j_H(Pd) = 1.2446 \times 10^{-8}\ kg/m^2.s$$

$$j_H = -\frac{D_{H-Pd}(0.351 - \rho_{H,1})}{1 \times 10^{-3}} = -\frac{6.4 \times 10^{-9} \times (0.351 - \rho_{H,1})}{1 \times 10^{-3}} = 1.2446 \times 10^{-8}\ kg/m^2.s$$

Solve the above equation for the unknown $\rho_{H,1}$. The solution is:

$$\rho_{H,1} = 0.353\ kg/m^3$$

Thus, the concentration of atomic hydrogen on the Pd side of Ni/Pd interface is $0.353\ kg/m^3$.

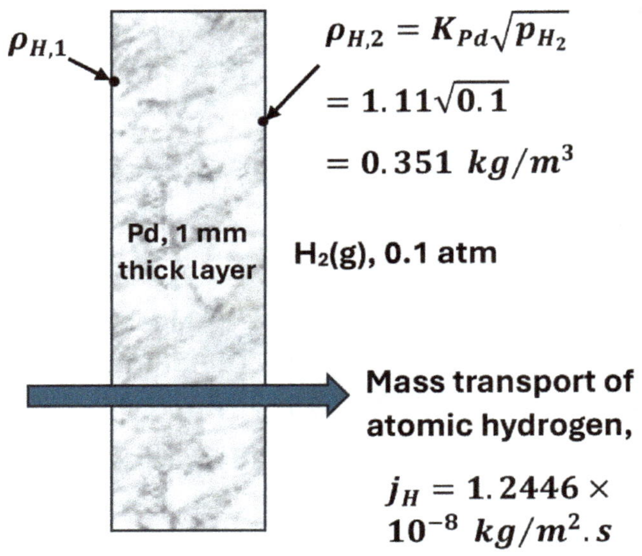

iii) Plot the concentration of atomic hydrogen in the composite wall

The plot of atomic hydrogen concentration in the composite wall is shown below.

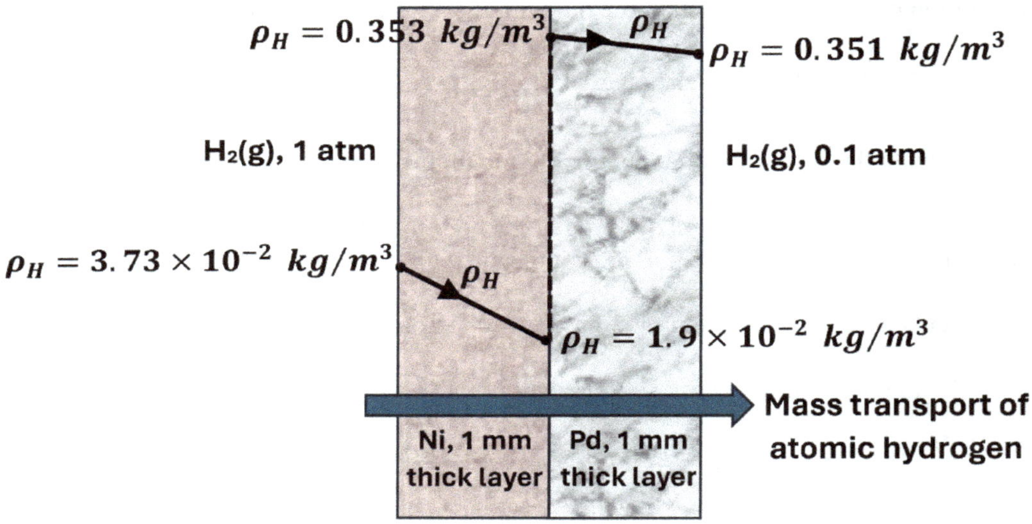

An important point to note is that at the interface between the two materials, nickel and palladium, there is no continuity of the concentration of diffusing species (atomic hydrogen). There occurs a sharp jump in the atomic hydrogen concentration at the interface.

16.8.9 Example 9

Aerobic fermentation is being considered to decompose organic matter into its components. The bacteria are uniformly distributed in organic matter. The oxygen to the bacteria is supplied from the atmosphere. Consider a slab of organic matter of thickness L supported on an inert material that is impermeable to oxygen (see figure below). The organic slab is exposed to air at the top. The concentration of oxygen (species A) in the slab at the surface exposed to air is maintained at $C_{A,0}$. The oxygen diffuses through the organic matter and is simultaneously consumed by the bacteria. The diffusivity of oxygen in organic matter is D_{AB} and the local volumetric rate at which oxygen is consumed by the bacteria is given in units of kmol/s.m^3 as $R_A = -k_1 C_A$, where k_1 is a rate constant. How does the concentration of oxygen $C_A(x)$ vary in the organic layer?

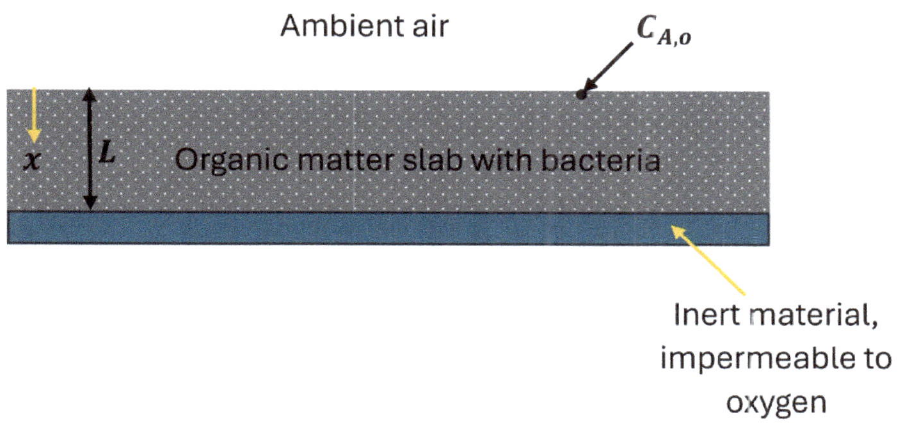

Solution

The equation governing concentration distribution of a diffusing species A in a medium with chemical reaction is as follows:

$$\frac{\partial C_A}{\partial t} = D_{AB}\nabla^2 C_A + R_A$$

Assuming steady state, one-dimensional diffusion in x direction, we get:

$$D_{AB}\,\frac{d^2 C_A}{dx^2} + R_A = 0 \text{ where } R_A = -k_1 C_A$$

We need to solve the following 2nd order, linear, homogeneous ODE with constant coefficients:

$$\frac{d^2 C_A}{dx^2} - \left(\frac{k_1}{D_{AB}}\right) C_A = 0$$

The boundary conditions are:

B.C. 1: at $x = 0,\ C_A = C_{A,o}$

B.C. 2: at $x = L,\ J_{A,x} = -D_{AB}\frac{dC_A}{dx} = 0,\ \frac{dC_A}{dx} = 0$

The solution for a 2nd order, linear, homogeneous ODE with constant coefficients, is $C_A = e^{mx}$. Upon substitution of $C_A = e^{mx}$ into the differential equation, we get:

$$m^2 - \frac{k_1}{D_{AB}} = 0$$

$$m = \pm\sqrt{\frac{k_1}{D_{AB}}}$$

Thus, the general solution of the differential equation under consideration is

$$C_A = ae^{(\sqrt{k_1/D_{AB}})x} + be^{-(\sqrt{k_1/D_{AB}})x}$$

Using the two boundary conditions, we get:

$$C_{A,o} = a + b$$

$$\left.\frac{dC_A}{dx}\right|_{x=L} = a\left(\sqrt{k_1/D_{AB}}\right)e^{(\sqrt{k_1/D_{AB}})L} - b\left(\sqrt{k_1/D_{AB}}\right)e^{-(\sqrt{k_1/D_{AB}})L} = 0$$

Solving the above two equations for a and b:

$$a = C_{A,o}\left[\frac{e^{-2(\sqrt{k_1/D_{AB}})L}}{1+e^{-2(\sqrt{k_1/D_{AB}})L}}\right]$$

$$b = \left[\frac{C_{A,o}}{1+e^{-2(\sqrt{k_1/D_{AB}})L}}\right]$$

Thus, the concentration distribution in the organic matter slab is:

$$\frac{C_A}{C_{A,o}} = \left[\frac{e^{(\sqrt{k_1/D_{AB}})(x-2L)}+e^{-(\sqrt{k_1/D_{AB}})x}}{1+e^{-2(\sqrt{k_1/D_{AB}})L}}\right]$$

16.8.10 Example 10

To harden the surface of mild steel parts, the part is exposed to carburizing gas in a furnace at high temperature for a certain period. Consider a steel part whose initial carbon content is 0.12 percent by mass. It is hardened in a furnace at 1150 K. The diffusivity of carbon in steel is 7.2×10^{-12} m²/s at the furnace temperature of 1150 K. We would like to continue the hardening process of the steel part until its mass fraction of carbon is raised to 0.32 percent at a depth of 0.7 mm from the exposed surface of the steel part. How long should we keep the steel part in the carburizing environment inside the furnace to achieve the desired hardening. You can assume that the surface of the steel part exposed to the carburizing gas is maintained at a carbon mass fraction of 0.011.

Solution

This problem involves transient diffusion. As carbon penetrates only a very thin layer at the surface of the steel part, it is reasonable to assume that the steel part is semi-infinite. Thus Equation (16.55), re-written in terms of mass fractions, is applicable.

$$\frac{\omega_A-\omega_{A,i}}{\omega_{A,s}-\omega_{A,i}} = 1 - erf\left[\frac{x}{2\sqrt{D_{AB}t}}\right]$$

where ω_A is mass fraction of diffusing species, that is, carbon.

Given that:

$$\omega_{A,i} = \frac{0.12}{100} = 0.0012$$

$\omega_{A,s} = 0.011$

$$\omega_A = \frac{0.32}{100} = 0.0032 \quad \text{at} \quad x = 0.70\ mm$$

$D_{AB} = 7.2 \times 10^{-12}$ m^2/s

Thus,

$$\frac{\omega_A - \omega_{A,i}}{\omega_{A,s} - \omega_{A,i}} = \frac{0.0032 - 0.0012}{0.011 - 0.0012} = 0.20408 = 1 - erf\left[\frac{x}{2\sqrt{D_{AB}t}}\right]$$

$$erf\left[\frac{x}{2\sqrt{D_{AB}t}}\right] = 0.79592$$

From error function tables,

$$\frac{x}{2\sqrt{D_{AB}t}} = 0.89838$$

$$t = \frac{x^2}{4D_{AB}(0.89838)^2} = \frac{(0.7 \times 10^{-3})^2}{4 \times 7.2 \times 10^{-12} \times (0.89838)^2} = 2.108 \times 10^4\ s = 5.856\ h$$

NOTATION

A	area of mass transfer
$A_{log-mean}$	log-mean area of a cylindrical shell
$A_{geometric-mean}$	geometric mean area of spherical shell
C	total molar concentration of species (including all species)
C_A	molar concentration of species A
$C_{A,i}$	initial molar concentration of species A
$C_{A,s}$	molar concentration of species A in the material medium at the surface
C_B	molar concentration of species B
D_{AB}	diffusivity of species A in species B
D_{AB}^o	diffusivity of species A in species B at infinitely dilute concentration of species A
erf	error function
F_A	force acting on the molecule of species A
$\hat{\imath}$	unit vector in x-direction of cartesian coordinates
$\vec{j}$	diffusional mass flux vector
$\vec{j}_A$	diffusional mass flux vector of species A
J_A	diffusional molar flux of species A
$J_{A,x}$	diffusional molar flux of species A in x-direction
$J_{A,y}$	diffusional molar flux of species A in y-direction
$J_{A,z}$	diffusional molar flux of species A in z-direction
$J_{A,r}$	diffusional molar flux of species A in r-direction
$J_{A,\theta}$	diffusional molar flux of species A in θ-direction
$J_{A,\varphi}$	diffusional molar flux of species A in φ-direction

$\vec{J_A}$	diffusional molar flux vector of species A
k	rate constant of a chemical reaction
k_B	Boltzmann constant
L	length or thickness
M	molecular weight
M_A	molecular weight of species A
M_B	molecular weight of species B
$\dot{M}_A$	molar flow rate of species A
N	Avogadro number
p	pressure
p_A	partial pressure of A
P	permeability
r	radial coordinate or distance in radial direction
$\hat{r}$	unit vector in radial direction
R_A	rate of production of species A per unit volume of the medium
R_m	resistance to diffusion mass transfer
S	solubility
t	time
T	temperature
T^*	dimensionless temperature, see Equation (16.7)
u_A	velocity of a molecule of species A
V_A	molar volume of species A
$\vec{V}$	velocity vector
x, y, z	space coordinates in cartesian coordinate system

Greek Letters

Δ	change in variable
ε	characteristic molecular interaction energy
ε_A	characteristic molecular interaction energy of pure A.
ε_B	characteristic molecular interaction energy of pure B.
ε_{AB}	characteristic energy of molecular interaction for the mixture
η	viscosity of solution
η_B	viscosity of solvent (species B)
θ	space co-ordinate
$\hat{\theta}$	unit vector in θ-direction
ρ_A	mass concentration of species A
σ	collision diameter
σ_A	collision diameter of species A
σ_B	collision diameter of species B
σ_{AB}	average collision diameter
φ	space co-ordinate in spherical co-ordinate system
$\hat{\varphi}$	unit vector in φ-direction
ψ_B	association parameter for the solvent, species B
ω_A	mass fraction of species A

Mathematical Operators

∇	del operator
∇^2	Laplacian operator

PRACTICE PROBLEMS

Problem 1

Using the Chapman-Enskog equation, estimate D_{AB} for the binary gaseous system of methane (species A) and propane (species B) at T=25 °C and P=1 atm.

Problem 2

Estimate D_{AB} for a binary liquid solution of species A and species B at T=25 °C. The species A is present in small amounts, that is, the solution is very dilute in A. Given that: $\eta = 0.90\ cP$, $V_A = 74\ cm^3/gmol$, $\psi_B = 2.6$, and $M_B = 18$.

Problem 3

It is being considered to use a polyethylene film for packaging of a pharmaceutical product. The film thickness is 0.12 mm. Determine the molar diffusion flux of O_2 from ambient air to inside the package. The partial pressure of O_2 outside is 0.21 atm and inside the package it is 0.008 atm. The permeability of O_2 through the film is 1.86×10^{-10} mol.m/(s.m^2.atm).

Problem 4

A rectangular vessel made of steel contains hydrogen gas. The thickness of the vessel walls is 13 mm. Hydrogen gas leaks through the walls of the vessel by a diffusion process. The molar concentration of hydrogen in the steel at the outside surface is negligible. On the inside surface of the steel wall, the molar concentration of hydrogen is 1.34 kmol/m^3.

Estimate the molar diffusion flux of hydrogen through the steel walls given that the diffusivity of hydrogen in steel is 0.26 x 10^{-12} m^2/s.

Problem 5

A spherical container made of nickel contains hydrogen gas. The outer diameter of the container is 5.15 m, and it is 8 cm thick. The molar concentration of hydrogen in the nickel at the outside surface is negligible. On the inside surface of the nickel shell, the molar concentration of hydrogen is 0.1 kmol/m^3. Estimate the mass flow rate at which hydrogen leaks from the nickel container through a diffusion process given that the diffusivity of hydrogen in nickel is 1.2 x 10^{-12} m^2/s and that the molecular weight of hydrogen is 2.

Problem 6

Consider diffusion of hydrogen gas across a 0.5 mm thick plastic membrane. On one side of the membrane, the pressure is 4.5 bars and on the other side of the membrane it is 1.2 bars. The permeability of H_2 through the film is 1.3×10^{-12} mol.m/(s.m^2.Pa). Determine the mass diffusion flux of hydrogen through the membrane? The molecular weight of hydrogen is 2.

Problem 7

The mass transport of oxygen through the lung tissue could be modelled approximately as a diffusion process in a slab (lung tissue) of thickness L. The molar concentration of oxygen (species A) in the tissue at $x = 0$ is maintained at $C_{A,o}$ due to inhalation of air from the outside. At the other end of the tissue at $x = L$, the assimilation of oxygen by blood maintains a constant molar concentration of oxygen equal to $C_{A,L}$. Due to metabolic processes, oxygen is also consumed by the lung tissue at a volumetric rate of $R = -k$, where R is in the units of mol/s.m^3.

i) Determine the distribution of oxygen concentration in the lung tissue

ii) Estimate the molar rate of oxygen transfer from the lung tissue to the blood.

Problem 8

We would like to develop a simple diffusion model for the estimation of the concentration distribution of gaseous NO_2 in the atmosphere. We are given that the molar flux of NO_2 at the ground level ($x = 0$) is $J_{A,o}$. The concentration of NO_2 as $x \to \infty$ is zero. In addition to diffusion, NO_2 undergoes a chemical reaction with unburned hydrocarbons in the atmosphere to produce peroxyacetyl nitrate. The volumetric rate at which NO_2 is consumed by a chemical reaction is given as $R = -k_1 C_A$ where R is in the units of mol/s.m^3 and C_A is the local molar concentration of NO_2 in units of mol/m^3. Determine the concentration distribution of NO_2 (species A) in the atmosphere above the ground level.

Problem 9

Consider combustion of fine coal particles in pure oxygen gas. The combustion reaction given below is carried out at 1175 °C and 1 atm.

$$C(s) + O_2(g) \to CO_2(g)$$

Estimate the rate of combustion of a spherical particle of radius 150 μm in units of gmol/s. You can assume that the combustion reaction is fast enough to keep the concentration of oxygen at the surface of the particle negligible. The diffusivity $D_{O_2-CO_2}$ is 1.69×10^{-4} m^2/s. The diffusion of oxygen to the particle surface is governed by the following differential equation:

$$\frac{d}{dr}\left(r^2 D_{AB} \frac{dC_A}{dr}\right) = 0$$

where D_{AB} is the diffusivity of oxygen in carbon-dioxide. You can treat oxygen as an ideal gas.

Problem 10

A spherical container made from fused silica (species B) is used to store helium gas (species A). The inside diameter of the container is 0.29 m, and its thickness is 3.5 mm. The initial pressure of helium in the container is 4.32 bars. Due to diffusion of helium through the container wall, the pressure of helium inside the container decreases. Determine the initial rate of decrease in pressure inside the container. You are given that: $D_{AB} = 0.4 \times 10^{-13}$ m^2/s and the solubility of helium in fused silica is $S = 0.45 \times 10^{-3}$ kmol/m^3.bar. The partial pressure of helium in the outside air is negligible. You can treat helium as an ideal gas.

Problem 11

Consider a thick slab of nickel containing atomic hydrogen at a uniform concentration of 10 gmol/m^3. One face of the slab is exposed to a vacuum to remove atomic hydrogen from the slab. The diffusivity of atomic hydrogen in nickel is 2.66×10^{-9} m^2/s.

i) How long does it take for the local atomic hydrogen concentration to decrease to 5 gmol/m^3 at a depth of 1.1 mm below the exposed surface?

ii) What is the molar flux of atomic hydrogen at the surface of the slab at time calculated in part (i)?

You can assume that the concentration of atomic hydrogen in the slab at the surface exposed to vacuum is zero?

FURTHER READING

1. T.L. Bergman, A.S. Levine, F.P. Incropera, and D.P. Dewitt. *Fundamentals of Heat and Mass Transfer*. 7th Edition, New York, Wiley & Sons, 2011.
2. R.B. Bird, W.E. Stewart, and E.N. Lightfoot. 2007. *Transport Phenomena*. 2nd Edition, Wiley & Sons, New York, 2007.
3. J.R. Welty, C.E. Wicks, R.E. Wilson, and G.L. Rorrer. *Fundamentals of Momentum, Heat, and Mass Transfer*. 5th Edition, Wiley & Sons, New York, 2008.
4. A. Bejan. *Heat Transfer*. Wiley & Sons, New York, 1993.
5. Y.A. Cengel. *Heat and Mass Transfer*, 3rd Edition, McGraw Hill, New York, 2007.
6. C.J. Geankoplis, A.A. Hersel, D.H. Lepek. *Transport Processes and Separation Process Principles*. 5th Edition, Prentice Hall, New Jersey, 2018.
7. W.L. McCabe, J.C. Smith, P. Harriott. *Unit Operations of Chemical Engineering*. 6th Edition, McGraw Hill, New York, 2001.
8. D.R.Gaskell. *An Introduction to Transport Phenomena in Materials Engineering*. Macmillan Publishing Company, New York, 1992.
9. R. Pal. *Electromagnetic, Mechanical, and Transport Properties of Composite Materials*. CRC Press, Boca Raton, Florida, 2015.
10. A.F. Mills. *Basic Heat and Mass Transfer*. 2nd Edition, Prentice Hall, New Jersey, 1999.
11. S. Middleman. *An Introduction to Mass and Heat Transfer*. Wiley & Sons, New York, 1998.

Chapter 17

Mass Transfer in a Moving Fluid (Convective Mass Transfer)

In the analysis and discussion presented in Chapter 16, it was assumed that the medium (mixture of species A and B) was stationary and that there was no bulk motion involved. Strictly speaking, this is generally not true. Even when species B is stationary and only species A is diffusing, the velocity of the medium (mixture) is non-zero. Therefore, the analysis and discussion presented in Chapter 16 is restricted to the situation where the bulk velocity of the medium is negligibly small. When the bulk velocity of the medium is non-negligible, the diffusional fluxes $\vec{J}_A$ (molar flux) or $\vec{j}_A$ (mass flux) are no longer the absolute fluxes, that is, fluxes with respect to a stationary observer. They are instead relative fluxes, that is, fluxes with respect to an observer moving with the velocity of the medium.

The transfer of mass from a high concentration region to a low concentration region of a fluid in the presence of bulk macroscopic motion of the fluid is referred to as "convective mass transfer". Mass transfer between a solid boundary and a fluid flowing adjacent to the solid boundary when there exists a concentration difference between the fluid and the solid surface is also referred to as "convective mass transfer". Thus, convective mass transfer operations involve bulk fluid motion as well as concentration variations within the fluid and between the fluid and contact surface.

In a convective mass transfer process, mass of some species is transported from one region of fluid to another by two mechanisms acting simultaneously. The mechanisms are "molecular diffusion" and "advection" (advection being caused by bulk macroscopic motion of fluid).

When the medium is in motion ($\vec{V} \neq 0$), then an advective mass flux of species A, given as $\rho_A\vec{V}$, is present in addition to the diffusional mass flux $\vec{j}_A$ at any given point in the medium. Thus, the total mass flux of species A at any given point in the medium consists of two parts:

$$\vec{n}_A = \vec{j}_A + \rho_A\vec{V} \tag{17.1}$$

where $\vec{n}_A$ and $\vec{j}_A$ are the vectorial total and diffusional mass fluxes of species A, respectively, and $\rho_A\vec{V}$ is the advective mass flux of species A. Note that $\vec{n}_A$ is the absolute mass flux of species A relative to a stationary observer whereas $\vec{j}_A$ is the relative mass flux of species A relative to the observer moving with velocity $\vec{V}$ of the medium.

The bulk or macroscopic motion in mixtures of different species can be caused either by external means such as a pump or by molecular-level diffusion process itself. In the latter case, the advective mass flux of species A (given as $\rho_A\vec{V}$) is often referred to as "bulk diffusion flux".

17.1 SOME DEFINITIONS AND CONCEPTS

17.1.1 Velocities

The mass-average velocity $\vec{V}$ of a multi-component mixture is defined as:

$$\vec{V} = \frac{\sum_i \rho_i \vec{V}_i}{\rho} \tag{17.2}$$

where ρ_i is the mass concentration of species i (mass per unit volume of mixture), ρ is the total concentration of the mixture ($\rho = \sum_i \rho_i$), and $\vec{V}_i$ is the absolute velocity of species i (relative to a stationary observer). Note that $\vec{V}_i$ at any point in the mixture is the average velocity of all the species i molecules contained in a small volume about the point.

For a binary mixture of species, A and B, the mass-average velocity becomes:

$$\vec{V} = \frac{\rho_A \vec{V}_A + \rho_B \vec{V}_B}{\rho} \tag{17.3}$$

where $\vec{V}_A$ and $\vec{V}_B$ are absolute velocities of species A and B, respectively.

The molar-average velocity ($\vec{V}^*$) of a multi-component mixture is defined as:

$$\vec{V}^* = \frac{\sum_i C_i \vec{V}_i}{C} \tag{17.4}$$

where C_i is the molar concentration of species i (moles per unit volume of mixture), and C is the total molar concentration of the mixture ($C = \sum_i C_i$). For a binary mixture:

$$\vec{V}^* = \frac{C_A \vec{V}_A + C_B \vec{V}_B}{C} \tag{17.5}$$

17.1.2 Fluxes

As already noted, mass transfer flux could be defined in terms of either mass or moles, and it is a vector quantity. In any given direction, it is the amount (either mass or molar units) of a particular species that passes through a unit area normal to the given direction per unit time.

Flux may be defined either with reference to co-ordinates fixed in space (stationary coordinates) or with reference to coordinates moving with either $\vec{V}$ (mass-average velocity of the medium) or $\vec{V}^*$ (molar-average velocity of the medium).

The absolute mass flux of species i with respect to stationary coordinates, $\vec{n}_i$, is defined as:

$$\vec{n}_i = \rho_i \vec{V}_i \tag{17.6}$$

The diffusional mass flux of species i with respect to mass-average velocity ($\vec{V}$) of the medium is given as:

$$\vec{j}_i = \rho_i (\vec{V}_i - \vec{V}) \tag{17.7}$$

where $\vec{V}_i$ is the absolute velocity of species i.

The absolute molar flux of species i, $\vec{N}_i$, and diffusional molar flux of species i, $\vec{J}_i$, are defined as:

$$\vec{N}_i = C_i \vec{V}_i \tag{17.8}$$

$$\vec{J}_i = C_i(\vec{V}_i - \vec{V}^*) \tag{17.9}$$

17.2 RELATIONSHIP BETWEEN FLUXES

17.2.1 Mass Fluxes

From Equations (17.2), (17.6), and (17.7), it follows that:

$$\vec{j}_i = \rho_i(\vec{V}_i - \vec{V}) = \rho_i\vec{V}_i - \rho_i \sum_{j=1}^{N} \frac{\rho_j \vec{V}_j}{\rho} = \vec{n}_i - \omega_i \sum_{j=1}^{N} \vec{n}_j \tag{17.10}$$

where N is the total number of species in the multi-component mixture and ω_i is the mass fraction of species i. Equation (17.10) can be re-written as:

$$\vec{n}_i = \vec{j}_i + \omega_i \sum_{j=1}^{N} \vec{n}_j \tag{17.11}$$

For a binary mixture of species, A and B, Equation (17.11) can be expressed as:

$$\vec{n}_A = \vec{j}_A + \omega_A(\vec{n}_A + \vec{n}_B) \tag{17.12}$$

$$\vec{n}_B = \vec{j}_B + \omega_B(\vec{n}_A + \vec{n}_B) \tag{17.13}$$

Using Fick's Law of diffusion (Equation (16.58)), Equations (17.12) and (17.13) become:

$$\vec{n}_A = -\rho D_{AB} \nabla \omega_A + \omega_A(\vec{n}_A + \vec{n}_B) \tag{17.14}$$

$$\vec{n}_B = -\rho D_{BA} \nabla \omega_B + \omega_B(\vec{n}_A + \vec{n}_B) \tag{17.15}$$

Interestingly,

$$\sum_{i=1}^{N} \vec{j} = 0 \tag{17.16}$$

that is, the sum of mass diffusional fluxes of all species present in the mixture is always zero. This result simply follows from Equation (17.10) when it is summed over N species:

$$\sum_{i=1}^{N} \vec{j} = \sum_{i=1}^{N} \vec{n}_i - \left(\sum_{j=1}^{N} \vec{n}_j\right)\left(\sum_{i=1}^{N} \vec{\omega}_i\right) = 0 \tag{17.17}$$

Note that:

$$\sum_{i=1}^{N} \vec{\omega}_i = 1 \quad \text{and} \quad \sum_{i=1}^{N} \vec{n}_i = \sum_{j=1}^{N} \vec{n}_j \tag{17.18}$$

Thus, in a binary mixture of species A and B, $\vec{j}_A$ is always equal to $-\vec{j}_B$:

$$\vec{j}_A = -\vec{j}_B \tag{17.19}$$

The addition of Equations (17.14) and (17.15) leads to another interesting result, that is,

$$D_{AB} = D_{BA} \tag{17.20}$$

17.2.2 Molar Fluxes

From Equations (17.4), (17.8), and (17.9), it follows that:

$$\vec{J}_i = C_i\vec{V}_i - C_i\vec{V}^* = \vec{N}_i - \frac{C_i}{C}\sum_{j=1}^{N} C_j\vec{V}_j = \vec{N}_i - x_i\sum_{j=1}^{N}\vec{N}_j \tag{17.21}$$

Or,

$$\vec{N}_i = \vec{J}_i + x_i\sum_{j=1}^{N}\vec{N}_j \tag{17.22}$$

where x_i is the mole fraction of species i in the mixture. For a binary mixture of species, A and B, Equation (17.22) gives:

$$\vec{N}_A = \vec{J}_A + x_A(\vec{N}_A + \vec{N}_B) \tag{17.23}$$

$$\vec{N}_B = \vec{J}_B + x_B(\vec{N}_A + \vec{N}_B) \tag{17.24}$$

Using Fick's law of diffusion (Equation (16.57)), Equations (17.23) - (17.24) can be re-written as:

$$\vec{N}_A = -CD_{AB}\nabla x_A + x_A(\vec{N}_A + \vec{N}_B) \tag{17.25}$$

$$\vec{N}_B = -CD_{BA}\nabla x_B + x_B(\vec{N}_A + \vec{N}_B) \tag{17.26}$$

The addition of Equations (17.25) and (17.26) leads to $D_{AB} = D_{BA}$. Furthermore, Equation (17.22) when summed over N species gives:

$$\sum_{i=1}^{N}\vec{J}_i = 0 \tag{17.27}$$

Thus, for a binary mixture, Equation (17.27) implies that:

$$\vec{J}_A = -\vec{J}_B \tag{17.28}$$

17.2.3 Conditions Under Which Absolute Flux and Diffusional Flux are Equal

Consider Equation (17.23) as follows:

$$\vec{N}_A = \vec{J}_A + x_A\vec{N}_A + x_A\vec{N}_B \tag{17.29}$$

When species B is stationary, Equation (17.29) gives:

$$\vec{N}_A = \vec{J}_A + x_A\vec{N}_A \tag{17.30}$$

If the diffusing species A is present in small amounts such that $x_A \ll 1$, Equation (17.30) gives:

$$\vec{N}_A = \vec{J}_A \tag{17.31}$$

Thus, the absolute flux and diffusional flux are equal under the conditions: Species B is stationary, and the concentration of diffusing species A is small. Under these conditions, the analysis presented in Chapter 16 is applicable. This is often the case with diffusion of solute in liquid solutions and solids.

17.3 MATERIAL BALANCE

To obtain the concentration distribution of species A in a moving medium (mixture of A and B), material balance of species A is required. Consider the differential control volume shown in Figure 17.1. Species A is entering and leaving the control volume through absolute molar fluxes as shown in the figure. The species A is also being generated within the control volume through a chemical reaction. Let the rate of production of species A per unit volume through chemical reaction be R_A. The material or mole balance over the control volume gives:

$$N_{A,x}\big|_x A - N_{A,x}\big|_{x+\Delta x} A + R_A(A\Delta x) = (A\Delta x)\frac{\partial C_A}{\partial t} \quad (17.32)$$

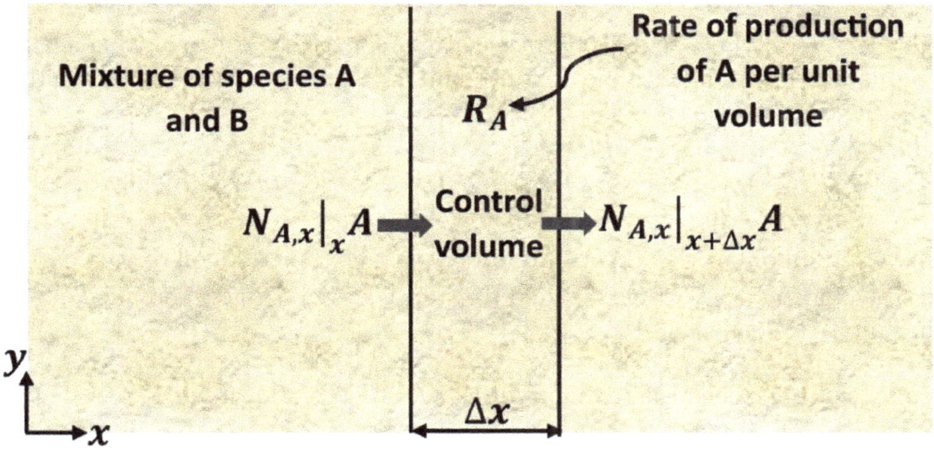

Figure 17.1 Material balance over a differential control volume.

Upon re-arrangement, Equation (17.32) becomes:

$$-\left[\frac{N_{A,x}\big|_{x+\Delta x} - N_{A,x}\big|_x}{\Delta x}\right] + R_A = \frac{\partial C_A}{\partial t} \quad (17.33)$$

Taking the limit $\Delta x \to 0$, we get:

$$-\frac{\partial N_{A,x}}{\partial x} + R_A = \frac{\partial C_A}{\partial t} \quad (17.34)$$

Equation (17.34) can now be generalized to a three-dimensional form as:

$$-\frac{\partial N_{A,x}}{\partial x} - \frac{\partial N_{A,y}}{\partial y} - \frac{\partial N_{A,z}}{\partial z} + R_A = \frac{\partial C_A}{\partial t} \quad (17.35)$$

Thus,

$$-\nabla \cdot \vec{N}_A + R_A = \frac{\partial C_A}{\partial t} \quad (17.36)$$

From Equation (17.8) and (17.9),

$$\vec{N}_A = \vec{J}_A + C_A\vec{V}^* \quad (17.37)$$

Upon substitution of $\vec{N}_A$ from Equation (17.37) into Equation (17.36), we obtain:

$$\frac{\partial C_A}{\partial t} = -\nabla \cdot \vec{J}_A - C_A(\nabla \cdot \vec{V}^*) - \vec{V}^* \cdot \nabla C_A + R_A \quad (17.38)$$

From Fick's law of diffusion:

$$\vec{J}_A = -C D_{AB} \nabla x_A \quad (17.39)$$

Upon substitution of $\vec{J}_A$ from Equation (17.39) into Equation (17.38), we obtain:

$$\frac{\partial C_A}{\partial t} = \nabla \cdot (C D_{AB} \nabla x_A) - C_A(\nabla \cdot \vec{V}^*) - \vec{V}^* \cdot \nabla C_A + R_A \quad (17.40)$$

This is a very general form of the material balance equation for species A in a moving medium. We can call it Fick's second law of diffusion for a moving medium. An equivalent relationship in terms of mass units (instead of molar units) is as follows:

$$\frac{\partial \rho_A}{\partial t} = \nabla \cdot (\rho D_{AB} \nabla \omega_A) - \rho_A(\nabla \cdot \vec{V}) - \vec{V} \cdot \nabla \rho_A + r_A \quad (17.41)$$

where ρ_A is mass concentration of A (kg/m^3), ρ is total mass concentration of mixture, ω_A is mass fraction of A, $\vec{V}$ is mass average velocity of the medium (mixture), and r_A is rate of production of A by chemical reaction in the units of kg/(s.m^3).

Under the assumptions of no chemical reaction, constant C and D_{AB}, Equation (17.40) simplifies to:

$$\frac{\partial C_A}{\partial t} = D_{AB} \nabla^2 C_A - C_A(\nabla \cdot \vec{V}^*) - \vec{V}^* \cdot \nabla C_A \quad (17.42)$$

For a stationary medium, $\vec{V}^* = 0$, Equation (17.42) gives:

$$\frac{\partial C_A}{\partial t} = D_{AB} \nabla^2 C_A \quad (17.43)$$

This equation, referred to as mass diffusion equation or Fick's second law of diffusion for a stationary medium, was derived in Chapter 16 (see Equation (16.22)).

17.4 DILUTE SYSTEMS

Consider an isothermal non-reacting binary mixture of species A and B. Let the mixture viscosity (η), mixture total mass concentration (ρ), mixture total molar concentration (C), and diffusion coefficient D_{AB} be constant. With these restrictions, the governing equations of convective mass transfer are as follows:

$$\nabla \cdot \vec{V} = 0 \qquad \text{(continuity equation)} \quad (17.44)$$

$$\rho \left[\frac{\partial \vec{V}}{\partial t} + \vec{V} \cdot \nabla \vec{V}\right] = -\nabla p + \rho \vec{g} + \eta \nabla^2 \vec{V} \qquad \text{(Navier-Stokes equation)} \quad (17.45)$$

$$\frac{\partial C_A}{\partial t} = D_{AB} \nabla^2 C_A - C_A(\nabla \cdot \vec{V}^*) - \vec{V}^* \cdot \nabla C_A \qquad \text{(material balance on A)} \quad (17.46)$$

For systems dilute in A: $\eta \approx \eta_B$, $\rho \approx \rho_B$, $C \approx C_B$, and $\vec{V}^* \approx \vec{V}_B$. Thus, the governing equations become:

$$\nabla \cdot \vec{V}_B = 0 \tag{17.47}$$

$$\rho_B \left[\frac{\partial \vec{V}_B}{\partial t} + \vec{V}_B \cdot \nabla \vec{V}_B\right] = -\nabla p + \rho_B \vec{g} + \eta_B \nabla^2 \vec{V}_B \tag{17.48}$$

$$\frac{\partial C_A}{\partial t} + \vec{V}_B \cdot \nabla C_A = D_{AB} \nabla^2 C_A \tag{17.49}$$

where fluid B is the solvent and species A is the solute present at a dilute level. These three equations, Equations (17.47) – (17.49), need to be solved simultaneously to determine the velocity, pressure, and concentration distribution in the fluid. Note that the hydrodynamic problem (solution of equations of motion, Equations (17.47) and (17.48)) is uncoupled from the mass transfer problem (solution of Equation (17.49)). Thus, Equation (17.49) can be solved after obtaining the velocity distribution from the equations of motion.

17.5 APPLICATIONS

17.5.1 Mass Transport of Species from the Free Surface into a Falling Liquid Film

Consider the steady laminar flow of a Newtonian liquid film flowing down a flat inclined plate due to gravity (see Figure 17.2). Species A is diffusing into the falling liquid film from the free surface. The plate is inclined at an angle α with the horizontal. The length and width of the plate are L and W, respectively. The thickness of the liquid film is δ. The concentration of the species A at gas/liquid interface ($x = 0$) within the liquid film is $C_{A,o}$. The plate is impermeable to A (and B). No chemical reaction is taking place and $C_A = 0$ at the entrance where $z = 0$. Species A is only sparingly soluble in B and the contact time between the liquid and species A, that is, L/V_{max} is small. Determine the concentration distribution $C_A(x, z)$ in the fluid. What is the molar flux of A at the gas/liquid interface?

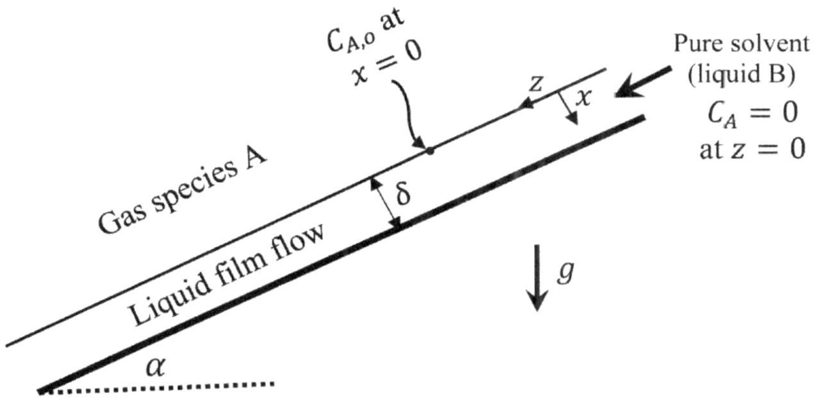

Figure 17.2 Mass transport of species from the free surface into a falling liquid film.

Solution

Velocity Distribution:
We assume that the amount of species A present in the liquid film is so small that any changes in the properties of liquid solvent B can be neglected. Thus, the liquid present in the film is treated as pure liquid B for solving the hydrodynamic problem.

As the flow is in z-direction, only V_z is non-zero. From continuity equation:

$$\frac{\partial V_z}{\partial z} = 0 \tag{17.50}$$

The z -component of the Navier-Stokes equation in Cartesian coordinates gives:

$$\eta \frac{d^2 V_z}{dx^2} = -\rho g_z \tag{17.51}$$

As $g_z = g \sin \alpha$, Equation (17.51) becomes:

$$\eta \frac{d^2 V_z}{dx^2} = -\rho g \sin \alpha \tag{17.52}$$

Upon integration:

$$V_z = -\left(\frac{\rho g \sin \alpha}{\eta}\right)\frac{x^2}{2} + \mathrm{a}x + \mathrm{b} \tag{17.53}$$

The integration constants a and b can be determined from the following boundary conditions:

B.C. 1: at $x = 0$, $\tau_{xz} = 0, \frac{dV_z}{dx} = 0$ (shear stress at liquid surface is zero) (17.54)

B.C. 2: $at\ x = \delta, V_z = 0$ (no-slip boundary condition) (17.55)

The solution is:

$$\mathrm{a} = 0\,, \quad \mathrm{b} = \left(\frac{\rho g \sin \alpha}{\eta}\right)\frac{\delta^2}{2} \tag{17.56}$$

$$V_z(x) = V_{max}\left[1 - \left(\frac{x}{\delta}\right)^2\right] \tag{17.57}$$

where

$$V_{max} = \frac{\rho g (\sin \alpha)\delta^2}{2\eta} \tag{17.58}$$

Concentration Distribution:
Assuming diffusion in the z −direction is negligible as compared with diffusion in the x −direction, that is, $\frac{\partial^2 C_A}{\partial z^2} \ll \frac{\partial^2 C_A}{\partial x^2}$, Equation (17.49) simplifies to:

$$V_z \frac{\partial C_A}{\partial z} = D_{AB}\frac{\partial^2 C_A}{\partial x^2} \tag{17.59}$$

Using Equation (17.57), Equation (17.59) gives:

$$V_{max}\left[1 - \left(\frac{x}{\delta}\right)^2\right]\frac{\partial C_A}{\partial z} = D_{AB}\frac{\partial^2 C_A}{\partial x^2} \tag{17.60}$$

Equation (17.60) needs to be solved subject to the boundary conditions:

B.C. 1: at $z = 0$, $C_A = 0$ (17.61)

B.C. 2: at $x = 0$, $C_A = C_{A,o}$ (17.62)

B.C. 3: at $x = \delta$, $\frac{\partial C_A}{\partial x} = 0$ (17.63)

The solution of Equation (17.60), subjected to the given boundary conditions, can be obtained using the separation of variable technique. The solution is in the form of an infinite series.

However, Equation (17.60) can be simplified further by assuming that the diffusing species A does not penetrate too far in the liquid film due to short contact time between species A and liquid (L/V_{max} is small) and low solubility of species A. Thus, the concentration variation of species A is restricted in a very thin liquid layer at the top of the film where $V_z = V_{max}$. Consequently, Equation (17.60) can be simplified to:

$$V_{max}\frac{\partial C_A}{\partial z} = D_{AB}\frac{\partial^2 C_A}{\partial x^2} \tag{17.64}$$

We need to solve Equation (17.64) subject to the boundary conditions: B.C. 1 (Equation (17.61)), B.C. 2 (Equation (17.62)), and the following third boundary condition:

$$\text{B.C. 3: at } x = \infty,\ C_A = 0 \tag{17.65}$$

The third boundary condition is modified as $C_A = 0$ for relatively large values of x when the diffusing species A does not penetrate too far in the liquid film.

Equation (17.64) can be solved using the similarity technique. According to the similarity `technique, the two independent variables (x and z) are combined into a single variable η (called the similarity variable) such that the original partial differential equation is converted into an ordinary differential equation. For the problem at hand, the similarity variable is:

$$\eta = \frac{x}{\sqrt{z}} \tag{17.66}$$

The partial differential equation (Equation (17.64)) can now be converted into an ordinary differential equation using η. Thus, Equation (17.64) becomes:

$$\frac{d^2 C_A}{d\eta^2} = -\frac{\eta V_{max}}{2D_{AB}}\frac{dC_A}{d\eta} \tag{17.67}$$

and the new boundary conditions are:

$$C_A|_{\eta=0} = C_{A,o} \quad \text{and} \quad C_A|_{\eta=\infty} = 0 \tag{17.68}$$

The solution of Equation (17.67) subjected to the given boundary conditions is as follows:

$$\frac{C_A}{C_{A,o}} = 1 - erf\left[\frac{x}{\sqrt{\frac{4D_{AB}z}{V_{max}}}}\right] \tag{17.69}$$

where erf is the error function.

Molar Flux of A at the Gas/Liquid Interface:

The molar flux $N_{A,x}$ is given as:

$$N_{A,x} = -D_{AB}\frac{\partial C_A}{\partial x} + C_A V_x = -D_{AB}\frac{\partial C_A}{\partial x} \tag{17.70}$$

Note that $V_x = 0$. From Equation (17.69), we get:

$$\left.\frac{\partial C_A}{\partial x}\right|_{x=0} = -C_{A,o}\sqrt{\frac{V_{max}}{\pi D_{AB} z}} \tag{17.71}$$

Thus, the molar flux $N_{A,x}$ at the gas/liquid interface is:

$$\left.N_{A,x}\right|_{x=0} = -D_{AB}\left.\frac{\partial C_A}{\partial x}\right|_{x=0} = C_{A,o}\sqrt{\frac{D_{AB}V_{max}}{\pi z}} \tag{17.72}$$

17.5.2 Mass Transport of Species from the Wall into a Falling Liquid Film

Consider the steady laminar flow of a liquid film (species B) down a flat inclined wall containing species A. Mass transfer of species A takes place from the wall to the liquid film. Species A is only sparingly soluble in the liquid and its concentration is $C_{A,o}$ at the solid/liquid interface on the liquid side (see Figure 17.3). At the entrance, the liquid is pure species B, that is, at $z = 0,\ C_A = 0$. The length and width of the wall are L and W, respectively. The thickness of the liquid film is δ. The concentration of species A at the surface of the film ($y = \delta$) is zero. We are interested in determining the concentration distribution in the film and the rate of mass transfer from the wall to the liquid film. Assume that the fluid properties are constant and that the concentration C_A changes are restricted to the region close to the wall due to short contact time between the fluid and the wall (L/V_{max} is small).

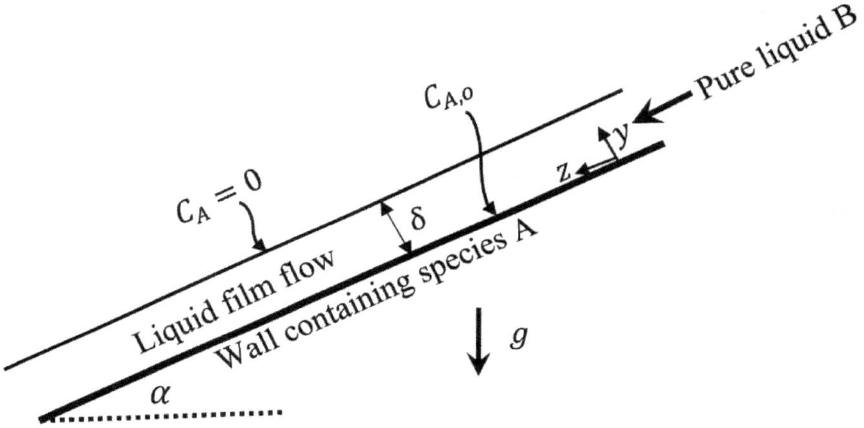

Figure 17.3 Mass transport of species from the wall into a falling liquid film.

Solution

Velocity Distribution:
The velocity distribution in the film was obtained in Chapter 12 (Equation (12.84)) as follows:

$$V_z(y) = V_{max}\left[1-\left(1-\frac{y}{\delta}\right)^2\right] = V_{max}\left[2\left(\frac{y}{\delta}\right)-\left(\frac{y}{\delta}\right)^2\right] \quad \text{where} \quad V_{max} = \frac{\rho g(\sin\alpha)\delta^2}{2\eta} \tag{17.73}$$

Close to the wall, $y/\delta \gg (y/\delta)^2$ and therefore, the velocity distribution is approximately linear in the wall region:

$$V_z = 2V_{max}\left(\frac{y}{\delta}\right) = \left(\frac{\rho g\delta\sin\alpha}{\eta}\right)y \tag{17.74}$$

Concentration Distribution:

Assuming diffusion in the z −direction is negligible as compared with diffusion in the y −direction, that is, $\frac{\partial^2 C_A}{\partial z^2} \ll \frac{\partial^2 C_A}{\partial y^2}$, Equation (17.49) simplifies to:

$$V_z \frac{\partial C_A}{\partial z} = D_{AB} \frac{\partial^2 C_A}{\partial y^2} \tag{17.75}$$

The diffusing species A does not penetrate too far in the liquid film due to short contact time between species A and liquid (L/V_{max} is small) and low solubility of species A. Thus, the concentration variation of species A is restricted to a very thin liquid layer close to the wall region where V_z is linear given by Equation (17.74). From Equations (17.74) and (17.75), we obtain:

$$\left(\frac{\rho g \delta \sin\alpha}{\eta}\right) y \frac{\partial C_A}{\partial z} = D_{AB} \frac{\partial^2 C_A}{\partial y^2} \tag{17.76}$$

Using re-arrangement, Equation (17.76) gives:

$$y \frac{\partial C_A}{\partial z} = \beta \frac{\partial^2 C_A}{\partial y^2} \tag{17.77}$$

where $\beta = D_{AB}\eta/\rho g \delta \sin\alpha$. To obtain the concentration distribution, Equation (17.77) needs to be solved subject to the following boundary conditions:

$$\text{B.C. 1: at } z = 0,\ C_A = 0 \tag{17.78}$$

$$\text{B.C. 2: at } y = 0,\ C_A = C_{A,o} \tag{17.79}$$

$$\text{B.C. 3: at } y = \infty,\ C_A = 0 \tag{17.80}$$

Equation (17.77) can be solved by using the combination of variables. Let ε be a new variable defined as:

$$\varepsilon = y/(9\beta z)^{1/3} \tag{17.81}$$

The partial differential equation in Equation (17.77) can be reduced to the following ordinary differential equation by substitution of the new variable ε:

$$\left(\frac{d^2\Upsilon}{d\varepsilon^2}\right) + 3\varepsilon^2 \frac{d\Upsilon}{d\varepsilon} = 0 \quad \text{where} \quad \Upsilon = \frac{C_A}{C_{A,o}} \tag{17.82}$$

The boundary conditions become:

$$\text{B.C.1: } \varepsilon = 0,\quad \Upsilon = 1 \tag{17.83}$$

$$\text{B.C.2: } \varepsilon = \infty,\quad \Upsilon = 0 \tag{17.84}$$

The following solution is obtained for Equation (17.82) subject to the given boundary conditions:

$$\Upsilon = \frac{C_A}{C_{A,o}} = \frac{\int_\varepsilon^\infty e^{-\lambda^3} d\lambda}{\int_0^\infty e^{-\lambda^3} d\lambda} = \frac{\int_\varepsilon^\infty e^{-\lambda^3} d\lambda}{\Gamma(4/3)} \tag{17.85}$$

where Γ is the gamma function.

Molar Flux of A at the Wall:

The molar flux $N_{A,y}$ is given as:

$$N_{A,y} = -D_{AB}\frac{\partial C_A}{\partial y} + C_A V_y = -D_{AB}\frac{\partial C_A}{\partial y} \tag{17.86}$$

Note that $V_y = 0$. From Equation (17.85), we get:

$$\left.\frac{\partial C_A}{\partial y}\right|_{y=0} = -\frac{C_{A,o}}{(9\beta z)^{1/3}\Gamma(4/3)} \tag{17.87}$$

Thus, the molar flux $N_{A,y}$ at the wall is:

$$\left.N_{A,y}\right|_{y=0} = -D_{AB}\left.\frac{\partial C_A}{\partial y}\right|_{y=0} = \frac{C_{A,o}D_{AB}}{(9\beta z)^{1/3}\Gamma(4/3)} \tag{17.88}$$

Note that $\beta = D_{AB}\eta/\rho g\delta \sin\alpha$.

The rate of mass transfer of species A from the wall (inclined plate) surface to the liquid film in molar units can be calculated as follows:

$$\dot{M}_A = W\int_0^L \left.N_{A,y}\right|_{y=0} dz = \frac{1.5C_{A,o}D_{AB}WL}{(9\beta L)^{1/3}\Gamma(4/3)} \tag{17.89}$$

17.6 EXAMPLES

17.6.1 Example 1

Consider one-dimensional equimolar counter diffusion in a binary gas mixture of non-reacting species A and B at constant temperature and pressure. Derive the expressions for the absolute molar fluxes of species A and B.

Solution

Refer to the figure below.

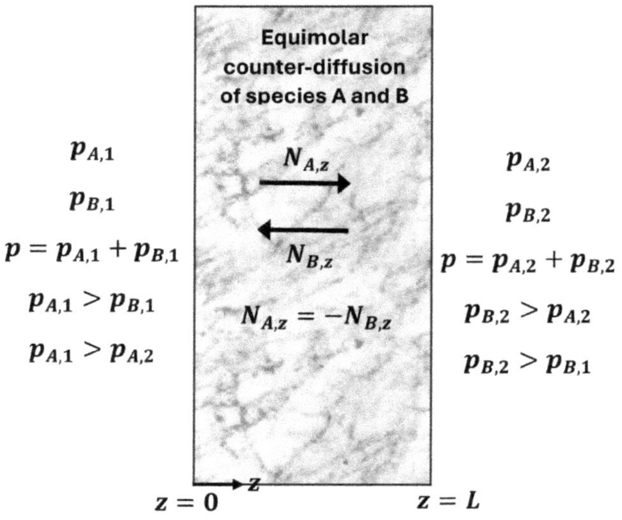

In equimolar counter diffusion, 1 mole of species A diffuses in one direction for each mole of species B diffusing in the opposite direction. This type of behavior is often observed in a binary distillation process where a mole of species A diffuses to the gas-liquid interface and condenses for every mole of species B that evaporates and moves away from the interface.

In such cases, $\vec{N}_A = -\vec{N}_B$ and the molar-average velocity $(\vec{V}^*)$ of the mixture is zero. From Equations (17.25) and (17.26):

$$\vec{N}_A = -CD_{AB}\nabla x_A \quad ; \quad \vec{N}_B = -CD_{BA}\nabla x_B$$

Let z be the direction of mass transfer. For one-dimensional diffusion in z-direction, we can write the above equations as:

$$N_{A,z} = -CD_{AB}\frac{dx_A}{dz} \quad ; \; N_{B,z} = -CD_{BA}\frac{dx_B}{dz}$$

Assuming C, D_{AB}, and D_{BA} to be constant, we get:

$$N_{A,z} = -CD_{AB}\frac{dx_A}{dz} = -D_{AB}\left[\frac{C_{A,2}-C_{A,1}}{L}\right] \; ; \; N_{B,z} = -CD_{BA}\frac{dx_A}{dz} = -D_{BA}\left[\frac{C_{B,2}-C_{B,1}}{L}\right]$$

Using ideal gas law,

$$C_A = \frac{p_A}{RT} \; ; \; C_B = \frac{p_B}{RT}$$

Thus,

$$N_{A,z} = -\frac{D_{AB}}{RT}\left[\frac{p_{A,2}-p_{A,1}}{L}\right] \; ; \; N_{B,z} = -\frac{D_{BA}}{RT}\left[\frac{p_{B,2}-p_{B,1}}{L}\right]$$

As $p = p_{A,1} + p_{B,1} = p_{A,2} + p_{B,2}$, it follows that:

$$p_{B,2} - p_{B,1} = \left(p_{A,1} - p_{A,2}\right)$$

Consequently,

$$N_{A,z} = -\frac{D_{AB}}{RT}\left[\frac{p_{A,2}-p_{A,1}}{L}\right] \; ; \; N_{B,z} = -\frac{D_{AB}}{RT}\left[\frac{p_{A,1}-p_{A,2}}{L}\right] = -N_{A,z}$$

As expected, $N_{A,z}$ and $N_{B,z}$ are equal and opposite in an equimolar counter diffusion process. Note that the diffusivity in binary mixtures is symmetric, that is, $D_{AB} = D_{BA}$.

17.6.2 Example 2

Consider two large containers containing gas mixtures of N_2 and CO_2 at 273 K and 1 atm (101.3 kPa), as shown in the figure below. The containers are connected to each other by a tube of diameter 75 mm and length 1.5 m. The partial pressure of CO_2 in the left container (see figure) is maintained at 15 kPa and in the right container is maintained at 6 kPa. Assuming equimolar counter diffusion, do the following:

i) Estimate the absolute molar fluxes of CO_2 and N_2

ii) Calculate the mass flow rates of CO_2 and N_2

The diffusivity of CO_2 in N_2 is $D_{CO_2-N_2} = 1.44 \times 10^{-5}$ m²/s. The molecular weights of CO_2 and N_2 are 44 and 28, respectively. You can further assume that the gas mixture behaves as an ideal gas mixture.

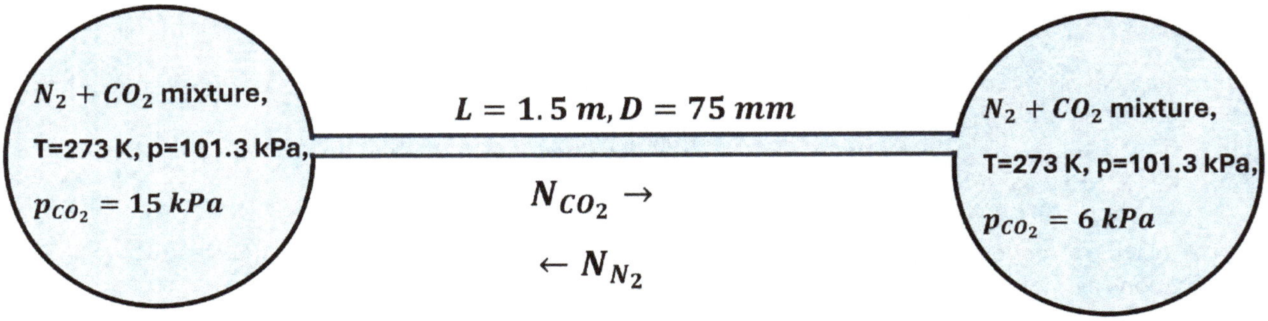

Solution

Let CO_2 be species A and N_2 be species B.

i) Estimate the absolute molar fluxes of CO_2 and N_2

Under the condition of equimolar counter diffusion,

$$N_{A,z} = -\frac{D_{AB}}{RT}\left[\frac{p_{A,2} - p_{A,1}}{L}\right] = -\frac{1.44 \times 10^{-5}}{8.314 \times 273}\left[\frac{(6 - 15) \times 10^3}{1.5}\right] = 3.8066 \times 10^{-5}\ \ gmol/m^2.s$$

$$p_{B,1} = p - p_{A,1} = 101.3 - 15 = 86.3\ kPa$$

$$p_{B,2} = p - p_{A,2} = 101.3 - 6 = 95.3\ kPa$$

$$D_{BA} = D_{AB} = 1.44 \times 10^{-5}\ m^2/s$$

$$N_{B,z} = -\frac{D_{BA}}{RT}\left[\frac{p_{B,2} - p_{B,1}}{L}\right] = -\frac{1.44 \times 10^{-5}}{8.314 \times 273}\left[\frac{(95.3 - 86.3) \times 10^3}{1.5}\right] = -3.8066 \times 10^{-5}\ \ gmol/m^2.s$$

As expected, $N_{A,z} = -N_{B,z}$ in equimolar counter diffusion. The justification for observing equimolar counter diffusion in the present case is that under conditions of constant temperature and pressure and fixed volumes of containers, the number of moles of the containers do not change. If some moles of species A leave a container, an equal number of moles of species B must enter the container to keep the total moles constant.

ii) Calculate the mass flow rates of CO_2 and N_2

$$\dot{m}_A = N_{A,z} \times A \times M_A = (3.8066 \times 10^{-5})\left(\frac{\pi}{4} \times 0.075^2\right) \times 44 = 7.4 \times 10^{-6}\ g/s$$

$$\dot{m}_B = N_{B,z} \times A \times M_B = (3.8066 \times 10^{-5})\left(\frac{\pi}{4} \times 0.075^2\right) \times 28 = 4.709 \times 10^{-6}\ g/s$$

17.6.3 Example 3

Consider the diffusion of gaseous species A to a solid surface where species A is instantly converted into gaseous species B according to the reaction:

$$A(g) \rightarrow 3B(g)$$

Thus, each mole of species A is converted into three moles of species B. The reaction occurs instantly resulting in the concentration of species A on the solid surface to be zero. The process is illustrated in the figure below.

Determine the mole fraction distribution of species A in the diffusion zone, $0 \leq z \leq \delta$, at steady state. The temperature and pressure can be assumed constant in the diffusion zone, $0 \leq z \leq \delta$. The absolute molar flux of species A in the diffusion zone is given to be N_o.

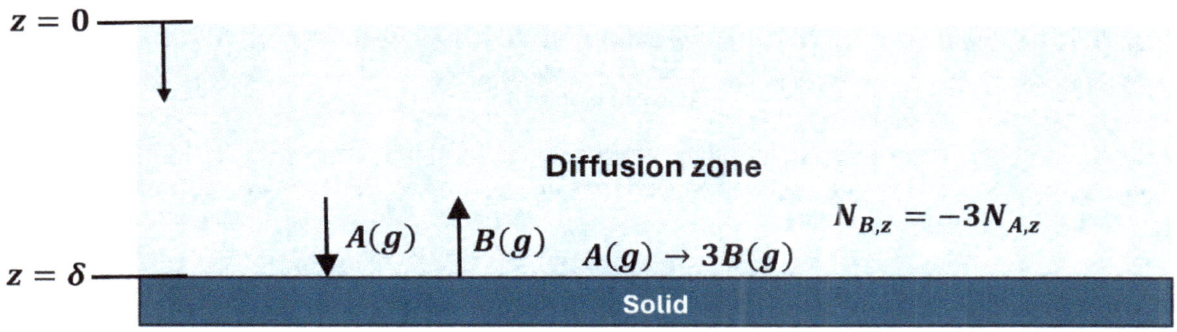

Solution

From Equation (17.25),

$$\vec{N}_A = -CD_{AB}\nabla x_A + x_A(\vec{N}_A + \vec{N}_B)$$

$$N_{A,z} = -CD_{AB}\frac{dx_A}{dz} + x_A(N_{A,z} + N_{B,z})$$

As 1 mole of A(g) produces 3 moles of B(g),

$$N_{B,z} = -3N_{A,z} = -3N_o$$

Thus,

$$N_o = -CD_{AB}\frac{dx_A}{dz} + x_A(N_o - 3N_o)$$

Upon re-arrangement, we get:

$$N_o(1 + 2x_A) = -CD_{AB}\frac{dx_A}{dz}$$

Upon integration,

$$\frac{1}{2}[ln(1 + 2x_A)] = -\frac{N_o z}{CD_{AB}} + a$$

where a is an integration constant. Using the boundary condition that at $z = \delta, x_A = 0$, we get:

$$a = \frac{N_o\delta}{CD_{AB}}$$

Upon substitution of a in the integrated expression and re-arrangement, we obtain:

$$x_A = \frac{1}{2}\left[exp\left\{\frac{2N_o}{CD_{AB}}(\delta - z)\right\} - 1\right]$$

17.6.4 Example 4

Consider one-dimensional diffusion of gaseous species A through stagnant gaseous species B at constant temperature and pressure. Derive the expression for the absolute molar flux of species A taking into consideration non-dilute concentrations of the diffusing species.

Solution

Refer to the figure below.

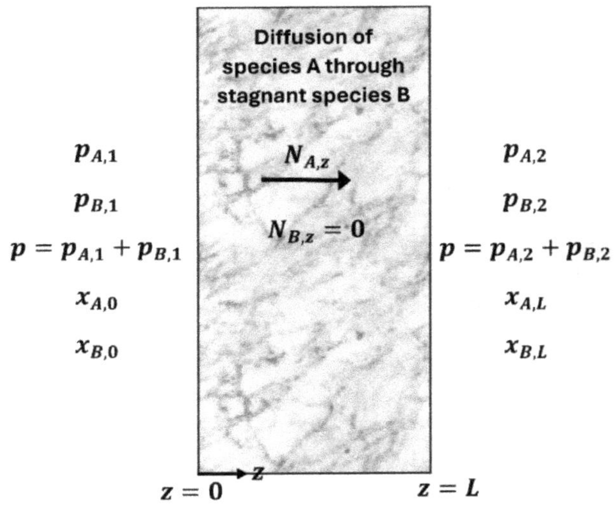

When the concentration of the diffusing species is not small, the bulk or advective effects cannot be neglected. In a binary system where A is diffusing through stagnant B, the absolute molar flux of A is given as:

$$\vec{N}_A = -CD_{AB}\nabla x_A + x_A\vec{N}_A$$

Considering one-dimensional diffusion in the z-direction, the above equation can be re-written as:

$$N_{A,z}(1 - x_A) = -CD_{AB}\left(\frac{dx_A}{dz}\right)$$

where $N_{A,z}$ is the absolute molar flux of A in the z-direction. Assuming steady state ($N_{A,z}$ =constant), and constant C (total concentration of mixture) and D_{AB}, the above equation can be integrated over the diffusion zone of thickness L as:

$$N_{A,z}\int_{z=0}^{L} dz = -CD_{AB}\int_{x_{A,0}}^{x_{A,L}} \frac{dx_A}{1 - x_A}$$

where $x_{A,0}$ is the mole fraction of species A at $z = 0$ and $x_{A,L}$ is the mole fraction of species A at $z = L$. Upon integration, we get:

$$N_{A,z} = \frac{-CD_{AB}}{L} ln\left[\frac{1-x_{A,0}}{1-x_{A,L}}\right]$$

Upon re-arrangement, the above equation could be re-cast as:

$$N_{A,z} = \frac{CD_{AB}}{(x_B)_{lm}}\left[\frac{x_{A,0}-x_{A,L}}{L}\right]$$

where $(x_B)_{lm}$ is log-mean mole fraction of B, defined as:

$$(x_B)_{lm} = \left[\frac{x_{B,0}-x_{B,L}}{ln\left(\frac{x_{B,0}}{x_{B,L}}\right)}\right]$$

We can also express $N_{A,z}$ in terms of partial pressures as shown below.

If species A and B are ideal gases (at low pressures, gases can be treated as ideal gases), it follows that:

$$C = \frac{p}{RT},\ C_A = \frac{p_A}{RT},\ C_B = \frac{p_B}{RT}$$

$$x_A = \frac{C_A}{C} = \frac{p_A}{p},\ x_B = \frac{p_B}{p},\ (x_B)_{lm} = \frac{(p_B)_{lm}}{p}$$

where

$$(p_B)_{lm} = \left[\frac{p_{B,0} - p_{B,L}}{ln\left(\frac{p_{B,0}}{p_{B,L}}\right)}\right]$$

Thus, the expression for $N_{A,y}$ in terms of partial pressures is:

$$N_{A,z} = \frac{D_{AB}}{RT}\frac{p}{(p_B)_{lm}}\left(\frac{p_{A,0} - p_{A,L}}{L}\right)$$

17.6.5 Example 5

Consider the evaporation of liquid water (species A) at the bottom end of a thin tube, as shown in the figure below. The water vapor diffuses from the liquid surface towards the open end of the tube. A stream of pure air flows slowly past the open end of the tube to maintain the mole fraction of water vapor at the open end of the tube to zero. The air is insoluble in water so there is no diffusion of air into or away from the liquid surface. In other words, air (species B) is stationary. Only species A (water vapor) is diffusing in the up-ward direction through stagnant species B.

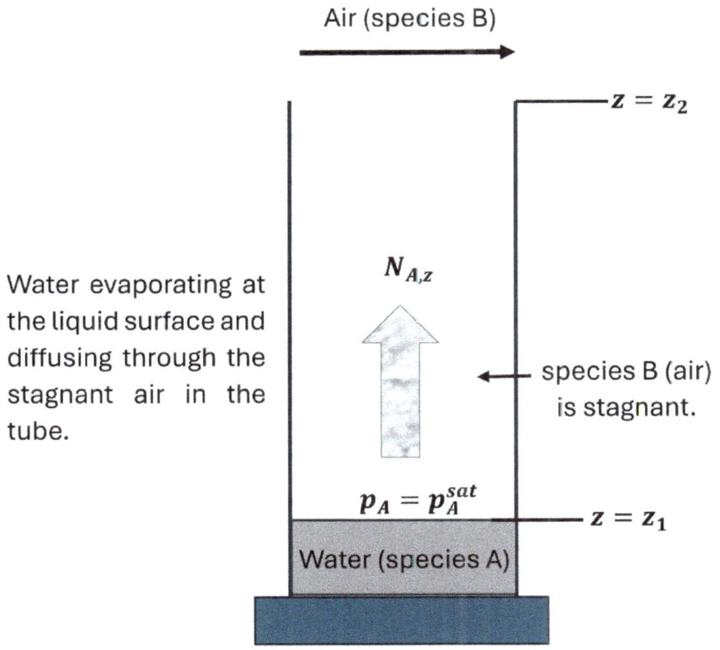

Estimate the absolute molar flux of water vapor at steady state under the following conditions:

Total pressure, $p = 1$ atm; $T = 25^oC$; vapor pressure of water, $p_A^{sat} = 23.8$ mmHg; tube length, $z_2 - z_1 = 1$ ft; diffusivity, $D_{AB} = 26.3 \times 10^{-6}$ m^2/s.
You can assume that species A and B are ideal gases and the liquid level in the tube is held constant.

Solution

We can apply the equation developed in the preceding example:

$$N_{A,z} = \frac{D_{AB}}{RT}\frac{p}{(p_B)_{lm}}\left(\frac{p_{A,0}-p_{A,L}}{L}\right)$$
$$p_{A,0} = 23.8/760 = 0.03132\ atm = 0.03132 \times 101.325 = 3.1735\ kPa$$
$$p_{A,L} = 0$$
$$p_{B,0} = p - p_{A,0} = 1 - 0.03132 = 0.96868\ atm$$
$$p_{B,L} = p - p_{A,L} = 1 - 0 = 1\ atm$$
$$(p_B)_{lm} = \left[\frac{p_{B,0} - p_{B,L}}{ln\left(\frac{p_{B,0}}{p_{B,L}}\right)}\right] = \frac{0.96868 - 1}{ln\left(\frac{0.96868}{1}\right)} = 0.98426\ atm = 0.98426 \times 101.325 = 99.73\ kPa$$

Thus,
$$N_{A,z} = \frac{D_{AB}}{RT}\frac{p}{(p_B)_{lm}}\left(\frac{p_{A,0}-p_{A,L}}{L}\right) = \frac{26.3\times10^{-6}}{8.314\times298.15}\left(\frac{101.325}{99.73}\right)\left(\frac{3.1735\times10^3}{1\times0.1524}\right) = 2.245 \times 10^{-4}\ \frac{mol}{m^2.s}$$

NOTATION

A	area
C	total molar concentration of mixture
C_A	molar concentration of species A
C_B	molar concentration of species B
C_i	molar concentration of species *i*
D_{AB}	diffusion coefficient of species A in B
D_{BA}	diffusion coefficient of species B in A
erf	error function
g	magnitude of acceleration due to gravity
g_z	component of gravity in z-direction
$\vec{g}$	acceleration due to gravity vector
$\vec{j}_A$	diffusional mass flux vector of species A
$\vec{j}_B$	diffusional mass flux vector of species B
$\vec{j}_i$	diffusional mass flux vector of species *i*
$\vec{J}_A$	diffusional molar flux vector of species A
$\vec{J}_B$	diffusional molar flux vector of species B
$\vec{J}_i$	diffusional molar flux vector of species *i*
L	length of plate
N	total number of species in the multi-component mixture
$\vec{n}_A$	absolute mass flux vector of species A
$\vec{n}_B$	absolute mass flux vector of species B
$\vec{n}_i$	absolute mass flux vector of species *i*
$\vec{N}_A$	absolute molar flux vector of species A
$\vec{N}_B$	absolute molar flux vector of species B
$\vec{N}_i$	absolute molar flux vector of species *i*

$N_{A,x}$	x-component of absolute molar flux vector of species A
$N_{A,y}$	y-component of absolute molar flux vector of species A
$N_{A,z}$	z-component of absolute molar flux vector of species A
p	pressure
r_A	rate of production of A by chemical reaction in the units of kg/(s.m^3).
R_A	rate of production of A by chemical reaction in the units of moles/(s.m^3).
t	time
$\vec{V}$	mass average velocity vector
$\vec{V}^*$	molar average velocity vector
$\vec{V}_A$	absolute velocity of species A
$\vec{V}_B$	absolute velocity of species B
$\vec{V}_i$	absolute velocity of species i
V_z	z-component of velocity vector
V_{max}	maximum velocity
W	width of a plate
x_A	mole fraction of species A
x_B	mole fraction of species B
x, y, z	space coordinates in cartesian coordinate system

Greek Letters

α	angle
β	parameter in Equation (17.77)
Γ	Gamma function
δ	film thickness
ε	parameter defined in Equation (17.81)
η	viscosity of fluid or parameter defined in Equation (17.66)
η_B	viscosity of solvent (species B)
ρ	density or mass concentration of mixture
ρ_A	mass concentration of species A
ρ_B	mass concentration of species B
ρ_i	mass concentration of species i
Υ	dimensionless concentration, Equation (17.82)
ω_A	mass fraction of species A
ω_B	mass fraction of species B
ω_i	mass fraction of species i

Mathematical Operators

∇	del operator
∇^2	Laplacian operator

PRACTICE PROBLEMS

Problem 1

Is $\vec{j}_1$ equal to $\vec{J}_1 M_1$? Explain.

Problem 2

Consider a binary gas mixture of species A and B at 490 K and 1.2 bar. The absolute mass fluxes of the species are: $\vec{n}_A = -0.79\,\hat{k}$ kg/m^2.s and $\vec{n}_B = +0.79\,\hat{k}$ kg/m^2.s, where $\hat{k}$ is a unit vector in the z-direction. Determine the following:

i) The mass average velocity $\vec{V}$ of the mixture

ii) The molar average velocity $\vec{V}^*$ of the mixture

iii) The diffusional mass fluxes of the species $\vec{j}_A$ and $\vec{j}_B$

iv) The diffusional molar fluxes of the species $\vec{J}_A$ and $\vec{J}_B$

Problem 3

Consider a hot carbon surface (see figure below) where the following reaction is taking place:

$$3C(s) + 2O_2(g) \rightarrow 2CO(g) + CO_2(g)$$

The oxygen diffuses to the carbon surface at a constant absolute molar flux of N_o. Under steady state condition,

i) Express the absolute molar fluxes of $CO(g)$ and $CO_2(g)$ in terms of N_o.

ii) What is the total absolute molar flux of all components, that is, the sum of absolute molar fluxes of all species involved.

You can assume that the components are present in accordance with the stoichiometry of the reaction.

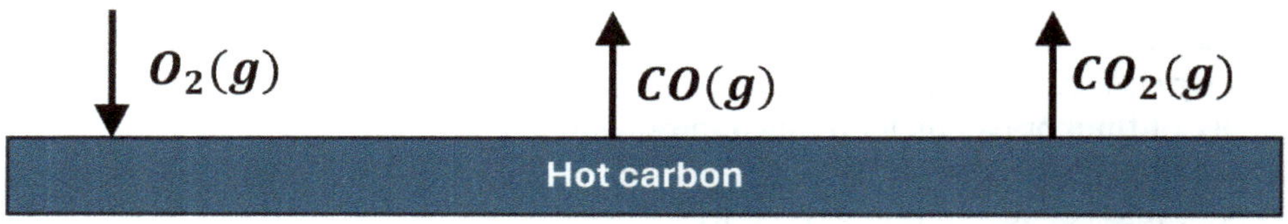

$$3C(s) + 2O_2(g) \rightarrow 2CO(g) + CO_2(g)$$

Problem 4

Consider the steady laminar flow of a Newtonian liquid film flowing down a flat inclined plate due to gravity (see figure below). The plate is inclined at an angle α with the horizontal. The length and width of the plate are L and W, respectively. The thickness of the liquid film is δ. Species A is diffusing into the falling liquid film from the free surface. Species A is also reacting with the solvent (species B) by a first order irreversible chemical reaction. The volumetric rate of reaction is as follows:

$$R_A = -k_1 C_A$$

where R_A is in units of moles/m^3.s. The concentration of the species A at gas/liquid interface ($x = 0$) within the liquid film is $C_{A,o}$. The plate is impermeable to A (and B). You can assume that $C_A = 0$ at the entrance where $z = 0$. Species A is only sparingly soluble in B and the contact time between the liquid and species A, that is, L/V_{max} is small. What is the molar flux of A at the gas/liquid interface?

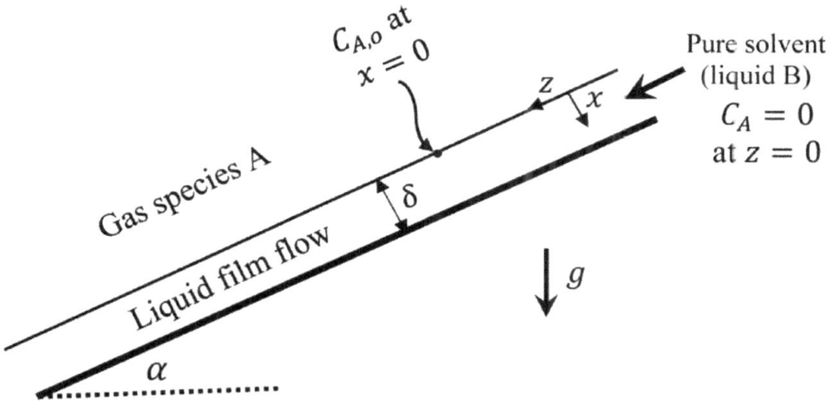

Problem 5

Gaseous species A and B are undergoing equimolar counter diffusion through a tube 1.85 ft long and 18 mm in inside diameter. The ends of the tube are connected to large vessels. The entire system is at 25 °C and 1 atm. The partial pressure of species A is maintained at 17 kPa in the left vessel and at 5.5 kPa in the right vessel. The diffusivity D_{AB} at 25 °C and 1 atm is 2.25×10^{-5} m^2/s.

Estimate the absolute molar fluxes and molar flow rates of species A and B. What are the mass flow rates of species A and B if the molecular weights are 17 and 28, respectively. You can assume that the gases are ideal in nature.

Problem 6

Consider a pool of equimolar binary mixture of liquid species A and B exposed to ambient air at 25 °C and 1 atm. Species B is non-volatile. The vapor pressure of pure species A at 25 °C is $p_A^{sat} = 20\ kPa$. The ambient air is insoluble in the liquid mixture. Species A evaporates from the surface of the liquid and diffuses through the stagnant ambient air. The partial pressure of species A 10 cm above the liquid surface is 2.5 kPa.

Estimate the absolute molar flux of species A if the diffusivity is, $D_{A-cir} = 3.45 \times 10^{-5}$ m^2/s.

You can assume that the liquid solution and gases are ideal in nature.

FURTHER READING

1. R.B. Bird, W.E. Stewart, and E.N. Lightfoot. 2007. *Transport Phenomena*. 2nd Edition, Wiley & Sons, New York, 2007.
2. J.R. Welty, C.E. Wicks, R.E. Wilson, and G.L. Rorrer. *Fundamentals of Momentum, Heat, and Mass Transfer*. 5th Edition, Wiley & Sons, New York, 2008.
3. C.J. Geankoplis, A.A. Hersel, D.H. Lepek. *Transport Processes and Separation Process Principles*. 5th Edition, Prentice Hall, New Jersey, 2018.
4. W.L. McCabe, J.C. Smith, P. Harriott. *Unit Operations of Chemical Engineering*. 6th Edition, McGraw Hill, New York, 2001.
5. D.R.Gaskell. *An Introduction to Transport Phenomena in Materials Engineering*. Macmillan Publishing Company, New York, 1992.
6. R. Pal. *Electromagnetic, Mechanical, and Transport Properties of Composite Materials*. CRC Press, Boca Raton, Florida, 2015.
7. A.F. Mills. *Basic Heat and Mass Transfer*. 2nd Edition, Prentice Hall, New Jersey, 1999.
8. S. Middleman. *An Introduction to Mass and Heat Transfer*. Wiley & Sons, New York, 1998.
9. T.L. Bergman, A.S. Levine, F.P. Incropera, and D.P. Dewitt. *Fundamentals of Heat and Mass Transfer*. 7th Edition, New York, Wiley & Sons, 2011.
10. Y.A. Cengel. *Heat and Mass Transfer*, 3rd Edition, McGraw Hill, New York, 2007.

Chapter 18

Mass Transfer Coefficient Correlations and Interphase Mass Transfer

In Chapter 17, analytical solutions of concentration distribution and mass flux in fluid were obtained for problems involving steady laminar flow of Newtonian fluids of constant properties. However, it is not always possible to obtain an analytical solution to the governing equations of motion and component material balance even in the case of laminar flows when the geometry is complicated. Furthermore, turbulent flows are completely out of reach as it is not at all possible to obtain any analytical solutions of turbulent flows although the governing equations of motion and component material balance are still valid. Fortunately, in many practical applications involving convective mass transfer, detailed concentration and mass flux distributions in the fluid are not required. Often, the quantity of practical interest is the mass flux at the wall on the flow geometry. The mass transfer rate from the solid boundary (wall) to the fluid and vice versa can be calculated from the knowledge of mass flux at the solid boundary and the mass transfer area. Two types of flow situations are encountered in practical applications of convective mass transfer: external flows (see Figure 18.1) and internal flows (see Figure 18.2). There is enough experimental evidence to suggest that the molar (or mass) flux at the wall $N_{A,w}$ is directly proportional to the characteristic concentration difference of species A between the solid boundary and the fluid moving adjacent to the solid boundary, that is,

$$N_{A,w} \propto (C_{A,w} - C_{A,c}) \tag{18.1}$$

where $C_{A,w}$ is the concentration of species A at the solid boundary on the fluid side and $C_{A,c}$ is the characteristic concentration of species A in the fluid.

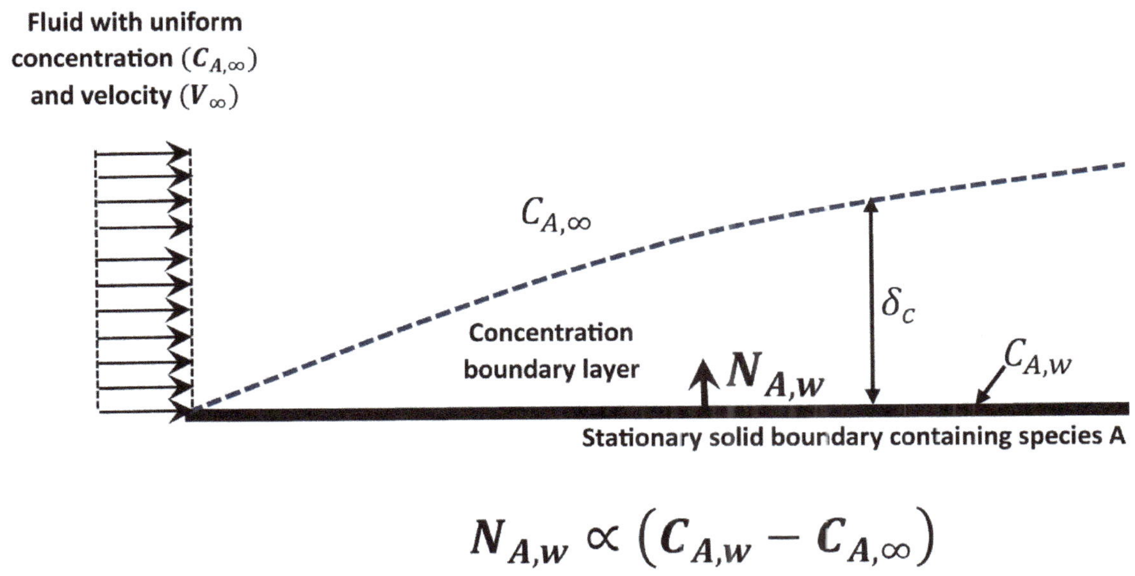

Figure 18.1 Convective mass transfer of species A from solid surface to fluid in external flow.

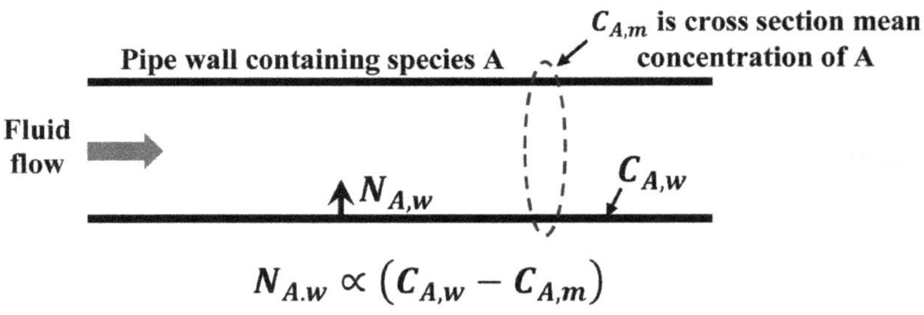

Figure 18.2 Convective mass transfer of species A from solid surface to fluid in internal flow.

For external flows:

$$C_{A,c} = C_{A,\infty} \tag{18.2}$$

where $C_{A,\infty}$ is the free-stream concentration of species A in the fluid. For internal flows:

$$C_{A,c} = C_{A,m} \tag{18.3}$$

where $C_{A,m}$ is the mean concentration of species over the cross-section of the pipe at any given axial location.

The mass transfer coefficient is defined as the proportionality factor between $N_{A,w}$ and $(C_{A,w} - C_{A,c})$. Thus, Equation (18.1) can be re-written as:

$$N_{A,w} = k(C_{A,w} - C_{A,c}) \tag{18.4}$$

For external flows,

$$N_{A,w} = k(C_{A,w} - C_{A,\infty}) \tag{18.5}$$

For internal flows,

$$N_{A,w} = k(C_{A,w} - C_{A,m}) \tag{18.6}$$

The mass or molar flux at the wall and hence the mass transfer rate between the solid boundary and the fluid can be calculated from Equations (18.4) - (18.6) provided that the mass transfer coefficient k for the given situation is known or could be estimated from correlations. Note that the units of k is m/s.

The mass transfer coefficient k is linked to concentration gradient in the fluid at the solid boundary through the following relationship:

$$k = \frac{-D_{AB}\left(\frac{\partial C_A}{\partial y}\right)_{y=0}}{C_{A,w} - C_{A,c}} \tag{18.7}$$

where $(\partial C_A/\partial y)_{y=0}$ is the concentration gradient in the fluid at the solid boundary and y is the distance from the solid boundary. Equation (18.7) follows from the equality of Fick's law of diffusion and Equation (18.4) at the solid boundary.

18.1 MASS TRANFER COEFFICIENT IN EXTERNAL FLOWS

In many practical applications, mass transfer takes place between a solid surface and the fluid moving over the solid surface. The total rate of mass transfer in molar units $\dot{M}_A$ can be expressed as:

$$\dot{M}_A = \int_A N_{A,w} dA \tag{18.8}$$

where $N_{A,w}$ is the local molar flux at any point on the surface of the solid, normal to the surface, and A is the surface area of the solid object over which mass transfer is taking place.

At a solid/fluid interface, the local molar flux normal to the solid surface is related to the difference between the local solid surface concentration $C_{A,w}$ and the free stream temperature of the fluid $C_{A,\infty}$ as given by Equation (18.5). Thus, the mass transfer rate can be expressed as:

$$\dot{M}_A = \int_A k\left(C_{A,w} - C_{A,\infty}\right) dA \quad (18.9)$$

For a given geometry (shape of the solid boundary) and concentration difference $\left(C_{A,w} - C_{A,\infty}\right)$, $\dot{M}_A$ can be calculated from Equation (18.9) provided that k is known. Note that k is by no means constant. It varies from point to point on the surface of the solid boundary. It is a complicated function of fluid properties, fluid motion, and geometry.

18.1.1 Flow Over a Flat Plate

The mass transfer coefficient is usually expressed in a dimensionless form as Sherwood number defined below:

$$Sh_x = \frac{kx}{D_{AB}} \quad (18.10)$$

where Sh_x is the local Sherwood number and x is the distance from the leading edge of the solid boundary (plate in the present case). Dimensional analysis indicates that Sherwood number is a function of Reynolds number and Schmidt number:

$$Sh_x = f(Re_x, Sc) \quad (18.11)$$

where Re_x is the local Reynolds number defined as $\rho V_\infty x/\eta$ and Sc is the Schmidt number defined as:

$$Sc = \nu/D_{AB} \quad (18.12)$$

The following relations are valid for local mass transfer coefficient in laminar and turbulent boundary layers over a solid flat plate:

$$Sh_x = \frac{kx}{D_{AB}} = 0.332 Sc^{1/3}(Re_x)^{1/2} \quad (18.13)$$

$$Sh_x = \frac{kx}{D_{AB}} = 0.0296(Re_x)^{4/5} Sc^{1/3} \quad (18.14)$$

Equation (18.13) is valid for laminar boundary layer and Equation (18.14) is valid for turbulent boundary layer. They are valid for $Sc > 0.60$. Equation (18.13) is an exact theoretical relationship obtained from the solution of boundary layer equations. The solution is given in Appendix E. Equation (18.14) is an empirical relation based on Chilton-Colburn analogy between momentum and mass transfer (see Appendix E).

18.1.2 Flow Across a Solid Cylinder

Another commonly encountered geometry in convective mass transfer applications is a solid cylinder. Consider mass transfer from a solid cylinder into fluid which is flowing normal to the axis of the cylinder. This type of mass transfer problem could be encountered in sublimation of solid into the moving fluid. Other applications may involve mass transfer of some species present at the surface of cylinder in the form of a coating.

Bedingfield and Drew developed the following correlation for mass transfer from solid cylinder into gas flowing normal to the axis of the cylinder:

$$Sh = \frac{kD}{D_{AB}} = 0.281(Re_D)^{1/2}Sc^{0.44} \qquad (18.15)$$

where D is diameter of the cylinder, and Re_D is the Reynolds number based on cylinder diameter ($=\rho D V_\infty/\eta$). This correlation is valid for $400 < Re_D < 25{,}000$ and $0.6 < Sc < 2.6$. For liquids the following correlation of Linton and Sherwood could be used

$$Sh = \frac{kD}{D_{AB}} = 0.281(Re_D)^{0.6}Sc^{1/3} \qquad (18.16)$$

This correlation is valid for $400 < Re_D < 25{,}000$ and $Sc < 3000$.

18.1.3 Flow Around a Solid Sphere

The Sherwood number (Sh) for forced convective mass transfer from a particle is a function of particle Reynolds number (Re) and Schmidt number (Sc). The general form of the Sherwood number relation for a single particle is as follows:

$$Sh = \frac{kD_p}{D_{AB}} = Sh_0 + C\,Re^m\,Sc^n \qquad (18.17)$$

where k is the mass transfer coefficient, D_p the particle diameter, D_{AB} the diffusion coefficient, and C, m, and n are constants. Sh_0 is the Sherwood number in the limit $Re \rightarrow 0$, corresponding to diffusion from a particle in an infinite stationary fluid. For a solid spherical particle, Sh_0 is 2. Note that the particle Reynolds number is defined as $\rho D_p V_\infty/\eta$, where V_∞ is the free-stream velocity of fluid, and ρ and η are fluid density and viscosity, respectively.

Froessling proposed the following relationship for mass transfer from a solid spherical particle:

$$Sh = \frac{kD_p}{D_{AB}} = 2 + 0.552\,Re^{1/2}\,Sc^{1/3} \qquad (18.18)$$

This equation correlates the data well for mass transfer from a spherical particle into gases with $2 < Re < 800$ and $0.6 < Sc < 2.7$.

For higher Re ($1500 < Re < 12{,}000$) and $0.6 < Sc < 1.85$), Steinberger and Treybal proposed the modification of the Froessling equation as:

$$Sh = \frac{kD_p}{D_{AB}} = 2 + 0.552\,Re^{0.53}\,Sc^{1/3} \qquad (18.19)$$

McCabe et al. recommend a slightly modified version of the Froessling equation for Re up to 1000.

$$Sh = \frac{kD_p}{D_{AB}} = 2 + 0.6\,Re^{1/2}\,Sc^{1/3} \qquad (18.20)$$

This equation is probably less restrictive on Sc and could be applied to liquid flow around the particle. However, for creeping flows ($Re \ll 1$) with high Peclet number ($Pe = ReSc$), this equation tends to under-predict the values of the Sherwood number. In such situations, the following equation is recommended:

$$Sh = \frac{kD_p}{D_{AB}} = \left(4.0 + 1.21Pe^{2/3}\right)^{1/2} \qquad (18.21)$$

This equation could be applied for Pe up to as high as 10,000. For $Pe > 10{,}000$, the following Levich equation could be used:

$$Sh = \frac{kD_p}{D_{AB}} = 1.01(Re\,Sc)^{1/3} \tag{18.22}$$

So far mass transfer from a solid spherical particle is considered. For bubble or droplets, the mass transfer rates are somewhat higher due to internal circulation effect. Assuming that no surfactants or impurities are present at the surface of bubbles or droplets, transmission of tangential stresses occur from the external matrix fluid to internal fluid of droplets and bubbles. The transmission of stresses causes internal circulation within the droplets/bubbles. The internal circulation, in turn, enhances the mass transfer rate. McCabe et al. recommend the following correlation for the estimation of mass transfer coefficient in the presence of internal circulation effect:

$$Sh = \frac{kD_p}{D_{AB}} = 1.13\,Re^{1/2}\,Sc^{1/2} \tag{18.23}$$

This equation should be modified to the following form to consider diffusion from a droplet/particle under no-flow condition:

$$Sh = \frac{kD_p}{D_{AB}} = 2.0 + 1.13Re^{1/2}Sc^{1/2} \tag{18.24}$$

In the equations discussed thus far, natural convection is assumed to be absent. If natural convection is present along with forced convection, the mass transfer correlation can be expressed in the following form:

$$Sh = \frac{kD_p}{D_{AB}} = Sh_{nc} + C\,Re^m\,Sc^n \tag{18.25}$$

where Sh_{nc} is the contribution of natural convection and the second term is the contribution from forced convection. For flow around a solid sphere, the correlation proposed by Steinberger and Treybal is as follows:

$$Sh = \frac{kD_p}{D_{AB}} = Sh_{nc} + 0.347\,Re^{0.62}\,Sc^{1/3} \tag{18.26}$$

This correlation is valid over $1 < Re < 30{,}000$ and $0.6 < Sc < 3200$ with natural convection contribution Sh_{nc} as follows:

$$Sh_{nc} = 2.0 + 0.569(GrSc)^{1/4} \quad \text{for} \quad GrSc < 10^8 \tag{18.27}$$

$$Sh_{nc} = 2.0 + 0.0254(GrSc)^{1/3}Sc^{0.244} \quad \text{for} \quad GrSc > 10^8 \tag{18.28}$$

where Gr is the Grashof number based on the particle diameter. The natural convection term becomes negligible at high Re.

18.2 MASS TRANFER COEFFICIENT IN INTERNAL FLOWS

Consider a cylindrical tube (radius R) whose surface is coated with some material (species A) that can dissolve and diffuse into the fluid moving inside the tube. The molar flux of species A at the fluid/solid interface ($r = R$) can be expressed in terms of the mass transfer coefficient as given in Equation (18.6).

The mean concentration $C_{A,m}$ of species A in the fluid over a given cross-section is:

$$C_{A,m} = \frac{2}{\bar{V}R^2}\int_0^R rC_AV_z dr \quad (18.29)$$

where $\bar{V}$ is the average velocity and V_z is the local velocity.

At the pipe wall, the molar flux normal to the solid surface can also be expressed as:

$$N_{A,w} = -D_{AB}\left(\frac{\partial C_A}{\partial y}\right)_{y=0} + (C_AV_z)_{y=0} = -D_{AB}\left(\frac{\partial C_A}{\partial y}\right)_{y=0} \quad (18.30)$$

Note that at the pipe wall, $V_z = 0$. In terms of radial co-ordinate ($r = R - y$), Equation (18.30) gives:

$$N_{A,w} = -D_{AB}\left(\frac{\partial C_A}{\partial y}\right)_{y=0} = D_{AB}\left(\frac{\partial C_A}{\partial r}\right)_{r=R} \quad (18.31)$$

From Equations (18.6) and (18.31), it follows that:

$$k = \frac{D_{AB}\left(\frac{\partial C_A}{\partial r}\right)_{r=R}}{C_{A,s}-C_{A,m}} \quad (18.32)$$

Thus, the mass transfer coefficient in internal flow can be obtained if both concentration and velocity distributions are known.

From material balance over a differential section of the tube, it can be readily shown that:

$$\bar{N}_A = \bar{k}(\Delta C_A)_{lm} \quad (18.33)$$

where $\bar{N}_A$ is the average molar flux of A over the entire length of the tube, $\bar{k}$ is the average mass transfer coefficient, and $(\Delta C_A)_{lm}$ is the log mean concentration difference of A defined as:

$$(\Delta C_A)_{lm} = \frac{\Delta C_{A,2}-\Delta C_{A,1}}{ln\left(\frac{\Delta C_{A,2}}{\Delta C_{A,1}}\right)} \quad (18.34)$$

In Equation (18.34), $\Delta C_{A,1} = C_{A,w} - C_{A,1}$, $\Delta C_{A,2} = C_{A,w} - C_{A,2}$, and $C_{A,1}$ and $C_{A,2}$ are average inlet and outlet concentrations of species A. The average mass transfer coefficient $\bar{k}$ in Equation (18.33) is given as:

$$\bar{k} = \frac{1}{L}\int_0^L kdz \quad (18.35)$$

where k is the local mass transfer coefficient and L is the length of the tube.

18.2.1 Laminar Flow

For hydrodynamically fully developed laminar flow in a pipe, the velocity distribution is given as:

$$V_z = V_{max}\left[1 - \left(\frac{r}{R}\right)^2\right] \quad (18.36)$$

where V_z is the local velocity in axial direction, and V_{max} is the maximum velocity at the centerline ($r = 0$). The material balance for species A under steady-state laminar flow conditions can be expressed as:

$$V_z \frac{\partial C_A}{\partial z} = D_{AB}\left(\frac{\partial^2 C_A}{\partial r^2} + \frac{1}{r}\frac{\partial C_A}{\partial r}\right) \quad (18.37)$$

Therefore, for hydrodynamically fully developed laminar flow the concentration distribution is given by the following equation:

$$\frac{\partial C_A}{\partial z} = \frac{D_{AB}}{V_{max}\left[1-\left(\frac{r}{R}\right)^2\right]}\left(\frac{\partial^2 C_A}{\partial r^2} + \frac{1}{r}\frac{\partial C_A}{\partial r}\right) \quad (18.38)$$

This equation needs to be solved subject to the given boundary conditions. As an example, consider the case where the wall concentration of species A is constant ($C_{A,w}$=constant) and the flow is hydrodynamically and concentration-wise fully developed. Note that concentration-wise fully developed flow is characterized by the following condition:

$$\frac{\partial}{\partial z}\left(\frac{C_{A,w}-C_A}{C_{A,w}-C_{A,m}}\right) = 0 \quad (18.39)$$

Under these restrictions, the solution to Equation (18.38) yields the following result:

$$Sh = \frac{kD}{D_{AB}} = 3.66 \quad (18.40)$$

where D is the pipe diameter. For hydrodynamically fully developed but concentration-wise developing flows, the solution to Equation (18.38) (assuming constant wall concentration) can be expressed in the form:

$$Sh_z = \frac{kD}{D_{AB}} = f(Gz_m) \quad (18.41)$$

where Gz_m is Graetz number for mass transfer, defined as:

$$Gz_m = \frac{\pi}{4}\left(\frac{D}{z}\right) Re\, Sc \quad (18.42)$$

where Re is the Reynolds number for pipe flow ($=\rho\bar{V}D/\eta$). For the calculation of average mass transfer coefficient ($\bar{k}$) or average Sherwood number ($S\bar{h}$), the following Hausen correlation is often used:

$$S\bar{h} = \frac{\bar{k}D}{D_{AB}} = 3.66 + \frac{0.0668(D/L)(Re\, Sc)}{1+0.04[(D/L)(Re\, Sc)]^{2/3}} \quad (18.43)$$

The combined entry length problem where the flow is hydrodynamically, and concentration-wise developing is much more difficult to solve as velocity and concentration both depend on r as well as z. For the estimation of the average Sherwood number in the case of combined entry length problem dealing with laminar flow through a pipe (assuming constant wall concentration), the following correlation is recommended:

$$S\bar{h} = \frac{\bar{k}D}{D_{AB}} = 1.86(ReSc)^{1/3}\left(\frac{D}{L}\right)^{1/3} \quad (18.44)$$

18.2.2 Turbulent Flow

The mass transfer coefficient in fully developed turbulent flows can be determined from the Chilton-Colburn analogy between momentum and mass transfer expressed as:

$$St_m\, Sc^{2/3} = \frac{f}{2} \tag{18.45}$$

where St_m is called Stanton number for mass transfer, defined as $Sh/(Re\, Sc)$. Thus, the mass transfer coefficient can be obtained if only the friction factor is known. For smooth pipes, the friction factor in turbulent flow can be estimated from the following empirical relation:

$$f = \frac{0.046}{Re^{1/5}} \tag{18.46}$$

This equation can be applied over the Reynolds number range of 2 x 10^4 to 10^6. Substitution of Equation (18.46) into Equation (18.45) results in the following equation for Sherwood number:

$$S\bar{h} = \frac{\bar{k}D}{D_{AB}} = 0.023(Re)^{4/5}\,(Sc)^{1/3} \tag{18.47}$$

This equation is valid for Sc >0.5.

Modified versions of Equation (18.47) have been proposed in the literature. For example, Gilliland and Sherwood proposed the following correlation based on experimental data on vaporization of different liquids (wetting the inner surface of the pipe) into air flowing through the tube:

$$\frac{\bar{k}D}{D_{AB}}\left(\frac{p_{B,lm}}{p}\right) = 0.023(Re)^{0.83}\,(Sc)^{0.44} \tag{18.48}$$

where p is the total pressure of the system and $p_{B,lm}$ is the log-mean partial pressure of the carrier gas, given as:

$$p_{B,lm} = \frac{p_{B,2}-p_{B,1}}{ln\left(\frac{p_{B,2}}{p_{B,1}}\right)} \tag{18.49}$$

where $p_{B,1}$ and $p_{B,2}$ are partial pressures of carrier gas at inlet and outlet, respectively.

Equation (18.48) is valid for gases under the conditions: $2000 < Re < 3.5x10^4$ and $0.6 < Sc < 2.5$. For dilute solutions ($x_A << 1.0$), $p_{B,lm}/p$ is nearly unity.

Linton and Sherwood proposed another correlation based on experimental data on dissolution of solid solute (coated on the inner surface of the pipe) in liquid solvent flowing through the tube.

$$S\bar{h} = \frac{\bar{k}D}{D_{AB}} = 0.023(Re)^{0.83}\,(Sc)^{1/3} \tag{18.50}$$

This equation is valid for liquids under the conditions: $2000 < Re < 3.5x10^4$ and $1000 < Sc < 2260$

18.3 INTERPHASE MASS TRANSFER

In the preceding sections, convective mass transfer was considered between a solid surface and a fluid moving adjacent to the solid surface. For convective mass transfer between a solid surface and a moving fluid, the concept of individual mass transfer coefficient (k) was introduced. In many mass transfer problems of practical interest, the transfer of some material species (say A) takes place between two immiscible fluid phases (gas/liquid, liquid/liquid) in intimate contact with one another. Also, the two phases are in motion and consequently, the transfer of species A from one phase to the interface and from interface to the second phase occurs by convection. Examples of interphase mass transfer processes include "absorption" and "stripping (desorption)". In absorption process, a solute gas (for example, ammonia) is absorbed from an inert gas (for example, air) into a liquid (for example, water) in which the solute is soluble. In desorption or stripping process, the solute is transferred from the solvent liquid phase to the gas phase. In the analysis of interphase mass transfer problems, the overall mass transfer coefficient and its relation to individual mass transfer coefficients are important concepts. Also, in most situations it is assumed that the interface itself does not offer any resistance to the transfer of the diffusing species across the interface. Thus, the rate of mass transfer between the phases is controlled by the individual mass transfer resistances in the two phases. Across the interface between the two phases, equilibrium is assumed to exist so that the adjacent concentrations of diffusing species across the interface are related by a thermodynamic equilibrium relation.

Figure 18.3 shows the concentration profile of solute A diffusing from gas phase to liquid phase. In the gas phase, the partial pressure of solute A is plotted as a concentration variable. In the liquid phase, the molar concentration of A is plotted. The bulk phase concentrations are $p_{A,G}$ (partial pressure of A in bulk gas phase) and $C_{A,L}$ (molar concentration of A in bulk liquid phase). At the interface, the partial pressure of A on the gas side is $p_{A,i}$ and the molar concentration of A on the liquid side is $C_{A,i}$. N_A is the molar flux of A from the gas phase to the liquid phase (see Figure 18.3). As equilibrium exists at the interface, the concentrations of species A across the interface are related by an equilibrium relation such as:

$$p_{A,i} = f(C_{A,i}) \tag{18.51}$$

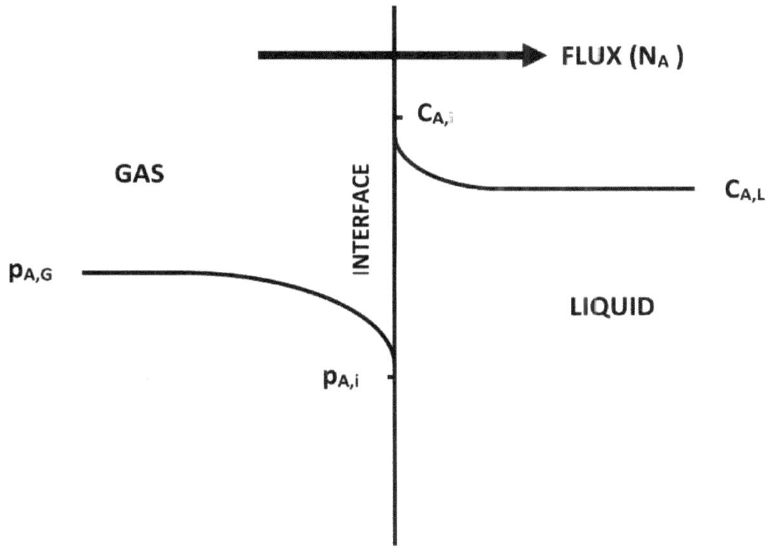

Figure 18.3 Concentration profile of solute A diffusing from gas phase to liquid phase.

18.3.1 Individual Mass-Transfer Coefficients and Interface Concentrations

Under steady state condition, the molar (or mass) flux in one phase must be equal to the molar flux in the second phase. Thus:

$$N_A = k_G(p_{A,G} - p_{A,i}) = k_L(C_{A,i} - C_{A,L}) \tag{18.52}$$

where k_G is the gas-phase mass transfer coefficient (S.I. units of k_G are moles/(m^2.s.Pa)), and k_L is the liquid-phase mass transfer coefficient (S.I. units of k_L are m/s).

Equation (18.52) can be rearranged as:

$$-\frac{k_L}{k_G} = \frac{p_{A,G}-p_{A,i}}{C_{A,L}-C_{A,i}} \tag{18.53}$$

This expression allows the evaluation of the unknown interface concentrations ($p_{A,i}$ and $C_{A,i}$) from the knowledge of bulk concentrations ($p_{A,G}$ and $C_{A,L}$) and the individual mass transfer coefficients, provided that the equilibrium relationship (Equation (18.51)) is known.

The procedure for determining $p_{A,i}$ and $C_{A,i}$ is shown graphically in Figure 18.4. From the knowledge of the bulk compositions ($p_{A,G}$ and $C_{A,L}$), point "O" is marked on the diagram. A straight line of slope $-k_G/k_L$ passing through the point "O" intersects the equilibrium curve at a point where $p_A = p_{A,i}$ and $C_A = C_{A,i}$.

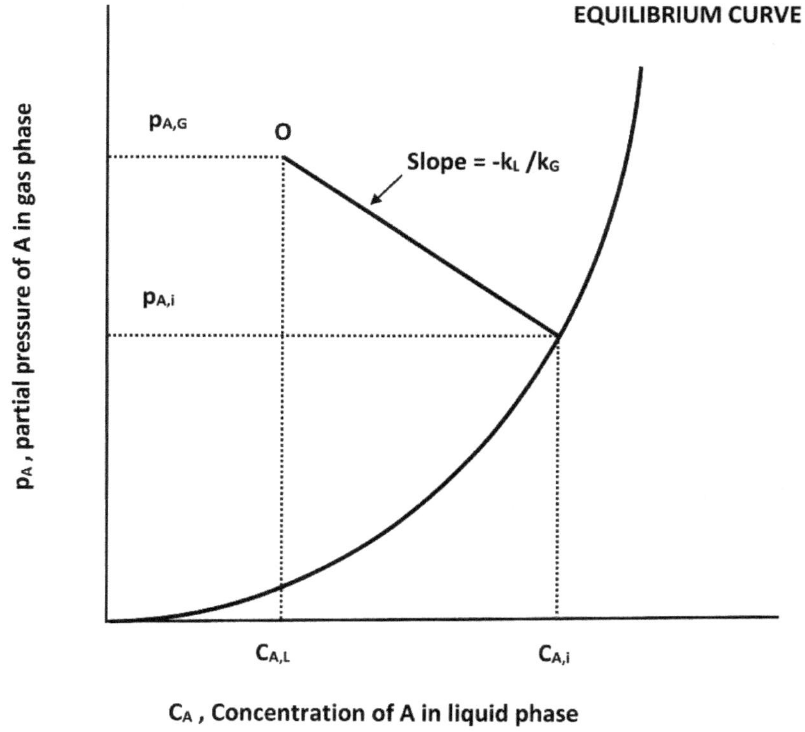

Figure 18.4 Graphical procedure for determining the interfacial quantities $p_{A,i}$ and $C_{A,i}$.

18.3.2 Concept of Overall Mass Transfer Coefficient

The concept of overall mass transfer coefficient is very useful in that it allows the evaluation of the mass flux without the knowledge of the interfacial concentrations ($p_{A,i}$ and $C_{A,i}$).

The overall mass transfer coefficient based on the gas phase (K_G) is defined as:

$$N_A = K_G\left(p_{A,G} - p_A^*\right) \tag{18.54}$$

where p_A^* is the partial pressure of A in the gas phase if the gas phase was in equilibrium with the bulk liquid phase (having concentration of A as $C_{A,L}$). Note that p_A^* is a fictitious quantity whose value is determined from the equilibrium relation (Equation (18.51)) at $C_A = C_{A,L}$. Thus p_A^* is a measure of the bulk liquid concentration $C_{A,L}$.

Similarly, the overall mass transfer coefficient based on liquid phase (K_L) is defined as:

$$N_A = K_L\left(C_A^* - C_{A,L}\right) \tag{18.55}$$

where C_A^* is the concentration of A in the liquid phase if the liquid phase was in equilibrium with the bulk gas phase (having partial pressure of A as $p_{A,G}$). Once again, C_A^* (like p_A^*) is a fictitious quantity whose value is determined from the equilibrium relation (Equation (18.51)) at $p_A = p_{A,G}$. Thus C_A^* is a measure of the partial pressure of A in the bulk gas phase. The quantities p_A^* and C_A^* are shown schematically in Figure 18.5.

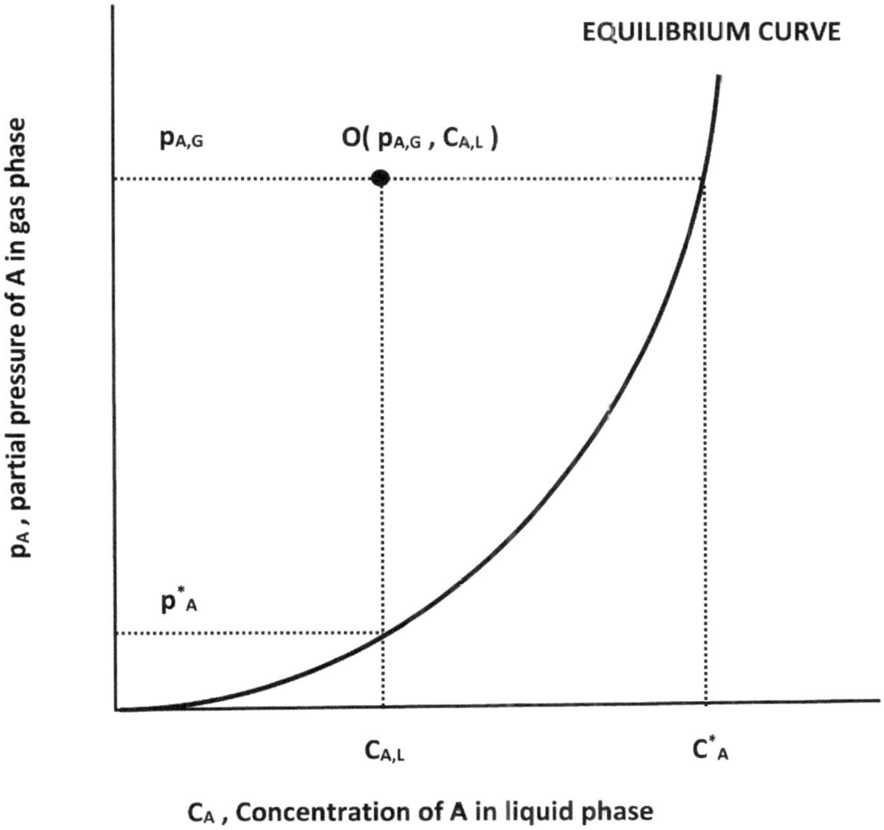

Figure 18.5 Graphical description of quantities p_A^* and C_A^*.

The overall mass transfer coefficient can be related to individual phase coefficients if the equilibrium relationship (Equation (18.51)) is known. Assuming a linear equilibrium relationship:

$$p_A = mC_A \tag{18.56}$$

where m is the equilibrium constant. This linear equilibrium relationship is generally valid at low concentrations and is referred to as "Henry's law". As equilibrium exists at the interface, Equation (18.56) can be expressed in terms of the interface concentrations:

$$p_{A,i} = mC_{A,i} \tag{18.57}$$

From Equation (18.54), it follows that:

$$\frac{1}{K_G} = \frac{p_{A,G}-p_A^*}{N_A} = \frac{p_{A,G}-p_{A,i}}{N_A} + \frac{p_{A,i}-p_A^*}{N_A} \tag{18.58}$$

Using the relations: $p_A^* = mC_{A,L}$and $p_{A,i} = mC_{A,i}$, Equation (18.58) can be re-written as:

$$\frac{1}{K_G} = \frac{p_{A,G}-p_{A,i}}{N_A} + \frac{m(C_{A,i}-C_{A,L})}{N_A} \tag{18.59}$$

From Equations (18.52) and (18.59), it can be readily shown that:

$$\frac{1}{K_G} = \frac{1}{k_G} + \frac{m}{k_L} \tag{18.60}$$

This equation expresses the relationship between overall mass transfer coefficient (K_G) and individual phase coefficients (k_G and k_L). A similar relation for K_L can be obtained as follows:

$$\frac{1}{K_L} = \frac{C_A^*-C_{A,L}}{N_A} = \frac{C_A^*-C_{A,i}}{N_A} + \frac{C_{A,i}-C_{A,L}}{N_A} \tag{18.61}$$

This equation follows from Equation (18.55). As $p_{A,G} = mC_A^*$ and $p_{A,i} = mC_{A,i}$, it follows that:

$$\frac{1}{K_L} = \frac{p_{A,G}-p_{A,i}}{mN_A} + \frac{C_{A,i}-C_{A,L}}{N_A} \tag{18.62}$$

From Equations (18.52) and (18.62):

$$\frac{1}{K_L} = \frac{1}{k_L} + \frac{1}{mk_G} \tag{18.63}$$

Equation (18.63) is the relationship between overall mass transfer coefficient K_L and individual mass transfer coefficients (k_G and k_L).

For systems with large values of m (solute with low solubility in liquid), Equation (18.63) gives:

$$\frac{1}{K_L} \approx \frac{1}{k_L} \tag{18.64}$$

that is, the main resistance to mass transfer lies in the liquid phase. This type of system is "liquid-phase controlled" from mass transfer point of view.

For systems with small values of m (solute with high solubility in liquid), Equation (18.60) gives:

$$\frac{1}{K_G} \approx \frac{1}{k_G} \tag{18.65}$$

In this case, the main resistance to mass transfer of solute lies in the gas phase. This type of system is "gas-phase controlled".

18.3.3 Mass Transfer Coefficients Using Mole Fraction Difference as the Driving Force

In the discussion so far on mass transfer coefficients, either molar concentration difference ΔC_A or partial pressure difference Δp_A were considered as the driving force for mass transfer, and the mass transfer coefficients were defined through the following equations:

$$N_A = k_L \Delta C_A \tag{18.66}$$

$$N_A = k_G \Delta p_A \tag{18.67}$$

$$N_A = K_G \Delta p_A \tag{18.68}$$

$$N_A = K_L \Delta C_A \tag{18.69}$$

It is not necessary to express the driving force for mass transfer in terms of either concentration difference (ΔC_A) or partial pressure difference (Δp_A). It is equally valid to express the driving force for mass transfer in terms of mole fraction difference. Consider mass transfer of solute A from gas phase to liquid phase, as shown in Figure 18.3. For the gas phase, one can express the molar flux as:

$$N_A = k_y \left(y_{A,G} - y_{A,i} \right) \tag{18.70}$$

where $y_{A,G}$ is the mole fraction of A in the bulk gas phase, $y_{A,i}$ is the mole fraction of A in the gas phase at the interface, and k_y is the gas-phase mass transfer coefficient with mole fraction difference as the driving force. For the liquid phase, the molar flux can be expressed as:

$$N_A = k_x \left(x_{A,i} - x_{A,L} \right) \tag{18.71}$$

where $x_{A,i}$ is the mole fraction of A in the liquid phase at the interface, $x_{A,L}$ is the mole fraction of A in the bulk liquid phase, and k_x is the liquid-phase mass transfer coefficient with mole fraction difference as the driving force.

In a similar manner, the overall mass transfer coefficients can be expressed in terms of mole fraction difference as the driving force. Thus,

$$N_A = K_y \left(y_{A,G} - y_A^* \right) \tag{18.72}$$

$$N_A = K_x \left(x_A^* - x_{A,L} \right) \tag{18.73}$$

where K_y and K_x are overall mass transfer coefficients based on gas-phase and liquid-phase, respectively with mole fraction difference as the driving force. $y_{A,G}$ is the mole fraction of A in the bulk gas phase and y_A^* is the mole fraction of A in the gas phase if gas phase was in equilibrium with the bulk liquid (having mole fraction of A as $x_{A,L}$). $x_{A,L}$ is the mole fraction of A in the bulk liquid phase and x_A^* is the mole fraction of A in the liquid phase if liquid phase was in equilibrium with the bulk gas (having mole fraction of A as $y_{A,G}$). It should be noted that mole fractions y_A^* and x_A^* are fictitious quantities evaluated from the equilibrium relationship.

Liquid-phase mass transfer coefficients (k_L and k_x)

The molar flux in the liquid phase is:

$$N_A = k_L(C_{A,i} - C_{A,L}) = k_x(x_{A,i} - x_{A,L}) \tag{18.74}$$

As $C_A = x_A C$, where C is the total molar concentration of the liquid mixture, Equation (18.74) gives:

$$k_L C(x_{A,i} - x_{A,L}) = k_x(x_{A,i} - x_{A,L}) \tag{18.75}$$

Thus,

$$k_x = k_L C \tag{18.76}$$

Gas-phase mass transfer coefficients (k_G and k_y)

The molar flux in the gas phase is:

$$N_A = k_G(p_{A,G} - p_{A,i}) = k_y(y_{A,G} - y_{A,i}) \tag{18.77}$$

As $p_A = y_A p$, where p is the total pressure of the gas mixture, it follows that:

$$k_G p(y_{A,G} - y_{A,i}) = k_y(y_{A,G} - y_{A,i}) \tag{18.78}$$

Thus,

$$k_y = k_G p \tag{18.79}$$

Overall mass transfer coefficients based on liquid-phase (K_L and K_x)

The molar flux of species A from gas phase to liquid phase is given by:

$$N_A = K_L(C_A^* - C_{A,L}) = K_x(x_A^* - x_{A,L}) \tag{18.80}$$

As $C_A = x_A C$, Equation (18.80) gives:

$$K_L C(x_A^* - x_{A,L}) = K_x(x_A^* - x_{A,L}) \tag{18.81}$$

Thus,

$$K_x = K_L C \tag{18.82}$$

Overall mass transfer coefficients based on gas-phase (K_G and K_y)

The molar flux of species A from gas phase to liquid-phase is:

$$N_A = K_G(p_{A,G} - p_A^*) = K_y(y_{A,G} - y_A^*) \tag{18.83}$$

As $p_A = y_A p$, where p is the total pressure of the gas mixture, it follows that:

$$pK_G(y_{A,G} - y_A^*) = K_y(y_{A,G} - y_A^*) \tag{18.84}$$

Thus,

$$K_y = pK_G \tag{18.85}$$

Relationship between overall mass transfer coefficient K_x and individual mass transfer coefficients k_x and k_y

From Equation (18.80), one can write:

$$\frac{1}{K_x} = \frac{x_A^* - x_{A,L}}{N_A} = \frac{x_A^* - x_{A,i}}{N_A} + \frac{x_{A,i} - x_{A,L}}{N_A} \tag{18.86}$$

Assuming that the equilibrium relationship in terms of mole fractions is linear:

$$y_A = m' x_A \tag{18.87}$$

where m' is the equilibrium constant, one can re-write Equation (18.86) as:

$$\frac{1}{K_x} = \frac{y_{A,G} - y_{A,i}}{m' N_A} + \frac{1}{k_x} \tag{18.88}$$

Equation (18.88) implies that:

$$\frac{1}{K_x} = \frac{1}{k_x} + \frac{1}{m' k_y} \tag{18.89}$$

Relationship between overall mass transfer coefficient K_y and individual mass transfer coefficients k_x and k_y

From Equation (18.83), one can write:

$$\frac{1}{K_y} = \frac{y_{A,G} - y_A^*}{N_A} = \frac{y_{A,G} - y_{A,i}}{N_A} + \frac{y_{A,i} - y_A^*}{N_A} \tag{18.90}$$

Using the equilibrium relation (Equation (18.87)), Equation (18.90) can be re-written as:

$$\frac{1}{K_y} = \frac{1}{k_y} + \frac{m'(x_{A,i} - x_{A,L})}{N_A} \tag{18.91}$$

Equation (18.91) implies that:

$$\frac{1}{K_y} = \frac{1}{k_y} + \frac{m'}{k_x} \tag{18.92}$$

18.4 EXAMPLES

18.4.1 Example 1

The local Sherwood number and mass transfer coefficient in laminar boundary layer over a flat plate are given as follows:

$$Sh_x = \frac{kx}{D_{AB}} = 0.332Sc^{1/3}(Re_x)^{1/2}$$

$$k = Sh_x\left(\frac{D_{AB}}{x}\right) = 0.332Sc^{1/3}(Re_x)^{1/2}\left(\frac{D_{AB}}{x}\right)$$

Derive the expressions for average Sherwood number and average mass transfer coefficient over a plate of length L if the boundary layer flow over the entire plate is laminar.

Solution

The average mass transfer coefficient is defined as:

$$\bar{k} = \frac{1}{L}\int_0^L k dx$$

Upon substituting the following expression for local mass transfer coefficient under laminar condition in the integral,

$$k = Sh_x\left(\frac{D_{AB}}{x}\right) = 0.332Sc^{1/3}\left(\frac{\rho V_\infty x}{\eta}\right)^{1/2}\left(\frac{D_{AB}}{x}\right) = 0.332Sc^{1/3}D_{AB}\left(\frac{\rho V_\infty}{\eta}\right)^{1/2}\frac{1}{x^{1/2}}$$

we get:

$$\bar{k} = \frac{1}{L}\int_0^L k dx = \frac{1}{L}\int_0^L 0.332Sc^{1/3}D_{AB}\left(\frac{\rho V_\infty}{\eta}\right)^{1/2}\frac{dx}{x^{1/2}} = \frac{0.332Sc^{1/3}D_{AB}}{L}\left(\frac{\rho V_\infty}{\eta}\right)^{1/2}(2\sqrt{L})$$

$$= 2 \times 0.332Sc^{1/3}\left(\frac{\rho V_\infty L}{\eta}\right)^{1/2}\left(\frac{D_{AB}}{L}\right)$$

$$\bar{k} = 2k|_{x=L}$$

Thus, the average mass transfer coefficient over a length L of the plate with laminar boundary layer is two times the local mas transfer coefficient evaluated at the end of the plate (at $x = L$).

It also follows from the above equations that the average Sherwood number is two times the local Sherwood number evaluated at the end of the plate (at $x = L$), that is:

$$S\bar{h} = 2Sh_x|_{x=L}$$

18.4.2 Example 2

For mixed laminar and turbulent boundary layer flow over a plate, derive the expressions for the average Sherwood number and average mass transfer coefficient over a plate of length L.

Solution

The average mass transfer coefficient for a mixed boundary layer flow can be calculated as:

$$\overline{k} = \frac{1}{L}\int_0^L k dx = \frac{1}{L}\left[\int_0^{x_c} k_{laminar} dx + \int_{x_c}^L k_{turbulent} dx\right]$$

Here x_c is the distance from the leading edge of the plate where transition occurs from laminar to turbulent regime.

Substitute the following expressions for $k_{laminar}$ and $k_{turbulent}$ in the above integrals:

$$k_{laminar} = 0.332 Sc^{1/3} D_{AB}\left(\frac{\rho V_\infty}{\eta}\right)^{1/2}\frac{1}{x^{1/2}}$$

$$k_{turbulent} = 0.0296 Sc^{1/3} D_{AB}\left(\frac{\rho V_\infty}{\eta}\right)^{4/5}\frac{1}{x^{1/5}}$$

$k_{laminar}$ follows from Equation (18.13) and $k_{turbulent}$ follows from Equation (18.14). Upon substitution of these expressions and integration, we get:

$$\overline{k} = \frac{1}{L}\left[\int_0^{x_c} k_{laminar} dx + \int_{x_c}^L k_{turbulent} dx\right]$$

$$= \frac{1}{L}\left[2 \times 0.332 Sc^{1/3} D_{AB}\left(\frac{\rho V_\infty x_c}{\eta}\right)^{1/2} + 0.037 Sc^{1/3} D_{AB}\left\{\left(\frac{\rho V_\infty L}{\eta}\right)^{4/5} - \left(\frac{\rho V_\infty x_c}{\eta}\right)^{4/5}\right\}\right]$$

$$= \frac{D_{AB}}{L}\left[0.664 Sc^{1/3}\left(Re_{x_c}\right)^{1/2} + 0.037 Sc^{1/3}\left\{(Re_L)^{4/5} - \left(Re_{x_c}\right)^{4/5}\right\}\right]$$

where $Re_{x_c} = \frac{\rho V_\infty x_c}{\eta}$ and $Re_L = \frac{\rho V_\infty L}{\eta}$ are critical Reynolds number and Reynolds number at the end of the plate.

We can re-write the above average mass transfer coefficient expression in terms of the average Sherwood number as follows:

$$S\overline{h} = \frac{\overline{k}L}{D_{AB}} = \left[0.664 Sc^{1/3}\left(Re_{x_c}\right)^{1/2} + 0.037 Sc^{1/3}\left\{(Re_L)^{4/5} - \left(Re_{x_c}\right)^{4/5}\right\}\right]$$

Note that the critical Reynolds number Re_{x_c} for flow over a plate is commonly taken as 5×10^5.

18.4.3 Example 3

Consider a large pool of water, 750 m in length, 400 m wide, 2 m in depth. The water temperature is 25 °C. Air at 25 °C blows in a direction parallel to the length of the pool (750 m side of the pool) at a free stream velocity of 1.95 m/s (see figure below).

Estimate the rate of evaporation of water from the pool.

Given that: $\nu_{air} = 1.65 \times 10^{-5}\ m^2/s$, $p_A^{sat} = 3.17\ kPa$, $D_{AB} = 2.58 \times 10^{-5}\ m^2/s$, $Re_{x_c} = 5 \times 10^5$, $M_A = 18$, water is species A and air is species B.

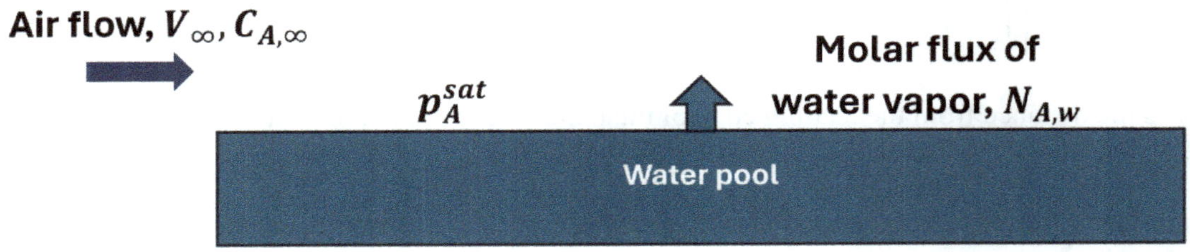

Solution

Calculate Reynolds number at the end of the pool, $L = 750\ m$

$$Re_L = \frac{\rho V_\infty L}{\eta} = \frac{V_\infty L}{\nu} = \frac{1.95 \times 750}{1.65 \times 10^{-5}} = 8.864 \times 10^7$$

As $Re_L \gg Re_{x_c}$, we have a mixed boundary layer, laminar at the beginning and turbulent at the end. The point of transition is:

$$x_c = Re_{x_c}\left(\frac{\nu}{V_\infty}\right) = 5 \times 10^5 \times \left(\frac{1.65 \times 10^{-5}}{1.95}\right) = 4.231\ m$$

For mixed boundary layer flow,

$$S\bar{h} = \frac{\bar{k}L}{D_{AB}} = \left[0.664Sc^{1/3}\left(Re_{x_c}\right)^{1/2} + 0.037Sc^{1/3}\left\{(Re_L)^{4/5} - \left(Re_{x_c}\right)^{4/5}\right\}\right]$$

$$= \left[0.664\left(Re_{x_c}\right)^{1/2} + 0.037\left\{(Re_L)^{4/5} - \left(Re_{x_c}\right)^{4/5}\right\}\right] Sc^{1/3}$$

$$= \left[0.664(5 \times 10^5)^{1/2} + 0.037\{(8.864 \times 10^7)^{4/5} - (5 \times 10^5)^{4/5}\}\right]\left(\frac{1.65 \times 10^{-5}}{2.58 \times 10^{-5}}\right)^{1/3}$$

$$= [469.519 + 0.037\{2.281 \times 10^6 - 3.624 \times 10^4\}] \times 0.8616 = 7.197 \times 10^4$$

Thus,

$$\bar{k} = S\bar{h}\left(\frac{D_{AB}}{L}\right) = 7.197 \times 10^4 \times \frac{2.58\times10^{-5}}{750} = 2.476 \times 10^{-3}\ \text{m/s}$$

The absolute molar flux of water vapor from the surface:

$$N_{A,w} = \bar{k}\left(C_{A,w} - C_{A,\infty}\right)$$

$$C_{A,w} = \frac{p_A^{sat}}{RT} = \frac{3.17 \times 10^3}{8.314 \times 298.15} = 1.2788\ mol/m^3$$

$$C_{A,\infty} = 0$$

$N_{A,w} = \overline{k}(C_{A,w} - C_{A,\infty}) = 2.476 \times 10^{-3}(1.2788 - 0) = 3.1664 \times 10^{-3}\ \ moles/s.m^2$

$\dot{M}_A = N_{A,w} \times A_w = 3.1664 \times 10^{-3} \times (750 \times 400) = 949.92\ moles/s$

where A_w is the contact area of the plate.

$\dot{m}_A = \dot{M}_A \times M_A = 949.92 \times 18 = 17098.6\ \ g/s = 17.099\ \ kg/s$

Thus, the rate of evaporation of water from the pool is 17.099 kg/s.

18.4.4 Example 4

Consider a solid sphere (diameter 30 mm) of naphthalene placed in the path of air flow. The air flows at a free stream velocity of 0.295 m/s. The naphthalene and air are at 25 °C and 1 atm pressure. The naphthalene evaporates (sublimates) into the moving air. The vapor pressure of naphthalene is 74 Pa.

Estimate the following:

i) The mass transfer coefficient for convection of naphthalene from sphere to air

ii) The absolute molar flux of naphthalene at the surface of the sphere

iii) The rate of mass transfer of naphthalene from sphere to air

Given that: $\eta_{air} = 1.9 \times 10^{-5}\ Pa.s,\ \ \rho_{air} = 1.1\ \ kg/m^3,\ \ D_{AB} = 6.9 \times 10^{-6}\ \ m^2/s,\ \ M_A = 128.174$, naphthalene is species A and air is species B.

Solution

i) The mass transfer coefficient for convection of naphthalene from sphere to air

Use the Froessling relationship, Equation (18.18), for mass transfer from a solid spherical particle:

$$Sh = \frac{kD_p}{D_{AB}} = 2 + 0.552\, Re^{1/2}\, Sc^{1/3}$$

This equation correlates the data well for mass transfer from a spherical particle into gases with 2< *Re*<800 and 0.6< *Sc*<2.7.

$$Re = \frac{\rho D_p V_\infty}{\eta} = \frac{1.1 \times 30 \times 10^{-3} \times 0.295}{1.9 \times 10^{-5}} = 512.37$$

$$Sc = \frac{\nu}{D_{AB}} = \frac{\eta}{\rho D_{AB}} = \frac{1.9 \times 10^{-5}}{1.1 \times 6.9 \times 10^{-6}} = 2.503$$

$$Sh = \frac{kD_p}{D_{AB}} = 2 + 0.552\,(512.37)^{1/2}(2.503)^{1/3} = 18.965$$

$$k = Sh\left(\frac{D_{AB}}{D_p}\right) = 18.965\left(\frac{6.9 \times 10^{-6}}{30 \times 10^{-3}}\right) = 4.362 \times 10^{-3}\ \ m/s$$

ii) The absolute molar flux of naphthalene at the surface of the sphere

$$N_{A,w} = k\left(C_{A,w} - C_{A,\infty}\right)$$

$$C_{A,w} = \frac{p_A^{sat}}{RT} = \frac{74}{8.314 \times 298.15} = 0.029853\ mol/m^3$$

$$C_{A,\infty} = 0$$

$$N_{A,w} = k\left(C_{A,w} - C_{A,\infty}\right) = 4.362 \times 10^{-3} \times (0.029853 - 0) = 1.302 \times 10^{-4}\ \text{moles/s.m}^2$$

iii) The rate of mass transfer of naphthalene from sphere to air

$$\dot{M}_A = N_{A,w} \times A_w = 1.302 \times 10^{-4} \times (\pi)(30 \times 10^{-3})^2 = 3.6813 \times 10^{-7}\ moles/s$$

where A_w is the surface area of the sphere.

$$\dot{m}_A = \dot{M}_A \times M_A = 3.6813 \times 10^{-7} \times 128.174 = 4.718 \times 10^{-5}\ g/s$$

18.4.5 Example 5

Consider convective mass transport of naphthalene from a naphthalene-coated pipe to air flowing in the pipe at an average velocity of 0.80 m/s. The air is at 44.85 °C and 1 atm pressure. The pipe inside diameter is 0.02 m and its length is 1.10 m.

Estimate the following:

i) the exit concentration of naphthalene in air

ii) the molar rate of evaporation (sublimation) of naphthalene from the surface of the pipe

Given that: $\eta_{air} = 1.932 \times 10^{-5}\ Pa.s$, $\rho_{air} = 1.114\ kg/m^3$, $D_{AB} = 6.92 \times 10^{-6}\ m^2/s$, $M_A = 128.174$, naphthalene is species A and air is species B.

Solution

Refer to the figure below.

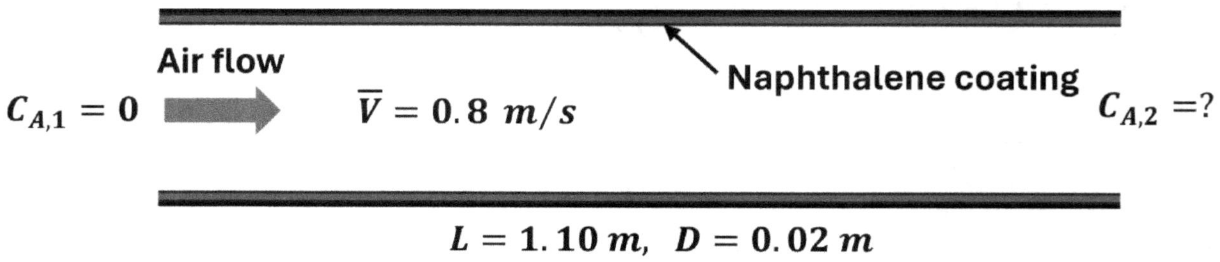

i) the exit concentration of naphthalene in air

$$C_{A,w} = \frac{p_A^{sat}}{RT} = \frac{74}{8.314 \times 318} = 2.799 \times 10^{-2}\ mol/m^3$$

$$\Delta C_{A,1} = C_{A,w} - C_{A,1} = 2.799 \times 10^{-2} - 0 = 2.799 \times 10^{-2}\ mol/m^3$$

From Equations (18.33) and (18.34), we get:

$$\dot{M}_A = \bar{k}A_w(\Delta C_A)_{lm} = \bar{k}A_w\left[\frac{\Delta C_{A,2}-\Delta C_{A,1}}{ln\left(\frac{\Delta C_{A,2}}{\Delta C_{A,1}}\right)}\right]$$

where $\dot{M}_A$ is the total molar flow rate of A over the entire length of the pipe, $\bar{k}$ is the average mass transfer coefficient, A_w is the inner surface area of the pipe and $(\Delta C_A)_{lm}$ is the log mean concentration of A.

From material balance,

$$\dot{M}_A = \left(\frac{\pi D^2}{4}\right)\bar{V}(C_{A,2} - C_{A,1}) = \left(\frac{\pi D^2}{4}\right)\bar{V}(\Delta C_{A,1} - \Delta C_{A,2})$$

Equating the above two expressions for $\dot{M}_A$, we get:

$$\bar{k}(\pi DL)\left[\frac{\Delta C_{A,2} - \Delta C_{A,1}}{ln\left(\frac{\Delta C_{A,2}}{\Delta C_{A,1}}\right)}\right] = \left(\frac{\pi D^2}{4}\right)\bar{V}(\Delta C_{A,1} - \Delta C_{A,2})$$

Upon rearrangement,

$$ln\left(\frac{\Delta C_{A,1}}{\Delta C_{A,2}}\right) = \frac{4\bar{k}L}{D\bar{V}}$$

We need to estimate the average mass transfer coefficient $\bar{k}$ to proceed further.

$$Re = \frac{\rho D\bar{V}}{\eta} = \frac{1.114 \times 0.02 \times 0.80}{1.932 \times 10^{-5}} = 922.5673$$

$$Sc = \frac{\eta}{\rho D_{AB}} = \frac{1.932 \times 10^{-5}}{1.114 \times 6.92 \times 10^{-6}} = 2.5062$$

Since $Re < 2100$, the flow is laminar. We can use the Hausen correlation, Equation (18.43):

$$S\bar{h} = \frac{\bar{k}D}{D_{AB}} = 3.66 + \frac{0.0668(D/L)(Re\,Sc)}{1+0.04[(D/L)(Re\,Sc)]^{2/3}}$$

$$S\bar{h} = \frac{\bar{k}D}{D_{AB}} = 3.66 + \frac{0.0668(D/L)(Re\,S\,c)}{1 + 0.04[(D/L)(Re\,S\,c)]^{2/3}}$$

$$= 3.66 + \frac{0.0668(0.02/1.10)(922.5673 \times 2.5062)}{1 + 0.04[(0.02/1.10)(922.5673 \times 2.5062)]^{2/3}} = 5.5528$$

$$\bar{k} = 5.5528\left(\frac{D_{AB}}{D}\right) = 5.5528\left(\frac{6.92 \times 10^{-6}}{0.02}\right) = 1.9213 \times 10^{-3}\ m/s$$

Now that $\bar{k}$ is known, we can do exit concentration calculation as shown below.

$$ln\left(\frac{\Delta C_{A,1}}{\Delta C_{A,2}}\right) = \frac{4\bar{k}L}{D\bar{V}} = \frac{4 \times 1.9213 \times 10^{-3} \times 1.10}{0.02 \times 0.80} = 0.52836$$

$$\frac{\Delta C_{A,1}}{\Delta C_{A,2}} = 1.69615$$

$$\Delta C_{A,2} = \Delta C_{A,1}/1.69615 = 2.799 \times 10^{-2}/1.69615 = 0.0165$$

$$\Delta C_{A,2} = C_{A,w} - C_{A,2} = 0.0165$$

$$C_{A,2} = C_{A,w} - 0.0165 = 2.799 \times 10^{-2} - 0.0165 = 0.01149\ mol/m^3$$

ii) the molar rate of evaporation (sublimation) of naphthalene from the surface of the pipe

$$\dot{M}_A = \left(\frac{\pi D^2}{4}\right)\bar{V}(C_{A,2} - C_{A,1}) = \left(\frac{\pi D^2}{4}\right)\bar{V}(\Delta C_{A,1} - \Delta C_{A,2})$$

$$= \left(\frac{\pi \times 0.02^2}{4}\right) \times 0.80 \times (2.799 \times 10^{-2} - 0.0165) = 2,8878 \times 10^{-6}\ \ mol/s$$

18.4.6 Example 6

Consider flow of air at 10 °C in a naphthalene-coated pipe at an average velocity of 45 ft/s. The inside diameter of the pipe is 1.2 inch. Estimate the mass transfer coefficient using two different correlations and compare the values.

Given that: $\eta_{air} = 1.2 \times 10^{-5}\ lb_m/ft.s$, $\rho_{air} = 0.078\ lb_m/ft^3$, $D_{AB} = 0.2\ ft^2/h$, naphthalene is species A and air is species B.

Solution

$$Re = \frac{\rho D\bar{V}}{\eta} = \frac{0.078 \times (1.2/12) \times 45}{1.2 \times 10^{-5}} = 2.925 \times 10^4$$

$$Sc = \frac{\eta}{\rho D_{AB}} = \frac{1.2 \times 10^{-5}}{0.078 \times (0.2/3600)} = 2.7692$$

As $Re \gg 2100$, the flow is turbulent.

Use Colburn equation, Equation (18.47):

$$Sh̄ = \frac{\bar{k}D}{D_{AB}} = 0.023(Re)^{4/5}(Sc)^{1/3} = 0.023(2.925 \times 10^4)^{4/5}(2.7692)^{1/3} = 120.804$$

$$\bar{k} = Sh̄\left(\frac{D_{AB}}{D}\right) = 120.804 \times \frac{0.2}{3600} \times \frac{1}{(1.2/12)} = 0.0671\ ft/s$$

Use Gilliland and Sherwood equation, Equation (18.48):

$$\frac{\bar{k}D}{D_{AB}}\left(\frac{p_{B,lm}}{p}\right) = 0.023(Re)^{0.83}(Sc)^{0.44} = 0.023(2.925 \times 10^4)^{0.83}(2.7692)^{0.44} = 183.3362$$

Assuming dilute system, $x_A \ll 1$,

$$\left(\frac{p_{B,lm}}{p}\right) \cong 1$$

$$\bar{k} = Sh̄\left(\frac{D_{AB}}{D}\right) = 183.3362 \times \frac{0.2}{3600} \times \frac{1}{(1.2/12)} = 0.10185\ ft/s$$

Thus, Gilliland and Sherwood correlation predicts a higher $\bar{k}$ value than that predicted by the Colburn equation. The percent difference is:

$$\%\text{difference} = \left(\frac{0.10185 - 0.0671}{0.0671}\right) \times 100 = 51.79\%$$

18.4.7 Example 7

Consider absorption of ammonia gas by water in a wetted wall tower where ammonia-air stream flows upward and water flows downward as a liquid film on the wall of the column (see figure below). The overall mass transfer coefficient K_G is 2.8×10^{-9} kmol/(m^2.s.Pa). At a certain location in the tower, the gas phase contains 7.7 mol% ammonia. The liquid phase at the same location has an ammonia concentration of 0.058 kmol/m^3. Under the operating conditions of the tower, the Henry law constant is 1.299×10^3Pa/(kmol/m^3). Nearly 90% of the total mass transfer resistance encountered in this interphase mass transport problem resides in the gas phase. The tower is operating at 20 °C and 1 atm pressure.

Estimate the following:

i) the individual mass transfer coefficients k_G and k_L

ii) the molar flux of ammonia at the specified location in the tower

iii) the interfacial composition (concentrations of ammonia) at the specified location in the tower

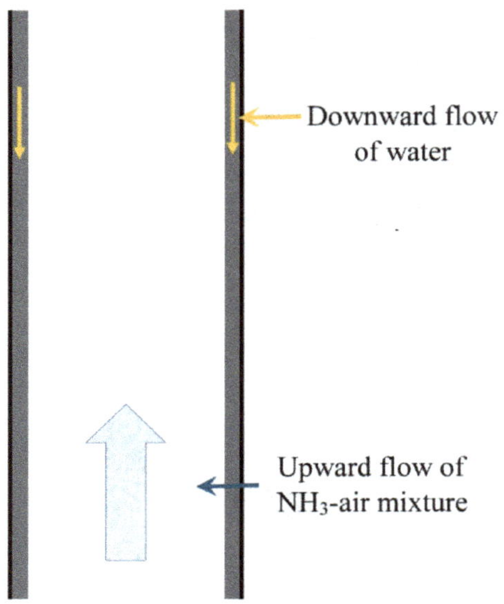

Solution

i) the individual mass transfer coefficients k_G and k_L

$K_G = 2.8 \times 10^{-9}$ kmol/(m^2.s.Pa)

$p_{A,i} = mC_{A,i}$ where $m = 1.299 \times 10^3$ Pa/(kmol/m^3).

Total resistance in both phases:

$$\frac{1}{K_G} = \frac{1}{2.8\times10^{-9}} = 3.5714 \times 10^8 \text{ m}^2\text{.s.Pa/kmol}$$

Resistance in the gas phase:

$$\frac{1}{k_G} = 0.90(3.5714 \times 10^8) = 3.2143 \times 10^8 \text{ m}^2\text{.s.Pa/kmol}$$

$$k_G = 3.111 \times 10^{-9} \text{ kmol/(m}^2\text{.s.Pa)}$$

$$\frac{1}{K_G} = \frac{1}{k_G} + \frac{m}{k_L}$$

$$\frac{m}{k_L} = \frac{1}{K_G} - \frac{1}{k_G} = 3.5714 \times 10^8 - 3.2143 \times 10^8 = 3.571 \times 10^7$$

$$k_L = \frac{m}{3.571\times10^7} = \frac{1.299\times10^3}{3.571\times10^7} = 3.6376 \times 10^{-5} \text{ m/s}$$

ii) the molar flux of ammonia at the specified location in the tower

$$N_A = K_G(p_{A,G} - p_A^*)$$

$$p_{A,G} = y_{A,G}p = 0.077 \times 101.325 = 7.802\ kPa$$

$$C_{A,L} = 0.058 \text{ kmol/m}^3$$

$$p_A^* = mC_{A,L} = 1.299 \times 10^3 \times 0.058 = 75.341\ Pa$$

$$N_A = K_G(p_{A,G} - p_A^*) = 2.8 \times 10^{-9}(7.802 \times 10^3 - 75.341) = 2.1635 \times 10^{-5} \text{ kmol/s.m}^2$$

iii) the interfacial composition (concentrations of ammonia) at the specified location in the tower

$$N_A = k_G(p_{A,G} - p_{A,i})$$

$$p_{A,i} = p_{A,G} - (N_A/k_G) = 7.802 \times 10^3 - (2.1635 \times 10^{-5}/3.111 \times 10^{-9}) = 847.64\ Pa$$

$$y_{A,i} = \frac{p_{A,i}}{p} = \frac{847.64}{101.325 \times 10^3} = 8.366 \times 10^{-3}$$

$$C_{A,i} = \frac{p_{A,i}}{m} = \frac{847.64}{1.299 \times 10^3} = 0.6525 \text{ kmol/m}^3$$

18.4.8 Example 8

Consider absorption of ammonia (NH_3) into water from an air-ammonia mixture in an absorption tower. The individual mass transfer coefficients are $k_L = 1.10$ ft/hr and $k_G = 0.25$ lb-mol/(hr.ft^2.atm). The Henry law for dilute solutions of ammonia (species A) in water is given as:

$$p_{A,i} = 0.25C_{A,i}$$

where $p_{A,i}$ is in atm and $C_{A,i}$ is in lb-mol/ft^3 of solution.

In the table given below, three values are given for each of the mass transfer coefficient. Circle the correct value corresponding to a given mass transfer coefficient.

k_y in lb-mol/(hr.ft^2)	0.0175	0.75	3.75
k for gas phase in ft/hr	1000	10.5	94.9
K_G in lb-mol/(hr.ft^2.atm)	100	116	0.236
K_y in lb-mol/(hr.ft^2)	0.71	0.021	17.1
K_L in ft/hr	65.23	1.234	0.059

Note that the universal gas constant R=0.73 atm.ft^3/(lb-mol.^{0}R) and ^{0}R=^{0}F+460

Solution

k_y in lb-mol/(hr.ft^2)	0.0175	0.75	3.75
k for gas phase in ft/hr	1000	10.5	94.9
K_G in lb-mol/(hr.ft^2.atm)	100	116	0.236
K_y in lb-mol/(hr.ft^2)	0.71	0.021	17.1
K_L in ft/hr	65.23	1.234	0.059

NOTATION

A area
C total molar concentration of mixture or constant in Equations (18.17) and (18.25)
$C_{A,c}$ characteristic molar concentration of species A
$C_{A,i}$ molar concentration of species A at interface on the liquid side
$C_{A,L}$ molar concentration of species A in bulk liquid
$C_{A,m}$ mean molar concentration of species A
$C_{A,w}$ molar concentration of species A in the fluid at the boundary/wall
$C_{A,\infty}$ free stream molar concentration of species A
C_A^* concentration of A in the liquid phase if the liquid phase was in equilibrium with the bulk gas phase (having partial pressure of A as $p_{A,G}$)
D_{AB} diffusivity of species A in species B
D diameter of pipe
D_p diameter of spherical particle
f friction factor
Gr Grashof number
Gz_m Graetz number for mass transfer
k mass transfer coefficient
$\bar{k}$ average mass transfer coefficient
k_G mass transfer coefficient for gas phase
k_L mass transfer for liquid phase
k_x liquid-phase mass transfer coefficient with mole fraction difference as the driving force
k_y gas-phase mass transfer coefficient with mole fraction difference as the driving force
K_G overall mass transfer coefficient based on gas phase
K_L overall mass transfer coefficient based on liquid phase
K_x overall mass transfer coefficient based on liquid phase with mole fraction difference as the driving force
K_y overall mass transfer coefficient based on gas phase with mole fraction difference as the driving force
L length of a tube
m constant in Equations (18.17) and (18.25) or Henry law constant (Equation (18.57))
m' Henry law constant (Equation (18.87))
$\dot{M}_A$ molar flow rate of species A
n constant in Equations (18.17) and (18.25)
N_A absolute molar flux of species A
$\bar{N}_A$ average absolute molar flux of species A
$N_{A,w}$ absolute molar flux of species at the wall
p pressure
$p_{A,G}$ partial pressure of species A in the bulk gas phase
$p_{A,i}$ partial pressure of species A at the interface on the gas side

$p_{B,lm}$	log-mean partial pressure of species B
p_A^*	partial pressure of A in the gas phase if the gas phase was in equilibrium with the bulk liquid phase (having concentration of A as $C_{A,L}$).
Pe	Peclet number for mass transfer ($= ReSc$)
r	radial coordinate or position
R	radius of tube
Re	Reynolds number
Re_D	Reynolds number based on cylinder diameter
Re_x	local Reynolds number, external flow
Sc	Schimdt number
Sh	Sherwood number
$\bar{Sh}$	average Sherwood number
Sh_{nc}	contribution of natural convection
Sh_o	Sherwood number in the limit $Re \to 0$
Sh_x	local Sherwood number, external flow
St_m	Stanton number for mass transfer
V_{max}	maximum velocity
V_z	component of velocity in z-direction
$\bar{V}$	average velocity
V_∞	free stream velocity of fluid
x, y, z	space coordinates in cartesian coordinate system
x_A	mole fraction of species A in liquid
$x_{A,i}$	mole fraction of species A at the interface on the liquid side
$x_{A,L}$	mole fraction of species in bulk liquid phase
x_A^*	mole fraction of species A in the liquid phase if liquid phase was in equilibrium with the bulk gas (having mole fraction of A as $y_{A,G}$).
y_A	mole fraction of species A in gas phase
$y_{A,G}$	mole fraction of species A in bulk gas phase
$y_{A,i}$	mole fraction of species A at the interface on the gas phase
y_A^*	mole fraction of A in the gas phase if gas phase was in equilibrium with the bulk liquid (having mole fraction of A as $x_{A,L}$).

Greek Letters

Δ	change in variable
η	shear viscosity of fluid
ν	kinematic viscosity of fluid
ρ	density of fluid

PRACTICE PROBLEMS

Problem 1

Consider the flow of dry air over a pool of liquid water, 11 m long and 5 m wide, at a free stream velocity of 2.95 m/s. Estimate the average mass transfer coefficient. You can take the critical Reynolds number to be 5×10^5.

Given that: $\nu_{air} = 1.55 \times 10^{-5}\ m^2/s$, $D_{AB} = 2.54 \times 10^{-5}\ m^2/s$, water vapor is species A and air is species B.

Problem 2

Liquid acetone is accidentally spilled over a solid smooth surface (impermeable to acetone), 1 m in length and 1 m in width. To dry the surface of the acetone layer, dry air at 25 °C is blown over the layer of acetone at a free stream velocity of 5.9 m/s.

Estimate the following:

i) the mass transfer coefficient at a mid-point of the solid surface (0.5 m downstream the leading edge of the solid surface)

ii) the rate of mass transfer of acetone from the surface

The transition from laminar to turbulent flow in the boundary layer occurs at $Re_{x_c} = 2 \times 10^5$.

Given that: $\nu_{air} = 1.55 \times 10^{-5}\ m^2/s$, $D_{AB} = 0.93 \times 10^{-5}\ m^2/s$, acetone vapor is species A and air is species B. $p_A^{sat} = 3.07 \times 10^4\ Pa$. Molecular weight of acetone is 58.08.

Problem 3

Consider a pool of liquid water, 3.5 m in length, 3.5 m in width, and 1 m in depth. The temperature of the water is 25 °C. Air at 25 °C flows over the water surface at a free stream velocity of 5.85 m/s. Estimate the following:

i) the local mass transfer coefficient at 1 m from the leading edge of the pool.

ii) the average value of mass transfer coefficient over the entire pool of length 3.5 m.

iii) the mass flux of water vapor and the mass rate of evaporation of water from the pool

The transition from laminar to turbulent flow in the boundary layer occurs at $Re_{x_c} = 2 \times 10^5$.

Given that: $\nu_{air} = 1.55 \times 10^{-5}\ m^2/s$, $D_{AB} = 2.55 \times 10^{-5}\ m^2/s$, water vapor is species A and air is species B. $p_A^{sat} = 3.21 \times 10^3\ Pa$. Molecular weight of water is 18.

Problem 4

Consider convective mass transfer from a flat boundary. The Reynolds number at the end of the boundary is $Re_L = 4.5 \times 10^6$. What percentage of the total mass transfer occurs in the laminar boundary layer over the boundary assuming the critical Reynolds number to be 5×10^5?

Problem 5

An air stream is flowing over a solid plate of naphthalene, 1.5 m in length and 1.5 m in width, at a free stream velocity of 28.5 m/s. The temperature and pressure are 25 °C and 1 atm. Estimate the rate of sublimation of naphthalene from the plate. The diffusivity of naphthalene in air is $6.9\text{x}10^{-6}$ m^2/s and vapor pressure at 25 °C is 24.5 Pa. The kinematic viscosity of air at 25 °C is $1.55\text{x}10^{-5}$ m^2/s. You can take critical Reynolds number to be 5×10^5.

Problem 6

A solid sphere (diameter 50 mm) of naphthalene placed in the path of air flow. The air flows at a free stream velocity of 0.20 m/s. The naphthalene and air are at 25 °C and 1 atm pressure. The naphthalene evaporates (sublimates) into the moving air. The vapor pressure of naphthalene is 74 Pa.

Estimate the following:

i) The mass transfer coefficient for convection of naphthalene from sphere to air

ii) The absolute molar flux of naphthalene at the surface of the sphere

iii) The rate of mass transfer of naphthalene from sphere to air

Given that: $\eta_{air} = 1.86 \times 10^{-5}\ Pa.s$, $\rho_{air} = 1.168\ kg/m^3$, $D_{AB} = 6.9 \times 10^{-6}\ m^2/s$, $M_A = 128.174$, naphthalene is species A and air is species B.

Problem 7

The stripping process is commonly used to remove organic compounds from wastewater using air as stripping gas. Consider a wetted wall type stripping column where wastewater is flowing downward along the walls of the column and air is flowing upwards in the core of the column. The organic solute to be removed from wastewater is trichloroethylene (TCE). TCE is a commonly used solvent in industrial applications and is often present in wastewater. The stripping process is carried out at 20 °C and 1 atm total pressure. The flow rate of air is 2650 cm^3/s. The inside diameter of the column is 45 mm.

i) Estimate the gas-phase mass transfer coefficient (k_G) using the Gilliland and Sherwood correlation

ii) Estimate the overall mass transfer coefficient K_L

Given that: $k_L = 2.55 \times 10^{-5}\ \mathrm{m/s}$, Henry law constant: $m = 3.05 \times 10^4$ atm/(kmol/m^3), $D_{AB} = 8.08 \times 10^{-6}\ m^2/s$, $\rho_{air} = 1.19\ kg/m^3$, $\eta_{air} = 1.84 \times 10^{-5}\ Pa.s$. TCE is species A and air is species B.

Problem 8

Consider absorption of H_2S from an air-H_2S mixture by water in a wetted wall column where water is flowing downward along the walls of the column and air-H_2S mixture is flowing upwards. The absorption process is carried out at 29 ^{0}C and 1.49 atm. At a certain location in the column, the partial pressure of H_2S in the bulk gas phase is 0.048 atm and the mole fraction of H_2S at the interface on the liquid side is 1.98×10^{-5}. Estimate the molar flux of H_2S at this location. Given that: $\mathrm{k\,(gas\ phase)} = 9.6 \times 10^{-4}\ m/s$ and the Henry law for this system is:

$$P_{A,i} = 609x_{A,i}$$

where $P_{A,i}$ is in atm (H_2S is species A)

Problem 9

Species "A" is being stripped from wastewater using a wetted wall column and air as the stripping gas. At a certain location in the column, the concentration of species "A" in the bulk aqueous phase is 3.95 kmol/m^3 and the particle pressure of species "A" in the bulk gas phase is 2.96 x10^5 Pa. The Henry law for the system is:

$$P_{A,i} = mC_{A,i}$$

where m is 1328 Pa/(kmol/m^3). The overall mass transfer coefficient is $K_G = 2.99 \times 10^{-9}$ kmol/(m^2.s.Pa).

If the gas phase offers 72% of the total resistance to mass transfer, calculate the following:

i) The gas-phase mass transfer coefficient, k_G

ii) The liquid-phase mass transfer coefficient, k_L

iii) The overall mass transfer coefficient, K_L

iv) The molar flux, N_A

Problem 10

Consider a mass transfer tower where wastewater stream containing an organic solute (species *A*) is introduced to the top. The air stream flows upward in the tower. At some location in the tower, the concentration of species "A" in the water stream is 1.2×10^{-3} mol/m^3 and the air is basically free of any "A". The individual phase mass transfer coefficients for this system are as follows:
$k_L = 4.8 \times 10^{-4}$ m/s and $k_G = 0.0098$ kmol/(m^2.s.atm). The Henry law for the system is:

$$P_{A,i} = mC_{A,i}$$

where m is 11.2 atm/(kmol/m^3).

Calculate the following: (a) the molar flux of species "A" at the specified location; (b) the overall mass-transfer coefficients, K_L and K_G.

FURTHER READING

1. J.R. Welty, C.E. Wicks, R.E. Wilson, and G.L. Rorrer. *Fundamentals of Momentum, Heat, and Mass Transfer*. 5th Edition, Wiley & Sons, New York, 2008.
2. C.J. Geankoplis, A.A. Hersel, D.H. Lepek. *Transport Processes and Separation Process Principles*. 5th Edition, Prentice Hall, New Jersey, 2018.
3. W.L. McCabe, J.C. Smith, P. Harriott. *Unit Operations of Chemical Engineering*. 6th Edition, McGraw Hill, New York, 2001.
4. D.R.Gaskell. *An Introduction to Transport Phenomena in Materials Engineering*. Macmillan Publishing Company, New York, 1992.
5. R.B. Bird, W.E. Stewart, and E.N. Lightfoot. 2007. *Transport Phenomena*. 2nd Edition, Wiley & Sons, New York, 2007.
6. R. Pal. *Electromagnetic, Mechanical, and Transport Properties of Composite Materials*. CRC Press, Boca Raton, Florida, 2015.
7. R.A. Greenkorn, D.P.Kessler. *Transfer Operations*. McGraw Hill, New York, 1972.
8. H. Bedingfield, H. and T.B. Drew. 1950. Analogy between heat transfer and mass transfer -A psychrometric study. *Ind. Eng. Chem.* 42: 1164–1173.
9. W.H. Linton, W.H. and T.K. Sherwood. 1950. Mass transfer from solid phase to water in streamline and turbulent flow. *Chem. Eng. Progress*. 46: 258-264.
10. E.R Gilliland, and T.K. Sherwood. 1934. Diffusion of vapors into air streams. *Ind. Eng. Chem.* 26: 516-523.
11. N. Froessling. 1938. Uber die Verdunstung fallender Tropfen. *Gerlands Beitr. Zur Geophyik* 52: 170-216.
12. R.L. Steinberger, R.E. Treybal. 1960. Mass transfer from a solid soluble sphere to a flowing liquid stream. *AICHE.J.* 6: 227-232.
13. V.G. Levich. 1962. *Physicochemical Hydrodynamics*. Prentice-Hall, Englewood Cliffs, NJ.
14. H. Hausen. 1943. Darstellung des Warmeuberganges in Rohren durch verallgemeinerte Potenzbeziehungen. *VDI Z.*, no. 4, p. 91

Chapter 19

Analysis and Design of Packed Columns for Mass Transfer

19.1 INTRODUCTION

In this chapter, we consider the analysis and design of *continuous* contacting devices for mass transport, namely the packed columns. In packed columns, the heavier phase (liquid) flows downward in the column over the surface of some packing material and the lighter phase (gas or vapor) flows counter-currently in the upward direction. As the contact between the two phases is continuous, that is, maintained throughout the column, the packed columns are also called *continuous* contactors. Packed columns are also called rate-based mass exchange devices as mass transfer of solute from one phase to another occurs throughout the column. No phase equilibrium is established between the two phases in the column. Packed towers are used for mass transfer processes such as absorption, stripping, distillation, and liquid-liquid extraction.

19.2 PACKINGS AND PACKING MATERIALS

The purpose of the packing in packed columns is to provide a large surface area for contact between the two phases. The liquid flows downward the column over the surface of the packing material in the form of thin films and the vapor flows upward in the free/void space available inside the packing. The packing should preferably have an open structure so as to provide low resistance to gas flow. The packing should also promote uniform distribution of liquid over the packing surface. Different materials are used for the generation of packings for columns. The popular materials are a) ceramics – ceramics provide good wettability and are corrosion resistant. However, they are brittle in nature; b) metals – metals have good strength and good wettability; and c) plastics – plastics have good strength and are relatively inexpensive. However, they may have wettability issues. Packings come in many different styles. They can be broadly classified into two categories: *random* packing and *structured* packing.

19.2.1 Random Packings

In random packing, the packing consists of discrete particles. The particles are simply dumped into the column and the arrangement of particles in the column is random. Particles used for random packings come in different shapes and sizes. Figure 19.1 shows some of the popular shapes of particles. They are Spheres, Raschig rings, Pall rings, and Saddles. Raschig rings are pieces of cylindrical tubes with diameter nearly the same as the length. Pall ring is a modification of Raschig ring. It has a more open structure to help maintain an even distribution of liquid and minimize plugging and fouling. Also, it offers lower resistance to gas flow. Saddle-shaped particles are also popular for packing of columns. In some applications, spherical particles are used. Figure 19.2 shows random packings of Raschig rings and spherical particles.

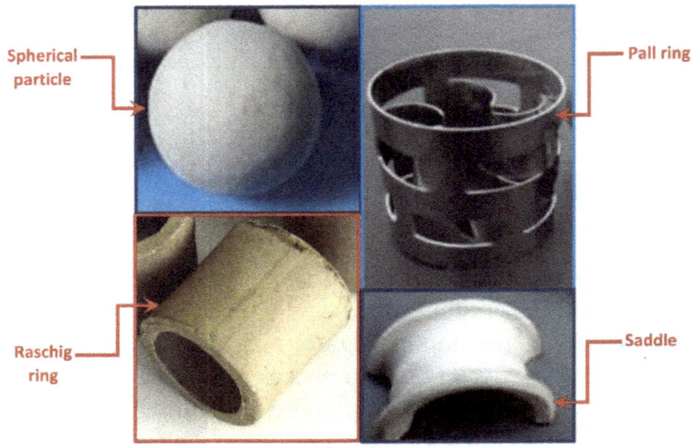

Figure 19.1 Different shapes of particles used in random packings.

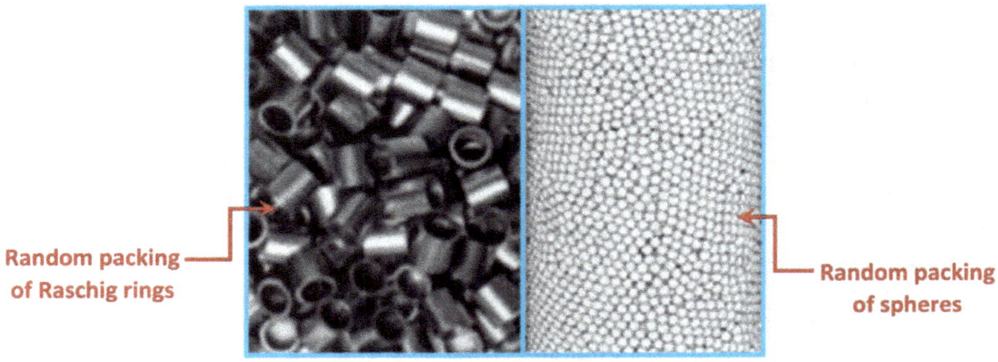

Figure 19.2 Random packings of Raschig rings and spheres.

19.2.2 Structured Packings

Structured packings are packings with regular geometry. They are usually made of corrugated sheets of perforated metal, as shown in Figure 19.3. The layers of corrugated metal sheets are stacked together in a crisscrossing manner to create flow channels for the vapor or gas phase. The liquid spreads uniformly over the packing surface and flows downward as a continuous thin film held by surface tension.

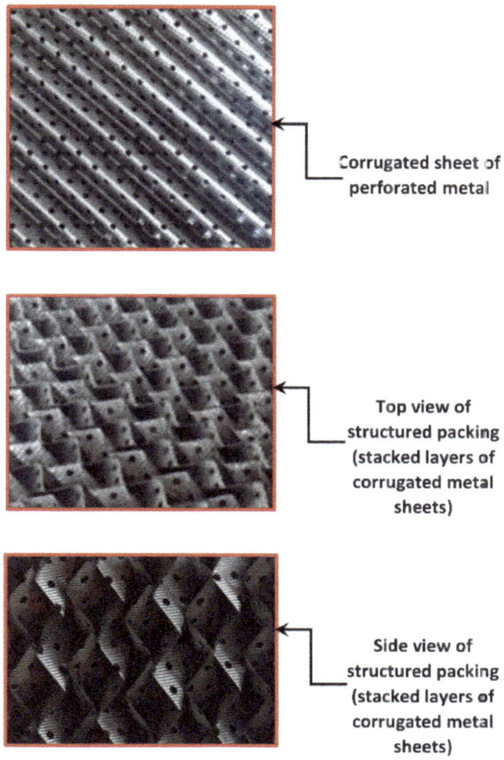

Figure 19.3 Intalox structured packing of corrugated sheets of porous metal.

19.2.3 Liquid Distribution

Even distribution of liquid over the entire cross-sectional area of the packed column is important. If the liquid is not distributed evenly over the entire cross-section of the packed column, the separation efficiency of the column is reduced significantly. Uneven distribution or channelling of liquid must be avoided. Ideally the entire packing surface area should be wetted by the liquid and the liquid should flow downwards in the form of thin films over the packing surface.

There are several different types of liquid distributors available commercially (see Figure 19.4). Some of the popular ones are a) pan distributor, b) pipe-arm distributor, and c) spray nozzle distributor. Pan distributor consists of a pan with cylindrical gas risers and orifices at the bottom of the pan. Liquid is distributed through the orifices and gas passage is provided by gas risers. Pipe-arm distributor consists of a row of perforated pipes. The liquid is distributed evenly over the packed bed through perforated holes under the pipes. Spray nozzle distributor consist of a row of pipes with spray nozzles. There are numerous small holes on the nozzles to distribute the liquid.

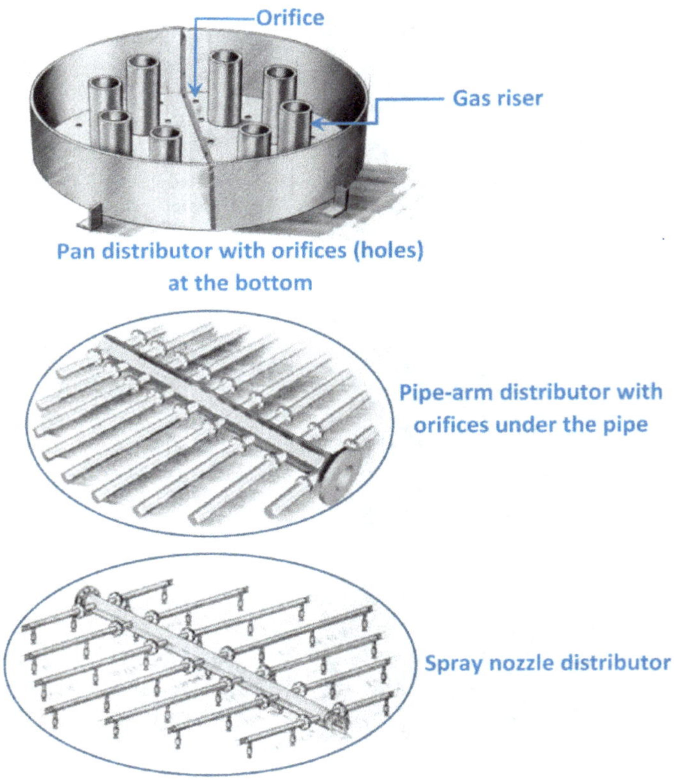

Figure 19.4 Examples of liquid distributors.

19.3 HEIGHT OF PACKED COLUMNS

19.3.1 HETP Method

Although packed column is a continuous contactor, it can still be treated like a staged column using the concept of *HETP* (height equivalent to a theoretical plate). In the *HETP* method, it is imagined that the total height of the packed column can be divided into layers or segments of packing of equal height, referred to as *HETP*, such that the gas and liquid phases leaving this imaginary segment are at equilibrium with each other. Thus, *HETP* is the height of packing needed to obtain the same change in vapor (gas) composition as that obtained in one equilibrium stage. The concept of *HETP* is illustrated schematically in Figure 19.5. If at any point in the column, x is the solute concentration in the liquid phase and y is the solute concentration in the gas phase at the same point, then *HETP* is the height of the packing above this point where the gas concentration is y^* corresponding to equilibrium with the liquid concentration x .
The column height h can be expressed as:

$$h = (HETP) \times N_{theoretical} \tag{19.1}$$

where $N_{theoretical}$ is the number of equilibrium or theoretical stages estimated from the McCabe-Thiele graphical method or Kremser equations.

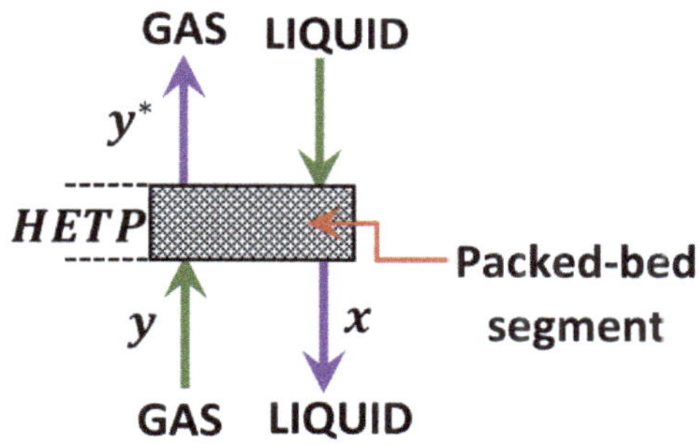

Figure 19.5 *HETP* of packing.

HETP can be readily determined experimentally for a given type of packing. *HETP* is related to the mass transfer area provided by the packing. Higher the surface contact area of the packing per unit volume of the column, more efficient is the packing with low value of *HETP*. Note that a low value of *HETP* implies a shorter column.

For randomly packed columns, the surface area of the particles per unit volume of the column is inversely related to the equivalent diameter d_p of the particles. Consequently,

$$HETP \propto d_p \tag{19.2}$$

One rule of thumb for the estimation of *HETP* is as follows:

$$HETP(m) = 0.018\, d_p(mm) \tag{19.3}$$

For small diameter columns ($D < 2$ ft), $HETP \approx D$ regardless of the particle size. The *HETP* value is generally assumed to be 1 ft when column diameter is less than 1 ft.

For structured packings, the following rule of thumb may be used to estimate the *HETP:*

$$HETP(m) = \frac{100}{a} + 0.10 \tag{19.4}$$

where a is the surface area of the packing per unit volume.

Although the key parameter affecting the *HETP* is the surface area per unit volume a, *HETP* also depends on the physical properties to some extent. For example, Strigle (R.F. Strigle, Packed Tower Design and Applications, 2nd Edition, Gulf Pub. Co, 1994) proposed the following *HETP* correlation for Intalox corrugated sheet type structured packings:

$$ln(HETP) = n - 1.01 - 0.187 ln(\sigma) + 0.213 ln(\eta_L) \tag{19.5}$$

where *HETP* is in m, σ is the surface tension in N/m, and η_L is the liquid viscosity in Pa.s. The parameter n is dependent on the type of Intalox structured packing (see Table 19.1).

Lockett (M.J. Lockett, Chemical Engineering Progress, Vol. 94, 1, 1998) proposed the following correlation for the estimation of *HETP* for high density corrugated sheet type structured packings:

$$HETP\ (m) = \frac{1.54 g^{0.5}(\rho_L - \rho_V)^{0.5} \mu_r^{-0.06}}{a[1+0.78(\rho_V/\ \rho_L)^{0.25} exp(0.00058a)]^2} \tag{19.6}$$

where g is the acceleration due to gravity ($9.8\ m/s^2$), $\mu_r = \mu_L(mPa.s)/0.978$, a is the specific area of packing in units of m^2/m^3, ρ is the density in kg/m^3.

Table 19.1 Values of n (Equation (19.5)) for different Intalox corrugated sheet type structured packings

Packing-type	Surface area per unit volume, m²/m³	n
1T, Intalox corrugated sheet structured packing	310	0.7683
2T, Intalox corrugated sheet structured packing	215	1.0128
3T, Intalox corrugated sheet structured packing	170	1.3868

19.3.2 HTU-NTU Method

In the HTU-NTU method of analysis and design of packed towers, it is imagined that the total height of the packed column can be divided into a number of layers of packing (referred to as number of transfer units, NTU) of equal height (referred to as height of transfer unit, HTU) such that the change in gas concentration over the HTU is equal to the average driving force for mass transfer over the whole column. The concept of *HTU* for gas absorption is illustrated schematically in Figure 19.6. If at any point in the column, x is the solute concentration in the liquid phase and y is the solute concentration in the gas phase at the same point, then *HTU* is the height of the packing above this point where the gas concentration is $y - (y - y^*)_{avg}$ where $(y - y^*)_{avg}$ is the average driving force for mass transfer of solute in the column from gas phase to liquid phase.

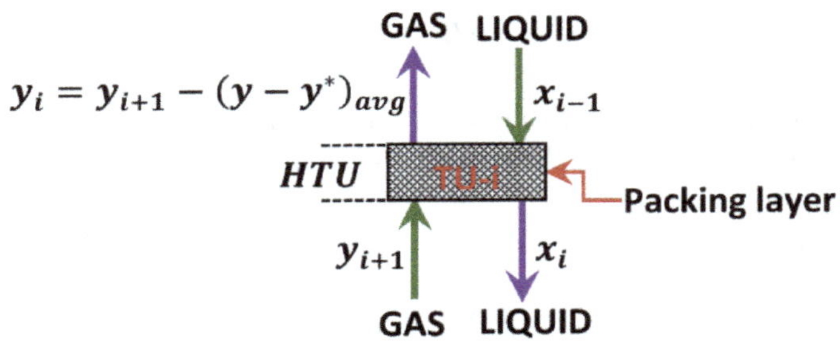

Figure 19.6 *HTU* of packing for gas absorption.

Figure 19.7 shows a schematic diagram of the packed column divided into N transfer units (NTU=N) of equal height, that is, HTU. Thus, the column height h can be expressed as:

$$h = (HTU) \times (NTU) \tag{19.7}$$

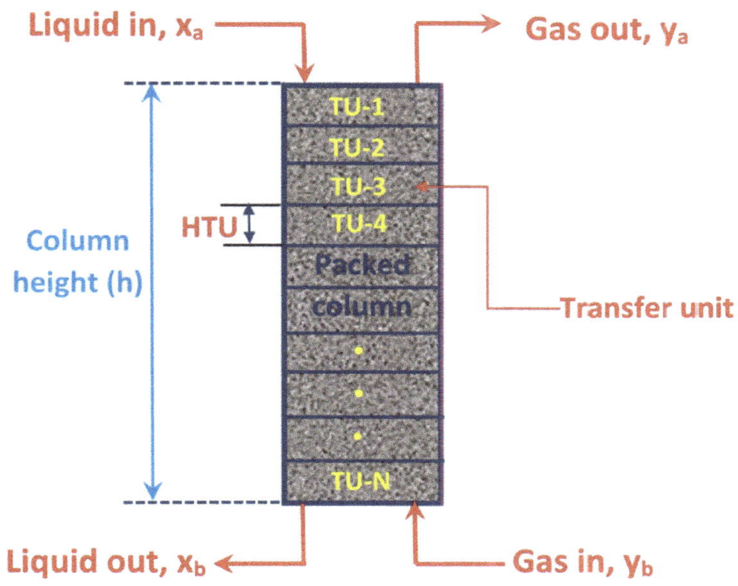

Figure 19.7 Packed column with imaginary transfer units.

The number of transfer units (NTU) depends on the required separation whereas the height of the transfer unit (HTU) depends on the efficiency of mass transfer. More efficient the mass transfer, smaller is the value of HTU.

In the discussion to follow, a detailed analysis is presented of counter-current *gas absorption* in a packed column where some solute, species A, present in the gas phase is being transferred to the liquid phase. The analysis of other mass transfer operations (stripping, binary distillation, liquid-liquid extraction, etc.) in packed columns can be treated in a similar manner.

Interphase Mass Transfer

In a packed column, interphase mass transfer of solute occurs from one phase to another by diffusion. Figure 19.8 shows the concentration profiles, in units of mole fraction, on both sides of the gas-liquid interface in gas absorption operation. The molar flux of solute N_A (moles/s.m^2) across the gas film on the left side of the gas-liquid interface can be expressed as:

$$N_A = k_y(y - y_i) \tag{19.8}$$

where k_y is the gas-phase mass transfer coefficient (moles/s.m^2). The mass transfer coefficient k_y is a measure of the ability of the system to transfer solute across the gas film. For the same driving force $(y - y_i)$, a higher k_y implies a higher rate of mass transfer of solute across the gas film, and hence, more efficient mass transfer.

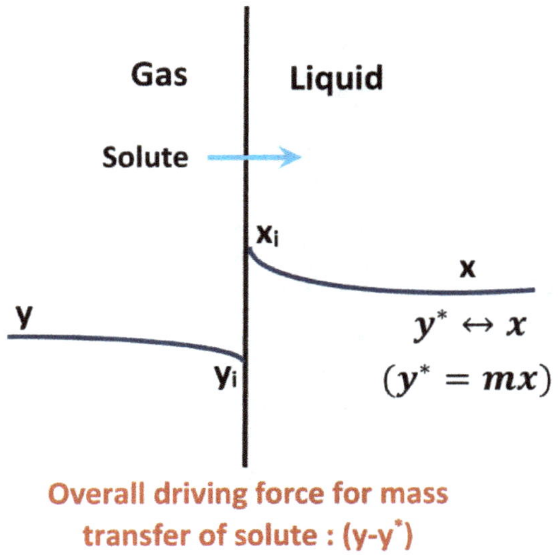

Figure 19.8 Interphase mass transfer of solute in gas absorption.

The molar flux of solute N_A (moles/s.m^2) across the liquid film on the right side of the gas-liquid interface can be expressed as:

$$N_A = k_x(x_i - x) \tag{19.9}$$

where k_x is the liquid-phase mass transfer coefficient (moles/s.m^2). The mass transfer coefficient k_x is a measure of the ability of the system to transfer solute across the liquid film. For the same driving force $(x_i - x)$, a higher k_x implies a higher rate of mass transfer of solute across the liquid film.

At steady state, there cannot be accumulation of solute at the interface and therefore,

$$N_A = k_y(y - y_i) = k_x(x_i - x) \tag{19.10}$$

The driving forces for mass transfer of solute included in the molar flux expressions, Equations (19.8) and (19.9), are the individual phase driving forces, $(y - y_i)$ for gas phase and $(x_i - x)$ for liquid phase. These driving forces contain the interface compositions (y_i, x_i) which are usually not known. In avoid the interface compositions, it is equally appropriate to express the molar flux of solute as:

$$N_A = K_y(y - y^*) \tag{19.11}$$

where K_y is the overall mass transfer coefficient based on gas phase, $(y - y^*)$ is the overall driving force for mass transfer of solute from gas phase to liquid phase, and y^* is the mole fraction of solute in the gas phase in equilibrium with the existing bulk liquid phase. Note that y^* is a fictitious quantity to be calculated from the equilibrium relation (Henry's law):

$$y^* = mx \tag{19.12}$$

where m is the equilibrium constant.

From Equations (19.10) - (19.12), it can be readily shown that:

$$\frac{1}{K_y} = \frac{1}{k_y} + \frac{m}{k_x} \tag{19.13}$$

If the solute is highly soluble in the liquid phase, m value is small, and the resistance to mass transfer in the liquid phase is small. Thus, the resistance to mass transfer of solute lies mainly in the gas phase. Consequently,

$$\frac{1}{K_y} \approx \frac{1}{k_y} \tag{19.14}$$

If the solute is only sparingly soluble in the liquid phase, m value is large, and the resistance to mass transfer of solute lies mainly in the liquid phase. Consequently,

$$\frac{1}{K_y} \approx \frac{m}{k_x} \tag{19.15}$$

Material Balance

Consider material balance of solute A over a thin section dz of the packed gas absorption column shown in Figure 19.9. Let the contact area of the packing per unit volume of packing be a. The interphase mass transfer rate of solute from gas phase to liquid phase over a section dz of the column is:

$$\text{Rate of mass transfer} = K_y a(y - y^*)Sdz \tag{19.16}$$

where S is the cross-sectional area of the column. The mass transfer rate of solute from gas phase to liquid phase over a section dz of the column is also given as:

$$\text{Rate of mass transfer} = Vy|_z - Vy|_{z+dz} \tag{19.17}$$

where V is the molar flow rate of gas. From Equations (19.16) and (19.17):

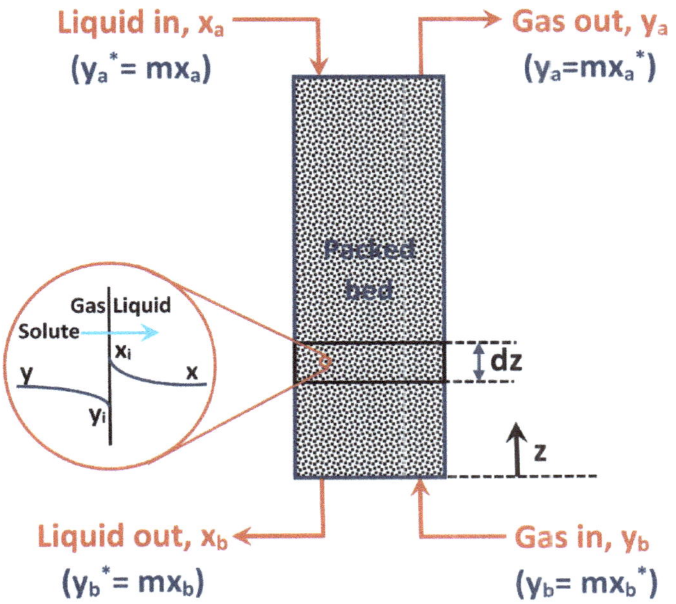

Figure 19.9 Packed gas absorption column.

$$Vy|_z - Vy|_{z+dz} = K_y a(y - y^*)Sdz \tag{19.18}$$

Thus,

$$-\frac{d(Vy)}{dz} = K_y a(y - y^*)S \tag{19.19}$$

Assuming the gas flow rate V to be constant, Equation (19.19) reduces to:

$$-V\frac{dy}{dz} = K_y a(y - y^*)S \tag{19.20}$$

Height of a Packed Column

Upon rearrangement and integration of the material balance equation, Equation (19.20), we obtain:

$$h = -\left(\frac{V/S}{K_y a}\right)\int_{y_b}^{y_a} \frac{dy}{y - y^*} \tag{19.21}$$

Upon inverting the limits of the integral:

$$h = \left(\frac{V/S}{K_y a}\right)\int_{y_a}^{y_b} \frac{dy}{y - y^*} \tag{19.22}$$

This equation could be recast as:

$$h = (HTU) \times (NTU) \tag{19.23}$$

where

$$HTU = \left(\frac{V/S}{K_y a}\right) \tag{19.24}$$

$$NTU = \int_{y_a}^{y_b} \frac{dy}{y - y^*} \tag{19.25}$$

The number of transfer units *(NTU)*, defined in Equation (19.25), depends on the degree of desired separation $(y_b - y_a)$ and the average driving force for mass transfer $(y - y^*)_{avg}$. It is simply the ratio of $(y_b - y_a)$ to $(y - y^*)_{avg}$. The height of the transfer unit (HTU), given by Equation (19.24), is a measure of the efficiency of mass transfer. More efficient the mass transfer, smaller is the value of HTU. A larger mass transfer coefficient K_y implies a faster rate of interphase mass transfer of solute and hence a lower HTU. Likewise, a larger contact area of the packing (per unit volume of the packing) implies a faster rate of interphase mass transfer of solute and hence a lower HTU.

Given the mass transfer coefficient K_y , the packing contact area a per unit volume of the packing, the gas molar flow rate V, and the column diameter D $(S = \pi D^2/4)$, Equation (19.24) can be used to calculate the HTU. The estimation of *NTU* is described in the following section.

Estimation of NTU

The number of transfer units is given by the integral in Equation (19.25). To evaluate this integral, we need to develop an equation for the operating line of the packed column based on material balance. The solute material balance over the dashed envelope shown in Figure 19.10 gives:

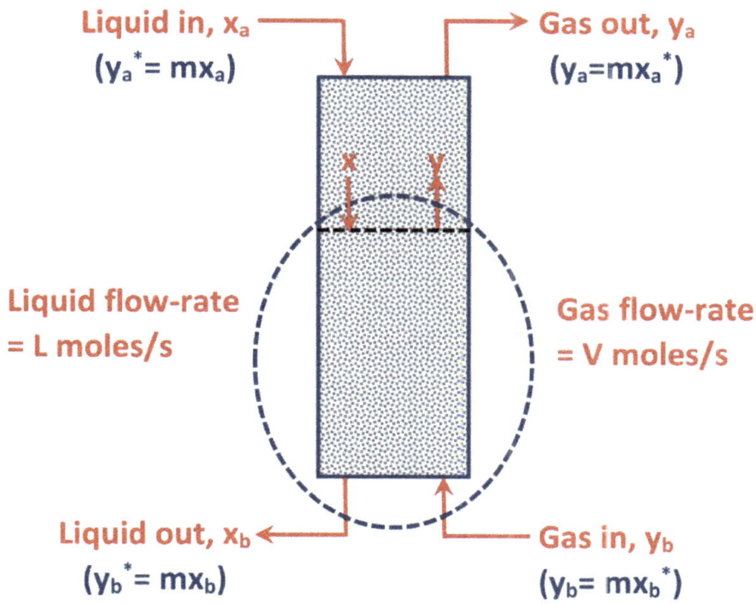

Figure 19.10 Material balance on a packed gas absorption column.

$$Lx + Vy_b = Lx_b + Vy \tag{19.26}$$

The liquid and gas flow rates are assumed to be constant. Upon rearrangement of Equation (19.26), we get:

$$y = y_b + (L/V)(x - x_b) \tag{19.27}$$

This is the equation for the "operating line" of the packed column. It relates the compositions of passing liquid and gas streams at any section of the column. The operating line has a slope of L/V and passes through the terminal compositions (x_a, y_a) and (x_b, y_b). Figure 19.11 shows the plots of the operating line, Equation (19.27), and the equilibrium relationship (assumed to be linear). The driving force for mass transfer of solute $(y - y^*)$ generally varies with y.

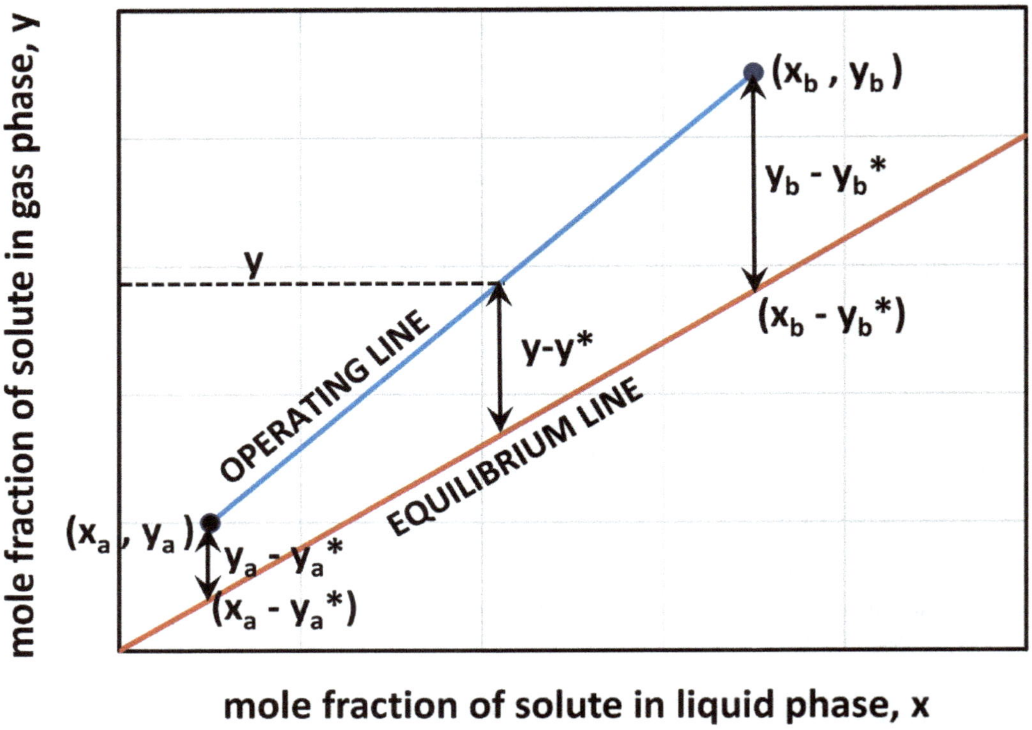

Figure 19.11 Plots of operating and equilibrium lines for counter-current absorption process.

For linear operating and equilibrium lines with different slopes, it can be readily shown that:

$$NTU = \int_{y_a}^{y_b} \frac{dy}{y-y^*} = \frac{y_b-y_a}{(y-y^*)_{log-mean}} \tag{19.28}$$

where

$$(y-y^*)_{log-mean} = \frac{(y_b-y_b^*)-(y_a-y_a^*)}{ln\left(\frac{y_b-y_b^*}{y_a-y_a^*}\right)} \tag{19.29}$$

In the special case when the slopes of the operating and equilibrium lines are the same, that is, the two lines are parallel and the driving force for mass transfer $(y-y^*)$ is constant,

$$NTU = \int_{y_a}^{y_b} \frac{dy}{y-y^*} = \frac{y_b-y_a}{(y_a-y_a^*)} = \frac{y_b-y_a}{(y_b-y_b^*)} \tag{19.30}$$

19.3.3 Relationship Between HETP and HTU

HETP and *HTU* are related to each other. Also, *NTU* is related to the number of equilibrium or theoretical stages $N_{theoretical}$. These relationships are derived here for counter-current gas absorption in packed columns.

The absorption factor, A, is defined as:

$$A = \frac{L/V}{m} \tag{19.31}$$

where m is the Henry law constant (assuming equilibrium relationship is linear, that is, Henry law is applicable).

In the special case of $A = 1$, the number of theoretical or equilibrium stages given by the Kremser equation is the same as *NTU* given by Equation (19.30). Thus,

$$N_{theoretical} = NTU = \frac{y_b - y_a}{(y_b - y_b^*)} = \frac{y_b - y_a}{(y_a - y_a^*)} \quad (19.32)$$

The total height h of a packed column can be expressed in two equivalent ways as:

$$h = HETP \times N_{theoretical} = HTU \times NTU \quad (19.33)$$

When $A = 1$, $N_{theoretical} = NTU$. Therefore,

$$HETP = HTU \quad (19.34)$$

In the general case when $A \neq 1$, the Kremser equation gives the following number of theoretical (equilibrium) stages:

$$N_{theoretical} = \frac{ln\left(\frac{y_b - y_b^*}{y_a - y_a^*}\right)}{lnA} \quad (19.35)$$

Starting from Equation (19.28), it can be shown, after some algebra, that:

$$NTU = \frac{y_b - y_a}{(y - y^*)_{log-mean}} = \left(\frac{A}{A-1}\right) ln\left(\frac{y_b - y_b^*}{y_a - y_a^*}\right) \quad (19.36)$$

From Equations (1.35) and (19.36), it follows that:

$$NTU = \left(\frac{A}{A-1}\right)(lnA)N_{theoretical} \quad (19.37)$$

Upon substitution of *NTU* from Equation (19.37) into Equation (19.33), we get:

$$HETP = \left(\frac{A}{A-1}\right)(lnA)(HTU) \quad (19.38)$$

19.4 DIAMETER OF PACKED COLUMNS

The estimation of the column diameter is based on consideration of "flooding phenomenon" in packed columns. To discuss flooding, it is necessary to define superficial gas and liquid mass velocities. They are defined as follows:

$$G_V = \frac{\dot{m}_V}{A_{column}} = \frac{4\dot{m}_V}{\pi D^2} = \left(\frac{4M_V}{\pi D^2}\right)V \quad (19.39)$$

$$G_L = \frac{\dot{m}_L}{A_{column}} = \frac{4\dot{m}_L}{\pi D^2} = \left(\frac{4M_L}{\pi D^2}\right)L \quad (19.40)$$

where G is the superficial mass velocity, $\dot{m}$ is the mass flow rate, A is the cross-sectional area of the column, D is the column diameter, M is the molecular weight, V is the molar flow rate of gas, and L is the molar flow rate of liquid. For a given column and fluids, the mass velocity is directly proportional to the molar flow rate of the phase (gas or liquid). With the decrease in column diameter, the mass velocity increases keeping other factors constant (same fluid, same molar flow rate). In the discussion to follow, the mass velocity is often referred to as simply velocity.

19.4.1 Flooding in Packed Columns

Consider a counter-current gas absorption column of known diameter, packed with a packing of known type and size. The liquid flows in the downward direction in the column at some fixed molar flow rate L (fixed liquid velocity G_L to be more precise). The gas flows in the upward direction. What happens if we increase the gas molar flow rate (gas velocity G_V to be more precise)?

At low gas velocities, the downward flow of liquid through the packing is not affected by the upward flow of gas. The quantity of liquid retained in the packing, that is, liquid holdup of packing is constant unaffected by the changing gas flow rate or velocity. With further increase in gas velocity, the interphase friction between gas and liquid streams begins to hinder the downward flow of liquid. Consequently, more liquid is held up in the packing and liquid holdup of the packing increases. The gas velocity where liquid holdup of the packing begins to rise is called "loading point" (see Figure 19.12). As the gas velocity is increased further, the accumulation of liquid within the packing increases. At some gas velocity, referred to as "flooding point" (see Figure 19.12), the liquid is no longer able to flow down the packing and is blown out of the column with the rising gas. Obviously, one must operate the column at gas velocities below the flooding point velocity.

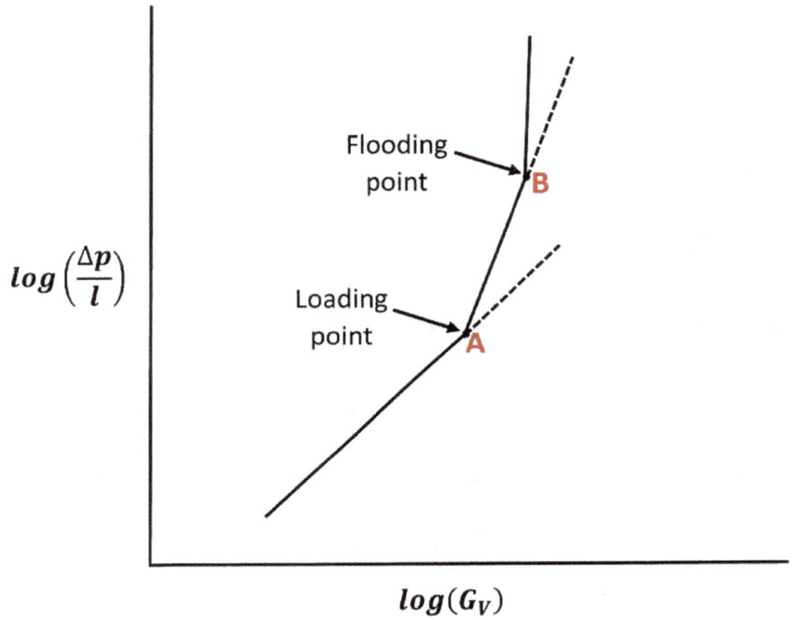

Figure 19.12 Increase of pressure gradient in a packed bed with the increase in gas mass velocity.

19.4.2 Flooding Curve and its Interpretation

The flooding velocity is a function of liquid velocity, fluid properties, and packing characteristics. The flooding curve for random packings is shown in Figure 19.13. The flooding curve is a plot of $(G_V^2\psi)$ versus F_{LV}. The parameters ψ and F_{LV} are defined as follows:

$$\psi = \frac{F_p\phi\eta_L^{0.2}}{\rho_V\rho_L g_c} \tag{19.41}$$

$$F_{LV} = \frac{\dot{m}_L}{\dot{m}_V}\sqrt{\frac{\rho_V}{\rho_L}} = \frac{G_L}{G_V}\sqrt{\frac{\rho_V}{\rho_L}} \tag{19.42}$$

where F_p is the packing factor that has units of area/volume, that is, ft^{-1}, $\phi = \rho_{water}/\rho_L$, g_c= 32.2 $lb_m(ft/s^2)/lb_f$, η_L is liquid viscosity in cP, ρ_V and ρ_L are densities in units of lb_m/ft^3, $\dot{m}$ is the mass flow rate in lb_m/s, and G is the mass velocity in units of $lb_m/(s.ft^2)$. The packing factor F_p depends on the type and size of packing. Smaller packing sizes of the same packing family have larger values of packing factor and specific contact area. The values of the packing factors for various packings are given in Table 19.2. Also note that the ordinate of Figure 19.13, that is, $(G_V^2\psi)$, is not dimensionless and we must use the units of various quantities as already indicated here.

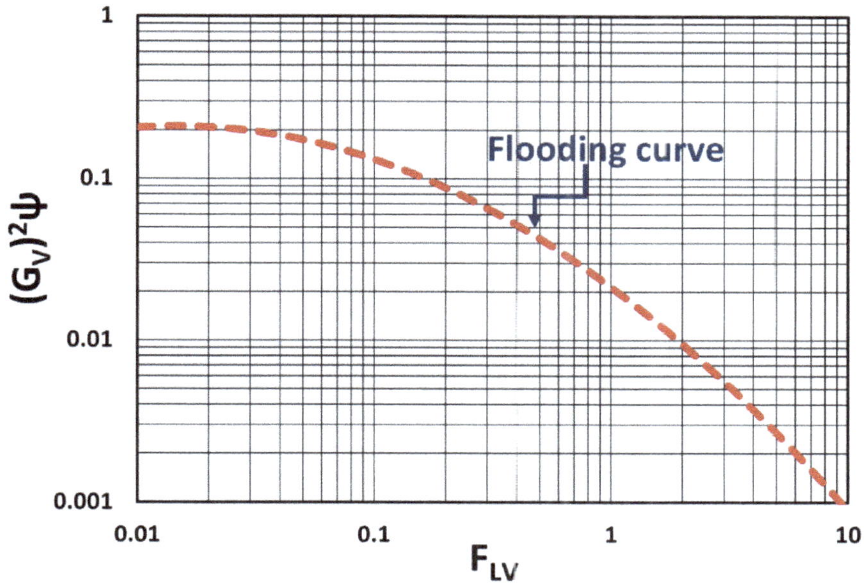

Figure 19.13 Flooding curve for random-packed columns.

The flooding curve shown in Figure 19.13 is generated from the following equation proposed by Kessler and Wankat (D.P. Kessler, P.C. Wankat, Chemical Engineering, Vol. 95, 72-74, 1988):

$$log_{10}(G_V^2\,\psi) = -1.6678 - 1.085log_{10}(F_{LV}) - 0.29655[log_{10}(F_{LV})]^2 \tag{19.43}$$

We can interpret the flooding curve in different ways as discussed below:

- Given gas and liquid mass flow rates, fluid properties, type and size of packing, and column diameter, would there be flooding in the column under the given conditions? To answer this question, we need to calculate F_{LV} and $G_V^2\psi$. Then locate the point on Figure 19.13. If the point falls below the flooding curve, there occurs no flooding at the given conditions. If the point falls on or above the flooding curve, the column will be flooded at the given conditions.

- Given liquid mass flow rate, fluid properties, type and size of packing, and column diameter, what happens if the gas mass flow rate is increased? As the gas mass flow rate is increased, keeping other factors constant, we move in the direction indicated by the arrow in Figure 19.14. At some gas velocity, the flooding point is reached.

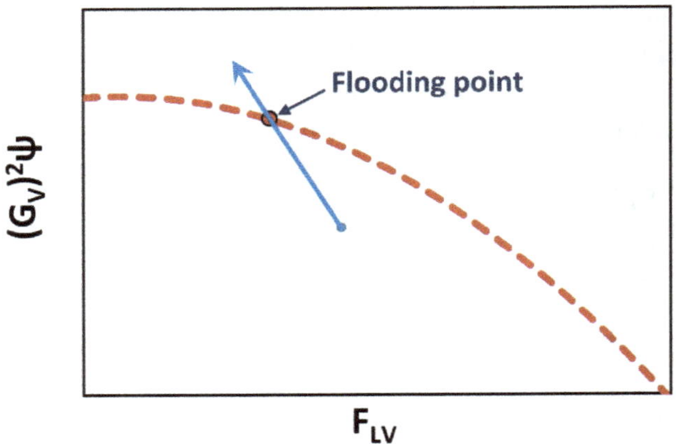

Figure 19.14 Flooding due to increase in gas velocity, keeping other factors constant.

- Both gas and liquid mass flow rates are increased in equal proportions, keeping other factors (fluid properties, type and size of packing, and column diameter) the same, would there be flooding in the column? As gas and liquid mass flow rates are increased in the same proportions, the flow parameter F_{LV} does not change. However, $G_V^2\psi$ increases due to an increase in gas velocity. Figure 19.15 shows the path followed. At some gas velocity (fixed F_{LV}) flooding occurs.

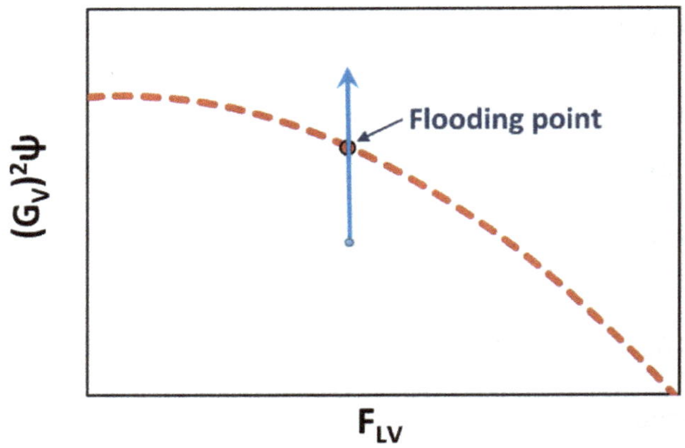

Figure 19.15 Flooding due to increase in gas and liquid mass flow rates in equal proportions, keeping other factors constant.

- Given gas mass flow rate, fluid properties, type and size of packing, and column diameter, what happens if the liquid mass flow rate is increased? As the liquid mass flow rate is increased, keeping other factors constant, we move in the direction indicated by the arrow in Figure 19.16. At some liquid velocity, the flooding point is reached.

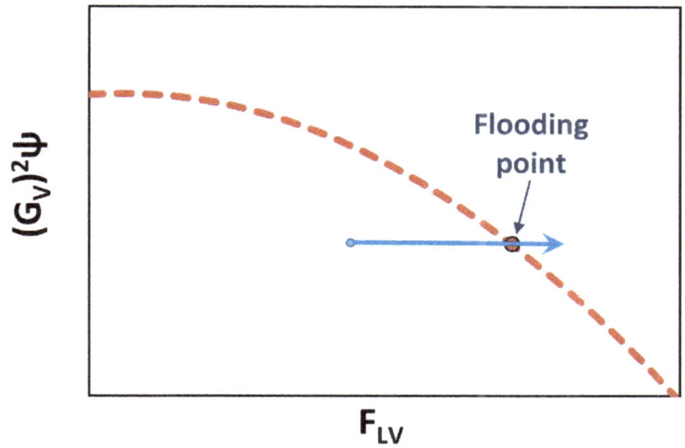

Figure 19.16 Flooding due to increase in liquid mass flow rate, keeping other factors constant.

- Given gas and liquid mass flow rates, fluid properties, type and size of packing, what happens if the column diameter is decreased? As the diameter of the column is decreased, keeping other factors constant, the flow parameter F_{LV} does not change. However, $G_V^2\psi$ increases due to an increase in the gas velocity. Figure 19.17 shows the path followed. At some column diameter (fixed F_{LV}) flooding occurs.

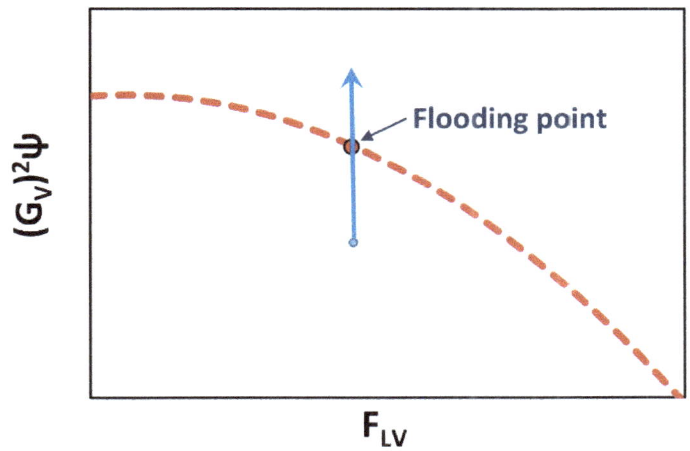

Figure 19.17 Flooding due to a decrease in column diameter, keeping gas and liquid mass flow rates and other factors constant.

Table 19.2 Packing factors for various packings

Packing -type	Nominal size (inches)	Packing factor, F_P (ft^{-1})
Raschig rings (metal, 1/16 in wall)	0.5	410
	1.0	137
	1.5	83
	2.0	57
Raschig rings (ceramic)	0.5	580
	1.0	155
	1.5	95
	2.0	65
Pall rings (metal)	0.625	70
	1.0	48

	1.5	28
	2.0	20
Pall rings (plastic)	0.625	97
	1.0	52
	1.5	32
	2.0	25
Intalox saddles (ceramic)	0.5	200
	1.0	98
	1.5	52
	2.0	40

Finally, it should be noted that the flooding curve shown in Figure 19.13 and the corresponding correlation, Equation (19.43), are valid for random packings. Flooding correlations have been proposed in the literature for structured packings as well. For example, Lockett (M.J. Lockett, Chemical Engineering Research and Design, Vol. 73, 379-384, 1995) proposed the following flooding correlation for high density structured packings:

$$U_V \rho_V^{0.5} = (\rho_L - \rho_V)^{0.5} \left[\frac{0.887 g^{0.25} (a)^{-0.25} \eta_r^{-0.03}}{1+\beta(\rho_V/\rho_L)^{0.25}(L/V)^{0.5}}\right]^2 \tag{19.44}$$

where β is given as:

$$\beta = 0.78 exp(0.00058a) \tag{19.45}$$

In these equations, U_V is the flooding superficial gas velocity (m/s), g is the acceleration due to gravity ($9.8\, m/s^2$), $\eta_r = \eta_L(mPa.s)/0.978$, a is the specific area of packing in units of m^2/m^3, ρ is the density in kg/m^3.

Equation (19.44) can be modified as follows:

$$G_V^2 \psi_S = \left[1 + \beta\sqrt{F_{LV}}\right]^{-4} \tag{19.46}$$

where ψ_S is given as:

$$\psi_S = \frac{1.615 a \eta_r^{0.12}}{\rho_V(\rho_L - \rho_V)g} \tag{19.47}$$

Figure 19.18 shows the plots of flooding curves generated from Equation (19.46) for two different values of specific area of packing. With the increase in specific area (m^2/m^3) of packing, the flooding gas velocity decreases.

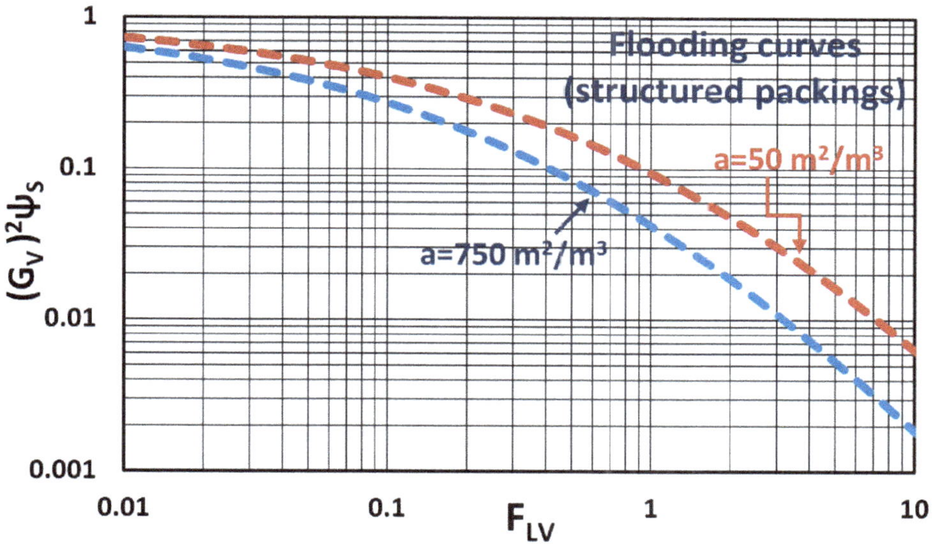

Figure 19.18 Flooding curves for structured packings, generated from Equation (19.46).

19.4.3 Estimation of Column Diameter using the Flooding Curve/Equation

The diameter of a packed column is selected such that the superficial mass velocity of gas G_V at which the column is operated is well below the flooding point. Thus:

$$(G_V)_{actual} = \alpha (G_V)_{flooding} \tag{19.48}$$

where $\alpha < 1$. For packed columns, the column is usually designed with α anywhere in the range: $0.50 < \alpha < 0.80$. Often the packed columns operate at about 70% flooding ($\alpha = 0.70$). Once the value of α is selected, the column diameter is calculated from:

$$D = \sqrt{\frac{4\dot{m}_V}{\pi \alpha (G_V)_{flooding}}} \tag{19.49}$$

The flooding velocity of gas $(G_V)_{flooding}$ is determined either from flooding curves or flooding correlations, Equation (19.43) or (19.46).

19.5 EXAMPLES

19.5.1 Example 1

A binary liquid mixture is to be distilled in a distillation column at 1 atm pressure. The feed is an equimolar of components. The more volatile component in the distillate product has a mole fraction of 0.98. The less volatile component in the bottoms product has a mole fraction of 0.95. The column is to be operated at a reflux ratio of 3. The number of theoretical (equilibrium) stages required is 10 (including the partial reboiler). The viscosity of feed liquid under column conditions is 0.5 cP. The average relative volatility (α) is 2.5.
Calculate the following:

a) the HETP and column height if the column is a packed tower consisting of 1.5-inch Raschig rings as packing particles. Use the rule of thumb, Equation (19.3), to estimate HETP.
b) the HETP and column height if the column is a packed tower consisting of 1T Intalox corrugated sheet structured packing. Use the rule of thumb, Equation (19.4), to estimate HETP.
c) the value of HETP for the structured packing of part (c) using the Strigle correlation, Equation (19.5). Take η_L=0.5 cP and σ =5 mN/m.
d) the value of HETP for the structured packing of part (c) using the Lockett correlation, Equation (19.6). Take η_L=0.5 cP, ρ_V =1.28 kg/m^3 and ρ_L =995 kg/m^3

Solution

a) Using the rule of thumb, Equation (19.3):

$$HETP(m) = 0.018\, d_p(mm) = 0.018 \times 38.1 = 0.686\, m$$
$$h = (10 - 1) \times 0.686 = 6.17\, m$$

Note that $d_p = 38.1\ mm$ and that the partial reboiler is excluded from the number of given theoretical stages to calculate the tower height h.

b) Using the rule of thumb, Equation (19.4):

$$HETP(m) = \frac{100}{a} + 0.10 = \frac{100}{310} + 0.10 = 0.42\, m$$

$$h = (10 - 1) \times 0.42 = 3.78\, m$$

Note that $a = 310\ m^2/m^3$ for 1T Intalox corrugated sheet structured packing (see Table 19.1).

c) Strigle correlation, Equation (19.5), gives:

$$ln(HETP) = n - 1.01 - 0.187 ln(\sigma) + 0.213 ln(\eta_L)$$
$$ln(HETP) = 0.7683 - 1.01 - 0.187 ln(5 \times 10^{-3}) + 0.213 ln(0.5 \times 10^{-3}) = -0.87$$
$$HETP = 0.419\, m$$

Note that $n = 0.7683$ is taken from Table 19.1

d) Lockett correlation, Equation (19.6), gives:

$$HETP\ (m) = \frac{1.54 g^{0.5}(\rho_L - \rho_V)^{0.5}\eta_r^{-0.06}}{a[1 + 0.78(\rho_V/\ \rho_L)^{0.25} exp(0.00058a)]^2}$$

$$HETP\ (m) = \frac{1.54(9.8)^{0.5}(995 - 1.28)^{0.5}(0.5/0.978)^{-0.06}}{310[1 + 0.78(1.28/\ 995)^{0.25} exp(0.00058 \times 310)]^2} = 0.37\, m$$

Note that $a = 310\ m^2/m^3$ for 1T Intalox corrugated sheet structured packing (see Table 19.1).

19.5.2 Example 2

A packed tower is to be used to absorb species "A" from a gas stream using pure water as a solvent. The gas entering the tower contains 2.5 mol% A and 95 percent of it is to be absorbed. The tower is to operate at 25 °C and 1 atm, and the liquid and gas molar flow rates on a unit area basis are to be 150 kmol/h.ft^2 and 30 kmol/h.ft^2, respectively.
The equilibrium and overall mass transfer coefficient data are as follows:

$$y^* = 3x\ ; \qquad K_y a = 7.5\ kmol/h.ft^3$$

Determine the following:

a) the number of transfer units, NTUs
b) the height of the transfer unit, HTU
c) the height of the packed tower

You can assume that the gas and liquid molar flow rates are constant.

Solution

Refer to the schematic diagram of the packed tower shown below.
We first need to calculate x_b. From Equation (19.27):

$$y_a = y_b + (L/V)(x_a - x_b)$$

Upon rearrangement,

$$x_b = x_a + (y_b - y_a)/(L/V) = 0 + (0.025 - 0.00125)/(150/30) = 4.75 \times 10^{-4}$$

Note that y_a is 5% of y_b, that is, $y_a = 0.05 \times 0.025 = 0.00125$
Thus all the mole fractions at inlet and outlet of the tower are now known.

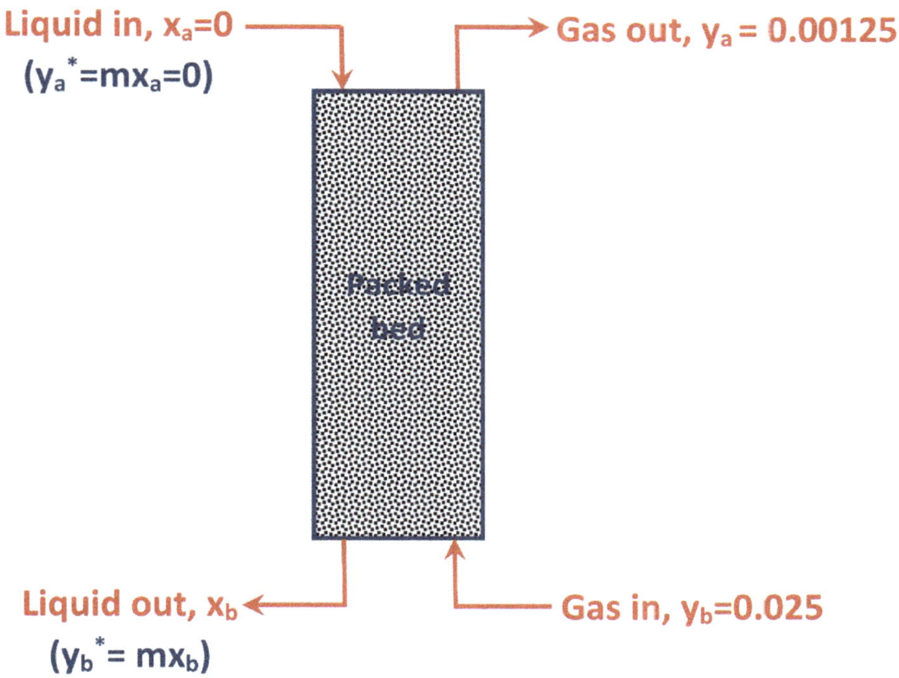

a) the number of transfer units, NTUs

For linear operating and equilibrium lines with different slopes, we can use Equation (19.28) to determine the NTUs:

$$NTU = \frac{y_b - y_a}{(y - y^*)_{log-mean}} = \frac{0.025 - 0.00125}{(y - y^*)_{log-mean}}$$

$$(y - y^*)_{log-mean} = \frac{(y_b - y_b^*) - (y_a - y_a^*)}{ln\left(\frac{y_b - y_b^*}{y_a - y_a^*}\right)} = \frac{(0.025 - 1.425 \times 10^{-3}) - (0.00125 - 0)}{ln\left(\frac{0.025 - 1.425 \times 10^{-3}}{0.00125 - 0}\right)}$$

$$= 7.6 \times 10^{-3}$$

$$NTU = \frac{0.025 - 0.00125}{7.6 \times 10^{-3}} = 3.125$$

Note that $y_a^* = 3x_a = 0$ and $y_b^* = 3x_b = 3(4.75 \times 10^{-4}) = 1.425 \times 10^{-3}$

b) the height of the transfer unit, HTU

From Equation (19.24):

$$HTU = \left(\frac{V/S}{K_y a}\right) = \frac{30}{7.5} = 4\ ft$$

c) the height of the packed tower

$$h = (HTU) \times (NTU) = 4 \times 3.125 = 12.5\ ft$$

19.5.3 Example 3

A packed tower is to be used in the treatment of a gas stream that enters the tower at a volumetric flow rate of 9.81 ft^3/s. The tower is packed with 1-inch ceramic Raschig rings. The mass flow rates of liquid and gas streams are the same.
Estimate the diameter of the tower such that the gas mass velocity is 70% of the flooding limit.
The properties of the fluids are:

$\rho_V = 0.062\ lb_m/ft^3$, $\rho_L = 62.4\ lb_m/ft^3$, $\eta_L = 1$ cP

For 1-inch ceramic Raschig rings, the packing factor $F_p = 155\ ft^{-1}$. Note that water is the liquid phase.

Solution

From Equation (19.42):

$$F_{LV} = \frac{\dot{m}_L}{\dot{m}_V}\sqrt{\frac{\rho_V}{\rho_L}} = 1 \times \sqrt{\frac{0.062}{62.4}} = 0.0315$$

Using Equation (19.43), we can estimate the flooding velocity:

$$log_{10}(G_V^2\,\psi) = -1.6678 - 1.085 log_{10}(F_{LV}) - 0.29655[log_{10}(F_{LV})]^2$$

$$log_{10}(G_V^2\,\psi) = -1.6678 - 1.085 log_{10}(0.0315) - 0.29655[log_{10}(0.0315)]^2 = -0.684$$

$$G_V^2\,\psi = 0.207$$

From Equation (19.41):

$$\psi = \frac{F_p \phi \eta_L^{0.2}}{\rho_V \rho_L g_c} = \frac{155 \times 1 \times 1^{0.2}}{0.062 \times 62.4 \times 32.2} = 1.244$$

Thus, the flooding mass velocity G_V is:

$$G_{V,flooding} = \sqrt{\frac{0.207}{1.244}} = 0.408\ \frac{lb_m}{s.ft^2}$$

Using Equation (19.49),

$$D = \sqrt{\frac{4\dot{m}_V}{\pi\alpha(G_V)_{flooding}}} = \sqrt{\frac{4 \times 9.81 \times 0.062}{\pi \times 0.70 \times 0.408}} = 1.65\ ft$$

19.5.4 Example 4

Ammonia is to be absorbed in a packed tower packed with 1T Intalox corrugated sheet structured packing using pure water as a solvent. The gas stream enters the tower at a mass flow rate of 600 kg/h. The mass flow rate of liquid is 1.75 times that of gas stream.
Estimate the diameter of the tower such that the gas mass velocity is 70% of the flooding limit.
The properties of the fluids are:

$\rho_V = 1\ kg/m^3$, $\rho_L = 1000\ kg/m^3$, $\eta_L = 0.001$ Pa.s

For 1T Intalox corrugated sheet structured packing, the surface area per unit volume is 310 m^2/m^3.

Solution

Refer to the schematic diagram of the packed tower given below.

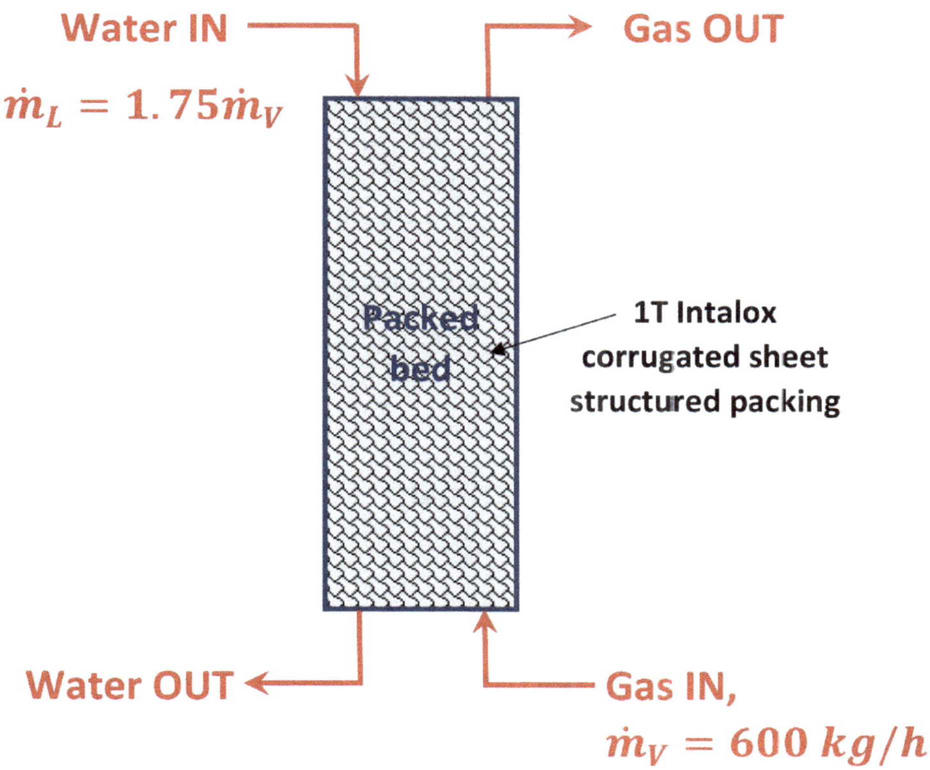

From Equation (19.42):

$$F_{LV} = \frac{\dot{m}_L}{\dot{m}_V}\sqrt{\frac{\rho_V}{\rho_L}} = 1.75 \times \sqrt{\frac{1}{1000}} = 0.055$$

Using Equation (19.46), we can estimate the flooding velocity:

$$G_V^2\psi_S = \left[1 + \beta\sqrt{F_{LV}}\right]^{-4}$$

where β is given as:

$$\beta = 0.78exp(0.00058a) = 0.78\exp(0.00058 \times 310) = 0.934$$

Upon substitution of F_{LV} and β, we get:

$$G_V^2\psi_S = \left[1 + (0.934)\sqrt{0.055}\right]^{-4} = 0.453$$

ψ_S is given by Equation (19.47) as:

$$\psi_S = \frac{1.615a\eta_r^{0.12}}{\rho_V(\rho_L - \rho_V)g} = \frac{1.615 \times 310 \times (1/0.978)^{0.12}}{1 \times (1000 - 1) \times 9.8} = 0.051$$

Thus,

$$G_{V,flooding} = \sqrt{\frac{0.453}{0.051}} = 2.98\ \frac{kg}{s.m^2}$$

Using Equation (19.49),

$$D = \sqrt{\frac{4\dot{m}_V}{\pi\alpha(G_V)_{flooding}}} = \sqrt{\frac{4 \times (600/3600)}{\pi \times 0.70 \times 2.98}} = 0.319\ m$$

19.5.5 Example 5

Develop the equations for height of transfer unit (HTU) and number of transfer units (NTU) for liquid stripping in a packed column.

Solution

As described in the case of packed column gas absorption (see section 19.3.2), it is imagined here that the total height of the liquid stripping packed column can be divided into a number of layers of packing (referred to as number of transfer units, NTU) of equal height (referred to as height of transfer unit, HTU) such that the change in liquid concentration over the HTU is equal to the average driving force for mass transfer from liquid phase to gas phase over the whole column (see Figure 19.7).

The concept of *HTU* for liquid stripping is further illustrated schematically in the figure below. If at any point in the column, x is the solute concentration in the liquid phase and y is the solute concentration in the gas phase at the same point, then *HTU* is the height of the packing above this point where the liquid concentration is $x + (x - x^*)_{avg}$ where $(x - x^*)_{avg}$ is the average driving force for mass transfer of solute in the column from liquid phase to gas phase.

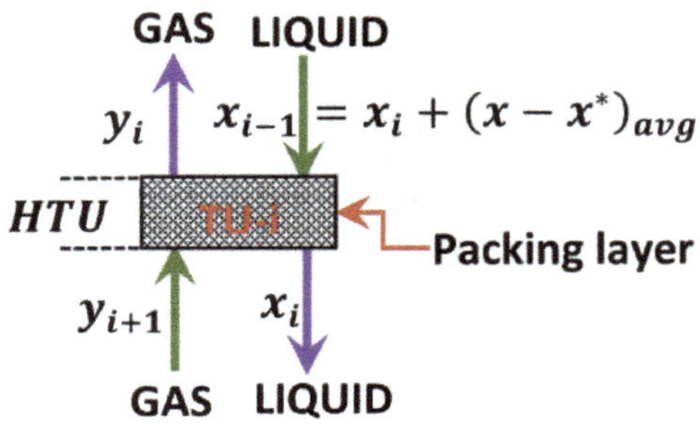

In a packed column, interphase mass transfer of solute occurs from one phase to another by diffusion. Figure given below shows the concentration profiles, in units of mole fraction, on both sides of the gas-liquid interface in liquid stripping operation. The molar flux of solute N_A (moles/s.m^2) across the liquid film on the left side of the gas-liquid interface can be expressed as:

$$N_A = k_x(x - x_i)$$

where k_x is the liquid-phase mass transfer coefficient (moles/s.m^2).

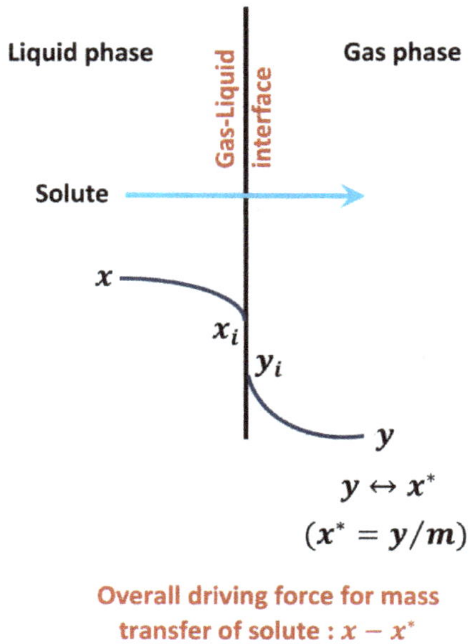

The molar flux of solute N_A (moles/s.m^2) across the gas film on the right side of the gas-liquid interface can be expressed as:

$$N_A = k_y(y_i - y)$$

where k_y is the gas-phase mass transfer coefficient (moles/s.m^2).

At steady state, there cannot be accumulation of solute at the interface and therefore,

$$N_A = k_x(x - x_i) = k_y(y_i - y)$$

To avoid the interface compositions, it is equally appropriate to express the molar flux of solute as:

$$N_A = K_x(x - x^*)$$

where K_x is the overall mass transfer coefficient based on liquid phase, $(x - x^*)$ is the overall driving force for mass transfer of solute from liquid phase to gas phase, and x^* is the mole fraction of solute in the liquid phase in equilibrium with the existing bulk gas phase. Note that x^* is a fictitious quantity to be calculated from the equilibrium relation (Henry's law):

$$y = mx^*$$

where m is the equilibrium constant. It can be readily shown that:

$$\frac{1}{K_x} = \frac{1}{k_x} + \frac{1}{mk_y}$$

$$\frac{1}{K_x a} = \frac{1}{k_x a} + \frac{1}{mk_y a}$$

We now consider material balance of solute over a thin section dz of the packed liquid stripping column shown in the figure below. Let the contact area of the packing per unit volume of packing be a. The inter-phase mass transfer rate of solute from liquid phase to gas phase over a section dz of the column is:

$$\text{Rate of mass transfer} = K_x a(x - x^*)Sdz$$

where S is the cross-sectional area of the column. The mass transfer rate of solute from liquid phase to gas phase over a section dz of the column is also given as:

$$\text{Rate of mass transfer} = Lx|_{z+dz} - Lx|_z$$

where L is the molar flow rate of liquid. Equating the two expressions, we get:

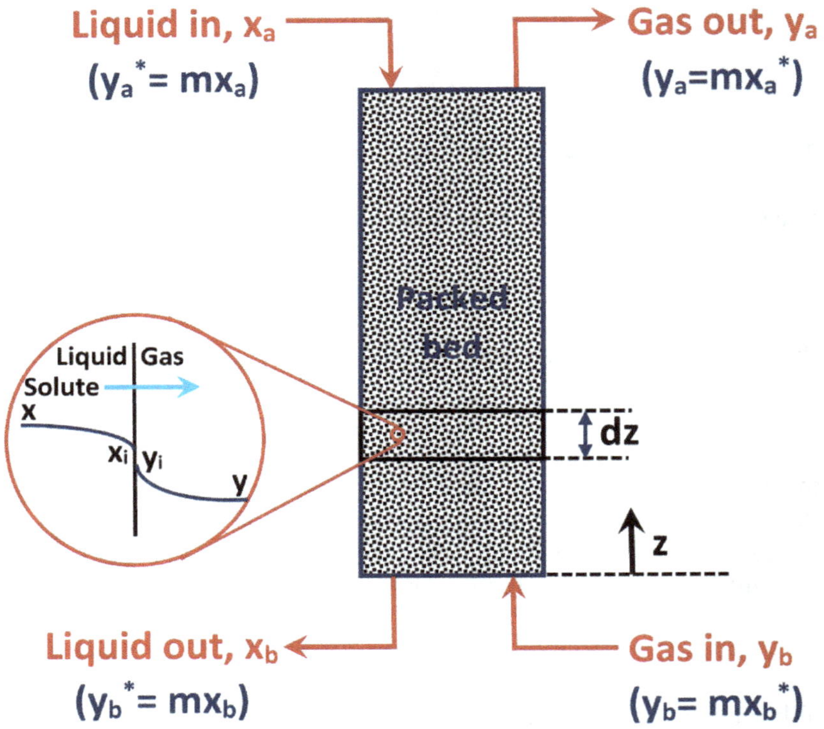

$$Lx|_{z+dz} - Lx|_z = K_x a(x - x^*)Sdz$$

Thus,

$$\frac{d(Lx)}{dz} = K_x a(x - x^*)S$$

Assuming the liquid flow rate L to be constant:

$$L\frac{dx}{dz} = K_x a(x - x^*)S$$

Upon rearrangement and integration,

$$h = \left(\frac{L/S}{K_x a}\right)\int_{x_b}^{x_a}\frac{dx}{x - x^*}$$

This equation could be recast as:

$$h = (HTU) \times (NTU)$$

where

$$HTU = \left(\frac{L/S}{K_x a}\right)$$

$$NTU = \int_{x_b}^{x_a}\frac{dx}{x - x^*}$$

The number of transfer units (*NTU*) depends on the degree of desired separation $(x_a - x_b)$ and the average driving force for mass transfer $(x - x^*)_{avg}$. It is simply the ratio of $(x_a - x_b)$ to $(x - x^*)_{avg}$. The height of the transfer unit (HTU) is a measure of the efficiency of mass transfer. More efficient the mass transfer, smaller is the value of HTU. A larger mass transfer coefficient K_x implies a faster rate of interphase mass transfer of solute and hence a lower HTU. Likewise, a larger contact area of the packing (per unit volume of the packing) implies a faster rate of interphase mass transfer of solute and hence a lower HTU. Given the overall mass transfer coefficient K_x , the packing contact area a per unit volume of the packing, the liquid molar flow rate L, and the column diameter D ($S = \pi D^2/4$), HTU can be calculated.

The NTU can be calculated from the following equation:

$$NTU = \int_{x_b}^{x_a} \frac{dx}{x-x^*} = \frac{x_a-x_b}{(x-x^*)_{log-mean}}$$

where $(x - x^*)_{log-mean}$ is the log-mean driving force for mass transfer in the packed column, given as:

$$(x - x^*)_{log-mean} = \frac{(x_a - x_a^*) - (x_b - x_b^*)}{ln\left(\frac{x_a - x_a^*}{x_b - x_b^*}\right)}$$

19.5.6 Example 6

A packed tower is used to remove a toxic chemical from wastewater. Pure air is used as the stripping gas. The wastewater entering the tower contains 5 mol% toxic chemical. It is desired to remove 98% of the toxic chemical. The equilibrium relation is $y = 0.85x$, where y and x are mole fractions of toxic chemical in air and water at equilibrium, respectively. The gas and liquid molar flow rates per unit cross-sectional area of the column are: 0.8 kmol/min.m^2 and 0.4 kmol/min.m^2, respectively. The mass transfer coefficients are:

$$k_x a = 6 \times 10^{-2}\ kmol/s.m^3\ ;\ k_y a = 4 \times 10^{-2}\ kmol/s.m^3$$

Determine the following:

a) the number of transfer units, NTUs
b) the height of the transfer unit, HTU
c) the height of the packed tower

You can assume that the gas and liquid molar flow rates are constant.

Solution

Refer to the diagram below

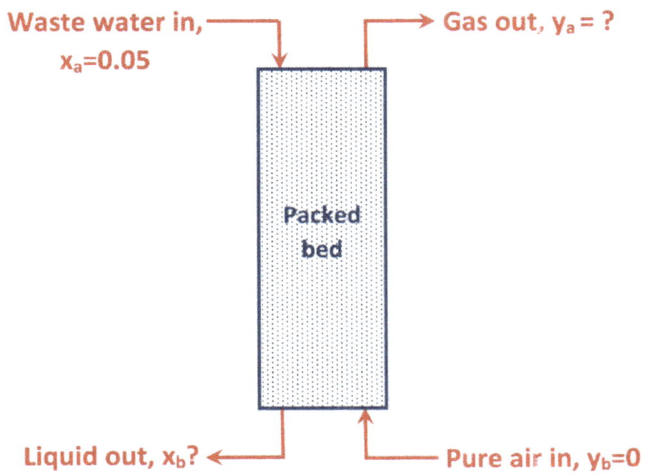

Let us first calculate the terminal compositions. As 98% of the toxic chemical is to be removed, $x_b = (1-0.98)x_a = 0.02 \times 0.05 = 0.001$. From the operating line equation,

$$y_a = y_b + (L/V)(x_a - x_b) = 0 + (0.4/0.8)(0.05 - 0.001) = 0.0245$$

a) the number of transfer units, NTUs

$$NTU = \int_{x_b}^{x_a} \frac{dx}{x - x^*} = \frac{x_a - x_b}{(x - x^*)_{log-mean}}$$

$$(x - x^*)_{log-mean} = \frac{(x_a - x_a^*) - (x_b - x_b^*)}{ln\left(\frac{x_a - x_a^*}{x_b - x_b^*}\right)} = \frac{(0.05 - 0.02882) - (0.001 - 0)}{ln\left(\frac{0.05 - 0.02882}{0.001 - 0}\right)} = 6.61 \times 10^{-3}$$

Note that: $x_a^* = y_a/m = 0.0245/0.85 = 0.02882$, $x_b^* = y_b/m = 0/0.85 = 0$. Thus, NTUs is:

$$NTU = \frac{x_a - x_b}{(x - x^*)_{log-mean}} = \frac{0.05 - 0.001}{6.61 \times 10^{-3}} = 7.413$$

b) the height of the transfer unit, HTU

We first need to calculate the overall mass transfer coefficient $K_x a$ as:

$$\frac{1}{K_x a} = \frac{1}{k_x a} + \frac{1}{m k_y a} = \frac{1}{6 \times 10^{-2}} + \frac{1}{0.85 \times 4 \times 10^{-2}} = 46.07843$$

$$K_x a = 0.0217 \text{ kmol/s.m}^3$$

The HTU is given as:

$$HTU = \left(\frac{L/S}{K_x a}\right) = \frac{0.4}{0.0217 \times 60} = 0.3071 \text{ m}$$

c) the height of the packed tower

$$h = (HTU) \times (NTU) = 0.3071 \times 7.413 = 2.277 \text{ m}$$

NOTATION

a surface area per unit volume of packing
A absorption factor, defined in Equation (19.31)
D column diameter
d_P equivalent diameter of particles of a packing
F_{LV} flow parameter, defined in Equation (19.42)
F_P packing factor of a packing, see Table 19.2
g acceleration due to gravity, 32.2 ft/s^2 or 9.8 m/s^2
g_c conversion factor, 32.2 lb$_m$.(ft/s^2)/lb$_f$
G_L superficial mass velocity of liquid
G_V superficial mass velocity of gas or vapor
$(G_V)_{flooding}$ superficial mass velocity of gas or vapor corresponding to flooding in a column

h height of a packed column
$HETP$ height equivalent to a theoretical plate
HTU height of transfer unit
k_x liquid phase mass transfer coefficient, see Equation (19.9)
k_y gas phase mass transfer coefficient, see Equation (19.8)
K_x overall mass transfer coefficient based on liquid phase
K_y overall mass transfer coefficient based on gas phase, see Equation (19.11)
L molar flow rate of liquid
m Henry law constant of solute defined in Equation (19.12)
M_L molecular weight of liquid phase
M_V molecular weight of vapor or gas phase
$\dot{m}_L$ mass flow rate of liquid
$\dot{m}_V$ mass flow rate of vapor or gas
n parameter in Equation (19.5)
NTU number of transfer units
N_A molar flux of solute
$N_{theoretical}$ total number of equilibrium or theoretical stages
S cross-sectional area of a column
U_V superficial flooding velocity of gas, see Equation (19.44)
V molar flow rate of vapor or gas
x mole fraction of solute in the liquid phase
x_a mole fraction of solute in liquid stream entering the column at the top, see Figure 19.9
x_b mole fraction of solute in liquid stream leaving the column at the bottom, see Figure 19.9
x_i mole fraction of solute at the gas-liquid interface on the liquid side (see Figure 19.8) or mole fraction of solute in liquid stream leaving ith stage (see Figure 19.6)
x_{i-1} mole fraction of solute in liquid stream entering ith stage (see Figure 19.6)
x_a^* mole fraction of solute in the liquid phase in equilibrium with the gas stream leaving the column, see Figure 19.9
x_b^* mole fraction of solute in the liquid phase in equilibrium with the gas stream entering the column, see Figure 19.9
y mole fraction of solute in the gas phase
y_a mole fraction of solute in gas stream leaving the column at the top, see Figure 19.9
y_b mole fraction of solute in gas stream entering the column at the bottom, see Figure 19.9
y_i mole fraction of solute at the gas-liquid interface on the gas side (see Figure 19.8) or mole fraction of solute in gas stream leaving ith stage (see Figure 19.6)
y_{i+1} mole fraction of solute in gas stream entering ith stage (see Figure 19.6)
y^* mole fraction of solute in the gas phase if the gas phase was in equilibrium with the existing liquid phase
y_a^* mole fraction of solute in the gas phase in equilibrium with the liquid stream entering the column, see Figure 19.9
y_b^* mole fraction of solute in the gas phase in equilibrium with the liquid stream leaving the column, see Figure 19.9
z column height variable, see Figure 19.9

Greek Letters

β parameter defined in Equation (19.45)
η_L liquid viscosity, in units of cP (Equation (19.41)) and mPa.s
η_r relative viscosity given as $\eta_L(mPa.s)/0.978$, see Equation (19.44)
ρ density
ρ_L density of liquid

ρ_V density of gas or vapor
ρ_{water} density of water
σ surface tension
ϕ density ratio, ρ_{water} / ρ_L
ψ parameter defined in Equation (19.41)
ψ_S parameter defined in Equation (19.47)

PRACTICE PROBLEMS

Problem 1

A packed tower is to be used to absorb SO_2 from air stream using pure water as a solvent. The gas entering the tower contains 2.0 mol% A and 85 percent of it is to be absorbed by water. The molar flow rates of gas and liquid streams are 0.05 kmol/s and 1.98 kmol/s, respectively. Under the tower conditions, the equilibrium relationship is as follows: $y = 35x$. The column height is 4 m and its diameter is 1.25 m. Estimate the value of the mass transfer capacity coefficient $K_y a$. You can assume that the gas and liquid molar flow rates are constant.

Problem 2

A packed tower is to be used to absorb cyclohexane from airstream. A non-volatile oil is used to absorb cyclohexane. The air entering the tower contains 1.5 mol% cyclohexane and 90 percent of it is absorbed by oil. The oil entering the column is cyclohexane-free. The tower is operated at 30 °C and 1 atm. The gas and liquid molar flow rates on a unit area basis are 0.5 lb-mol/min.ft^2 and 0.17 lb-mol/min.ft^2, respectively. The equilibrium and mass transfer coefficient data are as follows:

$$y^* = 0.15x \ ; \ k_x a = 0.30 \ lb - mol/h.ft^3 \ ; \ k_y a = 14.0 \ lb - mol/h.ft^3$$

Determine the following:

a) the number of transfer units, NTUs
b) the height of the transfer unit, HTU
c) the height of the packed tower

You can assume that the gas and liquid molar flow rates are constant.

Problem 3

A packed tower is to be designed to absorb ammonia from an air stream using pure water as a solvent. The gas enters the tower at a mass flow rate of 1500 lb$_m$/h. The mass flow rates of liquid is 3000 lb$_m$/h. The tower is packed with 1.5-inch metal Pall rings.
Estimate the diameter of the tower such that the gas mass velocity is 70% of the flooding limit.
The properties of the fluids are:

$\rho_V = 0.062$ lb$_m$ /ft^3 , $\rho_L = 62.4$ lb$_m$ /ft^3 , $\mu_L = 1$ cP

For 1.5-inch metal Pall rings, the packing factor $F_p = 32$ ft^{-1}.

Problem 4

Show that:

$$\frac{1}{K_y} = \frac{1}{k_y} + \frac{m}{k_x}$$

If m is small, which phase (gas or liquid) is controlling with respect to mass transfer? The controlling phase is one which offers the greatest resistance to mass transfer.

Problem 5

Show that:

$$\frac{1}{K_x} = \frac{1}{k_x} + \frac{1}{m k_y}$$

If m is large, which phase (gas or liquid) is controlling with respect to mass transfer? The controlling phase is one which offers the greatest resistance to mass transfer.

Problem 6

Show that for a packed gas absorption column,

$$NTU = \frac{y_b - y_a}{(y - y^*)_{log-mean}}$$

where

$$(y - y^*)_{log-mean} = \frac{(y_b - y_b^*) - (y_a - y_a^*)}{ln\left(\frac{y_b - y_b^*}{y_a - y_a^*}\right)}$$

What are the restrictions on this expression of NTU?

Problem 7

Show that for a liquid stripping packed column,

$$NTU = \frac{x_a - x_b}{(x - x^*)_{log-mean}}$$

where

$$(x - x^*)_{log-mean} = \frac{(x_a - x_a^*) - (x_b - x_b^*)}{ln\left(\frac{x_a - x_a^*}{x_b - x_b^*}\right)}$$

What are the restrictions on this expression of NTU?

Problem 8

A packed tower is used to remove a chemical from wastewater using pure air as the stripping gas. The wastewater entering the tower has a chemical mole fraction of 0.05. The chemical mole fraction in the exit water stream is 0.001. The equilibrium relation is $y = 1.5x$, where y and x are mole fractions of chemical in air and water at equilibrium, respectively. The gas and liquid molar flow rates per unit cross-sectional area of the column are: 0.9 kmol/min.m^2 and 0.6 kmol/min.m^2, respectively. The overall mass transfer coefficient based on the liquid phase is:

$$K_x a = 0.04 \;\; kmol/s.m^3$$

Determine the following:

a) the number of transfer units, NTUs
b) the height of the transfer unit, HTU
c) the height of the packed tower

You can assume that the gas and liquid molar flow rates are constant.

FURTHER READING

1. R. Pal. *Electromagnetic, Mechanical, and Transport Properties of Composite Materials*. CRC Press, Boca Raton, Florida, 2015.
2. R. Pal. *Introduction to Analysis and Design of Equilibrium Stages Separation Processes*. 2nd Edition, Central West Publishing, Australia, 2022.
3. J.R. Welty, C.E. Wicks, R.E. Wilson, and G.L. Rorrer. *Fundamentals of Momentum, Heat, and Mass Transfer*. 5th Edition, Wiley & Sons, New York, 2008.
4. C.J. Geankoplis, A.A. Hersel, D.H. Lepek. *Transport Processes and Separation Process Principles*. 5th Edition, Prentice Hall, New Jersey, 2018.
5. W.L. McCabe, J.C. Smith, P. Harriott. *Unit Operations of Chemical Engineering*. 6th Edition, McGraw Hill, New York, 2001.
6. S. Middleman. *An Introduction to Mass and Heat Transfer*. Wiley & Sons, New York, 1998.

7. P.C. Wankat. *Separation Process Engineering: Includes Mass Transfer Analysis*. 5th Edition. Prentice Hall, NJ, 2022

Chapter 20

Multicomponent Diffusion in Gases and Liquids

20.1 FICK'S LAW OF DIFFUSION

The diffusion in gas mixtures and liquid solutions is described by the Fick's law of diffusion, given as:

$$\vec{J}_A = -CD_{AB}\nabla x_A \tag{20.1}$$

where D_{AB} is called the Fick's diffusivity of species A in B. Strictly speaking, Fick's law is applicable only to ideal binary mixtures (gas or liquid). For ideal binary mixtures, D_{AB} is independent of the composition of the mixture and is symmetric in nature, that is, $D_{AB} = D_{BA}$.

To extend the applicability of the Fick's law (Equation (20.1)) to non-ideal binary mixtures (gas or liquid), we need to express it in terms of the chemical potential gradient as follows:

$$\vec{J}_A = -\frac{c_A D_{AB}^*}{RT}\nabla_{T,P}(\mu_A) \tag{20.2}$$

where μ_A is the chemical potential of species A in a binary mixture, $\nabla_{T,P}(\mu_A)$ is the spatial gradient of chemical potential of species A at constant T and P, and D_{AB}^* is the diffusion coefficient defined using chemical potential gradient as the driving force.

Equation (20.2) is valid for both ideal and non-ideal systems (gas mixtures and liquid solutions). It could be further re-written as:

$$\vec{J}_A = -C(D_{AB}^*\Gamma)\nabla x_A \tag{20.3}$$

where Γ is a thermodynamic factor given as:

$$\Gamma = 1 + x_A\left(\frac{\partial ln\hat{\varphi}_A}{\partial x_A}\right)_{T,P} \quad \text{(gas mixture)} \tag{20.4a}$$

$$\Gamma = 1 + x_A\left(\frac{\partial ln\gamma_A}{\partial x_A}\right)_{T,P} \quad \text{(liquid solution)} \tag{20.4b}$$

where $\hat{\varphi}_A$ is the fugacity coefficient of species A in a binary mixture and γ_A is the activity coefficient of species A in a binary liquid solution.

Upon comparison of Equation (20.3) with the Fick's law, Equation (20.1), we get:

$$D_{AB} = D_{AB}^*\Gamma \tag{20.5}$$

For ideal systems, the thermodynamic factor is unity and hence, $D_{AB} = D_{AB}^*$ and Equation (20.3) reduces to the Fick's law, that is, Equation (20.1). Note that for non-ideal binary mixtures (gas or liquid), the Fick's diffusivity D_{AB} is dependent on the composition of the mixture as the thermodynamic factor Γ is a function of the composition. Furthermore, the Fick's diffusivity is not symmetric.

20.2 MAXWELL-STEFAN (MS) DIFFUSION MODEL

Fick's law of diffusion is not the only law to describe diffusion in binary systems. Another very important model to describe diffusion in binary systems is the Maxwell-Stefan (MS) diffusion model. The MS model is based on the balance between the driving force of diffusion of species and the opposing friction force on the species. The MS model for diffusion in ideal binary mixtures is given as:

$$\nabla x_1 = -\frac{(x_2\vec{N}_1 - x_1\vec{N}_2)}{C\text{Đ}_{12}} = -\frac{(x_2\vec{J}_1 - x_1\vec{J}_2)}{C\text{Đ}_{12}} \tag{20.6}$$

where Đ_{12} is the MS diffusivity which is generally different from the Fick's diffusivity D_{12} although they have the same units of m^2/s. Also note that the fluxes in the MS diffusion model are not given directly as a function of concentration gradients as in the Fick's law of diffusion. Furthermore, the fluxes are coupled in that the concentration gradient of a species is a function of fluxes of all the components present in the system. Although the MS model for diffusion is derived for ideal gas mixtures (see Appendix F), it is often applied to ideal liquid solutions as well.

Interestingly, the MS equation for binary diffusion is equivalent to Fick's law of diffusion for binary mixtures. Since,

$$\vec{J}_1 = -\vec{J}_2 \tag{20.7}$$

the MS model reduces to:

$$\nabla x_1 = -\frac{(x_2\vec{J}_1 + x_1\vec{J}_1)}{C\text{Đ}_{12}} = -\frac{\vec{J}_1}{C\text{Đ}_{12}} \tag{20.8}$$

Upon rearrangement, Equation (20.8) gives:

$$\vec{J}_1 = -C\text{Đ}_{12}\nabla x_1 \tag{20.9}$$

This is Fick's law of diffusion for ideal binary mixtures (see Equation (20.1)). Clearly, for ideal binary mixtures the Fick's diffusivity is the same as the MS diffusivity:

$$\text{Đ}_{12} = D_{12} \tag{20.10}$$

As the Fick's diffusivity is symmetric for ideal binary mixtures ($D_{12} = D_{21}$), it follows from Equation (20.10) that the MS diffusivity is also symmetric: $\text{Đ}_{12} = \text{Đ}_{21}$.

20.3 MULTICOMPONENT DIFFUSION

20.3.1 Pseudo-Binary Model

One way to describe multicomponent diffusion in ideal systems involving three or more species is to generalize the Fick's law, Equation (20.1)), as follows:

$$\vec{J}_i = -CD_{im}\nabla x_i \tag{20.11}$$

where C is the total molar concentration of all species and D_{im} is the pseudo-binary diffusivity of species i in the mixture. In dilute systems where the concentrations of different species (other than the solvent) are small in comparison with the solvent (species B), the diffusivities D_{im} can be approximated as:

$$D_{im} \approx D_{iB} \tag{20.12}$$

where D_{iB} is the binary diffusivity of species i in the solvent B. In non-dilute systems, D_{im} can be estimated from empirical equations. For example, Fairbanks and Wilke equation could be used to estimate D_{1m} for gas mixtures containing n components as:

$$D_{1m} = \frac{1-y_1}{\left(\frac{y_2}{D_{12}}\right)+\left(\frac{y_3}{D_{13}}\right)+\cdots\cdots+\left(\frac{y_n}{D_{1n}}\right)} \tag{20.13}$$

where D_{1m} is diffusivity of component 1 in the gas mixture, D_{1n} is the binary diffusivity for the pair 1 and n, y_n is the mole fraction of component n in the gas mixture. Likewise, D_{2m} can be estimated from the following equation:

$$D_{2m} = \frac{1-y_2}{\left(\frac{y_1}{D_{21}}\right)+\left(\frac{y_3}{D_{23}}\right)+\cdots\cdots+\left(\frac{y_n}{D_{2n}}\right)} \tag{20.14}$$

The main drawback of Fick's law of the form given in Equation (20.11) is that it fails to describe certain important diffusion phenomena observed in multicomponent systems.

In multicomponent systems, it is not so uncommon to observe phenomena such as osmotic diffusion, reverse diffusion, and diffusion barrier. Figure 20.1 compares the variation of diffusional flux of species 1 with the negative concentration gradient of species 1 ($-\nabla x_1$) for binary and ternary systems. In the case of binary diffusion, the plot of $\vec{J}_1$ versus $-\nabla x_1$ is a straight line passing through the origin (see Figure 20.1a). This is the normal expected behavior where the movement of a species is always proportional to the negative of its concentration gradient. However, a very different picture emerges in a ternary system (see Figure 20.1b). Now the diffusional flux $\vec{J}_1$ is not zero even when $-\nabla x_1$ is zero, that is, the species diffuses even when its concentration gradient is zero. This behavior is referred to as "osmotic diffusion". It should also be noted that the flux in a ternary system $\vec{J}_1$ can be positive even when the concentration gradient $-\nabla x_1$ is negative, that is, the species diffuses in the direction of its increasing concentration. This phenomenon is referred to as "reverse diffusion" or "uphill diffusion". Furthermore, diffusion of a species in a ternary system may not occur at all even though there is present a concentration gradient in the species, that is, $\vec{J}_1 = 0$ when $-\nabla x_1 \neq 0$. This phenomenon is referred to as "diffusion barrier".

The complicated diffusion behavior observed in multicomponent systems is due to cross-diffusion or coupling effects, that is, the diffusion flux of a species is not just proportional to its own concentration gradient, it also depends on the concentration gradients of other species present in the system.

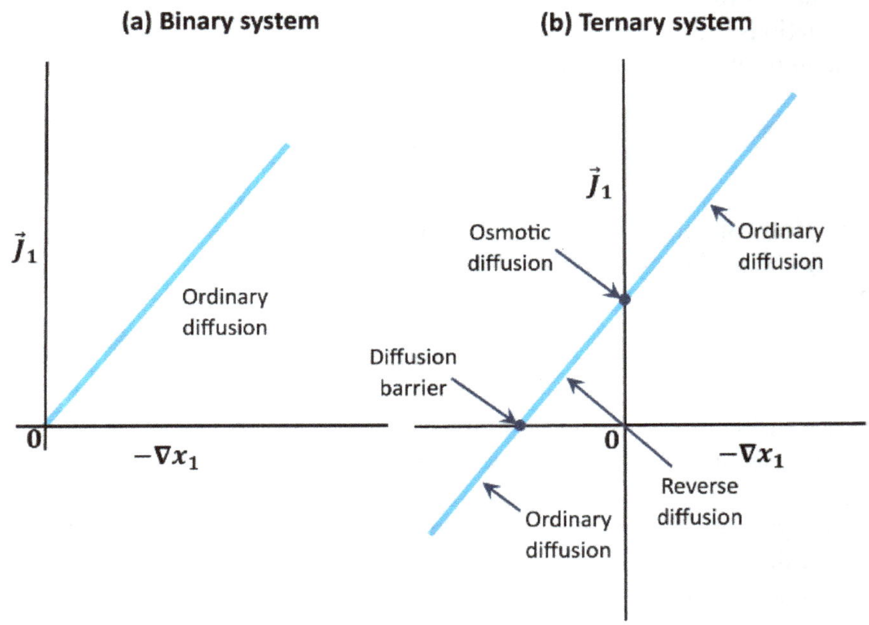

Figure 20.1 Variation of diffusion flux with composition gradient in binary and ternary systems.

20.3.2 Generalized Fick's Law

In order to take into account the coupling effects, the correct form of the generalized Fick's law is as follows:

$$\vec{J}_i = -C\sum_{j=1}^{n-1} D_{ij}\nabla x_j \qquad (i = 1,2, \ldots, n-1) \tag{20.15}$$

Note that there are $n-1$ independent fluxes as $\sum_{i=1}^{n}\vec{J}_i = 1$ where n is the number of components. According to the generalized Fick's law, the flux of the diffusing component i now depends on the concentration gradients of all components present in the mixture. There are $(n-1)^2$ independent diffusion coefficients involved. D_{ii} is the main diffusivity as it connects the flux of a component to its own concentration gradient and D_{ij} is the cross diffusivity as it connects the flux of a component to concentration gradients of other components.

It should be noted that the diffusion coefficients in a multicomponent mixture are not binary properties as they are affected by the presence of molecules of other species in the mixture. Thus, the values of D_{ij} in a multicomponent system are different from D_{ij} in a binary mixture of components i and j. Furthermore, the diffusion coefficients in a multicomponent mixture are generally composition-dependent and not symmetric ($D_{ij} \neq D_{ji}$).

We can re-write the generalized Fick's law in $n-1$ dimensional matrix form as:

$$\begin{pmatrix} \vec{J}_1 \\ \vdots \\ \vec{J}_{n-1} \end{pmatrix} = -C\begin{pmatrix} D_{1,1} & \cdots & D_{1,n-1} \\ \vdots & \cdots & \vdots \\ D_{n-1,1} & \cdots & D_{n-1,n-1} \end{pmatrix}\begin{pmatrix} \nabla x_1 \\ \vdots \\ \nabla_{n-1} \end{pmatrix} \tag{20.16}$$

In condensed matrix notation, Equation (20.16) can be re-written as:

$$(\vec{J}) = -C(\boldsymbol{D})(\boldsymbol{\nabla x}) \tag{20.17}$$

For a ternary 3-component system, the generalized Fick's law gives:

$$\begin{pmatrix}\vec{J}_1\\ \vec{J}_2\end{pmatrix} = -C\begin{pmatrix}D_{11} & D_{12}\\ D_{21} & D_{22}\end{pmatrix}\begin{pmatrix}\nabla x_1\\ \nabla x_2\end{pmatrix} \tag{20.18}$$

Thus,

$$\vec{J}_1 = -C[D_{11}\nabla x_1 + D_{12}\nabla x_2] \tag{20.19}$$

$$\vec{J}_2 = -C[D_{21}\nabla x_1 + D_{22}\nabla x_2] \tag{20.20}$$

Note that: $\vec{J}_3 = -(\vec{J}_1 + \vec{J}_2)$

20.3.3 Generalized Maxwell-Stefan Model

The Maxwell-Stefan model developed for ideal binary mixtures can be generalized to ideal multicomponent mixtures as follows:

$$\nabla x_i = -\sum_{j\neq i}^{n}\frac{(x_j\vec{J}_i - x_i\vec{J}_j)}{C Đ_{ij}} = -\sum_{j\neq i}^{n}\frac{(x_j\vec{N}_i - x_i\vec{N}_j)}{C Đ_{ij}} \tag{20.21}$$

The generalized MS model is fully capable of describing cross diffusion or coupling effects in multicomponent mixtures.

The MS diffusion coefficients are generally assumed to be symmetric for multicomponent mixtures, that is:

$$Đ_{ij} = Đ_{ji} \tag{20.22}$$

There are $n(n-1)/2$ independent MS diffusion coefficients. For ideal gas mixtures, $Đ_{ij}$ are independent of the mixture composition and multicomponent $Đ_{ij}$ are equal to binary $Đ_{ij}$.

It should be noted that there are $(n-1)$ independent MS equations as:

$$\sum_{i=1}^{n}\nabla x_i = 0 \tag{20.23}$$

For a ternary system, the independent MS equations are as follows:

$$\frac{dx_1}{dz} = -\left[\frac{x_2J_1 - x_1J_2}{CĐ_{12}}\right] - \left[\frac{x_3J_1 - x_1J_3}{CĐ_{13}}\right] \tag{20.24}$$

$$\frac{dx_2}{dz} = -\left[\frac{x_1J_2 - x_2J_1}{CĐ_{21}}\right] - \left[\frac{x_3J_2 - x_2J_3}{CĐ_{23}}\right] \tag{20.25}$$

Note that there are three independent diffusion coefficients as $Đ_{12} = Đ_{21}$. Also, there are two independent fluxes as $J_1 + J_2 + J_3 = 0$.

We can re-write the ternary MS equations in a matrix form as:

$$C\begin{pmatrix}\frac{dx_1}{dz}\\ \frac{dx_2}{dz}\end{pmatrix} = \begin{pmatrix}B_{11} & B_{12}\\ B_{21} & B_{22}\end{pmatrix}\begin{pmatrix}J_1\\ J_2\end{pmatrix} \tag{20.26}$$

where

$$B_{11} = \frac{x_1}{Đ_{13}} + \frac{x_2}{Đ_{12}} + \frac{x_3}{Đ_{13}} \quad (20.27)$$

$$B_{12} = -x_1\left(\frac{1}{Đ_{12}} - \frac{1}{Đ_{13}}\right) \quad (20.28)$$

$$B_{21} = -x_2\left(\frac{1}{Đ_{21}} - \frac{1}{Đ_{23}}\right) \quad (20.29)$$

$$B_{22} = \frac{x_2}{Đ_{23}} + \frac{x_1}{Đ_{21}} + \frac{x_3}{Đ_{23}} \quad (20.30)$$

In general, we can write the MS equations in $n-1$ dimensional matrix form as:

$$C\begin{pmatrix} \nabla x_1 \\ \vdots \\ \nabla x_{n-1} \end{pmatrix} = -\begin{pmatrix} B_{1,1} & \cdots & B_{1,n-1} \\ \vdots & \cdots & \vdots \\ B_{n-1,1} & \cdots & B_{n-1,n-1} \end{pmatrix}\begin{pmatrix} \vec{J}_1 \\ \vdots \\ \vec{J}_{n-1} \end{pmatrix} \quad (20.31)$$

where

$$B_{ii} = \frac{x_i}{Đ_{in}} + \sum_{j \neq i}^{n} \frac{x_j}{Đ_{ij}} \qquad \text{(Diagonal components)} \quad (20.32)$$

$$B_{ij} = -x_i\left(\frac{1}{Đ_{ij}} - \frac{1}{Đ_{in}}\right) \qquad \text{(Off-diagonal components)} \quad (20.33)$$

20.3.4 Relationship Between Maxwell-Stefan and Fick Diffusivities

Ideal Mixtures

In the condensed matrix notation, the MS equations for ideal mixtures can be written as:

$$C(\nabla \mathbf{x}) = -(\boldsymbol{B})(\vec{J}) \quad (20.34)$$

Upon re-arrangement, Equation (20.34) becomes:

$$(\vec{J}) = -C(\boldsymbol{B})^{-1}(\nabla \mathbf{x}) \quad (20.35)$$

The generalized Fick's law in matrix form is:

$$(\vec{J}) = -C(\boldsymbol{D})(\nabla \mathbf{x}) \quad (20.36)$$

Upon comparison of Equations (20.35) and (20.36), the Fickian diffusivity matrix can be expressed as:

$$(\boldsymbol{D}) = (\boldsymbol{B})^{-1} \quad (20.37)$$

For a ternary system, it can be readily shown that:

$$\begin{pmatrix} D_{11} & D_{12} \\ D_{21} & D_{22} \end{pmatrix} = \begin{pmatrix} BI_{11} & BI_{12} \\ BI_{21} & BI_{22} \end{pmatrix} \quad (20.38)$$

where

$$BI_{11} = \frac{Đ_{13}[x_1 Đ_{23} + (1-x_1)Đ_{12}]}{x_1 Đ_{23} + x_2 Đ_{13} + x_3 Đ_{12}} \quad (20.39)$$

$$BI_{12} = \frac{x_1 Đ_{23}(Đ_{13} - Đ_{12})}{x_1 Đ_{23} + x_2 Đ_{13} + x_3 Đ_{12}} \tag{20.40}$$

$$BI_{21} = \frac{x_2 Đ_{13}(Đ_{23} - Đ_{12})}{x_1 Đ_{23} + x_2 Đ_{13} + x_3 Đ_{12}} \tag{20.41}$$

$$BI_{22} = \frac{Đ_{23}[x_2 Đ_{13} + (1 - x_2) Đ_{12}]}{x_1 Đ_{23} + x_2 Đ_{13} + x_3 Đ_{12}} \tag{20.42}$$

Clearly, the Fickian diffusivities in multicomponent systems are composition dependent. Furthermore, they are not symmetric. In the special case where all the MS diffusivities are equal, that is, $Đ_{12} = Đ_{13} = Đ_{23} = Đ$, we get:

$$\begin{pmatrix} D_{11} & D_{12} \\ D_{21} & D_{22} \end{pmatrix} = \begin{pmatrix} Đ & 0 \\ 0 & Đ \end{pmatrix} \tag{20.43}$$

In this situation, there are no cross effects. Each species diffuses independently of other species present in the mixture and the fluxes are given as:

$$J_1 = -CĐ\frac{dx_1}{dz} \tag{20.44}$$

$$J_2 = -CĐ\frac{dx_2}{dz} \tag{20.45}$$

Non-Ideal Mixtures

For non-ideal mixtures, we can apply the following equation:

$$\left(\frac{x_i}{RT}\right)\nabla_{T,P}(\mu_i) = -\sum_{j\neq i}^{n}\frac{(x_j\vec{J}_i - x_i\vec{J}_j)}{CĐ_{ij}} \tag{20.46}$$

The chemical potential gradient $\nabla_{T,P}(\mu_i)$ for non-ideal mixtures is given as:

$$\nabla_{T,P}(\mu_i) = RT\nabla_{T,P}(ln\hat{a}_i) = RT\nabla_{T,P}[ln(x_i\gamma_i)] = \frac{RT}{x_i\gamma_i}\nabla_{T,P}(x_i\gamma_i) \tag{20.47}$$

Consequently,

$$\left(\frac{x_i}{RT}\right)\nabla_{T,P}(\mu_i) = \frac{1}{\gamma_i}\nabla_{T,P}(x_i\gamma_i) = \sum_{j=1}^{n-1}\Gamma_{ij}\,\nabla x_j \tag{20.48}$$

where

$$\Gamma_{ij} = \delta_{ij} + x_i\left(\frac{\partial ln\gamma_i}{\partial x_j}\right)_{T,P,x_{i\neq j}} \tag{20.49}$$

Combining Equations (20.46) and (20.48), we can write:

$$\sum_{j=1}^{n-1}\Gamma_{ij}\,\nabla x_j = -\sum_{j\neq i}^{n}\frac{(x_j\vec{J}_i - x_i\vec{J}_j)}{CĐ_{ij}} \qquad i = 1,2,\ldots\ldots(n-1) \tag{20.50}$$

Note that for ideal solutions,

$$\Gamma_{ij} = \delta_{ij} \tag{20.51}$$

where δ_{ij} is Kronecker delta ($\delta_{ij} = 1$ when $i = j$; $\delta_{ij} = 0$ when $i \neq j$).

Using matrix notation, Equation (20.50) can be expressed as:

$$C(\boldsymbol{\Gamma})(\nabla \mathbf{x}) = -(\boldsymbol{B})(\vec{J}) \tag{20.52}$$

Upon rearrangement, Equation (20.52) gives:

$$(\vec{J}) = -C(\boldsymbol{B})^{-1}(\boldsymbol{\Gamma})(\nabla \mathbf{x}) \tag{20.53}$$

Upon comparison with the generalized Fick's law in matrix form (Equation (20.36)), we get:

$$(\boldsymbol{D}) = (\boldsymbol{B})^{-1}(\boldsymbol{\Gamma}) \tag{20.54}$$

Equation (20.54) is the relationship that links Fickian and Maxwell-Stefan diffusivities for non-ideal systems. From the knowledge of thermodynamic factor matrix $(\boldsymbol{\Gamma})$ and $(\boldsymbol{B})^{-1}$, we can determine the Fickian diffusivities.

20.3.5 Prediction of Fickian and Maxwell-Stefan Diffusivities

Gas Mixtures

Binary Mixtures

For binary gas mixtures (assuming ideal), the MS diffusivity and the Fickian diffusivity are the same, that is, $Đ_{12} = D_{12}$. The Fick's diffusivity of a binary gas mixture at ordinary pressures can be predicted reasonably well using the kinetic theory of gases. The following Chapman-Enskog equation for diffusivity is based on the kinetic theory of gases:

$$D_{12} = 0.0018583 \frac{\sqrt{T^3\left(\frac{1}{M_1}+\frac{1}{M_2}\right)}}{p\sigma_{12}^2\Omega_{12}} \tag{20.55}$$

where D_{12} is in cm²/s, T is in K, P is in atm, σ_{AB} is the average collision diameter in A^o, and Ω_{12} is dimensionless collision integral based on the Lennard-Jones potential. The average collision diameter σ_{12} is given as:

$$\sigma_{12} = \frac{\sigma_1+\sigma_2}{2} \tag{20.56}$$

The collision integral Ω_{12} is a function of dimensionless temperature defined as:

$$T^* = T/(\varepsilon_{12}/k_B) \tag{20.57}$$

where ε_{12} is the characteristic energy of molecular interaction for the mixture and k_B is the Boltzmann constant. ε_{12}/k_B is given as:

$$\varepsilon_{12}/k_B = \sqrt{(\varepsilon_1/k_B)(\varepsilon_2/k_B)} \tag{20.58}$$

where ε_1 and ε_2 are the characteristic molecular interaction energies of pure 1 and 2. The collision integral Ω_{12} can be calculated from the following equation:

$$\Omega_{12} = 1.06036(T^*)^{-0.1561} + 0.193exp(-0.47635T^*) + 1.03587exp(-1.52996T^*) + 1.76474exp(-3.894117T^*) \tag{20.59}$$

Multicomponent Mixtures

For multicomponent gas mixtures (assuming ideal), the MS diffusion coefficients are independent of the mixture composition and multicomponent $Đ_{ij}$ are equal to binary $Đ_{ij}$. Furthermore, the multicomponent MS diffusion coefficients are symmetric, that is, $Đ_{ij} = Đ_{ji}$. The binary MS diffusivity $Đ_{ij}$ (= binary Fick's diffusivity D_{ij}) can be calculated from the Chapman-Enskog equation (Equation (20.55)).

Liquid Solutions

Infinitely Dilute Binary Solutions

The Fickian diffusivity of solute molecules in an infinitely dilute solution can be estimated from the Wilke and Chang correlation. The Wilke-Chang correlation is given as:

$$\frac{D_{AB}^{o}\eta}{T} = \frac{7.4\times10^{-8}}{V_A^{0.6}}(\psi_B M_B)^{1/2} \tag{20.60}$$

where D_{AB}^{o} is the Fick's diffusivity of solute molecules in an infinitely dilute solution in units of cm^2/s, η is solution viscosity in centipoises, T is the absolute temperature in K, V_A is molar volume of solute A at normal boiling point in units of cm^3/gmol, M_B is the molecular weight of solvent B, and ψ_B is the "association parameter" for the solvent. The recommended values of ψ_B for some common solvents are as follows: 2.26 for water; 1.9 for methanol; 1.5 for ethanol; and 1.0 for benzene, ether, heptane, and other unassociated solvents. The Wilke-Chang correlation generally gives good predictions of D_{AB}^{o}, usually within $\pm$10%. The MS diffusivity of a solute in an infinitely dilute solution ($Đ_{AB}^{o}$) is equal to the Fickian diffusivity (D_{AB}^{o}). Thus, $Đ_{AB}^{o} = D_{AB}^{o}$.

Non-Dilute Binary Liquid Solutions

For non-dilute binary liquid solutions, the MS diffusivity and the Fickian diffusivity are related as follows:

$$Đ_{AB} = \frac{D_{AB}}{\Gamma} \tag{20.61}$$

where Γ is the thermodynamic factor given as:

$$\Gamma = 1 + x_A\left(\frac{\partial \ln\gamma_A}{\partial x_A}\right)_{T,P} \tag{20.62}$$

There are several mixing rules proposed in the literature for the prediction of MS (and hence Fickian) diffusivities of non-dilute binary solutions:

a) Darken Rule:

$$Đ_{AB} = \frac{D_{AB}}{\Gamma} = (Đ_{AB}^{o})x_B + (Đ_{BA}^{o})x_A \tag{20.63}$$

b) Vignes Rule:

$$Đ_{AB} = \frac{D_{AB}}{\Gamma} = (Đ_{AB}^{o})^{x_B}(Đ_{BA}^{o})^{x_A} \tag{20.64}$$

c) Hartley and Crank Rule:

$$Đ_{AB} = \frac{D_{AB}}{\Gamma} = \frac{1}{\eta}[(Đ_{AB}^{o})x_B\eta_B + (Đ_{BA}^{o})x_A\eta_A] \tag{20.65}$$

d) Leffler and Cullinan Rule:

$$Đ_{AB} = \frac{D_{AB}}{\Gamma} = \frac{1}{\eta}[(Đ_{AB}^{o}\eta_B)^{x_B}(Đ_{BA}^{o}\eta_A)^{x_A}] \tag{20.66}$$

where η_A, η_B, and η are the viscosities of species A, species B, and solution, respectively.

Multi-Component Solutions

For multi-component liquid solutions, the Fickian diffusivities can be estimated from Equation (20.54) from the knowledge of MS diffusivities, solution composition, and thermodynamic factor. The MS diffusivities themselves depend on the liquid composition and can be predicted using the mixture rules proposed in the literature. For example, the Krishna-Vignes rule for ternary liquid solutions is as follows:

$$Đ_{ij} = \left(Đ_{ij}^{o}\right)^{z_j}\left(Đ_{ji}^{o}\right)^{z_i} \qquad i,j = 1,2,3 \tag{20.67}$$

where

$$z_i = \frac{x_i}{x_i+x_j}; \quad z_j = 1 - z_i \tag{20.68}$$

x_i and x_j are mole fractions of species i and j, respectively.

20.4 EXAMPLES

20.4.1 Example 1

Consider an ideal ternary gas mixture of hydrogen (1)-nitrogen (2)-carbon dioxide (3). The MS diffusivities of this system estimated from the kinetic theory of gases are given as: $Đ_{12} = 8.33 \times 10^{-5}$, $Đ_{13} = 6.8 \times 10^{-5}$, $Đ_{23} = 1.68 \times 10^{-5}$ m^2/s. The composition of the system is as follows: $x_1 = 0.25$, $x_2 = 0.50$, and $x_3 = 0.25$.

a) Estimate the elements of the matrix $(\boldsymbol{B})$ at the given composition.
b) Estimate the elements of $(\boldsymbol{B})^{-1}$.
c) Determine the matrix of Fick diffusivities $(\boldsymbol{D})$
d) Does the system show coupling effects?
e) Express the diffusional fluxes $(\vec{J})$ in terms of the concentration gradients.
f) Which species undergoes osmotic diffusion when $\Delta x_1 = 0.50$, $\Delta x_2 = 0$, and $\Delta x_3 = -0.50$ over the diffusion length of L m?

Solution

a) The elements of the matrix $(\boldsymbol{B})$ are as follows:

$$B_{11} = \frac{x_1}{Đ_{13}} + \frac{x_2}{Đ_{12}} + \frac{x_3}{Đ_{13}} = \frac{0.25}{6.8 \times 10^{-5}} + \frac{0.50}{8.33 \times 10^{-5}} + \frac{0.25}{6.8 \times 10^{-5}} = 0.1336 \times 10^{5}$$

$$B_{12} = -x_1\left(\frac{1}{Đ_{12}} - \frac{1}{Đ_{13}}\right) = -0.25\left(\frac{1}{8.33 \times 10^{-5}} - \frac{1}{6.8 \times 10^{-5}}\right) = 6.753 \times 10^{2}$$

$$B_{21} = -x_2\left(\frac{1}{Đ_{21}} - \frac{1}{Đ_{23}}\right) = -0.50\left(\frac{1}{8.33 \times 10^{-5}} - \frac{1}{6.8 \times 10^{-5}}\right) = 0.2376 \times 10^{5}$$

$$B_{22} = \frac{x_2}{Đ_{23}} + \frac{x_1}{Đ_{21}} + \frac{x_3}{Đ_{23}} = \frac{0.50}{1.68 \times 10^{-5}} + \frac{0.25}{8.33 \times 10^{-5}} + \frac{0.25}{1.68 \times 10^{-5}} = 0.4764 \times 10^{5}$$

Thus,

$$(\boldsymbol{B}) = \begin{bmatrix} 0.1336 & 0.0068 \\ 0.2376 & 0.4764 \end{bmatrix} \times 10^{5}$$

b) The elements of $(\boldsymbol{B})^{-1}$ are as follows:

$$(\boldsymbol{B})^{-1} = \begin{bmatrix} 7.68 & -0.11 \\ -3.83 & 2.15 \end{bmatrix} \times 10^{-5}$$

c) The matrix of Fick diffusivities $[\boldsymbol{D}]$ is as follows:

$$(\boldsymbol{D}) = (\boldsymbol{B})^{-1} = \begin{bmatrix} 7.68 & -0.11 \\ -3.83 & 2.15 \end{bmatrix} \times 10^{-5}$$

d) The system shows coupling effects as the off-diagonal components of matrix of Fick diffusivities $(\boldsymbol{D})$ are non-zero.

e) The diffusional fluxes $(\vec{J})$ in terms of the concentration gradients are given as follows:

$$\begin{pmatrix} J_1 \\ J_2 \end{pmatrix} = -C \begin{bmatrix} 7.68 \times 10^{-5} & -0.11 \times 10^{-5} \\ -3.83 \times 10^{-5} & 2.15 \times 10^{-5} \end{bmatrix} \begin{pmatrix} \dfrac{dx_1}{dz} \\ \dfrac{dx_2}{dz} \end{pmatrix}$$

$$J_1 = -C\left[7.68 \times 10^{-5}\left(\frac{dx_1}{dz}\right) - 0.11 \times 10^{-5}\left(\frac{dx_2}{dz}\right)\right]$$

$$J_2 = -C\left[-3.83 \times 10^{-5}\left(\frac{dx_1}{dz}\right) + 2.15 \times 10^{-5}\left(\frac{dx_2}{dz}\right)\right]$$

$$J_3 = -(J_1 + J_2) = C\left[3.85 \times 10^{-5}\left(\frac{dx_1}{dz}\right) + 2.04 \times 10^{-5}\left(\frac{dx_2}{dz}\right)\right]$$

We can re-write the expression for J_3 by replacing dx_1/dz as:

$$\left(\frac{dx_1}{dz}\right) = -\left[\left(\frac{dx_2}{dz}\right) + \left(\frac{dx_3}{dz}\right)\right]$$

$$J_3 = -C\left[3.85 \times 10^{-5}\left(\frac{dx_3}{dz}\right) + 1.81 \times 10^{-5}\left(\frac{dx_2}{dz}\right)\right]$$

f) Which species undergoes osmotic diffusion when $\Delta x_1 = +0.50$, $\Delta x_2 = 0$, and $\Delta x_3 = -0.50$ over the diffusion length of L m?

Let us calculate the three fluxes for the given concentration gradients.

$$J_1 = -\left(\frac{C}{L}\right)[7.68 \times 10^{-5}\Delta x_1 - 0.11 \times 10^{-5}\Delta x_2] = -\left(\frac{C}{L}\right)(7.68 \times 10^{-5} \times 0.5)$$
$$= -\left(\frac{C}{L}\right)(3.84 \times 10^{-5})$$

The concentration gradient and flux have opposite signs (Δx_1 is positive, J_1 is negative). Thus, diffusion of species 1 is downhill in the direction of decreasing concentration.

$$J_2 = -\left(\frac{C}{L}\right)[3.83 \times 10^{-5}\Delta x_1 + 2.15 \times 10^{-5}\Delta x_2] = -\left(\frac{C}{L}\right)(-3.83 \times 10^{-5} \times 0.5 + 2.15 \times 10^{-5} \times 0)$$
$$= \left(\frac{C}{L}\right)(1.915 \times 10^{-5})$$

In this case, the flux J_2 is non-zero even though the corresponding concentration gradient is zero ($\Delta x_2 = 0$). Thus species 2 undergoes osmotic diffusion.

$$J_3 = -\left(\frac{C}{L}\right)[3.85 \times 10^{-5}\Delta x_3 + 1.81 \times 10^{-5}\Delta x_2] = -\left(\frac{C}{L}\right)[3.85 \times 10^{-5} \times (-0.50)]$$
$$= \left(\frac{C}{L}\right)(1.925 \times 10^{-5})$$

The concentration gradient and flux have opposite signs (Δx_3 is negative, J_3 is positive). Thus, diffusion of species 1 is downhill in the direction of decreasing concentration.

20.4.2 Example 2

Consider a non-ideal ternary liquid solution of acetone (1)-benzene (2)-carbon tetrachloride (3) at 25°C. The MS diffusivities of this system at the given composition are: $Đ_{12} = 3.4 \times 10^{-9}$, $Đ_{13} = 2.5 \times 10^{-9}$, $Đ_{23} = 1.7 \times 10^{-9}$ m^2/s. The composition of the system is as follows: $x_1 = 0.35$, $x_2 = 0.35$, and $x_3 = 0.30$.

a) Estimate the elements of the matrix $(\boldsymbol{B})$ at the given composition.
b) Estimate the elements of $(\boldsymbol{B})^{-1}$.
c) Given the following matrix of thermodynamic factor $(\boldsymbol{\Gamma})$, estimate the elements of the matrix of Fick diffusivities $(\boldsymbol{D})$.

$$(\boldsymbol{\Gamma}) = \begin{bmatrix} 0.69 & -0.13 \\ 0.07 & 1.05 \end{bmatrix}$$

d) Does the system show coupling effects?
e) Express the diffusional fluxes $(\vec{J})$ in terms of the concentration gradients.

Solution

a) The elements of the matrix $(\boldsymbol{B})$ are as follows:

$$B_{11} = \frac{x_1}{Đ_{13}} + \frac{x_2}{Đ_{12}} + \frac{x_3}{Đ_{13}} = \frac{0.35}{2.5 \times 10^{-9}} + \frac{0.35}{3.4 \times 10^{-9}} + \frac{0.30}{2.5 \times 10^{-5}} = 0.363 \times 10^9$$

$$B_{12} = -x_1\left(\frac{1}{Đ_{12}} - \frac{1}{Đ_{13}}\right) = -0.35\left(\frac{1}{3.4 \times 10^{-9}} - \frac{1}{2.5 \times 10^{-9}}\right) = 0.0371 \times 10^9$$

$$B_{21} = -x_2\left(\frac{1}{Đ_{21}} - \frac{1}{Đ_{23}}\right) = -0.35\left(\frac{1}{3.4 \times 10^{-9}} - \frac{1}{1.7 \times 10^{-9}}\right) = 0.103 \times 10^9$$

$$B_{22} = \frac{x_2}{Đ_{23}} + \frac{x_1}{Đ_{21}} + \frac{x_3}{Đ_{23}} = \frac{0.35}{1.70 \times 10^{-9}} + \frac{0.35}{3.4 \times 10^{-9}} + \frac{0.30}{1.7 \times 10^{-9}} = 0.485 \times 10^9$$

Thus,

$$(\boldsymbol{B}) = \begin{bmatrix} 0.363 & 0.0371 \\ 0.103 & 0.485 \end{bmatrix} \times 10^9$$

b) The elements of $(\boldsymbol{B})^{-1}$ are as follows:

$$(\boldsymbol{B})^{-1} = \begin{bmatrix} 2.82 & -0.215 \\ -0.598 & 2.11 \end{bmatrix} \times 10^{-9}$$

c) The matrix of Fick diffusivities $[\boldsymbol{D}]$ is as follows:

$$(\boldsymbol{D}) = (\boldsymbol{B})^{-1}(\boldsymbol{\Gamma}) = \begin{bmatrix} 2.82 & -0.215 \\ -0.598 & 2.11 \end{bmatrix} \begin{bmatrix} 0.69 & -0.13 \\ 0.07 & 1.05 \end{bmatrix} \times 10^{-9} = \begin{bmatrix} 1.931 & -0.592 \\ -0.265 & 2.293 \end{bmatrix} \times 10^{-9}$$

d) The system shows coupling effects as the off-diagonal components of matrix of Fick diffusivities $(\boldsymbol{D})$ are non-zero.

e) The diffusional fluxes $(\vec{J})$ in terms of the concentration gradients are given as follows:

$$\begin{pmatrix} J_1 \\ J_2 \end{pmatrix} = -C \begin{bmatrix} 1.931 \times 10^{-9} & -0.592 \times 10^{-9} \\ -0.265 \times 10^{-9} & 2.293 \times 10^{-9} \end{bmatrix} \begin{pmatrix} \dfrac{dx_1}{dz} \\ \dfrac{dx_2}{dz} \end{pmatrix}$$

$$J_1 = -C\left[1.931 \times 10^{-9}\left(\frac{dx_1}{dz}\right) - 0.592 \times 10^{-9}\left(\frac{dx_2}{dz}\right)\right]$$

$$J_2 = -C\left[-0.265 \times 10^{-9}\left(\frac{dx_1}{dz}\right) + 2.293 \times 10^{-9}\left(\frac{dx_2}{dz}\right)\right]$$

$$J_3 = -(J_1 + J_2) = C\left[1.666 \times 10^{-9}\left(\frac{dx_1}{dz}\right) + 1.701 \times 10^{-9}\left(\frac{dx_2}{dz}\right)\right]$$

We can re-write the expression for J_3 by replacing dx_1/dz as:

$$\left(\frac{dx_1}{dz}\right) = -\left[\left(\frac{dx_2}{dz}\right) + \left(\frac{dx_3}{dz}\right)\right]$$

$$J_3 = -C\left[1.666 \times 10^{-9}\left(\frac{dx_3}{dz}\right) - 0.035 \times 10^{-9}\left(\frac{dx_2}{dz}\right)\right]$$

NOTATION

$\hat{a}_i$	activity of species i in non-ideal mixture
$(\boldsymbol{B})$	matrix in MS equation, see Equations (20.31) and (20.34), components of matrix depend on mixture composition and MS diffusivities of species
B_{ii}	diagonal component of matrix $(\boldsymbol{B})$,
B_{ij}	off diagonal component of matrix $(\boldsymbol{B})$
$(\boldsymbol{B})^{-1}$	inverse of matrix $(\boldsymbol{B})$
C	total molar concentration of all species
C_A	molar concentration of species A
D_{12}	Fick diffusivity of species 1 in species 2
D_{1n}	Fick diffusivity of species 1 in species n
D_{AB}	Fick diffusivity of species A in species B
D_{iB}	Fick diffusivity of species i in species B
D_{ii}	Fick's main diffusivity of species i in a mixture, see Equation (20.15)
D_{ij}	Fick's cross diffusivity of species i with respect to species j in a mixture, see Equation (20.15)
D_{im}	Fick diffusivity of species i in mixture, see Equation (20.13)
D_{ji}	Fick's cross diffusivity of species j with respect to species i in a mixture, see Equation (20.15)
D_{AB}^{o}	Fick diffusivity of species A in Species B in a dilute binary liquid solution

D_{AB}^*	diffusion coefficient of species A in species B defined using chemical potential gradient as the driving force
D_{BA}^o	Fick diffusivity of species B in Species A in a dilute binary liquid solution
$(\boldsymbol{D})$	Fickian diffusivity matrix, see Equation (20.17)
$Đ_{12}$	MS diffusivity of species 1 in species 2
$Đ_{21}$	MS diffusivity of species 2 in species 1
$Đ_{ij}$	MS diffusivity of species i in species j
$Đ_{ji}$	MS diffusivity of species j in species i
$Đ_{ij}^o$	MS diffusivity of species i in species j in a dilute binary liquid solution
$Đ_{ji}^o$	MS diffusivity of species j in species i in a dilute binary liquid solution
J_1, J_2, J_3	one-dimensional diffusion molar fluxes of species 1,2, and 3
$\vec{J}_i$	diffusional molar flux vector of species i
$(\vec{\boldsymbol{J}})$	matrix of diffusion molar flux vectors of all species
k_B	Boltzmann constant
M_i	molecular weight of species i
$\vec{N}_i$	absolute molar flux of species i
R	universal gas constant
T	absolute temperature
T^*	dimensionless temperature, see Equation (20.57)
V_A	molar volume of species A
x_i	mole fraction of species i
y_i	mole fraction of species i in a gas mixture
z_i	mole fraction defined in Equation (20.68)
z_j	mole fraction defined in Equation (20.68)

Greek Letters

γ_i	activity coefficient of species i
Γ	thermodynamic factor, see Equations (20.4a) and (20.4b)
Γ_{ij}	component of thermodynamic factor matrix
$(\boldsymbol{\Gamma})$	thermodynamic factor matrix
ε_1	characteristic molecular interaction energy of pure species 1
ε_2	characteristic molecular interaction energy of pure species 2
ε_{12}	characteristic energy of molecular interaction for the binary mixture
η	viscosity of liquid solution
η_A	viscosity of species A
η_B	viscosity of species B
μ_A	chemical potential of species A
μ_i	chemical potential of species i
σ_1	collision diameter of species 1
σ_2	collision diameter of species 2
σ_{12}	average collision diameter
$\hat{\varphi}_A$	fugacity coefficient of species A
ψ_B	association parameter of solvent (species B)
Ω_{12}	collision integral

Mathematical Operators

∇	del operator
$\nabla_{T,P}$	del operator evaluated at constant T and P

PRACTICE PROBLEMS

Problem 1

Consider an ideal ternary gas mixture of species (1)-(2)-(3). The MS diffusivities of this system estimated from the kinetic theory of gases are given as: $Đ_{12} = 7.33 \times 10^{-5}$, $Đ_{13} = 5.8 \times 10^{-5}$, $Đ_{23} = 1.48 \times 10^{-5}$ m^2/s. The composition of the system is as follows: $x_1 = 0.18$, $x_2 = 0.55$, and $x_3 = 0.27$.

a) Estimate the elements of the matrix $(\boldsymbol{B})$ at the given composition.
b) Estimate the elements of $(\boldsymbol{B})^{-1}$.
c) Determine the matrix of Fick diffusivities $(\boldsymbol{D})$
d) Does the system show coupling effects?
e) Express the diffusional fluxes $(\vec{J})$ in terms of the concentration gradients.
f) Which species, if any, undergoes osmotic diffusion when $\Delta x_1 = 0.50$, $\Delta x_2 = 0$, and $\Delta x_3 = -0.50$ over the diffusion length of L m?

Problem 2

Consider a non-ideal ternary liquid solution of species (1)-(2)-(3) at 25°C. The MS diffusivities of this system at the given composition are: $Đ_{12} = 3.1 \times 10^{-9}$, $Đ_{13} = 2.35 \times 10^{-9}$, $Đ_{23} = 1.62 \times 10^{-9}$ m^2/s. The composition of the system is as follows: $x_1 = 0.31$, $x_2 = 0.31$, and $x_3 = 0.38$.

a) Estimate the elements of the matrix $(\boldsymbol{B})$ at the given composition.
b) Estimate the elements of $(\boldsymbol{B})^{-1}$.
c) Given the following matrix of thermodynamic factor $(\boldsymbol{\Gamma})$, estimate the elements of the matrix of Fick diffusivities $(\boldsymbol{D})$.

$$(\boldsymbol{\Gamma}) = \begin{bmatrix} 0.73 & -0.15 \\ 0.08 & 1.15 \end{bmatrix}$$

d) Does the system show coupling effects?
e) Express the diffusional fluxes $(\vec{J})$ in terms of the concentration gradients.

FURTHER READING

1. R.B. Bird, W.E. Stewart, and E.N. Lightfoot. 2007. *Transport Phenomena*. 2nd Edition, Wiley & Sons, New York, 2007.
2. R. Krishna, T.A. Wesselingh. 1997. The Maxwell-Stefan approach to mass transfer. *Chem. Eng. Sci.* 52: 861-911.
3. R.B. Bird, D.J. Klingenberg. 2013. Multicomponent diffusion – a brief review. *Adv. in Water Res.*62: 238-242.
4. A. Bardow, E. Kriesten, M.A. Voda, F. Casanova, B. Blumich, W. Marquardt. 2009. Prediction of multicomponent mutual in liquids: model discrimination using NMR data. *Fluid Phase Equilibria.* 278: 27-35.
5. X. Liu, T.J.H. Vlugt, A. Bardow. 2011. Maxwell-Stefan diffusivities in liquid mixtures: using molecular dynamics for testing model predictions. *Fluid Phase Equilibria.* 301: 110-117.
6. C.F. Curtiss, R.B. Bird. 1999. Multicomponent diffusion. *Ind. Eng. Chem. Res.* 38: 2515-2522.
7. X. Liu, T.J.H. Vlugt, A. Bardow. 2011. Predictive Darken equation for Maxwell-Stefan diffusivities in multicomponent mixtures. *Ind. Eng. Chem. Res.* 50:10350-10358.
8. T. Allie-Ebrahim, Q. Zhu, P. Brauer, G.D. Moggridge, C. D'Agostino. 2017. Maxwell-Stefan diffusion coeficeint estimation for ternary systems: an ideal ternary alcohol system. *Phys. Chem. Chem. Phys.* 19: 16071-16077.
9. R. Haase, H.J. Jansen. 1980. Diffusion in binary nonelectrolyte solutions. *Z. Naturforsch.* 35a: 1116-1117.
10. R. Krishna. 2015. Uphill diffusion in multicomponent mixtures. *Chem. Soc. Rev.* 44: 1812-2836.

11. G.L. Standard, R. Taylor, R. Krishna. 1979. The Maxwell-Stefan formulation of irreversible thermodynamics for simultaneous heat and mass transfer. *Chem. Eng. Commun*. 3: 277-289.

Appendix A

Boundary Layer Equations

(pertaining to Chapters 6 and 13)

A.1 Concept of a Boundary Layer

When fluid flows over a solid surface, it experiences significant velocity changes. The fluid velocity at the solid surface is reduced to zero because of the no-slip condition. With increasing distance y normal to the surface, the velocity of the fluid increases until it approaches the free stream velocity V_∞. The velocity change from zero at the solid surface to the free stream velocity occurs in a thin region near the solid surface. This thin layer of fluid moving adjacent to the solid surface where the fluid velocity is significantly influenced by the presence of the solid boundary is called "boundary layer" (see Figure A.1). The flow in the boundary layer is characterized by large velocity gradients and shear stresses. In the fluid moving outside the boundary layer, velocity gradients and shear stresses are negligible. The boundary layer thickness δ is defined to be that distance normal to the solid surface where $V = 0.99V_\infty$. The boundary layer thickness δ increases with increasing x-distance from the leading edge of the solid surface.

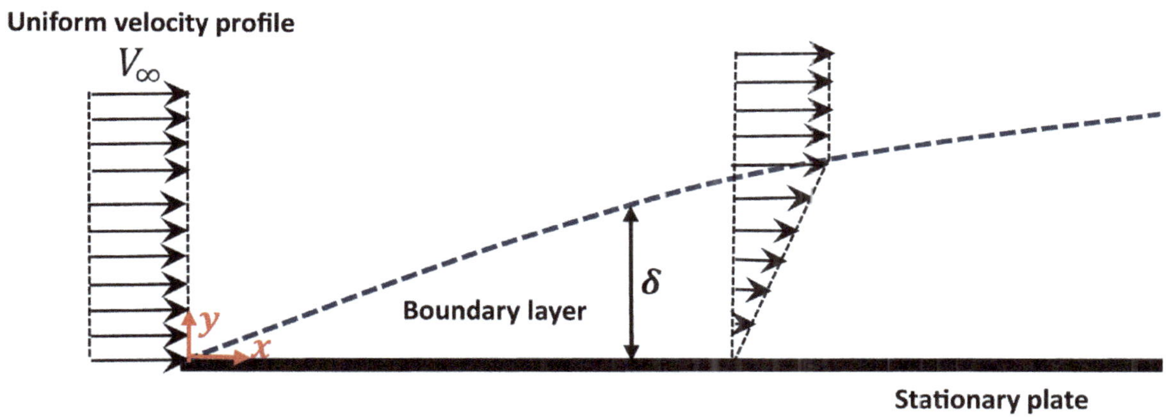

Figure A.1 Development of hydrodynamic boundary layer over a flat plate.

It should be noted that at relatively small values of x (leading portion of the solid surface), flow within the boundary layer is laminar. However, some distance downstream the leading edge, transition to turbulent flow begins to occur. In the transition region, the flow fluctuates between laminar and turbulent regimes. Finally, the boundary layer becomes completely turbulent when x exceeds a certain value. The transition from laminar to turbulent flow is governed by the Reynolds number defined as:

$$Re_x = \frac{\rho V_\infty x}{\eta} \tag{A.1}$$

For flow over a flat surface, the flow in the boundary layer is usually laminar for $Re_x < 2 \times 10^5$. The transition from laminar boundary layer to turbulent boundary layer begins at $Re_x = 2 \times 10^5$ and is completed at $Re_x = 3 \times 10^6$. The boundary layer is fully turbulent for $Re_x \geq 3 \times 10^6$.

The discussion thus far was about hydrodynamic boundary layer. Just as a hydrodynamic boundary layer develops when fluid flows over a solid surface, a thermal boundary layer development occurs if the free stream temperature T_∞ is different from the solid surface temperature T_w. The thermal boundary layer is a thin layer of fluid moving adjacent to the solid surface where the fluid temperature is significantly influenced by the presence of a solid boundary. Significant temperature gradients exist in the thermal boundary layer. In the fluid moving outside the thermal boundary layer, temperature gradients are negligible.

Figure A.2 shows the thermal boundary layer development on an isothermal flat solid surface. The thermal boundary layer thickness (δ_T), defined as the distance normal to the solid surface where the temperature ratio $(T_w - T)/(T_w - T_\infty)$ is 0.99, increases as we travel downstream the solid boundary. It is important to note that the thermal boundary layer thickness (δ_T) is not necessarily equal to the hydrodynamic boundary layer thickness δ. For laminar boundary layers, δ_T is related to δ by a dimensionless group called Prandtl number (Pr) as:

$$\frac{\delta_T}{\delta} \approx (Pr)^{-n} \tag{A.2}$$

where Pr (Prandtl number) is defined as the ratio of momentum diffusivity to thermal diffusivity, and n is a positive exponent. For gases, $Pr \approx 1$ and $\delta_T \approx \delta$; for non-metallic liquids, $Pr \gg 1$ and hence $\delta_T \ll \delta$; for metallic liquids, $Pr \ll 1$ and therefore $\delta_T \gg \delta$. For turbulent boundary layers, $\delta_T \approx \delta$.

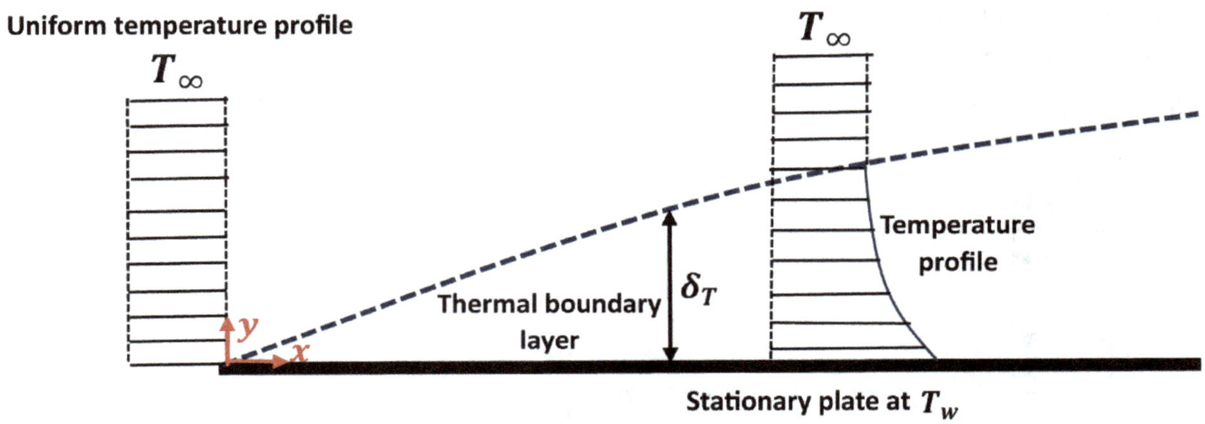

Figure A.2 Development of thermal boundary layer over a flat plate.

A.2 The Boundary Layer Equations

For two-dimensional flow of a Newtonian fluid (with constant properties), the x- and y- components of the Navier-Stokes equation are:

$$\rho\left(\frac{\partial V_x}{\partial t} + V_x\frac{\partial V_x}{\partial x} + V_y\frac{\partial V_x}{\partial y}\right) = \rho g_x - \frac{\partial P}{\partial x} + \eta\left[\frac{\partial^2 V_x}{\partial x^2} + \frac{\partial^2 V_x}{\partial y^2}\right] \tag{A.3}$$

$$\rho\left(\frac{\partial V_y}{\partial t} + V_x\frac{\partial V_y}{\partial x} + V_y\frac{\partial V_y}{\partial y}\right) = \rho g_y - \frac{\partial P}{\partial y} + \eta\left[\frac{\partial^2 V_y}{\partial x^2} + \frac{\partial^2 V_y}{\partial y^2}\right] \tag{A.4}$$

and the continuity equation is:

$$\frac{\partial V_x}{\partial x} + \frac{\partial V_y}{\partial y} = 0 \tag{A.5}$$

In boundary layer flow, certain order-of-magnitude simplifications of the Navier-Stokes equations (Equations (A.3) and (A.4)) are possible as the boundary layer thickness is rather small (few thousandths of a mm). Assuming that the solid surface is nearly flat with x-coordinate parallel and y-coordinate perpendicular to the solid surface, one can say that:

$$V_x >> V_y \tag{A.6}$$

$$\frac{\partial V_x}{\partial y} >> \frac{\partial V_x}{\partial x}, \frac{\partial V_y}{\partial y}, \frac{\partial V_y}{\partial x} \tag{A.7}$$

$$\frac{\partial^2 V_x}{\partial y^2} >> \frac{\partial^2 V_x}{\partial x^2}, \frac{\partial^2 V_y}{\partial y^2}, \frac{\partial^2 V_y}{\partial x^2} \tag{A.8}$$

These boundary layer approximations indicate that the velocity component in the direction along the surface is significantly larger than that perpendicular to the surface, and that the gradients perpendicular to the surface are significantly larger than those along the surface. With the foregoing approximations, the Navier-Stokes equations (Equations (A.3) and (A.4)) reduce to:

$$\rho\left(\frac{\partial V_x}{\partial t} + V_x\frac{\partial V_x}{\partial x} + V_y\frac{\partial V_x}{\partial y}\right) = \rho g_x - \frac{\partial P}{\partial x} + \eta\left[\frac{\partial^2 V_x}{\partial x^2}\right] \tag{A.9}$$

$$0 = \rho g_y - \frac{\partial P}{\partial y} \tag{A.10}$$

Assuming that the effect of gravity is negligible in thin boundary layers, Equation (A.10) yields $\partial P/\partial y = 0$. Thus, pressure is a function of x alone and does not vary in the direction normal to the surface. Furthermore, the pressure gradient in x-direction is related to the free stream velocity gradient as follows:

$$\frac{dP}{dx} = -\rho V_\infty \frac{dV_\infty}{dx} \tag{A.11}$$

Equation (A.11) follows from the application of Bernoulli equation to the free stream. Combining Equations (A.9) and (A.11) gives:

$$\left(\frac{\partial V_x}{\partial t} + V_x\frac{\partial V_x}{\partial x} + V_y\frac{\partial V_x}{\partial y}\right) = V_\infty \frac{dV_\infty}{dx} + \nu\left[\frac{\partial^2 V_x}{\partial x^2}\right] \tag{A.12}$$

This equation together with the continuity equation (Equation (A.5)) are known as the boundary layer equations.

The thermal energy equation, neglecting the viscous dissipation term, for two-dimensional flow under consideration is:

$$\left(\frac{\partial T}{\partial t} + V_x\frac{\partial T}{\partial x} + V_y\frac{\partial T}{\partial y}\right) = \alpha\left[\frac{\partial^2 T}{\partial x^2} + \frac{\partial^2 T}{\partial y^2}\right] \tag{A.13}$$

Using the boundary layer approximations

$$\frac{\partial T}{\partial y} >> \frac{\partial T}{\partial x} \tag{A.14}$$

$$\frac{\partial^2 T}{\partial y^2} >> \frac{\partial^2 T}{\partial x^2} \tag{A.15}$$

The thermal energy equation reduces to:

$$\left(\frac{\partial T}{\partial t} + V_x \frac{\partial T}{\partial x} + V_y \frac{\partial T}{\partial y}\right) = \alpha \left[\frac{\partial^2 T}{\partial x^2}\right] \tag{A.16}$$

Note that the term $V_x(\partial T/\partial x)$ cannot be neglected in comparison with the term $V_y(\partial T/\partial y)$as $V_x >> V_y$.

The boundary layer equations (Equations (A.5), (A.12) and (A.16)) are valid for both laminar and turbulent flows. In turbulent flow, however, the velocities and temperature are instantaneous quantities.

A.3 Integral Analysis of the Boundary Layer

In practical applications involving fluid flow, one is often interested in the calculation of the wall stress or drag force exerted by the fluid on a solid boundary. In practical applications dealing with convective heat transfer between a solid surface and fluid, the quantity of most interest is the heat flux at the solid surface. One can develop the relationships for wall stress and wall heat flux using the von Karman integral analysis.

Momentum Integral Analysis

As an example, consider steady flow over a flat plate (Figure A.3). We will carry out momentum integral analysis of the control volume shown in the figure. The control volume is of unit depth, Δx length, and δ thickness. According to Newton's second law of motion:

$$\sum \vec{F} = \frac{\partial}{\partial t} \iiint_{CV} \rho \vec{V}\, dV + \iint_{CS} \rho(\hat{n} \cdot \vec{V})\vec{V} dS \tag{A.17}$$

where CV refers to control volume, CS refers to control surface, $\hat{n}$ is outward unit normal, V is volume, and $\vec{V}$ is fluid velocity vector. Note that the right-hand side of Equation (A.17) is simply the time rate of change of linear momentum of a fluid instantaneously occupying the given control volume. The application of Equation (A.17) in x-direction gives:

$$\sum F_x = \frac{\partial}{\partial t} \iiint_{CV} \rho V_x dV + \iint_{CS} \rho(\hat{n} \cdot \vec{V})V_x dS \tag{A.18}$$

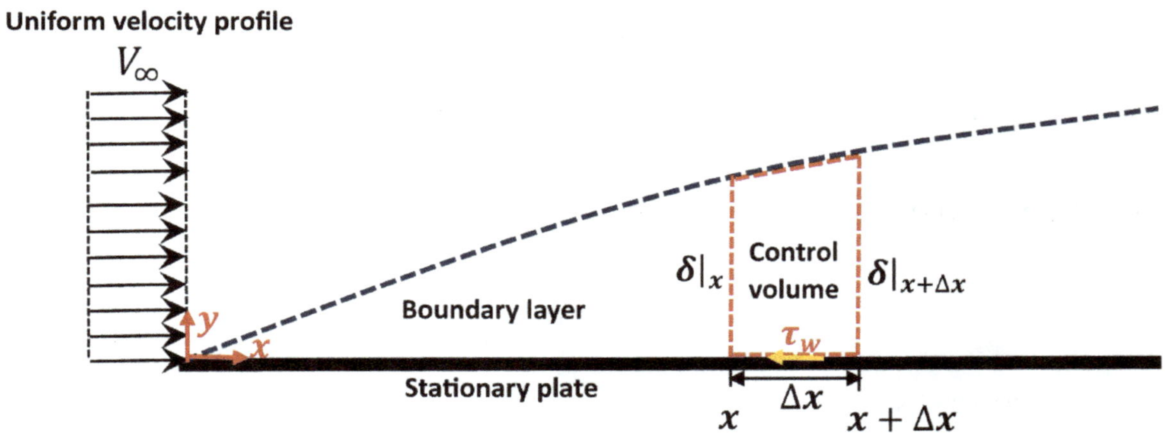

Figure A.3 Control volume for momentum integral analysis.

The forces acting on the control volume (of unit depth) in x-direction are:

$$\sum F_x = P\delta|_x - P\delta|_{x+\Delta x} + P_{top}(\delta|_{x+\Delta x} - \delta|_x) - \tau_w \Delta x \tag{A.19}$$

where τ_w is the wall stress acting on the control volume. As $dP/dx = 0$ and $P_{top} = P$ for flow over a flat plate, the pressure terms in Equation (A.19) cancel each other. Thus

$$\sum F_x = -\tau_w \Delta x = \frac{\partial}{\partial t}\iiint_{CV} \rho V_x dV + \iint_{CS} \rho(\hat{n}\cdot\vec{V})V_x dS \tag{A.20}$$

The first integral term on the right-hand side of Equation (A.20) is exactly zero under steady state condition. The second integral term can be written as:

$$\iint_{CS} \rho(\hat{n}\cdot\vec{V})V_x dS = \int_0^\delta \rho V_x^2\, dy\Big|_{x+\Delta x} - \int_0^\delta \rho V_x^2\, dy\Big|_x - \dot{m}_{top}V_\infty \tag{A.21}$$

where $\dot{m}_{top}$ is the rate of mass entering the control volume from the top surface. From mass balance under steady state condition:

$$\iint_{CS} \rho(\hat{n}\cdot\vec{V})dS = \int_0^\delta \rho V_x\, dy\Big|_{x+\Delta x} - \int_0^\delta \rho V_x\, dy\Big|_x - \dot{m}_{top} = 0 \tag{A.22}$$

Combining Equations (A.20), (A.21) and (A.22) gives:

$$\sum F_x = -\tau_w \Delta x = \int_0^\delta \rho V_x^2\, dy\Big|_{x+\Delta x} - \int_0^\delta \rho V_x^2\, dy\Big|_x - V_\infty \int_0^\delta \rho V_x\, dy\Big|_{x+\Delta x} + V_\infty \int_0^\delta \rho V_x\, dy\Big|_x \tag{A.23}$$

Upon dividing Equation (A.23) with Δx and taking the limit $\Delta x \to 0$, we get:

$$\tau_w = \rho\frac{d}{dx}\int_0^\delta V_x V_\infty dy - \rho\frac{d}{dx}\int_0^\delta V_x^2\, dy \tag{A.24}$$

Note that ρ is assumed to be constant (incompressible fluid) and V_∞ is constant for flow over a flat plate. Equation (A.24) can be re-written as:

$$\frac{\tau_w}{\rho} = \frac{d}{dx}\int_0^\delta V_x (V_\infty - V_x)dy \tag{A.25}$$

Equation (A.25) is referred to as von Karman momentum integral expression. The momentum integral expression can be used to determine the boundary layer thickness, wall shear stress, and friction factor, provided that the velocity profile in the boundary layer is known.

Energy Integral Analysis

Energy integral analysis involves the application of energy balance on a control volume inside the thermal boundary layer. As an example, consider steady flow over a flat plate. Under steady state condition, the application of the first law of thermodynamics to a control volume gives:

$$\iint_{CS} \rho\left(u + gz + \frac{1}{2}V^2\right)(\hat{n}\cdot\vec{V})dS = -\iint_{CS} \vec{q}''\cdot d\vec{S} + \iint_{CS}(\vec{\pi}_{(\hat{n})}\cdot\vec{V})dS \tag{A.26}$$

Neglecting potential and kinetic energies as well as viscous dissipation of energy, Equation (A.26) simplifies to:

$$\iint_{CS} \rho\, u(\hat{n}\cdot\vec{V})dS = -\iint_{CS} \vec{q}''\cdot d\vec{S} - \iint_{CS}(\hat{n}P\cdot\vec{V})dS \tag{A.27}$$

As enthalpy $h = u + (P/\rho)$, we can re-write Equation (A.27) as:

$$-\iint_{CS}\vec{q}''\cdot d\vec{S}=\iint_{CS}\rho\, h(\hat{n}\cdot\vec{V})dS \qquad \text{(A.28)}$$

The heat is being added to the control volume of unit depth through conduction from the solid boundary (plate). Neglecting conductive fluxes at other surfaces of the control volume (surfaces not in contact with the solid boundary), energy balance, Equation (A.28), becomes:

$$q''_w\Delta x=-k\Delta x\left(\frac{\partial T}{\partial y}\right)_{y=0}=\iint_{CS}\rho\, h(\hat{n}\cdot\vec{V})dS \qquad \text{(A.29)}$$

Equation (A.29) indicates that the rate of thermal energy carried away by the fluid is equal to that conducted through the wall. The energy efflux term (integral on the right-hand side of Equation (A.29)) can be expressed as:

$$\iint_{CS}\rho\, h(\hat{n}\cdot\vec{V})dS=\int_0^{\delta_T}(\rho V_x\, h)dy\Big|_{x+\Delta x}-\int_0^{\delta_T}(\rho V_x\, h)dy\Big|_{x}-\dot{m}_{top}h_\infty \qquad \text{(A.30)}$$

where h_∞ is the enthalpy of free-stream fluid (fluid entering from the top surface of the control volume). The mass flow rate from the top surface is:

$$\dot{m}_{top}=\int_0^{\delta_T}\rho V_x\, dy\Big|_{x+\Delta x}-\int_0^{\delta_T}\rho V_x\, dy\Big|_{x}=\left[\frac{d}{dx}\int_0^{\delta_T}\rho V_x dy\right]\Delta x \qquad \text{(A.31)}$$

From Equations (A.29) - (A.31), we get:

$$q''_w\Delta x=-k\Delta x\left(\frac{\partial T}{\partial y}\right)_{y=0}=\int_0^{\delta_T}(\rho V_x\, h)dy\Big|_{x+\Delta x}-\int_0^{\delta_T}(\rho V_x\, h)dy\Big|_{x}-\left[\frac{d}{dx}\int_0^{\delta_T}(\rho V_x h_\infty)dy\right]\Delta x \qquad \text{(A.32)}$$

Divide Equation (A.32) by Δx and take the limit $\Delta x\rightarrow 0$, we get:

$$q''_w=-k\left(\frac{\partial T}{\partial y}\right)_{y=0}=\frac{d}{dx}\int_0^{\delta_T}\rho V_x\,(h-h_\infty)dy=\frac{d}{dx}\int_0^{\delta_T}\rho V_x c_p\,(T-T_\infty)dy \qquad \text{(A.33)}$$

This is von Karman energy integral expression. The energy integral expression can be used to determine the thermal boundary layer thickness, wall heat flux, and heat transfer coefficient, provided that both the velocity and temperature profiles are known.

A.4 Laminar Flow Over a Flat Plate

Hydrodynamic Boundary Layer Thickness and Friction Factor

To derive the expressions for boundary layer thickness and friction factor from integral analysis, the first step is to assume velocity profile in the boundary layer. Consider steady laminar flow over a flat plate with a free stream velocity of the fluid V_∞.

The velocity profile can be assumed to be a four-term polynomial as:

$$V_x=a+by+cy^2+dy^3 \qquad \text{(A.34)}$$

The constants in Equation (A.34) can be obtained from the boundary conditions:

$$y=0, V_x=0 \qquad \text{(A.35)}$$

$$y = \delta, V_x = V_\infty \tag{A.36}$$

$$y = \delta, \frac{\partial V_x}{\partial y} = 0 \tag{A.37}$$

$$y = 0, \frac{\partial^2 V_x}{\partial y^2} = 0 \tag{A.38}$$

The last boundary condition in Equation (A.38) follows from the boundary layer equation (Equation (A.12)) as $\partial V_x/\partial t = 0$, $dV_\infty/dx = 0$, and at $y = 0$, $V_x(\partial V_x/\partial x) = 0$, $V_y(\partial V_x/\partial y) = 0$. Using the boundary conditions (Equations (A.35) - (A.38)), it can be easily shown that:

$$a = 0,\ b = 3V_\infty/2\delta,\ c = 0,\ d = -V_\infty/2\delta^3 \tag{A.39}$$

Thus, the velocity profile in the boundary layer becomes:

$$\frac{V_x}{V_\infty} = \frac{3}{2}\left(\frac{y}{\delta}\right) - \frac{1}{2}\left(\frac{y}{\delta}\right)^3 \tag{A.40}$$

The momentum integral expression (Equation (A.25)) can now be written as:

$$\frac{\tau_w}{\rho} = \frac{d}{dx}\int_0^\delta V_x\,(V_\infty - V_x)dy = \frac{d}{dx}\int_0^\delta V_\infty^2\left(\frac{V_x}{V_\infty}\right)\left(1 - \frac{V_x}{V_\infty}\right)dy = (V_\infty^2)\frac{d}{dx}\int_0^\delta \left[\frac{3}{2}\left(\frac{y}{\delta}\right) - \frac{1}{2}\left(\frac{y}{\delta}\right)^3\right]\left[1 - \frac{3}{2}\left(\frac{y}{\delta}\right) + \frac{1}{2}\left(\frac{y}{\delta}\right)^3\right]dy = (V_\infty^2)\left(\frac{39}{280}\right)\frac{d\delta}{dx} \tag{A.41}$$

From Newton's Law of viscosity

$$\tau_w = \mu\left(\frac{\partial V_x}{\partial y}\right)_{y=0} = \eta\left(\frac{3}{2}\frac{V_\infty}{\delta}\right) \tag{A.42}$$

Combining Equations (A.41) and (A.42), one can obtain:

$$\delta\frac{d\delta}{dx} = \left(\frac{140}{13}\right)\frac{\eta}{\rho V_\infty} \tag{A.43}$$

Upon integration with the limit $\delta = 0$ at $x = 0$ (leading edge of the plate), we obtain:

$$\frac{\delta}{x} = \frac{4.64}{Re_x^{1/2}} \tag{A.44}$$

The local friction factor can now be determined as:

$$f = \frac{\tau_w}{\frac{1}{2}\rho V_\infty^2} = \frac{3\eta}{\rho\delta V_\infty} = \frac{0..647}{(Re_x)^{1/2}} \tag{A.45}$$

These results derived from the momentum integral analysis are approximate. The exact results for δ and f (obtained from the solution of the boundary layer equations) are:

$$\frac{\delta}{x} = \frac{5}{(Re_x)^{1/2}} \tag{A.46}$$

$$f = \frac{0..664}{(Re_x)^{1/2}} \tag{A.47}$$

The exact solution for the boundary layer equations (Equations (A.5) and (A.12)) was obtained by Blasius using the concept of stream function and the similarity technique.

Thermal Boundary Layer Thickness and Heat Transfer Coefficient

Thus far we have dealt with the fluid mechanics aspect of the convective heat transfer problem involving steady laminar flow over a flat plate. Now we would like to develop a relationship for the heat transfer coefficient using the energy integral analysis. Consider steady laminar flow over an isothermal plate ($T = T_w$ =constant). The free stream velocity and temperature of the fluid are V_∞ and T_∞, respectively.

The temperature profile is assumed to be similar in form to the velocity distribution:

$$T = a + by + cy^2 + dy^3 \tag{A.48}$$

The constants in Equation (A.48) can be obtained from the boundary conditions:

$$y = 0, T = T_w \tag{A.49}$$

$$y = \delta_T, T = T_\infty \tag{A.50}$$

$$y = \delta_T, \frac{\partial T}{\partial y} = 0 \tag{A.51}$$

$$y = 0, \frac{\partial^2 T}{\partial y^2} = 0 \tag{A.52}$$

The last boundary condition follows from the boundary layer thermal energy equation (Equation (A.16)) as at $y = 0$, $\partial T/\partial t = 0$, $V_x(\partial T/\partial x) = 0$, $V_y(\partial T/\partial y) = 0$. Using the boundary conditions (Equations (A.49) - (A.52)), it can be shown that:

$$a = T_w, \; b = 3(T_\infty - T_w)/2\delta_T, \; c = 0, \; d = (T_\infty - T_w)/2\delta_T^3 \tag{A.53}$$

Thus, the temperature profile in the boundary layer becomes:

$$\frac{T-T_w}{T_\infty - T_w} = \frac{3}{2}\left(\frac{y}{\delta_T}\right) - \frac{1}{2}\left(\frac{y}{\delta_T}\right)^3 \tag{A.54}$$

To utilize the energy integral expression (Equation (A.33)) knowledge of both velocity and temperature distributions is required. When $\delta_T \ll \delta$ (non-metallic liquids, $Pr \gg 1$), the velocity distribution is given by Equation (A.40). For the case where $\delta_T >> \delta$ (metallic liquids where $Pr \ll 1$), it is reasonable to assume that the velocity distribution is uniform ($V_x \approx V_\infty$).

For the case where $Pr \gg 1$, substitution of temperature distribution (Equation (A.54)) and velocity distribution (Equation (A.40)) into the integral expression (Equation (A.33)) and carrying out the integration yields the following results:

$$\frac{\delta_T}{\delta} = \frac{Pr^{-1/3}}{1.025} \tag{A.55}$$

$$Nu_x = \frac{hx}{k} = 0.331(Re_x)^{1/2}(Pr^{1/3}) \tag{A.56}$$

where Nu_x is the local Nusselt number.

For the case where $Pr \ll 1$ (metallic liquids) substitution of temperature distribution (Equation (A.54)) and $V_x \approx V_\infty$ into the integral expression (Equation (A.33)) and carrying out the integration yields the following results:

$$\frac{\delta_T}{\delta} = 0.61 Pr^{-1/2} \tag{A.57}$$

$$Nu_x = \frac{hx}{k} = 0.530(Re_x)^{1/2}\left(Pr^{1/2}\right) \tag{A.58}$$

The exact solution of the boundary layer equations yields the following results:

a) When $Pr \gg 1$

$$\frac{\delta_T}{\delta} = Pr^{-1/3} \tag{A.59}$$

$$Nu_x = \frac{hx}{k} = 0.332(Re_x)^{1/2}\left(Pr^{1/3}\right) \tag{A.60}$$

b) When $Pr \ll 1$

$$Nu_x = \frac{hx}{k} = 0.564(Re_x)^{1/2}\left(Pr^{1/2}\right) \tag{A.61}$$

Clearly the agreement between the approximate integral analysis and exact results is good.

The heat transfer equations derived in this section assumed that the solid surface (flat plate) was isothermal. In a similar manner, equations can be derived for constant wall-flux problems. Constant surface heat flux problems are equally important from a practical point of view. The results for $Pr \gg 1$ are given here without going into the derivations:

$$Nu_x = \frac{hx}{k} = 0.417(Re_x)^{1/2}\left(Pr^{1/3}\right) \quad \text{(integral analysis)} \tag{A.62}$$

$$Nu_x = \frac{hx}{k} = 0.453(Re_x)^{1/2}\left(Pr^{1/3}\right) \quad \text{(exact analysis)} \tag{A.63}$$

Analogy Between Momentum and Heat Transfer

There exists a close similarity between momentum and heat transfer equations. For example, the friction factor equation (Equation (A.47)) can be re-written as:

$$\frac{f}{2} = \frac{0..332}{(Re_x)^{1/2}} \tag{A.64}$$

The heat transfer equation, Equation (A.60), can be re-written as:

$$Pr^{2/3}\left[\frac{Nu_x}{Re_x Pr}\right] = \frac{h}{\rho c_p V_\infty} = \frac{0.332}{(Re_x)^{1/2}} \tag{A.65}$$

The right-hand sides of the Equations (A64) and (A6) are the same. Thus,

$$St_x \, Pr^{2/3} = \frac{f}{2} \tag{A.66}$$

where St_x is called Stanton number, defined as $Nu_x/(Re_x \, Pr)$. Equation (A.66) is known as Chilton-Colburn analogy between momentum and heat transfer. It expresses the relationship between heat transfer and fluid friction for steady laminar flow over a flat solid surface. It is valid over a Prandtl number range of 0.6< Pr <50.

A.5 Turbulent Flow Over a Flat Plate

As turbulent flows are quite complex in nature, it is not possible to carry out any exact analysis. Based on experimental data, the local friction factor in turbulent flow over a flat plate is given as:

$$f = \frac{0..0592}{(Re_x)^{1/5}} \tag{A.67}$$

Equation (A.67) is valid over the Reynolds number range of $5 \times 10^5 < Re_x < 10^7$. The heat transfer co-efficient in turbulent flow over a flat plate of constant surface temperature can be determined from the Chilton-Colburn analogy (Equation (A.66)) once the friction factor relationship is known. From Equations (A.66) and (A.67), one can show that:

$$Nu_x = \frac{hx}{k} = 0.0296(Re_x)^{4/5}\left(Pr^{1/3}\right) \tag{A.68}$$

This relationship is generally valid over the Pr range of 0.6 to 60 and the Re_x range of $5 \times 10^5 < Re_x < 10^7$. Equation (A.68) could also be applied to turbulent flow over a flat plate with constant surface heat flux. The local Nusselt number for constant heat flux condition is only slightly higher than the isothermal surface condition. A better correlation for turbulent flow under constant heat flux condition is:

$$Nu_x = \frac{hx}{k} = 0.0308(Re_x)^{4/5}\left(Pr^{1/3}\right) \tag{A.69}$$

The boundary layer thickness in turbulent flow over a flat plate is approximately given by the relation:

$$\delta_T \approx \delta \approx \frac{0.37x}{(Re_x)^{1/5}} \tag{A.70}$$

Appendix B

Heisler and Grober Charts for Cylinder and Sphere

(pertaining to Chapter 11)

Heisler Charts for Cylinder

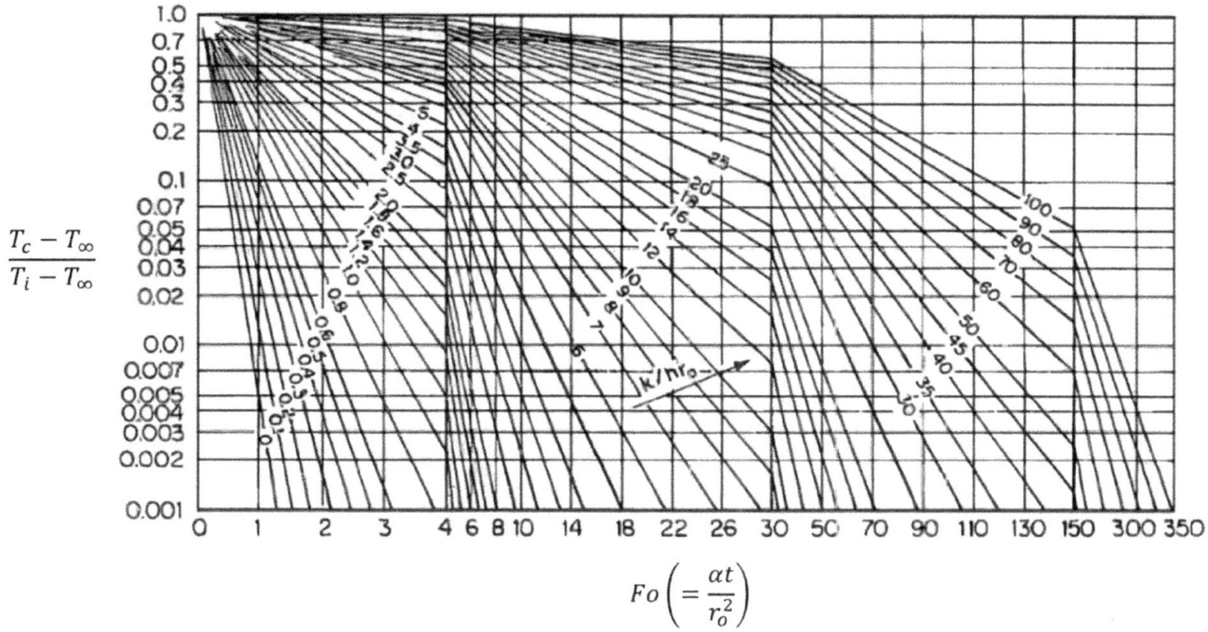

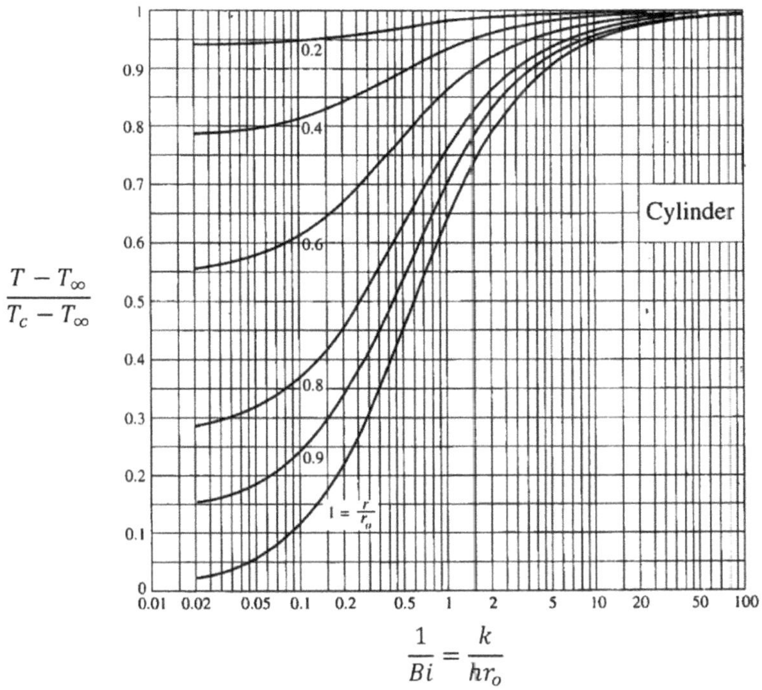

Grober Chart for Cylinder

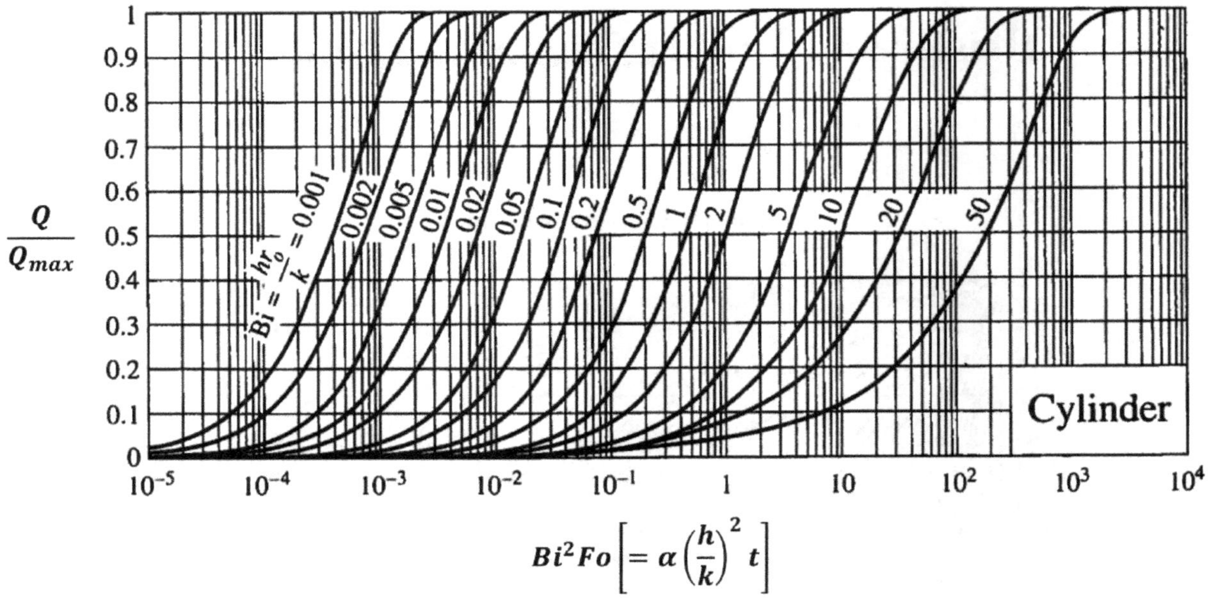

Heisler Charts for Sphere

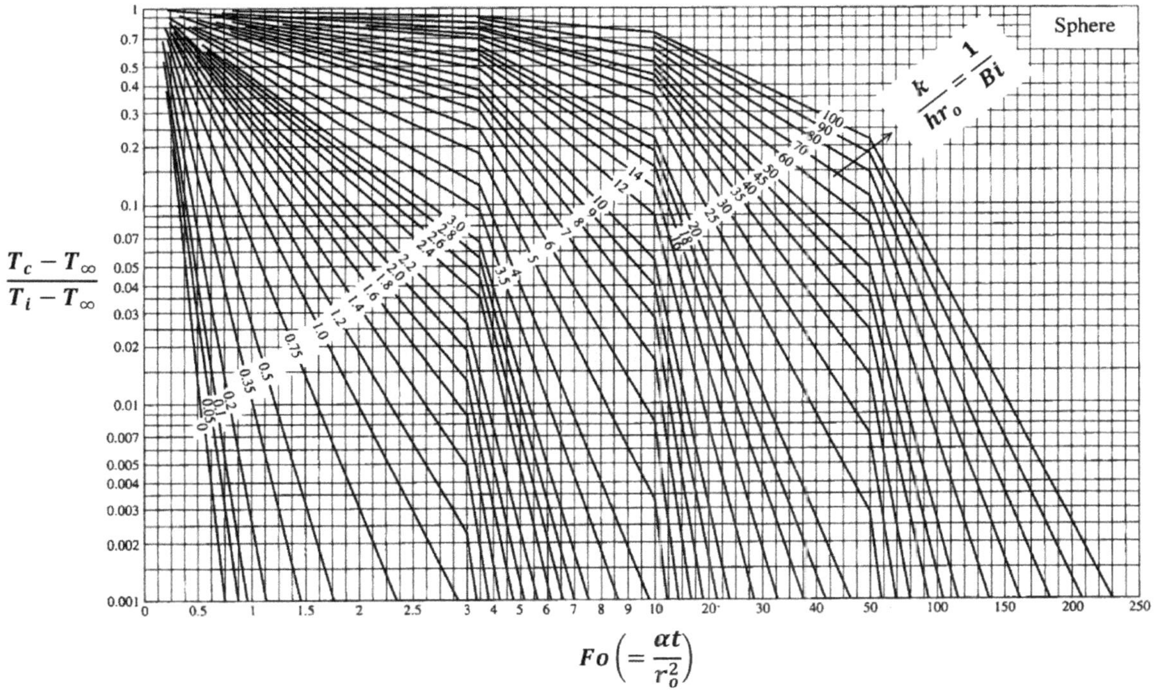

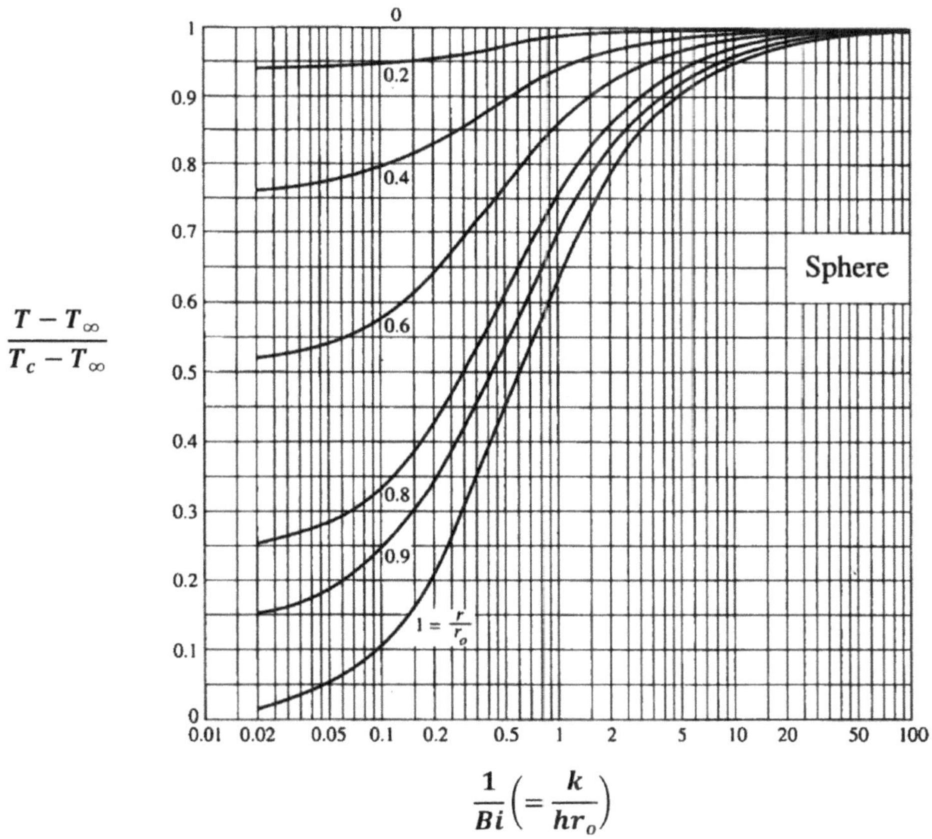

Grober Chart for Sphere

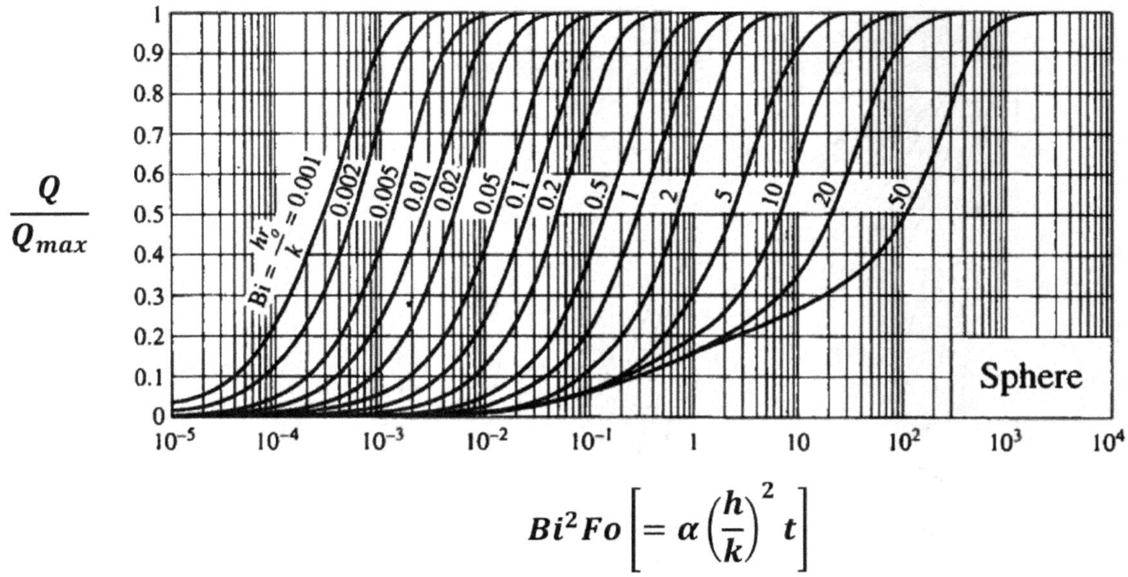

Appendix C

Basic Concepts of Convective Heat Transfer in Internal Flows

(pertaining to Chapter 13)

C.1 Heat Transfer Coefficient

Internal flows refer to flow inside conduits such as pipes or ducts. In internal flows the fluid is confined or surrounded by a solid boundary. At the solid surface (such as pipe wall) where $y = 0$, the local heat flux normal to the solid surface is given by the following form of the Newton's law of cooling:

$$q''|_{y=0} = h(T_w - T_m) \tag{C.1}$$

where T_w is the wall temperature and T_m is the mean temperature (also referred to as mixing-cup or bulk temperature) of the fluid at any cross-section of the conduit, defined as:

$$T_m = \frac{1}{\bar{V}A}\int_A V_z T dA \tag{C.2}$$

where $\bar{V}$ is the average velocity of the fluid, V_z is the local velocity, T is the local temperature, and A is the cross-sectional area. For flow in a circular pipe, the expression for T_m reduces to:

$$T_m = \frac{2}{\bar{V}R^2}\int_0^R rT(r)V_z(r) \tag{C.3}$$

where R is the pipe radius. The defining expression for T_m is derived using the reasoning that the thermal energy transported by the fluid through any cross-section in actual flow is equal to the thermal energy transported through the same cross-section if the fluid was at a mean temperature T_m.

At the solid surface (pipe wall), the local heat flux normal to the solid surface can also be obtained from the Fourier law of heat conduction:

$$q''|_{y=0} = -k\left(\frac{\partial T}{\partial y}\right)_{y=0} \tag{C.4}$$

where $\left(\frac{\partial T}{\partial y}\right)_{y=0}$ is the temperature gradient at the solid surface. For circular pipes, $\left(\frac{\partial T}{\partial y}\right)_{y=0} = -\left(\frac{\partial T}{\partial r}\right)_{r=R}$.

Using Equations (C.1) and (C.4), one can write:

$$h = \frac{-k\left(\frac{\partial T}{\partial y}\right)_{y=0}}{(T_w - T_m)} = \frac{k\left(\frac{\partial T}{\partial r}\right)_{r=R}}{(T_w - T_m)} \tag{C.5}$$

Thus, the heat transfer coefficient in internal flow can be obtained if both temperature and velocity distributions are known.

C.2 Basic Concepts

Consider the flow of fluid in a circular cross-section pipe. At the pipe entrance, the velocity is uniform. As the fluid moves along the pipe, the velocity distribution changes from uniform at the entrance to fully developed some distance downstream. Note that the velocity profile is called fully developed when it no longer changes with increasing axial distance, that is, $dV_z/dz = 0$. The length of the pipe required for the velocity profile to become fully developed is called hydrodynamic entry length (L_h). Figure B.1 shows the development of the hydrodynamic boundary layer when the fully developed flow is laminar. The fully developed flow is laminar when the Reynolds number is less than about 2100. The pipe Reynolds number is defined as:

$$Re = \frac{\rho D \bar{V}}{\eta} \tag{C.6}$$

where $\bar{V}$ is the average velocity given as:

$$\bar{V} = \frac{Q}{A} = \frac{\int_A V_z dA}{A} = \frac{2}{R^2}\int_0^R rV_z dr \tag{C.7}$$

The hydrodynamic entry length for fully developed laminar flow is approximately given as:

$$\frac{L_h}{D} \approx 0.05Re \tag{C.8}$$

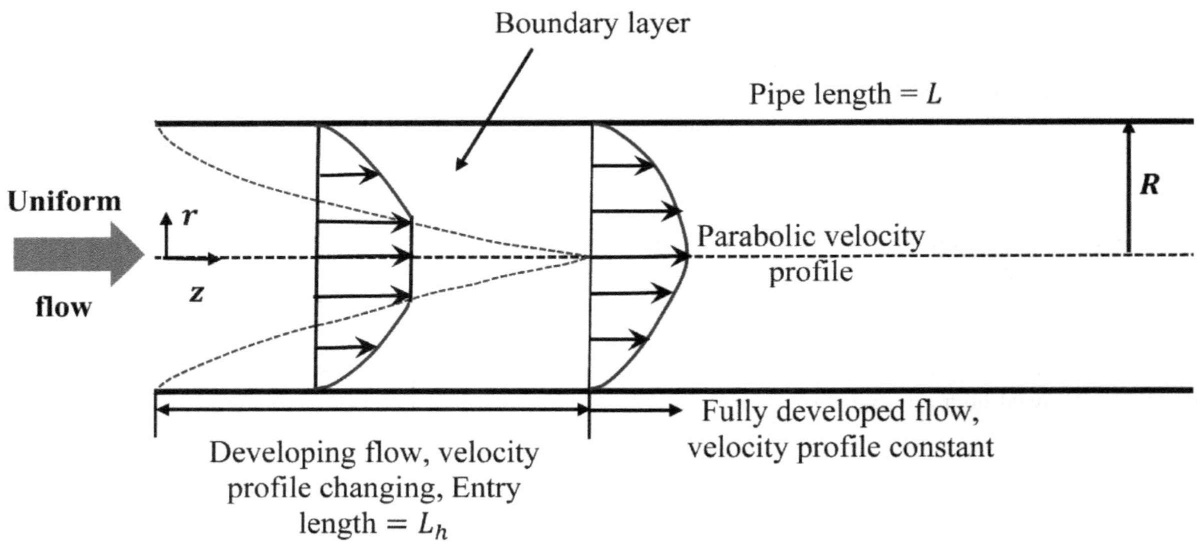

Figure C.1 Development of the hydrodynamic boundary layer for flow in a pipe.

When Re is high (Re >2100), the fully developed flow is turbulent. However, the flow in the boundary layer at the entrance of the pipe is always laminar. It becomes turbulent before the fully developed region is reached. The entry length for fully developed turbulent flow is approximately in the range (nearly independent of the Re): $10 \le L_h/D \le 60$

Figure C.2 shows the development of a thermal boundary layer in a heated pipe. The fluid enters the pipe at a uniform temperature. As the fluid moves along the pipe, it gets heated up and the temperature profile changes. After some distance downstream the pipe, the flow becomes thermally fully developed where:

$$\frac{\partial}{\partial z}\left[\frac{T_w - T}{T_w - T_m}\right] = 0 \tag{C.9}$$

That is, the dimensionless temperature $(T_w - T)/(T_w - T_m)$ no longer changes with increasing axial distance. The length of the pipe required for the dimensionless temperature to become fully developed is called thermal entry length (L_t).

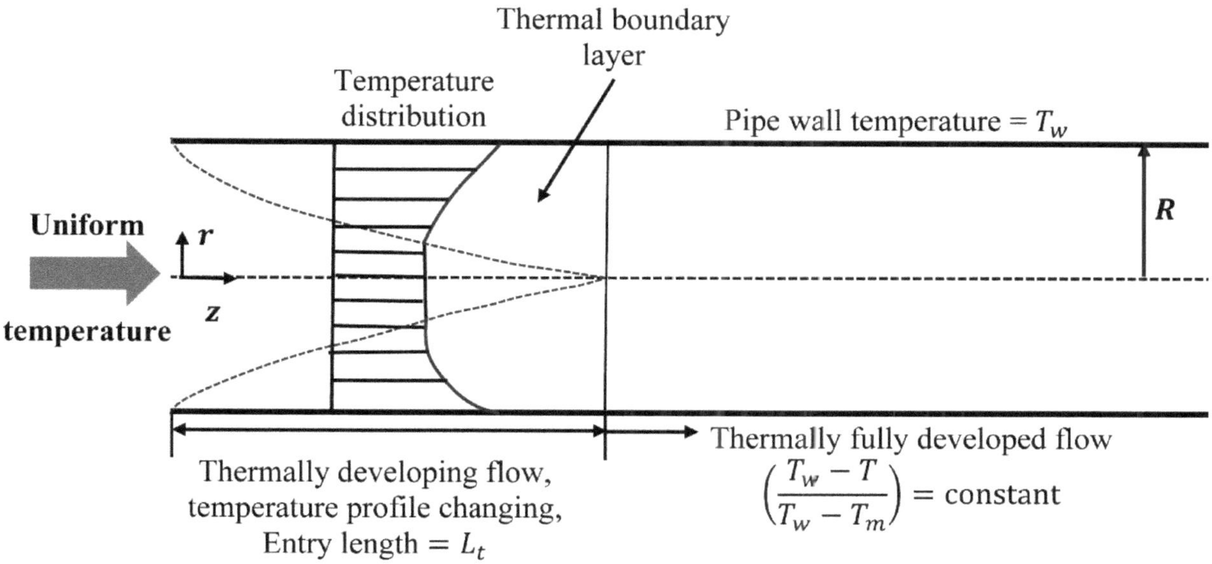

Figure C.2 Development of the thermal boundary layer for flow in a pipe.

For the case where the hydrodynamically fully developed flow is laminar, the thermal entry length is given as:

$$\frac{L_t}{D} \approx 0.05RePr \tag{C.10}$$

When the hydrodynamically fully developed flow is turbulent, the thermal entry length is approximately given as L_t/D=10 (nearly independent of Re and Pr).

C.3 Hydrodynamically and Thermally Fully Developed Flows

For flow to be hydrodynamically and thermally fully developed, the conditions to be satisfied are:

$$\frac{\partial v_z}{\partial z} = 0 \tag{C.11}$$

$$\frac{\partial}{\partial z}\left[\frac{T_w - T}{T_w - T_m}\right] = 0 \tag{C.12}$$

From Equation (C.12), it follows that the derivative of $(T_w - T)/(T_w - T_m)$ with respect to r is also independent of z. Thus, the following relation is independent of z:

$$\frac{\partial}{\partial r}\left(\frac{T_w - T}{T_w - T_m}\right)\Bigg|_{r=R} = -\frac{(\partial T/\partial r)_{r=R}}{T_w - T_m} \tag{C.13}$$

From Equations (C.5) and (C.13), it follows that the heat transfer coefficient h becomes constant independent of z in thermally fully developed region. The convection coefficient is very large at the pipe entrance due to a large temperature gradient $(\partial T/\partial r)_{r=R}$. As the thermal boundary layer grows h decreases rapidly until a constant value is reached in thermally fully developed flow.

Axial Temperature Gradient at Constant Wall Heat Flux

For the case where the wall heat flux is constant, the temperature difference $(T_w - T_m)$ is constant in thermally fully developed region. This follows from the Newton law of cooling:

$$T_w - T_m = q_w''/h \tag{C.14}$$

This equation also implies that in thermally fully developed region:

$$\frac{dT_w}{dz} = \frac{dT_m}{dz} \tag{C.15}$$

From Equations (C.12) and (C.15), one can further show that:

$$\frac{\partial T}{\partial z} = \frac{dT_w}{dz} = \frac{dT_m}{dz} \neq f(r) \tag{C.16}$$

Thus, the axial temperature gradient $\partial T/\partial z$ in thermally fully developed flow with constant wall flux is independent of radial location. From energy balance over a differential section of the pipe, it follows that:

$$\dot{m}c_p dT_m = q_w'' P dz \tag{C.17}$$

or,

$$\frac{dT_m}{dz} = \frac{q_w'' P}{\dot{m}c_p} \tag{C.18}$$

where P is the perimeter of the pipe ($= \pi D$ for a circular cross-section pipe). This energy balance assumes that the fluid is incompressible, and the flow work is negligible $d(P/\rho) \ll d(c_p T_m)$. It is valid for both developing and developed flows. For constant wall flux, this equation upon integration gives:

$$T_m(x) = T_{m,i} + \frac{q_w'' P}{\dot{m}c_p} z \tag{C.19}$$

Thus, the mean temperature varies linearly with z along the pipe.

From Equations (C.16) and (C.19), it also follows that:

$$\frac{\partial T}{\partial z} = \frac{dT_w}{dz} = \frac{dT_m}{dz} = \frac{q_w'' P}{\dot{m}c_p} = \text{constant} \tag{C.20}$$

Equation (C.20) implies that in thermally fully developed flow in a pipe subjected to constant wall flux, the temperature gradient $\partial T/\partial z$ is constant, independent of z and r. However, T a function of z and r. It can also be concluded that T, T_w and T_m all vary linearly with z.

Axial Temperature Gradient at Constant Wall Temperature

For the case where the wall surface temperature is constant (T_w = constant), $dT_w/dz = 0$. Using Equation (C.13) with $dT_w/dz = 0$, it follows that:

$$\frac{\partial T}{\partial z} = \left(\frac{T_w - T}{T_w - T_m}\right)\frac{dT_m}{dz} \tag{C.21}$$

As T is a function of z and r, $\partial T/\partial z$ now depends on both z and r.

Temperature Distribution and Nusselt Number in Fully Developed Flows

For steady laminar flow in a pipe, the thermal energy equation is:

$$V_z \frac{\partial T}{\partial z} = \alpha\left[\frac{1}{r}\frac{\partial}{\partial r}\left(r\frac{\partial T}{\partial r}\right) + \frac{\partial^2 T}{\partial z^2}\right] \tag{C.22}$$

Equation (C.22) assumes that the fluid properties are constant, and that viscous dissipation of energy is negligible.

Consider the situation where the wall heat flux imposed on the pipe is constant, that is, q''_w=constant. In thermally fully developed region, $\partial T/\partial z$ = constant and therefore $\partial^2 T/\partial z^2 = 0$. Thus, Equation (C.22) reduces to:

$$V_z \frac{\partial T}{\partial z} = \alpha\left[\frac{1}{r}\frac{\partial}{\partial r}\left(r\frac{\partial T}{\partial r}\right)\right] \tag{C.23}$$

Assuming hydrodynamically fully developed flow, the velocity distribution is given as:

$$V_z = V_{max}\left[1 - \left(\frac{r}{R}\right)^2\right] \tag{C.24}$$

Consequently, Equation (C.23) becomes:

$$\frac{1}{r}\frac{\partial}{\partial r}\left(r\frac{\partial T}{\partial r}\right) = \frac{V_{max}}{\alpha}\left[1 - \left(\frac{r}{R}\right)^2\right]\left(\frac{dT_m}{dz}\right) \tag{C.25}$$

Note that $\partial T/\partial z = dT_m/dz$ = constant. Solving Equation (C.25) subject to the boundary conditions:

$$dT/dr = 0 \quad \text{at} \quad r = 0 \tag{C.26}$$

$$T = T_w \quad \text{at} \quad r = R \tag{C.27}$$

we obtain:

$$T = T_w - \frac{V_{max}R^2}{\alpha}\left(\frac{dT_m}{dz}\right)\left[\frac{3}{16} + \frac{1}{16}\left(\frac{r}{R}\right)^4 - \frac{1}{4}\left(\frac{r}{R}\right)^2\right] \tag{C.28}$$

where the wall temperature T_w does vary with z.

The bulk or mean temperature of the fluid, defined in Equation (C.3), can now be obtained using the velocity distribution (Equation (C.24)) and temperature distribution (Equation (C.28)). Thus, Equations (C.3), (C.24), and (C.28) yield:

$$T_m = T_w - \left(\frac{11}{96}\right)\left(\frac{V_{max}R^2}{\alpha}\right)\frac{dT_m}{dz} \tag{C.29}$$

From Equation (C.20):

$$\frac{dT_m}{dz} = \frac{q''_w P}{\dot{m} c_p} = \frac{8 q''_w}{\rho c_p D V_{max}} \tag{C.30}$$

Combining Equations (C.29) and (C.30), we obtain:

$$T_m = T_w - \left(\frac{11}{48}\right)\left(\frac{q''_w D}{k}\right) \tag{C.31}$$

From Newton's law of cooling and Equation (C.31), it follows that:

$$Nu = \frac{hD}{k} = \frac{48}{11} = 4.36 \tag{C.32}$$

Thus, in a cylindrical pipe subjected to constant wall flux, the Nusselt number in laminar fully developed (hydrodynamically and thermally) flow is constant, equal to 4.36.

Let us now consider the case where the wall temperature of the pipe is held constant (T_w= constant). Although $\partial^2 T/\partial z^2$is not zero in this situation, it is reasonable to assume $\partial^2 T/\partial z^2 << \partial^2 T/\partial r^2$. Thus, the thermal energy equation is the same as that in the case of constant wall flux (Equation (C.23)). Using Equations (C.21), (C.23) and (C.24), it can be shown that:

$$\frac{1}{r}\frac{\partial}{\partial r}\left(r\frac{\partial T}{\partial r}\right) = \frac{V_{max}}{\alpha}\left[1-\left(\frac{r}{R}\right)^2\right]\left(\frac{dT_m}{dz}\right)\left[\frac{T_w - T}{T_w - T_m}\right] \tag{C.33}$$

Equation (C.33) is to be solved subject to the boundary conditions:

$$\partial T/\partial r = 0 \quad \text{at} \quad r = 0 \tag{C.34}$$

$$T = T_w \quad \text{at} \quad r = R \tag{C.35}$$

Equation (C.33) is more complicated to solve than Equation (C.25). Without going into the details, we simply state the result that the Nusselt number in fully developed region (thermally and hydrodynamically) of laminar flow in pipe (subjected to constant wall temperature) is equal to 3.66, that is,

$$Nu = \frac{hD}{k} = 3.66 \tag{C.36}$$

C.4 Developing Flows

First consider the thermal entry length problem where the flow is hydrodynamically fully developed but thermally developing. The velocity profile is given by Equation (C.24) and the thermal energy equation is given by Equation (C.23) (assuming $\partial^2 T/\partial z^2 << \partial^2 T/\partial r^2$). Thus, the governing equation for thermal entry length problem is:

$$\frac{1}{r}\frac{\partial}{\partial r}\left(r\frac{\partial T}{\partial r}\right) = \frac{V_{max}}{\alpha}\left[1-\left(\frac{r}{R}\right)^2\right]\left(\frac{\partial T}{dz}\right) \tag{C.37}$$

Equation (C.37) is to be solved subject to the appropriate boundary conditions.

If the pipe wall temperature is held constant, the boundary conditions are:

$$T = T_i \quad \text{at} \quad z = 0 \tag{C.38}$$

$$\partial T/\partial r = 0 \quad \text{at} \quad r = 0 \tag{C.39}$$

$$T = T_w = \text{constant} \quad \text{at } r = R \tag{C.40}$$

If the pipe wall is subjected to constant heat flux, the boundary condition at the pipe wall becomes:

$$q''_w = k(\partial T/\partial r)_{r=R} = \text{constant} \tag{C.41}$$

Thus, the thermal entry length problem can be solved analytically by solving Equation (C.37) subject to the stated boundary conditions. We simply state without going into the details of the solution to this thermal entry length problem that the result can be expressed in the form:

$$Nu_z = \frac{hD}{k} = f(Gz_z) \tag{C.42}$$

where h is the local heat transfer coefficient and $Gz_x$is the local Graetz number, defined as:

$$Gz_z = \frac{\pi}{4}\left(\frac{D}{z}\right) RePr \tag{C.43}$$

Note that the product $RePr$ is also referred to as Peclet number.

The combined entry length problem where the flow is both hydrodynamically and thermally developing is much more difficult to solve analytically as velocity and temperature both depend on r as well as z. Empirical correlations are available in the literature to predict heat transfer coefficient for combined entry length problem.

Appendix D

Fraction of Total Black-Body Radiation Emitted Over a Range of 0 to λT

(pertaining to Chapter 15)

$\lambda T\ (\mu m.K)$	$\int_0^{\lambda T} \frac{E_{\lambda,b} d(\lambda T)}{\sigma T^5}$	$\lambda T\ (\mu m.K)$	$\int_0^{\lambda T} \frac{E_{\lambda,b} d(\lambda T)}{\sigma T^5}$
200	0.000000	6,200	0.754140
400	0.000000	6,400	0.769234
600	0.000000	6,600	0.783199
800	0.000016	6,800	0.796129
1,000	0.000321	7,000	0.808109
1,200	0.002134	7,200	0.819217
1,400	0.007790	7,400	0.829527
1,600	0.019718	7,600	0.839102
1,800	0.039341	7,800	0.848005
2,000	0.066728	8,000	0.856288
2,200	0.100888	8,500	0.874608
2,400	0.140256	9,000	0.890029
2,600	0.183120	9,500	0.903085
2,800	0.227897	10,000	0.914199
2,898	0.250108	10,500	0.923710
3,000	0.273232	11,000	0.931890
3,200	0.318102	11,500	0.939959
3,400	0.361735	12,000	0.945098
3,600	0.403607	13,000	0.955139
3,800	0.443382	14,000	0.962898
4,000	0.480877	15,000	0.969981
4,200	0.516014	16,000	0.973814
4,400	0.548796	18,000	0.980860
4,600	0.579280	20,000	0.985602
4,800	0.607559	25,000	0.992215
5,000	0.633747	30,000	0.995340
5,200	0.658970	40,000	0.997967
5,400	0.680360	50,000	0.998953
5,600	0.701046	75,000	0.99713
5,800	0.720158	100,000	0.999905
6,000	0.737818		

Appendix E

Convective Mass Transport from a Flat Solid Surface

(pertaining to Chapter 18)

E.1 Boundary Layer Equations

For steady flow of a Newtonian fluid (with constant properties) over a flat solid surface, the boundary layer mass and momentum equations are:

$$\frac{\partial V_x}{\partial x} + \frac{\partial V_y}{\partial y} = 0 \tag{E.1}$$

$$V_x \frac{\partial V_x}{\partial x} + V_y \frac{\partial V_x}{\partial y} = V_\infty \frac{dV_\infty}{dx} + \nu \frac{\partial^2 V_x}{\partial y^2} \tag{E.2}$$

where V_x and V_y are x and y components of fluid velocity, respectively, V_∞ is the free stream velocity of the fluid, and ν is the kinematic viscosity ($= \eta/\rho$).

The concentration boundary layer equation is as follows:

$$V_x \frac{\partial C_A}{\partial x} + V_y \frac{\partial C_B}{\partial y} = D_{AB} \frac{\partial^2 C_A}{\partial y^2} \tag{E.3}$$

Equations (E.1) - (E.3) need to be solved, subject to the given boundary conditions, for the three unknowns V_x , V_y, and C_A.

E.2 Integral Analysis of the Boundary Layer

In practical problems dealing with convective mass transfer between a solid surface and moving fluid, the quantity of most interest is the mass or molar flux at the solid/fluid interface. The following von-Karman mass integral expression (given here without proof) could be used to determine the molar flux at the solid/fluid interface:

$$N_{A,w} = -D_{AB} \left(\frac{\partial C_A}{\partial y}\right)_{y=0} = \frac{d}{dx} \int_0^{\delta_c} (C_A - C_{A,\infty}) V_x dy \tag{E.4}$$

where $N_{A,w}$ is the absolute molar flux of species A at the solid surface, $C_{A,\infty}$ is the free-stream concentration of A, and δ_c is the thickness of the concentration boundary layer. The integral expression (Equation (E.4)) is valid for dilute systems with low mass transfer rate and small concentration of diffusing species A. It can be used to determine the molar flux at the solid/fluid interface provided that both the velocity and concentration profiles are known.

E.3 Mass Transfer Coefficient

For external flows, the absolute molar flux at the solid/fluid interface (normal to the interface) can be expressed in terms of the mass transfer coefficient as:

$$N_{A,y}\big|_{y=0} = k\left(C_{A,w} - C_{A,\infty}\right) \tag{E.5}$$

where k is the mass transfer coefficient, $C_{A,w}$ is the molar concentration of species A in the fluid at the fluid/solid interface ($y = 0$), and $C_{A,\infty}$ is the free-stream molar concentration of species A in the fluid. At the fluid/solid interface, the molar flux normal to the interface is also given by the Fick's law of diffusion:

$$N_{A,y}\big|_{y=0} = J_{A,y}\big|_{y=0} = -D_{AB}\left(\frac{\partial C_A}{\partial y}\right)_{y=0} \tag{E.6}$$

From Equations (E.5) and (E.6), it follows that:

$$k = \frac{-D_{AB}\left(\frac{\partial C_A}{\partial y}\right)_{y=0}}{C_{A,w} - C_{A,\infty}} \tag{E.7}$$

E.4 Laminar Flow Over a Flat Plate

Consider laminar flow over a solid flat plate. The concentration of species A in the fluid at the fluid/solid interface ($y = 0$) is $C_{A,w}$. The free-stream concentration of A ($y > \delta_c$) is $C_{A,\infty}$. The velocity profile for laminar flow over a flat plate is given as:

$$\frac{V_x}{V_\infty} = \frac{3}{2}\left(\frac{y}{\delta}\right) - \frac{1}{2}\left(\frac{y}{\delta}\right)^3 \tag{E.8}$$

where δ is the hydrodynamic boundary layer thickness. From momentum integral analysis (see Appendix A), it is found that:

$$\frac{\delta}{x} = \frac{4.64}{Re_x^{1/2}} \tag{E.9}$$

where x is the distance from the leading edge of the plate and Re_x is the local Reynolds number defined as $\rho V_\infty x/\eta$. The concentration profile can be assumed to be a four-term polynomial as:

$$C_A - C_{A,w} = a + by + cy^2 + dy^3 \tag{E.10}$$

The constants in Equation (E.10) can be obtained from the boundary conditions:

$$y = 0, C_A - C_{A,w} = 0 \tag{E.11}$$

$$y = \delta_c, C_A - C_{A,w} = C_{A,\infty} - C_{A,w} \tag{E.12}$$

$$y = \delta_c, \frac{\partial}{\partial y}\left(C_A - C_{A,w}\right) = 0 \tag{E.13}$$

$$y = 0, \frac{\partial^2}{\partial y^2}\left(C_A - C_{A,w}\right) = 0 \tag{E.14}$$

The last boundary condition follows from the concentration boundary layer equation (Equation (E.3)) as $V_x(\partial C_A/\partial x) = 0$ and $V_y(\partial C_A/\partial y) = 0$ at $y = 0$. Note that $C_{A,w}$ is assumed to be constant. Using the boundary conditions (Equations (E.11) - (E.14)), it can be easily shown that:

$$\frac{C_A - C_{A,w}}{C_{A,\infty} - C_{A,w}} = \frac{3}{2}\left(\frac{y}{\delta_c}\right) - \frac{1}{2}\left(\frac{y}{\delta_c}\right)^3 \tag{E.15}$$

To utilize the mass integral expression (Equation (E.4)) knowledge of both velocity and concentration distribution is required. Assuming that $\delta_c < \delta$, the velocity distribution is given by Equation (E.8). Substitution of concentration distribution (Equation (E.15)) and velocity distribution (Equation (E.8)) into the integral expression (Equation (E.4)) and carrying out the integration yields the following results:

$$\frac{\delta_c}{\delta} = \frac{0.976}{Sc^{1/3}} \tag{E.16}$$

$$Sh_x = \frac{kx}{D_{AB}} = 0.331(Re_x)^{1/2} Sc^{1/3} \tag{E.17}$$

where Sc is the Schmidt number, defined as ν/D_{AB}, Sh_x is the local Sherwood number, and Re_x is the local Reynolds number.

These results (Equations (E.16) – (E.17)) are valid for $Sc > 0.60$ and they are very close to exact solution of the boundary layer equations (Equations (E.1) – (E.3)). The exact solution gives:

$$\frac{\delta_c}{\delta} = \frac{1}{Sc^{1/3}} \tag{E.18}$$

$$Sh_x = \frac{kx}{D_{AB}} = 0.332(Re_x)^{1/2} Sc^{1/3} \tag{E.19}$$

Using Equation (E.19) and the following friction factor equation for laminar flow over a flat plate:

$$\frac{f}{2} = \frac{0.332}{Re_x^{1/2}} \tag{E.20}$$

it can easily be shown that:

$$St_m Sc^{2/3} = \frac{f}{2} \tag{E.21}$$

where St_m is the mass transport Stanton number, defined as:

$$St_m = \frac{Sh}{ScRe} \tag{E.22}$$

Equation (E.21) is referred to as the Chilton-Colburn analogy between mass transfer and momentum transfer. It is valid for $Sc > 0.6$.

E.5 Turbulent Flow Over a Flat Plate

The local friction factor in turbulent flow over a flat plate is given by the following empirical equation:

$$f = \frac{0.0592}{Re_x^{1/5}} \tag{E.23}$$

The mass transfer coefficient in turbulent flow over a flat plate can be determined from the Chilton-Colburn analogy (Equation (E.21)) using the above friction factor relation. Thus:

$$Sh_x = \frac{kx}{D_{AB}} = 0.0296(Re_x)^{4/5} Sc^{1/3} \tag{E.24}$$

Appendix F

Derivation of Maxwell-Stefan Diffusion Model

(pertaining to Chapter 20)

The Maxwell-Stefan (MS) model is based on the balance between the driving force of diffusion of species and the opposing friction force on the species.

Consider the diffusion of species 1 in a binary mixture of gaseous species 1 and 2. The driving force of diffusion of species 1 per unit volume of the mixture is given as follows:

$$f_1 = -\frac{dP_1}{dz} \tag{F.1}$$

where P_1 is the partial pressure of species 1 and $(-dP_1/dz)$ is the gradient in partial pressure of species 1. Assuming the gas mixture to be ideal, the molar concentration of species 1 is:

$$C_1 = \frac{P_1}{RT} \tag{F.2}$$

Thus, the driving force of diffusion per unit mole of species 1 is:

$$F_1 = \frac{f_1}{C_1} = \frac{1}{C_1}\left(-\frac{dP_1}{dz}\right) = \frac{RT}{P_1}\left(-\frac{dP_1}{dz}\right) \tag{F.3}$$

Upon re-arrangement, Equation (F.3) becomes:

$$F_1 = RT\left(-\frac{dlnP_1}{dz}\right) \tag{F.4}$$

We can now express the diffusion force per mole of species in terms of the chemical potential gradient. The chemical potential of a species in an ideal gas mixture is given as:

$$\mu_1^{ig}(T,P) = \mu_{1,pure}^{ig}(T,P) + RTlnx_1 \tag{F.5}$$

where x_1 is the mole fraction of species 1. At constant temperature,

$$\mu_{1,pure}^{ig}(T,P) = \mu_1^o(T,P^o) + RTln\left(\frac{P}{P^o}\right) \tag{F.6}$$

Thus,

$$\mu_1^{ig}(T,P) = \mu_1^o(T) + RTln\left(\frac{x_1P}{P^o}\right) = \mu_1^o(T) + RTln\left(\frac{P_1}{P^o}\right) \tag{F.7}$$

If we take the spatial derivative of $\mu_1^{ig}(T,P)$ at constant temperature, we get:

$$F_{1,T} = -\nabla_T(\mu_1) = -\left(\frac{d\mu_1}{dz}\right)_T = RT\left(-\frac{dlnP_1}{dz}\right) \tag{F.8}$$

Thus, the diffusion driving force on species 1 at constant temperature is simply $\nabla_T(\mu_1)$. For convenience, we have dropped the superscript "*ig*" from the chemical potential. Since $P_1 = x_1 P$, we can further write:

$$F_{1,T} = -\nabla_T(\mu_1) = RT\left(-\frac{dlnP_1}{dz}\right) = RT\left[-\frac{d\ln(x_1P)}{dz}\right] \tag{F.9}$$

Assuming total pressure to be constant, we get:

$$F_{1,T,P} = -\nabla_{T,P}(\mu_1) = -\frac{RT}{x_1}\left(\frac{dx_1}{dz}\right) \tag{F.10}$$

The opposing frictional force on species 1 is directly proportional to the difference in the average velocities of components $(\vec{V}_1 - \vec{V}_2)$ and the mole fraction of species 2. Thus,

$$F_2 \propto x_2(\vec{V}_1 - \vec{V}_2) \tag{F.11}$$

The proportionality constant is the drag coefficient ξ_{12}. Consequently,

$$F_2 = \xi_{12}x_2(\vec{V}_1 - \vec{V}_2) \tag{F.12}$$

This equation can be re-written as:

$$F_2 = \left(\frac{RT}{Đ_{12}}\right)x_2(\vec{V}_1 - \vec{V}_2) \tag{F.13}$$

where $Đ_{12}$ is Maxwell-Stefan diffusivity related to ξ_{12} as $Đ_{12} = RT/\xi_{12}$. Upon equating the diffusion driving force with the opposing frictional force ($F_{1,T,P} = F_2$), we get:

$$-\nabla_{T,P}(\mu_1) = -\frac{RT}{x_1}\left(\frac{dx_1}{dz}\right) = \left(\frac{RT}{Đ_{12}}\right)x_2(\vec{V}_1 - \vec{V}_2) \tag{F.14}$$

Upon simplification,

$$-\left(\frac{dx_1}{dz}\right) = \left(\frac{x_1x_2}{Đ_{12}}\right)(\vec{V}_1 - \vec{V}_2) \tag{F.15}$$

Thus,

$$\nabla x_1 = -\left(\frac{x_1x_2}{Đ_{12}}\right)(\vec{V}_1 - \vec{V}_2) \tag{F.16}$$

Note that we have made the following assumptions in arriving at this result: temperature and pressure are constant, and the mixture is an ideal gas mixture.

We can re-write Equation (F.16) in terms of absolute molar fluxes of species 1 and 2 as shown below:

$$\vec{N}_1 = C_1\vec{V}_1 \tag{F.17}$$

$$\vec{N}_2 = C_2\vec{V}_2 \tag{F.18}$$

$$\nabla x_1 = -\left(\frac{x_1x_2}{Đ_{12}}\right)\left(\frac{\vec{N}_1}{C_1} - \frac{\vec{N}_2}{C_2}\right) \tag{F.19}$$

Using the relations,

$$C_1 = x_1C \quad \text{and} \quad C_2 = x_2C \tag{F.20}$$

we get,

$$\nabla x_1 = -\frac{(x_2\vec{N}_1 - x_1\vec{N}_2)}{C Ð_{12}} \tag{F.21}$$

Using the following relations,

$$\vec{N}_1 = \vec{J}_1 + x_1\vec{N} \tag{F.22}$$

$$\vec{N}_2 = \vec{J}_2 + x_2\vec{N} \tag{F.23}$$

it can be readily shown that:

$$x_2\vec{N}_1 - x_1\vec{N}_2 = x_2\vec{J}_1 - x_1\vec{J}_2 \tag{F.24}$$

Thus,

$$\nabla x_1 = -\frac{(x_2\vec{N}_1 - x_1\vec{N}_2)}{C Ð_{12}} = -\frac{(x_2\vec{J}_1 - x_1\vec{J}_2)}{C Ð_{12}} \tag{F.25}$$

This is the Maxwell-Stefan model for binary diffusion. It can be generalized to multicomponent mixtures as follows:

$$\nabla x_i = -\sum_{j\neq i}^{n}\frac{(x_j\vec{J}_i - x_i\vec{J}_j)}{C Ð_{ij}} = -\sum_{j\neq i}^{n}\frac{(x_j\vec{N}_i - x_i\vec{N}_j)}{C Ð_{ij}} \tag{F.26}$$

Index

A

B

C

D

E

F

G

H

I

J

K

L

M

N

O

P

R

S

T

U

V

www.ingramcontent.com/pod-product-compliance
Lightning Source LLC
LaVergne TN
LVHW082012200926
848481LV00010B/2845